HIGH-SPEED MARINE CRAFT

This book details the efforts to build a large naval vessel capable of traveling at 100 knots. It is the first book to summarize this extensive work from historical and technical perspectives. It explores the unique principles and challenges in the design of high-speed marine craft. This volume explores different hull form concepts, requiring an understanding of the four forces affecting the lift and the drag of the craft. The four forces covered are hydrostatic (buoyancy), hydrodynamic, aerostatic, and aerodynamic. This text will appeal to naval researchers, architects, graduate students, and historians, as well as others generally interested in naval architecture and propulsion.

Peter Mantle's long career as a naval architect and aerospace engineer includes positions as a Chief Engineer and Test Pilot for the first surface effect ship (an aerodynamic air cushion craft); Technical Director and Program Manager of the US Navy 100-ton displacement surface effect ship, "SES-100B," that set a world speed record of 91.90 knots; Director of Technology Assessment, Office of Secretary of Navy (SECNAV), and Chief of Naval Operations (OPNAV) in The Pentagon for all R&D on aircraft, ships, submarines, missile systems and classified projects; Chairman, NATO Studies on Air, Land and Sea Battles; and Chairman, US Delegation to NATO Industrial Advisory Group, on defense matters for NATO. He is the author of numerous research articles and three books: *A Technical Summary of Air Cushion Craft Development*, *Air Cushion Craft Development*, and *The Missile Defense Equation: Factors for Decision Making.*

High-Speed Marine Craft

ONE HUNDRED KNOTS AT SEA

Peter J. Mantle

CAMBRIDGE
UNIVERSITY PRESS

CAMBRIDGE
UNIVERSITY PRESS

32 Avenue of the Americas, New York, NY 10013-2473, USA

Cambridge University Press is part of the University of Cambridge.

It furthers the University's mission by disseminating knowledge in the pursuit of education, learning, and research at the highest international levels of excellence.

www.cambridge.org
Information on this title: www.cambridge.org/9781107090415

First published 2015

A catalog record for this publication is available from the British Library.

Library of Congress Cataloging in Publication Data
Mantle, Peter J.
High-speed marine craft : one hundred knots at sea / Peter J. Mantle.
pages cm
1. Warships – Design and construction – History – 21st century.
2. Warships – Design and construction – History – 20th century.
3. Warships – United States – Technological innovations. 4. Ground-effect machines – United States – Design and construction. 5. Hydrofoil boats – United States – Design and construction. I. Title. II. Title: One hundred knots at sea.
V800.M24 2015
623.825–dc23 2014048657

To Dr. Scott Carson Rethorst, who with his original work on the Surface Effect Ship "Columbia," *almost got us there to a practical one hundred knots at sea*

Contents

About the Author

Peter J. Mantle has been closely involved in hovercraft and surface effect ship development since Sir Christopher Cockerell first developed the hovercraft with Saunders Roe, Ltd (now British Hovercraft Corporation) in England in 1956. Mr. Mantle began his career at Saunders Roe, Ltd in 1951. In 1958, after graduating from the College of Aeronautics, Cranfield, England with a Master's Degree in Aeronautical Science (DCAe), he emigrated to Quebec, Canada to conduct upper atmosphere research with Dr Gerald Bull at the Canadian Armament Research & Development Establishment (CARDE) in Valcartier, Quebec. In Canada, he received a Master's Degree in Aerodynamics and Mathematics (*magna cum laude*) from Laval University, Montreal in 1959. He immigrated to the US in 1960, receiving his Engineer's Degree (Ae. E.) in Aeronautics from California Institute of Technology (CALTECH), Pasadena, California in 1964.

From 1960 to 1964 he designed, built and was the test pilot of the US Maritime Administration (MARAD) first surface effect ship (SES) test craft (*VRC-1),* based on Dr Scott Rethorst's *"Columbia"* channel flow surface effect ship which was the chosen high speed ship concept by (MARAD) in 1961, to find a new solution to the declining US Merchant Marine.

Peter Mantle continued his career in hovercraft and surface effect ship development by becoming the Technical Director and Program Manager of the US Navy Surface Effect Ship *"SES-100B"* at Bell Aerosystems (now Bell Aerospace) in 1969. He holds the patent for the unique propulsion system (a semi-submerged, supercavitating, controllable pitch propeller system) that allowed the *"SES-100B"* under US Navy trials to achieve the world speed record of 91.9knots on 30 June 1976 displayed in the *Guinness Book of Records* of that year.

In 1975, he wrote the first technical compendium of the technology of hovercraft and surface effect ship work, with a major update in 1980 called *"Air Cushion Craft Development"* published by the US Navy.

From 1976 to 1978, Peter Mantle was the Deputy Project Officer and Technical Director of the US Navy's Advanced Naval Vehicle Concepts Evaluation (ANVCE) Project, a major effort of analyses, model testing and full scale testing of several candidates including aircraft, ships, hydrofoils, hovercraft, SWATH ships and surface effect ships, to help the Navy and the Office of the Secretary of Defense (OSD) make informed decisions on the proper course to follow for future Navy high speed ship development.

From 1978 to 1984, he was the Director of Technology Assessment in the Office of the Secretary of the Navy (SECNAV) responsible for advising on the Basic Research, Technology and Advanced Development of all Navy vehicles (ships, submarines, and aircraft). This included Special Access and stealth programs.

In 1984, he joined Lockheed Marine Systems, part of Lockheed Corporation (now Lockheed Martin Corporation) in Seattle (and later in Sunnyvale, California), as Director of Strategic Planning and after different assignments became Program Director of the (now declassified) *Sea Shadow* stealth ship.

From 1989 to 2000, Mr. Mantle was the Chairman of several NATO studies on sea, land and air defense, chairing what became the basis for the current NATO missile defense plans to protect both Europe and the US from missile attack. From 2003 to 2007, Peter Mantle was the Chairman (now Chairman Emeritus) of the US Delegation to the NATO Industrial Advisory Group (NIAG) to advise NATO on a variety of industrial and defense matters before the twenty-six (now twenty-eight) nation Alliance. His book "*The Missile Defense Equation – Factors for Decision Making*" based on that work was published by the American Institute of Aeronautics and Astronautics (AIAA) in 2004. Peter Mantle is an aerospace consultant and resides with his wife Kathleen on Vashon Island, Washington.

Preface

The design principles of *high speed marine craft* are much less established than the well-established techniques honed over centuries of practice in the design of *low speed marine craft* with their displacement hull origins. *High speed marine craft* design, on the other hand, involves study of different hull form concepts, each requiring an understanding of four basic forces resulting in the lift and the drag of the craft. These four forces are: hydrostatic (buoyancy); hydrodynamic; aerostatic and aerodynamic. Each of these four forces scale by different laws of physics making scaling difficult from small models to large ship sizes. The combination of those forces applies differently in each case depending on the choice of concept being considered. In a broad sense, for the *hydrofoil*, the hydrodynamic forces dominate; for the *amphibious air cushion craft*, the aerostatic forces dominate; for the *wing-in-ground-effect craft (WIG)*, the aerodynamic forces dominate. Coupled with these different forces, the high speed marine craft must also contend with the physics of subcavitating and supercavitating flows in both the hull hydrodynamics and in the propulsion schemes envisaged.

This influence of the four forces has a significant impact on the size and speed of the craft and its use or mission. This intrinsic triad of "*size-speed-mission*" is a key consideration when asking what is achievable in attaining high speed at sea. This relationship is expanded upon throughout the book. Because of these complex interactions between the various forces and choice of craft concept, the programmatic history of developing high speed marine craft has been somewhat sporadic with isolated successes among various setbacks caused by both technology issues and programmatic stumbles.

Upper limits of *low speed marine craft* speeds using displacement hulls have remained relatively unchanged over several centuries with typical values of 25–35 knots, depending on ship size and sea conditions. The "speed limits" for *high speed marine craft* vary widely depending on the concept selected but speeds from 50 to 250 knots covers the experience base under discussion.

Two major thrusts in the US for high speed marine craft were by the US Maritime Administration (MARAD) for commercial shipping, and by the US Navy for military ships and craft. MARAD conducted two major thrusts; first a *surface piercing hydrofoil* with planned speeds of 60–100 knots and subsequently, a combination *hovercraft* and *WIG* design for 100–150 knots. The US Navy also had two major thrusts; first, for an *amphibious air cushion craft* for amphibious assault missions with speeds of 40–50 knots (other nations have developed craft with higher speeds). The second thrust was with a *non-amphibious air cushion craft* for open ocean naval missions of speeds of 80–100 knots. Each of these programs had both successes and setbacks that are detailed in the various chapters of the book.

It is important to note that speed alone was never the aim of either the MARAD or the US Navy programs. The world speed record over water today is 276 knots, so obtaining speed alone over water is relatively "easy" to obtain for special purpose boats. The problem is how to achieve *high speed* in a commercially viable or military useful ship under water conditions more likely to be found at sea than on a calm water lake under controlled conditions speeding in a straight line. History documents many false starts over many centuries in high speed ship development to conquer, or at least tame, the "cruel sea" as various ship designs have been tried, tested and failed to get much beyond the 25–35 knot capability of even today's most modern ship designs. In the last four decades, serious effort had been embarked upon by many nations to achieve some significant improvement over this 25–35 knot "barrier". In the 1960s and 1970s as the hovercraft and hydrofoil principles and other forms of dynamic lift in ship design gained a foothold, the idea of a "100 knot Navy" started to take on a major thrust in ship development. Even though there was no direct rationale for the specific value of "100 knots" (economically or militarily) this number did provide an important impetus to aeronautical engineers, naval architects and military planners alike to seek answers in this new territory of "high speed" marine transportation.

The US Navy's 105 ton displacement surface effect ship *"SES-100B"* achieved a world speed record of 91.9 knots for "warships" on 30 June 1976. This was the pinnacle in the decade long US Navy pursuit of "100 knots at sea" that ended on 9 January 1980 with the cancelation of the frigate size 3000 ton displacement surface effect ship *"3KSES"* program without attaining that "100 knot" goal in a practical design. The end of this major thrust by the US Navy was not the result of the fuel crisis of the 1970s (a common misconception), it was due to unresolved technological solutions to complex issues in the design of such craft. This book takes a look at the history of the failures and many successes in the critical two decades of development (1960–1980) chasing the "*one hundred knots at sea*" capability; first as a means to document the history for later researchers to use appropriately, and secondly to identify some possible avenues to explore in this, as yet, unrealized dream. Along the way, many valuable offshoots of technology and designs produced significant findings in other mission areas and in many areas of hydrodynamics and aerodynamics basics for high speed marine craft design across a broad speed spectrum other than "100 knots". These are documented in the book. The book contains much previously unpublished original work on technical solutions, theories, analyses and test data.

The book has twelve self–contained chapters (with ample cross-referencing) for ease of documentation for both the historical record and the technical features in the development of high speed marine craft.

Chapter 1: *The Goal of One Hundred Knots.* This chapter covers the key developments that have been pursued on hydrofoils, hovercraft, and wing-in-ground-effect craft including the *ekranoplan* variants and outlines the successful technology features used in each of the operational high speed marine craft.

Chapter 2: *History of High Speed Ship Development.* More technical detail on each craft type and programmatic history of the key concepts under discussion is provided in this chapter.

Chapter 3: *The First Surface Effect Ship.* The history and technical design characteristics of the US Maritime Administration's program of the *Columbia* Surface Effect Ship Project is given in this chapter.

Chapter 4: *US Maritime Administration "Large Surface Effect Ship Program".* This chapter documents the entire program by MARAD for the nuclear ship *N.S. Savannah*; the hydrofoil *HS Denison* and the *Columbia* project to seek improvements in the US Merchant Marine. It documents a critical review by independent experts on the merits of the various candidates for high speed ships for ocean commerce.

Chapter 5: *History of US Navy "Large High Speed Surface Effect Ship" Program.* This chapter gives a detailed chronological history of all programmatic, political and technical achievements in the US Navy program to develop a high speed frigate size ship for military missions. It contains a postscript of the Navy's subsequent plans on high speed ships for ocean transport and combatants.

Chapter 6: *SES-100A and SES-100B Test Craft and the Three Thousand Ton SES.* The detailed characteristics of the two test craft, and their Trials data is provided in this chapter. It covers both performance and stability characteristics as well as subsystem descriptions. It also provides details on how these two craft contributed to the design of the 3,000 ton class frigate size surface effect ship (SES).

Chapter 7: *Economic Considerations.* How the economic requirements of a ship design are related to the technical characteristics is provided in this chapter.

Chapter 8: *Technical Considerations.* This chapter delves into the many technical characteristics of high speed marine craft. It provides detailed derivation of many of the key characteristics of high speed ships on both performance and stability parameters. It includes many experimental results from multiple sources for comparison with the developed theories. It provides design tools for determining optimum geometries for efficient cruise speeds under varying conditions. It treats both hydrodynamic aspects and aerodynamic aspects of the different hull forms applicable to high speed marine craft.

Chapter 9: *Navy Military Operations Considerations.* An important element in the design of high speed marine craft for military use is the need to incorporate features driven by military operations' needs. This chapter develops the military considerations for amphibious assault missions and for stealth missions and outlines guidelines for future designs.

Chapter 10: *The US Navy ANVCE Program.* This chapter documents the US Navy Advanced Navy Vehicles Concepts Evaluation (ANVCE) Project which was a major effort to evaluate many surface and air concepts for high speed missions. These concepts included air loiter aircraft, airships, seaplanes, wing-in-ground-effect craft including *ekranoplan* variants, monohull ships, planing hull craft (including supercritical planing hulls), SWATH ships, hovercraft, surface effect ships and hydrofoils. Much original technical work pertaining to these vehicles is documented and serves as a substantial data base for future work.

Chapter 11: *Aerodynamic Air Cushion Craft.* The chapter covers *basic wing-in-ground-effect (WIG)* craft; *channel flow* craft (a combination of hovercraft and

WIG) and *ekranoplan* craft. Detailed development of the key equations on performance and other characteristics are provided together with much experimental data obtained from multiple sources of wind tunnels and hydrodynamic tow tanks to form a reliable set of design tools.

Chapter 12: *Lessons Learned and Where to Next.* This chapter pulls together the key results from the earlier chapters and summarizes the current efforts to guide future developments of this fruitful area of high speed marine craft for many different uses, both commercially and militarily.

Acknowledgments

Sir Christopher Cockerell, of course, deserves especial acknowledgment with his seminal work in originating and developing both the amphibious and the sidehull forms of air cushion craft. Both forms of which have received significant development in the UK (country of origin) and in the US as well as elsewhere in the world. His sage advice is deeply acknowledged, especially as the author struggled with the later studies by the US Navy to seek the proper forms of high speed ships. Without the work by such stalwarts in the hovercraft industry initiated in Saunders-Roe on the Isle of Wight, England (later British Hovercraft Corporation) this book could not have been written.

All the folks at Saunders-Roe were pillars in the industry. Dick Stanton Jones (Chief Aerodynamicist and later Managing Director) is especially acknowledged together with Jack Lloyd (Dick's right arm); Bill Crago and Dave Perry of the Saunders-Roe Tow Tank; Ray Wheeler in the Stress Department (who later took over from Stanton-Jones as Chief Designer); Peter Crewe (Chief of the all-important Project Office where original ideas including Christopher Cockerell's hovercraft were nurtured in the company) and the many other engineers and designers at Saunders-Roe whose names are unfortunately forgotten in the passage of time but their help and nurturing of the author as an apprentice at Saunders-Roe in those early years is gratefully acknowledged.

Dr Scott Rethorst and all the engineers and technicians at Vehicle Research Corporation, Pasadena, California (Tor Strand, Toshio Fujita, Helge Norstrud, Dr Win Royce, and many others) deserve special thanks for originating and developing the *first* surface effect ship, the "*Columbia*" and the test craft "*VRC-1*" for the US Maritime Administration. The experience gained by the author in designing, building and as test pilot for the "*VRC-1*" proved invaluable in gaining insight into the sophisticated characteristics of wing-in-ground-effect vehicles (WIG) specifically and in high speed marine vehicles generally.

All of the 40 engineers who moved with the author from Buffalo, New York to New Orleans, Louisiana to design, build and test the world record breaking Surface Effect Ship, "*SES-100B*" deserve special thanks. The cultural shock of moving from Buffalo to New Orleans was offset by the enthusiasm and excitement of designing, developing and testing such a unique ship. The long and friendly association with Ed Butler, the liaison officer from the US Navy/MARAD SES Joint Project Office, while the author managed and directed the SES-100B Program, made many of the difficult decisions a pleasant experience.

During the US Navy ANVCE Project, continued discussions on key aspects of unique marine craft with such stalwarts as Baron Hanns von Schertel, Dott. Ing. Leopoldo Rodriquez, Peter Payne, Peter Crewe, Dr Alexander Lippisch, Sir Christopher Cockerell, and many other highly capable engineers and scientists will be forever appreciated. The team in the ANVCE Project Office: Capt Tom Meeks (Project Director); Cdr "Corky" Graham; Dave Gicking performed herculean tasks keeping such a wide ranging project under control; and their direction, guidance and support is deeply appreciated.

Especial acknowledgement is due to Bill Ellsworth at the US Navy David Taylor Research & Development Center (DTRDC) who sponsored the author's first two books on air cushion craft technology and design in 1975 and in 1980, that laid the ground work for this new book on high speed marine transportation and the pivotal subject of "100 knots at Sea".

1 The Goal of One Hundred Knots

"Those who cannot remember the past are condemned to repeat it"
George Santayana, (1863–1952)

This book has a primary focus. It is to document the history, with all its successes and failures, of the US Navy's efforts to achieve the "100 knot Navy". It includes the critical decade 1969–1979 when the Navy spent more than $650 million (and closer to $1 billion if all related technologies are included) to develop a 3,000 ton displacement (frigate size) ship capable of conducting navy missions at 100 knots. The Navy program was canceled on 9 January 1980 after failed technological development and the Navy turned its attention to ships with much lower speeds.

This book also has a second major focus which is to examine the various hydrodynamic and aerodynamic theories and substantiated data of various forms of marine craft designed for high speed other than just "100 knots". This fall-out from that intensive effort produced a wealth of data and information on innovative forms of high speed marine craft. This has been an important byproduct that is included and expanded upon in the various chapters of this book.

The average speed of naval fleets using conventional displacement ships is about 25 knots. The goal to *quadruple* that speed to 100 knots proved to be a "bridge too far". Immediately following the cancelation in 1980 of those intensive efforts, the Navy re-directed its attention to advanced design ships with speeds closer to 50 knots – a mere *doubling* of the current fleet speeds! But even that "goal" has not yet gained a foothold for either commercial or naval fleets in any sustainable manner. The Navy, after some sketchy and limited beginnings in 1965, initially in consort with the US Maritime Administration, had concluded that the best way of achieving a "100 knot Navy" was to select one form of *non-amphibious air cushion craft* out of several other advanced marine vehicle concepts available at that time to achieve such a capability. A significant amount of development, both before, during and after the decade in question was available on the various other available concepts – some of which might have been a better choice! Where necessary, reference will be made to these other developments to form a backdrop and experience base from which the US Navy efforts extracted the necessary engineering, operational and cost data to guide its efforts to achieve such an ambitious goal. This book is not another compilation of all the various vehicle types that have appeared on the scene, from early historical efforts dating back to the 18th century to today but rather, a focused effort on the Navy's choice of "100 knots" and what was needed to obtain that speed in *normally experienced sea states*, at *minimum risk* and in a *practical platform* capable of *conducting real military missions*. The reader is referred to the many historical summaries on advanced forms of *air cushion vehicles, surface effect ships, hydrofoils*, etc that already exist. They form a useful backdrop to this book. In particular, the reader is referred to the well documented general texts such as *Jane's Surface Skimmer Systems*, and *Jane's Surface Skimmers* and *Jane's High Speed Marine Transportation* from the first issue in 1967 to the present day issue (2009–2010) to become familiar with the wide range of advanced concept ships and craft from around the world for both military and commercial purposes. Some of the advanced concepts have been directed at specific purposes (for example, *amphibious hovercraft* for across the beach landings of Marines) and some directed for more multipurpose high speed vehicles missions. More in-depth technical summaries of these advanced concept ships may be found in the technical summaries by this author, *"A Technical Summary of Air Cushion Craft Development"* (1975)[1] and updated book, *"Air Cushion Craft Development"* (1980).[2]

Those two books contain numerous references to all the earlier developments around the world contributing to the knowledge base on *hovercraft* and *surface effect ships* including hitherto lost information on the emerging *aerodynamic versions* of the air cushion craft concept integrating the advantages of the air cushion to ships. Other authors have summarized their work on advanced naval vehicles in the *Naval Engineers Journal*, Special Edition, *"Modern Ships and Craft"*, (1985)[3] and various issues of *Jane's* (that are referred to in the text). Each of these technical summaries contains the results of various craft that have been built, tested, modified, and even rejected as the technology evolved. This documentation is an important backdrop for the US Navy's pursuit of a *one hundred knot, large tonnage ship for naval missions* and other variants and is recommended reading as adjuncts to this current book.

1.1 A Brief Outline of the Key Types of Advanced Marine Vehicles

As a snapshot of the types of "vehicles" that have formed the "set" under discussion, some typical examples of advanced marine craft are shown here. Some of these craft no longer operate and other types have replaced them but they serve here as representative of the types of craft that were under consideration in seeking a naval platform to achieve sustained operational speeds of 100 knots. There are always variations in designs in particular cases. These concepts and types will be discussed as necessary throughout the book, and they serve as a reference point for the main discussion. There have been and continue to be variants of the types of ships and craft depicted here in development and operation throughout the world but it helps to have a "base" for meaningful discussion. They are presented in no particular order but are provided with the generic names used throughout the book to capture the types of vehicles that were under consideration by the US Navy as it made its decision.

1.1.1 Fundamental Types of Hydrofoils

The **Rodriquez RHS 70** (Rodriquez Hydrofoil Ship) hydrofoil represents the simplest form of the *surface piercing foil* system which uses the basic principle of variable area stabilization. As the ship pitches or rolls, the foil gains or loses

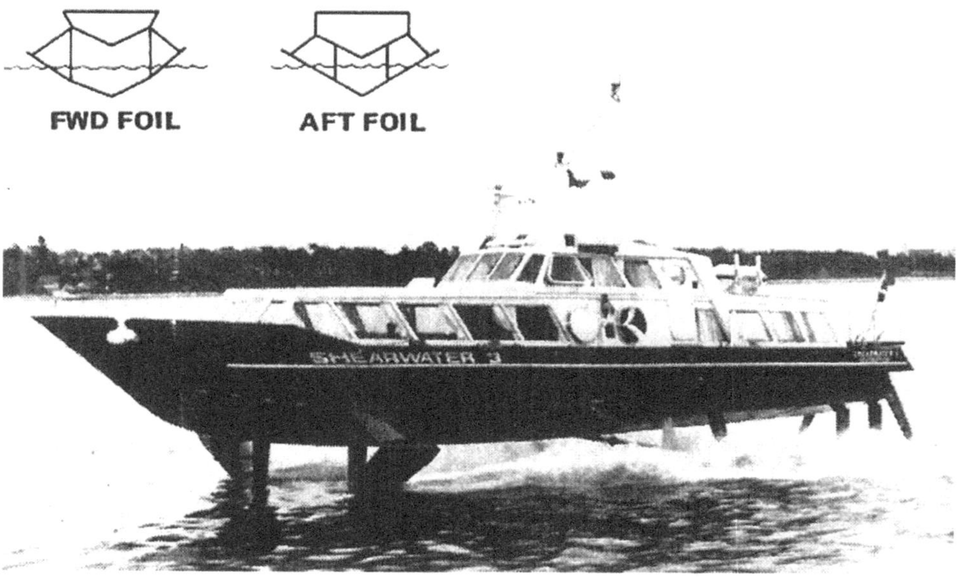

Figure 1.1: Surface Piercing Hydrofoil (1982)
Reprinted by permission of SNAME[a]

supporting foil area and provides the necessary restoring and stabilization force. During operation, the trim angle of the bow foil can be adjusted (within narrow limits) by an hydraulic actuator. The RHS 70 shown here (Figure 1.1) is the *Shearwater 3*, operated in regular passenger service by the Red Funnel Steamers Company between Southampton and Cowes, Isle of Wight, England during the period 1972–1982. It was an improved design over that previously used which had been the type of hydrofoil that had seen successful service between the Channel Islands and France in 1964.

It was replaced by later models of the *Shearwater* series up through 1999. Such craft were simple in design and relatively inexpensive both in procurement cost and in operating cost with low maintenance of their simple static foil systems. They operated at about 40 knots and were quite suited for the choppy waters between Southampton and the Isle of Wight. It cut in half the time usually taken by the existing standard displacement ship car ferries that operated over the same route.

The second type of surface piercing hydrofoil is that designed and built by the Supramar company in Lucerne, Switzerland. Figure 1.2 shows the Supramar PTS 150 Mk III, the *"Queen of the Waves"*.

This novel form of hydrofoil represents a combination of the *surface piercing foil* shown above in the Shearwater craft and the *fully submerged foil* system (see later). In this design, the bow foil is of the Schertel-Sachsenberg surface-piercing foil design and carries 60 per cent of the craft displacement. The rear foil is fully submerged, carries 40 percent of the craft displacement and is stabilized through use of the Schertel-Sachsenberg air stabilization system. The basic principle in this scheme recognizes that the subcavitating foil achieves the majority of its lift through

[a] See SNAME permission statement at end of chapter

Figure 1.2: Air Stabilized Surface Piercing Foil (1974)
Reprinted by permission of SNAME

the low pressure region on its upper surface and that if air is introduced into this low pressure region the lift is reduced. The *"Queen of the Waves"* typically operated at about 36 knots.

The Schertel-Sachsenberg air stabilization scheme comprises a series of ducts connecting the free atmosphere air to the foil upper surface through small air exit openings. As the craft heaves, pitches or rolls such that as the foils lifts up, a gyro-sensed valve opens the ducting to atmosphere, air is drawn into the ducting and into the low pressure region reducing lift and causing the foil (and the craft) to settle lower, whereupon the valve is closed and lift is regained. In this manner the craft motions are minimized and a smooth ride in rough water is achieved. This particular photograph shows the hydrofoil *"Queen of the Waves"* operating the Malmo-Copenhagen run between Sweden and Denmark in 1974.

The two main surface piercing hydrofoil companies of Rodriquez and Supramar characterize the type of *surface piercing hydrofoil* systems operated extensively in Europe and other parts of the world. Both Baron Hanns von Schertel (Supramar) and Leopoldo Rodriquez (Rodriquez Shipyards) were brought on board as consultants to the US Navy Advanced Naval Vehicles Concepts Evaluation (ANVCE) Project in 1976 to help in the evaluation of these types of hydrofoil craft as the US Navy evaluated high speed ships. See Chapter 2 and Chapter 10 for more discussion of these gentlemen and their pioneering work.

In the US, interest was centered on the *fully submerged foil* system. Figure 1.3 shows one basic form exemplified by the Grumman-Lockheed AG(EH)-1[b] or *"Plainview"* hydrofoil. The *Plainview* is the third example of hydrofoil form and has a *fully submerged foil* system of conventional airplane configuration stabilized by incidence control using an autopilot system. The (split) bow foils carry 90 per cent of the craft displacement while the aft foil carries 10 per cent. Foil lift variation, controlled automatically, is by incidence variation which controls both heave and roll of the craft. The single aft foil controls pitch.

The **AG(EH)-1 *Plainview*** was the largest hydrofoil in the world (at 367 short tons displacement) until it was sold to a private buyer in 1979, and is shown here operating in Puget Sound off Seattle for the US Navy at a speed of 50 knots. It was sold for "pennies on the dollar" ($128,000 sale price against an

[b] Auxiliary General Experimental Hydrofoil, AG(EH)

original procurement cost of $21 million!). It is now scrapped and laying in the yard in pieces in Chinook, Washington – a sad end for such a sleek craft.

Figure 1.3: US Navy Hydrofoil AG(EH)-1 *"Plainview"* (1969)
Reprinted by permission of SNAME

The insert in Figure 1.3 shows the forward foils and the aft foil, each of which retract (not shown here) for hullborne operation, which is of value for operation alongside existing docks throughout the world. Each tapered foil pivots about the quarter chord point to vary foil incidence and lift. No flaps are incorporated. This serves in simplifying the design to offset the complexities of a dynamic stabilization system essential to any fully submerged foil system.

The **Boeing *Jetfoil***, (Figure 1.4) along with its big sister, the Boeing NATO PHM (see photo, Chapter 2), represents the most recent development of the fully submerged foil system utilizing trailing edge flap control for automatic stabilization. The foil system is a canard arrangement, with approximately 32 per cent of the dynamic lift provided by the single bow foil, and the remaining 68 per cent by the single aft foil. The trailing edge flaps provide both automatic control in pitch, heave and roll while underway, and lift augmentation during take-off. These foils also retract for hullborne operation in restricted

Figure 1.4: Boeing *"JetFoil"* (1974)
Reprinted by permission of SNAME

waterways and alongside standard docks. The foils are simple constant chord, retractable foils.

The *Jetfoil* can operate at high speeds (around 45–50 knots) and has an interesting safety feature built in (as does the US Navy PHM) of relatively soft bolts in the upper struts such that in the event of a major collision, with say a surface floating log, the soft bolt will extrude through the opening allowing a "graceful" folding of the strut without structural damage as the craft settles into the water after the collision. Because of the relatively high cost of such dynamic craft, this is a serious concern to minimize the maintenance costs.

The *Jetfoil* used to operate on the waters between the Hawaiian Islands, but the peculiar swells in those waters were not favorable to the craft as it kept "falling off" the foils. The Jetfoil still operates today between Hong-Kong and Macao. An interesting cost item is that while the cost of such dynamically stabilized, fully submerged hydrofoil craft is high, the gambling casino operators had computed that any cost of the craft was more than offset by the *increased time* it gave gamblers at the tables with the corresponding increased monies earned by the casino!! Allegedly, gambling is the main source of income for Macao. Not the usual cost trade-off analysis embarked upon when determining the cost-effectiveness of any particular high speed ship!

The Soviet Union began development of the hydrofoil in 1945 at the Krasnoye Soromovo Shipbuilding Co., in Gorky, Russia. That work, under the leadership of its Chief Naval Architect Rostislav Alekseyev[c], concentrated on the known phenomenon (from Baron von Schertel's original work) of the immersion depth effect or *shallow draft foil* effect. This effect is essentially the inherent characteristic of a fully submerged foil that loses its lift as the foil approaches the surface.

After some earlier models, the ***Raketa*** (Rocket) shallow draft foil system was introduced in 1957 and is shown in Figure 1.5. This immersion depth or surface effect has been applied in several *Soromovo* craft. The system, as it has been applied to the ***Raketa*** comprises two main horizontally fully-submerged foils, one forward and one aft, with virtually no dihedral, and each carrying 50 per cent of the craft displacement and operating at low lift coefficients close to the surface (about one chord depth, submerged). The loss of lift as the foils approach the surface is the basic self-stabilizing feature of the foils and thus eliminates the need for automatic control systems for low Sea State operations. The low lift coefficients preclude generation of the necessary high lift for take-off and thus an auxiliary set of subfoils in the form of planing ski surfaces are added near the bow to generate this take-off lift. There were many other craft that followed the ***"Raketa"*** by the Soromovo company, based on the same principle. These included the ***"Meteor"*** *(1960)*, "***Sputnik***" (1961) and others.

The insert in Figure 1.5 illustrates the form of the submerged foil system that typically operates at a depth of one chord or less. The auxiliary subfoil which is more in the form of an arrowhead planing ski, is shown in position behind the bow foil. The

[c] See later and in Chapter 11 for description of the work by Rostislav Alekseyev (or Alekseeva) on *aerodynamic air cushion craft* and specifically those craft known as "*ekranoplan*"

Figure 1.5: Krasnoye Soromovo *"Raketa"* Shallow Draft Hydrofoil (1957)
Reprinted by permission of SNAME

Raketa is operating in England as well as seeing extensive river traffic usage on the Volga River in Russia since its debut in 1957. Typical speeds on the river are around 35–40 knots. The shallow draft foil system of Alekseyev functioned well in calm water but was not suitable for operation in any significant sea states or rough waters. Hence, the design was used almost exclusively for use on rivers which were in abundance throughout the country and provided an economic form of transport on inland waterways.

Over the years there have been many variants of foil systems from the basic types shown here. These include the development of *supercavitating foils* to provide a greater than 50 knot continuous operation capability, but these have not been proven in any sustainable operation. Other foil configurations include tandem foils, midship foils and *superventilating foils* to name but a few and each of these variants provide some unique feature in lowering cost or improved maintenance but all operate in the 30–50 knot speed range (see Chapter 10 for more discussion of the use of *supercavitating foil* systems to achieve about 70 knots) but none have approached a "100 knot" capability in any reliable form.

1.1.2 Fundamental Types of Air Cushion Craft

In addition to the hydrofoil concepts, the US Maritime Administration, and subsequently the US Navy, explored the possibility of the new form of craft that was just beginning to appear on the scene at that time which was the various forms of *air cushion supported craft*. The history of the development of such craft as it pertains to the "100 knot" goal is expanded on in more detail in Chapter 2 and just the basic forms are described here. As these novel forms of craft became established, it became clear that they settled into three basic or generic forms, which are:

1. Amphibious Air Cushion Craft
2. Non-Amphibious Air Cushion Craft
3. Aerodynamic Air Cushion Craft

Many of the different types have early beginnings and have disappeared only to appear again in slightly different forms and yet others have appeared on the scene and seen steady improvements as operational experience was gained. A brief summary of the main types follows.

Amphibious Air Cushion Craft Basic Form Figure 1.6 shows the basic forms of the *amphibious air cushion craft*. The pressurized air is forced beneath the craft by some means (usually a fan system of some type) that then lifts the craft. The efficiency of the pressure system varies with the mechanism to control the air flow into the "chamber" beneath the craft. In Figure 1.6, can be seen the "simple plenum" and the "simple peripheral jet". How the peripheral jet was invented is described in Chapter 2. Such a scheme however needed large amounts of lifting power to generate any appreciable ground clearance. The invention of flexible skirts (see later) allowed the craft to be raised to good ground and wave clearance (for the same power) as shown in the middle sketches of Figure 1.6. As the flexible skirt systems matured, most craft designs settled into forms of a modified plenum system with skirts as shown in the bottom sketch of Figure 1.6.

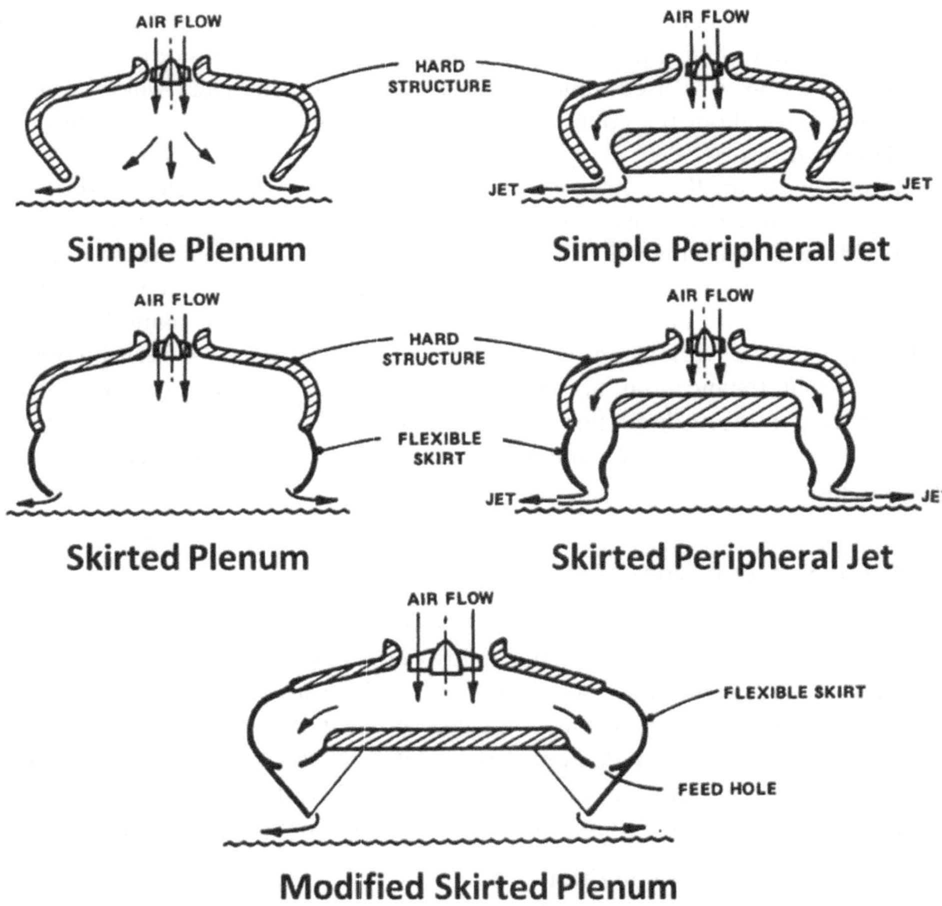

Figure 1.6: Amphibious Air Cushion Craft Basic Forms

Figure 1.7: British Hovercraft Corporation, "*SRN.4 Mountbatten*" (1969)
Reprinted by permission of SNAME

The first practical *hovercraft* (using the name given to it by its inventor Christopher Cockerell) made its maiden voyage across the English Channel on 25 July 1959 (on the 50th anniversary of Bleriot's first flight across the same waters in 1909). This was the hovercraft "*SRN.1*" which is discussed in more detail in Chapter 2. After ten years of successful development of such air cushion craft in progressive increases in size, speed and operational ruggedness, the largest hovercraft in the world, the "*SRN.4*" was launched, a short ten years later, in 1969. Figure 1.7 shows the ***"SRN.4 Mountbatten"*** on one of its regular runs across the English Channel and is a good example of this amphibious form of air cushion craft.

Most of the key features of such craft are seen in the insert in Figure 1.7. A fan system pressurizes the drawn-in ambient air and expels that pressurized air underneath the craft to lift it above the surface. In this amphibious form, the pressurized air cushion is contained by a peripheral flexible skirt (of the type shown in Figure 1.6 with the modified skirted plenum skirt) that "gives" as the craft moves along over waves at sea or over obstacles on land. In the case of the "SRN.4" (and other air cushion craft) the lift system uses centrifugal fans while the generic sketches in Figure 1.6 show an axial fan. Such fan design variations do not change the basic form of such craft. The "*SRN.4*" could achieve operational speeds in relatively calm seas of 70 knots. The references already cited above provide details of all the many forms of the *amphibious hovercraft* but the essential features are as embodied in the "*SRN.4*".

Non-Amphibious Air Cushion Craft Basic Form The non-amphibious form of air cushion craft has its beginnings much before the advent of the amphibious form as it began life as a means to pump air under existing boat forms to either lubricate the hull (and reduce the friction resistance from the water) or to lift the boat clear of the water altogether. The basic forms that have been considered are shown in Figure 1.8.

The simplest form is shown in Figure 1.8 as in the upper sketch (A) called the *hydrokeel*. It has its origins in antiquity as various inventors have pumped air under their boats to reduce resistance. It is actually a fundamental form that has appeared in such craft as Von Thomamhul's boat (see Chapter 2); in the 1962 US Navy's *Hydrokeel* landing craft (LCVP(K); and is even a basic underpinning of the Russian *Ekranoplan* set of *aerodynamic air cushion craft* discussed later. The second main type of the non-amphibious air cushion craft is shown in sketch (B) in Figure 1.8 and is called the *captured air bubble* form. This form has its "reduction to practice" beginnings in 1929 when Douglas Warner invented such a scheme to give his boat a high speed capability sufficient for him to win the boat races on Lake Compounce, Connecticut of that year. More details of Warner's *captured air bubble* boat are provided in Chapter 2.

The third basic form is as shown in sketch (C) of Figure 1.8. It is this form that has seen the greatest success in the US Navy's quest to achieve "100 knots at sea" and the two forms shown as sketch (B) and (C) will receive more in depth analysis in the following chapters to seek the best hull forms.

The first practical craft in recent history using this concept of the *non-amphibious air cushion craft* was based on the design by Christopher Cockerell

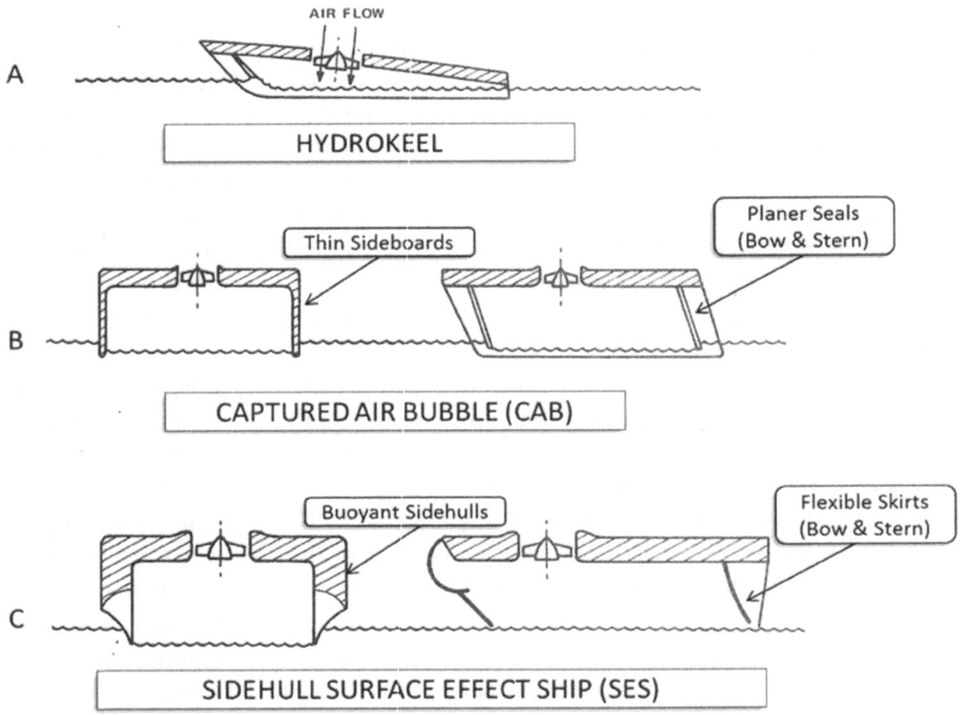

Figure 1.8: Non-Amphibious Air Cushion Craft Basic Forms

Figure 1.9: Hovermarine *"HM.2"* (1967)
Reprinted by permission of SNAME

(see Chapter 2) and was the Denny Company's *"D.1"* that entered service in 1960 in Scotland. Chapter 2 provides more details on this first craft and its role in the development of such craft. The Chief Engineer working for Christopher Cockerell on the Denny craft was Edward (Ted) Gunston Tattersall. Later, Ted Tattersall formed Hovermarine Ltd in September 1965 in Southampton, Hants, England to develop the *"HM.2"* series of craft based on these early developments of the Denny *"D.1"*. The *"HM.2"* entered service in 1967 and is shown in Figure 1.9 to illustrate the basic features that form the prototype configuration of such concepts that influenced the designs of the successful *sidehull surface effect ship* designs of the US Navy discussed later. As in the amphibious form of air cushion craft, the pressurized air flow lifting the craft is generated by simple passive fan(s) shown schematically in Figure 1.9, feeding the cushion directly. The air leakage is minimized by the sealing action of the sidehulls in the water which typically support 5–10 percent of the ship's displacement.

The advantage of this concept is that the air cushion support power is lessened over its amphibious sister described above and also that the various types of available marine propulsors could now be used. The concept loses its ability to come up onto land of course, but from the perspective of "large tonnage ships" sought by the US Navy, this was not a problem. The restriction to waterborne operation with the increased water resistance of the sidehulls meant that the speeds of such craft were lowered over that expected from the amphibious air cushion craft. The *Hovermarine HM.2* operated at speeds of approximately 35–40 knots.

As the technology advanced and the US Navy sought higher speeds, new technologies over those used in the simple passive concepts embodied into the *"Hovermarine HM.2"* had to be incorporated. This then led to an additional problem of "ride quality" while operating at very high speeds over "bumpy" water. There was a need to reduce the high acceleration levels that would be unacceptable to crew, passengers, cargo and equipment alike that came from operating at high speed over waves. Also, such high speeds (to obtain the 100 knot goal) also dictated that the sidehulls had to be treated differently. They had to be thin to minimize water resistance yet thick enough to provide sufficient hydrodynamic forces to maintain transverse stability of such high metacentric height craft. There were many other unknowns related to supercavitating flow associated with high speed that also had to be addressed before such concepts could be considered to be an acceptable platform for Navy missions.

A significant milestone along the way to develop the "100 knot" capability for the US Navy occurred in 1972 with the development of the two 100 ton displacement test craft of the *non-amphibious air cushion craft* concept. Because of the uncertainty of the technologies mentioned above to achieve this speed, two radically different designs were contracted for to explore and seek the best solution. The first was to explore the possibilities of the *captured air bubble* form (sketch B in Figure 1.8) where the design used (a) minimal buoyancy sidewalls or "thin sideboards" and (b) mechanical rigid planer seals. This was embodied in the Aerojet General *"SES-100A"* (see Figure 1.10). The second was to explore the possibilities of the *sidehull surface effect ship* form (see sketch C in Figure 1.8) where the design used (a) thicker buoyant sidehulls for stability and (b) flexible skirts as developed on the amphibious air cushion craft. This was embodied in the Bell Aerospace *"SES 100B"* (see Figure 1.11).

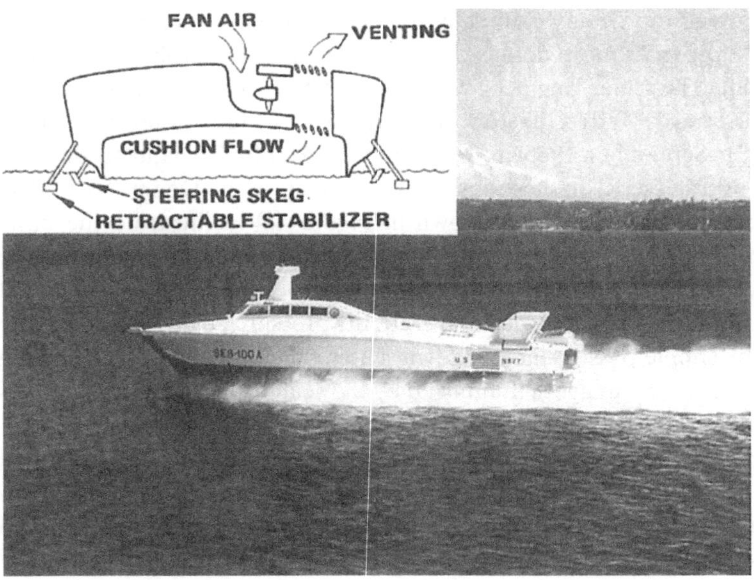

Figure 1.10: Aerojet General *"SES-100A"* (1972)
Reprinted by permission of SNAME

Also, because it was unclear as to which type of propulsion system should be used; the *"SES-100A"* used fully submerged waterjets while the *"SES-100B"* used semi-submerged marine propellers.

Figure 1.11: Bell Aerospace *"SES-100B"* (1972)
Reprinted by permission of SNAME

The features of the "SES-100A" were that the sidehulls were to be narrow (to reflect the CAB principle) with the transverse stability to be obtained through retractable stabilizers and steering skegs (see insert to Figure 1.10). The ride quality was to have been obtained through the use of a lift fan system made up of an axial flow fan together with active venting activated by pressure sensing. The seals at the bow and stern were to be of a mechanical set of "rigid planer seals" as called for in the CAB patents.

This latter feature was a preferred feature by the US Navy to avoid use of the British patented flexible skirt system. The propulsion system was a fully submerged pod inlet waterjet. This was later retrofitted to a flush waterjet inlet in its final form because of technical limitations of the fully submerged system.

The Bell Aerospace *"SES-100B"* design took on a more conservative approach where the key features were hydrodynamically shaped sidehulls that were narrow but of sufficient width and deadrise to provide the necessary hydrodynamic restoring forces without resort to appendages like the *"SES-100A"*. The "seals" at the bow and stern were the flexible skirt designs used successfully in all the operational British hovercraft and sidehull air cushion craft described earlier.

The unique feature that allowed the SES-100B craft to meet the high speed record of 91.9 knots was the surface piercing supercavitating controllable pitch marine propellers. The ride quality was obtained by the use of centrifugal fans and a passive venting system. Figure 1.11 shows the SES-100B operating near Panama City, Florida at 50 knots in a slight chop. The simple venting system for improved ride quality can be seen in the sides of the ship (in sketch and photo).

Chapters 6, 8 and 10 provide an in-depth technical evaluation of all the different unique features developed on both the *"SES-100A"* and the *"SES-100B"* and how they influenced the final designs considered by the US Navy for its planned "3000 ton, ocean going 100 knot frigate".

Aerodynamic Air Cushion Craft Basic Forms The third basic form of air cushion craft, namely the *aerodynamic air cushion craft*, has not received the in-depth or systematic development that the other two forms have received. This is unfortunate in that such aerodynamic forms offer the greatest possibility for sustained high speed over water especially if "100 knots" is the objective over anything other than dead calm seas.

The phenomenon of "ground effect" has been a welcome feature to aircraft pilots since the first days of the Wright Brothers development of airplanes in 1903. All aircraft pilots in their basic training learn how to conduct, first *"normal take-off & landings"*, then they migrate in their training to *"short field take-off & landings"* and *"soft field take-off & landings"*. It is the *"soft field take-off & landing"* where the pilot, at the end of his final approach, will flare out the aircraft and gently adjusts the angle of attack, the flap angle and throttle causing the aircraft to settle on to the "air cushion" forming between the ground and the aircraft so that he can "float" down the soft airfield until the speed has slowed sufficiently that he can gently lower the aircraft wheels on to the soft soil or snow or slush. The pilots of seaplanes enjoy the same characteristic as they gently bring their floatplanes into a "landing" on the surface of the water. Birds use a similar approach in their landings. As is discussed later on the various crashes of *aerodynamic air cushion craft*, the gentle contact with the waves is to be preferred to any sudden contact at high speed with a medium that is 840 times as dense as the air that is supporting the craft!!

This "air cushion" ground effect is caused by two basic aerodynamic characteristics of wings as they approach the surface, they are:

(1) ***Lift Increase***

The first characteristic is the increase in the bound circulation as the wing approaches the surface that distorts the streamlines. The stagnation point moves aft on the under surface of the wing and the resulting increased curvature of the streamlines gives increased circulation and increased lift.

(2) ***Drag Decrease***

The second gain is that due to the partial destruction of the wing tip vortices as they interact with the ground (or wave surface). This feature reduces the induced drag of the wing and of the aircraft.

Both these features result in not only an increase in lift but an increase in the *lift-drag ratio (L/D)* of the aircraft. It is this latter gain that has attracted pilots and designers in the design and operation of both conventional aircraft and the aerodynamic air cushion craft. While the designer of *conventional aircraft* uses these beneficial characteristics of "ground effect'" at the two ends of the flight (both take-off and landing); the designer of *aerodynamic air cushion craft* seeks to take advantage of the "ground effect" throughout the entire duration of the "flight" for both economic and military advantage.

The most quoted early example of flight in "ground effect" is the transatlantic flight of the largest "flying boat" (seaplane) at the time in 1931 of the Dornier DO-X

Figure 1.12: Dornier Seaplane "DO-X" on Lake Constance (1929)

aircraft (Figure 1.12). It made the crossing while flying just a few feet above the surface of the waves to take advantage of the gain in lift-drag ratio and thus gain in range in order to make the crossing.

This was more of a necessity than any demonstration of "ground effect" as the aircraft was known to be a "gas guzzler" with its twelve engines and any gain in range capability was welcomed as it made the transatlantic crossing. While the Do-X had a chequered life and was eventually broken up in October 1934, it did generate interest in such phenomenon as ground effect. The remaining pieces of the Do-X were on display in the Deutsche Luftfahrtsammlung (German Aviation Collection) in Berlin until the museum's destruction in WW II. Other known examples of pilots taking advantage of ground effect were during WW II where bomber pilots returning from bombing raids over Europe and running low on fuel flew across the English Channel to their bases in England just a few feet above the waves, again to gain every last inch of range from the increased lift-drag ratio (L/D) gained from flying in "ground effect".

As *aerodynamic air cushion craft* designers sought to use this phenomenon to advantage it was found that four basic forms or Categories emerged for such craft. Figure 1.13 shows the four basic forms of aerodynamic air cushion craft that have received development of varying degrees by several nations over the past four decades. Many examples exist of these variations on taking advantage of ground effect. Some designs are simply wings that fly close to the ground (water surface) for improved lift-drag ratio (L/D) and other designs augment the aerodynamic lift through various techniques such as the ram wing (in various design variants) and some designs exploit both the lift increase and the drag decrease characteristics. While there are design variations in all cases, the four basic Categories are described below.

CATEGORY A: WING-IN-GROUND-EFFECT. The first type shown in sketch (A) of Figure 1.13 is the basic *"wing-in-ground-effect"* version of the aerodynamic air cushion craft. It has its origins in the known increase in lift experienced by all

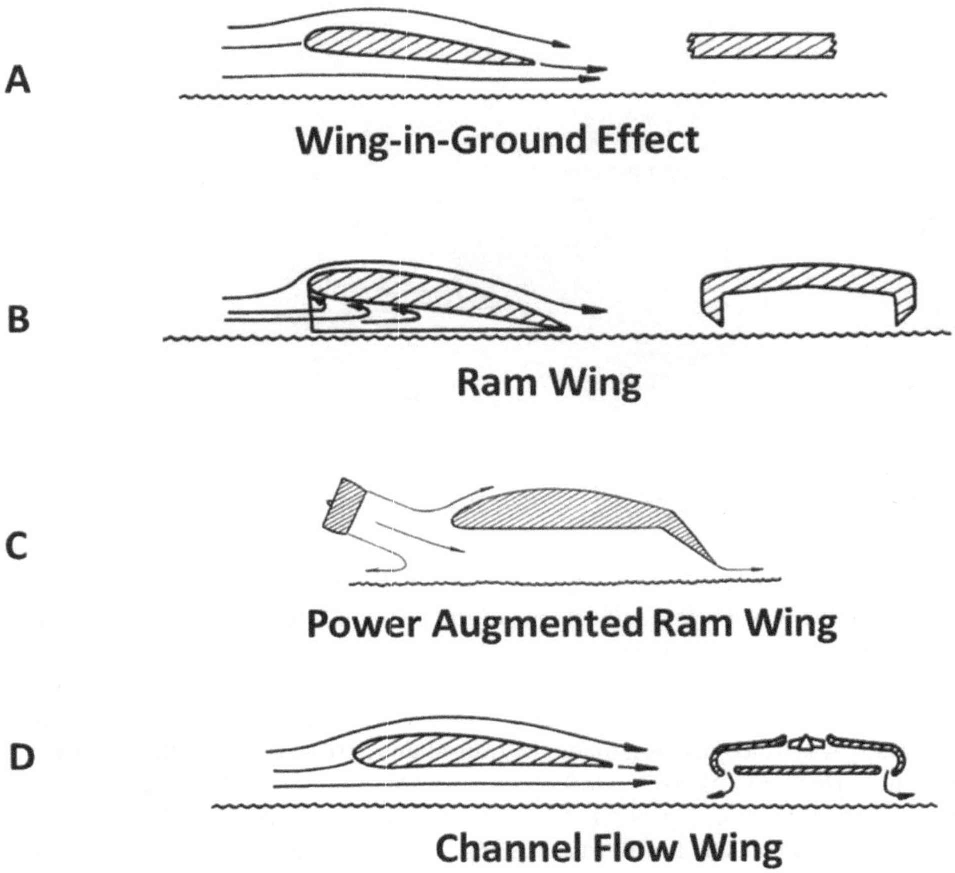

Figure 1.13: Aerodynamic Air Cushion Craft Basic Forms

wings as they approach the ground (or water surface) together with a decrease in (induced) drag as outlined above. Standard texts on aerodynamic design provide details on how the air flow is distorted by the wing and any ground or surface near the wing.

In the 1960s, the German aeronautical engineer Alexander Lippisch (famous for his original work on low aspect ratio wings) introduced practical examples of craft designed specifically to fly in ground effect (as distinct from a conventional aircraft like the Do-X). Dr Lippisch (see Figure 1.14) was a pioneer of aerodynamics and was known for his understanding of flying wings and of ground effect. He designed a series of delta wing gliders and rocket powered aircraft. He joined the Messerschmidtt company in Munich, Germany in 1939 and was responsible for such pioneering efforts as the Me 163 *Komet*.

After WW II, Dr Lippisch emigrated to the US and joined Collins Radio Company (1950–1964) working on various projects including several wing-in-ground-effect craft. Later, in 1966, he formed his own company, the Lippisch Research Corporation to develop various forms of *aerodynamic air cushion craft* using the delta wing as a fundamental platform. One of the known problems with wings flying in ground effect is that the center of lift is further aft than when flying out of ground effect. This has the unfortunate characteristic of incurring a destabilizing lift force and moment as the craft pitches up or climbs up out of ground effect. This is the basic reason why all the wing-in-ground-effect aircraft produced (the Russian *ekranoplan* shown later is a good example) have been built with large tailplanes. Dr Lippisch, found that if the delta wing is used "in reverse", that is with the apex of the delta facing aft, this problem was minimized. There are other ways to control the center of lift as discussed in Chapter 11 but the Lippisch designs all have this "reversed delta planform" characteristic.

Figure 1.14: Dr Alexander Lippisch (1894–1976)
Photo permission Special Collections Department, Iowa State University Library

Figure 1.15 illustrates an example of the principle of the *"wing-in-ground-effect air cushion craft" (WIG)*. The particular WIG shown is the "X-113Am" shown flying in 1971. It is a follow-on to earlier Lippisch designs, the *"X-112"* in 1963. A later version was the X-114 which first flew in 1977, which had a gross weight of 1.65 tons and a cruise speed of 80 knots. Shown in Figure 1.16 is Lippisch's design for a 300 ton craft *("X-300")*. The winglets in both the actual *"X-113Am"* and in the *"X-300"* sketch

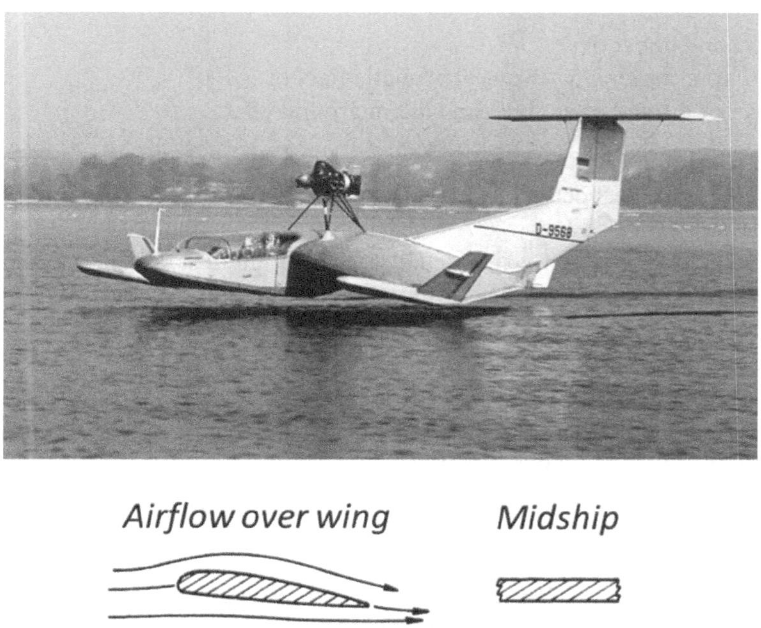

Figure 1.15: Lippisch "X-113Am" WIG in Flight (1971)
Photo permission from Special Collections Department, Iowa State University Library

Figure 1.16: Lippisch "X-300" Design
Permission from Special Collections Department, Iowa State University Library

provide increased span and a mechanism for decreasing the induced drag (see Chapter 11 for span effects). In the basic *"wing-in-ground effect"* version of such aerodynamic air cushion craft, there are different variations that are pursued by different designers.

Regulatory Categories of Aerodynamic Air Cushion Craft Without changing the concept, the WIG craft could be designed to:

(a) always fly in ground effect or
(b) have the ability to occasionally fly out of ground effect or
(c) normally fly out of ground effect but can land in ground effect.

The International Maritime Organization (IMO) have classified such variations as **Type A, Type B** and **Type C** respectively and these design distinctions are covered in more detail in Chapter 11: Aerodynamic Air Cushion Craft. Such variations influence whether the vehicle is treated (from a regulatory viewpoint of safety) as either a "ship" or an "aircraft". **Type C** versions are more in the category of conventional aircraft with assisted air cushion landing (and take-off) features. **Type A** and **Type B** are treated under maritime rules.

The Lippisch designs were primarily of the IMO **Type A** categorization, with the occasional ability to fly out of ground effect in some designs for maneuverability reasons such as to turn without contacting the waves (see Chapter 11 on turning issues).

CATEGORY B: RAM WING. The second classification of "aerodynamic air cushion craft" is the "ram wing" which was shown illustratively in Figure 1.13 (middle sketch B). Not many such craft exist but Douglas Warner was prolific in developing the "ram wing" type and variants. Figure 1.17 shows one model of Warner's ram wing that he demonstrated in the early 1940s and patented (US Patent 2390859) on 11 December 1945[4].

Photograph of Warner's RAM WING Flying Model (1945)

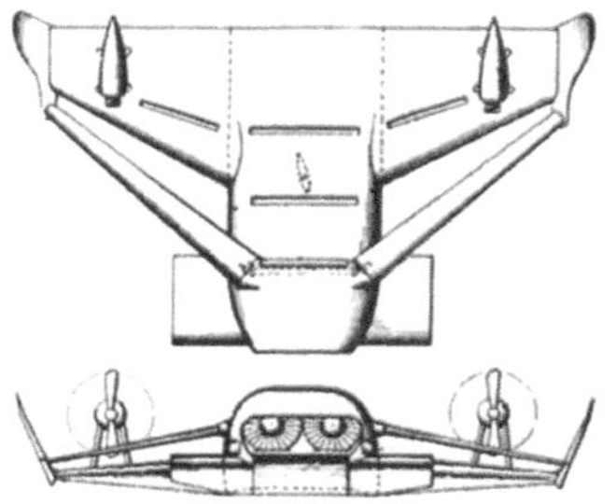

Drawing of RAM WING from Warner Patent (1945)

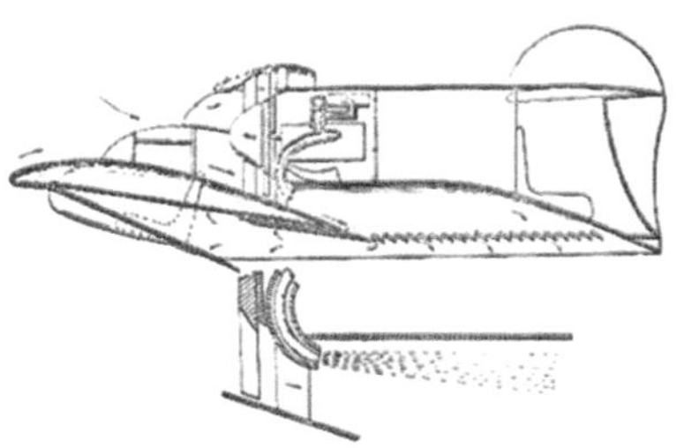

Drawing from Warner Patent showing Air Injection Beneath Main Body

Figure 1.17: Douglas Warner's Ram Wing-in-Ground-Effect (1945)

An examination of Warner's Ram Wing in Figure 1.17 will show that he had included many features that anticipated solutions to problems that have occurred in later wing-in-ground-effect vehicles. The design shows wing tip winglets to further reduce the induced drag. The engines are buried and back from the salt spray environment in any sea application. The ram air is directed beneath the wing to provide a hovering (or at least an air lubrication) capability. The design also includes forward stub wings mounted at a high angle of attack to facilitate planing and take-off assistance to alleviate hump drag problems. Clearly, Warner had both anticipated and experienced the characteristics of wings flying close to the water with all the sea's unforgiving features.

There is an overlap between the pure ram wing and one where the "ram effect" increasing the air pressure beneath the wing is augmented by an on board power source. Warner's Ram Wing included this extra *air injection* effect (Figure 1.17). This leads to the third category of aerodynamic air cushion craft which has become known (in the US) as the Power Augmented Ram Wing-in-Ground-Effect.

CATEGORY C: POWER AUGMENTED RAM WING-IN-GROUND-EFFECT. The third classification of *aerodynamic air cushion craft* from Figure 1.13 (sketch C) is a relative newcomer and is exemplified by the Russian *"ekranoplan"* series. The Russian word *"ekranoplan"* loosely translates as "screen plane[e]" which unfortunately does not capture the distinctive feature of the concept since it refers more to a generic wing-in-ground-effect rather than the unique feature to augment the lift with a "ram effect" by air injection. The many confusing names used for these various concepts are found to be a common problem in sorting out the difference between true "concepts" and the many "design variants" of the same concept that will appear often in the discussion. This particular form of *aerodynamic air cushion craft* is the product of the Soviet Central Hydrofoil Design Bureau (CHDB) led by Chief Designer Rostislav E. Alekseyev[f], at the time, mentioned earlier in the design of the *shallow draft hydrofoil* design concept that was used extensively throughout the FSU on rivers and waterways. Figure 1.18 is a photo of Rostislav Alekseyev taken in the 1950s when he was involved in the Soromovo Shipyards shallow draft hydrofoils of his design.

Figure 1.18: Rostislav Alekseyev (1916–1980)
From Hovering Craft & Hydrofoil, 1964[d]

In the 1960s Alekseyev directed his attention to another form of high speed marine vehicle, this time using a form of the wing-in-ground-effect principle discussed above. The unique feature of the Alekseyev design was the incorporation of *air injection* beneath the wings using the jet efflux from forward mounted engines (a variant of the Warner ram wing approach).

Figure 1.19 and Figure 1.20 show two of the more recognized craft that appeared on the scene in the 1960s and 1980s. Figure 1.19 shows the earliest craft called the Caspian Sea Monster by the US because it was spotted by intelligence sources operating in the Caspian Sea. The large size led the US Intelligence to refer to it as "monster" alluding more to its size than any other reason. The Russian name is simply "KM" (*корáбль моДéль*) (*korabi maket*) or "ship prototype".

[d] *Hovering Craft & Hydrofoil:* The International Review of Air Cushion Vehicles and Hydrofoils, published by Kalerghi Publications from 1961 to 1979

[e] The Russian word *ekranoplan* has its origins in the French word *écran* meaning "screen"

[f] Alternatively spelled *Alekseev* and *Alekseeva* in the Russian literature

Figure 1.19: "*Caspian Sea Monster*" or "KM" (1966)

Figure 1.20: Ekranoplan "*Lun*" with Missiles (1989)

The US tends to use the words "power augmented" for this type of ram wing. The Russians use the descriptor "air injection" which relates more directly to the feature that gives the craft a capability, like that in the Warner concept, to provide a take-off ability before the benefits of flying in ground effect can be fully realized.

The basic principle is that the "stubby" wing, chosen to minimize the footprint of such a large aircraft, gets its lift force augmented by the efflux from the forward mounted engines. In the case of the Caspian Sea Monster, the craft is approximately 540 tons weight and eight VD-7 turbojets mounted forward to direct their efflux under the wing for augmented lift and two more engines mounted on the tail fin for cruise power. The maximum speed is advertised as 270 knots (and cruise at 230 knots) in ground effect – certainly more than the "100 knot" goal sought by the US Navy.

Figure 1.20 is the *"Lun"* (Hen Harrier or Loon) (NATO name "Utka") and is a missile carrier version of the same concept as the Caspian Sea Monster. It has a take-off weight of 400 tons and an advertised maximum speed of 300 knots and a cruise speed of approximately 250 knots in ground effect. It uses eight NK-87 turbofans (each 28,660 lb thrust) for both lift and propulsion. The (six) missile launchers for the anti-ship missiles can be seen on the upper body of the craft.

The Russian *ekranoplan* designs were essentially waterborne vehicles except that in certain cases they were given an "amphibious" capability by means that has not been explained completely by the craft designers. There are videos (on the web) that show the Caspian Sea Monster leaving the water and briefly coming up the beach amid great amounts of spray and sand being dispersed. These photos appear to be more of a publicity stunt than any reliable mission capability. The wear on the structure and the ingestion in the engines would be brutal.

There is evidence however, that in 1972 a new version of the *ekranoplan*, the *A.90.150* or *Orlyonok* (Eaglet), was built with an amphibious capability deliberately designed into the craft. Details of how the amphibious capability is attained is not provided by the designers but Jane's[5] notes that in smaller craft (*Volga-2*) designed by the same Bureau, a soft balloon-like structure was integrated into the lower hull to give it an amphibious capability.

There is no question that some form of "amphibious" capability was built into the *Orlyonok* as seen in Figure 1.21 where the craft is seen leaving the hard surface ramp and entering the water. While not visible in most photos of the *Orlyonok*, it is believed that it was fitted with a retractable "landing gear" of 10 aircraft type wheels with shock absorbing struts and a retractable nose gear to allow it to traverse solid ramps such as seen in Figure 1.21 but would have limited capability on soft sandy beaches. Hence, the term "amphibious" for the Russian *ekranoplan* craft must be viewed in a generous light when compared to, say, *amphibious air cushion craft* (discussed earlier) that have the capability to operate overland with large variations in terrain and landscapes as well as operate at sea.

Figure 1.21: *"Orlyonok"* leaving ramp "on cushion" (c. 1990)

The ability of the *ekranoplan* designs to negotiate

Figure 1.22: *"Orlyonok"* on beach demonstrating its cargo unloading capability (c. 1990)

terrain is limited to small variations of less than six inches or so in height and limited to a surface nature not prone to inflicting wear on the craft. The photo of the *Orlyonok* coming down the ramp (Figure 1.21) illustrates this limited "amphibious" feature. A demonstration of the *Orlyonok* unloading cargo on the beach can be seen in Figure 1.22 taken sometime in the mid-1990s. Again, the craft is shown unloading on a smooth sandy beach just a few feet from the water. If such craft were planned to be used for amphibious landings of marines on enemy coastline, the choices of landing spots would be quite limited (see Chapter 9 for more discussion).

The *Orlyonok* has a take-off weight of 125 tons and a cruise speed of 216 knots in ground effect. It uses two NK-87 turbofans (each 28,660 lb thrust) for augmented lift and one Kuznetsov NK-12 (11,000 kW) turboprop for sustained cruise. In its passenger version, the *Orlyonok* could carry 100–150 passengers (single deck version) or 350 passengers (twin deck version) over a range of 1,080 n. miles. A cargo version was designed to carry 30 tons of cargo over 540 n. miles. Five craft were built with the first flight in 1972. One crashed in 1975 and one crashed in 1992. The remaining three were in service by the Soviet Navy until 1993.

The major development that was enjoyed by the *ekranoplan* designers was severely hampered with the collapse of the Soviet Union and later attention has been turned, by many groups, to applying the same principles to much smaller craft and mainly in the recreational and small airplane community. The *Aquaglide-5*, for example, is typical of this new thrust in the use of "PAR WIG" technology. The *Aquaglide-5*, launched in 1995, is a private company designed and built small 2.4 ton, **Type A** *ekranoplan*, designed for use as a "five passenger marine recreational passenger" craft. According to its manufacturer, the Arctic Trade and

Transport Company (ATTK), Moscow, Russia it has a crew of 1 (pilot) and 4 passengers. The cruising speed is advertised as 81–92 knots over relatively calm water. It is powered by a Mercedes-Benz V8 gasoline engine driving two 4½ foot diameter air propellers that direct their airstream under the wings in the fashion discussed for the *ekranoplan* design of aerodynamic air cushion craft. Most of the variations of these "derivatives" of the Russian *ekranoplan* design that appear in such countries as China, US, Germany and Japan and other countries are similar to this craft. Only design sketches appear in the literature for any return to much larger sizes. The size limitation and other characteristics of the PAR WIG designs are given in Chapter 11.

CATEGORY D: CHANNEL FLOW WING-IN-GROUND-EFFECT. The fourth and final category of aerodynamic air cushion craft is as shown in Figure 1.12 (sketch D) which is the multi-mode *channel flow air cushion craft*. This was the first type to be explored by the US Maritime Administration (MARAD) in its exploration of a "100 knots at sea" capability. Figure 1.23 shows the test craft *"VRC-1"* piloted by the author under test at Edwards Air Force Base, California in 1964. Its maiden flight was on 17 March 1964.

This version of this *aerodynamic air cushion craft* takes advantage of two basic forms: (a) the "conventional" amphibious form exemplified in the "hovercraft" designs pioneered by Christopher Cockerell and (b) the wing-in-ground effect craft. The "*VRC-1*" test craft (a man carrying model of the *"Columbia"* project, see Chapter 3) is shown operating in full peripheral jet mode across the Rogers Dry Lake, California. The inserts show the two main modes of operation, which are: (a) "hover mode" with full peripheral jets operating to generate the lifting air

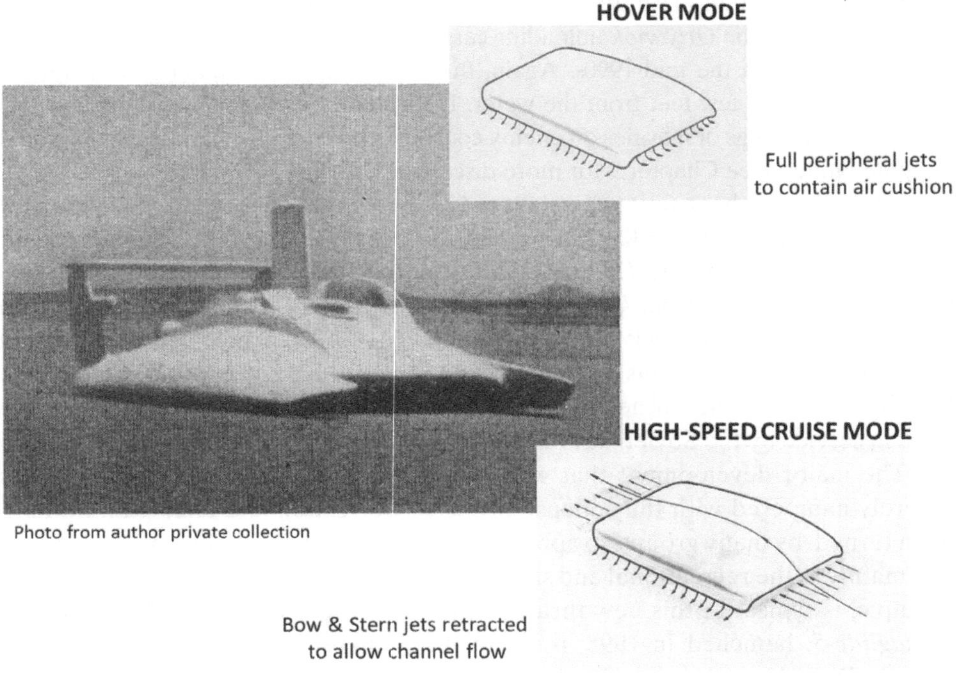

Figure 1.23: Multi-Mode *"VRC-1"* (Under Test at Edwards Air Force Base, 1964)

Figure 1.24: MARAD *"Columbia"* Channel Flow Craft (1960)
Art by VRC artist J. Bryant

cushion and (b) the cruise mode where the bow and stern skirts (or seals) have been retracted allowing the air flow to channel through and be sealed on the sides by the remaining peripheral jets to reduce induced drag. The craft can also operate in an intermediate mode of "ram wing" by retracting just the bow seal and leaving the stern seal operational. No hard structure is in contact with the land or sea during operation.

One problem with any "wing-in-ground effect" designs is the limitation in turn capability because of the close proximity of the wing to the surface of the water and waves. This problem is eliminated by the channel flow design by the addition of a dorsal fin (shown in Figure 1.23) that can generate the required turning force without the need to "bank" the wing.

For clarity, in the inserts of Figure 1.23, the means to generate lift (lift fan systems or equivalent) and the propulsion mechanism (propellers, jets, etc) have been omitted. In the case of the *"VRC-1"* shown, centrifugal fans provided the lift, and air propellers provided the forward thrust. The *"VRC-1"* was a man carrying test craft of the MARAD ship "*Columbia*" designed to provide economic transportation at 100+ Knots. Chapter 3 provides a technical description of the *"VRC-1"* and its characteristics.

A sketch of the proposed channel flow craft *"Columbia"* is shown in Figure 1.24. It was designed to carry 150 passengers (or 40 tons freight) over distances of 500 nautical miles at a low direct operating cost to fill a particular "gap" in the transportation spectrum. Its design speed was 150 knots. A technical description of the *"Columbia"* is provided in Chapter 3. The economic considerations by MARAD that led to the choice of the 100 knot "*Columbia*" design are provided in Chapter 7.

1.2 Main Marine Vehicles considered by the US Navy for High Speed

While the set of *hydrofoil, amphibious hovercraft* and *sidehull surface effect ships* were the main contenders considered by the US Navy in its quest for a "100 knot

Navy", there were other serious candidates that were available throughout the world at that time in various stages of development but received only sporadic treatment, which in retrospect was a missed opportunity in the quest for "100 knots at sea". Some of these concepts especially the *aerodynamically supported air cushion craft* summarized here provide a basic reference point for the technical treatments that occur in later chapters. Other forms such as *advanced monohulls, planing craft, supercritical planing hull* and *small waterplane area twin hull (SWATH)* were also considered by the US Navy but these were more in the lines of incorporating other attractive features and not just speed as it was expected that none of these hull forms would be able to attain speeds of 100 knots without an exorbitant expenditure of power.

Discussion of all these types is included throughout the book where necessary to illustrate a point but the main concentration is on achieving the objective of "100 knots at sea" in an economic and (in particular) a military useful ship. Treatment of the economic considerations in achieving such high speeds is treated in more detail in Chapter 7: Economic Considerations. Treatment of the technical considerations in the analysis and design of the various candidate concepts is provided in Chapter 8: Technical Considerations.

Photo permissions

Figures 1.1 to 1.5, 1.7, and 1.9 to 1.11 originally appeared in the paper *"Cushions and Foils"* by Peter J. Mantle at the SNAME Spring Meeting, 2–5 June 1976, Philadelphia, PA. They are reprinted here by permission of the Society of Naval Architects and Marine Engineers (SNAME) and cannot be reprinted without written permission from the Society, 601 Pavonia Ave., Jersey City, NJ 07306.

Figures 1.14, 1.15 and 1.16 are provided under permission by the Special Collections Department of the Iowa State University Library.

References

1. Mantle, Peter J., *"A Technical Summary of Air Cushion Development"*, US Navy DTNSRDC Report No. 4727, October 1975, Stock Number 008-047-00196-7
2. Mantle, Peter J. *"Air Cushion Craft Development"*, US Navy Report No. DTNSRDC-80-012, January 1980, Stock Number 008-047-00291-2
3. *"Modern Ships and Craft"*, Naval Engineers Journal, Special Edition (Chmn, Editing Committee William Ellsworth), February 1985
4. Warner, Douglas K., *Compression Airplane*, US Patent No. 2390859, issued 11 December 1945
5. *"Jane's High Speed Marine Transportation 2009–2010"*, Ed. Stephen Phillips, 42nd edition, published 2009

2 History of High Speed Ship Development

"I wish to have no connection with any ship that does not sail fast, for I intend to go into harm's way"

John Paul Jones, 1778

It is almost impossible to provide a clear historical basis for all the many ways that "high speed" has been introduced into the marine world. Inventors have tumbled over themselves in conceptualizing, building, testing and otherwise trying out their designs. History has been constantly repeating itself as new designs appear that are frequently nothing more than a reappearance of some earlier design but have gained status because the earlier designs were either lost in the patent offices or abandoned by the original inventor for say, financial burdens or personal crises and other reasons. Sometimes the "new" design came into being simply because the "time was right" and the original inventor is lost to antiquity. In some cases, the speed improvement is overshadowed by the introduction of a new and unexpected technology. The "invention" of the Clipper Ship that was so successful in the 1840s was soon eclipsed by the invention and application of the steam engine for marine use. In these days of seeking "alternative energy sources" perhaps the use of sails may come back!

There are many books already written that document the ways that "high speed" has been introduced in the marine field and it is not the purpose here to repeat such documentation. Christopher Dawson's book, *"A Quest for Speed at Sea"*[1] published in 1972 provides an excellent treatment of advances in sail, engine propulsion, hull form and the start of the "dynamic lift" ships such as hydrofoils and air cushion craft. Dawson shows how the conflicting requirements for speed, load carrying, seaworthiness, endurance, economy and reliability have greatly influenced each of the various designs from historical times to the present day. Although Dawson's book was published over 40 years ago, not much has changed in these conflicting requirements since that time. Frequently, in modern day developments, time and money has been spent on re-living this history.

Another respected historian is H.F.King, a noted former editor of (now defunct) *Flight International* and *Air Cushion Vehicles.* King in his 1966 book *"Aeromarine Origins"*[2] documents the results of his research into many of the designs used today. It is worth quoting from the flyleaf of his book:

"One of the most remarkable facts to emerge is that the hydrofoil boat appears not only to have been a British invention, dating back over a hundred years, but that British ideas and techniques were used by some of the earliest experimenters abroad. The planing boat, or hydroplane, is seen as the invention of a British clergyman, who experimented with rocket-propelled craft of this type about a century ago. This same man proposed fearsome "rocket-rams", and one of these, intended to weigh 140 tons, is described and illustrated. Such weapons were to "sweep away existing navies" and to "render war at sea no longer possible".

A vessel with an air lubricated hull, of the type claimed by recent American inventors as something novel, is shown to have been in regular public service on the New York-New Jersey run over a century ago. Well before the First World War, the Wright brothers were experimenting not only with hydrofoils, but with air lubricated floating bodies also. The author (Dawson) undertakes the fullest enquiry to date into man's earliest consciousness of the phenomenon of "ground effect", upon which the air cushion craft of today depends."

Much of this current book is concerned with the application of the "air cushion" to ship design because enough evidence now exists that suggests that for the large tonnage ship class of consideration for oceanic use that is required to carry large tonnage payloads for either commercial or military use, that variations of hovercraft, air cushion craft, surface effect ships or similar variants, will be the most likely solution. Hydrofoil ships in their various forms (surface piercing or fully submerged) will be more applicable for short haul routes and littoral waters use and less so for transoceanic use (although this last point deserves re-examination, see Chapter 10 on the US Navy ANVCE Project). Work on hydrofoils will be discussed as needed in this overall context. For more detailed treatment of hydrofoils and their history, the reader is referred to the excellent treatments by Capt Robert Johnson, Chapter V, *Hydrofoils* in the Special Edition of Naval Engineers Journal, *"Modern Ships and Craft"* published February 1985[3]; and Bill Ellsworth's book[4], *"Twenty Foilborne Years"*, published by DTNSRDC in 1986.

Leslie Hayward, in his book[5], *"The History of Air Cushion Vehicles"*, published in 1963 provides an insightful documentation of the various designs and hardware examples of vehicles that have employed the principle of air cushion technology dating back to 1716 and to the present day (of then 1963) that includes many of the foundation designs of today's craft. His in-depth research of designs and patents reveals that "not much is new under the sun" and that much of the success of today's craft owes as much to "being in the right place at the right time" as it does to new thought.

In the book[6], *"Air Cushion Craft Development"* by this author, published by the US Government Printing Office in 1980, documents in some detail the major developments of the past using the above references liberally and takes the reader further into the technology of air cushion craft generally and in the technical detail of Surface Effect Ship (SES) developments around the world and specifically in the development of the high speed SES developments by the US Navy.

In the two decades (1960–1980) since the major thrusts in the US and around the world, the developments have slowed down considerably (in hydrofoil, hovercraft and SES development) and it is with this wealth of historical data base as "ground zero" that a perspective can now be gained as to issues, successes and failures that have occurred in this quest for high speed at sea. This book will pull heavily on the wealth of previous work to provide a useful perspective for further development.

2.1 One Hundred Knots Under What Conditions?

The author keeps a display of photos on his office wall as a constant and gentle reminder that, as he either thinks up new ship designs or is presented new designs by enthusiastic engineers, the real world does not treat kindly those designs that are not rugged enough to pass muster on the high seas of the world. Figure 2.1, taken from that display of photos, shows that from smaller ocean going ships to the largest military ships in the world have to not only survive but to function *usefully* in their mission under such conditions. Watching waves break over the deck (even from the protection of the captain's bridge) of an aircraft carrier – a virtual floating city – already towering some 30–40 feet above the sea is a sobering experience and a reminder that any "new ship design concept" had better recognize this reality of the Earth's oceans.

These particular ships shown (in Figure 2.1) are the DD-688 (USS *Remey*) at approximately 3000 tons displacement and the CVS-9 (USS *Essex*) at approximately 44,000 tons displacement. Operating in these conditions the calm water speeds of these ships normally taken as 35–40 knots would be considerably reduced to speeds much less than 20 knots. Attempting these standard operating procedures at 100 knots or even 50 knots would be foolhardy – yet the literature is replete with engineering designs suggesting just that! This is especially true if the proposed designs are not as structurally rugged as today's conventional displacement ship. Note that the US Navy's program on SES development was geared to producing a 3000 ton displacement ship (of about the same displacement as the DD *Remey* shown in Figure 2.1) to operate at speeds of at least 80 knots in Sea State 3.

In the nautical world there are several ways to describe conditions at sea. First, it must be recognized that "seas" do not act in regular wave patterns and steady conditions. They are most often irregular and non-repetitive. Oceanographers and naval engineers have many "standards" that are used from such well known "Beaufort Scale" treatments frequently used by weather forecasters and yachtsmen to more mathematical treatments such as Pierson-Moskowitz (and more recently Bretschneider, JONSWAP, Ochi and others) used more in controlled testing environments. Figure 2.2 shows the typical percentage of time certain seas are experienced in two of the most traveled sea routes; the North Atlantic and South East Asia. It is usually taken that if ships can be designed to operate as required in the North Atlantic in Winter, they will probably be able to handle most of the rough seas in the world.

As will be discussed further in the book, the US Navy were looking for a ship that could travel at 100 knots in calm seas; 80 knots in Sea State 3; and 50 knots in Sea State 6. From Figure 2.2, this implies that the high speed ship would have to be designed to accommodate Sea State 3 seas that could occur about 50% of the time in the North Atlantic and about 30% of the time in South East Asia. Similarly, Sea State 6 is expected to occur about 15% of the time in the North Atlantic and 10% of the time in South East Asia. Another point of concern for the ship designer is that the height of the waves can be deceiving. It is standard practice to talk about "significant wave heights" which is one way of categorizing the highly irregular

3,000 ton Destroyer Refueling in Rough Seas US Navy photo

44,000 ton Aircraft Carrier in Rough Seas US Navy photo

Figure 2.1: Small and Large Ships must operate at an unforgiving interface on the Earth's oceans

wave forms with which the ship designer must contend. In Sea State 6, the significant wave height varies from about 11 feet to 15 feet, yet the "Average $1/10^{th}$ highest" wave varies from 14 feet to 21 feet. *Woe betide* the ship designer who does not recognize this real life probability when designing the ship both structurally and from the viewpoint of stability and control.

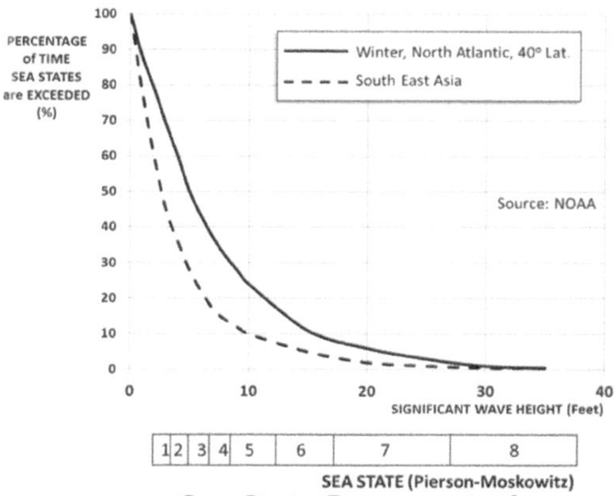

Sea State Representation

Actual Seas

Figure 2.2: Seas Around the World

The origins of the improvements in ships (or boats) to make them go faster are lost in antiquity. Archaeological excavations have uncovered dugout boats from 6000 B.C. History books provide ample descriptions of the ship developments by the Phoenicians and Greeks (circa 1200 B.C.); the Romans (circa 100 B.C.); the Vikings (circa 1000 A.D.) and up into the sixteenth and seventeenth centuries. The first major technological improvement in sail to gain more speed can be tied to the introduction in about 1840 of the "clipper ships". These were designed with magnificent masts of sails reaching several stories high. These ships were designed to "clip along" at speed, hence the name Clipper Ships, at speeds approaching 10–15 knots which was more than *twice* the speed of the merchant sailing vessels of the time. In 1492 Christopher Columbus sailed across the Atlantic in the *Santa Maria* on the outbound leg at speeds of 4–8 knots.

2.2 High Speed at Sea

Concentrating on just speed for the moment, it is seen from Figure 2.3, that the world water speed records have been grouped into two categories. The first is Turbojet Propelled Boats and the worlds' speed record is held by Ken Warby in his *"Spirit of Australia"* on 8 October 1978 in Blowering Dam, Australia with a speed of 276 knots. The second category is Propeller Propelled Boats and that record is held by Dave Villwock in *"Miss Budweiser"* on 13 March 2004 on Lake Oroville, California with a speed of 191.73 knots.

The success of the propeller driven boats in achieving ever higher speeds was due primarily to the introduction of what was known in the high speed ship community as "prop riders". This meant that the propeller was "riding half out of the water" and today, these are referred to as surface piercing propellers. The "prop riders" first saw success with the hydroplane *"Slo-Mo-Shun"*, designed by Stanley Sayres in Seattle and driven by Ted Jones to a world speed record (in 1950) of 138.90 knots on Lake Washington near Sand Point. The author, after studying the hydroplane speed records and designs by these "prop riders", designed the SES 100B with a unique surface piercing supercavitating controllable pitch propeller system that

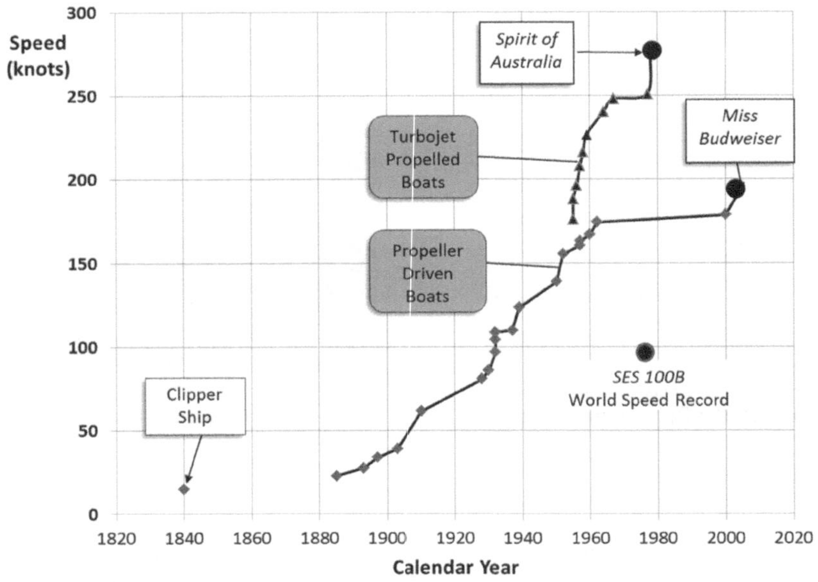

Figure 2.3: World Water Speed Records

allowed it to achieve a speed of 91.90 knots on 30 June 1976 in the waters off Panama City, Florida. These water speed record holders are displayed in Figure 2.4. Chapter 6 provides more details on the SES 100B design.

The Clipper ship *"Flying Cloud"* and the hydroplanes *"Spirit of Australia"* and *"Miss Budweiser"* are shown in Figure 2.4 together with the US Navy *SES-100B*. The significant difference between both the Clipper Ships and SES 100B and the later "world water speed record holders" is that, high speed was attained in both the clipper ships and the SES 100B *while still maintaining all the requirements of a ship for its mission at sea*. Seekers of just world water speed records on the other hand, are not concerned with ship mission requirements and operational and environmental factors and simply try to *attain high speed under very limited conditions of calm water with no payload (other than the driver)*.

Speed alone is an insufficient condition for efficient transport at sea and this led to the approach by the US Navy to develop a high speed ship that could still conduct a mission. The SES 100B at 91.90 knots, achieved in 1976, was an important step toward achieving that goal.

"High speed" is a relative term and does not have a precise definition. This is expanded upon in more detail in Chapter 8 under Technical Considerations. From a safety point of view it is interesting to note that the International Maritime Organization (IMO) classify "high speed" to be almost the speed of most conventional displacement ships today. This conservative approach can be seen in the formula used by the IMO[7] that defines "high speed craft" as one that has a maximum speed equal to or exceeding a speed given by:

$$V = 3.7(\Delta)^{1/6} \tag{2.1}$$

where the speed (V) is expressed in meters per second and the craft displacement (Δ) is in cubic meters. In water (salt or fresh), this converts, to a close approximation, to:

Clipper Ship, *Flying Cloud* (1851)

Spirit of Australia
Turbojet Propelled
276 knots on October 8, 1978

Miss Budweiser
"Prop Rider"
191.73 knots on March 13, 2004

SES 100B
Surface Piercing Controllable Pitch Propeller Propelled
91.90 knots on June 30, 1976
US Navy photo

Figure 2.4: World Water Speed Records

$$V = 7.19(W)^{1/6} \tag{2.2}$$

where now the speed is expressed in knots and the displacement in tonnes. For a 3000 tonne displacement ship (the size of interest for the "100 knot Navy"), this IMO formula gives a speed of about 27 knots which clearly is not in the same context as intended by the US Navy. The IMO interests are more to consider any type of craft (other than perhaps a slow barge) from the viewpoint of safety at sea and thus set the "limit" fairly low. Chapter 8 develops definitions of "high speed" of interest more in tune with the intent of "high speed" missions based more on scaling laws and related physics.

Before concentrating on the modern day development to achieve "100 knots at sea" it is useful to have a brief background of the developments in the two main types of craft: the hydrofoil and the air cushion craft.

2.3 Brief History of Hydrofoils

The first operational hydrofoil craft is considered to be Enrico Forlanini's *fully-submerged ladder foil* craft which operated in 1906 on Lake Maggiore in Italy. An earlier craft by Comte de Lambert operated on the River Seine, France in 1897, but used planing skiis and is not strictly a member of the hydrofoil family. Also excluded is the Englishman Thomas Moy's experiment in 1861 when he fitted "wings" beneath his boat to study aerofoil patterns and found his boat rise out of the water!

Forlanini's hydrofoil weighed 1.65 tons and achieved approximately 40 knots with 75 horsepower. It was propelled by tandem air propellers, and its ladder foil system was constructed with zero dihedral. Other work followed closely on Forlanini's heels; notably among these were Crocco, Ricaldoni and Meacham, but it was not until 1919 did Alexander Graham Bell (aided by Frederick (Casey) Baldwin) successfully demonstrate his *"Hydrodrome"*, the *"HD-4"*, (Figure 2.5) which displaced 5.5 tons and reached a record breaking 61.5 knots on Braddeck Bay in Nova Scotia on 9 September 1919[8]. This was the most successful of the Bell and Baldwin hydrofoils that started with their *HD-1* in 1911 and continued until the successful *HD-4* in 1919.

The *"HD-4"* also employed a ladder foil system (that Alexander Graham Bell incorporated based on his discussions with Forlanini), except that some dihedral was now added for improved stability and control. The craft was powered by two Liberty engines with a total horsepower rating of 700 horsepower driving two air propellers.

The work of Forlanini also inspired a young engineer, Baron von Schertel, who began refining some of Forlanini's (and his collaborator General Arturo Crocco) designs of a dihedral surface piercing foil system and developed the first successful "vee-foil" boat that had solved some of the stability problems of the earlier Forlanini-Crocco designs. Baron von Schertel designed and demonstrated a successful hydrofoil that caught the attention in the late 1930s of the

Figure 2.5: Alexander Graham Bell's *"Hydrodrome, HD-4"* (1919) Photo permission by Parks Canada, Alexander Graham Bell National Historical Site

Köln-Düsseldorfer Shipping Line for a commercial operation along the Rhine River. In 1939 this same successful hydrofoil caught the attention of the German Navy and early in World War II, the surface piercing (area stabilized) hydrofoils received extensive development under urging of Adolf Hitler[9] who was seeking ways of transporting military supplies to Rommel's army in Africa at high speed over the rough Mediterranean waters from Sicily. Baron von Schertel (see Figure 2.6) had to put aside his plans for commercial hydrofoil operation and concentrated on the German Navy needs. This resulted in the hydrofoil with the Sachsenburg shipyards called the *"VS-6"* hydrofoil (Figure 2.7) and is representative of this early development. It had a displacement of 17 tons and reached a speed of 47 knots in 1941. It was propelled by two Hispano-Suiza gasoline engines (each 1,560 hp) driving two subcavitating marine propellers. The foils were surface piercing and used trailing edge flaps for control.

Figure 2.6: Baron Hanns von Schertel (1902–1985) From Hovering Craft & Hydrofoil 1968

Figure 2.7: Schertel-Sachsenburg "VS-6" underway (1939) From Hovering Craft & Hydrofoil 1965

Because of Baron Hanns von Schertel's expansive experience in such craft, the author personally sought him out and brought him on board as an advisor to the US Navy's Advanced Naval Vehicle Concepts Evaluation Project (ANVCE) in 1976. This is discussed in Chapter 10.

The surface piercing hydrofoil achieved a significant milestone in 1961 with the Grumman built, Maritime Administration (MARAD), *"Denison"* (see later). This craft of 90 tons displacement achieved a speed of 62 knots in 1962. It was powered by a single 14,000 hp GE LM 1500 gas turbine. This speed record held until Boeing's experimental 16.5 ton *"FRESH-1"* operating on fully-submerged superventilating foils achieved 80 knots on 10 July 1963. As discussed later in this Chapter, this proved to be a fateful short-lived achievement with an accident that curtailed the US Navy's pursuit of such high speeds with such craft. Fully submerged hydrofoils received impetus in 1963 with the PCH-1, *"Highpoint"*, which was the first hydrofoil to be commissioned by the US Navy. After a history of trials and tribulations this craft accumulated significant operational and technical data before being deactivated on 1 December 1984 and turned over to the US Navy's DTNSRDC (operated by Boeing) for R&D purposes. The *"Highpoint"* was sold privately in 2002 and subsequently its foils were removed and it was used briefly as a conventional boat.

The largest hydrofoil in the world was the Grumman designed, Lockheed built AG(EH)-1, *"Plainview"*. The *"Plainview"* was launched on 28 June 1965 and entered trials with the US Navy on 1 March 1969. The full load displacement was 328 tons and achieved foilborne speeds in excess of 50 knots. It was powered by two GE LM 1500 gas turbines (29,000 hp total) driving two supercavitating propellers. Like the *"Highpoint"* it had an unglamorous end in 1979 (see Chapter 1).

After the *PCH-1* and *Plainview* programs, attention was given to the fully submerged hydrofoils: the *JETFOIL* (for commercial use) and the *PHM* (for NATO forces use). The *JETFOIL* (see Chapter 1) has a displacement of 106 tons, a foilborne speed of 45 knots and is powered (foilborne) by two Allison 501-K20A gas turbines, each developing 3300hp. The *JETFOIL*, launched on 29 March 1974, entered commercial service in Hawaii on 15 June 1975 for the Pacific Sea Transport Company and in Hong-Kong on 25 April 1975 for the Far-East Hydrofoil Company.

The *PHM* has a displacement of 230 tons, is designed to operate at speeds in excess of 48 knots and is powered foilborne by a single GE LM 2500 gas turbine developing 18,000hp. For hullborne operation the PHM can achieve 12 knots with two Mercedes-Benz marine diesels at 1600hp. Both the *JETFOIL* and the *PHM* employ fully-submerged foils with hydraulically activated, automatically controlled trailing edge foils to provide both control and stabilization. Figure 2.8 shows the PHM in operation foilborne (1977). The shape of the forward foil can be clearly seen in the photograph. The US Navy Patrol Combatant Hydrofoil-Missile (PHM) was commissioned on 9 July 1977 the first of the class was the *Pegasus* (with only six built compared to an original projected production run of 100 craft for NATO navies) and was decommissioned on 30 July 1993. The PHM was the result of the Chief of Naval Operations (CNO) Admiral Elmo Zumwalt's desire to expand the US Navy's operations with high speed vessels in a cost-effective manner. A common theme that will be returned to many times in this book is that the reality was that such craft were found to be too expensive to acquire and operate. The PHM program lost its funding upon Admiral Zumwalt's retirement and the monies used for other Navy priorities. The participating NATO navies in the program, namely Canada, the UK Royal Navy, the Italian Navy and the German Navy all canceled their participation in the program.

Figure 2.8: US Navy "PHM" (*Taurus*) US Navy photo

A more inexpensive hydrofoil system was that employed by the Russian *Raketa* (see photograph in Chapter 1). The *Raketa* (rocket) was the first commercial Russian hydrofoil and was launched in June 1957. It has a displacement of 25.3 tons, has a maximum speed of 33 knots and is powered by a single 1200 hp M-50 diesel driving a subcavitating propeller. Over 400 (at time of writing) of these craft had entered service mainly in Russia and in April 1974 began regular passenger service on the River Thames in England. The *Raketa* has proven to be highly successful and economical to operate and may be found operating on regular commercial service on many inland rivers in Russia, Bulgaria and other nations.

Table 2.1 summarizes the leading particulars of these representative types of hydrofoils. It shows that the state of the art has provided a substantial data base of hydrofoils with displacements up to about 370 tons and consistent operational speeds in moderate Sea States of around 50 knots (with a few limited special purpose craft with higher speeds).

With the above examples (admittedly a snapshot) it can be seen that hydrofoils in various configuration forms are a successful means of high speed marine transportation in the 30–50 knot speed range and up to about 370 tons displacement provided the waterways are relatively protected. Extension of this form of craft into higher displacements (of the frigate size) and at speeds in excess of 50 knots requires special treatment and this is considered in later chapters (see Chapter 10 on the US Navy's concerted efforts to explore the performance boundaries of ocean going versions of hydrofoils for naval missions).

One hundred knots does not appear to be a realistic goal for such craft for operational missions.

2.4 Brief History of Air Cushion Craft

Turning now to the air cushion craft, a similar pattern emerges in its development. While sketches in 1716 by Emmanuel Swedenborg, a Swedish scientist and mystic, of a hand-powered air cushion machine and the (unsuccessful) experiments by Gustaf de Laval, a Swedish inventor in 1882, are often cited as early starts to the air cushion craft, it was not until 1916 did a specific craft appear using the air cushion principle. In 1916, Herr von Thomamhul (see later for photograph and description) built an air cushion supported torpedo boat for the Austrian Navy. This craft reached a speed of 40 knots, was approximately 40–50 feet long, and was powered by four engines delivering 480 hp to two marine propellers. After Thomamhul's craft, several ideas and craft appeared on the scene, all seeking to place an air cushion beneath a craft to either reduce friction or to lift the craft entirely out of the water.

The frustrated attempts by A. U. Alcock in Perth, Australia to demonstrate the air cushion principle preceded Herr von Thomamhul's by four years in 1912 with his clearly demonstrated experiments to reporters of the Australian *Sunday Times* (1912) and to officials of the Australian Government. His 4ft x 4ft model had an electric motor that drove a compressor that pumped air beneath the platform. A separate propeller provided the necessary thrust. Alcock was issued provisional patents (Patent number 14309/1914) but he did not proceed further due, apparently, to lack of funding. This event sadly passed into history without recognition and Alcock, who subsequently moved to Devon, England, did not pursue his invention further. Alcock coined the phrase *"floating traction"* for his air cushion principle which is quite descriptive.

Fifteen years later, in 1929, Douglas K. Warner (from Warner Research Laboratories in Sarasota, Florida) won the boat races on Lake Compounce,

Table 2.1: *Leading Particulars of Representative Hydrofoils*

Hydrofoil	Displ. (s.ton)	No. of pass. or payload (tons)	Range (n.m)	Length (ft)	Beam (ft)	Foil Span (ft)	Strut Length (ft)	Foilborne Power (hp)	Max. Speed (kts)	Sig. Wave height (ft)	Type
CANADA											
Bras d'Or	263.2	20 crew		151	21.5	66	17	22,000	60	12	(1)(4)
ITALY											
Rodriquez											
RHS 70	35.28	71 pass	300	72.5	19.2	24.25	~6.5	1x1350	36.5		(1)(4)
RHS 110	60.48	110 pass	300	84		30.2	~8.5	2x1350	40		(1)(4)
RHS 140	72.8	140 pass	900	94.1		35.2	~9.0	2x1350	36		(1)(4)
RHS 160	91.84	160 pass	300	101.5		41.3	~10.5	2x1920	39		(1)(4)
SWITZERLAND											
Supramar											
PT20 Mk II	67.2	74 pass	140	67.5	17.05	26.4	~8.0	1x1100	34	4.25	(1)(4)
PT50 Mk II	370.9	140 pass	325	91.6	20.05	35.4	~9.0	2x1100	36.5		(1)(4)
PTS75 MkIII	89.6	170 pass	180	98.5	19.1	38.1	~11.0	2x1650	39		(1)(4)
PTS150 Mk III	184.8	250 pass	250	124.2	24.6	52.45	~16.0	2x3400	36.5	10.0	(1)(4)(5)
USA											
Tucumcari	71.6	18.7t		80.3	19.5	36		1x3200	>50		(2)(3)(7)
High Point	141.1			115.5	30	36.5		2x3800	>50	12	(2)(3)(7)
PHM	263.2			132.7	28.2	47.6		1x25,000	>40		(2)(3)(7)
JETFOIL	118.7	250 pass	>500	90	30	30	10	2x33,000	>40	8–12	(2)(3)(7)
Flagstaff	75.6		>600	73	21.6	37.16		1x3,550	>40		(2)(4)(6)
Plainview	367.4	5.3t		212	40.4	70		2x14,500	50		(2)(4)(6)
Fresh-1	18.5			47	22.5	<beam		18,000Lbt	80		(2)(3)(4)
FSU											
Taifun	72.8	105 pass		103.2	18.25	32.9		2x1750	42	6.5	(2)(4)(7)
Kometa	62.72	102 pass		115.5	31.5			2x900	34	4	(1)(4)
Raketa	28.34	64 pass		88.4	16.5			1x1200	33	2.7	(2)(4)(8)
Vikhr	131.6	268 pass	240	156	29.6			4x1200	43	5	(2)(4)(8)
Strela	54.5	94 pass	450	96	26.4			2x970	40	4	(1)(4)(7)

Type Notes: (1) Surface piercing (2) Fully submerged (3) Canard (4) Airplane (5) Air stabilized (6) Incidence control (7) Flap control (8) Surface Effect

Connecticut, with the first *"captured air bubble"* craft where the air "bubble" or cushion is trapped between the two sidehulls and the rigid, planing bow and stern seals. The bow and stern seals were mechanical spring loaded "planer seals". More details of the design are provided later and shown later in Figure 2.14. This design became the forerunner of later variants of the same concept that were the driving concepts in the US Navy surface effect ships that are discussed in more detail in later chapters. Also, as will be discussed later, the prolific inventor Warner also designed variants of *ram wings* and other *aerodynamic forms of air cushion craft* discussed in detail in Chapter 11 on *aerodynamic air cushion craft.* A few years later, in 1935, another prolific inventor, this time from Finland, Toivo Kaario, an engineer at Valmet Corporation, Linnavouri, Finland, made several models of several different types of air cushion but primarily versions of *aerodynamic air cushion craft.* Kaario presented his ideas and evidence of his models at the Princeton Symposium in 1959 along with Douglas Warner, mentioned above, who was also present. Both Warner and Kaario were aware of the stability problems of air cushion craft, regardless of the type, if the craft "sits" too high on the air bubble. Warner presented his findings to the US Navy[a] based on stability and "plow-in" problems experienced with his *captured air bubble* craft, and Kaario presented his thoughts on the subject at the same Princeton Symposium. See later discussion on details of Warner's CAB boat design.

In the early 1950s, the form of the air cushion craft started to take on the familiar lines of today's craft. In 1955, about the time that Christopher Cockerell was exploring his concept and filing patents, other inventors were working on similar designs. In particular, Renalto Alves De Lima filed a provisional patent (in August 1955) in Brazil for a design of a machine that drew in air by counter-rotating fans (to avoid vehicle precession) that compressed the air beneath the craft through peripheral jets that captured the air beneath the craft to lift it.

De Lima showed how this concept could be added to an aircraft to give it an *"air cushion landing device"* for use over rough terrain. Unfortunately, like many similar ideas, no interest was shown by any authority and De Lima's patent languished. De Lima's work on air cushion landing gears did re-emerge in November 1970 when Bell Aerospace[b], under a USAF contract, added such a device to a Lake LA-4 light amphibian aircraft and subsequently to a de Havilland CC-115 Buffalo air craft (Figure 2.9). This feature was different to that originally suggested by De Lima in that instead of the flexible skirt under the air craft using peripheral jets, for the LA-4 and the de Havilland Buffalo (re-designated XC-8A) the skirts were perforated with about 6,800 small holes to allow air leakage into the doughnut hole of the inflated air cushion landing system (ACLS) bag. This design allowed the aircraft to land and take-off on unprepared surfaces, including water, ice, snow, sand, marsh and many

[a] Documentation of the US Navy (circa 1930) efforts in analyzing the stability problems of the *captured air bubble* (CAB) craft of the type described by Warner was not found by this author. Perhaps, if Warner's findings and concerns about the stability of the *captured air bubble* (CAB) had been documented and made more readily available, some of the later problems with this concept could have been avoided.

[b] Patented by T. Desmond Earl and Wilfred J. Eggington at Bell (US Patent 3,275,270 issued 27 Sept 1966).

Figure 2.9: Air Cushion Landing System *("ACLS")* on de Havilland Buffalo XC-8A (1975)
USAF photo

other natural open field terrain. The first "cushion landing" took place on 11 April 1975 at Wright Patterson Air Force Base, Dayton, Ohio.

The ACLS shown in Figure 2.9 is not really an air cushion craft in the contextual sense discussed under the general heading of "air cushion craft" but it is of importance first to recognize the prescience of De Lima but also to lay the ground work for design possibilities in the *aerodynamic air cushion craft* treated in more detail in Chapter 11. It was found during the extensive test program in the early 1970s that the ACLS could be used on a variety of landing fields. Unique features, such as "pneumatic pillows", were incorporated into the inflated skirt that when inflated provided a breaking mechanism for the landing mode. The ACLS was designed in such a way that the elasticity of the entire system allowed it to retract smoothly into the contour of the aircraft when deflated and the aircraft was in normal flight, thus keeping a minimal drag profile. Funding limitations on this program curtailed further development. While it is difficult to ascertain from the literature, it is conjectured that a similar design feature was explored (but not used?) in the Russian design of the *Orlyonok* and later smaller craft at the same time to give it its "amphibious" capability[c].

In 1953, Christopher Cockerell perfected the idea of hovercraft and improved on the peripheral jet concept by ensuring that the peripheral jet was inclined inward (De Lima's peripheral jet was directed vertically) such that the centrifugal forces generated by the turned jet would impinge on the ground (or surface) such that the pressure force acting on the vehicle would lift it. Cockerell's work is expanded on further in the next section of this Chapter. With the jets inclined inwards at about 45 degrees instead on straight down as in De Lima's patent, the centrifugal forces increased and the supporting air cushion pressure would also be increased by an additional 40% or more. The next section expands on the history of the development of the modern day air cushion craft but Table 2.2 summarizes the backdrop of representative air cushion craft that have demonstrated a capability up to about 300 tons displacement and about 70 knots operational speeds in nominal Sea States. It was soon realized that the principle of the air cushion craft required small air gaps to be efficient and such unique craft would not have progressed further were it not for the incorporation of the skirt or flexible extension beneath the craft which allowed the gap to be small but allow the hard structure of the craft to be sufficiently high to avoid impact with either waves or hard ground. This idea came into being in England in 1958 with the work of Cecil Hugh ("Chookie") Latimer-Needham (see

[c] This must be left as a conjecture because despite frequent references to such schemes in the literature no reliable data has appeared to document such designs in the Russian *ekranoplans* (see Chapter 11).

Table 2.2: *Leading Particulars of Representative Air Cushion Craft*

Air Cushion Craft	Displ. (tons)	No. of pass. or payload (tons)	Range (n.m)	Length (ft)	Beam (ft)	Cushion Depth (ft)	Prop'n Power (hp)	Lift Power (hp)	Max. Speed (kts)	Sig. Wave height (ft)	Type
CANADA											
Bell Voyageur	45	25t	300	64.8	36.7	4.0	2x1300	inc	47		(1)(2)
Viking	16.25	6–7t	680	44.5	26.0	4.0	1x1300	inc	50	6	
FRANCE											
N.300	29.75	120 pass	180	78.75	34.5	6.6	2x1500	inc	62		(1)(2)
N.500	237	400 & 41 cars		177.16	78.75				76	13	(1)(2)
GERMANY											
Lippisch											
X-113	0.38	1 crew		27.68	19.34	n/a		n/a	108[a]		(5)
X-114	1.5		1160	42	23	n/a	200	n/a	81[b]	5	(5)
JAPAN											
Mitsui											
MV-PP15	56	155 pass	250	80.1	41.66	5	2x1950	inc	60		(1)(2)
MV-PP5 Mk II	19.3	7.2t	160	59.5	28.16	4	1x1050	inc	76		(1)(2)
United Kingdom											
SRN.5	6.6	18 pass	200	38.75	23	4	1x900	inc	66	~5	(1)(2)
SRN.6	9.0	38 or 3t	200	48.4	23	4	1x900	inc	60	~5	(1)(2)
SRN.4 Mk 2	200	200t	100	130.1	78.0	8	4x3400	inc	70		(1)(2)
Super 4	300	92t		185	76	9	4x3800	inc	65+	~8	(1)(2)
BH.7 Mk 2	56		140	76.5	44	6	1x4250	inc	58		(1)(2)
HM.2	16	60 pass	300	51.0	20.0	3	2x320	1x185	35	3	(4)
VT.2	105	32		99.0	43.5	5.5	2x4500	inc	60+		(1)(2)
USA											
JEFF A	166.5	60t	200	96.1	48.0	5	6x2800	inc	50+	8	(1)(2)
JEFF B	162.5	60t	200	86.75	47.0	5	6x2800	inc	50+	8	(1)(2)
LCAC	153.5	54.3t		85.76	45.76	5	4x5304	inc	50	8	(1)(2)
SES 100A	123	10t		81.9	41.9	6	4x3500	inc	76	6	(1)(3)
SES 100B	105	10t		77.67	35.0	6	3x3260	3x500	91.9	6	(2)(4)
SES 200	152	12 crew	1550	109.88	39.03	6	2x1804	1x349	30		(2)(4)
SES 200 stretch	207		3850	159.74	39.03	6	1649	436	28		(2)(4)

Footnotes: (a) At altitude; (b) In ground effect

Type Notes: (1) Amphibious (2) Passive ride control (3) Active ride control (4) Side hull, non-amphibious (5) Aerodynamic later) and has now developed into many forms of flexible skirts, all with different characteristics, but all aimed at providing wave or surface clearance and craft stability. It is not the intent to expand on all the many skirt geometries that have been explored and used successfully on many operational craft. The reader is referred to the already cited references for more detail including skirt development in other countries besides the country of origin (UK).

2.5 Modern Day Developments to Achieve 100 knots

The modern day thrust into the pursuit of "one hundred knots at sea" has its roots in the efforts by the US Maritime Administration in the late 1950s and early 1960s to find ways to solve the problems of the ailing US maritime industry. According to the 1963 SNAME paper by Dr Scott Rethorst of Vehicle Research Corporation and his co-author, W.T. "Taylor" Potter, of the US Department of Commerce:[10]

> *"The Department of Commerce in 1960, established an Office of Research & Development in the Maritime Administration. This office was charged with finding, through applied research, practical methods of making American merchant marine competitive with the merchant fleets of other countries. Today (1963), one third of America's ships at sea are aided by the operating differential and construction subsidy programs. Under the Merchant Marine Act of 1936, these programs were instituted to keep the American sea transportation system alive. Without these programs the face of the marine industry would undergo severe changes...... A new framework must be found if America is to maintain her international transportation system. One statistic points up this urgent need: 90%[d] of America's imports and exports is carried in the ships of other nations."*

The first efforts by the US Maritime Administration (MARAD) to tackle this problem was first (a) the development of the nuclear powered ship *Savannah;* and (b) the second major thrust was the development of a 100 knot hydrofoil in the form of the HS *DENISON*; and (c) the third major thrust was in the development of a 100 knot surface effect ship (*Columbia*). The *Savannah* (Figure 2.10) was the first nuclear-powered cargo-passenger ship, launched on 21 July 1959, as part of President Eisenhower's "Atoms for Peace" initiative. The *NS Savannah* had several built-in features that foretold its demise that emphasized the problems facing any new ship design. It was small in size (its cargo holds were several times smaller than conventional ships). Its streamlined hull, designed for

Figure 2.10: NS Savannah approaching Golden Gate Bridge (1962)
US MARAD photo

[d] The share of world's shipping by US ships has declined even further in the intervening years to today.

speed, made loading of the forward holds difficult for the dock crews. Because of the nature of the unique (nuclear) propulsion plant, its crews were about one third greater in size than those of conventionally powered ships and further required specialized training that caused significant union problems and increased cost.

Further, because of the use of a nuclear power plant, several ports around the world denied entry of the ship. Because of these "built-in" difficulties, MARAD eventually was forced to decommission the ship on 10 January 1972. Such features serve to emphasize that in any new ship design one must consider all factors. Not just the one feature that appears on the surface to be an advantage.

The second initiative pursued by MARAD was the hydrofoil program. This was introduced (after several internal studies) in 1958 by the then Chief of Research & Development of MARAD, Charles R. Denison. On 28 January 1962, an existing experimental ocean-going hull was outfitted with a set of surface piercing subcavitating foils of similar design characteristics of the successful European designs. The ship was christened HS *DENISON* in the memory of Charles Denison whose sad and untimely death during the early design stages of the program precluded his seeing the success of his work. The HS *DENISON* is shown in Figure 2.11.

HS *Denison* foilborne **HS *Denison* hullborne**

Figure 2.11: US MARAD HS *DENISON* Hydrofoil (1962)
US MARAD photos

The HS *DENISON* was outfitted first with subcavitating surface piercing foils with a speed capability of about 60 knots. The plan was to replace these foils later with supercavitating surface piercing foils to give the ship a speed capability of 100 knots. This speed of 100 knots was driven by the MARAD studies for commercial advantage (see later). While the *HS DENISON* was an experimental ship to determine the characteristics of hydrofoils, MARAD was keenly aware of the need to keep as close to conventional (displacement) configuration as possible so as to be able to use the same port and dock facilities as conventional ships based on their experience with the *N. S. Savannah*.

The Annual Report of MARAD in 1964 had the following entry:

"During July and August of 1963, the 95 ton, 60 knot experimental hydrofoil ship, HS Denison, completed 14 months of performance testing, including a run in the open sea during which she maintained speeds over 55 knots in waves more than 10feet high.

A demonstration program, aimed at displaying the hydrofoil concept and showing the craft in action to the marine industry, was inaugurated in September 1963. The demonstrations also provided extended experimental operating experience and kept the craft in service in the winter. Beginning in Portland, Maine, the Denison traveled south to Miami, touching at most of the major ports along the Atlantic seaboard. During this time several thousand people from the marine industry, Government, and business were given demonstration rides. The response was generally favorable.

In March 1964 the demonstration program was discontinued and the ship ordered to the reserve fleet in Wilmington, N.C. While proceeding up Cape Fear River, N.C., the Denison ran aground on a sand bar, sustaining substantial damage to the hull and main foils. She was undergoing repairs at Grumman Aircraft Engineering Corp., Bethpage, N.Y., at the end of the year."

This occurred unfortunately at about the same time that MARD was rethinking its entire program complete with funding cutbacks and decided to co-develop its work with the US Navy. The US Navy wished to pursue, based on its own studies, the fully submerged foil design (rather than the simpler surface piercing foil design favored by MARAD), with the ill-fated US Navy *FRESH-1* experimental craft shown in Figure 2.12.

The hydrofoil *FRESH-1*, had achieved speeds of about 80 knots in trials, but during the Navy Acceptance Trials on 18 July 1963 off Dilworth Point, Vashon Island, Washington, during a high speed test run at 80 knots slowing to 70 knots, it lost lift through ventilation of the main foils (due to improper activation of the trailing edge flaps). The craft flipped over causing major damage to the hull and foils but fortunately no major injuries to those on board. All three occupants escaped through the broken front window and door of the upside down cabin. A set of "transition foils" (built by Grumman) to smooth transition from subcavitating flow to fully cavitating flow at the higher speeds had been constructed but were never installed on *FRESH-1*. The *FRESH-1* was "mothballed" following this incident and caused the Navy to re-direct its efforts to concentrate on lower speeds closer to 50 knots, and US Navy plans to pursue 100 knots with hydrofoils were suspended from that time forth.

Based on the new thrust by MARAD, to pursue the surface effect ship programs based on its own studies, this ended MARAD's interest in hydrofoil development. This was especially so in the light of those studies that showed that hydrofoils (fully submerged or surface piercing) did not meet the tonnage requirements deemed essential for oceanic commerce. This decision was revisited by the US Navy in 1976 as part of the US Navy's Advanced Navy Vehicles Concepts Evaluation (ANVCE) Project and is described in Chapter 10.

Figure 2.12: Boeing-US Navy fully submerged hydrofoil FRESH-1 (1963)
US Navy photo

The third initiative by MARAD to solve its declining American marine industry was the investigation into surface effect ship development. Based on internal studies, MARAD had determined that the high speed of "100 knots" was best met by an *aerodynamically supported surface effect ship* and on 2 August 1961 gave a contract to Vehicle Research Corporation (VRC) in Pasadena, California to develop a 100 ton prototype craft, called "*Columbia*". The complete story on that effort is provided in Chapter 3.

Later, under new leadership at MARAD while the SES *Columbia* project was still underway, a new MARAD study was initiated to examine other forms of high speed craft that were just entering the scene. This effort was initiated in a Memorandum on 6 July 1965, when the Undersecretary for Transportation, Department of Commerce, Mr. Alan S. Boyd, requested that a study be made of the surface effect ship (SES) concept in all its forms. He addressed his request to Dr. J. Herbert Holloman, Assistant Secretary for Science and Technology, Department of Commerce and Chairman of the Commerce Technical Advisory Board. The Committee to study Surface Effect Ships for Ocean Commerce (SESOC Committee) was formed and started work on 9 September 1965. The SESOC Committee was headed by Dean George Maslach, College of Engineering, University of California. The MARAD Undersecretary for Transportation, Mr Alan Boyd assigned Ed M. McCutcheon, the new Chief of Research & Development, to be his full time representative of this study.

While the SESOC Committee study into various forms of Surface Effect Ships was a noble effort, it was somewhat flawed in that the Committee accepted the proposals from both Government laboratories and industry alike on their "point designs" that in many cases were nothing more than marketing brochures. If the presenter said his concept could travel at 100 knots in 8foot seas – *independent of ship size* – he was initially believed and the committee members had to rely on their own experience base and to make their selections based on how well the presenter presented his material.

The assumptions going into the study were also quite constraining, even if they sounded somewhat logical on first glance.

The MARAD requirements (for the ship) were:

1. Be able to operate and survive in its ocean environment
2. Be able to utilize existing facilities, e.g. canals, channels, pier facilities and harbors to the greatest extent possible
3. Be of such size, shape and arrangement to substantially reduce the required turnaround time and minimize the handling of cargo in port.
4. Offer a substantially increased, more flexible service to the shipper at a cost at least equivalent to present costs, and at a lower cost if possible.
5. Be as insensitive to construction and operating labor cost as possible
6. Be capable of being operated and maneuvered with the experienced merchant seamen existing in our fleets today with a proper retraining course.

One can see how the experience with the MARAD nuclear ship NS *Savannah* had influenced these requirements. Unfortunately, because of the state of the art of surface effect technology in 1965, this placed all participants into a competitive environment. But it did launch the surface effect ship program as we know it today. As will be covered in a later chapter (Chapter 4), MARAD first formed a Joint Surface Effect Ship Project Office (JSESPO) on 27 February 1967 with the US Navy shortly after the conclusion of the SESOC Committee work and then, shortly afterward on 1 July 1971, pulled out completely leaving the development of Surface Effect Ships (in the US) with the US Navy.

2.6 Early Developments (1916–1930)

As said earlier, the first use of air cushions to either lubricate or partially support vehicles can be traced back to 1716, but the first "reduction to practice" invention can be directly related to several inventors and practitioners of the art in the early 1900s. In 1916, Herr Dagobert Muller von Thomamhul designed and built for the Austrian Navy a torpedo boat that was based on the principle of pumping air from on-board blowers, between the sidewalls. This unusual boat is shown in Figure 2.13.

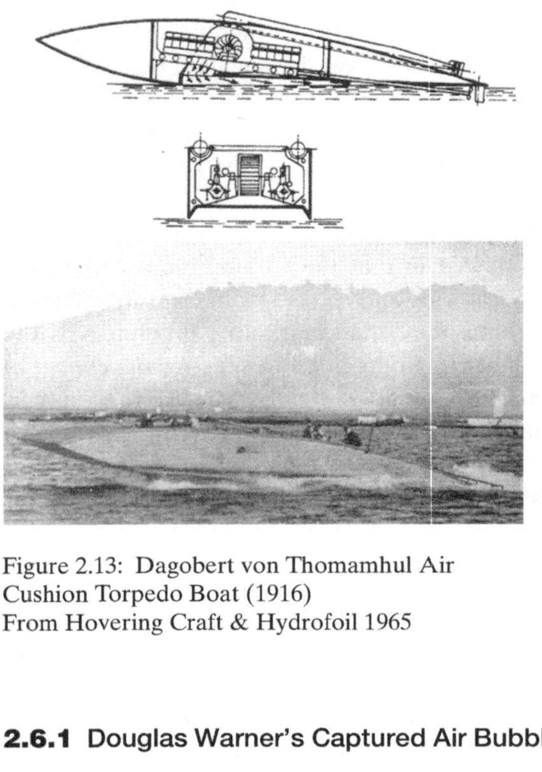

Figure 2.13: Dagobert von Thomamhul Air Cushion Torpedo Boat (1916)
From Hovering Craft & Hydrofoil 1965

According to the official records, von Thomamhul's *"trapped air boat"* achieved a speed of 40 knots. This boat was built for the Austrian Navy as a fast torpedo boat.

A torpedo can be seen in the upper drawing in Figure 2.13. The air blower can be seen as well in the upper sketch as the mechanism for pumping air beneath the boat and between the two sidewalls (or sidehulls). It appears from the drawing, that the rear "seal" is the trailing edge of the boat. Later in Chapter 11 when *aerodynamic air cushion craft* are discussed, it will be seen that this technique of containing the air beneath the boat was also used by the Russian designers in their *ekranoplan* craft.

2.6.1 Douglas Warner's Captured Air Bubble Boat (1929)

There were several other patents after 1916 describing different versions of trapping the air between the sidewalls and other features of new technology but the most relevant work to today's design of surface effect ships is the work done by a prolific inventor with a boat built as described in the documented work of Douglas Kent Warner of Bristol, Connecticut when he introduced the idea of a *captured air bubble* underneath a speed boat[11] in 1929. Warner[e] was seeking a way to win the boat races

[e] Warner called his boat by a more generic term of *air cushion boat* but the principle used was *captured air bubble*.

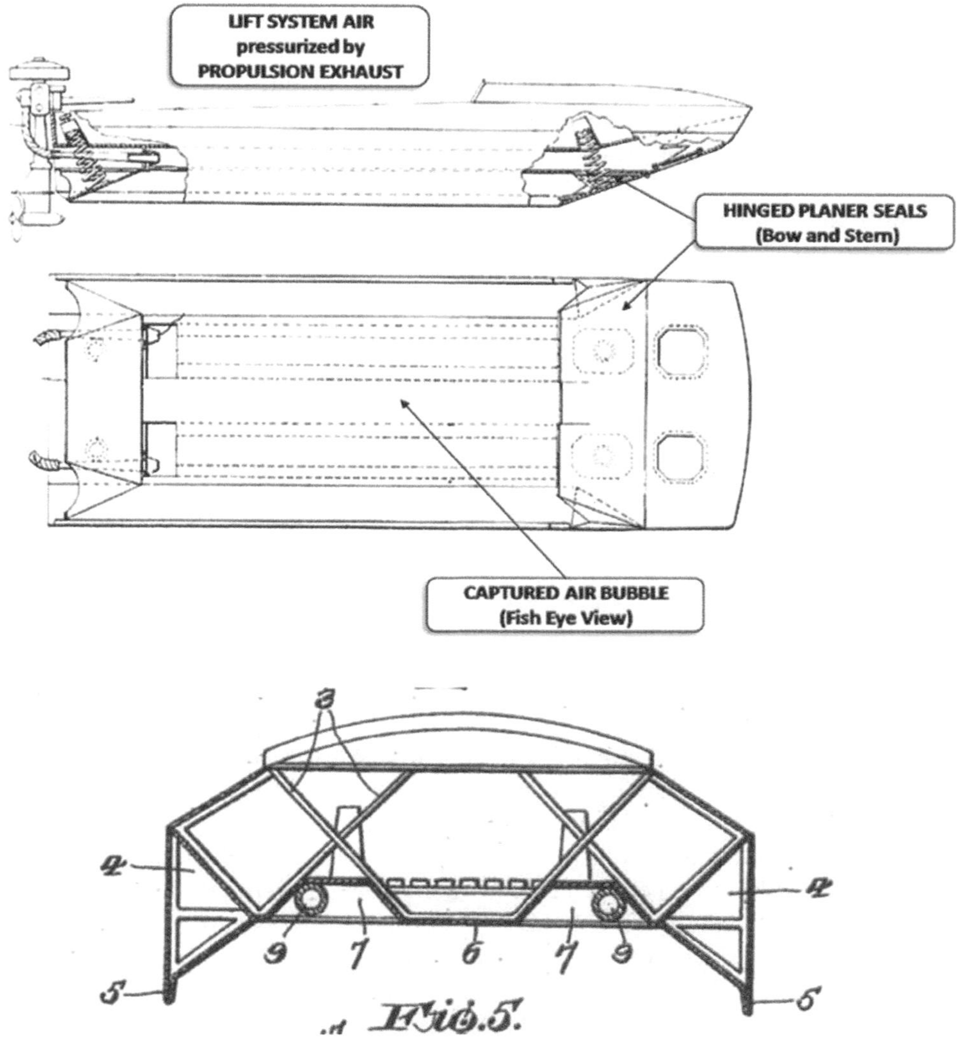

Figure 2.14: Douglas K. Warner's *Captured Air Bubble (CAB)* Boat (1929)

and did exactly that with his captured air bubble craft shown in his patent drawings, Figure 2.14.

Warner built this boat and won the boat races at Middleton on Lake Compounce, Connecticut in 1929. As can be seen from the drawing, the Warner CAB design is an integrated lift and propulsion system in that the lift system air is pressurized from the propulsion engine exhaust and the free stream ram air then ducted to the underneath of the boat as seen in the "fish eye view" in the middle part of the drawing. The air is trapped in the bubble between the twin sidehulls by spring loaded flat "planer seals" in the bow and stern of his boat. The springs were strong enough to help in resisting any restoring forces from the water as it ran at speed and contributed to providing pitch stiffness and stability to the boat. Not only did this allow Warner to win the races it also was adopted later in river boats that operated commercially on the New York-New Jersey run as documented by H.F. King in his "*Aeromarine Origins*" described above.

Warner, after his success, delivered one of his *captured air bubble* craft to the US Navy for test. In the various write-ups on his work, Warner describes the problems with the craft as wave action tended to scoop the air out from under the boat causing it to stop abruptly in the water. He redesigned his boat to use the mechanical "planer seals" as non-return valves for the trapped air which seemed to solve the problem and allowed him to win his races. Douglas Warner was a prolific inventor and builder of his unique boats. Many of his features can be found in his US Patent 1819216 that clearly shows the *sidewalls* and the *planer seals* and the air system (see Figure 2.14).

As the US Navy later pursued various versions of planer seals in its SES designs, it is useful to document some of the key features that Warner explored in his work. Much of that work was prescient of the later problems in planer seal design. From the Warner patent:

> *".... provide a boat with spring resisted bow and stern flaps to permit the tops of high waves to pass under the boat without disturbing the force of the waves and to prevent any suction from being created under the boat by the waves...."*
>
> *".... In very high speed boats, this flap may be reduced to only two narrow flexible strips which plane over the water and hold the bow a predetermined distance above the average water surface....."*
>
> *".... At still higher speeds, the stern flap may also be replaced by narrow skids thus reducing the variation in air pressure caused by large waves and holding the boat even steadier...."*

The patent contains many other features, where Warner has included ducts and valves to redirect pressurized air into the cushion as conditions dictate when traveling at high speed over waves. The stability of the boat was assured by sufficient width buoyancy hulls as shown in the boat cross section in Figure 2.14. An important difference between D K Warner's original *captured air bubble* design (1929) and that of the later US Navy version (1964) was Warner's inclusion of thickness on the inside of the hulls to provide the necessary buoyant force and roll stability to the craft (as shown in the cross-sectional view in Figure 2.14). Compare this buoyancy addition with the US Navy concept in Figure 2.25 that does not have this stability feature.

In the later discussion of the US Navy developments in the 3KSES Program to obtain a successful design of planer seals, first for the SES-100A and then, for the larger displacement high speed SES, these findings by Warner will be revisited.

Warner went on to patent, design and build many versions of hovercraft and sidehull SES-like craft that have application to many of the designs used today and proposed by others in future work including the *aerodynamic SES* discussed and evaluated later in this book (Chapter 11).

After Warner's "reduction to practice" of the *captured air bubble* boat in 1929, other inventors appeared on the scene exploring the principle of air cushion support. Each inventor was motivated by different aims but the end result was the introduction of various types of air cushion supported craft, some using static lift and some using dynamic lift, each with a different purpose. In January 1935, the Finnish engineer, Tovio Kaario[12] pioneered his first air supported vehicle that operated in ground effect that was actually a forerunner of the aerodynamically supported air cushion craft. This is discussed more in Chapter 11 on *"aerodynamic air cushion craft*". Kaario built other machines in the late 1940s and presented some of his ideas at the Princeton Symposium in 1959.[13]

2.7 Dr. William R. Bertelsen's Hovercraft (1958+)

In the mid-1950s, Dr William R. Bertelsen (1920–2009), a medical doctor and inventor was exploring ways of rapidly getting to his patients in outlying rural areas of Illinois in all kinds of weather. He explored various innovations in VSTOL flight and worked closely with NACA on his designs. Dr Bertelsen concluded that for "all weather" operation to get to his patients, that helicopters were not the answer for cost and safety reasons and explored the use of the air cushion. He designed and built a small air cushion craft that resembled the later work by Sir Christopher Cockerell. Bertelsen called it the *Aeromobile (A-35)*. It was powered by a twin-cylinder 35hp Righter drone engine. Dr Bertelsen was quite prolific in his inventions and the pursuit of unique forms of air cushion craft. Figure 2.15 shows the *Popular Science* magazine cover of July 1959 recording his early "flights" of the *Aeromobile A-35* on 6 April 1959, together with a photograph (provided by his son William D. Bertelsen to this author) of Dr Bertelsen flying his *Aeromobile* on a country road in Neponset, Illinois. Dr Bertelsen also demonstrated this hovercraft with a larger engine (the *Aeromobile-72*) at the Princeton Symposium[14] on Ground Effect Phenomena on 21–23 October 1959. The flight of the *Aeromobile (A-35)* on 6 April 1959 preceded the 25 July 1959 flight across the English Channel of the Saunders Roe SRN.1 (see later).

Dr Bertelsen's developments cover a wide range of innovative vehicles including novel forms of VTOL aircraft and air cushion craft that include the *Arcopter VTOL* that follows principles of directing propulsion air beneath the craft for air cushion support; the *gimbal fan air cushion craft* that integrates propulsion and lift units;

***Popular Science* cover**
(July 1959)

Photo courtesy of William D. Bertelsen

Dr. William R. Bertelsen flying his Aeromobile (A-35)
on country road in Neponset, Illinois

Figure 2.15: Dr William R. Bertelsen's Air Cushion Craft (1959) and other forms yet to be exploited or developed by the air cushion craft community. See Mantle, *Air Cushion Craft Development* for more description of Dr Bertelsen's *gimbal fan Aeromobiles* and projections for larger craft up to 300 ton displacement size. Many of Dr Bertelsen's designs and inventions had direct applicability to some of the US Navy's objectives in seeking a fast yet economical platform to be used in a variety of terrain (or sea) conditions.

2.8 Sir Christopher Cockerell's Developments (1955–1999)

About a quarter of a century after Douglas K. Warner's development of the *captured air bubble* boat, in 1953 Christopher Cockerell (1910–1999) (Figure 2.16), an electrical engineer, who owned a small boatyard in England, was looking into ways of making ships go faster. He expanded on the earlier work of Douglas K. Warner, by building, first a 16 foot boat and then a 20 foot boat that employed the same *captured air bubble* concept as Warner's boat except that now a powered air blower fan was installed in the boat to provide the air in much the same manner as today's hovercraft and surface effect ships.

He set up a small company called Ripplecraft in 1953 to develop his ideas and received his first patent[15] on hovercraft inventions on 12 December 1955 and was later knighted in 1969 for his many achievements in this new form of transportation. His name for all forms of air cushion supported craft whether amphibious or sidehull sealed was "hovercraft" which was officially included in the English language as a word in the *Collins English Dictionary* in 1959.

Unfortunately, while Christopher Cockerell intended the word to cover all forms of air cushion supported craft, the general dictionary definition is for *"a vehicle that travels over land or water supported by air thrust downwards. . .".* Such a definition would rule out those craft that use sidewalls or sidehulls. Yet the hovercraft organization, *"The International Hovercraft Society"* and others clearly includes both types of craft!! This confusion will come up at several points in the discussion but the reader should be able to distinguish from the context at any particular juncture which form is being discussed.

Figure 2.16: Sir Christopher Cockerell (on left) and the author in discussion, Brighton, England, 1974
Photo from author private collection

The author prefers the broader term *"air cushion craft"* that meets the intention for all forms of amphibious and non-amphibious types of air supported craft including the aerodynamic lift versions that have also appeared in the total mix of air cushion supported craft throughout the world and are discussed in more detail in Chapter 11. Because of Sir Christopher's vast experience that started from these humble beginnings, the author brought him on board as an advisor to the US Navy's Advanced Naval Vehicles Concepts Evaluation Project (ANVCE) in 1976 (see Chapter 10).

Cockerell described his approach and techniques in various papers at the time (circa 1959) and his papers were published as part of the *Symposium on Ground Effect Phenomena* held at Princeton University on 21–23 October 1959 in cooperation with the US Army[16]. Those Princeton papers describe some of his work in his own words. It was a well attended symposium with papers from experts around the world (UK, US, Canada, France, etc) and included over 250 experts together with interested people from both government and industry.

In the words of Christopher Cockerell, describing his second *captured air bubble* boat:

> *"Then a large centrifugal blower was installed in a 20 foot, 20 knot launch, and the air supply was taken over the bow and fed out at the rear of a hollow step....... The next scheme was for a craft with deep side walls, hinged trailing doors fore and aft, with the center pumped up with a blower"*

Christopher Cockerell pursued many alternate solutions for the *rigid planer seals* that Douglas K. Warner (and Cockerell) used in his boats. Cockerell also explored the use of "water curtains" and "air curtains" instead of the rigid hinged, planer seals, as he was aware of the inherent problems of such seals if the intention was to go fast over water. This was prophetic, as will be discussed later on the trials and tribulations in the development of the US Navy large, high speed surface effect ship program (see Chapter 6) This developmental work occupied him throughout the period of 1953–1956 (as described in his notes and shared at the symposium) that included both *sidewall craft* and *amphibious craft* with various ways to contain the air cushion and also retain a reliable "seal" to the trapped air cushion beneath the boat.

In August 1957 Cockerell managed to secure a contract from the UK Ministry of Supply to be given to Saunders-Roe, Ltd in Cowes, Isle of Wight, England for more research and development of his ideas. There was an annoying bureaucratic classifying of his work in the beginning by the Ministry of Supply that held up meaningful debate, but the work was finally declassified and this allowed Christopher Cockerell to set up his Hovercraft Development Ltd (HDL) in Hythe, near Southampton, England under the auspices of the National Research Development Corporation (NRDC). This was to develop his patents and to license various entities who wished to build both his *side hull captured air bubble craft* and his *amphibious "hovercraft"* to use the word coined by Christopher Cockerell to cover his work.

Over the years, various writers and developers have made a distinction between "*sidehull SES*" to mean hovercraft *with* sidehulls, and "*hovercraft* or *air cushion vehicle*" to mean amphibious hovercraft *without* sidehulls. The reader will frequently find different categorizations of the basic principle depending on the need or sometimes through simple confusion or misunderstanding. This book will try to keep the confusion to a minimum! Even in those early days, it is interesting to note that Christopher Cockerell was well aware of the problems of a ship hull sitting high out of the water on a cushion of air to avoid rolling over. Accordingly, at the same

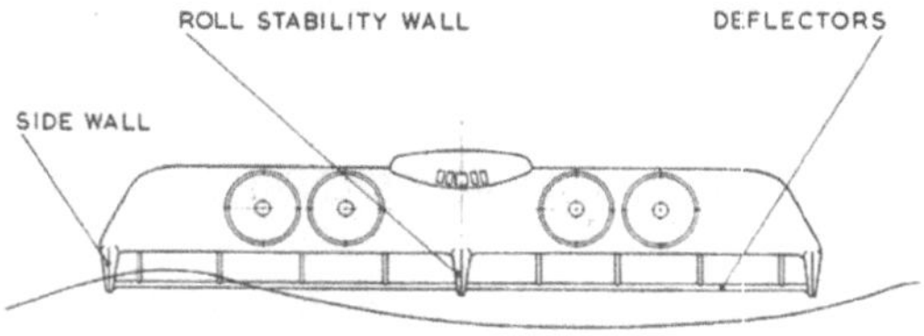

Figure 2.17: Christopher Cockerell's *Captured Air Bubble* Ship (Mid-ship section)(1959)

symposium in 1959, but in a different paper, Cockerell presented what the cross section of what a *captured air bubble* ship might look like with air captured between the two sidewalls of the ship by adding a "roll stability wall" running fore and aft down the center of the ship (see Figure 2.17 taken from the 1959 Princeton symposium) to give the necessary transverse or roll stability to such a craft. Note the low height-to-beam ratio (h_c/B) of the ship as an essential contribution to roll stability. The "deflectors" shown are another feature in Cockerell's design (as a means to control flow and raise the height of the air bubble if the cushion air is supplied by peripheral jets) but that detail is not discussed further here. Figure 2.17 shows the transverse sea wave that would cause leakage from the air bubble and further contributing to the destabilization of the sidewall designed ship.

2.9 Hovercraft Development Ltd Sidewall Hovercraft (1963+)

The first sidewall hovercraft, that were built under the Cockerell (or HDL) patents, were the HD-1 and the HD-2 built under license by J. Samuel White shipyard in Cowes, Isle of Wight, England in 1963 for HDL research. Both HD-1 and HD-2 were built as research craft to explore the various ship designs of Christopher Cockerell. These craft were also used to explore the new concept of *flexible skirts* introduced by a well-known English aircraft designer and inventor, C. H. ("Chookie") Latimer-Needham (1900–1975)[f]. Figure 2.18 shows the original sidewall craft HD-1 *after* it had been retrofitted with *flexible skirts* along the sides. Its length overall was 50 feet, with a beam of 23 feet. Its maximum speed was 35 knots.

Figure 2.18: HDL "*HD-1*" (1963) (Original sidewalls removed & replaced with skirts)
From Jane's Surface Skimmers 1967–1968

Then, based on the success of the *sidewall captured air bubble* craft HD-1 and HD-2 test craft, Cockerell, through his newly established Hovercraft

[f] Latimer-Needham subsequently sold his patents in October 1961 to Saunders Roe Ltd, that were the basis for all the flexible skirt and seal work that has appeared on both amphibious and sidehull SES around the world since that time.

Development Ltd company (HDL), licensed a shipyard in Scotland, Denny Shipbuilding to build the first such *capture air bubble ships* for commercial use. They were the Denny D-1 and the Denny D-2. Figure 2.19 shows the Denny D-2 on a regular passenger run in Scotland.

Figure 2.19: Denny Sidewall Hovercraft *"D-2"*, (1963) From Jane's Surface Skimmer Systems 1967–1968

The characteristics of such a craft can be seen in Figure 2.19, with the thin sidewalls and the seal across the bow, trapping the air cushion. In the designs of these sidewall craft, the seals were now variants of the rubberized fabric skirts that Latimer-Needham had introduced instead of the rigid planer seals. Unfortunately, due to reasons unrelated to hovercraft work, William Denny & Company went into voluntary liquidation in 1963.

2.10 Saunders-Roe Developments (1959–2000)

The Denny sidewall hovercraft *"D-2"* transported fare paying passengers on regular runs at speeds of 40 knots. Christopher Cockerell continued his work on both the *sidewall hovercraft* and most energetically on the *amphibious form of hovercraft.* Immediately upon forming his Hovercraft Development Ltd (HDL) in 1958 to control his patents, the Ministry of Supply had issued a contract to Saunders-Roe, in October 1958, to build his first amphibious form of hovercraft. Work started immediately on the Saunders-Roe hovercraft the *"SR-N1"*. This craft is shown in its original configuration and later with its four foot skirt attached in Figure 2.20. The hard ground clearance for the SRN.1 can be seen clearly in the left hand photograph. It was in this original configuration that it made its historic maiden voyage from Calais to Dover on 25 July 1959. The approximately 4.25 ton initial weight

Original *SRN.1* without skirts (1959) **SRN.1 outfitted with 4 foot skirt (1962)**

Figure 2.20: First Hovercraft *"SRN.1"*
Photos permission by GKNAerospace©

displacement craft crossed the calm sea of the English Channel at an average speed of 13.8 knots propelled by a 450 hp Alvis Leonides engine(with sprints to 20 knots depending on the wind and wave conditions during the crossing).

Christopher Cockerell was on board for that Channel crossing. The craft in its original configuration was somewhat limited with its low efficiency air jet propulsion and low air clearance but later improvements including upgraded propulsion (a Bristol Siddeley Viper engine of 1500lb thrust) and the addition of the 4 foot deep skirts allowed the craft to reach speeds of up to 60 knots in calm seas and about 25 knots in 4–5 foot seas. As the skirt and hovercraft design principles were refined, later versions of the Saunders-Roe hovercraft designs made significant leaps in performance and utility – but the concept of air cushion support had been established. The earlier cited references provide a wealth of descriptive material on all the British Hovercraft Corporation and other manufacturers' hovercraft.

The date of 25 July 1959 was chosen to be on the 50th anniversary of Bleriot's flight across the English Channel on 25 July 1909. As Christopher Cockerell said this was "to commemorate Bleriot's flight" (albeit traveling in the opposite direction). The description on this first hovercraft trip can be found in the 1959 Princeton Symposium in papers by Christopher Cockerell; by Lt Cdr Peter Lamb (Chief Test Pilot of Saunders-Roe) who "flew" the *SRN.1*; and by Richard Stanton-Jones (Chief Designer of the *SRN.1*). Richard Stanton-Jones (1926–1991) (Figure 2.21) had an illustrious career after joining Saunders-Roe in 1949 and moving up with increasing responsibility through the ranks first as Chief Aerodynamicist (1956), then Chief Designer (1959), then Technical Director of British Hovercraft Corporation (1966) and finally as Managing Director (1968) of British Hovercraft Corporation as Saunders-Roe passed through its corporate history from Saunders-Roe Ltd to being a division of Westland Aircraft Ltd and finally as British Hovercraft Corporation (BHC) that was formed from the companies of Westland Aircraft Ltd, Vickers Aviation and NRDC on 6 December 1966. The name was changed to Westland Aerospace in 1984 and design and manufacture of hovercraft ceased at that time. GKN Aerospace, a leading manufacturer of composite and metallic aerostructures and engine products, acquired Westland in 1994.

Figure 2.21: Richard Stanton-Jones (1926–1991)
Photo permission by GKN Aerospace©

Richard Stanton-Jones was the designer of the *SRN.1*, based on the patented work of Christopher Cockerell. He then went on to design or be responsible for most of the hovercraft built by the British Hovercraft Corporation up to and including the largest hovercraft in the world the *"SRN.4"* which was launched on 4 February 1968 (and entered commercial service on 1 August

Figure 2.22: *"Super 4"* (1978–2000)
Photo permission by GKN Aerospace©

1968); and then, on 6 April 1978 was launched in its stretched version as the *"SRN.4 Mark 3, The Princess Anne (the Super 4)"*.

Figure 2.22 shows the "Super 4". It made its last run on 1 October 2000. It was determined that by that time that the craft were too expensive to operate and maintain. The stretched SRN.4 has a displacement of 300 tons and can carry 416 passengers and 60 cars over a distance of 150 miles. Its speed is typically 60–65 knots in relatively calm seas (2 foot seas with 5 knot winds) and 35–45 knots in rough seas (8 foot seas and 27 knot winds)

The vast design and operating experience of all the British Hovercraft Corporation products provided significant background and design criteria for all forms of hovercraft, amphibious and sidewall type on such key matters as skirt (and seal) design. This commercial operating data over a variety of operating conditions in real world environments was most useful to augment the US Navy in its developmental work to determine the characteristics of high speed *sidehull surface effect ships* in the US Navy plan to develop an ocean going 3000 ton SES.

The author owes a debt of gratitude to Richard Stanton-Jones on at least two accounts. First, he encouraged and sponsored the author, after completing his apprenticeship at Saunders-Roe, to enter the College of Aeronautics, Cranfield, England to obtain his Masters Degree in Aeronautical Science; and second, he taught the author the value of "rules of thumb". This latter skill proved invaluable in later years as "common sense tests" based on physics, on claims of new high speed ship designs as the US Navy pursued the goal of a "100 knot navy". The first example the author experienced on the "rules of thumb" lesson was sometime in 1955 when he had been tasked to compute the take-off performance of the Saunders-Roe mixed propulsion unit (rocket and jet engine) fighter SR.53 (and its brother the P.177). After laboring for several hours on a "Marchant calculator" and a detailed spread sheet, produced a take–off time of 40 seconds. As the author stood alongside Richard Stanton-Jones' desk as the spread sheet was checked, he noticed that Stanton-Jones calmly looked over the sheet and jotted a few short formulae on a side notepad along with some quick calculations. After a few minutes he looked up and said, *"Would you mind going over these calculations again because I get about 30 seconds?* Dick was always polite. Since it was a complex set of calculations of thrust, drag and other parameters to do the spread sheet, the author was surprised but after redoing the calculations, did find an error and the final answer was indeed closer to 30 seconds instead of 40 seconds![g]

The second example was even more revealing. British Intelligence had sent a message through channels to the Aerodynamic Department at Saunders-Roe that said that one of their agents would have the opportunity to photograph a Soviet aircraft and the question was: *"Which would be the most useful viewpoint?"* Again, the task fell to the author with a lot of help from Stanton-Jones. The eventual answer after a lot of analysis was *"a three quarter rear view"*. With that photo, an analyst could determine the length of the fuselage (through co-ordinate geometry). This would lead to an estimate of the weight of the aircraft. The same view would give a measure of the wing span that would help in the estimation of the induced drag. Finally, the three-quarter rear view would give the diameter of the tail pipe that would lead to an estimation of the thrust. A knowledge of the state-of-the-art and projections of the state-of-the-art of aerodynamics, propulsion and structures coupled with those three dimensions would allow a reasonable estimate to be made of the aircraft performance. Years later, while serving in The Pentagon, the author had the opportunity to check the results of the Stanton-Jones derived "rules of thumb" with the actual values of the Soviet aircraft and found an amazing agreement. The use of simple rules based on physics and state-of-the art projections provide useful tools for evaluation of new concepts. It does not invalidate new claims (necessarily) but it forces the new claimant to be forthright and explain why his concept was different to the state-of-the-art and offers improved performance.

2.11 US Navy Amphibious Air Cushion Craft Development (1965–today)

While the commercial use of hovercraft has not expanded greatly in the intervening years, the example of military use of the amphibious air cushion craft is worthy of mention as it shows how speed has offered a distinct military advantage in at least one mission area. The US Navy in the 1965–1970 period developed a capability to get Marines across the beach at high speed to avoid all previous experiences in World War I and II where the Marines were exposed to enemy action in a most vulnerable manner.

The initial development resulted in prototypes called JEFF A and JEFF B (named after one of the Navy Program Manager's sons) that were launched in 1977 (JEFF B) and 1978 (JEFF A). The program resulted in an operational craft based on the JEFF B called the Landing Craft Air Cushion (LCAC) being delivered to the US Navy in 1984 and achieved an Initial Operating Capability (IOC) in 1986. Full production approval went into effect in 1987 and to date 91 craft have been delivered. The last craft, LCAC 91 was delivered to the US Navy in 2001. The program is now in an advanced stage of Service Life Extension Program (SLEP) which is expected to be completed in 2018. A new competitive program called the

[g] The numbers of "30 seconds" and "40 seconds" are arbitrary here as the elapsed period of over half a century from that event has dulled the author's memory a little!

LCAC-50 **operating in rough water**

LCAC-81 **unloading 60 ton tank on beach**

Figure 2.23: US Marine Corps Landing Craft Air Cushion (LCAC) in operation (circa 1990s) US Navy photos

Ship to Shore Connector (SSC) program is the new *air cushion landing craft* which are planned to be delivered in 2019. Figure 2.23 shows the LCAC operating with US Marine units.

The LCAC is a landing craft with a displacement of 170–182 tons measuring 87ft 11 inches long, 47 ft beam. It is powered by four Lycoming TF-40B gas turbines in an integrated (propulsion and lift) power arrangement. In the SLEP program these engines are upgraded to the ETF-40B versions. The range of the craft is about 200 miles at a speed of 40 knots carrying the payload of one 60 ton tank. It can carry 24 troops or one main battle tank (MBT). The speed of the LCAC is 40 knots+ with the full load and is capable of speeds in excess of 70 knots with a light load in relatively calm sea conditions.

The speed of 70 knots has been the "norm" for both the commercial and the military vehicles using the *amphibious air cushion concept* from the SRN-4 to the LCAC. The state of the art does not allow these craft to approach anything like the sustained speed of "100 knots" capability sought by the US Navy in its broader mission goals. The main advantage is clearly its unique ability to operate across the interface between the beach and the sea – an essential military need in a well defined military mission.

In the case of amphibious landings, the US Navy has opted for transporting these high speed (40–70 knot) craft to the area of operations in the slower speed "landing ship docks". Figure 2.24 shows the prototype JEFF (B) entering the USS *Spiegel Grove* in tests off Panama City, Florida in 1979. The LCACs have been transported to their areas of operation in recent years (1991+) in several types of ships such as the Multi-purpose amphibious assault ships such as the 23 knot USS *Wasp* that can carry different "load-outs" of amphibious assault mission make up of helicopters, VTOL aircraft and LCAC. Typically the LHD would carry three LCAC plus the aircraft and troops.

While Christopher Cockerell was experimenting with his "coffee can experiment" with the air cushion peripheral jet and noticing the increased pressure under the can, it was Richard Stanton-Jones[17] (with a little help from Gordon Elsley[18] in the Saunders-Roe Aerodynamics Department) who developed the simple theory relating the cushion pressure required to lift the vehicle to the pressurized air jet exhaust pressure and the geometry of the jets in what became known as the

Figure 2.24: JEFF (B) entering USS *Spiegel Grove* (LSD-32) (1979)
US Navy photo

exponential theory. With some simplification assumptions about the variation of the total pressure head across the peripheral jet it became possible to express the key parameters in a simple and concise form. Until the hovercraft and surface effect ship designs moved more into the plenum chamber configurations with various skirt and seal designs, the exponential theory for peripheral jet supported vehicles proved a useful design tool. This simple relationship is given by:

$$\frac{p_c}{H_j} = 1 - e^{-2x} \tag{2.3}$$

where (p_c) is the air cushion pressure to lift the craft; (H_j) is the pressure source total head and (x) is the peripheral jet geometry term,

$$x = \frac{t}{h}(1 + \cos\theta) \tag{2.4}$$

where (t) is the thickness of the jet, which is inclined inwards at an angle (θ). The term (h) is the actual clearance beneath the jet exit and not the clearance of the hard bottom of the craft. These simple relationships were confirmed with experimental results by Peter Payne[19] for the US Army in 1965. A discussion of Peter Payne's invaluable contributions is provided in Chapter 10 on the US Navy's Advanced Navy Vehicles Concepts Evaluation Project (ANVCE).

Several other authors have developed similar simple expressions at about the same time. Some approached the problem from the aeronautical world of ground effect of aircraft and helicopters and some from those directly involved in designing and building "hovercraft". These include Boehler[20], Chaplin[21] and Barratt[22] among others. A comparison of the simple exponential theory (given above) with various experimental results and also the comparison between peripheral jet theories and plenum theories (more applicable to today's hovercraft and surface effect ships) can be found in Mantle (1975).[23]

2.12 US Navy Developments in Sidehull SES (1960 –1980)

Following the flurry of development of the many different kinds of "air cushion craft", "hovercraft", "captured air bubble craft", and "ram wing air cushion craft" that had spanned the three decades from the Warner boats in 1929 to Cockerell's modern day developments starting in 1953 to the first hovercraft by Saunders-Roe in England in 1959; the US Navy officially entered the development process in 1963 when it began its research program into "sidewall" types of air cushion craft. The US Navy built its first sidewall air cushion craft (the XR-1) in 1963, that closely resembled Warner's captured air bubble boat of 1929 (see Figure 2.14) except that the XR-1 incorporated air blowers similar to those first used by Sir Christopher Cockerell in his original designs and test craft in 1959 (see earlier).

This design, patented by Allen G Ford[24] a year after the XR-1 launch, is shown in Figure 2.25.

This version of the captured air bubble concept may be compared with Warner's patent (Figure 2.14). The key features (quoting from the Ford patent) to contain the *captured air bubble* are;

(a) "..... a pair of thin sideboards...."
(b) ".....and a ski assembly fore and aft....."
(c) "..... a substantially airtight-watertight airspace...."

The first feature of this US Navy CAB patent is the (non-buoyant) *thin sideboards.* There is no discussion of the stability of the craft in the patent and the assumption is that it will be stable laterally without the benefit of restoring hydrodynamic forces from any *buoyant sidehulls.* Indeed, the assumption is that lateral stability will be provided by the two laterally divided "captured air bubbles" (between sideboards and skiis). Apparently, the center sideboard is not essential as, further in the patent, it states:

> *"...... removal of it (the center sideboard) and use only one assembly fore and one assembly aft identical in configuration to the one described but twice as wide would not detrimentally affect the basic principle of operation of the present invention....."*

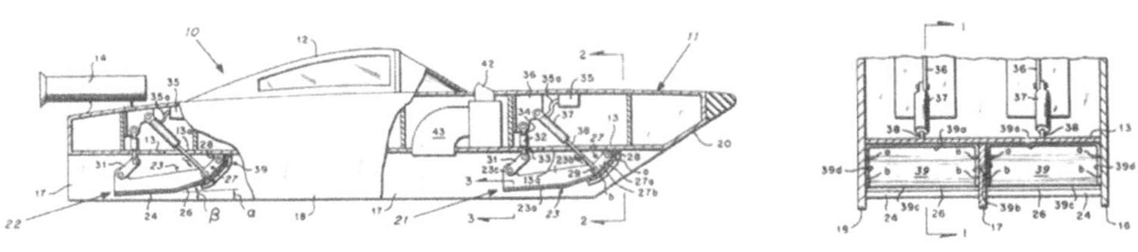

Figure 2.25: Allen G. Ford's *Captured Air Bubble* Boat *("XR-1")* (1964)

Clearly, the assumption is that sitting atop a bubble of air was stable, without any need for a restoring mechanism of any kind. This was an unfortunate feature of the Navy design which resulted in a catastrophic capsize of the XR-1 during a test run on 8 December 1964, one year after its construction in 1963. In the December 1964 Annual Report of the Naval Air Engineering Center (NAEC) which was the Navy lab owner and builder of the XR-1, the Commanding Officer, Capt A. H. Clancy, Jr reported the following on the XR-1 capsize[h]:

> *"..... On 8 December 1964, the XR-1 overturned during a starboard turn which was started at a speed of approximately 35 knots....... The basic cause of the capsizing of the CAB XR-1 test vehicle has not yet been substantiated firmly. Based on analysis conducted to date, it is believed that the cause was lateral instability of the particular configuration of the XR-1, due to a dynamic interaction between roll, pitch, and yaw..."*

The XR-1 had been conceived in 1960, less than one year after the Princeton Symposium where Warner, Cockerell and others had specifically cautioned against such a design (see earlier).

Capt Clancy continues:

> *"..... The initial theoretical studies indicated that CAB offered great promise of providing efficient high speed naval ships over a wide range of sizes and classes..... Discussions with BUSHIPS and David Taylor Model Basin working level personnel were held to ascertain interest in sponsoring a research program on CAB. Reaction was generally negative......."*

Despite this negative reaction, the program went ahead as the CAB program was transferred from NAEC to DTNSRDC, a sister Navy laboratory. Even if a basic stability analysis had not been made, the tow tank tests should have unearthed the instability of such a CAB concept. Unfortunately, the limitations of the NAEC linear Tow Tank restricted the test to only "head on" wave conditions, and no turning or maneuvering tests were conducted. The restraining of the model for head-on motion only, even in calm water conditions, precluded any uncovering of a basic "pitch-roll-yaw" interaction that was well known for any such marine vehicle. In addition to the Warner and Cockerell warnings in Princeton in 1959, some of the early tests with the Saunders-Roe hovercraft (SRN-1 and others) that involved related stability problems with "pitch clicks" and "plow-ins" should have raised a red flag.

What is surprising was the inclusion in the same Annual Report by NAEC, the comment:

> *"... The instability exhibited by the XR-1 is not considered inherent in the CAB concept, and can be corrected with appropriate design changes....."*

After that time, the XR-1 did indeed go through a radical design change and the craft became a work horse for trying out various ideas especially as the US Navy tried unsuccessfully to solve the problem of the second feature of the cited CAB patent, namely the seal mechanism or *"mechanical ski assembly"* used in the bow and stern of the craft. The intention was to design a type of seal mechanism that did not follow the tried and proven flexible skirt system patented by the British. Solving this seal

[h] The fatality of an onboard crew member that occurred as a result of this overturning was not reported in the NAEC Report. See Chapter 8 for a more complete description of conditions of this overturning of the XR-1.

problem proved to be a stumbling block in the US Navy program of the high speed SES initiative (1969–1979).

In 1965, the US Maritime Administration (MARAD) were still examining the form of the ship that would best give them the "100 knots" capability and a Surface Effect Ships for Ocean Commerce (SESOC) Committee was formed in September of that year. The concepts included in the evaluation were: the *amphibious air cushion craft* (based on the British hovercraft); *aerodynamic air cushion craft* (WIG and channel flow that MARAD was exploring); the air lubricated hull or *hydrokeel*, and the new US Navy *CAB concept* described above. Chapter 4 describes the MARAD program in more detail. The results of their analyses and recommendations were published in 1966[25].

Because the CAB concept was put forward with the features identified of "*thin sideboards*" as described above it should not be surprising to find that the predicted hydrodynamic resistance of the "design" was very low, which allowed the US Navy to put forward a *CAB design of 10,000 ton displacement with a maximum speed in calm water of 129 knots!* This result was in spite of all the documentation in the same report of the aerodynamic air cushion craft analyzed, that, with no contact with the water in cruise mode, projected no more than 100 knots capability. There is no mention in the 1965 SESOC report or subsequent US Navy report that initiated the US Navy SES program, of the failed XR-1 test craft in 1964 with the incurred lateral stability problem of "*thin sideboards*".

2.13 The Joint MARAD and US Navy Program on Surface Effect Ships (1967–1970)

The US Maritime Administration now wished to proceed with the 10,000 ton CAB design with its projected *129 knot speed over the sea* capability as (on paper) it presented a solution to their problem of a declining American shipping market. Based on these findings, both the US Maritime Administration and the US Navy redirected their efforts as they formed their Joint Surface Effect Ships Project Office (JSESPO) in 1967 to develop the CAB design and dropped all further consideration of the *aerodynamic forms of air cushion craft*. Based on the US Navy findings, MARAD canceled its ongoing contract with the "*Columbia*" project (see Chapter 1 and 3) and redirected its efforts to *captured air bubble* and *sidehull SES* concepts to attain its "100 knots at sea" goal. *Aerodynamic air cushion craft* were not reintroduced until 1976 when the US Navy evaluated, in the ANVCE Project (see Chapter 10), the successful Russian efforts on the *Ekranoplan* series of aerodynamic air cushion craft that had demonstrated speeds greater than 200 knots, and included them in the comparison of alternatives to the 3KSES as a result of the DDR&E directive in 1975.

The *aerodynamic air cushion craft* including US, Russian and other nations' operational craft, is discussed in detail in Chapters 3, 10 and 11.

2.14 US Navy Surface Effect Ship Program Office Developments (1969–1980)

The US Maritime Administration (MARAD) pulled out of the joint effort in 1970–1971 and the US Navy continued on its own program. A description of the MARAD efforts prior to leaving the development is described in Chapter 4. The major effort to seek a "100 knot at sea" capability was then pursued by the US Navy over the period 1969–1980 and a detailed treatment of that effort is provided in Chapter 5.

2.15 Postscript

From the historical assessments it is found that the *hydrofoil* and *amphibious air cushion craft* developments around the world had settled into a 50–70 knot capability. The development of 100 knot capability evolved from the MARAD (originally) and eventually with the US Navy efforts of different versions of those *air cushion craft* concepts that used *different forms of sidehulls* to both help contain the air cushion and also to provide a marine propulsion scheme. This billion dollar development effort with its trials and tribulations is discussed in detail in the following Chapters.

References

1. Dawson, Christopher, *"A Quest for Speed at Sea"*, Hutchinson & Co, London, England, 1972
2. King, H.F., *"Aeromarine Origins"*, Putnam & Co, London, England, 1966
3. Johnson, Robert (Capt retd), *Chapter V: Hydrofoils*, Naval Engineers Journal (Special Edition), February 1985
4. Ellsworth, William M., *"Twenty Foilborne Years"*, published by US Navy DTNSRDC, 1986
5. Hayward, Leslie, *"The History of Air Cushion Vehicles"*, Kalerghi-McLeavy Publications, 1963
6. Mantle, Peter J., *"Air Cushion Craft Development"*, published by US Government Printing Office January 1980 (ISBN 008-047-00291-2)
7. International Maritime Organization, *"IMO High Speed Craft Code"* (Part 0, Chapter 6), July 1998
8. Parkin, J. H., *"Bell and Baldwin (their development of aerodromes and hydrodromes in Baddeck, Nova Scotia)"* published by University of Toronto Press, 1964
9. Hanns von Schertel, *"Hitler's Hydrofoils"*, published in Sea Classics, Vol 7, No.1, pp 12–17, January 1974
10. Rethorst, Scott and Potter, W.T., *"Maritime Administration Surface Effect Ship"*, SNAME National Aero-Nautical Meeting, Paper Number 697D, April 8–11, 1963
11. Warner, Douglas K., *"Air Cushion Boat"*, US Patent 1,819,219 filed 20 August 1929, patented 18 August 1931
12. Kaario, Tovio, "Process for Eliminating Friction between a Surface Vehicle and the Surface", Finnish Patent 18630, January 1935
13. Kaario, Tovio, *"The Principles of Ground Effect Vehicles"*, Symposium on Ground Effect Phenomena, Princeton University, Paper 20, 21–23 October 1959
14. *Symposium on Ground Effect Phenomena*, at the Princeton University, Department of Aeronautical Engineering, and the US Army TRECOM, 21–23 October 1959
15. Cockerell, Christopher S., "Improvements in or Relating to Vehicles for Traveling Over Land and/or Water", British Patent 854,211 issued 12 December 1955
16. *Symposium on Ground Effect Phenomena*, at the Princeton University, Department of Aeronautical Engineering, and the US Army TRECOM, 21–23 October 1959
17. Stanton-Jones, Richard, *"The Development of the Saunders-Roe Hovercraft SRN.1"*, Symposium on Ground Effect Phenomena, Princeton, Paper 13, 21–23 October 1959
18. Elsley, Gordon. H., *"Two-Dimensional Hovercraft Lift"*, Report A/A/53, Saunders-Roe Ltd, November 1958.

19. Payne, Peter R., *"Curved Jet Flows",* US Army Materiel Laboratory Technical Report 65–20, May 1965
20. Boehler, Gabriel D., *"The Airborne Ground Vehicle, a new device for improving Tactical Mobility of Ground Based Military Units"*, paper submitted to US Army Ordnance Research Office, January 1957
21. Chaplin, Harvey R., *"Theory of the Annular Jet in Proximity to the Ground"*, DTMB Aero Report No. 923, July 1957
22. Barratt, M. J., *"Simple Curtain Theory and the Split Jet,"* Technical Memorandum HDL/61/115, HDL Ltd, August 1961
23. Mantle, Peter J., *"A Technical Summary of Air Cushion Craft Development",* US Navy DTNSRDC Report 4727, October 1975.
24. Ford, Allen G., *"Captured Air Bubble Vehicle"*, US Patent 3,146,752 filed 28 August 1962, patented 1 Sept 1964
25. *Surface Effect Ships For Ocean Commerce (SESOC),* US Department of Commerce report published February 1966

3 The First Surface Effect Ship

"Columbia, the GEM of the ocean"

Dr. Scott C. Rethorst, 1959

There have been many historical examples of adding either "*air lubrication*" or "*air cushion support*" to ships since very early times to provide mainly greater speed, but also to accomplish other mission objectives for efficient marine transportation, but the name "*surface effect ship*" was a modern creation.

The name "*surface effect ship*" was coined during a car ride down California Boulevard in Pasadena, California in 1962. In the car were Dr Scott Rethorst, President of Vehicle Research Corporation (driving); Ed MacCutcheon, Chief of Research & Development, US Maritime Administration; his Deputy Jim Higgins; and this author. A photograph of Scott taken at CALTECH for his PhD program in 1956 is in Figure 3.1.

Figure 3.1: Dr Scott Carson Rethorst (1915–2011)
Photo permission by CALTECH

A few years earlier in 1959, Scott had patented his new high speed ship design that incorporated air cushion support and had successfully won a contract from the US Maritime Administration to design and build a prototype of a *channel flow* air cushion craft. The "*hovercraft*" had been invented earlier in England by Christopher Cockerell in 1955 (later knighted for his efforts) and all such vehicles, regardless of the type, are called "hovercraft" in England since that time. In the US, much of the parallel development work, which was in the aircraft industry, had been influenced by the recognized increased lift or "*ground effect*" experienced by helicopters (for example) as they approached the ground or hovered close to the ground. Because of this, it was common practice to refer (in the US) to any vehicle incorporating this effect to be called a "*ground effect machine*" or GEM. Scott, with his impish sense of humor, had accordingly dubbed his unique design (with apologies to Shaw and Becket) as, *"Columbia, the GEM of the Ocean"!*

During that car ride coming back from lunch and returning to Scott's offices on California Boulevard, Ed said that he was uncomfortable with the word "ground" because in the maritime business "we deal with the sea and not the ground". After some light hearted discussion we all settled on the word "surface" instead of "ground" because that allowed us to retain the feature that certain designs could indeed traverse both "ground" and "water or sea". The next thing that Ed did not like was calling this a "machine" which again did not fit the world of the maritime community and clearly "ship" was better. Nobody liked the "machine" label anyway. Ed conceded that the "effect" was still valid and so as we pulled into the driveway, the name "*surface effect ship*" was agreed and decided upon. Scott grumbled a little about losing his great joke to posterity but still called his design *"Columbia"* which appeared in all the documentation and the Press from that point forward.

3.1 Basic Theory of Channel Flow

The basic concept of the *channel flow craft* is the marriage of a conventional *hovercraft* for efficient operation at low speeds and the *wing-in-ground-effect* for efficient operation at high speeds. The particular speeds involved depend upon the size of craft and desired height above the ground or surface. At "low speeds" the channel flow craft operates exactly like a hovercraft (using Christopher Cockerell's name for amphibious air cushion craft) with pressurized air generated by an on-board fan system that is funneled into peripheral air jets to contain the pressure beneath the craft. At "high speeds", defined where the shape of the body can generate sufficient aerodynamic lift to support the full weight of the craft, the channel flow craft operates in the wing-in-ground-effect mode. The uniqueness of the design incorporates how the air jets contain the underneath pressure at the low speeds and also provides the side sealing to lower induced drag in the high speed mode of operation. Figure 3.2 shows the basic concept.

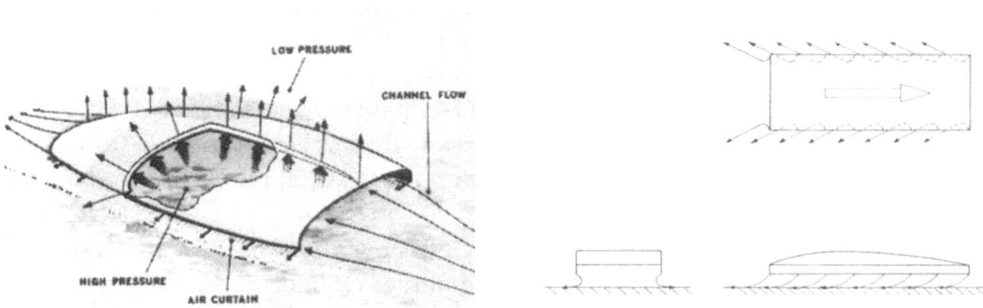

(a) Channel Flow System (b) Channel Flow Analytical Model

Figure 3.2: Channel Flow Concept

In Figure 3.2 (a) the craft is moving to the right and is in the cruise channel flow mode and the lift is being generated by the shaping of the body. The free stream air flow is channeled beneath and over the craft where the lift is generated on the upper surface through the low pressure and by the high pressure on the lower surface. All the usual features of wings or mounds close to the surface come to play in the generation of both lift and drag. While not shown in this diagram, it is easy to visualize that if air jets in the bow and stern were in operation, the high pressure underneath the craft would be generated by an on-board lift system (as in hovercraft) and the craft would be sustained by aerostatic lift. In the mode shown, the air jets are maintained along the sides of the craft to form an effective air seal that will (in the ideal case) eliminate the induced drag caused by any vortex formation from lower to upper surface flow. How well the air jets seal is a measure of the efficiency of the design.

In the right hand sketch of Figure 3.2 (b) is the analytical model for the air jet flow in the full channel flow operation. As in the first sketch, the craft is moving toward the right and the air jets are inclined both rearward and into the channel. The jets are inclined inwards to generate the necessary pressure rise and inclined rearwards to eliminate or minimize the ram drag that is inherent in any on-board pressurized air system. The rearward direction of the side air jets also serve to reduce the shearing forces between the sealing air and the adjacent free stream air.

The basic theory for channel flow craft may be found in the original work by Dr Scott Rethorst and Dr Winston Royce of Vehicle Research Corporation, Pasadena, California[1] under the sponsorship of the Office of Naval Research, US Navy. Dr Royce was a consultant to Vehicle Research Corporation during the early days of the channel flow development and was an Assistant Professor at the California Institute of Technology (CALTECH) in Pasadena, California. Dr Royce was also one of the author's thesis advisors at CALTECH in 1960. Another key person in the development of the concept was Tor Strand, who was the Director of Research at the company. Much of the basic aerodynamics of the inviscid incompressible flow of the air jets was developed by Tor.[2]

While the complete analysis of the channel flow craft as it translates along the surface of the ground (or sea), moving from the aerostatic mode (or conventional hovercraft mode) into the aerodynamic mode, is somewhat complex, it can be represented here using the work of Royce and Rethorst. In their 1961 paper, an example calculation is provided that describes all the essential features for a simple case of a simple rectangular vehicle. The resulting curves of power required and the lift-drag ratio for different operating heights provides an excellent visualization of how the concept works. Figure 3.3 (a) and (b) show the results.

Various assumptions have been made to facilitate the calculations. A lift coefficient of $C_L = 0.50$ has been assumed for the craft which is a conservative value for a body in "mound flow". For this example the craft can sustain itself in aerodynamic flight at a cruise speed of V = 120 knots.

The upper curves in Figure 3.3 (a) show the power required for the craft as it moves from the aerostatic mode (called "annular jet regime" in the chart); through the transition mode where the bow and stern jets are gradually retracted to form the channel; and into the fully aerodynamically supported mode (called "channel regime" in the chart) when the bow and stern air jets are fully retracted. The

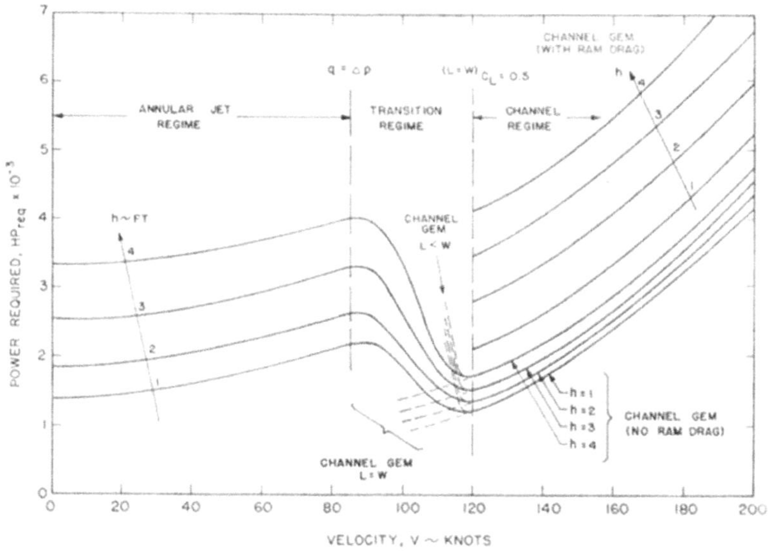

(a) Channel Flow Power Required from Aerostatic to Aerodynamic Mode

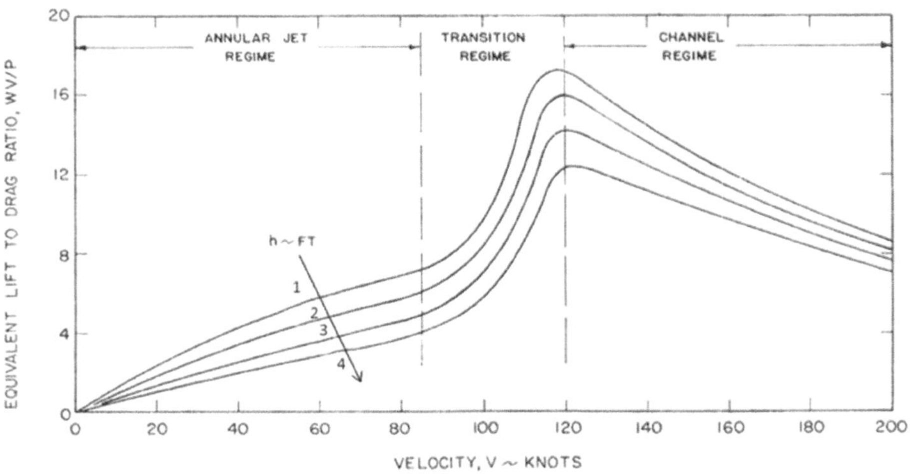

(b) Channel Flow Equivalent Lift-Drag Ratio through all Modes

Figure 3.3: Channel Flow Performance (sample calculation)

power required has been computed for a range of “hover heights” from 1–4 ft of clearance. As can be seen in the sample calculation, there is a transition regime where the dynamic pressure ($\frac{1}{2}\rho V^2$) of the free air stream exceeds the required pressure to support the craft but the aerodynamic lift when operated as a channel flow craft is insufficient to support the craft’s weight. This transition regime would require a careful opening of the bow and stern air jets as the supporting mechanism transitions. While not examined in the project it is suspected that retracting the bow air jets first would allow a "ram pressure" and lift to provide a smooth transition and then the rear air jets would be retracted.

In the lower set of curves in Figure 3.3 (b), the calculations have been made for the sample case for the equivalent lift-drag ratio. For ease of calculation the authors Rethorst and Royce assumed a propulsive efficiency (η = 100%) so that the transport efficiency (WV/P) and lift drag ratio are the same numerically. Obviously, in a real case the values shown for transport efficiency would be about half the values shown. The chart shows (i) that the transport efficiency decreases as height above the surface increases and (ii) the transport efficiency is significantly improved in the aerodynamic or channel flow mode over that in the aerostatic mode.

The original work for the *"Channel Flow GEM"* – a member of the *aerodynamic air cushion craft* family – was done under the sponsorship of the Office of Naval Research (ONR) during the period of 1958–1961. The US Maritime Administration became actively interested in *surface effect craft* in the Spring of 1961. After an industry wide competition, Vehicle Research Corporation (VRC) was awarded a contract on 2 August 1961 to undertake a *surface effect ship* (SES) (née GEM) design and development program sponsored by the US Maritime Administration. Extensive wind tunnel tests and experiments (in four different facilities in California and elsewhere) provided the necessary background to conduct a Preliminary Design of the ship, which was done in the period 1961–1963. These tests included both static wind tunnel tests and three degree of freedom tests to tackle static and dynamic stability issues as well as performance issues in the so-called *static mode* (as a conventional amphibious air cushion craft); and in *channel flow mode* of a wing-in-ground effect and in *transition mode* between the two modes of operation. The stability characteristics are discussed in Chapter 4 and 11. The specifics of size were not rigidly set as the development program progressed through a four Part program although the Class of Surface Effect Ship envisaged for the preliminary phases was the 100 ton Class. Chapter 7: *Economic Considerations* provides the projected Direct Operating Cost (DOC) for the *Channel Flow Surface Effect Ships* of 100 to 1000 ton displacement class. Preliminary designs showed a 40 ton payload and 10 ton fuel load. In response to the US Maritime Administration requirements, the design was to include a cargo compartment sufficient to carry four 8 x 8 x 20 ft containers of 10 tons each. Alternatively, the cargo compartment was to be altered to carry 150 passengers, their luggage, and break-bulk premium cargo. Operating speed for best performance was to be 100–120 knots, with a maximum speed of 150 knots. Over rough water it was estimated that the craft would operate at a nominal height of 9 ft (4 ft to the bottom of the jet nozzles and 9 ft to the body structure), which would enable the craft to negotiate large waves (to be specified after model testing). In 1961, the technology of the flexible extensions or "skirts" was not very mature (see Chapters 1 and 2), and only rough estimates could be made on performance. Preliminary work showed that the air gap (height that the nozzles would clear the water) would be about 1 ft. The suspension system's normal extension of 6 ft plus the hull depth of 5 ft would place the main body of the craft 12 ft above the water. Based on early analyses at Vehicle Research Corporation, the power would be provided to two mixed flow fans and two thrusting propellers by six engines having a total power of 13,620 HP for the 100 ton displacement ship. Calculations showed that 9,350 HP would be required for a cruise speed of 102 knots.

3.2 The MARAD Surface Effect Ship "*Columbia*"

The key features of the MARAD SES *"Columbia"* Program were given to the joint meeting of the Society of Automotive Engineers (SAE) and the American Society of Naval Engineers (ASNE) on 8 April 1963 by Dr Scott Rethorst (inventor) and a senior official from the US Maritime Administration in the US Dept. of Commerce, W. Taylor Potter, in a paper called *"Maritime Administration Surface Effect Ship"*.[3] The general concept of the *Columbia* is shown in Figure 3.4.

Earlier, a more in-depth technical analysis of the "*channel flow GEM*" had been published by the Institute of the Aerospace Sciences (IAS), forerunner of the AIAA, in 1962. Tor Strand, who was Director of Research at Vehicle Research Corporation, published the paper "*150 knot GEM Cruise*"[4].

The details are provided in the Strand paper but the overall design parameters predicted provide a useful guide as to what could be expected from such a design of an aerodynamic air cushion craft (see Table 3.1) The design point Equivalent Lift Drag Ratio (see Chapter 11) was selected, based on the wind tunnel test data and analysis, as $(L/D)_{eq} = 8.3$ for a 240 ton ship.

As the US Maritime Administration program continued at Vehicle Research Corporation in the period 1962–1963, the form of the expected design settled on the "100 ton Class" design and the "*Columbia*" took shape. The general description of the 100 ton displacement MARAD SES "*Columbia*" is shown in Figure 3.5 which is a sketch based on the engineering drawings at that time. The ship is a catamaran hull with the powered air cushion lift systems incorporated into the side hulls for sealing the air. At zero and low speeds the ship can operate as (a) a *conventional catamaran displacement ship*. Once leaving the dock, the air cushion systems are activated and the air cushion between the hulls is pressurized between the forward and rear seals. In this mode the ship is operating as (b) a *conventional hovercraft*.

Figure 3.4: MARAD Surface Effect Ship *"Columbia"*
Sketch by VRC artist J. Bryant

Table 3.1: *Preliminary Leading Particulars of 240 ton Class Channel Flow GEM*

Gross Weight	540,000 lb
Total Installed Power	37,600 HP
Side Jet Ground Clearance	6 ft
Side Jet width	1.1 ft
Air Volume Flow	84,600 ft^3/sec
Cushion Pressure at Cruise Speed	75 lb/ft^2
Lift Power	24,300 HP
Propulsion Power	13,300 HP
Cruise Speed	150 knots

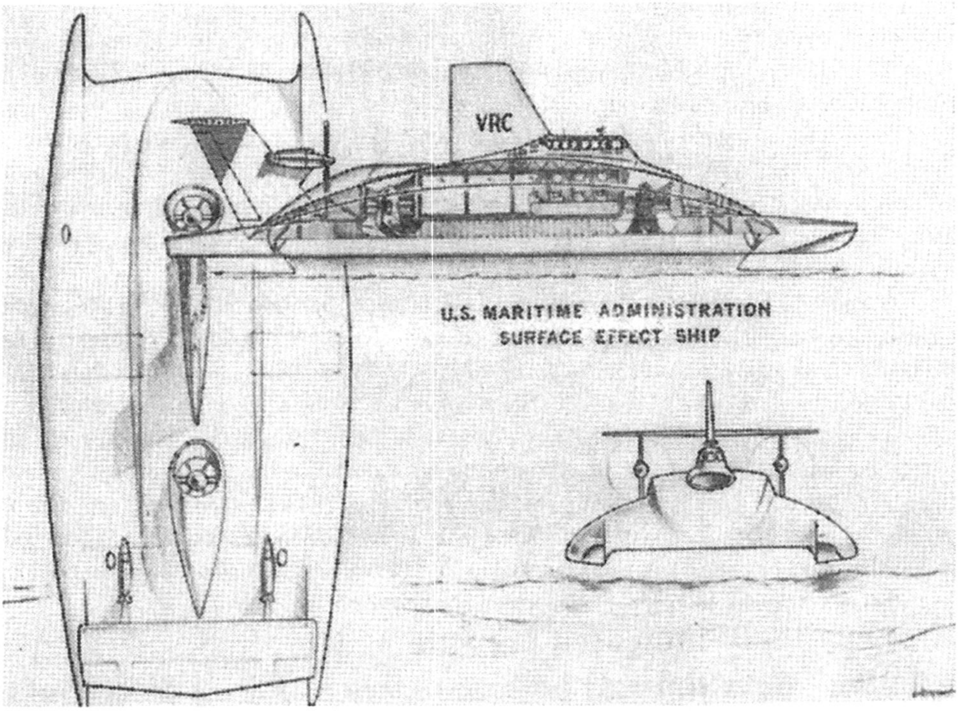

Figure 3.5: General Arrangement of MARAD Surface Effect Ship *"Columbia"*

At the design cruising speed, the forward and rear seals are retracted (as shown in Figure 3.3) and the main support is from the aerodynamic lift generated by the main body. In this third mode the craft is now operating as (c) a *form of wing-in-ground effect.* The side air jets remain activated for improved performance (reduced induced drag) and also for transverse stability and control. It is this *tri-mode design feature* that would allow the *channel flow SES* to operate in the various sea conditions necessary for a practical ship that can function both in port and in the open sea.

Much of the technical details of the *Columbia* design are available in the records of the US Maritime Administration but some descriptive material may be found in the open literature in articles in Flight International (1962)[5] and also the magazine

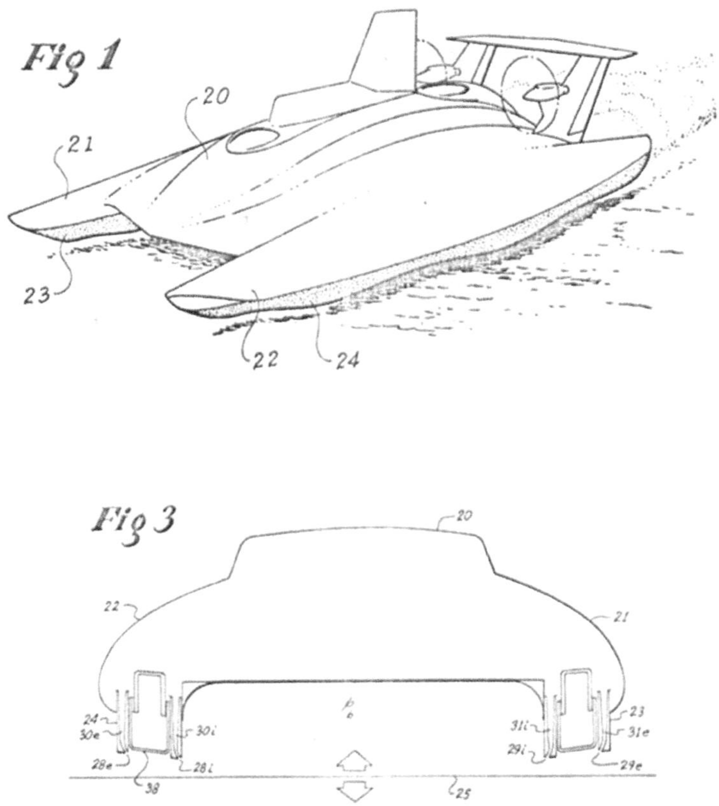

Figure 3.6: Channel Flow Craft Flexible Suspension System

Hovering Craft and Hydrofoil[6] (1961) where the discussion includes the addition of a flexible skirt system for improved operation over waves at the lower speeds.

The design of the flexible skirt system or suspension system for the Channel Flow ship described in the 1961 and 1962 articles is described in more detail in the patent by Scott Rethorst filed in 1963 and issued in 1965.[7] Figure 3.6, taken from the patent drawings, show the general features of the flexible suspension system mounted to the underneath of the two catamaran hulls. The basic intention of the suspension system is to act as a "linear tire" (to use the inventor's words) that can provide adjustable stiffness as needed depending on the surface to be traversed. Because the main intent is to provide sealing on the pressurized channel air in the cruise mode the additional skirt extensions for across the ship for low speed operation are not addressed in the patent. In the Test Craft (see later) such extensions were incorporated in the test program.

A detailed explanation of how the various parts of the "linear tire" are activated by variations in the base pressure may be found in the patent. The main intent here is merely to show just one example of how the flexible skirt systems that were developed for the amphibious air cushion craft under development in England (and elsewhere) at that time evolved into possible schemes for aerodynamic air cushion craft.

3.3 MARAD Test Craft *"VRC-1"*

On 2 August 1961, the Maritime Administration gave a contract to Vehicle Research Corporation in Pasadena, California to proceed with the Preliminary Design of the Surface Effect Ship *"Columbia"*. As part of that contract, based on the success of the

Preliminary Design Reviews, authorization was given on 30 August 1962 to proceed with the construction and test of a man carrying Test Craft (subsequently named *"VRC-1"*). This Test Craft was built as planned and testing started at Edwards Air Force Base, California on the Rogers Dry Lake Bed, Mojave Desert in early 1964. Several test runs were made and Figure 3.7 shows the author (who was the Chief Designer and Test Pilot) exiting the test craft at the South Base, Edwards Air Force Base, after one of the test runs on 17 March 1964. Several test runs were made and speeds of 55–70 knots were achieved.

Figure 3.8 shows the Test Craft cruising on Rogers Dry Lake Bed during one of its test runs in early 1964. During the design stages of *"Columbia"* and of the Test Craft *"VRC-1"*, the technology of hovercraft skirts was not well advanced

Figure 3.7: The author exiting cockpit after test run of MARAD SES Test Craft. *"VRC-1"* USAF photo

Figure 3.8: MARAD Surface Effect Ship Test Craft *"VRC-1"* Cruising at 50 knots Author photo

Figure 3.9: MARAD SES Test Craft *"VRC-1"* cruising outfitted with dorsal fin (1964)
Author photo

(see Chapter 2) and care had to be made in making banked turns because of the low ground clearance. To offset this problem, the basic design of the craft included the addition of a dorsal fin as shown in the artist's renderings in Figures 3.4 and 3.5.

The MARAD SES Test Craft "*VRC-1*" is seen cruising (author at the controls) in Figure 3.9 over the Mojave desert outfitted with the scaled dorsal fin for flat turns (minimal roll angle). Unfortunately, these photographs, complete with thumb prints, were taken by the mechanics with the author's Polaroid camera from the moving Jeep "chase car" but the essential features are clear! The author can personally testify that very comfortable coordinated flat turns could be performed through coordinated application of both dorsal fin and the rudders without the need to increase the lift power and bank the craft.

The design of the *"VRC-1"* was to be a scale model of the 100 ton *Columbia*. This simple statement brings with it the usual concerns of scaling which include the incompatibility of scaling in all key areas. This was especially important when it is noted that the key feature of the channel flow design was for a craft (ship) that could operate in three distinct modes of operation, each governed by different sets of hydrodynamic, aerostatic and aerodynamic forces. The high speed envisaged for the design brought with it the need to correctly scale the dynamic behavior of the craft as well as the normal performance scaling (Froude Number, Reynolds Number, etc)[8]. Normally, the largest scale possible would be the aim of the designer to minimize the incompatibilities.

Because it was intended to operate the *"VRC-1"* craft in different areas (desert and lakes) it was also necessary to ensure that the Division of Highways rulings were adhered to for any trailing of the craft around the State to meet width and weight constraints. Once all the necessary "trade-offs" had been made, the final selection of the *"VRC-1"* was to be a 1:4.5 geometric scale of a 100 ton (224,000 lb) ship. The Leading Particulars of the *"VRC-1"* are given in Table 3.2.

Table 3.2: *Leading Particulars of MARAD Test Craft* "VRC-1" *(as built)*

Gross Weight	4,505 lb
Total Installed Power:	275 HP
Airesearch gas turbine (GTCP 85–98) driving four separate hydraulic pumps (2 for lift, 2 for propulsion, each pump capable of 60 HP each	
Lift fans	2 x 39 in dia mixed flow fans
Propulsion	2 x 60 in dia FP tractor
Air Volume Flow (hovering air cushion craft mode)	37,600 CFM
Cushion Pressure (hover mode)	18.3 lb/sq ft
Length Overall (sidehull keel length)	31 ft
Width Overall	14 ft
Cushion Area	246 sq ft
Hover height beneath the sidehull keels (and air jets)	2 in
Height beneath the sidehull keels at 45–55 knots	6 in
Height beneath the sidehull keels at cruise speed (73 kts)	8–10 in
Crew (Pilot)	1
Passenger (optional)	1

The performance values were approximate pending test results and various Test Craft features were incorporated. For example, the incorporation of the trailing edge Jet Flap for stability and control requirements in the aerodynamic (or channel flow) mode changed the pressure and flow values slightly.

In order to keep the weight down for this aerodynamic craft, the structure was basically an aircraft space frame structure (with some use of shear panel framework) made of high grade aluminum (5086-H32 and 6061-T6 chosen for their good resistance to atmospheric and chemical corrosion). No loads were taken by the fiberglas skin which served to give both the external aerodynamic shape and to enclose the plenum chamber for the pressurized air from the fans. Figure 3.10 shows the space-frame construction. The upper photograph shows the general view of the structure on its construction stand in the manufacturing facility at Vehicle Research Corporation in Pasadena, California. The lower photograph in Figure 3.10 is a close-up of the structure after the gas turbine engine and piping was installed.

The fiberglas skin was attached with quick release fasteners to facilitate the maintenance of the craft both in the field and in the hangar as shown in Figure 3.11 where the *"VRC-1"* is being moved from the hangar at South Base, Edwards Air Force Base for start of an engine run-up prior to a test run. The removed fiberglas skin can be seen on the forward starboard section of the

Figure 3.10: Space-Frame Structure of MARAD "*VRC-1*"
Photos by author

Figure 3.11: MARAD "*VRC-1*" being towed out of South Base hangar
Photo by author

main body. The transporting dolly wheels seen in Figure 3.11 are removed prior to flight.

The unique feature of the "*channel flow SES*" is its ability to operate in three modes:

(a) Conventional *catamaran* mode over water
(b) Conventional "*amphibious hovercraft*" mode over land and sea
(c) "*Aerodynamic air cushion craft or wing-in-ground-effect*" mode over land and sea

The variable geometry feature that allowed all these modes to be accommodated in one design can be seen in Figure 3.12, 3.13 and 3.14.

3.3.1 Bow Flap

The early schematic drawings provide a clearer picture of the arrangement than the actual construction drawings. These particular schematics were drawn up during the lift and propulsion arrangement trade-off studies in 1961–1962. The mechanical drive shown in the schematics in Figure 3.12, 3.13 and 3.14 were replaced with an all hydraulic drive for the actual construction in 1963 but the jet configurations remained as shown.

It can be seen clearly in Figure 3.12, that in the conventional catamaran (hullborne) mode these jets would be retracted smoothly into the hull of the ship for a watertight seal. The craft would then operate as a conventional catamaran,

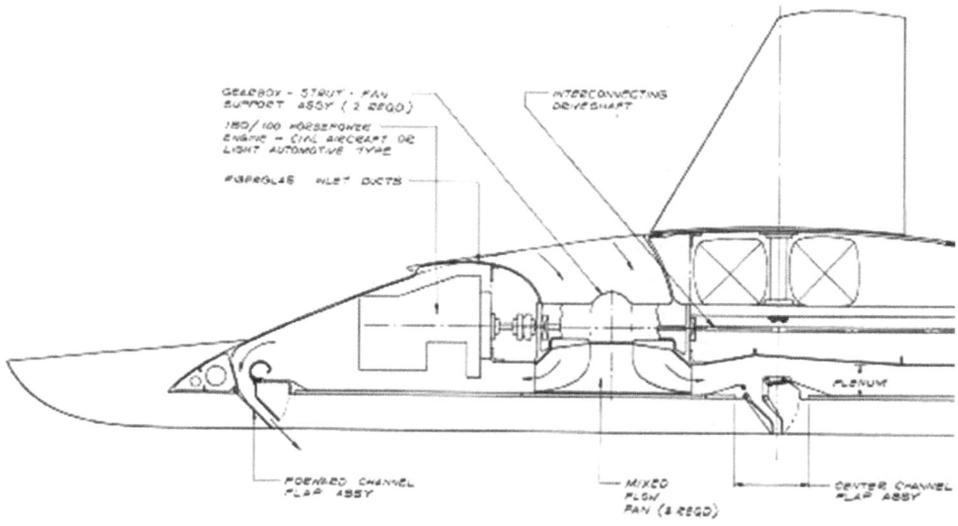

Figure 3.12: Schematic of Bow Flap and Center Channel Flap for "*VRC-1*"

say for motoring in harbors[a]. For the conventional amphibious hovercraft mode the flaps (bow, center and stern) would be lowered and pressurized air from the fans (shown) would provide the necessary lift. During later tests as information on the "skirts" from the UK became available, flexible extensions would be added to these mechanical jets to provide a greater "ground" clearance.

In the final "as built" configuration of the "*VRC-1*", the leading edge of the main body was rounded (see Figure 3.10, upper photograph) to provide the necessary aerodynamic shaping. Also, the main body was mounted on the catamaran hulls at a small positive (1°) angle for improved aerodynamic lift and moment arrangement based on extensive wind tunnel testing. Such a shaping aided the aerodynamic mode of operation and had no effect on the conventional amphibious air cushion craft mode of operation.

3.3.2 Center Channel Flap

The concept of a "channel flow SES" is that it is capable of operating in both aerostatic mode (like an amphibious air cushion craft) and also as a wing-in-ground-effect mode when operating in the aerodynamic mode. In broad terms, this means that the Center of Pressure (C.P.) will move from the 50% chord point to a point much more forward of the order of 25–33% chord as the craft transitions from one mode to the other. As part of the Test Program for the "*VRC-1*" a jet of air dividing the forward and rear sections of the craft was included in the configuration such that as the craft transitioned from the aerostatic mode to the aerodynamic mode various operational techniques could be explored to ease

[a] It is believed that the noise problem associated with "hovercraft" when in port or at the dock would be eliminated by this feature of the channel flow concept.

the transition as the Center of Pressure moved. This Center Channel Jet can be seen in Figure 3.12 and in Figure 3.13.

3.3.3 Rear Jet and Jet Flap

The combination of two flaps at the rear of the craft also aided in the transition from the aerostatic mode to the aerodynamic mode of operation. Figure 3.13 shows the rear jet arrangement that operated in much the same manner as the bow and center flaps to provide the necessary air flow and sealing for the air cushion mode of operation. From the early days of wing-in-ground effect designs (see Chapters 1, 2, 8 and 11) it was known of the potential hazard of such craft to "pitch up" as the craft operated at high speed where small disturbances could cause the main body to continue to rise as the aerodynamic Center of Pressure (C.P.) moved forward when the craft came out of ground effect. Observers of hydroplane races are all too familiar with some of the more spectacular examples of this tendency!! Accordingly, designers of such craft will normally incorporate large horizontal tailplanes to control this tendency to pitch up. The horizontal tailplane for the "*VRC-1*" can be seen in Figure 3.13.

Considerable research was available around the world, notably in England[9], France[10] and Germany[11] on the features of a jet flap and also a jet flap operating in ground effect. There is much more in the literature that was explored as the "*VRC-1*" augmented jet flap was studied prior to construction. In the design, use was made of a blowing jet flap to alleviate the size of the horizontal tailplane necessary to provide the restoring pitching moment. Extensive model tests with jet flaps were performed in 1961–1962[12] at Vehicle Research Corporation and in the Merrill Wind Tunnel at CALTECH to fully understand the characteristics of wing-in-ground-effect craft as they pitched both close to the ground (surface) and also as they rose to higher heights. The pitch stability of such craft varies with the height above ground and at different angles of attack. The degree of stabilizing pitch moment varies with both flap angle and the amount of air blown over the flap. It can be seen in Figure 3.13 that immediately above the T.E. flap is a second device called a *jet flap throttle* in the diagram. This nozzle allowed the pressurized air from the fans to provide the correct

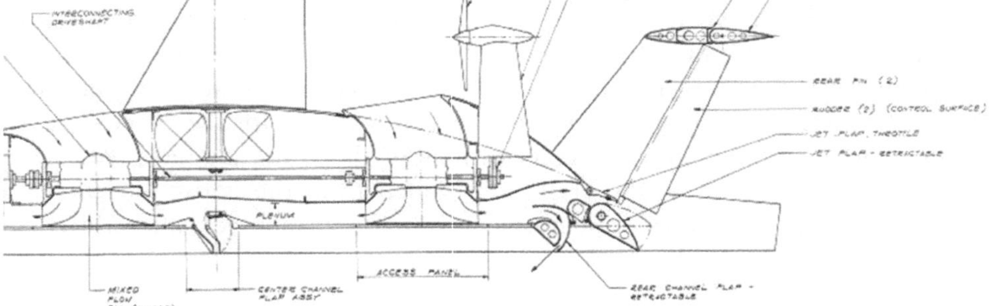

Figure 3.13: Schematic of Rear Flap and Jet Flap for "*VRC-1*"

amount of air flow to give the desired characteristics. A typical set out of the many variations explored is shown in Figure 3.14. The data shown is of the recorded pressure distribution on the upper surface and on the lower surface for the two cases of (a) no jet flap blowing and (b) with jet flap blowing. In both cases, the model is operating in the channel flow mode.

Over a wide range of tests, it was found that the upper pressure distribution was relatively unaffected by the jet flap (provided the flow remained attached) and that the main effect was on increasing the lower positive pressure on the craft (Figure 3.14). The use of the jet flap was found to both increase the lift and the location of the aerodynamic center of pressure. From the cited model tests, it was found that for the *"VRC-1"* configuration, a 5% trailing edge mechanical flap augmented by blown cushion air, provided the ability to shift the aerodynamic center aft approximately 4%. This is shown in Figure 3.15. The pitching moment coefficient, which is the rate of change of moment with angle of attack ($C_{M\alpha}$) versus the Center of Pressure change (expressed as a percentage of the chord or main body length) was measured for various angles of attack and for various amounts of blowing over the trailing edge (T.E.) flap.

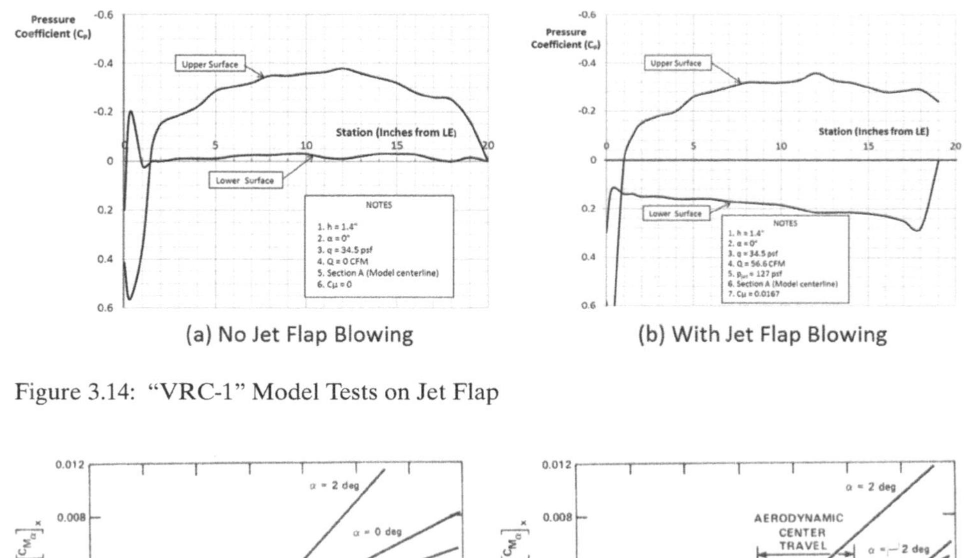

(a) No Jet Flap Blowing (b) With Jet Flap Blowing

Figure 3.14: "VRC-1" Model Tests on Jet Flap

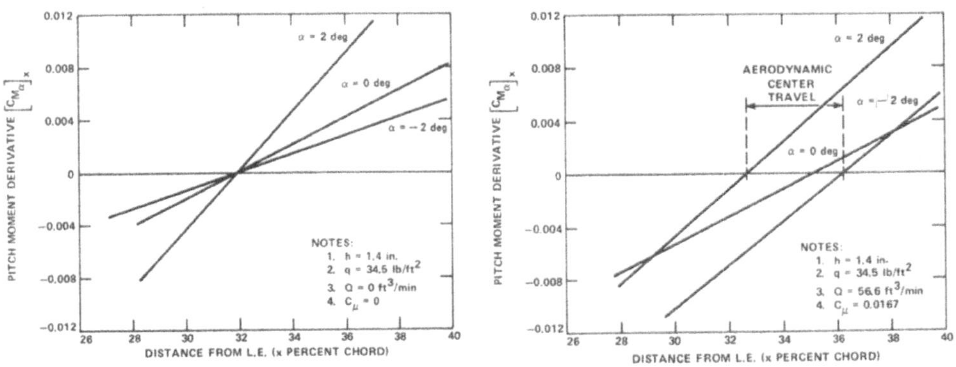

(a) No Jet Flap Blowing (b) With Jet Flap Blowing

Figure 3.15: Test Results of Jet Flaps in Ground Effect

In (a) the left-hand plot of Figure 3.15, the pitching moment is shown for the case of no jet blowing. For the particular configuration selected for the *"VRC-1"* main body, the Center of Pressure (C.P.) was at the expected value of 32% back from the leading edge (L.E.) of the main body. In (b) the right hand plot of Figure 3.15, it can be seen that for the particular case of blowing shown, the Center of Pressure (C.P.) moved back to about 36% of the chord, which is a significant savings in tail plane area. The effect of varying the angle of attack in both cases can also be seen from the plots from the wind tunnel data. The amount of air blowing required is measured by the jet blowing coefficient,

$$C_\mu = \frac{m_j v_j}{1/2 \rho V^2 S} \tag{3.1}$$

where m_j is the mass of air blown over the flap; v_j is jet velocity of the air over the flap; ρ is the air density; V is the craft speed over the ground and S is the reference area taken to be the plan area of the main body. The jet velocity (v_j) is determined by the normal laws of fluid dynamics and controlled by the opening of the jet nozzle (shown in Figure 3.13). Because the jet flap on the "*VRC-1*" was full span, then Equation 3.1 can be written:

$$C_\mu = 2\left(\frac{\rho_j}{\rho}\right)\left(\frac{v_j}{V}\right)^2\left(\frac{t}{c}\right) \tag{3.2}$$

For simplicity, it was taken in the analysis that the air density in the jet was the same as the free stream air density. In Equation 3.2, it can be seen that control of the opening of the jet flap would control the required value of C_μ. The opening of the jet flap was also carefully designed (by intruding slightly into the flow) to ensure that the flow remained attached by the Coanda effect and to maintain attached flow over the wing.

For the particular wind tunnel test data shown the jet blowing coefficient, C_μ = 0.017. The actual configuration of the jet flap for the "*VRC-1*" can be seen in the photo of the craft during construction in Figure 3.16. In that photo the jet nozzle had not yet been installed but the lower surface formed by the mechanical T.E. flap can be seen. This particular photo also shows part of the removable fiberglass skin used to cover the side catamaran hulls and also an access opening panel to facilitate attaching and removing of the tail fins and rudders. The horizontal tail plane and elevator control can be seen in Figure 3.16 in the upper right hand corner.

3.4 Stability & Control Jets and Induced Drag Reduction Mechanism

One of the features that distinguishes the "*channel flow SES*" from other more familiar forms of "*wing-in-ground-effect*" craft is the use of the air flow to seal the edges of the planform. The use of such peripheral jets to "seal" the channel has multiple purposes. As can be seen in the schematic drawing in Figure 3.17 there are two jets on either side of the craft. Each jet (port and starboard) can be individually opened or closed upon command by the pilot. From a "stability & control" function it can be seen that the "pads" of pressurized air between the two jets on either side of the craft can be set to either trim the craft laterally as well as to be used for slight banking of the craft to aid in any turning. This last feature was not used extensively because of the low ground clearance prior to any installation of flexible extensions or "hovercraft skirts".

Figure 3.16: *VRC-1* Jet Flap during Construction Phase (1962–1963)
Photo by author

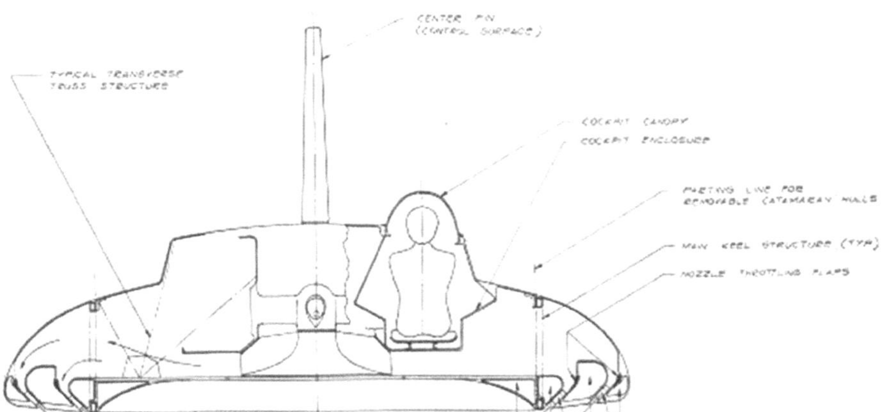

Figure 3.17: Mid-ship Cross Section (showing Lateral Stability & Control and Edge Sealing Jets)

The two sets of peripheral jets came together at the bow and stern of the craft for better transition with the bow and stern flaps to contain the air cushion. This resulted in "lenticular-like" pads of pressurized air to provide the lateral stability when on cushion. This can best be seen in Figure 3.18 which shows the "*VRC-1*" on the weigh scale in the Weight & Balance hangar in the North Base of Edwards Air Force Base

Figure 3.18: "*VRC-1*" on Weight & Balance Scale for weight and C.G. determination
Photo by author

in early 1964. While the jets openings are in the shadow in the bottom of the sidehulls, the lenticular geometry of the jets can be seen. These jets can be closed when the craft is operated as a conventional catamaran in hullborne operation or open when used in the air cushion modes (aerostatic or aerodynamic).

In addition to the side air jets being used for lateral stability and control of the craft a second and major purpose is to provide the edge sealing and thus reduction in the induced drag of low aspect ratio wings operating in ground (or surface) effect. A concern of the maritime world when considering the application of the wing-in-ground-effect principle is the problem of "span". Normally, wings (at least in subsonic flight) have spans that are much larger than the chord (or length) of the craft. This would give major problems when operating such craft into most ports around the world where the ports already have been designed and built for ships that are longer than they are wide.

This real life problem was recognized in the design of the "*Columbia*" Class of channel flow aerodynamic air cushion craft. The problem lies in the nature of lift generation on conventional wings that must contend with the vortex generation at the wing tips as the high pressure on the under surface of the wing flows into the low pressure on the upper surface of the wing. This energy dissipation is commonly referred to as "vortex drag" or more frequently as "induced drag".

Conventional aerodynamic theory expresses the drag of wings as the sum of the "profile drag" caused by the resistance of the wing moving through the air and the "induced drag" caused by the vortex creation discussed above. A common technique in wing design is to express the efficiency of any particular wing design by an "effectiveness term" (e) that measures how well an actual wing performs compared to the ideal wing[b]. This is expressed most simply through the standard expression for the total drag of the wing (in coefficient form) as:

$$C_D = C_{D_o} + \frac{{C_L}^2}{\pi(eAR)} \tag{3.3}$$

In this conventional equation, lift (L) and drag (D) are expressed in coefficient form as:

$$C_L = \frac{L}{1/2\rho V^2 S} \tag{3.4}$$

$$C_D = \frac{D}{1/2\rho V^2 S} \tag{3.5}$$

In the above expressions for the lift coefficient (C_L) and drag coefficient (C_D) of the wing, the other terms (ρ, V, S) are as given before. In equilibrium flight, L = W.

In Equation 3.3, the profile drag in coefficient form (C_{Do}) of the wing which, to keep the analysis simple in this example, assumes no contact with the water. (See Chapter 8 for inclusion of hydrodynamic drag term). The second term in Equation 3.2 is the induced drag coefficient (C_{Di}). The term (eAR) is the *effective aspect ratio* of the craft with operating side air jet seals, which provides the same induced drag as a wing without the side air jet seals and a geometric aspect ratio (AR = span/chord). The physical or geometric aspect ratio of the "*VRC-1*" is 0.5. Wind tunnel tests of the "*VRC-1*" configuration were conducted at the CALTECH wind tunnel facilities in 1961–1962[13] and a set of representative results for lift and drag are shown in Figure 3.19.

Induced drag varies inversely proportional to speed and the lift coefficient (C_L) from the data shown in Figure 3.19 at the height tested (h/c = 0.07) shows that the lift curve slope was 0.12 per degree and the drag varied with the angle of attack as shown, increasing at a higher rate than the square of the angle of attack (α). The profile drag coefficient (C_{Do}), in this configuration was about 0.02 with a zero lift angle of –4.2 degrees. Only representative test data has been shown here for brevity.

Using this data and the basic equation for drag (Equation 3.3) it is possible to calculate the "effective aspect ratio" which was about eAR = 14. When compared to the actual aspect ratio AR= 0.5, it provided an encouraging effective sealing and vortex destruction. In any full scale craft it is likely that this ratio would be reduced significantly from this value because of factors not easily simulated in the wind tunnel but it would still be a good value demonstrating the ability to reduce physical span for such craft. This is a significant decrease in the span required for a

[b] The reader is referred to standard aerodynamic texts where discussion can be found on pressure distributions on ideal wings and elliptical pressure distributions on elliptical wings and other phenomena. A practical measure is the comparison of the induced drag of the short span actual wing (low aspect ratio) with sealing jets compared to that with a much longer span (high aspect ratio) wing without the sealing air jets.

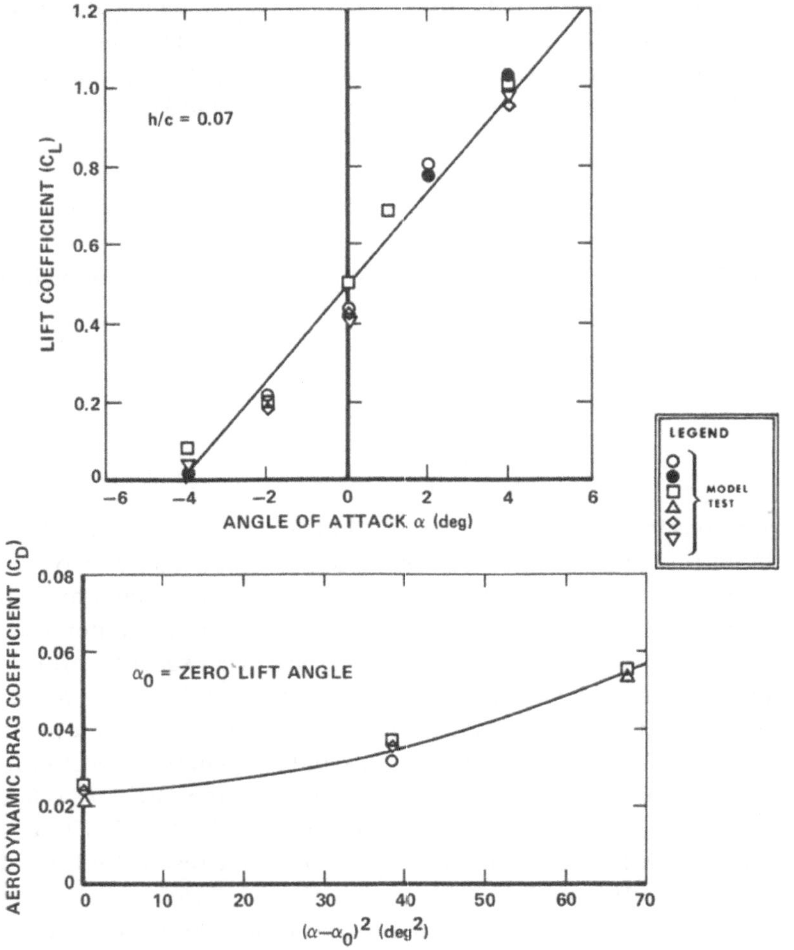

Figure 3.19: Channel Flow Aerodynamic Air Cushion Craft Lift & Drag

wing-in-ground-effect to have the same low drag characteristics. This result that was repeated in many of the other tests of various configurations for the channel flow craft was encouraging and showed that it was possible to design an aerodynamic air cushion craft with dimensions much more compatible with ports and maritime environments around the world. Chapter 11 discusses this and other mechanisms for reducing the span and induced drag of aerodynamic air cushion craft.

The "*VRC-1*" continued the Test Program into 1965 collecting valuable data for the MARAD Surface Effect Ship Program using the *channel flow aerodynamic form* of air cushion craft until the US Maritime Administration decided to cancel the program and switch its attention to the "new" concept (see Chapter 2 for more discussion) just introduced by the US Navy which was the *captured air bubble* concept. The history of the US Maritime Administration interests in high speed ships is summarized in Chapter 4 together with the description and results of the Surface Effect Ships for Ocean Commerce (SESOC) Study that was commissioned by the US Department of Commerce at that time to examine a wide range of concepts for high speed craft for transoceanic missions including *channel flow craft, captured air bubble* craft and other candidates before proceeding further on the program.

The US Administration dropped all work into such high speed hull forms in 1970. The cancelation of the *Columbia* project in 1965 essentially curtailed the US lead in aerodynamic air cushion craft at about the same time as development work on the Former Soviet Union *ekranoplan* aerodynamic air cushion craft was underway on the Caspian Sea in 1963 (see Chapters 1, 2 and 11).

The US Navy, who continued the pursuit of "100 knots at sea", did not revisit the aerodynamic form of air cushion craft (except for some limited laboratory work) until 1976 after the Russian activities in *wing-in-ground-effect craft* became more widely known. The history of the US Navy's activities into *wing-in-ground-effect craft* is described in Chapter 10 and Chapter 11.

References

1. *"Translational Characteristics of Ground Effect Machines"*, by Winston W. Royce and Scott C. Rethorst, Institute of Aeronautical Sciences (IAS) 29th Annual Meeting, New York, New York January 23–25, 1961; Paper No. 61–79
2. *"Inviscid Incompressible Flow Theory of Static Peripheral Jets in Proximity to the Ground"*, by Tor Strand, Journal of the Aerospace Sciences, January 1961, Vol 28, No. 1.
3. *"Maritime Administration Surface Effect Ship"*, by Scott Rethorst (Vehicle Research Corporation) and W. Taylor Potter (US Dept. of Commerce), National Aero-Nautical Meeting (Society of Automotive Engineers and American Society of Naval Engineers) Washington DC, 8–11 April 1963.
4. *"150 knot GEM Cruise"*, by Tor Strand, Vehicle Research Corporation, reprint Aerospace Engineering, Volume 21, No. 4, April 1962.
5. "*The 100-Knot Columbia"*, article in Air Cushion Vehicles, FLIGHT International Supplement 25 October 1962
6. "*Columbia in Detail"*, Abstracts from National Meeting on Hydrofoils and Air Cushion Vehicles sponsored by IAS, Hovering Craft & Hydrofoil Magazine, Kalerghi-McLeavy Publications, 1961
7. *"Suspension System for Surface Effect Ships"*, US Patent No. 3,211,247 by Scott Rethorst, filed 25 March 1963, issued 12 October 1965
8. "*On Dynamically Similar Motions for Sea-Going GEMs*", by Peter Mantle, Vehicle Research Corporation, Working Paper No. 24, July 1962
9. "*The Pressure Distribution on Two-Dimensional Wings Near the Ground*", unpublished notes by J. A. Bagley, Ministry of Supply, London, England, 1959
10. *"Recherches theorique et experimentale sur le controle de circulation par sufflage appliqué aux ailes d'avions"*, by L. Malavard, Ph. Poisson-Quiton and P. Jousserandet, ONERA Note Technique 37, 1956
11. *"Einfluß der Bodennahe auf die Auftriebeverteilung einer ebenen Platte in inkompressibler ebener Stromung."* By C. Braunss and W. Lincke, Bericht des Inst. F. Luftfahrttechnik der TH Darmstadt, Germany, 1961
12. "*Wind Tunnel Tests on Jet Flaps in Ground Effect*", by Helge Norstrud, Vehicle Research Corporation, Working Paper No. 50, November 1962
13. "*Thrust and Drag Estimate for the Channel Flow GEM"*, by Peter Mantle, Vehicle Research Corporation, Working Paper No. 36, September 1962

4 History of US Maritime Administration "Large Surface Effect Ship" Program

Never promise more than you can perform

Publilius Syrus, 1st century B.C.

The US Maritime Administration (MARAD) must be credited with taking bold initiatives in the late 1950s and early 1960s to seek novel forms of ships aimed at revitalizing the ailing American Merchant Marine. In the late 1950s the US Maritime Administration (MARAD) embarked upon several programs that included different approaches to the ship construction subsidy program; the use of standardized containers; improved data processing for handling ship operating information on board ship; more efficient shoreside operations; improved cargo handling means; and other ways of reducing costs and improve efficiency of American ships. The decline of the American Merchant Marine had been underway since the end of World War II. By 1964, the world merchant shipping fleet was moving around 200 million deadweight tons of cargo and goods. At that same time, the cargo and goods being shipped by the US fleet was only about 15% of that world value at a modest 35 million deadweight tons. In addition to the wide ranging initiatives being pursued by the US Maritime Administration there were three specific programs designed to seek new hull forms for the transport of goods over transoceanic routes at both reduced cost and with more efficient operation. These programs were:

(1) Nuclear Powered Ship (*N.S.Savannah*)	(1959–1972)
(2) High speed Hydrofoil (*HS Denison*)	(1958–1964)
(3) High speed Surface Effect Ship (*Columbia* and *VRC-1* Project)	(1961–1965)

The first program (*N.S. Savannah*) was aimed at reducing the operating costs of ships by switching from fossil fuel to nuclear power for propulsion. The second two programs (*hydrofoil* and *surface effect ship*) were aimed at incorporating vastly superior high speeds into ocean commerce with a goal of 100 knots or more. This goal for the *hydrofoil* and *surface effect ship* programs was based on in-house economic analyses that showed that such ships from 100 tons to 3,000 tons displacement traveling at 100 knots over transatlantic distances (approx 3,600 nautical miles) would offer Direct Operating Costs (DOC) between those experienced by conventional displacement ships and those of transport aircraft. The aim was to move cargo and goods faster "to market" at speeds much greater than the 20–30 knots of conventional displacement ships but at costs much less than transporting by air.

4.1 Nuclear Ship *N.S. Savannah*

In addition to being the first nuclear powered commercial cargo carrying ship, the *N.S. Savannah* was also built to be a showcase for American ingenuity as the ship traveled around the world. Under President Eisenhower's "Atoms for Peace" initiative, the *N.S. Savannah* began construction in 1956 and was launched 21 July 1959. First Lady Mamie Eisenhower christened the ship at the New York Shipbuilding Corporation, Camden, New Jersey. The maiden voyage of the *N.S. Savannah* was made on 20 August 1962. The *N.S. Savannah* ship was a joint program between the US Maritime Administration and the Atomic Energy Commission (AEC). In 1965, the AEC ended its participation in the project and transferred the liability of the reactor to the US Maritime Administration. The ship was designed to carry 9,400 tons of cargo, 60 first class passengers and 124 crew at a speed of 21 knots over 336,000 nautical miles before refueling. The *N.S. Savannah* was not intended to be a fast ship, the *Queen Mary* in 1936 for example had a cruise speed of 29 knots, but it was a luxury ship fitted out as a showcase of American ingenuity. Figure 4.1 shows the *N.S. Savannah* leaving amid fanfare on its first trip to Europe.

Figure 4.1: Nuclear Ship *Savannah* leaving for Europe
US MARAD photo

As part of the "showcase" nature of the design of *N.S. Savannah*, while it was intended to be a cargo ship, there were 30 air conditioned staterooms for the first class passengers each with their own bathroom. The luxurious dining facility could seat 100 passengers and a lounge that could also be used as a movie theater. The ship was also equipped with a swimming pool and a fully stocked library. The sleek lines of the ship were more in line with that expected for a luxury yacht rather than a cargo ship. Indeed, it was found that the hull lines complicated the cargo unloading at many docks around the world including the US.

The crew of 124 men far surpassed the number of a normal crew of about 30 men for a conventional cargo ship of the same size. The specialized requirements for a crew required to operate and maintain a nuclear powered ship caused problems with the labor unions. Labor-management disputes in 1963 resulted in the cancelation of further scheduled visits to foreign ports and cancelation of the original contract with States Marine Lines as general agent for the operation of the ship and on 23 July 1963, American Export Isbrandtsen Lines, Inc was selected as the new general agent. While the technical advantages of nuclear power over use of fossil fuel were analyzed in depth, the problems of taking such sips into commercial – and public – ports around the world were not fully appreciated in the design process. Considerable negotiations with foreign governments were needed before the *N.S.Savannah* was allowed to enter their ports. By 1963, negotiations to allow entry for the *N.S. Savannah* were concluded with 10 nations, including UK, Ireland, Sweden and Spain. By late 1963, negotiations were underway with France, Italy, Portugal and Turkey. By 5 May 1964, the N.S. Savannah, after a period of immobilization because of various disputes, sailed again from Galveston, Texas for US and foreign ports (Bremerhaven and Hamburg). The *N.S. Savannah* had been maintained at the nuclear facility base at Todd Shipyards Corp. in Galveston while new crews were being trained. The designer of the nuclear power plant was Babcock & Wilcox Co. and that company provided interim caretaker service and the training of the personnel of Government, Todd Shipyards and the new agent (American Export Isbrandtsen Lines) on manning the ship, operate the reactor and perform shoreside nuclear engineering services. All of which contributed to the high acquisition cost of the ship and its direct operating cost. The cost of the ship was $46.9 million in 1964 dollars. This cost included $28.3 million for the reactor alone. In retrospect, it is curious that a ship planned to improve the American Merchant Marine shipping and provide more efficient transport of cargo would waste valuable cargo space on first class cabins, lounge, and swimming pool, etc. Some reports indicate that the *N.S. Savannah* cost approximately $2 million dollars a year *more* in ship subsidies than a comparable conventionally powered (oil-fired steam plant) cargo ship. These concerns and inefficiencies eventually led to the cancelation of the program and the withdrawal from service of the *N.S. Savannah* on 10 January 1972 – a scant decade after its maiden voyage amid much fanfare.

The US Maritime Administration commissioned a few studies after 1964 into economically competitive applications of nuclear power for merchant ships but no commercial ships were ever pursued or built after the *N.S. Savannah*. Since the layup of the ship a new problem appeared over the period from 1972 until today which is the disposition of the nuclear fuel rods and other radioactive material. On 8 May 2008, the *N.S. Savannah* was towed from Norfolk to Baltimore for the removal of the radioactive material. At time of writing, it is believed once this is done satisfactorily, it is planned to exhibit the ship as a museum ship for nuclear powered ships. The *N.S. Savannah* is now a registered National Historic Landmark.

Currently, nuclear powered ships are the domain of US Navy ships under military control.

4.2 MARAD Hydrofoil Ship *Denison* Program

At about the same time as the nuclear ship *N.S. Savannah* was being designed, the US Maritime Administration began its research into high speed ships with speeds of the order of 50–100 knots. The history of hydrofoils is covered in more detail in Chapter 2 but it can be summarized that after early research beginnings at the turn of the twentieth century, hydrofoils became an established means of high speed marine transport starting in the 1930s in Europe and elsewhere. All of these hydrofoils were of the *surface piercing design type* (see Chapter 2) and used *subcavitating* foils. Such designs provided well established ships capable of traveling at speeds of the order of 30 knots or more in rough seas. By 1960, the technology was beginning to emerge on the incorporation of *supercavitating* foil cross sections into the surface piercing designs. The use of *supercavitating* foils would, it was projected, give increased speeds of the order of 100 knots. It was this recognition of possible high speed operation at 100 knots that brought the US Maritime Administration into the world of high speed ships.

The US Maritime Administration placed a contract in January 1960 to Dynamic Developments, Inc for the design of an experimental surface piercing hydrofoil craft. The program was to be in two phases. The first phase was to build and test a subcavitating foil design capable of speeds of about 60 knots. The second phase was to replace the subcavitating foils with supercavitating foils with projected speeds of about 100 knots.

Grumman Aircraft Engineering shortly after the initial work, acquired Dynamics Development and on 5 June 1962, the converted experimental displacement hull after being outfitted with the foil system began sea trials at Oyster Bay, Long Island, New York (see Figure 4.2). The displacement of the hydrofoil ship varied during its test program but was in the "95 ton" class, with an overall length of 104.6 ft and maximum hull beam of 23 ft. With the foils down, the draft was 15.4 ft but much less when in its hullborne mode and the foils raised – a useful feature when maneuvering in close harbors. The foil system was unique in that while it is called a member of the surface piercing foil family, only the forward foils that carried 85% of the weight were surface piercing. The aft foil carrying the remaining 15% of the weight was fully submerged which was different from the vast number of hydrofoils that operated in Europe and the (Former) Soviet Union. An innovation in the design was its greater control system over that of the "normal" surface piercing foil design. The main forward foils were fitted with trailing edge flaps controlled by a stability augmentation system. The aft foil had the added feature of incidence control. This design gave MARAD confidence that as the ship operated at high speed and in rough sea conditions there was a greater chance of an improved – and passenger acceptable – ride quality. The ship was powered by a single 14,000 HP General Electric gas turbine driving a supercavitating propeller through a right angle gear train.

HS *Denison* operating hullborne

HS *Denison* operating foilborne

Figure 4.2: MARAD Hydrofoil Ship *H.S.Denison* (1962)
US MARAD photos

Speeds attained by the ***H.S.Denison*** were typically 60 knots or more in test using the basic *subcavitating foil* system described above. Plans to change to *supercavitating foil design* to give the "100 knots or more" capability did not materialize as MARAD began to rethink what was the correct form for large tonnage ships to travel efficiently at those speeds. Studies were showing that it was doubtful that practical designs of 1000 tons or more operating at 100 knots could be achieved with a hydrofoil design. MARAD discontinued the test program of ***H.S.Denison*** in March 1964. As summarized in Chapter 2, it was during the last run to deliver the ship to the Reserve Fleet, the ***H.S. Denison*** ran aground sustaining considerable damage to both hull and the foils.

This ended the MARAD efforts into exploring the use of hydrofoils for their planned "100 knots at sea" program for oceanic commerce with large tonnage ships. The consideration of high speed hydrofoil ships of large tonnage (1000 tons or more) did not reappear until the US Navy studies in 1976 (see Chapter 10).

4.3 MARAD Aerodynamic Surface Effect Ship (*Columbia* and *VRC-1*) Program

The third leg of MARAD's effort in improving the American Merchant Marine and specifically to seek the incorporation of high speed (of the order of 100 knots or more) was the development of a variation of *conventional hovercraft*. The conventional hovercraft (using pressurized air for lift) was already well underway in England after the first craft operated in 1959 (see Chapters 1 and 2). This new form that MARAD studied incorporated *aerodynamic lift* into the design under the realization that such features were essential if such high speeds over water were to be realized in any practical design. There are various forms of aerodynamic lift air cushion craft and these are covered in more detail in Chapter 11. The fundamental types of air cushion craft are discussed in Chapter 1.

The particular form selected by MARAD as having the greatest promise was the *channel flow air cushion craft* which was the basis of the *Columbia* design. The complete description of the *Columbia* project is provided in Chapter 3. Economic studies conducted by MARAD showed that the *channel flow SES* would have low Direct Operating Costs for ship displacements up to 1000 tons (the largest size examined by MARAD at that time). The history of the MARAD Columbia Surface Effect Ship Program is provided in Chapter 3 and it was on this program that the term *"Surface Effect Ship (SES)"* was coined in 1962. Prior to that date all craft using various forms of pressurized air cushion lift were called *"Ground Effect Machines (GEM)"* in the US to distinguish them from the original work in England where they were called *"hovercraft"* by its inventor, Sir Christopher Cockerell in 1955. The aerodynamic form of air cushion craft called by its inventor Dr Scott Rethorst, *"Channel Flow GEM or SES"* was patented in 1959. The US Maritime Administration, based on studies underway at the Office of Naval Research gave a contract to Vehicle Research Corporation (Pres. Dr Scott Rethorst) on 2 August 1961 to undertake a design of a 100 ton prototype of the channel flow craft to demonstrate the capabilities of carrying cargo (four standardized 8x8x20ft containers) at speeds of 100 knots or more over rough seas in a safe and economical manner. As part of that program, the MARAD authorized (on 30 August 1962) the construction and test of a 2 ton craft called the "MARAD VRC-1". Tests were to be conducted first overland (at the Edwards Air Force Base) over Rogers dry lake bed in California and then on Meade Lake for overwater tests. Figure 4.3 shows the author (chief designer and test pilot) listening with an attentive ear to a test engineer and mechanic in front of the *"MARAD VRC-1"* in early 1964. More complete description of the craft, its form of operation and its test program is provided in Chapter 3.

Figure 4.3: MARAD *VRC-1* being readied for test (1964)
US MARAD photo

By 1965, the US Maritime Administration was continuing to explore high speed hull forms and early in 1965 halted the test program of the *MARAD Columbia* and *VRC-1* project while MARAD studied other hull forms as a result of new direction from the US Department of Commerce. The MARAD studies relied heavily on analyses conducted over the period 1963–1965 by Booz, Allen Applied Research, Inc. Based on these studies, on 6 July 1965, the Undersecretary for Transportation in the Department of Commerce, Mr. Alan S. Boyd, requested that a study be made of the Surface Effect Ship (SES) concept. In response to that request, on 14 September 1965, the Assistant Secretary for Science and Technology (J. Herbert Hollomon) directed that a broad based study be conducted to evaluate all forms of *surface effect ship* (the name now being used by MARAD to cover all forms of air cushion supported ships, especially when of large tonnage) and their application to high speed ocean transportation.

The study was led by Prof. George Maslach, Dean, College of Engineering, University of California. Mr. Edward M. MacCutcheon, Chief, Office of Research & Development of the Maritime Administration was assigned by Mr. Boyd to be the full time manager of this project and work started on 9 September 1965.

4.4 US Department of Commerce "Surface Effect Ships for Ocean Commerce (SESOC)" Study

The Committee work, directed by the US Department of Commerce, was very short in length of effort with the direction to form a committee issued in

September 1965 and a Final Report[1] issued by the Committee in February 1966 – a short three to four months[a] of study aimed at wide variety of basic concepts with fundamental differences in concept and purpose. The availability of concepts in any one type was limited with unconstrained prototypes or early forms available for study. The scope of the SESOC study was limited to commercial transoceanic applications of the surface effect ship concept. Military applications and feeder services by SES were not studied.

By direction, the study was to be constrained to examining:

(a) the *channel flow* concept (see above on the *Columbia* and *VRC-1* Project) and
(b) a concept put forward by the US Navy in 1964 known as the *captured air bubble (CAB)*

Other hull forms, in the words of the Department of Commerce directive, were to be included for the "sake of completeness". The *captured air bubble (CAB)* concept was originated by Douglas Warner (private inventor) in 1929 and again in 1964 by Allen Ford of the US Navy.

The history of the CAB concept is provided in Chapter 2 including the work on similar designs in England (during 1955–1963) of such craft by Christopher Cockerell and others.

The mission of the study was given as:

> *To determine the research and engineering problems associated with employing the surface effect ship concept in commercial ocean transportation systems.*
>
> *To recommend a research and development program competent to resolve such technological problems.*

The goals set for the Committee were:

1. *To identify those technological problems which are important to SES development.*
2. *To identify and define those technological problems which are crucial to SES development.*
3. *To rank the crucial problems in order of criticalness to the success of the SES program.*
4. *To rank the crucial problems in order of difficulty of solution.*
5. *To recommend the magnitude, time phasing, and scope of an effective R&D program to resolve the problems.*

As it started its work, the Committee added the following items:

1. *Consideration would be restricted to cargo liner systems handling a typical spectrum of high value containerized cargo.*
2. *The study would include all concepts of SES.*
3. *The hypothetical SES would have a gross weight of about 5,000 long tons and would cruise at a speed of about 100 knots.*

[a] While the SESOC Program Office began scoping the study in September 1965, the actual Panel sessions began in October 1965 and the entire month of January 1966 was consumed in writing the Final Report that was issued in February 1966.

While the physics of the problem would say otherwise, the SESOC Committee ascertained before embarking on its work that there were (only) two basic types of SES, described in their words as:

1. *One type is supported by a cushion of pressurized air which is constrained beneath the vehicle. This type has been designated as a "Category I" vehicle (aerostatic lift).*
2. *The second type of SES is mainly supported by aerodynamics as the vehicle transits at sufficient speed to obtain aerodynamic lift similar to an aircraft. The significant difference from aircraft is that the proximity of the sea surface provides and added degree of lift. This type is called a "Category II" vehicle (aerodynamic lift).*

Unfortunately, this overly simplistic categorization of the available concepts masked many of the very features that SESOC were tasked to study. Many of the candidates put forward for the "set" to be studied were not designed to any common set of requirements and in some cases were simply examples built by their inventors as prototypes or "brass board" test vehicles and not broad based candidates for careful evaluation. There were five basic craft types included in the SESOC study. These were identified (and named) by the Committee as:

1. Hydroskimmer
2. Captured Air Bubble (CAB)
3. Hydrokeel
4. VRC Channel Flow
5. Weiland Craft

As can be seen from the names, that some were named according to their basic concept of operation and some were simply names of existing craft and not the generic name of a concept of *surface effect ship* in the context of the study. This would prove to be an unfortunate feature in the aim of the study which was the selection for large surface effect ships capable of operating over the sea at 100 knots or more.

A short description of each of these "candidates" for the MARAD surface effect ship for transoceanic commerce follows.

4.4.1 Hydroskimmer

The *Hydroskimmer* was an amphibious air cushion craft designed and built for the US Navy Marine Corps and designated as SKMR-1 as a high speed means to deliver troops across the beach in an amphibious landing mission. The SKMR-1 was launched in 1963. It was the largest "*hovercraft*" in the US at that time with a gross weight of 22 to 28 tons, depending on payload and a calm water speed of 70 knots. It is surprising that because the Dept of Commerce were seeking solutions for high speed ships for commercial application that the SESOC Committee did not include in their "set" of candidate designs the impressive ongoing work in the UK, France and elsewhere in actual commercial operation of several forms of air cushion supported craft such as the SRN.5, SRN.6 and other craft that were operating on commercial routes at speeds up to 66 knots and operating under the more applicable maritime rules of construction and operation. In any event, the military *Hydroskimmer* gave valuable data, albeit at an R&D level. Figure 4.4 shows, in the upper photograph, the actual SKMR-1 outfitted with the four foot deep "skirts"[b] for operation in rough water and for over the beach obstacle clearance. The lower sketch[c] shows the scaled version (done by Bell for the SESOC Committee study) for a 1000 ton displacement version. The other types of amphibious air cushion craft built and operated in England, France, Japan and elsewhere were not added to the "set". Each of these other craft could have provided valuable operational data on different skirt systems, propulsion schemes and other design variations that would be applicable to larger ship sizes of interest to the US Maritime Administration mission.

Bell Aerosystems *Hydroskimmer* (*SKMR-1*) (1963)

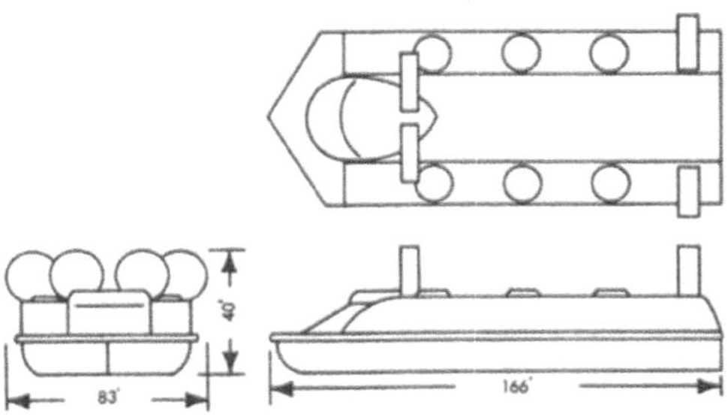

Hydroskimmer scaled to 1,000 tons

Figure 4.4: *Hydroskimmer* Candidate
US MARAD photo and sketch

The *Hydroskimmer* was a good example of the class of *amphibious air cushion craft* for military missions. It is unfortunate that the large amount of data available from the development and operation of amphibious air cushion craft (or *hovercraft*) underway in England and elsewhere at that time (1959–1965) for commercial applications was not used in the study.

4.4.2 Captured Air Bubble (CAB)

The US Navy contributed heavily to the MARAD SESOC Study with the work done on the Captured Air Bubble (CAB) concept. Chapter 2 provides a detailed history of the Captured Air Bubble concept from the original work by Douglas Warner in 1929 to the US Navy program in the early 1960s. Figure 4.5 shows, in the upper photograph, the US Navy CAB Test Craft "XR-1". The lower sketch shows

[b] Based on original development in England in 1961 and already in use on commercial hovercraft
[c] All sketches of the five "candidates" have been taken directly from the SESOC Final Report without editing

US Navy CAB *XR-1* (1964)

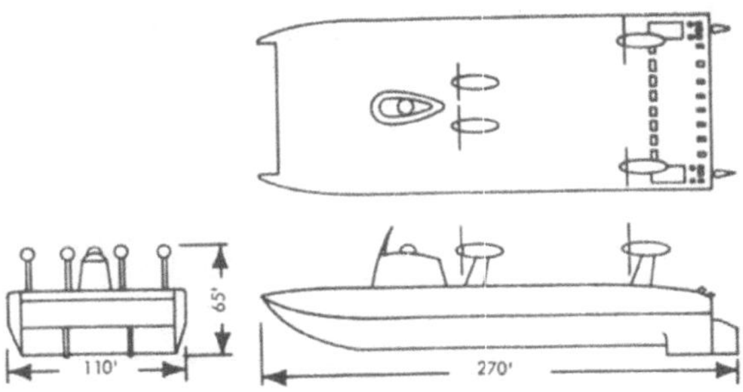

CAB Scaled to 1,000 tons

Figure 4.5: CAB Candidate
US MARAD photo and sketch

the scaled version (done by the US Navy for the study) for a 1000 ton displacement version.

As described in the US Navy patent, the CAB concept was designed to have two key features for a high speed hull design. First, it had minimum resistance at high speed through the water by the use of *thin non-buoyant sideboards* (to use the words from the 1964 patent). Second, the use of *rigid planing seals in bow and stern* provided a sealing or "capturing" of the pressurized air bubble while planing on the water, again to minimize hydrodynamic resistance. These two features were the subject of much development as the program unfolded first at MARAD and subsequently in the US Navy as it continued with the program. The concept of having no buoyancy or other restoring mechanism in the sideboards was of concern to the reviewers at SESOC.

It is unfortunate that in one of its first tests, the "*XR-1*" suffered a catastrophic roll-over in the water while exercising a modest turn because of this fundamental flaw in its pitch-roll-yaw stability. This accident occurred in December 1964 and a detailed report by the US Navy[2] is provided in Chapter 2. Surprisingly, if the accident was reported to SESOC, it was not mentioned or included in the Final Report on the findings relative to the CAB concept.

As discussed later in this Chapter, many of the engineering experts at SESOC correctly pointed out the potential for unstable operations of the CAB concept and gave their concern for such incidents in their Panel Reports to SESOC.

4.4.3 Hydrokeel

While the details of the CAB overturning incident may not have been provided to the SESOC Panels of experts, there was sufficient concern based on simple analyses that the "*thin sideboards*" of the CAB concept were insufficient to provide the ship with any acceptable roll stability. For this reason, the *Hydrokeel* was added to the list of candidates to be considered. In the words of the SESOC Final Report:

"..... The designs (for the Hydrokeel) use a spring or air loaded flap to close the bow as in the CAB, but the sidewalls are of appreciable thickness to give buoyant and planing lifts to stabilize the craft in roll....."

Figure 4.6 shows in the upper photograph, the US Navy Landing Craft the LCVP(K). The lower sketch is that drawn up for the SESOC Committee for a design of a 1000 ton displacement *Hydrokeel*. While at first glance, the *Hydrokeel* would not be a candidate for a high speed ship design it did have geometrical features that would appear later in high speed ship designs such as the ram wing versions of aerodynamic lift craft. The air was "rammed' into the forward openings and leakage eliminated by the two sealing buoyant sidehulls and a trailing edge "flap seal". Again, in comparison with the CAB concept, the SESOC experts concluded:

US Navy *Hydrokeel LCVP(K)* (1962)

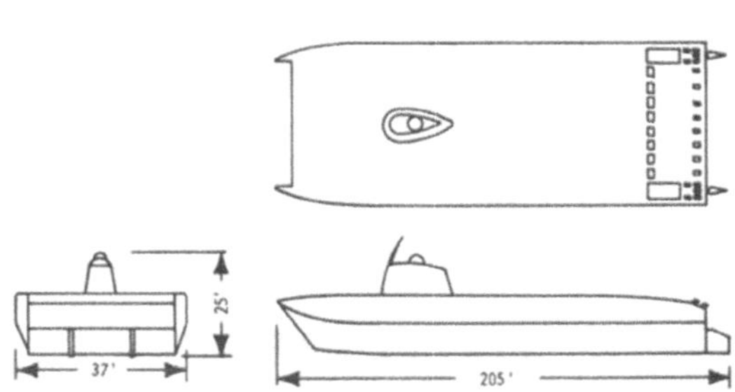

Hydrokeel scaled to 1,000 tons

Figure 4.6: *Hydrokeel* Candidate
US MARAD photo and sketch

"..... the aft planing surface (of the trailing edge seal flap) causes the stern of the craft to follow nearly the contour of the waves, producing greater accelerations than occur with the CAB concept. However, the present Hydrokeel designs appear to be stable under any sea condition...."

Very little work was done on the accelerations of rigid planing surfaces moving over waves in those early years and the large magnitude of those forces (some measured at 1000 g's or more) and the brutal treatment and maintenance nightmare had yet to be appreciated. Rigid seal or flap design proved to be one of the Achilles Heels in the design of any craft planned to operate at speeds of even 50 knots let alone 100 knots or more at sea. The irregularity of rough sea waves hitting such mechanisms only served to complicate the design and attempts in avoidance of such devices from tearing themselves apart in operation.

4.4.4 VRC Channel Flow

The SESOC Committee was specifically charged to examine the two main candidates for SES design, which were (a) the *channel flow wing* and (b) the *CAB* concepts. It is unfortunate that insufficient homework was done to understand the basic differences in features of the concepts to be evaluated. The misinterpretation of the *channel flow concept* can be seen in Figure 4.7 which shows the concept as interpreted by SESOC Committee and the actual concept as designed.

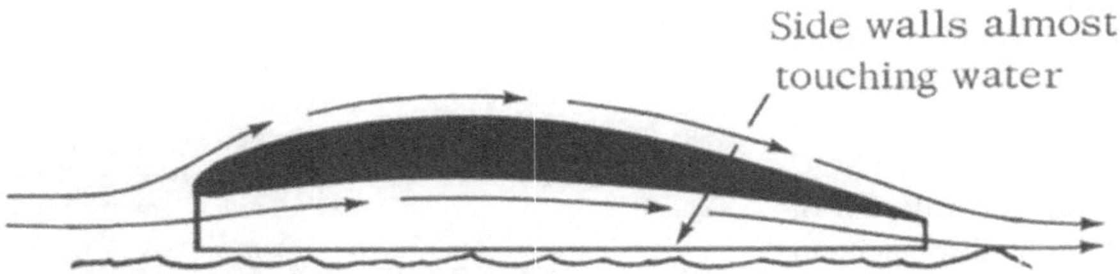

INCORRECT INTERPRETATION of Channel Flow Concept
(taken from SESOC Report)

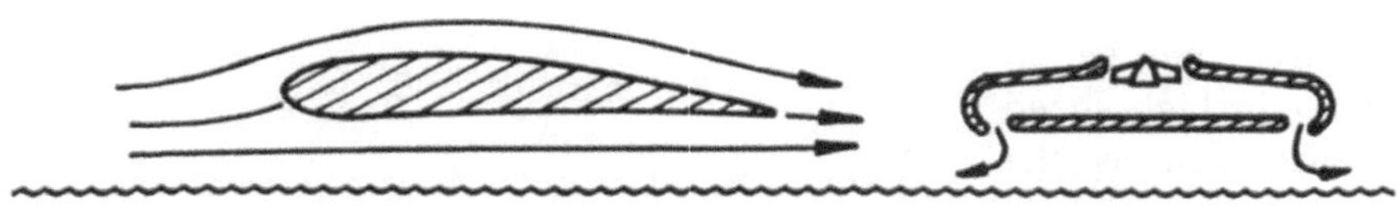

CORRECT DESCRIPTION of Channel Flow Concept

Figure 4.7: Incorrect and Correct Descriptions of *Channel Flow Concept*

The SESOC Committee incorrectly assumed that the *Channel Flow Concept* was simply a sidewall contained wing-in-ground-effect without the trailing edge flap normally associated with ram wings. As seen in the upper sketch (taken directly from the SESOC Final Report) shows (a) there are sidewalls "almost touching the water" and (b) there is no pressurized air flow incorporated in the design as found in amphibious air cushion craft. These SESOC shown features are in error and are a misrepresentation and understanding of the concept. Chapter 3 provides a detailed description of the *channel flow* that had been provided Booz-Allen and MARAD as part of the material supplied to SESOC by all the craft designers.

The lower sketch is the correct drawing that clearly shows that the sealing of the air flow on the sides of the craft come directly from the onboard pressurized air system (shown illustratively in this sketch as being an on-board fan system). The elimination of "hard" sidewalls that would come into contact with waves is part of the channel flow concept and is specifically intended to avoid what could be a catastrophic contact of hard sidewalls with waves at high speed. This very feature of the concept is discussed in the treatment of aerodynamic air cushion craft in Chapters 1, 2, 8, 10 and 11.

Chapter 3 shows the actual designs and "as built" configuration of the MARAD *Channel Flow VRC-1*. These designs and test data was provided to the SESOC Committee at the beginning of their study. This misinterpretation by Booz-Allen and SESOC of the channel flow concept led to unfortunate conclusions in their Final Report on Surface Effect Ships for Ocean Commerce.

The application of skirt systems to aerodynamic air cushion craft had not been attempted before the MARAD Columbia Project. Detailed analysis had been accomplished and a rudimentary skirt was applied to the VRC-1 during its Test Program in 1964. Fabric manufacturers had not yet begun their research in the US and suitable material was not readily available and only short duration speed runs were possible before significant wear occurred.

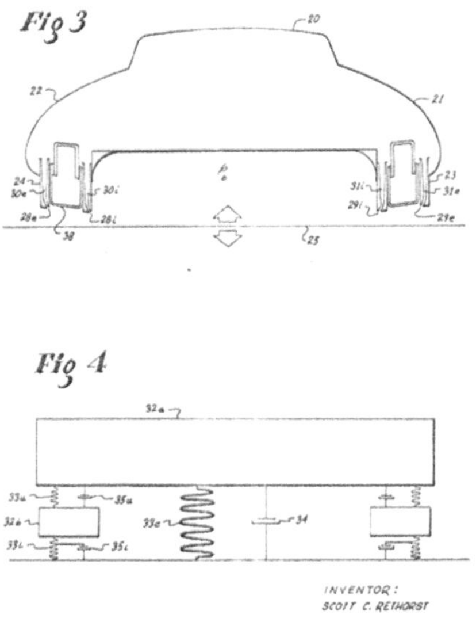

Figure 4.8: Flexible Skirt Design for Channel Flow Craft

The design of the skirts for *Columbia*-like craft may be seen in the patent drawings by the channel flow inventor Dr Scott Rethorst in the US Patent[3] and the sketches from the patent show the main features (see Figure 4.8). The patent called, *"Suspension System for Surface Effect Ships"* was filed on 25 March 1963 and design information made available to SESOC as part of their deliberations. The final patent was issued later on 12 October 1965. The first rudimentary version was installed on the MARAD VRC-1 on 18 June 1964. An entry in the contractual Monthly Report to MARAD for June 1964 reads:

> *"..... on 18 June 1964 the rubber jet extensions were installed and a tethered test was made..... and on 24 June 1964 two more flights were made..... one 10 minute flight at 55 mph at cruise heights between 4–6 inches with the flexible skirt jet extensions....."*

No further work was accomplished on flexible skirts for channel flow craft as other test priorities came into play in 1964 and it was decided to explore improved materials with material suppliers before the full patented suspension system be fabricated and tested on the *VRC-1*. Unfortunately, shortly after these preliminary efforts on skirts for channel flow craft, the MARAD *Columbia* Project was curtailed in response to the new Department of Commerce directive to study alternative concepts for SES with the formation of the SESOC Committee. These various features for *channel flow SES* were not examined by the SESOC Committee. The design concept for a 1000 ton displacement version of the *channel flow concept* as done by MARAD complete with an artist's rendition of such a ship is shown in Figure 4.9. The artist has attempted to show the air flow on the side of the ship (and not hard sidewalls) that was not considered in the SESOC evaluation and sketch shown in Figure 4.7. Neither the air flow nor any skirt system was included in the SESOC analysis. The MARAD *Columbia* Project that included the Channel Flow Test Craft *VRC-1* was not far enough along in its test program to show the results of

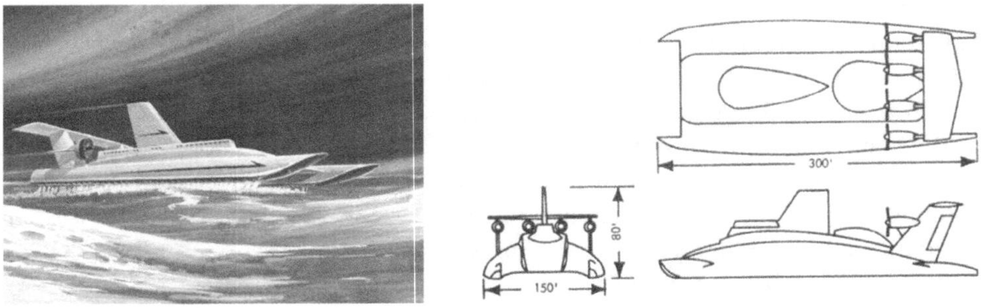

Figure 4.9: Channel Flow 1000 ton Candidate
US MARAD art

the *hovercraft-like* "skirts" that were tested in preliminary form during the Test program in 1964 at Edwards Air Force Base. The program was canceled early in 1965 as the SESOC Committee began its study work and the use of skirts in channel flow designs was not evaluated.

4.4.5 Weiland Craft

The fifth and last candidate for study by the SESOC Committee was the Weiland Craft. Carl Weiland was a Swiss inventor who resided in the US in the early 1960s. He consulted for a short while with Vehicle Research Corporation and later, at Douglas Aircraft Co., designed and built the *tandem wing-in-ground-effect craft* shown in Figure 4.10. The tandem wing feature was to provide improved altitude and pitch stability to single wing-in-ground-effect vehicles that had a tendency to pitch up as altitude increased (see Chapters 3 and 11 for further discussion on wing-in-ground-effect craft).

As for some of the other candidates selected by the SESOC Committee for evaluation, the Weiland Craft was just one example of aerodynamic air cushion craft and was not representative of the generic class of *aerodynamic air cushion craft*. The Weiland Craft was specifically built by its designer to explore stability features alone and was not intended to be representative of all aerodynamic wings operating in

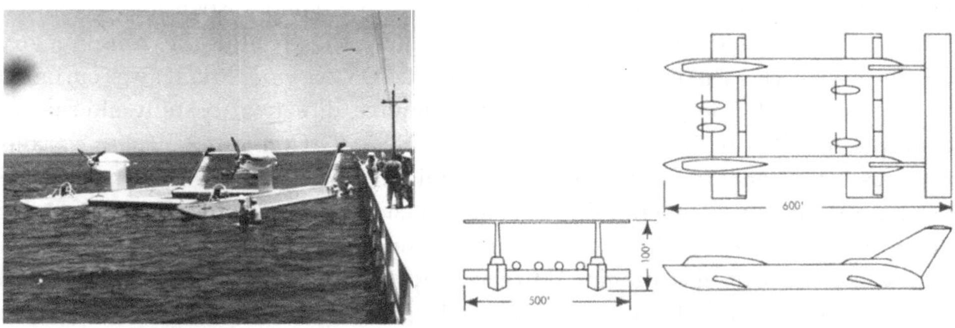

Weiland craft (Man-carrying prototype) (1963)

Weiland craft scaled to 1,000 tons

Figure 4.10: Weiland Craft Candidate
US MARAD photo and sketch

ground effect. Because of this, the span of the craft was not an issue and gives a false impression as to the possible smaller spans that are possible. The *channel flow concept* as shown above in the *Columbia* project sought to limit span by use of side air jets. The other solutions might be found with the Lippisch designs where *delta low aspect ratio wings* were used. Discussion of the Lippisch designs is given in Chapter 2 and again in Chapter 11. An example of the Lippisch design that is in the "same family" as the Weiland Craft is shown in Chapter 1 (Figure 1.16) with the "X-300" design. Such possibilities were not considered by the SESOC Committee and the scaled span of a 1000 ton design of the Weiland prototype would be 500ft – which would seriously limit it as a candidate to operate at normal ports around the world – but this would be an unfair conclusion of aerodynamic air cushion craft based on the above discussion.

4.5 Booz-Allen and MARAD Designs for Surface Effect Ships

MARAD with its contractor Booz-Allen Applied Research Inc. compiled a series of designs based on the above five candidates for the SESOC Committee to evaluate. There were several designs from 100 tons up to 1000 tons compiled. None of the companies involved, because of the state-of-the-art of the technologies involved, projected beyond 1000–2000 tons displacement. The sole departure was the US Navy which projected designs up to 10,000 tons displacement for the CAB concept.

It is not appropriate to summarize in detail all the designs done by Booz-Allen and MARAD that were then reviewed quite extensively by the various panels of SESOC. Only the 1000 ton designs have been highlighted in the above sketches taken from the SESOC report as examples. It is appropriate, however, in the context of understanding *"how is the best way to obtain a ship with a 100 knots at sea capability?"* to highlight some of the key findings from the complete set of evaluations. There are many technical and operational factors to consider such as performance (in all its complex facets); stability and control; powering; and the complex subject of operational feasibility including port and dock arrangements. The SESOC Panels were organized to tackle each of these key characteristics separately.

As discussed in Chapter 7 on Economical Considerations, there are a few key characteristics of the ship and its operations that feed directly into the Direct Operating Cost (DOC) as a measure of determining whether any particular candidate is suitable. These include the speed, both maximum (V_{max}) and cruise speeds (V_{cr}); the disposable load capability (W_D/W); propulsion efficiency (η_P) and power requirements measured by specific power (HP/ton) and the ability to transport its weight at speed for a given amount of power expenditure measured by transport efficiency (WV/P).

4.5.1 Booz-Allen Adjustments to Candidate Designs

Unfortunately, several adjustments were made to the submitted designs by the companies that seriously affected the outcome and results. For example, using the above mentioned key features for economic evaluation <u>all</u> of the designs submitted were downgraded in their speed capability, that is, <u>except</u> for the US Navy

CAB designs. The *hydroskimmer, channel flow* and *Weiland craft* were all downgraded (without explanation) to 100 knots top speed even though several of the submitted designs showed maximum speeds of 120 to 160 knots while the CAB designs were the only ones shown with speeds greater than 100 knots. For example, the top speed shown for the 5,000 ton CAB was 112 knots, and 129 knots for the 10,000 ton design!!

The *hydroskimmer* (as representative of the amphibious air cushion craft) was adjusted in power level such that at the 100 ton size it was adjusted to a specific power of 350 HP/ton while the actual Bell SKMR-1 (at 29 tons gross weight) exhibited only 148 HP/ton. Normally one would expect the specific power (HP/ton) to decrease slightly with increase in displacement, not increase and certainly not more than double! The rationale for this change in power by Booz-Allen for the *hydroskimmer* was to increase the air gap under the skirts to 4 ft which is totally unrealistic and has not been used in any *hovercraft* before or since. This "adjustment" raised the installed power for the 100 ton *hydroskimmer* design to a staggering 35,000 HP – which is almost three times the installed power in the 165 ton SRN.4 with 13,600 HP installed power!! Apparently, the rationale for this "adjustment" was that more power was required to allow the *hydroskimmer* to traverse 8 ft seas. This reasoning, unsubstantiated by any analysis or data from such craft showed a misunderstanding of the role of skirts in hovercraft design. The development of skirt technology had been underway since 1960 as British Hovercraft Corporation (BHC) and others were developing the capability. Figure 4.11 shows the early history of

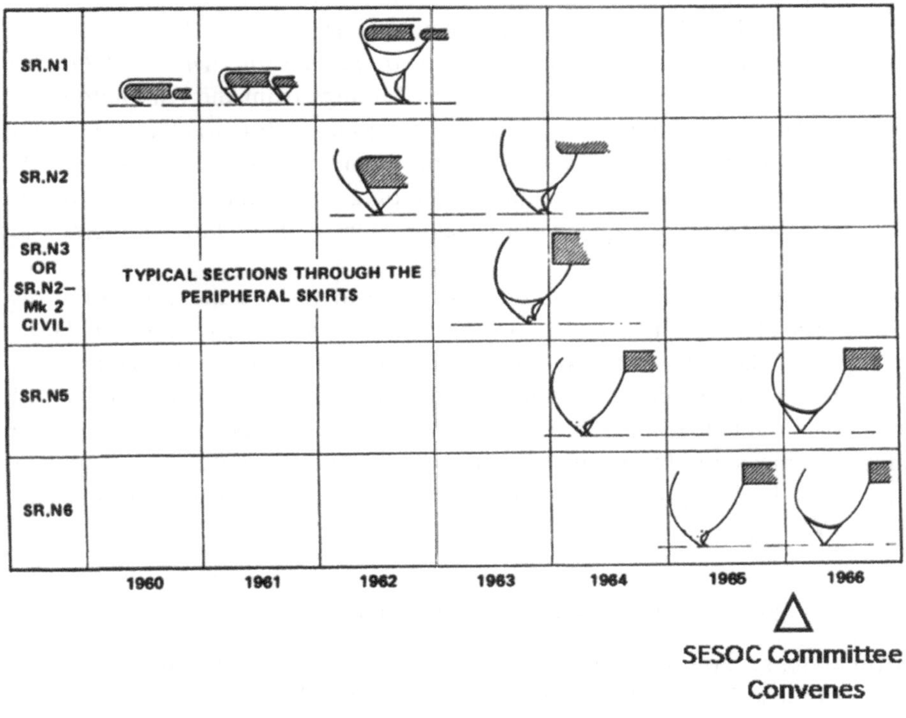

Figure 4.11: Hovercraft Early Flexible Skirt Development History

BHC skirt development and application to a range of hovercraft from the *SRN.1* in 1960 to the *SRN.6* in 1965.

A more complete history of skirt development through later developments may be found in the 1980 report[4] but Figure 4.11 shows the history that was certainly available to the SESOC Committee in 1965. The value of the skirt system is that height can be added to the hard structure of the craft *without additional power expenditure*! Because the Bell SKMR-1 with its skirt (modeled after the BHC designs) shown in Figure 4.4 had such skirts installed it is curious that this was not recognized. Booz-Allen went further in the adjustment to the *hydroskimmer* by an additional reduction in payload capability (because of the added power). Such distortions gave a decidedly unfair evaluation of the *hydroskimmer* or amphibious air cushion craft concept. All of the power requirements and the disposable load fractions for the amphibious air cushion craft designs in SESOC were seriously in error as a result of these "adjustments".

Other adjustments were made that were also very difficult to justify. The structural weight of the CAB design was adjusted down to 20% of the ship displacement which is a value that has never been attainable in any such dynamic lift craft. Such an unrealistic value led the adjusted design of the 10,000 ton CAB to boast a 73% disposable load fraction (W_D/W)!! Only large bulk container ships supported by full buoyancy forces and traveling at low speeds have approached such values.

These flawed adjusted designs formed the basis for the SESOC Panels of experts' evaluation. Fortunately, the SESOC Panels tackled each of these adjustments in a professional manner as described in the following summary, frequently coming to different conclusions than the Booz-Allen analysis claimed.

4.6 Conclusions and Recommendations of the SESOC Committee

The Surface Effect Ships for Ocean Commerce (SESOC) Committee work, while short in duration, was a key milestone along the path to find a practical and *efficient high speed ship of large tonnage* that was capable of operating on transoceanic routes and provide an economical advantage to the US Merchant Marine (and subsequently to the US Navy pursuits in a similar venture). The Panel divided its work into five panels each concentrating on specific areas. These were: (1) Aero-Hydro Dynamics and Control Panel, (2) Speed, Resistance, and Seakeeping Panel, (3) Propulsion Panel, (4) Hull Panel and (5) Operations Panel.

It is not the intent to summarize all the SESOC Committee work but some key results related to the two main candidates for consideration (by the US Department of Commerce) provides a firm basis for what followed over the next decade and a half from 1965 through 1980 where over $1 billion was spent attempting to produce a Surface Effect Ship capable of operating at 100 knots. These two candidates as noted earlier were the *channel flow concept* and the *CAB concept.*

4.6.1 SESOC Conclusions and Recommendations on Channel Flow and CAB Concepts

As stated above, the experts on the five panels were presented with a very difficult problem with only a matter of weeks allowed to generate supportable conclusions and recommendations. In many areas, over-simplified assumptions were made by Booz-Allen Applied Research in an attempt to develop broad conclusions. Some of these assumptions masked both advantages and fatal flaws in the concepts. Some extracts from the five Panels provide a general set of results on SES and how the follow-on efforts by MARAD (and subsequently the US Navy) proceeded.

Aero-Hydro Dynamics and Control Panel Findings In most cases, the SESOC Panel experts correctly spotted the Booz-Allen assumptions that affected the outcome. Because the stability and control of high speed ships is of a fundamental nature it was unfortunate that the Booz-Allen assumptions in this key area were lumped into a generality as can be seen from the SESOC Final Report that states:

> *"..... The Booz-Allen technological assumptions of concern to this Panel (Aero-Hydro Dynamics and Control Panel) are implicit in the economic studies. They can be reduced to the generality that the various craft examined will be stable and controllable about all axes in the specified sea states. If not, artificial means can be used and these are within the state-of-the-art......."*

This assumption is remarkable and ignores the very features that needed to be explored and evaluated. The SESOC Committee correctly did not accept this assumption. Indeed, the SESOC Final Report contains this conclusion:

> *"..... The Panel is unanimously of the opinion that all of the concepts have aero-hydro dynamic stability, control, and response problems potentially serious enough to preclude the actual development of a large prototype SES at the present time....."*

This fundamental conclusion proved prophetic in that as the program to develop a high speed surface effect ship unfolded it was hampered to a significant degree by such stability and control issues as correctly highlighted by the SESOC Committee.

Some specific comments by the SESOC Committee on the two basic candidates relative to their stability and control issues are:

Channel Flow Configuration

> *"..... It seems to offer some promise of superior performance and should be pursued on a research basis until it can be defined more closely and evaluated more rationally....."*

Other Configurations Here, the SESOC Panel grouped their findings as follows:

> *"..... The remaining three aerostatic concepts (air-lubricated hull, CAB, and annular jet) are all known to have performance, stability and control, and response problems that can be serious. It is believed that as each concept is improved, it will take on some of the better features of the other concepts and they will develop toward one common type embodying the best features of each concept. Consequently, the Panel has reviewed the various problems in an overall sense and has not differentiated among the three schemes, considering all of them as forerunners of a common descendant....."*

Here, it is believed, the SESOC Panel slipped into the same form of generality that it had just accused Booz-Allen which was of being over-simplified. The difference between *amphibious air cushion craft* and *sidehull craft* (see Chapter 1 on the development of the various types of aerostatic craft and others) is quite marked. The fundamental lack of roll stability in the CAB concept, for example, as demonstrated in its accident and roll over is difficult to ignore. The stability of the air lubricated hull with its trailing edge flaps was more than adequate. It is to the SESOC Committee's credit that it recognized that any program on a high speed SES should proceed slowly (with regard to stability and control) and to start with several 5–25 ton class man-carrying vehicles first with a carefully designed test program. As discussed in Chapter 5 these sage words of advice were quickly ignored as the program moved ahead.

Speed, Resistance, and Seakeeping Panel Findings In the early days of surface effect ship development there was very little data on which to make reliable predictions on resistance and seakeeping and the specifics are not explored further here but Chapter 8 discusses many of the characteristics especially those that developed after the SESOC Committee had completed its work. In general, the SESOC Committee believed that Booz-Allen had overstated the performance capabilities of many of the candidates, specifically the Report states:

> *"...... Without such research, it is very probable that early designs of stable craft will have lift-to-drag ratios substantially less than the Booz-Allen predictions....."*

This point is repeated throughout the SESOC Panel's findings in the assessment of the performance predictions. The Panel correctly pointed out the overly optimistic treatment of the effect of stability features in the overall resistance prediction with the conclusion:

> *"..... The one area that was found wanting in the drag projections was the effect of stabilizing schemes in degrading performance....."*

The statement that "artificial means" of achieving stability would not increase ship resistance was simply not accepted by the Panel and rightfully so. On the subject of stability and control and seakeeping the concern about CAB stability at sea (a key topic in the evaluation) was unfortunately sidestepped by:

> *"..... The Booz-Allen study does not treat the seakeeping problems of CAB-type vehicles...."*

On the subject of *aerodynamic air cushion craft*, the Panel unfortunately dismissed both the *channel flow* craft (which they were charged to evaluate) and the Weiland craft with the statements:

> *".....The channel flow machines were excluded because;*
>
> 1. *They are essentially very low aspect ratio wings which must operate within what appear to be impractically small distances from the water surface to overcome the inherent inefficiency of low aspect ratio wings.*
> 2. *They must be low density machines*
> 3. *They are very large compared with aerostatic machines of equivalent payload capacities...."*

".... The WIG machine was excluded (by the Panel) because:

1. *It is evidently much too large when practical payloads are considered*
2. *Relative to the other types of machines, there are no unknown or uncertain technical factors*
3. *If, for any reason, such machines find an application, working designs can be prepared on the basis of presently available information and data........"*

The later discussion of other forms of *aerodynamic air cushion craft* and the discussion of *hoverports* (see later discussion and Figure 4.14) show this to have been a missed opportunity by the Panel.

Further, the Panel took the position that if a feature was not acceptable for one reason or another it simply changed the configuration to suit the conclusion. For example, the fundamental differences between a flexible skirted craft (like the *hydroskimmer*) and a rigid sidewall craft (like the *CAB*) was dismissed by the statement;

"..... For example, the air lubricated hull is simply a CAB with a rigid rear seal; the annular jet machine with skirts has features of a CAB with flexible sidewalls and bubble leakage....."

Such generalities eliminate any possible treatment of the hydrodynamics of rigid versus flexible structures and the resultant differences in both performance and stability.

Propulsion Panel Findings While there were several detailed comments in areas of engines, transmissions and propulsors there were no major differences found between the five candidates. This is disappointing in that the schemes for propulsion for high speed ships vary considerably and has indeed hampered the development of the high speed surface effect ship. This subject is discussed in more detail in Chapter 8 on Technical Considerations. The Panel did make the general comment that the assumptions of propulsive efficiency assumed by Booz-Allen in their studies (for all candidates) were optimistic and that

".... the economics would be less attractive than those shown by Booz-Allen...."

The propulsion Panel also noted prophetically that

"..... Since the influence of cavitation will become evident at ship speeds above 50 knots, it is expected that appendage drag will become increasingly important....... The design of the intakes for the water jet system is identified as the most significant single problem......"

Later, this latter finding proved to be the Achilles Heel in the design of the US Navy CAB SES-100A that struggled with its water jet system design as it changed from a fully submerged pod inlet with its high drag to a subsequent retrofit with a flush inlet that was plagued by air ingestion from the air cushion bubble that supported the craft (see Chapter 6 on the history of the SES-100A and SES-100B Programs).

Hull Panel Findings The results of this Panel were probably the most prophetic in their deliberations. Much of the detail assessment was based on a review of the 5,000 ton CAB design provided by the US Navy CAB designers shown in Figure 4.12.

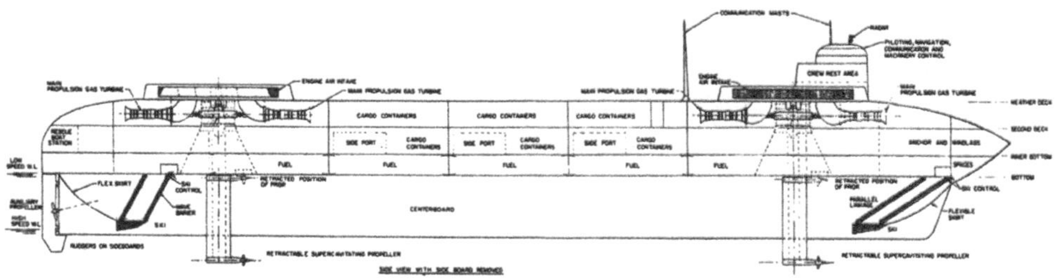

Figure 4.12: MARAD 5000 ton CAB Concept (US Navy design)

While a little difficult to see in the drawing, the details of the articulated skis shown at the bow and stern received particular criticism from the SESOC Committee. The design as shown included (undefined) ski control at the hinges; a wave barrier on the aft ski that was not included in the drag analysis; and large "spinnaker-like" sheets mounted on the articulated skis to contain the pressurized cushion air. The assessments by the Panel were almost entirely related to the problems of the CAB concept as can be seen from the following extracts from the SESOC Final report:

> *"..... It is generally agreed that a successful CAB design must employ fore and aft skis which contain the bubble with relatively small clearance at the waterline. Simple calculations quickly show that, for operations in a seaway at speeds up to 100 knots, the potential water loads on a hinged, but otherwise rigid, ski are enormous. In addition to intolerable structural loads, large longitudinal (fore and aft) accelerations would be imparted to the craft in even modest seaways"*
>
> *"..... Even if an ingenious and inventive concept for ski design is proposed in the near future, the current level of technical knowledge of CAB behavior in seaways would preclude an intelligent evaluation of the suitability of the concept....."*
>
> *"..... The Panel believes this entire problem area to be a major hurdle which must be overcome before serious CAB development can be undertaken. The practicality of the entire project may well hinge on whether reasonable solutions can be found to the problems of ski design....."*

While not generally known at the time, the extreme accelerations imparted to both flexible skirt ends as in the *hydroskimmer* like skirts (see Figure 4.11) and in the "feather" ends of the SES-100A CAB seals (see Chapter 6) caused extreme wear called flagellation. Under such conditions the "spinaker-like" sheets on the projected 5,000 ton CAB would have most certainly failed in operation.

In addition to the review of the structural design, the estimates of structural weight for the CAB concept in the light of the high loads predicted at 100 knots were questioned. The panel expressed serious concern about the Booz-Allen assumption of a structural weight of 25% of the ship displacement. The concern was also expressed that while the value of 25% given in the design analysis was not realistic, it was further reduced to 20% in the Booz-Allen economic analysis!

The specific comments by the Panel on these items were:

> *"...... The Panel is of the unanimous opinion that the figure of 20% is far too optimistic for use as a basis for economic studies. The Panel also doubts that the 25% level for structural weight can be achieved in practice....."*

"..... The Panel believes that this entire problem area (ski design) to be a major hurdle which must be overcome before serious CAB development can be undertaken......."

"..... the Panel is of the unanimous opinion that the CAB concept is not yet ready for Contract Definition (Project Definition) action....."

The Panel also examined, in lesser detail, the structural weights for the other candidate concepts and were less critical of the values given. Specifically the Panel concluded:

"..... The design of WIG-type SES, which are essentially large aircraft operating in ground-effect region, is within the current state-of-the-art as regards structural design. Structural weights in the vicinity of 25% of gross weight can probably be achieved in large aircraft of this type providing emergency survival in heavy seas in the displacement condition is not a design requirement. If survival in emergency landings at sea is required, the structural weight ratio is estimated to be 30% or more....."

Operations Panel Findings The Operations Panel raised serious objections to the analyses performed by Booz-Allen on the economic model used. In the words of the Panel:

"...... there are notable voids in the assumptions relative to economic analysis and justification of SESOC concepts......"

".... the identification of SESOC support facilities as a percentage of direct operating cost is considered inappropriate......"

The Panels comments on the CAB Candidate are summarized here:

CAB Concept

"....The selection of 7,000 hours per year utilization for the 5,000 ton CAB appears unrealistic.."

"..... Booz-Allen assumed a cargo load factor of 85 per cent for the 5,000 ton CAB. This figure is not specifically supported...."

There were additional comments about the economic analysis that caused the Panel concern that suggested that more realistic treatment of the key features needed to be reevaluated.

4.7 SESOC Committee Conclusions and Recommendations

Despite the paucity of data made available to the SESOC Committee and the short time available for study the SESOC Committee managed to capture many of the "sticking points" in the determination of which concept offered the best opportunity for achieving a "100 knots at sea capability". Essentially, the SESOC Committee concluded more research be tackled before any decision be made of which concept to pursue. Specifically, the SESOC Committee stated in their Final Report Conclusions:

"The commercial feasibility of the SES concept is critically dependent upon the achievement of optimum performance in the following areas: stability and control, effective lift-drag ratio, structure, propulsive efficiency, and fuel economy. At the present time, the

state of knowledge is not adequate to determine the optimum degree of performance in these areas, either as to what is required or what can be realized.

An improved state of knowledge for the solution of crucial technological problems can be achieved if immediate studies emphasize research rather than the development of scaled prototype vehicles. By the end of about 3 years and with an expenditure of about $10 million, it would be possible, through a coherently organized research program, to know whether the SES concept is technically feasible for ocean operation."

The SESOC Committee recommended that a two phase program be put together that would first explore the technology and then build a 500 ton prototype. It is important to note that the SESOC Committee did not have a recommended candidate, stating that much more research be done before any decision could be made. Critical evaluation of the various concepts has been summarized in the Panel reports.

It is interesting to note that the SESOC Committee did not use any of the information that was well underway in England and around the world where the air cushion principle was already being used in commercially operating vehicles. The air cushion principle originated in England, yet there were no Panel members or consultants from the UK hovercraft industry involved in the SESOC Committee deliberations.

No *hovercraft* or *wing-in-ground-effect craft* that were already "flying" were included in the mix of candidates being evaluated. Perhaps a different set of conclusions would have been reached if that ongoing experience had been made available. A brief summary of that ongoing industry is provided to serve as a back drop for discussion in this and other Chapters.

4.8 Ongoing SES Experience While SESOC Committee Was Deliberating

Before and during the SESOC Committee deliberations there were significant operations underway in the US, England, Russia and other countries on several of the concepts included in the SESOC Committee set but not included in the analyses. This was unfortunate as the conclusions might have been a little different if this real world experience had been included.

4.8.1 Amphibious Air Cushion Craft

A full two years before the SESOC Committee was convened, British Hovercraft Corporation (BHC)[d] was designing, building and selling to commercial operators several *amphibious air cushion craft* that more than embodied the *hydroskimmer* concept used by the SESOC Committee. Figure 4.13 shows just three key vehicles, the *SRN.5*, the *SRN.6* and the *SRN.4*.

The first hovercraft, the SRN.1, crossed the English Channel in 1959 – a full six years before the SESOC Committee convened. In that intervening period several designs were built and developed but the *SRN.5* (upper left hand photograph in Figure 4.13) went into a 10 craft production program on 6 April 1964 and began

[d] British Hovercraft Corporation ceased production of hovercraft in 1984 and became part of Westland Aerospace, which in 1994 became part of GKN Aerospace.

SRN.5 (6.7 tons; 66 knots) (1964)

SRN.6 (10.3 tons; 60 knots) (1966)

SRN.4 (165 tons; 70 knots) (1968)

Figure 4.13: Existing and Projected Commercial Hovercraft (circa early 1960s)
Photos permission by GKN Aerospace©

commercial operation (together with an earlier built craft the *SRN.2*) with regular passenger service across The Solent in England between Southsea on the mainland and Ryde on the Isle of Wight. The *SRN.5*, shortly after the Solent operations, was then operated by the British Interservice Hovercraft Trials Unit (BIHTU) for military evaluation. Trials were underway in the Far East, Japan, Norway, Germany and Brunei. Three were exported to Bell Aerosystems in the US (designer and builder of the Bell SKMR-1 that was unfortunately misrepresented in the SESOC study) in 1965 which started the *SK.5* series of military hovercraft, forerunner of the successful LCAC hovercraft for the US Marines.

In 1966, as the SESOC Committee was wrapping up its report, the *SRN.6* (shown upper right hand photograph in Figure 4.13) began a trial passenger service across the English Channel under the operation of the Hoverlloyd Co. formed by Swedish Lloyd and the Swedish American Steamship Company in the summer of 1966. This was a trial program that led to the even more impressive operations by the *SRN.4* hovercraft (see lower photograph in Figure 4.13) in 1968. While the *SRN.4* had not yet been launched at the time of the SESOC deliberations extensive model testing, tow tank experiments, analysis, design and construction work was indeed ongoing in the design offices of BHC and could have been made available to the SESOC Committee.

Advice from the British Hovercraft Corporation engineers to the SESOC Committee would have been invaluable in understanding what is possible for such craft. In particular, the Operations Panel would have benefited considerably if the passenger and cargo features of the *SRN.4* had been made known to SESOC together with all the design and maritime regulatory considerations that were included in the final configuration. Figure 4.14 shows the passenger compartment and the cargo deck of the *SRN.4*.

SRN.4 Passenger Deck (256 Seats) *SRN.4* Drive On Cargo Deck (30 Cars)

Figure 4.14: SRN.4 Typical Car and Passenger Configuration
Photos permission by GKN Aerospace©

In addition to the design features of the craft themselves there are a number of issues that must be worked out before commercial operation can go into effect. The criticism of the Booz-Allen assumptions in their economic analytic model by the SESOC Committee might have been alleviated if those concerned with putting together actual commercial operation and design of the port facilities and been consulted. The setting up of an operation that includes the design of the facilities for such unique craft (an issue with the SESOC Committee) was well underway for the *SRN.4* operations in 1965. In April 1965, the Swedish Lloyd Line and the Swedish American Steamship Company signed a contract for a five year charter for two *SRN.4* hovercraft for use as a passenger and car ferry between England and France operating between Ramsgate (England) and Calais (France). A new *"hoverport"* was constructed at Pegwell bay, Ramsgate and a new French Terminal established in the Calais harbor. The service was to begin in 1968 after the construction was complete. Figure 4.15 shows a photograph of a later *hoverport* built at Dover, England which had a similar layout to the first *hoverport* described for Ramsgate. Figure 4.16 shows a similar arrangement at Ryde, Isle of Wight in Southern England operated by Hovertravel Limited. The photograph shows the SRN.5 and the AP.1–88, a craft designed to carry 100 passengers.

Figure 4.15: British Rail's Hoverport at Dover, England (1968) (Not Shown: Similar Hoverlloyd's Hoverport at Ramsgate, England, 1965)
Photo permission by GKN Aerospace©

Figure 4.16: Hovercraft Terminal at Ryde, Isle of Wight
Photo permission by GKN Aerospace©

Several key points can be made about the setting up of the *hoverports* in England and France for the *SRN.4* operations. Many of the inputs essential to any direct operating cost (DOC) analysis involve knowledge of the facility costs (maintenance, crew costs, fueling, etc) would have been included in the work that ensued between the craft designers (BHC), the operators (Swedish companies) and the facilities managers that would have alleviated many concerns by the SESOC Operations Panel rather than rely on the assumptions in the Booz-Allen analysis of which SESOC Committee was most critical. There is an additional set of information that would have been useful to the SESOC Panel that essentially dismissed the aerodynamic air cushion craft (*channel flow* and *wing-in-ground-effect craft*) because of ship span. A quick look at Figure 4.15 would suggest that such aerodynamic craft could operate quite well from such *hoverports* – an unanticipated feature by the SESOC Committee.

Included in Figure 4.15 are the *SRN.4* in its original configuration and displacement of 165 tons on the right, and its larger sister craft, the stretched 300 ton *SRN. 4* (the "Super 4") on the left that went into operation later in 1978. The Original *SRN.4* had a top speed of 70 knots and the stretched *SRN.4* had a top speed of 60 knots. While the design and building of such craft were underway in the early 1960s the UK were also looking into transoceanic hovercraft. In 1966 the UK Ministry of Technology gave a contract to British Hovercraft Corporation for a 3000–4000 ton *hovership*. Figure 4.17 shows a model of a 4000 ton *hoverfreighter*. The design work on this large tonnage SES that was ongoing in 1964–1965 would have been an invaluable input to the SESOC study.

Figure 4.17: BHC 4,000 ton *Hoverfreighter* model
Photo permission by GKN Aerospace©

The BHC *hoverfreighter* was a design of an amphibious air cushion craft except that it was propelled by marine propellers. The reasonable assumption was that such large ships need not be "amphibious" in all aspects. It would have been useful if the preliminary design information available at the time on a ship the size of interest (around 4000 tons) to the SESOC Committee and the Maritime Administration could have been incorporated into the SESOC study.

There is a legitimate question as to the cost of maintenance of such extensive hovercraft skirts as shown in Figure 4.17 on the hoverfreighter, and a later craft such as the HM.2 and the SES-100B (see later) among others used partial sidehull and partial skirts in their design. The "partial sidehull" concept (see Chapter 6) where a trade-off has been made as to "how much sidehull and how much skirt" is examined for both stability and performance considerations. The additional question would be on the relative maintenance issues between the fully skirted amphibious hovercraft as shown in Figure 4.17 and a partial sidehull version. These considerations did not enter into the SESOC Committee deliberations and is discussed more fully in Chapter 6 and in Chapter 8.

4.8.2 Non-Amphibious Air Cushion Craft

British Sidewall Hovercraft In addition to the extensive development work that was ongoing in the world on the amphibious form of air cushion craft discussed above, extensive work was also ongoing on the *sidewall hovercraft* started by Christopher Cockerell in 1955 – a full ten years before the SESOC Committee met. Chapter 2 summarizes the early developments of sidehull craft first by Douglas Warner in 1929, and Christopher Cockerell later in 1955 and later throughout the 1960s. Christopher

Cockerell's work led to the set up of Hovercraft Development Ltd (HDL) and the National Research Development Corporation (NRDC) to explore both *amphibious hovercraft* and the *sidewall hovercraft*. The development work on sidewall hovercraft led to several companies who designed and built such craft for commercial operation with fare paying customers. Two craft that encapsulate the degree of success of such craft are the Denny Bros *D.1* and *D.2* craft and later the Hovermarine *HM.2* series of craft. Figure 4.18 shows the *D.2* and *HM.2* in operation.

Denny Bros was an established shipbuilding company since the 1840s in Dumbarton, Scotland on the River Clyde. In its time it was an aggressive shipbuilder with several innovations including its own tow tank. After Christopher Cockerell introduced Denny Bros to the sidewall hovercraft idea Denny built two craft the *D.1* which was more or less a research craft and the *D.2*, called the *"hoverbus"*, was used in commercial operation in Scotland and England. The left hand photograph in Figure 4.18 shows the *D.2* operating on the River Thames. For reasons not related to hovercraft work, Denny Bros went into voluntary liquidation in 1963. Fortunately, the work on *sidewall hovercraft* did not die there as the company spawned the new entity Hovermarine Transport Ltd and after some initial developmental work produced the *HM.2 sidewall hovercraft* shown in the right hand photograph in Figure 4.18. The successful development of the HM.2 was a significant milestone in that it formed a firm basis for the work later on the US Navy's SES development program. The SES-100B design was the next evolutionary step in *sidehull SES* that used the *HM.2* experience. The detailed description of the features of the HM.2 and how they were used in the US Surface Effect Ship Program is described in Chapter 6 but Figure 4.19 highlights some of the key features taken from the author's 1980 report.

Denny *D-2* (32 tons; 30+ knots (1962)
From Jane's Surface Skimmer Systems 1967 - 1968

Hovermarine *HM-2* (22 tons; 40 knots) (1966)
Reprinted by permission of SNAME

Figure 4.18: *British Sidewall Hovercraft in Commercial operation (circa early 1960s)*

The Hovermarine *HM.2* was launched about one year after the SESOC Committee had convened but the developmental work on the *D.1* and *D.2* and the model testing in different tow tanks of the *HM.2* was already complete when the SESOC Committee began its deliberations. Many of the stability problems of the "*thin sideboards*" of the CAB concept had already been resolved with the UK *sidewall hovercraft* work. Figure 4.19 shows several key features. The three view drawing on the left hand side shows the partial sidehull feature and the shaped and thickened sidewalls for stability. The photographs on the right hand side of

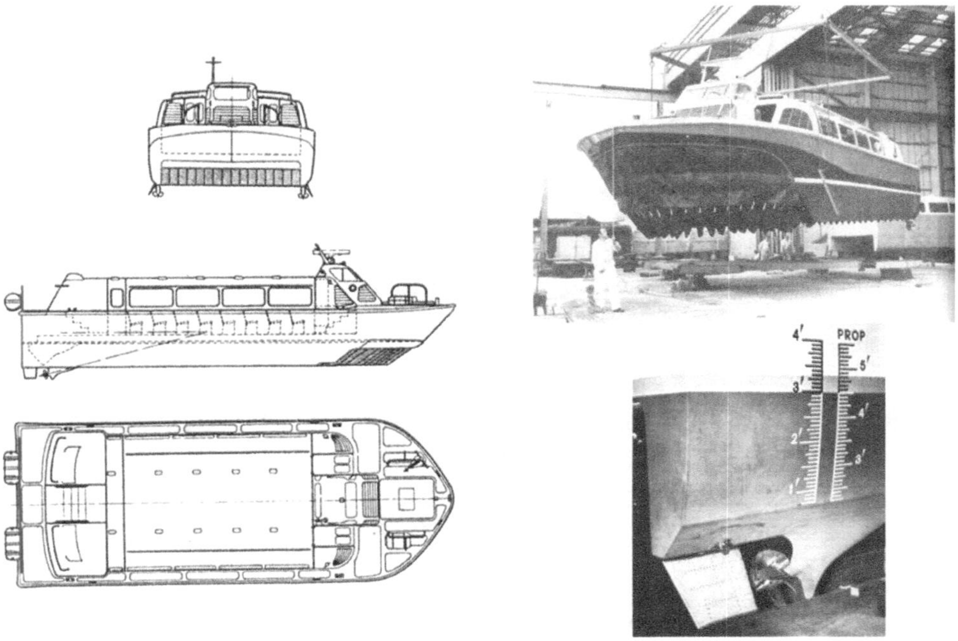

Figure 4.19: Hovermarine Sidewall Hovercraft *"HM-2"* (1966)

Figure 4.19 show the well developed flexible skirt system and the stability appendage at the stern of the ship.

Because of the speed regime of the *HM.2* (about 40 knots) a subcavitating propeller was sufficient. Supercavitating propellers came into play for the 100 knot SES (see Chapter 6 for more details). Again, unfortunately, none of this information was made available to the SESOC Committee.

Soviet "Sidewall" Craft Another set of valuable data that was not used in SESOC was the Soviet Union work on a variant of the British *sidewall hovercraft*. It is difficult to precisely categorize the Soviet work because of the uniqueness of design. There are two craft designs of interest and they are displayed in Figure 4.20.

Two years before the SESOC Committee began its work, the Krasnoye Sormovo Shipyard, in Gorky, Russia launched its "glissyor", the sidewall craft, the *"Opytnyi"* that operated on the Russian rivers and shallow waterways. The sidewalls were actually skegs and the mode of operation was for the nose of the craft to rise as ram air forced its way beneath the craft and between the skegs. The *Opytnyi* was waterjet propelled and it was common practice for the craft to run bow on into shallow banks for the embarking of passengers without the need for any docks or other loading mechanisms.

In this sense, the *Opytnyi* was a combination of the ram wing and sidewall concepts considered by the SESOC Committee. It was a unique application and it is difficult to see how the concept could be extended for large tonnage high speed operation in rough seas, but the knowledge base was there and should have been included in the SESOC Committee set. It could be likened to the *hydrokeel* concept that was included in the SESOC Committee set. The design is a good example of how

Soviet Krasnoye Sormovo *OPYTNYI* (Experimental) (22 tons; 22 knots) (1963)

Soviet Krasnoye Sormovo *SORMOVICH* (22 tons; 75 knots) (1965)

Figure 4.20: Soviet Air Cushion Craft in Commercial Operations
From Jane's Surface Skimmer Systems 1967–1968

the Soviet designers worked the concept to fit with their mission needs as the Former Soviet Union is laced with thousands of miles of rivers that were used successfully for commercial operation. In 1964, the *Opytnyi* started several experimental trips down various Russian rivers and in 1967 traveled the Sukhona River successfully traversing debris in the water including several large size logs. The unique bow shape and ram air, skegs and waterjet propulsion allowed the craft to operate without incident.

The lower photograph in Figure 4.20 is the faster design, the "*Sormovich*" which was also a passenger ferry used on Russian rivers but had a top speed of 75 knots and entered service in 1965 about six months before the SESOC Committee deliberations.

It was common practice for the Soviets to call their craft "amphibious" if they could run up on the beach, which does not fit the definition used by SESOC Committee and others or in this book. It does influence the Soviet designs however and it appears again in their wing-in-ground-effect craft or *ekranoplan* in that the craft are designed to be much more rugged structurally and thus have much larger structural weight fractions (W_S/W) than the US or UK designs (see Chapter 8 and Chapter 11).

None of these designs were added to the set to be studied by the SESOC Committee.

4.8.3 Aerodynamic Air Cushion Craft

The only two types of aerodynamic air cushion craft considered by the SESOC Committee were the *channel flow* and the *Weiland craft*. As has been already identified earlier both craft were misunderstood by the Committee and thus did not receive an adequate treatment although as stated by the Aero-Hydrodynamics

Panel, the *channel flow* concept did appear to "offer superior performance." However, these two types were just two examples of a broader class of aerodynamic air cushion craft. There are others. Carl Weiland was quite upset (in discussion with this author) when his experimental tandem wing craft was being used as an example of wing-in-ground-effect craft in general. He was specifically exploring the stability issues of WIGs (see Chapter 8 and Chapter 11) and was not putting the design forward as a generic form.

There were other researchers who were tackling the oft quoted "span problem" of WIGs and two examples were already in development and operation when the SESOC Committee convened its work. These were the Lippisch designs and the *ekranoplan* in Russia designs.

Lippisch Wing-in-Ground-Effect Craft A brief history of the German designer Alexander Lippisch is provided in Chapter 1 on the broader history of high speed development. Dr Lippisch, a renowned airplane designer at Messerschmidt, Munich, Germany emigrated to the US and joined Collins Radio in 1950 to develop a unique form of WIG. With his knowledge of the delta planform in aircraft design at Messerschmidt Lippisch discovered that if the delta planform was used in WIG – but reversed, with the apex pointed aft, the tendency to pitch up of WIGs was minimized. Dr Lippich built several prototypes to demonstrate this feature, many of which were flying one to two years before the SESOC Committee began its work. Figure 4.21 shows some of the Lippisch designs. In the upper photograph can be seen

Lippisch *X-112* Wing-in-Ground-Effect (1963)

X-113Am flying in ground effect

X-113Am flying out of ground effect
(shows reverse delta planform)

Lippisch *X-113Am* Wing-in-Ground-Effect (1964)

Figure 4.21: Early Lippisch Wing-in-Ground-Effect (WIG) Craft
Photos permission from Special Collections Department Iowa State University Library

the first prototype, the "*X-112*" flying in ground effect over the water in 1963. A clearer picture of the concept can be seen in the two photographs of his second design the "*X-113Am*" that was operating in 1964, about the same time as the *channel flow* craft, the *VRC-1* was starting its initial tests in California.

In the lower left hand photograph, the "X-113" is seen flying in ground effect (or surface effect) over water. The photograph in the lower right hand corner shows the "X-113" flying out of ground effect. In this photograph the *reverse delta planform* can be seen. As can be seen from these photographs it is not correct to assume that all WIGs have large span. There are various technical solutions to minimize span that are already well understood.

It is unfortunate that these fine examples were not made available to the SESOC Committee in its deliberations of the value of aerodynamic air cushion craft that clearly can operate at speeds of 100 knots or more without the severe drag penalties of other candidate designs.

Soviet *Ekranoplan* Development While the SESOC Committee may not have been exposed to the complete work on the Soviet *ekranoplan* development because of security reasons there was enough provided in the popular press at the time that should have peeked their interest. Figure 4.22 shows the basic concept. The upper

Artist's Sketch of Caspian Sea Monster in Popular Press (1966)

Actual *ekranoplan KM* in Sea Trials on Caspian Sea

Figure 4.22: Soviet *Caspian Sea Monster* (1965)

sketch shows how the popular press described the design. Essentially, the concept is that while the shorter span wing might suffer a lift penalty compared to its longer span conventional airplane brothers, it offsets this disadvantage by forward mounted engines whose efflux flowing under the wings augment the lift sufficient to support the craft.

The lower photograph in Figure 4.22 shows the actual craft once the photos were released. A more detailed description of the various versions of the *ekranoplan* (surface effect) craft is given in Chapter 1 and Chapter 11. It is shown here to emphasize that the "span criticism" levied by the SESOC Committee is not valid as shown by the *ekranoplan* (and the Lippisch design and the *channel flow* design). Designers had already worked the problem. A glance at the hoverport designed for the amphibious air cushion craft (see Figure 4.14 would easily accommodate these versions of *aerodynamic air cushion craft* also.

4.9 MARAD Plans Post SESOC

After the SESOC Committee had completed its work it is interesting to see how MARAD responded. The SESOC Committee Final Report said the following in their summary of the two concepts that they were charged to evaluate:

1. Channel Flow	*"promise of superior performance and should be pursued on a research basis"*
2. CAB	*"the Panel is of unanimous opinion that the CAB concept is not yet ready for Contract Definition (Project Definition) Action"*
3. Research Program	*"By the end of about 3 years...... it would be possible, through a coherently organized research program, to know whether the SES concept is technically feasible for ocean operation"*

In response to the SESOC Committee Conclusions and Recommendations in 1966, the US Maritime Administration (MARAD) took the following actions:

1. In 1965, MARAD canceled the *channel flow* program, and the MARAD *VRC-1* was relegated to a storage yard in Los Angeles, where it languished for about ten years and then sold for scrap.
2. In 1966, MARAD initiated the Preliminary Design of a 5,000 ton *CAB*[5].
3. In 1967, MARAD formed a Joint Program with the US Navy to start the development of CAB with a planned 500 ton ship within 5 years followed by a 5000 ton ship after a further 6 years.
4. In 1970, MARAD pulled out of the Joint Program and left all further development of SES to the US Navy (see Chapter 5).

Despite the SESOC Committee's strong recommendation NOT to proceed with the CAB design until its fundamental flaws were resolved; in April 1966, a bare two months after the SESOC Final Report was published, the Maritime Administration announced to the public[6] that it was proceeding with a CAB design of 5,000 tons displacement ship with a design speed of 114 mph (99 knots) using a US Navy supplied design. Further, the Maritime Administration projected, quoting from that Wall Street Journal article, this ship *"might take 10 years and cost more than $500 million"*. The article included a sketch of an XR-1 like CAB scaled up to 5,000 tons displacement. One month later (see above), in May 1966, the US Maritime Administration presented a paper at the 1966 National Transportation on their 5000 ton CAB design stating that based on their studies *"the CAB operating over transoceanic routes has the greatest potential of surface effect types for the American merchant marine at the present time"*

But early in 1970, MARAD terminated its participation in the joint MARAD/ US Navy SES development program and no further work was done by MARAD on the CAB or other SES concept in any form.

References

1. *"Surface Effect Ships for Ocean Commerce (SESOC), Final Report on a Study of the Technological Problems"*, by the SESOC Advisory Committee, issued by US Department of Commerce and published by US Government Printing Office, February 1966
2. *"Foundational Research Program"*, (Annual Report) by A.H. Clancy, Jr. Commanding Officer, US Naval Air Engineering Center (NAVAIRENGCEN), Philadelphia (inc report on *Research on Captured Air Bubble Test Vehicle*), 27 August 1965
3. *"Suspension System for Surface Effect Ships"*, by Scott Rethorst, US Patent 3,211247, filed 25 March 1963, issued 12 October 1965
4. *"Air Cushion Craft Development"*, by Peter J. Mantle, US Navy Report DTNSRDC-80-012, January 1980, published by GPO, Stock Number 008-047-00291-2
5. *"A Surface Effect Ship for the American Merchant Marine"* by James A. Higgins (MARAD) and Peter Fielding (Booz-Allen Applied Research), 1966 National Transportation Symposium, San Francisco, CA, 1–6 May 1966
6. *"Zippy Ships: US Planners See Speedy, Ocean Skimming Craft"*, Article in *The Wall Street Journal*, 26 April 1966

5 History of US Navy "Large High Speed Surface Effect Ship" Program

"Damn the torpedoes – full speed ahead!"

RADM David G. Farragut, 1864

The rise and fall of the US Navy's "large high speed SES" program covered a decade from 1969 to 1979 at a cost of more than $650 million (and closer to $1 billion if all related R&D is included). The original goal was to achieve a ship of "large tonnage" sufficient for transoceanic operation (2000–3000 tons) at an operational speed in Sea State 3 of 80 knots or more, with a "desired goal" of 100 knots.

The US Navy officially began its Surface Effect Ship (SES) activities when it jointly formed, in 1967, with the US Maritime Administration (MARAD) what was called the Joint Surface Effect Ship Project Office (JSESPO) headed by Marvin Pitkin, a respected industrial executive and Assistant Administrator for Commercial Development at MARAD. In the beginning, it was believed that both MARAD and the US Navy had common interests of developing a *high speed ship of high tonnage.* The MARAD interests based on its studies, believed that large ships (of up to approximately 15,000 ton displacement), capable of delivering some 500–5,000 tons of containerized cargo at speeds of 100 knots in calm seas was an attractive commercial possibility to revitalize the American marine industry (see Chapter 4 for the MARAD program).

The US Navy were more circumspect of their mission needs but saw the value of high speeds in the range of 80–100 knots for military advantage and in the 2,000–5,000 ton displacement class. Earlier visions were of a frigate size ship. Various missions such as ASW and Sea Control were being evaluated at that time. Both MARAD and the US Navy were aware of the significant progress that was already underway in England since the late 1950s with developments in both amphibious and sidehull type surface effect ships (although they were known under different names in England, such as *hovercraft* and *sidewall hovercraft* based on the original work by Sir Christopher Cockerell).

Based on these common interests, both MARAD and the US Navy agreed to jointly fund a program to explore the various applicable technologies and to develop a prototype ship and set up the Joint Surface Effect Ship Program Office (JSESPO) in 1967. From the very beginning, there was disagreement on the type of program that needed to be laid down. MARAD was very ambitious and wanted a 500 ton prototype within 5 years followed by a 5,000 ton ship 6 years later. The US Navy was more cautious and laid down a plan that would produce a 100 ton test craft within 5 years, followed by a 500 ton ocean capable ship 7 years after that, followed by a 2000 ton operational ship for some selected Navy mission 6 years after that. The US Navy projected 18 years to get to 2,000 tons and MARAD projected 11 years to get to 5,000 tons, which was a marked difference in viewpoint.

5.1 MARAD and US Navy Schedules for 100 Knot SES

Figure 5.1 shows the various schedules that describe the program to develop a high speed surface effect ship. In Figure 5.1, the two plans by MARAD and the US Navy can be seen and also the final schedule that ultimately ended with the program being canceled and the contractor receiving the termination notice on 9 January 1980.

MARAD pulled out of the program in 1970 after a scant 3 years joint effort and the US Navy continued under its own funding after that time. As can be seen from Figure 5.1, the original cautious approach by the US Navy to develop that ship in three "easy" steps from 100 tons to 500 tons to 2000 tons was quickly dropped and replaced by a more ambitious program that dropped the 500 ton test craft stage altogether and going directly to 3000 ton with an Initial Operational Capability (IOC) of 1979–1985. The IOC dates in the various planning documents fluctuated in the early stages. The first stage was well managed and the US Navy did indeed develop two test craft of 100 ton displacement and accomplished that successfully within 3 years of the start of the program with one of the test craft, the SES-100B, attaining 91.9 knots on 30 June 1976. The problems occurred after the 100 ton phase.

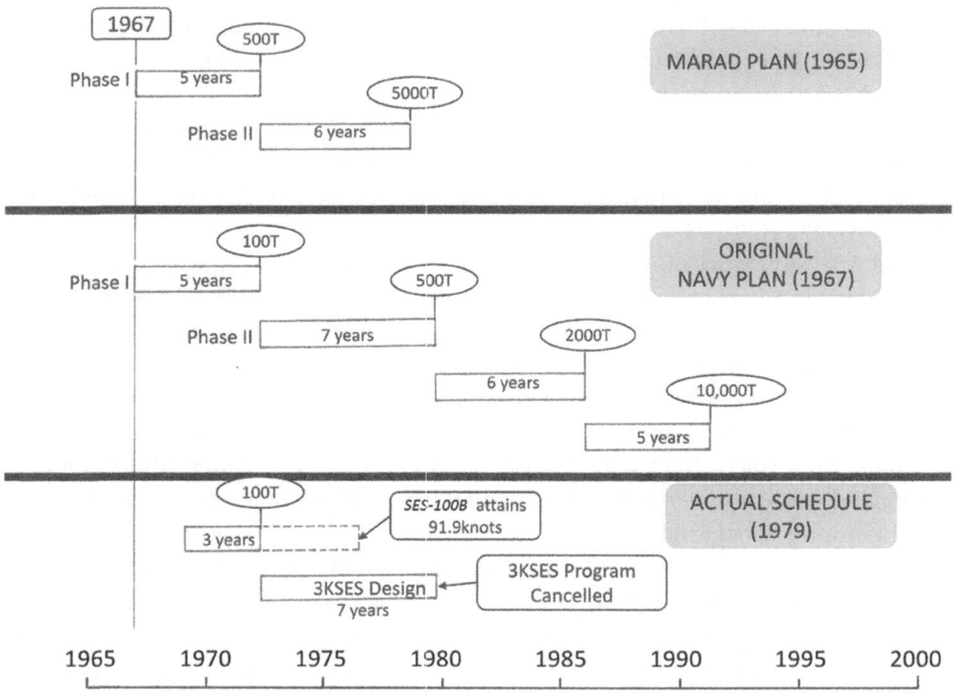

Figure 5.1: MARAD and US Navy Schedules for Development of Large High Speed SES

5.2 Chronological History of 100 Knot SES Program

The following is an encapsulated history of the Large High Speed SES Development that started with the successful SES 100 test craft program initiated in 1969. For ease of documentation of this development process (with its various successes, failures and frustrations) the key events are arranged chronologically.

1969

On 29 January 1969, the following joint news release was issued by MARAD and the US Navy:

> *"The Department of the Navy and the Maritime Administration of the US Department of Commerce announce the award of Maritime Administration contracts for the implementation of the joint 100-ton surface effect ship (SES) Testcraft Program to both Aerojet-General Corporation, El Monte Calif., and the Bell Aerosystems Co., Buffalo, N.Y.*
>
> *These Cost Plus Incentive Fee (CPIF) type Research and Development contracts will be incrementally funded. Funding for each contractor in the current Fiscal Year is $1.55 million and will cover detailed Engineering Design.*
>
> *The contract will be managed by the Joint Surface Effect Ships Program Office (JSESPO). This Office reports to both the Navy and Commerce Departments. The 100-ton Testcraft programs are a major phase of JSESPO's long-range efforts to determine the feasibility of building and operating large, high speed surface effect ships of 4,000 to 5,000 tons and capable of 80 knots or higher speed.*
>
> *These Testcraft will provide developmental testing on a major scale in actual sea conditions. In addition, they will provide vitally required information on alternate design configurations, structural, lift, propulsion and flexible seal systems. This type of surface effect ship rides on a "cushion" or "bubble" of air rather than cutting through water like displacement ships. This bubble of air will be contained by rigid side walls and by bow and stern seals of flexible material similar to that used by today's air cushion vehicles. These configurations will allow the testcraft to achieve very high water speeds of 80 knots or more.*
>
> *The Bell craft will be propelled by supercavitating propellers driven by gas turbine engines, whereas the Aerojet craft will use waterjet propulsion units driven by gas turbines"*

Work began on the two testcraft almost immediately at Aerojet General in Tacoma, Washington and Bell Aerosystems in New Orleans, Louisiana.

1970

In 1970, MARAD pulled out of the joint surface effect ship development to concentrate on other issues of higher priority in the American marine industry and the US Navy proceeded alone under what was now called the Surface Effect Ship Project Office or SESPO with the Navy designation PM-17.

1971

This initial effort by the US Navy in 1971 was funded at $37.4M. At that time, the US Navy were conducting preliminary studies on ship sizes of about 2000 ton displacement that seemed a reasonable goal, given the state of the art of the technology and the preliminary mission needs. A program plan was laid out that showed a progression in ship size from (a) 100 tons displacement first, then (b) 500 tons and finally (c) 2000 tons all spread over an 18 year period. The size of 2000 tons was the minimum size deemed appropriate to demonstrate both "open ocean" and a supportable military mission payload capability for the US Navy.

It was difficult to gain enthusiastic support in Congress for such a ship with only a vaguely defined mission and the Navy knew that it had better put together a prudent and conservative "risk avoidance" program if it were to gain the necessary support as the program passed through the various approval steps. Because the high speed of 80–100 knots was beyond the experience base in the England hovercraft activities, the US Navy decided upon a test program that would cover all the technology issues likely to befall such a program. Because of the uncertainty of many of the technologies, it was deemed prudent to stay with the original approach agreed with MARAD to build two ships each embodying the different technologies expected to be used in the "final" ship size of 2000 tons.

If the final ship design had been a conventional displacement ship and if the speeds were much lower, then simple Froude scaling from 100 knots down to much lower speeds (say 30 knots) for testing of any 100 ton "models" would have been sufficient. However, since the desired speed range was in the higher speed 80–100 knot class and included a complex hull and air cushion interaction this meant that normal displacement ship Froude scaling would not apply and it would be necessary to test the smaller "models" at the full scale speeds of 80–100 knots. In addition to hydrodynamic scaling issues, another concern expressed by the Navy experts at the time of the SESOC Committee in 1965 concerned the structure and use of aluminum alloys. The SESOC Committee Report stated:

> *"The cruciality of structural weight makes aluminum alloys seem desirable, but they present an undetermined jeopardy of catastrophic rupture, stress corrosion, and cavitation erosion for vehicles of such large size traveling at such high speeds over or through the sea".*

The Navy experts at the time probably had a similar photo to Figure 2.1 (Chapter 2) on their office walls!

This led to the first step of developing a 100 ton ship capable of speeds of 80–100 knots so such issues could be tested under full scale conditions before proceeding with any large scale ship construction. At the time of signing the contracts the requirement was to develop a high speed ship capable of 80 knots or more. The idea of a "100 knot Navy" came later (although 100 knots at sea was the idea of MARAD based on economical considerations). The driving force behind the world speed record achieved by the SES 100B of 91.9 knots in 1976 (see later and Chapter 6) was driven originally by competitive spirit and the fact that it was a CPIF contract to achieve 80 knots or more!!

The SES 100B was launched on 22 July 1971 in New Orleans, a short 31 months after signing the contract to begin Detailed Design. The SES 100A was launched a little later on 1 September 1971. The history and description of the two 100 ton Test Craft is provided in more detail in Chapter 6.

The day after the launching of the SES-100B, on 23 July 1971, the new Chief of Naval Operations, Admiral Elmo R. Zumwalt Jr (Figure 5.2) together with US Senator Allen J. Ellender (Chairman, Senate Appropriations Committee, SAC)

Figure 5.2: Inspecting SES-100B (New Orleans)
Front row, L to R: Congressman Hale Boggs; author; ADM Elmo Zumwalt; Senator Allen Ellender US Navy photo

and Majority Leader, Congressman Hale Boggs (D-Louisiana) inspected the SES-100B in New Orleans. Capt Robert L. von Gerichten (not shown), the new chief of the Surface Effect Ship Project Office (SESPO) was also in attendance.

ADM Zumwalt became the driving force behind the new initiative for a "100 knot Navy". He later (in 1973) took the helm of the SES-100B at a speed in excess of 55 knots, commenting that *"We weren't ever able to do this kind of speed in the destroyer Navy"*.

1971–1973

During this period, buoyed by the successes of the SES 100A and SES 100B Test craft, the US Navy continued to explore supporting technologies in various Navy laboratories and industry. Plans were also laid down to proceed faster into the 2000 ton displacement surface effect ship and to bypass the need for a 500 ton prototype.

Again, because of risk issues, the US Navy decided to change the original plans and now build two ships of 2000 ton displacement with each ship exploring different design choices and technologies.

This was the start of the US Navy problems with Congress.

1972

The SES 100B gets underway "on cushion" on 4 February 1972 (see Figure 5.3) for its first test mission down the Intracoastal Waterway and into the Mississippi River Gulf Outlet in New Orleans. At the same time, the SES-100A was having problems with its novel *mechanical CAB planer seal* designs and also with stability problems with its *thin sideboard hulls*, both features of the US Navy patent. These issues are discussed more fully in Chapter 6 which describes the SES-100A and SES 100B programs and their different ship characteristics.

There were many reviews and heated discussions on how many ships should be authorized during these first years of the program but based on reviews by the Defense Systems Acquisition Review Council (DSARC)[a], the Director of Defense Research and Engineering (DDR&E), John S. Foster Jr., authorized the Navy on 2 November 1972 to award four Preliminary Design contracts for 2000 ton prototypes.

Figure 5.3: SES-100B on First Underway Run (with Coast Guard Cutter escort)
4 February 1972
US Coast Guard photo

[a] After the Vietnam War, the US Budget came under intense scrutiny and the Defense Systems Acquisition Review Council (DSARC) was established in 1969. The DSARC required (at that time) that all decisions be made by the Office of Secretary of Defense (SecDef). Since 1969, the DSARC process has changed in detail with major changes occurring in 1981 and in 2000 with different levels of approval authority required and changes in the names of the Milestones, but the essential elements have remained unchanged. The 1972 approval authorized the Navy to proceed with the Demonstration and Validation Phase of the SES program (Milestone I in 1969 parlance or Milestone B in 2000 parlance).

On 8 November 1972, contracts were awarded to Aerojet General (designer and builder of SES 100A); Bell Aerospace (designer and builder of SES 100B); Lockheed Corporation and Litton Industries. Immediately upon award, Litton Industries novated its contract to its main subcontractor, Rohr Marine Industries (RMI).

In authorizing this "go-ahead", DepSecDef (DDR&E) John Foster was clear that the design work should now be directed toward developing *a ship with a definite military purpose* (and not a continuation of non-mission related R&D as the US Navy requested).

The 1972 Decision Concept Paper (DCP) specifically called for:

a. *the potential military advantages to be achieved by high speed naval surface ships in general and as ocean escorts, specifically the nominal 2000–3000 ton class SES,*
b. *the advantages of one versus two testcraft construction program,*
c. *the weapon/sensor suite candidates including a space and weight option, and*
d. *the 2000 ton test and evaluation plan with test milestones.*

In the same DCP, the US Navy were cautioned not to proceed with any commitments or to exclude options prior to the next milestone in the DSARC process which would be DSARC II (in 1969 parlance), which is the milestone for the authorization to proceed with Full Scale Development and Limited Rate Production.

On 4 June 1972, the SES 100B during trials on Lake Pontchartrain, New Orleans achieves 40 knots in choppy water.

On 8 September 1972, the SES 100B during trials on Lake Pontchartrain, New Orleans achieves speeds of 50 knots in choppy water (see Figure 5.4).

On 26 December 1972, DepSecDef clarified the DCP direction, stating that the funding was for the design, construction and test of two 2000 ton ships. No funds were to be allocated for long lead items, or weapons and sensors. It did authorize continued testing of the test craft, SES-100A and SES-100B. During this time, considerable debate ensued between the Committees, and funding was alternatively added and removed as the fiscal planning process continued.

Figure 5.4: SES-100B at 50 knots in choppy water
8 September 1972
US Navy photo

During this same period, the SES 100A operating in the Puget Sound, Seattle area, was still operating at low speed, due to problems of its *mechanical CAB planer seal* design that caused the craft to "plow in" at 45 knots in calm water. It would require a radical change in seal design away from the inadequately designed *mechanical planer seals*, and with re-designed *thicker sidehulls* along the lines of the SES-100B and a switch from a *fully submerged (pod) waterjet inlet* to a *flush waterjet inlet*, before the SES-100A could continue with its test program successfully (see Chapter 6 for detailed discussion).

1973

The House Armed Services Committee (HASC) passed its Authorization Bill including $50.1 million for SES development. The Senate Armed Services Committee (SASC) reduced that amount by $18 million questioning the need to build *two* large ships given that all the technologies were to have been "sorted out" during the two 100 ton testcraft phase.

The SASC, in its deliberations, stated:

> *"When the 2000 ton ship program is initiated, this should involve only one ship. Under this concept... if the single 2000 ton ship selected did not prove out, the Navy could then propose the second 2000 ton ship as an alternative solution".*

Notice the key word "propose" without any guarantees in the SASC language. There were many briefings and discussions on "the Hill" during this period with the main thrust being that the Navy (in the prevailing SASC view) should be very sure *before* embarking on a 2000 ton ship design, rather than continuing with concurrent Research & Development activities on two ships.

On 18 June 1973, VADM Frank Price (Director of Navy Acquisition) sent a note to ADM Zumwalt, saying that the HASC Staff (Mr Hy Fine) essentially agreed with the Navy approach, but they do not agree with the Navy's ability to complete Preliminary Design, make contractor selections, etc in time to award contracts in January 1974. This delayed the program start.

Meanwhile, the Preliminary Design work (awarded in 1972) was now essentially complete on 30 August 1973, except that now one contractor, Aerojet General declined to submit a "Final Design Report" as expected and instead issued a report challenging the schedule of the program based on the risks identified in SES development. The risks cited were on (a) *seal design*, (b) *waterjet flush inlets* (as used on the Aerojet General modified SES-100A), and (c) *ride control.* In parallel investigations, the General Accounting Office (GAO) based on its own analyses agreed with the Aerojet General position and had issued their own report cautioning about how to proceed without first removing these identified risks.

In addition, the new DDR&E, Dr Malcolm Currie, based on his staff's in-house studies also issued a concern about the pace of the program. Dr Currie reiterated the SecDef position that one ship should be pursued (not two) and that a thorough review be made of the Research & Development needs for all alternative types of advanced types of ocean-going vehicles including SES. The funding was reduced for the SES and this effectively delayed any start on the program from FY 75 to FY 76.

On 1 March 1973, the SES-100B on trial runs on Lake Pontchartrain, achieves 70+knots over a slight chop (see Figure 5.5).

Figure 5.5: SES-100B at 70+knots on Lake Pontchartrain, New Orleans, 1 March 1973
Photo courtesy of Textron Marine & Land Systems and Textron Inc.

Despite the DDR&E position, the US Navy, on 1 November 1973, selected a DCP option that called for the detail design and subsystem development for two prototypes of about 2200 tons displacement. Each prototype ship was to be capable of being equipped with appropriate weapons and sensors. Each contract should include selective advance procurement of construction materials. The intent was that at DSARC II (Milestone B), construction of two ships would be sought. On the same date, a baseline weapons and sensors suite was approved. ADM Zumwalt's plan was to begin initial testing of the two 2200 ton prototype SES in 1977.

On 9 November 1973, Admiral Elmo Zumwalt takes the helm during a test run of the SES-100B in excess of 55 knots in the Gulf of Mexico, off Panama City. Admiral Means Johnson (Inspector General of the Navy) and Capt Carl Boyd, the new SESPO Project Manager (PM-17) were also on board during this test run.

1974

The fight between the US Navy and Congress over whether to build one ship or two ships continued. The US Navy continued its efforts to proceed with two ships of 2000 ton displacement and the SASC reversed its position of 1973 and authorized the full $50.1 million. This time, however, the House Appropriations Committee (HAC) had since conducted more detailed studies and reduced the request by $18 million stating the Navy should continue with one not two 2000 ton ships.

On 23 January 1974, DDR&E Malcolm Currie again redirects the US Navy's plans, directing that a risk reduction effort should be pursued with minimum parallel efforts to ensure compatibility of the various subsystems development activities. The DDR&E specifically directed that risk reduction efforts focus on earlier identified SES unique subsystems such as seals, lift systems, ride control and propulsion (waterjets).

In late January 1974, the General Accounting Office (GAO) sent a report to Secretary of Defense (SecDef) James R. Schlesinger, stating the following:

"sufficient data had not been obtained to enable the program office to proceed with confidence into the detailed design phase"

"insufficient data is available to predict high speed-high sea state load and motion requirements for structures, particularly for strength and durability of air cushion seal materials, for low drag flush waterjet inlets and for stability and control"

"the test program is only partially finished and that neither test craft had operated continuously at 70 knots or more, even in calm water, for more than 15 minutes"

In April 1974, the US Navy solicited proposals for the work and contracts were awarded in June 1974 to Bell Aerospace and Rohr Marine Industries (RMI). Lockheed lost and Aerojet General had declined to bid based on the identified risks in the program.

On 19 June 1974, a revised DCP was issued that reflected the DDR&E Currie's direction but retained the pursuit of two (not one) 2000 ton class SES. The label "2KSES" now crept into the jargon to mean the 2000 ton SES. The DCP included expanded testing of the 100 ton testcraft, development of critical subsystems of the 2KSES and other risk reduction activities.

The plan to include work on the weapon systems was not part of the DCP. The program cost was placed at $628million with a slipped schedule (because of the risk reduction effort) to complete the ship by 1979 (and not 1977).

In August 1974, Congress reduced the budget from a planned $58 million to $46 million. The Senate Armed Services Committee (SASC) issued language:

"The Committee reiterates its position of previous years that the Navy should plan to build only one 2000 ton ship initially, because it is the Navy plan to build two".

DDR&E Malcolm Currie, in his detailed guidance, specifically called out action for the US Navy to do the following:

"Accommodate impact of the FY75 HAC budget action. Sequential rather than parallel prototype construction. In addition to opting for sequencing rather than parallel construction, consider deferring all prototype construction and reducing technology effort".

In November 1974, during the FY76 budget process, the US Navy decided finally to acquiesce to the DDR&E direction and changed its plans and proceeded with a one ship program for the 2KSES and reduced its budget request accordingly. Because of all the re-planning and budget changes, the US Navy now predicted that the ship would be completed in 1982 (and not 1979).

1975

On 13 February 1975, DDR&E Currie directed the Navy, in a memo to Assistant Secretary of the Navy (R&D) (H. Tyler Marcy), entitled; *"Summary of Understanding Concerning the SES Program"* that the SES program needed to be slowed down until the technology was more mature and the mission requirements clearly established. In the memo, the Service Budget Guidance was given as (paraphrased):

> *"SES R&D efforts were to be continued during FY 76 with the objective of readiness to proceed to contract design and construction of a large scale prototype when fully ready, after DSARC review and DDR&E approval. Emphasis was to be placed on technology development, testcraft experiments and subsystem development and test in support of the large scale prototype, as well as design studies to determine the optimum size and configuration for the prototype.*
>
> *Prior to requesting approval for contract design and construction of a large scale prototype, the DSARC was to be furnished with:*
>
> a. *Full test data and analysis results on all factors relating to technical risk;*
> b. *The Navy's plans and estimated costs for achieving a full SES operational capability should the large scale prototype meet its objectives;*
> c. *An analysis of the worth of that SES operational capability, in terms of additional capabilities gained or cost avoided, in an overall defense planning context;*
> d. *An assessment of vehicle alternatives, reflecting the results of the initial stages of the studies described below;*
> e. *An analysis of the desirability of various alternative sizes and configurations for the large scale prototype in terms of the impact on the ultimate cost, risk and IOC of the first follow SES class.*
>
> *R&D effort and studies were to be conducted as needed to evaluate advanced naval vehicles which have potential for application to naval missions. The evaluation was to be made primarily in the context of a DPPG Case 1 contingency and would weigh all alternatives in terms of the unique capabilities of each vehicle type. Vehicles were to be evaluated on the basis of performance in relationship to mission value, development effort necessary to achieve this performance with acceptable risk and costs. Where insufficient information existed for evaluation, analysis and testing of models of small scale prototypes were to be performed as necessary to provide it. Navy was to initiate participation by appropriate agencies, including NASA and USAF, to ensure that all available, applicable technology was incorporated. Vehicles to be considered included long range aircraft with airborne and/or seaborne loiter capability, wing-in-ground effect aircraft, airships, high L/B and low L/B SES, hydrofoil ships and small water area ships. The evaluation would be directed by a single manager, who would coordinate evaluation plans with DDR&E to ensure that DSARC information needs would be met; DDR&E acting as DSARC Chairman, was to effect necessary further coordination with other DSARC Principals. These efforts were to be integrated under unified management into a cohesive whole with parallel efforts supported under other 6.2 and 6.3 programs to provide a basis for sound overall naval vehicles R&D decision making….."*

This responsibility was assigned to the Chief of Naval Operations, Systems Analysis Division (OP-96). The Project Officer assigned was Capt Thomas

L. Meeks and the Technical Director & Deputy Project Manager was Peter J. Mantle (author). The project was called the Advanced Naval Vehicles Concept Evaluation (ANVCE) (pronounced "AN-Vee") Project, The work started on July 1975 and completed and reported in CNO documents on schedule in 1978 (Final Report issued by OP-96 in December 1979)[1]. A Technical Review Panel Chaired by Dr. Alan Berman, Naval Research Laboratory (NRL) provided an independent review of the work.

In October 1975, the House Appropriations Committee (HAC) recommended a major reduction in the SES program budget stating that the US Navy had not made a case for the mission requirement for the SES. The US Navy responded with affirmation of mission analysis and war gaming that gave the recommended mission for the SES as ASW with additional missions of AAW and SUW. The HAC continued with the cut but the Senate Appropriations Committee (SAC) restored the funds and the Joint Committee upheld the restoration but later cuts hampered the program and additional slippage was integrated into the schedule.

Also during 1975, DDR&E Currie indicated that the current ship size needed to be increased from 2000 tons to 2600–2800 tons so as to be more compatible with follow-on ships and weapons systems needs. Throughout 1975, many discussions between the US Navy and DDR&E ensued arguing about the ship size.

1976

In May 1976, DDR&E directed the US Navy that the operational prototype be increased in size from 2000 tons to 3000 tons in order that the ship can serve as:

> *"combined technology prototype and operational practicality prototype for a frigate-type operational SES".*

Unfortunately, this change in direction necessitated a design change from the planned LM 2500 engines to the FT-9 engines which would not be available in time and a further slip (of four months) was incurred in the schedule. Further, DDR&E directed that the ship be subjected to two years of at-sea testing including fleet operation simulation tests in realistic environments. This moved the completion date of the (now) 3KSES to 1983 (and not 1982).

On 8 April 1976, the SES-100B, underway at 60.80 knots in Sea State II conditions near Panama City, Florida, successfully launched a Standard Missile (SM-1) off its stern. The SM-1 missile hit its target ship some five nautical miles away.

The purpose of the test was to determine the feasibility of launching a missile vertically from a high speed ship that would be experiencing motion while traversing the waves. It was the first firing of an operational weapon from a high speed advanced marine vehicle and the first vertical launch from any Navy ship (see Figure 5.6).

No adverse effects were found during the firing. A small (~½ degree) excursion in ship pitch was experienced as the missile was launched. The exhaust from the SM-1 missile caused no noticeable ingestion into the ship's engine inlets and no noticeable heating of the ship aluminum structure or fabric stern seals.

The tests were conducted in the test range of the Armament Development and Test Center of the US Air Force Systems Command at Eglin Air Force Base, Florida.

Figure 5.6: SES-100B launches SM-1 missile (8 April 1976)
US Navy photo

On 30 June 1976, the SES-100B set a world speed record for a "warship" (US Navy's words) of 91.9 knots. This was entered into the *Guinness Book of World Records* after the necessary independent verification and authentication (using an Air Force radar). Earlier, the SES-100B had demonstrated 55 knots in 6 foot waves (see Chapter 6 for discussion of SES-100B speeds). This record breaking high speed run (at 91.9 knots) occurred during US Navy trials off Panama City, Florida in relatively calm water (Sea State II) with US Navy Lt. Tom Breitinger at the controls. On board was the Secretary of the Navy (SecNav) J. William Middendorf II (who had steered the ship earlier in the run at 45–50 knots in three foot waves).

The US Navy was so enthralled with the idea of "100 knots"–and the psychological impact of the number "100" – that the Navy broke with Navy tradition of only quoting their speeds in "knots" and started quoting this record speed by the SES-100B in miles per hour, and various official Navy papers in the literature will be seen stating this record speed as 105.7mph!!

After the run, SECNAV Middendorf, after being told that he was the fastest Secretary of the Navy ever, said: *"That record should stand for quite a while. That's an order!"* With the cancelation of the SES Program later (in 1980) that SECNAV order unfortunately came quite prescient.

The SES-100A with its retrofitted flush waterjet inlets achieved a top speed of 76 knots in 1976. The various features of both the SES-100A and the SES-100B that helped determine their two different top speeds (76 knots and 91.9 knots respectively) are covered in Chapter 6.

In November 1976, the preliminary President's Budget canceled all funding for the SES program beyond FY 77, citing the lack of a defined Navy mission as the reason for termination. The US Navy objected, citing its studies on operational use of the large SES for ASW with supplementary missions of AAW and SUW. This reclama by the US Navy was successful and funding was restored.

On 9 December 1976, after competitive bidding, a CPFF contract was awarded to Rohr Marine Industries (RMI) to design, conduct subsystem testing and procure long lead items for the 3KSES. The planned delivery date for the ship (in the contract) was October 1981. The contractor's cost estimate was higher than the funded amount and after adjusting the schedule to meet the available funding the delivery date was moved to the end of 1982. Also, to meet the funding available only the ASW weapons suite would be installed.

After the award and funding negotiations, a further reduction in the available funding occurred and a renegotiated contract with RMI on 30 November 1977, dropped the construction part and also reduced much of the weapons suite in the ship design.

1978

President Jimmy Carter, soon after assuming office, declares the surface effect ship *"interesting but unnecessary"* and drops it from the FY 79 President's Budget (January 1978). The President referred to the SES as the *"quickship"* program.

Throughout 1978, a series of discussions and renegotiations between the Congress and the US Navy and RMI that reflected new budget values and revised ship design. The revised program gave a new delivery date of November 1983, a slip of 25 months from the original contract date.

During this time, the OSD Budget process requested that the Services submit their budgets in groupings (or Bands) of priority. The highest level (meaning most likely to be cut in case of budget limitations) was called the Enhanced Level. Late in 1978, the Secretary of the Navy (SecNav) John Lehman submitted his budget with the following comment on the large high speed SES program:

> *"The SES represents one of the few radically new technology efforts of great potential, and the only major surface advanced hull and propulsion development now underway but I have included SES at the Enhanced Level. In my judgement, it should have higher priority but, given our past experience of losing both program and funding, I could not jeopardize other important development efforts to place it in a lower band".*

In December 1978, the Office of the Secretary of Defense (SecDef) canceled the SES program. The new Undersecretary for Defense Research & Engineering (USDR&E) Dr William J. Perry, in his FY80 Posture Statement said:

> *"After a thorough examination of all the issues involved, the Navy recommended against continuing with development of the 3000 ton SES test ship in this budget. The Navy's analysis showed that the payoff from this program was distant and very uncertain, and that SESs of the type being developed were very unlikely ever to have a significant positive impact on our naval posture. After conducting my own review, I reluctantly concurred in the Navy's recommendation. It is our intention to try to utilize the balance of the funds appropriated in FY79 to pursue vigorous development of technology and system concepts in the hope of being able to present a sound advanced ship program in the FY 81 budget".*

This essentially ended the large high speed SES program, with the remaining months dedicated to funding RMI at a low level to complete the Contract Design. No construction or purchase of long lead items had begun.

1979

The US Navy Advanced Naval Vehicles Concepts Evaluation (ANVCE) Project had completed its work and issued detailed recommendations for research & development of advanced naval vehicles and their key technologies and, in particular, gave specific recommendations relative to the troubled 3KSES Program.

Specifically, for the continuation of SES development, the following recommendations were given:

1. *"..... severely restrict the operating envelope to no more than 50 knots in any sea state until all engineering tests in that portion of the planned operating envelope are complete........"*
2. *"...... study a 50 knot SES in the 3000 tonne class for the same missions as the 80 knot SES to determine cost-effectiveness. The results of the analyses performed by the ANVCE Project indicate that by using high L/B platforms, the investment cost of the platform can be cut in half".*
3. *"..... as it is presently designed the 3KSES concept represents a high technology investment. It requires the use of sophisticated structures, high power propulsion systems, developmental lift fan systems to provide acceptable sea-keeping, and developmental seals to operate reliably in a seaway. All of these factors combine to increase the cost of the SES (both the initial investment and the operating cost). Accordingly, options within the SES concept that can lower those costs must be sought..... Based on the ANVCE Project findings, the development of features..... such as more slender configurations (high L/B); use of propeller propulsion; and a reduction in operating speeds are essential elements in producing a cost-effective SES....."*

The technical reasons behind these recommendations are included in Chapters 8 and 10.

In addition to the above recommendations for the SES Program, the ANVCE Project provided recommendations for the other "alternative concepts" in amphibious air cushion craft; hydrofoils (surface piercing and fully submerged); SWATH ships; planing craft (conventional and supercritical); wing-in-ground-effect (WIG) vehicles (in all their forms); long range aircraft; seaplanes; and semi- and fully buoyant airships. Detailed recommendations were also provided in the Key Technology areas of sea-keeping; marine propulsion; structural design; hull and foil design; skirts and seal development; lift systems; lightweight structures and efficiency. Recommendations were also provided in the important areas of Cost (both Investment and Operating) and Mission Usefulness.

1980

On 9 January 1980, RMI was notified by the US Navy that the 3KSES contract was terminated. RMI then set about closing up all operations and under a residual contract with the US Navy drew up plans for disposing of contract inventory and to settle outstanding subcontracts. This task was completed in December 1981.

The Secretary of the Navy (Edward Hidalgo) sent a letter in March 1980 to Chairman Melvin Price of the House Armed Services Committee (HASC) saying that the Navy would now pursue a different goal for surface effect ships. This new goal was to pursue a design of a Fast Logistics Ship of 10,000 to 15,000 ton displacement SES with a top speed of approximately 50 knots.

The goal of a "100 knot Navy" was no longer an option.

In 1980 the captured air bubble (CAB) design SES-100A was scrapped, with many of the problems of *high speed waterjet flush inlets* and the *CAB planer seals* unresolved.

The SES-100B made its last run on 17 January 1980 at the Navy Test Facility at Patuxent River, Maryland (see Figure 5.7) with LCDR Jim Boland (USN) as ship commander. On that day it conducted routine runs collecting valuable test data. The author was on board and at the helm (for a few minutes!) at a speed of 83 knots. During the last mission of that day (Mission 495), the SES-100B achieved a speed of 89.5 knots. Although an attempt was made during that last day to set a new speed record, a zinc anode detached from the hull and hit the propeller blade precluding breaking the earlier attained world speed record of 91.9 knots. After the last run, preparations were begun to remove essential equipment and eventually move the SES-100B from Patuxent River to DTNSRDC, Annapolis in 1983 for permanent display on the seawall.

In August 1980, the Vice Chief of Navy Operations established an Advanced Ship Development Working Group (ASDWG) to study requirements and operational concepts for advanced ships and to prepare a development plan. In September 1980, the US Navy, based on the recommendations from the OP-96 ANVCE Project, purchased the Bell-Halter BH-110 surface effect ship to pursue the technologies needed for this lower speed (50 knots) and higher length-to-beam (L/B) ratio SES designs.

Figure 5.7: SES-100B at 83 knots on Patuxent River, 17 January 1980 (author at helm)
US Navy photo

Chapter 8 and 10 provide more detailed technical discussion of these design parameters for a range of high speed surface effect ships (50–100 knots).

1981

In January 1981, the Advanced Ship Development Working Group (ASDWG) recommends pursuing an ocean-going low technology SES of approximately 1000 tons displacement to augment existing naval forces and to provide a technological and operational data base for such advanced ships based on the surface effect principle.

1982

In June 1982, the Commander of Naval Sea Systems Command disestablished the Surface Effect Ship Project Office (PMS 304) and reassigned all personnel and functions to the Navy Ship Design and Integration Directorate (NAVSEA 03).

The SES-100B was towed from Patuxent River to DTNRSDC Annapolis in December 1982.

1983

The SES-100B, having met its contractual requirements and completed its successful test program of demonstrating that a reliable high speed capable surface effect ship was possible and practicable, was installed next to the seawall of the Annapolis Laboratory of the US Navy David Taylor Ship Research and Development Center (DTNSRDC) in Maryland for permanent display. The ribbon cutting ceremony for the SES-100B permanent display was held on 10 June 1983.

Figure 5.8 shows the SES-100B installed at the seawall at Annapolis. The propulsion system, including the ramps for the flow into the semi-submerged propeller, has been removed, as were other key systems. The flexible bow bag and fingers of the seal system were left intact on the ship as part of the display. The ride control system vents in the ship side can be clearly seen.

1999

As part of the 1995 Congress decreed Base Realignment and Closure (BRAC) process, the Annapolis Center of DTNSRDC was closed in 1999 and after many years of being on display, the SES-100B was dismantled and sold for scrap along with the other assets of the Navy Annapolis R&D Center.

The US Navy program to pursue 100 knots at sea had ended.

5.3 Postscript

A review of the events over the ten year period (1969–1979) by the US Navy, following on from the earlier work by the US Maritime Administration, of developing a large (2000–3000 ton) surface effect ship (SES) capable of operating in a military mission at 100 knots, indicates that the "case had not been made" and that reaching for "100 knots" in one step was a "bridge too far". Probably it would have been better to have followed the original US Navy plan (see Figure 5.1) to develop

Figure 5.8: SES-100B Installed for Permanent Display at DTNSRDC Annapolis (1983–1999)
US Navy photo

the capability in smaller steps: a 100 ton ship first, then a 500 ton ship and then, after all technological problems had been reduced to minimum risk, proceed with a 2000–3000 ton displacement military capable ship. The technical issues had been recognized early by the engineers and designers who were working the problem but the political program enthusiasm overcame a prudent development process. An underlying reason throughout the development was the realization that if "100 knots" was the ultimate aim then probably the wrong concept had been selected. The concept selected was more practical for slower speeds. The history of the funding for the "Large SES Program" is shown in Figure 5.9 that showed that in excess of $650 million had been spent developing this capability.

Unfortunately much of the literature, that followed after the program was canceled, has been of a glowing revisionist history nature that has tended to mask the underlining technical issues of "large, high speed SES development". This has left much of the good technology and engineering work that had been accomplished on the program hidden and not available to new researchers.

Much of the funding expenditure had produced a significant amount of technical data that has been found useful in the later pursuit of surface effect ships even though the sights had been lowered to a much more attainable 50 knots. Also, many of the "false starts" and bad design choices have been hidden, so again new researchers and designers will not get the benefit of the hard work of those that had been involved in designing a high speed surface effect ship for military purposes.

The SES-100B was built by Bell Aerosystems (later called Bell Aerospace Textron) in New Orleans starting in 1969 and launched in 1971 as has been discussed earlier in this Chapter. The aim of the US Navy program was specifically geared to

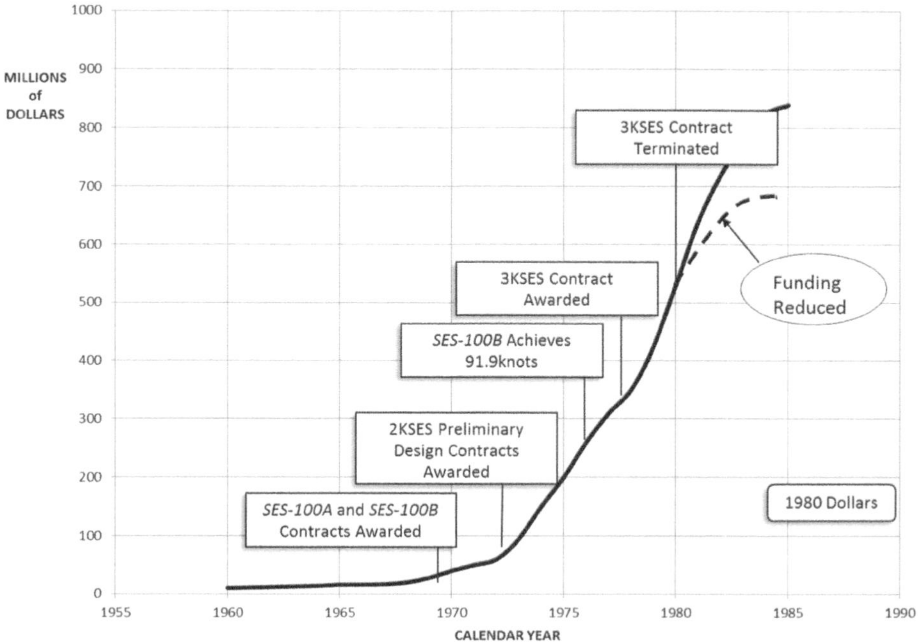

Figure 5.9: History of Funding of Large High Speed SES Program

use of high technology systems to achieve the specific program goals of a "100 knot capability" for the US Navy. While the SES-100B program was underway, Bell Aerospace began to explore applying the same basic features embodied in the SES-100B but to apply them in a more commercially accepted shipbuilding standards and environment to a much slower speed ship.

To this end, a joint venture with an experienced local shipbuilder, Halter Marine, was established in August 1977 to design and build such a ship. The same basic concept used in the successful SES-100B of a *sidehull air cushion craft* with the "*tried and true flexible skirts*" would be used but built to commercial marine standards. The use of conventional marine equipment, standard materials and construction methods in the shipbuilding industry would also be used. Diesel engines and subcavitating fixed pitch marine propellers would replace the gas turbines and supercavitating propellers of the SES-100B. This resulted in the surface effect ship BH-110, with construction started in January 1978 and within the year a maiden voyage on 20 December 1978. It is shown operating in the Gulf of Mexico in Figure 5.10. Bell-Halter built the BH-110 as a demonstration project to explore the "commercial shipbuilding practice" application of high speed sidehull SES. This venture saw several successful outgrowths with the Coast Guard and Navy. The leading particulars of the BH-110 are provided in Table 5.1.

More detailed description of the Bell-Halter[b] surface effect ship may be found in Jane's Surface Skimmers (1979 edition). The demonstration SES BH-110 led to several purchases by the US Coast Guard, US Army and the US Navy. While not expanded upon here in any detail in the context of seeking the "100 knot Navy", it

[b] Due to several mergers, the last in 1994, the company is now known as Textron Marine & Land Systems

Table 5.1: *Leading Particulars of BH-110*

LOA	110 ft	
Beam overall	39 ft	
All Up Weight	107 tons	
Normal payload	18 tons	
Propulsion:		
Two diesels @ 1335HP each		
Two 32 inch dia subcavitating propellers		
Lift:		
Two diesels @ 445HP each		
Two centrifugal fans		
Skirt system:		
Bow: Eight coated fabric fingers		
Stern: Similar to SES-100B of three inflated lobes of coated fabric material		
Cruise speed:	Sea State 0	Sea State 3
On cushion:	40 knots	33 knots
Hull borne:	19 knots	15 knots

Figure 5.10: Bell-Halter BH-110 (1978)
US Navy photo

was important to recognize the acceptance of such a design with its lower costs and high reliability as a working vessel for the US Coast Guard and others. The US Coast Guard ships that were launched throughout the period 1980–1983 were the *USCG Dorado* (WSES-1) in 1980; and later, the *Speed Command (1981), Swift Command (1981), Sea Hawk (1982), Shearwater (1982)* and *Petrel* (1983) The experience with these craft as used by the US Coast Guard has not been without the usual problems and the reader is referred to the 1998 report on the history of the US Coast Guard Surface Effect Ship Performance Improvements[2] for a detailed description of the development.

Following the cancelation of the 3KSES program in 1980 and the recommendations from the ANVCE Project (see above chronological listing of events, 1979), the US Navy purchased this "commercial practice SES", the BH-110, in September 1980 (with acceptance from Bell-Halter on 9 May 1981) with plans to obtain a *high length to beam ratio, lower speed SES* as soon as possible. Figure 5.11 shows the BH-110 cut in two at the Bell-Halter facilities in New Orleans prior to the addition of a 50 foot extension to convert the craft into what was now called the US Navy SES-200.

(a) *BH-110* cut in two ready for 50 foot insert

(b) *SES-200*

Figure 5.11: BH-110 Conversion to US Navy SES-200
US Navy photos

This means of obtaining a high L/B SES was a most cost-effective way to achieve such a capability. The SES-200 was accepted by the US Navy from Bell-Halter on

Table 5.2: *SES-200 Leading Particulars*

Displacement (AUW)	205 tons	
Empty Weight (Light Ship)	128 tons	
LOA	159.1 ft	
Beam overall	39 ft	
Effective cushion length	33.1 ft	
Effective cushion beam	31.6 ft	
Cushion height	6 ft	
Cushion length-to-beam ratio	4.25	
Propulsion:		
Two GM 16V-149TI diesel engines	1600hp each	
Two three bladed 40 in dia (48 in pitch) subcavitating marine propellers		
Lift:		
Two GM 8V-92TI diesel engines	435 hp each	
Two Bell 42 in dia centrifugal fans		
Skirt system:		
Bow: Eight coated fabric fingers		
Stern: Similar to SES-100B of three inflated lobes of coated fabric material		
Cruise speed:	Sea State 0	Sea State 3
On cushion:	32 knots	27 knots
Hull borne:	16 knots	14 knots

24 September 1982. The modified BH-110, or the SES-200 as it was now called, had the leading particulars as shown in Table 5.2.

The speed of the SES-200 with its higher L/B (= 4.25) hull is not as fast as the shorter L/B (= 2.65) BH-110. The SES-200 provided the necessary craft for US Navy evaluation of such designs. A more complete compilation of some of the key results that were obtained with the US Navy tests of the SES-200 may be found in the report by Adams and Beverly in the 1984 Naval Engineers Journal.[3]

5.3.1 SESPO's Designs for Larger SES in 50 knot Class

In the above chronological history of the US Navy's efforts to pursue a successful design of the surface effect ship, it was noted that in 1981 the US Navy's Advanced Ship Development Working Group (ASDWG) recommended that the US Navy explore first, the application of the concept to a 1000 ton displacement low technology SES. The Surface Effect Ship Project Office (PMS 304) then produced in 1981 a design called the Medium Displacement Combatant Surface Effect Ship (MDC) (see Figure 5.12). In this design by PMS 304, the strong resemblance to the SES-100B and SES-200 can be seen. The design uses buoyant sidehulls and bag and finger skirts for the bow seal and flexible inflated lobes for the stern seal. It also uses the surface piercing propellers design similar to the SES-100B (except that a design change was made to use an inclined propeller shaft). The leading particulars are as shown in Table 5.3. These characteristics are admittedly those produced in a very short time as PMS 304 was being disestablished but it provides a reasonable characterization of a possible "50 knot class" surface effect ship using all the established lessons learned from the SES-100B and the SES-200 experience.

An earlier effort by PMS-304 to explore alternative solutions was accelerated soon after the announced cancelation of the program in 1980 and detailed work

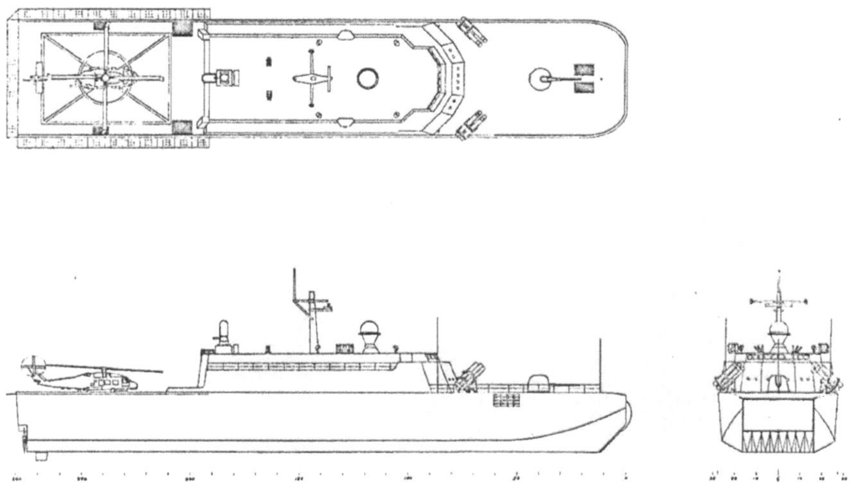

Figure 5.12: Medium Displacement Combatant (MDC) Surface Effect Ship

Table 5.3: *Medium Displacement Combatant SES (MDC) Leading Particulars*

Displacement (AUW)	1154 s. tons	
Empty Weight (Light Ship)	754 s. tons	
LOA	280 ft	
Beam overall	55 ft	
Effective cushion length	252 ft	
Effective cushion beam	33.5 ft	
Cushion Height	17 ft	
Cushion length-to-beam ratio	7.52	
Cushion Pressure	239 psf	
Propulsion(1):		
Four Allison 570-K gas turbine engines	6445 hp each (max cont)	
Two 12½ ft dia subcavitating controllable pitch semi-submerged marine propellers		
Lift(1):		
Three SACM 195V12 GZSHR diesel engines	2200 hp each (max cont)	
Six Aerophysics RD fans		
Skirt system (similar to SES-100B):		
Bow: Bag and Finger		
Stern: Three inflated lobes of coated fabric material		
Cruise speed:	Sea State 0	Sea State 3
On cushion:	53 knots	46 knots
Hull borne:	18 knots	17 knots

(1) The design is a CODOG in that the SACM diesels may be used for propulsion when off cushion and for lift when on cushion.

accomplished on designing a large tonnage SES. This effort was more in line with the original aims of the Maritime Administration in 1960 when the first thoughts were being put together of a *high tonnage* ship capable of transporting a payload at *high speed* over the ocean, except that the speed was now in the "50 knot class".

Edward A. Butler (1932–2004) who was the US Navy (SESPO) liaison official to the SES-100B effort, led this new design work on MDC and MPS. Ed Butler is shown with friend, on-board the SES-100B during the last run at Patuxent River, on 17 January 1980 (Figure 5.13).

Figure 5.13: Ed Butler (L) and author (R) onboard SES-100B during its Last Run, 17 January 1980
Author private collection

Figure 5.14 shows the general arrangement that again contains all the successful features obtained during the SES-100B program using *buoyant sidehulls, flexible skirt systems* and *surface piercing marine propellers.* The aim of this particular design was to produce a *multi-purpose ship* as a result of earlier criticism of designing a ship with only either a single purpose mission or a not well defined mission.

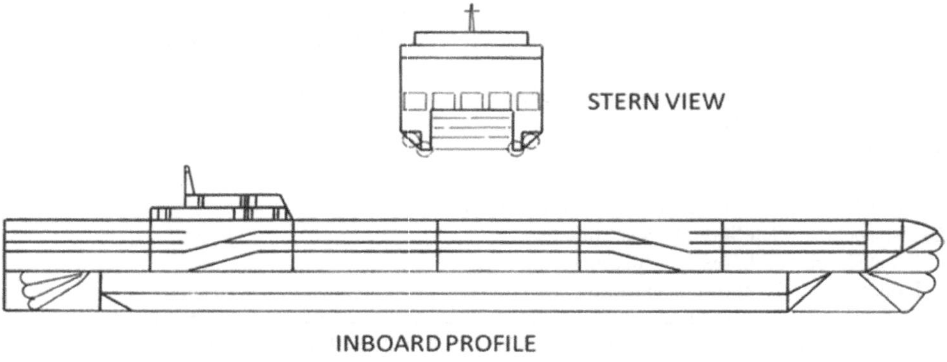

Figure 5.14: 15,000 ton Multi-Purpose Surface Effect Ship (MPS) (Transport Application)

The particular design shown in Figure 5.14 is for the transport version. The leading particulars are as shown in Table 5.4. The large size of the MPS brings into play the significant effect of a dynamic craft that is weight critical. For example, "on cushion" the speed of the ship (at maximum continuous power) is about 25 knots in calm water but that increases to about 70 knots at light ship weight. Hence, for a transoceanic voyage the normal cruise speed could vary from 25 knots to 70 knots as

Table 5.4: *Multi–Purpose Ship (MPS) Leading Particulars*

Displacement (AUW)	16,800 s. tons	
Empty Weight (Light Ship)	6,079 s. tons	
LOA	686 ft	
Beam overall	105 ft	
Effective cushion length	639 ft	
Effective cushion beam	75 ft	
Cushion height	33 ft	
Cushion length-to-beam ratio	8.52	
Cushion Pressure	680 psf	
Propulsion(1):		
Four LM 2500 or 2 SACM diesels	27,000 hp each turbine (max cont)	
	7,000 hp each diesel	
Four 12 ft dia. supercavitating controllable pitch semi-submerged marine propellers		
Lift(1):		
Six SACM diesel engines	7,000 hp each	
Six mixed flow fans		
Skirt system (extension of SES-100B bag and finger design):		
Bow: Compartmented Bag and Finger		
Stern: Three inflated lobes of coated fabric material		
Cruise speed:	Sea State 0	Sea State 3
On cushion:	25 knots @ FLD	23 knots @ FLD
	70 knots @ light ship weight	68 knots @ light ship weight
Hull borne:	23 knots @ FLD	21 knots @ FLD

(1) The design is a CODOG in that the SACM diesels may be used for propulsion when off cushion and for lift when on cushion.

the fuel is consumed. Obviously, there would be many variations of payload-range characteristics, that were detailed in the PMS-304 work but clearly the ship can be said to be in the "50 knot class" of high speed surface effect ships.

After the failed CAB rigid planer seal design, PMS 304 returned to the successful flexible skirt system as applied on the SES-100B and the SES-200 that was based on the earlier successful development of such flexible skirt systems on the *sidewall hovercraft HM* series and the *amphibious air cushion craft* of the *SRN* series in the UK over the preceding two decades.

5.3.2 "US Navy Successfully Meets All Objectives of High Speed SES"

The above headline should have appeared at the conclusion of the 100 ton phase of the SES Program but the US Navy had overlooked the success of one of their Test Craft, the SES-100B in their eagerness to pursue a concept that had failed repeatedly throughout the course of the program. The SES-100B had accomplished all the objectives of the program as laid down in the original contract release in January 1969 (see opening paragraphs of this Chapter). As discussed in detail in Chapter 7 on Economic Considerations and in the deliberations of the MARAD Surface Effect Ships for Ocean Commerce (SESOC) Committee in 1965 (see Chapter 4) there are fundamental requirements that must be met in any design. This is summarized well in the SESOC Committee Final Report[4] that says:

> *"The commercial feasibility..... is critically dependent upon the achievement of optimum performance in the following areas: stability and control, effective lift-drag ratio, structure, propulsive efficiency, and fuel economy."*

These features are as equally important in a military ship as they are in a commercial ship. Cost is paramount in both acquisition cost and direct operating cost. The demise over the years of many high speed ships including hydrofoils, air cushion ships and other types are a sobering testament to this real world fact. The SES-100B was successful in each of the areas of concern about any new venture. The successes can be grouped under first, technical performance and second under economic characteristics as identified in the SESOC Committee Final Report quoted. The achieved characteristics are in Table 5.5.

Table 5.5: *SES-100B Achieved Successes*

SES-100B Technical Characteristics	**Achieved**
Speed	
Contract speed (80 knots)	80 knots
Predicted speed (in 1969)	91 knots in Sea State 2
Actual speed (in 1976)	91.9 knots in Sea State 2
Stability & Control and Ride Quality	Stable in all axes in calm and rough seas Met all US Navy Habitability Criteria in ride quality
Turn radius: (no more than 8000 ft)	Less than 7000 ft
Acceleration: 55 knots in 60 secs	65 knots in 60 secs
Deceleration: Stop within 3000 ft	Stopped in 2750 ft in 60–70 secs
SES-100B Economic Characteristics	
Lift-drag Ratio (L/D)	High efficiency of 22–29 at 35 knots
Effective Lift-Drag Ratio $(L/D)_{EFF}$	High efficiency of 13.8 at 50 knots
Transport efficiency (WV/P)	High efficiency of 10 at 40 knots
Structure	Structural Weight (W_S/W) of 34%, well within SOA
Propulsive efficiency (η_P)	High efficiency at 62% at design speed
Fuel Economy (Range of 1500nm)	>1600nm at 40 knots

In addition to the above technical and economic characteristics, the SES-100B demonstrated that the life of the flexible skirt system (a key ingredient in any economic analysis) even at the high operating speeds was compatible with the best experience from the commercial hovercraft operating in England. A value of 400 hours was achieved with the actual operating speed profile of the SES-100B Test Program in the three test areas on Lake Pontchartrain, Panama City and Patuxent River (see Chapter 6, Figure 6.31 and Figure 6.48).

5.4 Summary of the Two Decades of Development

In summary, the two decades of development (1960–1980) by first, the US Maritime Administration (MARAD) and secondly, mainly the US Navy, has resulted in many successes and some failures in the goal of achieving a "100 knots at sea" capability in a technically sound (minimal risk) ship with a well defined military operational usefulness.

The goal of a 3,000 ton frigate size SES capable of cruising at 80 knots in calm seas was not met. The literature is replete with reasons as to why this goal was not met. Much literature points to political reasons and some even place the reason on the fuel crisis of the 1970s! In reality, the reasons were much more technical and design decision oriented in nature. There were many "red flags" along the way where one group after another (SESOC, DDR&E, SECNAV, GAO and others) would highlight specific technical issues that were not solved at the 100 ton test craft program stage where it was considered imprudent to proceed further to much larger size ships before the risks had been eliminated or at least significantly reduced. Each of these technical risk areas have been identified in the chronological listing of events above in this Chapter.

The one successful test craft, the SES-100B, did meet all the objectives for the program and actually demonstrated, on a routine test schedule, the capability to operate safely at 80 knots or more. It achieved a world record speed of 91.9 knots. There were no risks exhibited by any of the systems (structure, seal systems, propulsion system, etc). The earlier expressed concern of operating in supercavitating flows was alleviated by proper design and the testing of the *model scaled* SES-100B at the *full scale* speeds of a 3,000 ton SES at the supercavitating speeds of 80 knots and more. Despite this success, the US Navy pursued for its frigate size SES, the use of lower efficiency waterjets with severe limitations of cavitated flow (and air ingestion) into flush inlets, and pursuit of flawed CAB rigid planer seals instead of the well proven flexible skirt seals that resulted in a ship design with too much risk. This led most directly to the cancelation of the program.

It is difficult to understand why, that despite these successes, the US Navy elected to proceed into the 3KSES Phase of the program with the SES-100A with its problematic waterjet propulsion system and its flawed rigid planer seal. Because of these problems, the SES-100A did not achieve the basic contract speed of 80 knots and had not resolved the performance and stability issues of the rigid planer seal, and the captured air bubble concept of thin sideboards was found to be a flawed concept. The designer of the SES-100A (Aerojet Corporation) had corrected the stability problem after the "plow-in" of the SES-100A with the original CAB planer seal by replacing the seal with a flexible skirt system. Aerojet General then went on and declined to participate in the US Navy 3KSES Program because of these and other (identified) problem areas stating that more research and development was required. In a surprising move, the US Navy turned the SES-100A over to the new contractor (RMI) who then removed the successful flexible skirt system and installed again a CAB rigid planer seal. These actions of high technical risk led directly to the cancelation of the program by the Secretary of Defense in 1980.

If the successful SES-100B design had been pursued into the 3KSES Phase the headlines of the program might have been announcing the next step in the program rather than its cancelation.

The attempts to redirect the program to lower speed ships came too late as the SES Program Office was disestablished in 1982. It is interesting to note that in 1960, the US Maritime Administration was seeking a 15,000 long ton SES capable of delivering 500–5000 tons of containerized cargo at 100 knots in calm seas. In 1965, MARAD believed (with a US Navy design) that it could produce a 10,000 ton SES based on the *captured air bubble (CAB)* concept that would achieve 129 knots in calm seas[5]!!

Two decades later, as summarized above, the SES design by PMS-304 had now (in 1981) settled on a design, using a *sidehull SES with flexible seals and super-cavitating, semi-submerged marine propeller propulsion*, that has a full load displacement of 16,800 s. tons (15,000 long tons) with 6000 tons of payload, capable of transoceanic ranges more in line with the original goal for a high speed ship – but with a more modest speed of 50 knots (which was still twice the speed of conventional displacement ships) with 70 knot capability at lighter disposable loads, and a well-defined mission.

The high speed ship community today has now moved into a realm where many different hull forms are being tried, built, tested and operated throughout the world at these more "modest" speeds. It is left to other writings to summarize the expanding world of advanced ships and novel designs in this 21st Century, and the remaining chapters in this book will return to the possibility of achieving a "100 knots at sea" capability, and as will be shown, it will require a *change in basic ship concept* to achieve that goal.

References

1. *"Advanced Naval Vehicles Concept Evaluation (ANVCE) Project"*, Volumes 1,2,3; CNO (OP 96) Report No. CNO LTR 96/C194136, December 1979
2. " *History of Coast Guard Surface Effect Ship Performance Improvements*" by Gary Larimer, Joe McCollum, Cdr Benton Schaub, Donald van Liew and Charles Whipple; Naval Engineers Journal, Volume 100, Issue 3, 18 March 1998.
3. *"Technical Evaluation of the SES-200 High Length-to-Beam Surface Effect Ship"*, by John D. Adams and Walter F. Beverly III; Naval Engineers Journal, May 1984
4. *ibid*
5. *"Surface Effect Ships for Ocean Commerce (SESOC)"*, US Department of Commerce Final Report on a Study of the Technological Problems (of SES), pub by GPO February 1966

6 SES-100A and SES 100B Test Craft and the "THREE THOUSAND TON SES"

"There are two sides to every question"

Protagoras (485–410 BCE)

As described in Chapter 5, the US Navy program began to explore the value of ships capable of achieving speeds of 80 knots or more, with the issuance of contracts for two test craft in 1969. A key clause in the contract notification was:

> *"..... These Testcraft will provide developmental testing on a major scale in actual sea conditions. In addition, they will provide vitally required information on alternate design configurations, structural, lift, propulsion and flexible seal systems. This type of surface effect ship rides on a "cushion" or "bubble" of air rather than cutting through water like displacement ships. This bubble of air will be contained by rigid sidewalls and by bow and stern seals of flexible material similar to that used by today's air cushion vehicles. These configurations will allow the testcraft to achieve very high water speeds of 80 knots or more....."*

The complete text for the joint release by the US Maritime Administration (MARAD) and the US Navy is given in Chapter 5. To ensure that the craft were not designed for speed alone (see Chapter 1) the US Navy put in several "mission related" requirements to focus the design efforts for a military useful ship. Each were required to provide for a 10 ton payload (no specific mission payload defined); a range of 1500nm and a maximum required speed in calm water of 80 knots with an incentivized desired maximum speed of 100 knots. There were also other common requirements such as: stopping distances (less than 3000ft from all speeds); turn radius (less than 8000 ft at above hump speeds); mission duration of 24 hours and other features to keep the designs of useful capability rather than just speed. It was intended in the original program to capitalize on the well-grounded experience in theory, design and operations in both commercial and military operations that existed with the British Hovercraft industry at that time. The underline to this effect, in the above contract requirement, was added by the author. The US Navy Program Office departed from this specific direction by inserting a new untried design into one of the test craft of non-buoyant *thin sideboards* and *rigid planer seals* as used in the early Warner (1929) and later Ford (1964) designs on *captured air bubble (CAB)* air cushion craft.

6.1 Evolution of *Captured Air Bubble (CAB)* and *Sidehull SES*

The differences between the two test craft, called SES-100A and SES-100B, were distinct and evolved from: (a) the *captured air bubble* work in the US and (b) the established *sidewall hovercraft* work in the UK and in the vast amount of global development work on operational *amphibious air cushion craft.* This evolution is shown on Figure 6.1. The distinct differences between the two test craft provided a

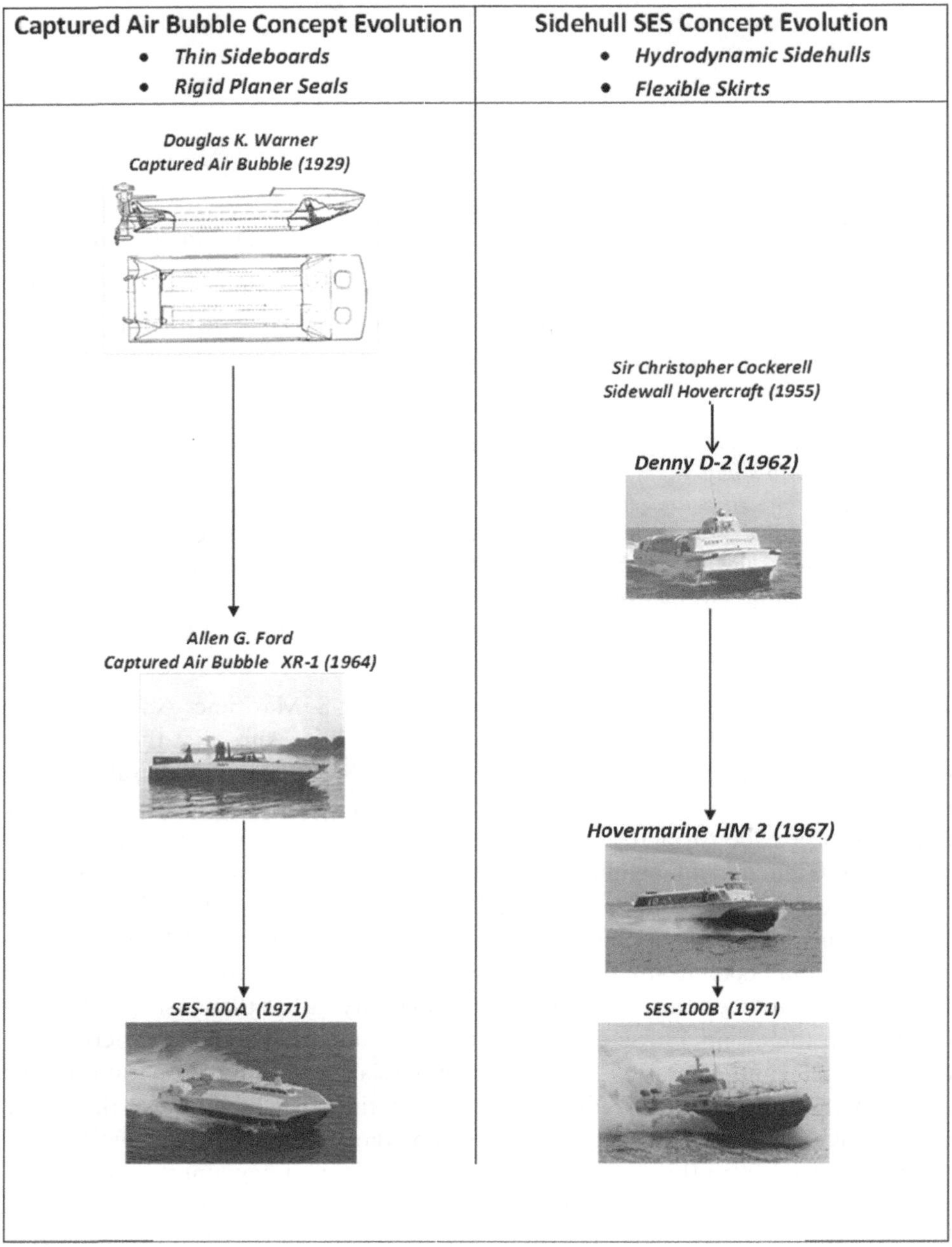

Figure 6.1: Evolution of *Captured Air Bubble* and *Sidehull SES* Concepts

wealth of data with which to design the ultimate goal of a 3,000 ton displacement frigate class SES.

Because of the high speed requirement, both test craft were designed for a low length-to-beam ratio of 2.0. (The actual value for the SES-100A, L/B = 1.95 and for the SES-100B, L/B = 2.2 based on cushion geometries). The different technologies that were to be explored with each craft were the result of extensive analyses and model testing in different facilities around the nation's industrial and governmental towing tanks and water tunnels. Prior international experience with operational *amphibious* and *sidewall hovercraft* from England proved invaluable. The evolutionary experience that was brought to bear on both the CAB SES-100A and the Sidehull SES-100B can be seen in Figure 6.1.

The first evolutionary trail led from the original *captured air bubble concept* of Douglas K. Warner (1929) and the later work by Allen G. Ford (1964) with the XR-1 that led to the key features of the SES-100A. The goal for the CAB SES-100A was to develop a craft that would indeed "capture" the air bubble between "thin sideboards" and "rigid planer seals" so that zero or minimum air leakage (and thus minimum lift power requirements) for an SES could be achieved. As problems occurred on the SES-100A performance, radical design changes were made later to change the thickness and shaping of the original sidehulls, to remove the stability appendages, to redesign the seals and to change the waterjet propulsion scheme (from submerged inlets to flush inlets). These changes will be discussed later as the features necessary to achieve the goal of "80 knots or more" are discussed.

The second evolutionary trail led from the historical development chain from Sir Christopher Cockerell's original work on *sidewall hovercraft* (1955) through the operational sidewall hovercraft, the *Denny D-2* (1962) and the *Hovermarine HM-2* (1967) that led to the key features embodied in the SES-100B (1971). These particular lower speed craft (the D-2 and the HM-2) will be referred to again throughout the book as each key technology feature is discussed in context of providing the SES-100B its characteristics for a "100 knot class" surface effect ship and how such features could be used in any larger tonnage displacement, high speed surface effect ship.

6.2 Different Key Technologies of the SES-100A and SES-100B

Each test craft was designed to explore different key technologies that previous studies had determined offered distinct advantages in one area or another associated with achieving high speed at sea and especially the ambitious goal of 100 knots. As identified above in the directive to the two contractors, these key technology areas are grouped under the headings of *structure, lift, propulsion* and *flexible seal systems.* It is useful in the discussion of the problems and developed solutions throughout the development of the two test craft to encapsulate the main features which has been done in Table 6.1.

As might be expected, several of these key areas resulted in an overlapping of design features in each test craft to accomplish the aims of the program. Most but not all of the problem areas were resolved at the time of the program cancelation in January 1980. The result of the various key features and technologies incorporated

Table 6.1: *Differentiating Key Technologies in the SES-100A and the SES-100B*

CAB SES-100A	Sidehull SES-100B
Sidehulls: Full length, thin width High deadrise angle (60°) Constant width sidehulls, bow to stern Forward mounted stabilizers & skegs This sidehull design is modeled after the CAB design seeking minimum resistance "thin side-boards" modified to provide the essential stability requirements.	*Sidehulls:* Partial length, thicker width for stability Low deadrise angle (30°) Sidehull widens toward stern (by 50%) No additional stabilizers required This sidehull design is modeled after previous UK sidewall hovercraft without the need for external stability devices.
Seal system: Bow seal: Bag and GRP rigid planer elements Stern seal: Upper panel hinged to nine separate lower rigid planing segments This "rigid planer seal" system is modeled after the CAB design patent on rigid elements to explore an alternative to the UK flexible skirt system	*Seal system:* Bow seal: Flexible fabric Bag and Finger Stern seal: Three flexible lobes This seal system is modeled after the UK hover-craft flexible skirt system adapted to high speed operation
Structure: All welded 5086 aluminum with GRP panels on wet deck to lower structural weight	*Structure:* All welded 5456 aluminum for high strength, low weight and simple design.
Engines: Integrated gas turbines for lift and propulsion for efficient use of power	*Engines:* Separate gas turbines for lift and propulsion for flexibility of control and also to facilitate the collection of test data for larger ships
Lift system: Axial fan system and automated ride control system. The pitch of the axial fans and inlet guide vanes are controlled by fan RPM automatically as cushion pressure varies through craft motion through the waves.	*Lift system:* Centrifugal fan system and simple venting ride control system that relieves cushion pressure as the craft moves through the waves.
Propulsion: Waterjets with submerged pod mounted inlets	*Propulsion:* Semi-submerged, supercavitating, controllable pitch marine propellers

into the two designs resulted in the configurations shown for the SES-100A in Figure 6.2 and the SES-100B in Figure 6.3.

The leading particulars of the *CAB SES-100A* and the *Sidehull SES-100B* are shown in Tables 6.2 and 6.3.

There were several detail differences between to the two test craft but those features most directly related to achieving high speed in a safe and useful manner are identified in Table 6.1 and may be seen in Figure 6.1 and Figure 6.2. As problems developed on the CAB SES-100A, several design changes were made from this version shown that was launched on 1 September 1971 in Seattle, Washington and tested in preliminary trials in Puget Sound before radical design changes were made. Retrofitted seal systems, thickened sidehulls, lowered deadrise angle (from 60° to 45°), flush inlets, and other changes were done 1974–1978 and the new configuration of the CAB SES-100A began testing in 1975 under Navy crews at the Surface Effect Ship Test Facility (SESTF) in Patuxent River, Maryland.

The redesigns of the CAB SES-100A will be discussed as appropriate throughout this Chapter. Each of the changes made to the SES-100A were made mainly to

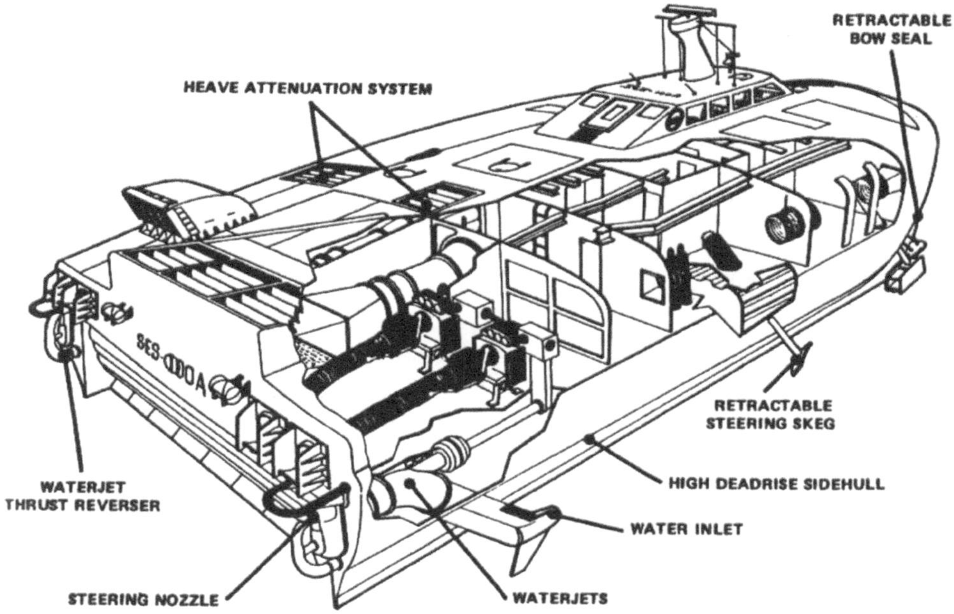

Figure 6.2: Captured Air Bubble SES-100A (as originally built) cutaway

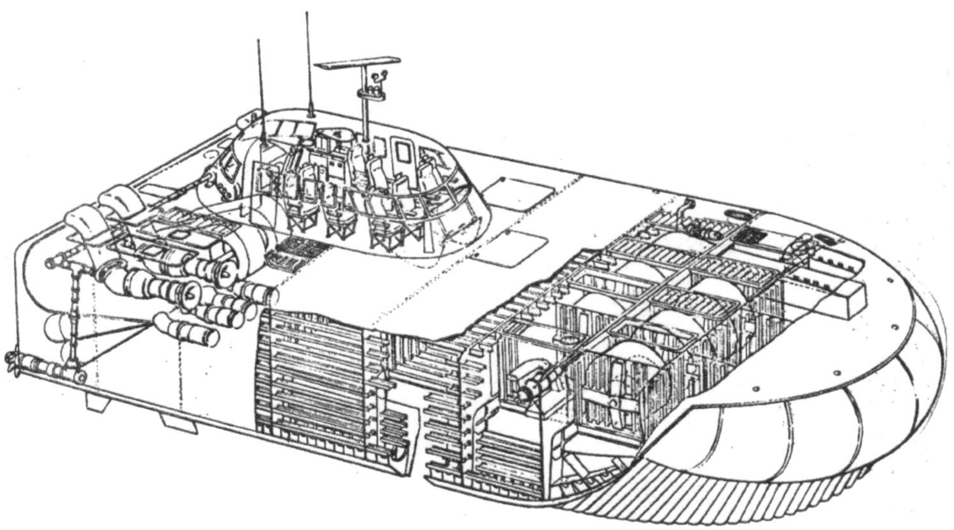

Figure 6.3: Sidehull SES-100B cutaway

Table 6.2: *CAB SES-100A Leading Particulars*

Dimensions	
Length Overall	81 ft 11 in
Beam Overall	41 ft 11 in (max)
Height Overall on cushion	23 ft
Height (Keel to Deck)	17 ft
Cushion Depth (Keel to Wet Deck)	6 ft 2 in
Cushion Area	1,998 sq ft
Cushion Pressure (p_c)	95 psf
Cushion Density (p_c/L)	1.40 psf/ft
Speeds (achieved in test)	
Maximum speed (Calm Water)	76 knots
Speed in Sea State 3 (6 ft waves)	Unknown
Weights	
Normal Gross Weight	230,400 lb (115 s. tons)[a]
Payload	10 s. tons (20 s. tons overload)
Fuel	n/a
Weight Empty	n/a
Integrated Lift System & Propulsion System	
Four Avco Lycoming TF-35 gas turbines	4 x 3,750 bhp max (15,000bhp)
Three axial flow lift fans	
Two Aerojet General two stage axial-centrifugal waterjets	Fully submerged, pod inlets

[a] The weight of the SES-100A varied as redesigns went into effect making it difficult to track the actual weight at any given time. Some documentation suggests the maximum weight in its final version was 244,342 lb (122 s.ton).

Table 6.3: *Sidehull SES-100B Leading Particulars*

Dimensions	
Length Overall	77 ft 8½ in
Beam Overall	35 ft
Height Overall (Keel to top of radar mast)	26 ft 11 in
Height (Keel to Deck)	19 ft 6 in
Cushion Depth (Keel to Wet Deck)	6 ft 2in
Cushion Area	1,905 sq ft
Cushion Pressure (p_c)	105 psf
Cushion Density (p_c/L)	1.70 psf/ft
Speeds (achieved in test)	
Maximum speed (Calm Water)	91.90 knots
Speed in Sea State 3 (6 ft waves)	55 knots
Weights	
Normal Gross Weight	210,000 lb (105 s. tons)
Payload	10 s.tons (20 s.tons overload)
Fuel	38.1 s. tons
Weight Empty (inc 1.3 s. tons operational load)	56.9 s. tons
Lift System	
Three Pratt & Whitney ST6J-70 marine gas turbines	3 x 620 bhp (1,860 bhp)
Eight directly coupled centrifugal fans	7 x 3.70ft dia; 1 x 4.15ft dia
Propulsion System	
Three Pratt & Whitney FT 12A-6 marine gas turbines	3 x 4500 bhp max (13,500 bhp)
Two Supercavitating, semi-submerged CRP marine screws	Each 3.5 ft dia

make the design functional because of various CAB features that were not working. The original *thin sidehulls* were not stable and needed to be made thicker and to have a lower deadrise angle for stability of the ship; the *CAB rigid planer seals* were not functioning properly and had to be replaced with the flexible bag and finger design so that pitch stability was restored to the craft; and the attempt to overcome the large drag of the fully submerged pod inlet was made by switching to a flush inlet.

The Sidehull SES-100B has retained its original configuration throughout its entire test program. The SES-100B was launched at Michoud Facility, New Orleans, Louisiana on 22 July 1971 and began underway testing on 4 February 1972 on Lake Pontchartrain, New Orleans. It was moved to the Naval Coastal Systems Laboratory, Panama City, Florida in May 1973 for rough water testing in the Gulf of Mexico. Testing at Patuxent River began in 1975 until the last run of the SES-100B on 17 January 1980.

Figure 6.4 captures the team spirit that accompanied the successful trials of the SES-100B. The author with a team of 40 engineers transferred from the Bell Aerosystems home offices in Buffalo, New York to New Orleans, Louisiana and the Michoud Facility on the Inland Coastal Waterway to begin the design, build and test of the SES-100B. At the time of launching on 22 July 1971, Vice President Bruce Horton, Bell Aerosystems was in charge of the New Orleans Division. Figure 6.4 (a) shows Bruce and the author (Technical Director and Program Manager) a few minutes after the launch standing on the main deck aft of the cargo hold (cover not yet in place). Less than one year later an enthusiastic dock crew prepared the SES-100B for its first cruise into the waterways surrounding New Orleans. Figure 6.4 (b) shows the team that made that happen. A smaller crew of 8 from that preparation dock team actually participated on-board for the first trial on 4 February 1972 and these have been identified in Figure 6.4 (c).

Table 6.1 summarized the Key Technologies incorporated into the SES-100A and SES-100B test craft. The resulting configurations for the two Test Craft are as shown in Figure 6.2 and Figure 6.3 in their "as built" versions. As discussed later, the SES-100A went through many design changes as the test program proceeded to solve its various captured air bubble (CAB) concept problems. The weights shown for the SES-100A (Table 6.2) are approximate values because of the various design changes that were made. The following discusses each of these Key Technology areas and how these have influenced the configurations shown. A complete description of all the performance characteristics are not addressed here, only those that impact designing an SES capable of achieving "100 knots" in a useful and practical manner.

6.3 Sidewall, Sidehull and Sideboard

The shaping of the catamaran hulls for surface effect ships has been a point of contention between the various designers and different names have been used to describe the basic characteristics. Figure 6.5 shows the names *sidewall, sidehull* and *sideboard* as they have been applied to the various designs.

In 1955, Christopher Cockerell began his experimentation of his hovercraft concept including both amphibious and non-amphibious designs. In his work on the non-amphibious hovercraft he called those designs *sidewall hovercraft*, a name that stayed in the community in England and elsewhere. He found early on, that the shaping of the *sidewall* (see left sketch in Figure 6.5) was critical to both performance

(a) Vice President Bruce Horton (L) and author (R)
Launch of *SES-100B* on 22 July 1971

(b) *SES-100B* First Cruise Dock Team (4 February 1972)

(c) *SES-100B* On-Board First Cruise Crew

1 Program Manager Peter Mantle; **2** Ship Commander Gordon MacClane;
3 First Officer Charlie Bernard; **4** Test Engineer Waide Galle;
5 Test Engineer John Wakefield (not shown); **6** Photographer Francis Jacob;
7 Mechanic Charles Lester; **8** Transmission Engineer James Pelkey

Figure 6.4: Launch and First Cruise Crew of SES-100B
Photos courtesy of Textron Marine & Land Systems and Textron Inc.

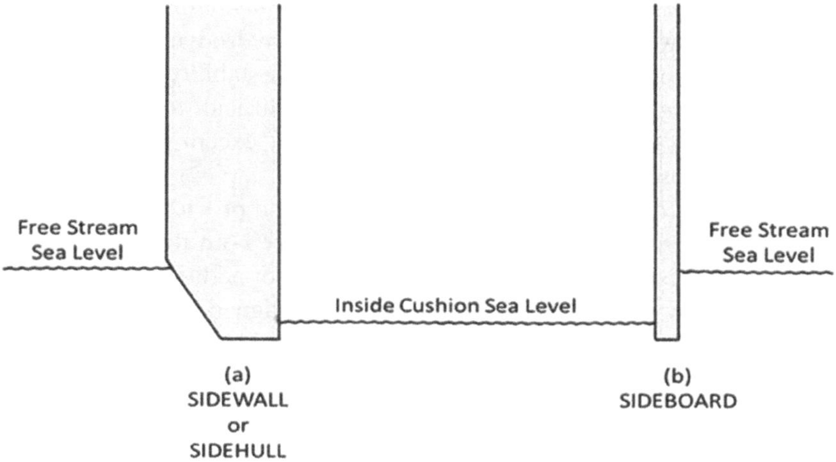

Figure 6.5: Comparison between Sidewall, Sidehull and Sideboard (Simplified)

and stability. The idea that the air cushion alone would provide the stability of the craft was not credible. He experimented with small boats including a 20 ft boat where he used *thin sideboards* (see right sketch in Figure 6.5) first and then moved into more stable forms of hull shape. Christopher Cockerell prepared many papers on his experiments and summaries may be found in his presentations to the broad community at the Princeton Symposium in 1959.[1] A discussion of his presentation and sketches of his *sidewall hovercraft* designs is given in Chapter 2.

It was quickly learned that the shaping of both the cross section and the length of the sidehull (compared to the length of the ship) was fundamental to both the performance and the stability characteristics. Extensive research and model tests were conducted in the facilities of the Hovercraft Development Limited (HDL) in Hythe, near Southampton, England and at the National Research Development Corporation (NRDC) Offices.

Later, in 1965, Hovermarine Ltd was formed in Southampton, England specifically to continue Cockerell's work and *"undertake the development and construction of low-cost, rigid sidewall hovercraft"* under license from Christopher Cockerell's company NRDC. At Hovermarine, Ltd, even more expansive analyses, model tests and full scale trials were pursued as the correct shaping of the sidehulls were developed. Much of the development of the *sidewall hovercraft* was done under the direction of Ted Tattersall, Technical Director of Hovermarine and a brief synopsis of the Hovermarine pursuits of such craft may be found in Jane's Surface Skimmers (1979 edition).[2] The flagship of the Hovermarine fleet of craft was the Hovermarine HM.2 (see Chapter 2). The extensive testing of the HM.2 in both tow tanks and full scale was found invaluable as the design of the Sidehull SES-100B unfolded.

The name of *sidewall* or *sidehull* has been retained to describe the hydrodynamically shaped hulls to provide the necessary performance and stability features. The fundamental difference between *sidewall* or *sidehull* and the *sideboard* described in the CAB patents is that the *sideboard* provides no roll or pitch stability (although it will contribute to yaw stability) to the craft and that the required stability features must be attained by some other means. CAB designers frequently refer to these other means as *"artificial means"*. The purpose of the *thin sideboard* was to seal the cushion air pressure between the two hulls with minimum hydrodynamic drag. The premise was that these "other means" or artificial means for stability would not add drag to the craft!! This is a contentious point in that it is difficult to visualize how stability could be added to the ship without adding drag – except perhaps by the installation of large gyroscopes in the center of the ship!

The *sidewall hovercraft*, HM.2 was launched in 1966 and provided an excellent data point as the design of the SES-100B proceeded where both the performance and the stability was invested in the sidehulls without need for additional or artificial means as was required by the CAB SES-100A which was designed to be the next step in CAB development. There are important differences, of course, as the different regimes of Froude Number and Cavitation Number between the HM.2 and the SES-100B were handled.

6.3.1 Sidehull Shaping of SES-100A and SES-100B

A fundamental feature that differentiated the two craft, SES-100A and SES-100B, was the choice to use thin, high deadrise (60°) ("*thin sideboard*") sidehulls for the CAB SES-100A and thicker, low deadrise (30°) sidehulls for the Sidehull SES-100B. The capsizing accident with the thin, high deadrise (60°) sideboards of the XR-1 in 1964 was still a vivid memory in 1969 as the 100 ton craft were designed. The approach used by the SES-100A designers was to use the thin, high deadrise hull features of the 1964 captured air bubble (CAB) concept – *but as a safety measure* – to augment the craft stability with forward mounted movable appendages to provide the needed stabilization in pitch, roll and yaw. These would be the "other means" described in the 1964 CAB patent.

The comparison between the two sidehulls is seen in Figure 6.6 that shows the distinct difference in design concept. The particular drawing shown in the upper left of Figure 6.6 shows the SES-100A in its original structural configuration with its *rigid planer seals* as called for in the CAB patent. The design concept for the SES-100A was to capture the concept of *"thin sideboards"* also as called for in the CAB patent, except that the SES-100A contractor designers made a prudent design decision to add stability augmentation devices (SAD) because of the realization that without some form of stabilization the craft would be unstable and prone to capsize like the XR-1. To the same scale, the SES-100B bow view is shown in the upper right of Figure 6.6.

In the two lower drawings (also drawn to scale) of Figure 6.6, are shown the actual cross sections of the two different designs of sidehull for the 41 ft 11 in beam SES-100A and the 35 ft beam SES-100B. The SES-100A had thin sidehulls and a high deadrise angle (60°) that maintained this constant cross section shown from bow to stern. The SES-100B sidehull, on the other hand, was thicker at the waterline (not shown) and with a lower deadrise angle (30°). A distinct difference between the two designs is that the width of the sidehull for the SES-100B thickened toward to the stern of the craft as shown by the shaded cross section while the SES-100A sidehull remained at constant (narrow) thickness.

There were two reasons for this thickening of the hull of the SES-100B toward the stern. First, based on stability reasons and learned from the HM.2 experience,

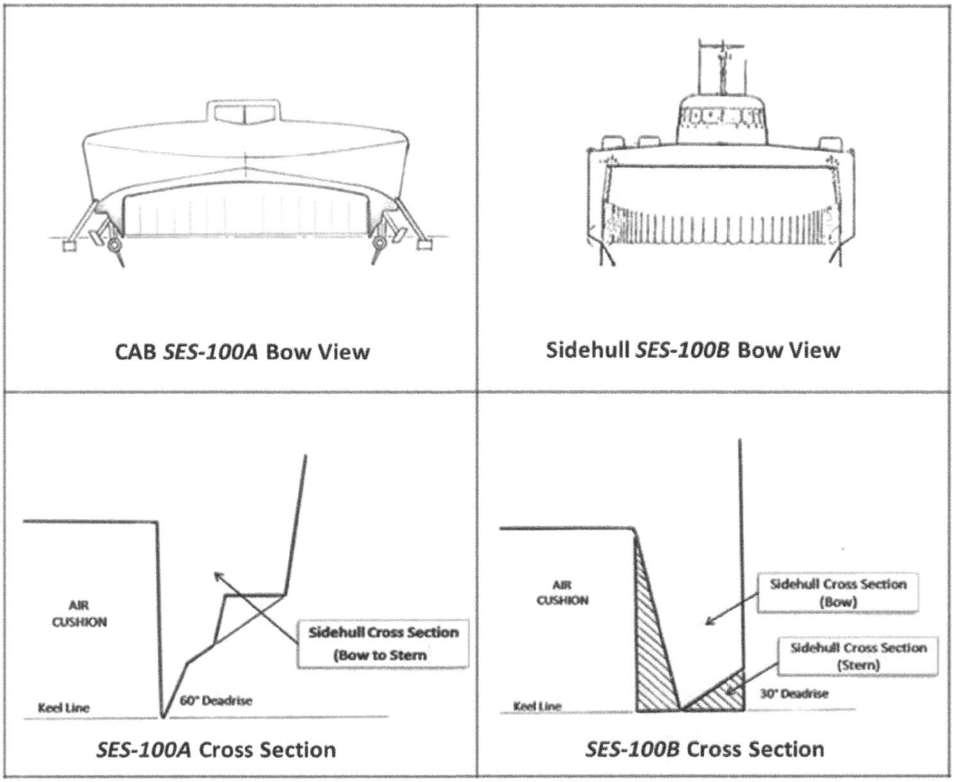

Figure 6.6: Sidehull Shaping of SES-100A and SES-100B (to same scale)

that exhibited a similar sidehull shaping, that it was recognized that the directional stability of the sidehull craft was improved considerably if buoyancy was added aft of the craft Center of Gravity (C.G.). The second and main reason was to shape the stern portions of the sidehulls to provide a smooth water flow entry into the surface piercing propellers. The sidehull width was increased by approximately 50% to give a >3.5 foot width sidehull in the aft and stern region (shown cross-hatched in Figure 6.6) to accommodate the 3.5 foot diameter propellers. More discussion of the propulsion technology differences between the SES-100A and the SES-100B is provided later in this Chapter.

The SES-100A was designed with the CAB concept in mind with thin sideboards to minimize hydrodynamic drag. While in actuality the thin sideboards did have some shaping as seen in Figure 6.6 the shaping was minimal and provided insufficient restoring forces for stability. The SES-100A designers were concerned that there would be insufficient stability mechanisms in the CAB thin sideboard design and went ahead and added the "artificial means" of stability with the addition of "stability augmentation devices" (SAD) as shown in Figure 6.7.

In the upper drawing drawn to scale in Figure 6.7, is the bow view of the SES-100A showing the additional means of stability in the form of stabilizers (fixed and retractable), skegs and fins. In the drawing, the retractable stabilizers are shown in their extended position. For clarity the rigid planer seals (also an essential feature of the CAB concept) are not shown. These are discussed later.

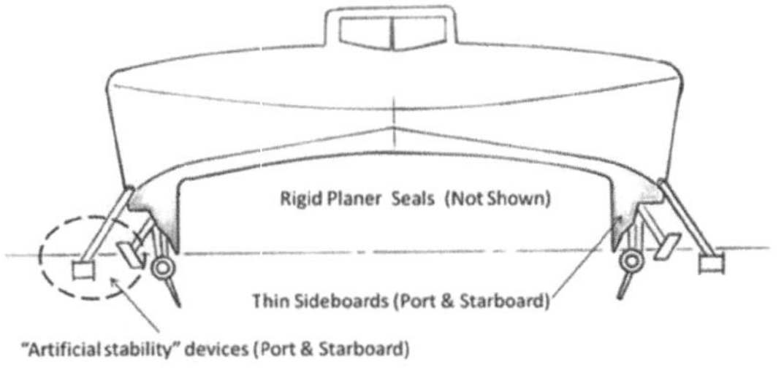

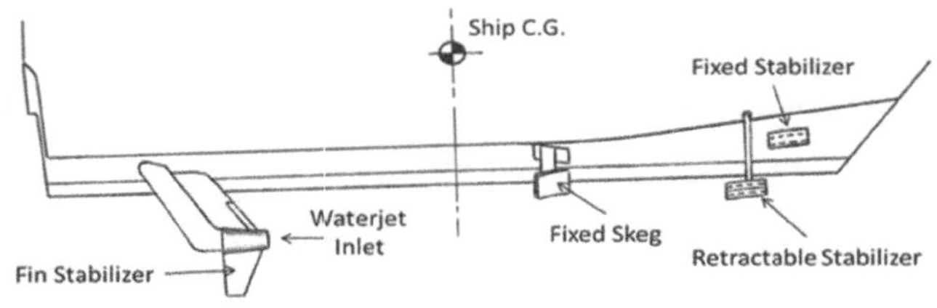

SES-100A at sea
with "artificial stability" devices extended

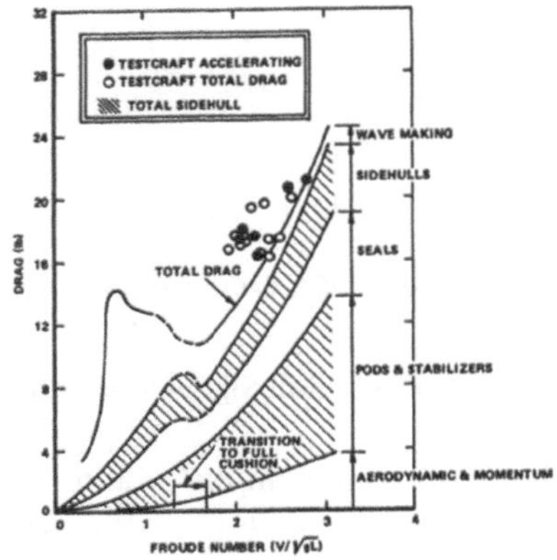

SES-100A Measured drag components

Figure 6.7: CAB Features of SES-100A

In the middle drawing is shown the sideview of the SES-100A that shows the various appendages in their actual locations along the hull and where these features fit in relation to the ship's center of gravity. Note that the directional stability of the ship is enhanced by the waterjet pod struts that also serve as fin area aft of the ship's C.G. Steerable rudders are then mounted beneath the waterjet pod intakes. The two bottom parts of Figure 6.7 show (on the left) a photograph of the SES-100A operating that clearly show these stability appendages and on the bottom right of Figure 6.7 can be seen the drag build-up for the test craft SES-100A as measured in trials and how these various components contribute to the drag of the ship.

As can be seen from the drag curves in Figure 6.7 that while the *thin sideboard* hulls of the SES-100A did indeed exhibit a low drag, the drag of the additional or artificial stability devices required, all but swamped the other drag components. The basic intent of the CAB concept was to design a craft that had much less lift power (because of the sideboards) than amphibious hovercraft and thus would be more "efficient" but it paid the price of the required stability features (and their high drag) needed to overcome the basic instability of thin sideboards.

It is interesting to note that the installed horsepower per ton of ship displacement (HP/ton) for the CAB SES-100A ended up no different from that of the SES-100B that designed the stability into the basic *sidehull* design without the added complexity and vulnerability of the extended stability devices. The reader might glance at conditions at sea from Figure 2.1 in Chapter 2 and visualize the CAB SES-100A operating with these devices as shown here in Figure 6.7 and wonder about the mechanical nightmare such a design would entail in rough seas.

From the data given in Table 6.2 it is seen that in the "as built" configuration the SES-100A had a value of HP/ton of 146HP/l.ton and the SES-100B had a value of 163 HP/l.ton. That is, both craft had about the same amount of HP/ton installed (about 12% difference) but considerably different stability and performance characteristics. How these values compare to other craft types is discussed in Chapter 8.

The Sidehull SES-100B design pulled heavily on the British experience from the original work by Christopher Cockerell and work at Hovermarine Ltd in the design of *sidewall hovercraft*. The design of the SES-100B hulls involved the integration of a high speed, low resistance shape that provided the required stability and the housing of the propulsion system. This latter requirement was the result of recognition of the high price paid by extending any propulsion mechanisms into the free stream. The high drag of the SES-100A waterjet pod strut (see Figure 6.7) is an example of such a large drag penalty.

Figure 6.8 shows the comparison between the HM.2 hull lines and those of the SES-100B. The two hull designs were designed to operate in very different flow regimes. These flow regimes are characterized by two "numbers". First, the Froude Number that relates the inertia forces acting on the hull to the gravity forces from the wave action. Second, the Cavitation Number that relates the difference between absolute pressure and vapor pressure of the water to the kinetic energy of the water

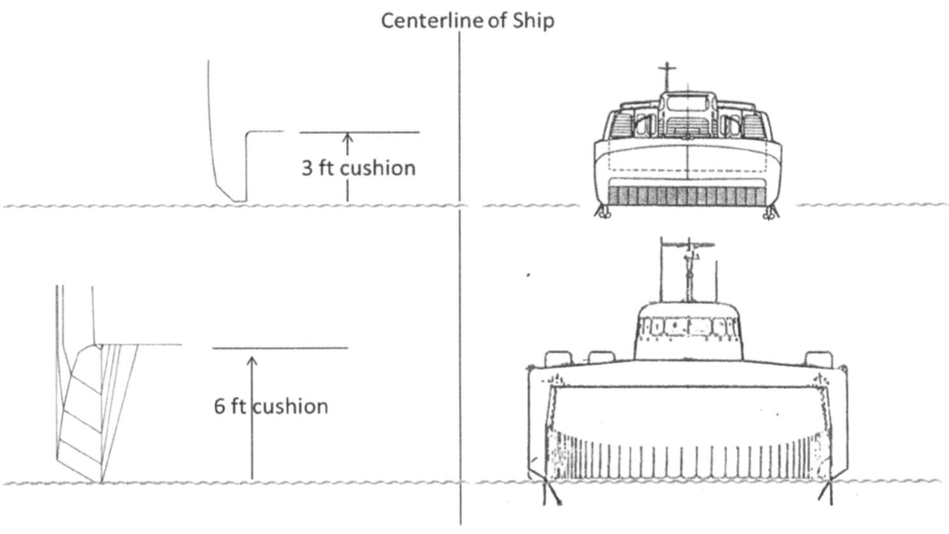

Figure 6.8: Comparison of HM.2 and SES-100B sidehull configurations

stream. That is, it measures the tendency or potential of the flow to cavitate. Designing hulls to operate in cavitating flows is quite different from designing hulls to operate in non-cavitating flow. These two numbers are given here for convenience of discussion:

$$\text{Froude Number (F. No.)} = \frac{V}{\sqrt{gL}} \tag{6.1}$$

$$\text{Cavitation Number } (\sigma) = \frac{p_{st} - p_v}{1/2\rho_w V^2} \tag{6.2}$$

In the above equations, (g) is the acceleration due to gravity; (p_{st}) is the static pressure acting on the sidehull; (p_v) is the vapor pressure of water; (ρ_w) is the density of water and (V) is the craft speed through the water and (L) is the length of the sidehull at the waterline.

The HM.2 hull was a high Froude Number hull design designed to operate at Froude Numbers of 1.0 to 2.0 but operating at a relatively high Cavitation Number (σ = 0.60–2.0). The SES-100B, on the other hand was designed to operate at very high values of Froude Number (> 2.5) but at a low value of Cavitation Number (σ < 0.15). The stability characteristics of such hulls is a complex set of considerations but experience in England operating under strict Maritime Safety Rules provided a good starting point for the SES-100B design. Figure 6.8 shows the comparison between the HM.2 and the SES-100B on some basic stability parameters. The left hand side of Figure 6.8 shows the HM.2 and the SES-100B drawn to the same scale with a view of the starboard side of the ship relative to a common center line.

Two simple parameters that contribute to the stability in roll are first, the ratio of the sidehull width or beam to the width of the ship (B_{SH}/B) and second, the ratio of the cushion height to the cushion beam (h_C/B_C). As can be seen from Figure 6.7, the SES-100B was designed to be conservative with these basic parameters because of the high speed nature of the design and also because of problems experienced with the XR-1 CAB design. Specifically, the values are shown in Table 6.4.

These values are tabulated from the values of the parameters at the chine as an indication of the stability mechanism. The value of B_{SH}/B_C for the SES-100B would be the larger value (0.18) if the increased sidehull width at the stern of the craft were taken into effect. Although the HM.2 was known to be very stable, a degree of conservatism was built in to the SES-100B design because of the greatly increased speed of the design (100 knots versus 30 knots).

The high speed requirement of the SES-100B demanded that there be minimal hull or appendages immersed in the water at high speed. To this end a unique propulsion scheme was conceived and integrated into the design of the SES-100B. A more complete description of this unique propulsion scheme follows later in this Chapter as the propulsion schemes for the SES-100A and the SES-100B are

Table 6.4: *Stability Parameters of HM.2 and SES-100B*

Stability Parameter	HM.2	SES-100B
B_{SH}/B_C	0.07	0.07–0.18 (at stern)
h_C/B_C	0.17	0.19

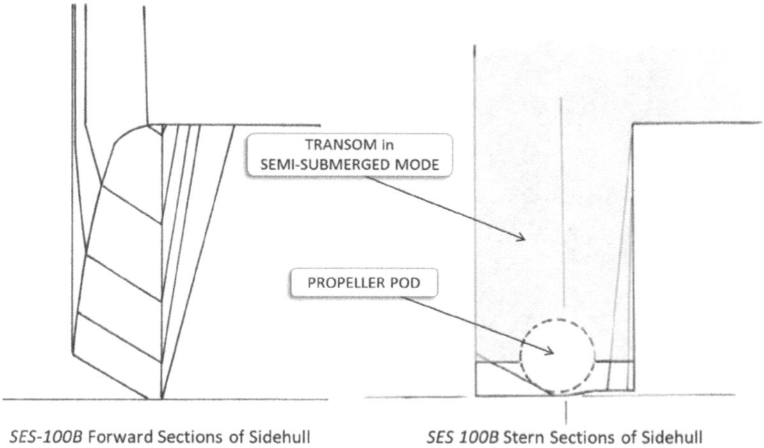

Figure 6.9: Bow and Stern Sidehull Lines of SES-100B

discussed. Figure 6.9 shows the sidehull lines of the SES-100B as the forward lines in left hand side of Figure 6.9 fair into the widened sidehull at the stern of the ship to accommodate the propulsion system.

The forward sections shown show a smooth entry into the water. A photograph of the actual hull may be seen in Figure 6.35 shown later in this Chapter as the structural design differences are discussed. The lines also show the gradual flaring into the cushion area toward the stern as the propulsion pod is approached. This flaring also contributes to giving the ship some extra buoyancy when in the hullborne mode. The right hand side of Figure 6.9 shows the stern sections. The shaded area shows the lines when the propulsion system ramps are in their down position for cruise speed when the propellers operate in their semi-submerged mode. At low speeds, the ramp raises and the hull lines "move up" to accommodate the water flow into the propeller for fully submerged propeller operation (not shown). More discussion follows later on this system. From a sidehull lines viewpoint, Figure 6.9 shows that the water lines of the sidehull are relatively narrow in the bow and then flare out significantly toward the stern of the ship.

Figure 6.10 is a photograph of the stern area in a partially assembled state (the ramps are not installed). The flaring out of the sidehulls as the stern of the craft is reached is clearly seen in Figure 6.10, which shows the SES-100B on its supporting rig during outfitting at the Michoud Facility. In the foreground is the port dihedral fin stabilizer and immediately behind the fin can be seen the port vertical rudder. Behind the rudder is the water flow opening into the surface piercing propeller. The sidehull shaping shows the general smooth widening of the hulls to provide both stabilizing hull form and flow lines into the propellers.

6.3.2 Some Additional Comments on Stability

For the CAB SES-100A, in addition to stability augmentation devices (SAD), the SES-100A was fitted with deployable foils or "retractable steering skegs" (shown in the upper left of Figure 6.5 and Figure 6.6). The intent was for these skegs to provide

Figure 6.10: Stern Shaping of SES-100B Sidehulls
Photo courtesy of Textron Marine & Land Systems and Textron Inc.

a lifting force to the outside of the turn (on cushion). The steerable nozzles of the waterjets would provide the turning moment at low speeds and when hullborne.

After the "plow-in" by the SES-100A, these SAD features were removed and the sidehulls were thickened more along the lines of the SES-100B. In the contractor test program, the CAB rigid *bow planer seal* was replaced with the flexible "*bag and finger*" skirt as seen in Figure 6.11 (and later in Figure 6.19) shown when operating over calm water.

The issue of high deadrise, thin hulls as used on the CAB SES-100A and the low deadrise, thicker hulls as used on the sidehull SES-100B is a further key distinction between the two technologies used in their respective designs. There is no hard and fast rule as to the shaping of the hulls for high speed SES but the potential for catastrophic capsize as a result of a "pitch-roll-yaw" interaction insists on a rigorous analysis and test program to ensure that the configuration is a safe one. The features contributing to the "pitch-roll-yaw" interaction can be reviewed separately for discussion.

First, in any small perturbation in yaw, the sidehull area ahead of the craft center of gravity (CG) will tend to generate a side force that will be destabilizing in yaw. If the craft pitches or rolls at the same time it is easy to visualize the possibility for a rapid build-up in sidehull forces that in general are destabilizing. To this end, it is advisable to keep the sidehull area ahead of the CG to a minimum. From a configuration viewpoint this results in a "more skirt, less hull" design, or a more wrap-around skirt feature. Such configurations are often called "*partial sidehull*" forms.

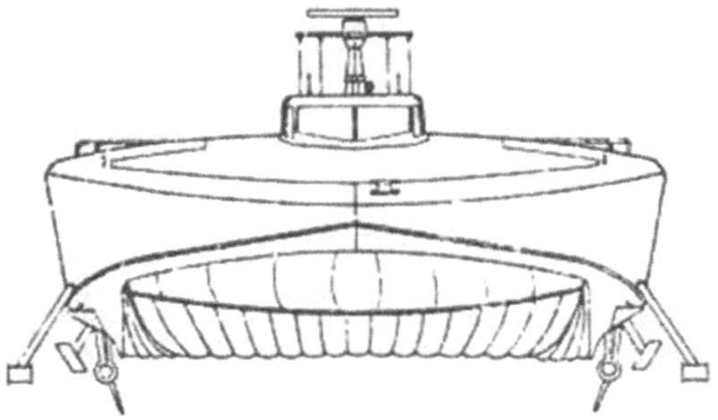

Figure 6.11: Stability Augmentation Devices (SAD) on SES-100A

Figure 6.12 shows some patent work on this concept by the author and Arnold Hall in 1966 at General Dynamics, Electric Boat in Connecticut, with model tests conducted at the General Dynamics (at that time) tow tank in San Diego, California.[3,4] The main intent of this design feature is to minimize the forward wetted area, and to properly position the centers of (air) cushion pressure lift and of (hydrodynamic) planing lift, for optimization of both stability and ship resistance.

Much later, in 1989, such a partial sidehull design appeared on the scene with the 30 knot Mitsui, *Sumidagawa* for the Tokyo Metropolitan Government Bureau of Construction (see issue of Jane's High Speed Marine Transportation, 2009–2010)[5]. The *Sumidagawa* had about a 50% partial sidehull design. The exact percentage of sidehull and skirted periphery for any particular design will depend on several factors of stability, and resistance. In the case of the SES-100B, the sidehull length was found to be optimum at about 75% of the ship length (see Figure 6.3). For the SES-100A, the original sidehull was about 90% of the total ship length, but after the craft was modified to incorporate the CAB planer seal the sidehull length became almost 100% of the craft length for the CAB SES-100A.

Secondly, if the high deadrise sidehulls are such that any hydrodynamic force is generated by that same large wetted area sidehull ahead of the CG, an additional concern comes into play relative to the righting moment of the craft in the roll or lateral plane. This is a complex interaction that needs more detailed treatment in any specific design but the following sketch (Figure 6.13) illustrates the main point.

The issue of how "thin" should the sidehull be, and in addition, how much deadrise the sidehull should have, must be orchestrated such that the craft is stable

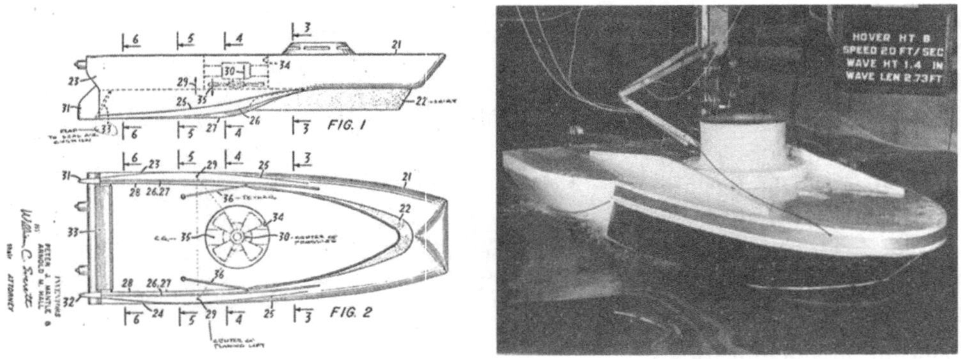

Figure 6.12: Partial Sidehull SES Patent and Model Test (1966)
Photo reprint by permission of NY Acad. Sciences

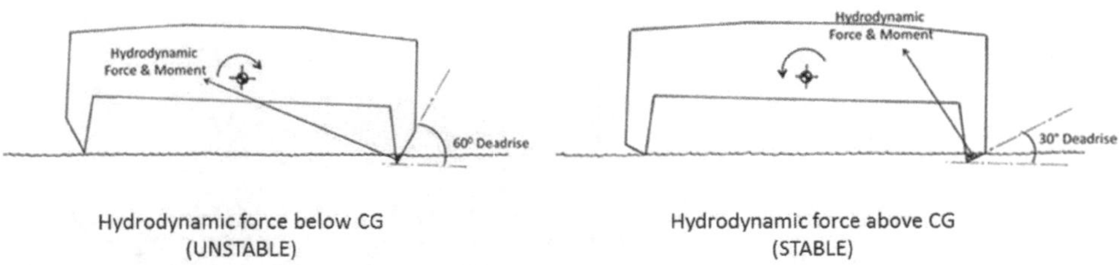

Figure 6.13: Possible Destabilizing & Stabilizing Forces from Deadrise Angle

in all modes of pitch, roll and yaw. Figure 6.13 illustrates both destabilizing and stabilizing moments that could be generated depending on the sidehull thickness, position of the craft's CG, and on the sidehull's deadrise angle. In the case of the SES-100A, the intent was to have the captured air bubble provide the lateral stability (with some help from the SAD features), whereas in the case of the SES-100B, the restoring moment was to come almost completely from the hydrodynamic forces generated by the thicker buoyant sidehull and lower deadrise angle.

In the hypothetical case illustrated in Figure 6.13, the ship has rolled to starboard causing a hydrodynamic force to be generated from the starboard rolled sidehull. In the case of a high deadrise angle (left hand sketch) the hydrodynamic force passes beneath the craft center of gravity and the resulting hydrodynamic rolling moment is destabilizing. In the same situation, for the lower deadrise angle (right hand sketch), the generated hydrodynamic force passes above the CG and the resulting moment is stabilizing. Clearly, the geometry of the craft, and the relative thickness of sidehulls to craft beam, height on cushion and where the craft CG and metacentric height is, all come into play to determine whether the craft would be stable in roll, marginally stable in roll or unstable in roll.

The capsizing experience with the CAB XR-1 in 1964 and the "plow-in" during early underway tests in 1973 with the CAB SES-100A in its original "as built" configuration, showed that such instabilities were real. During the 1973 test, with the original configuration of *thin sidehulls* and *high deadrise angle*, the SES-100A experienced such strong roll and yaw instabilities that the Stability Augmentation Devices (SAD) (see Figure 6.4) were overcome and the craft pitched forward enough to put "green water" over the deckhouse! Fortunately no one was injured in that test incident.

6.3.3 Lateral Stability Rules for Air Cushion Craft

As a set of "rules of thumb" on craft stability, the following charts encapsulate the overall ship geometry parameters as they influence the craft stability in pitch-roll-yaw interaction. In Chapter 8, where these Technical Considerations are expanded, more specifics of these features as they apply to designing for high speed are provided. The reader is cautioned that what follows are indeed "rules of thumb" and guidance values only because the nature of a high speed craft traversing either calm or rough water especially those craft that incorporate an air cushion is not readily reducible to precise engineering calculations. There is no substitute for a well-structured and scaled model testing in dynamic conditions for any reliable predictions for the stability of such craft.

The high speed non-amphibious air cushion craft, in particular, must contend with both hydrodynamic forces and air cushion related forces plus their interaction and the additional complication of cavitating and ventilating flow from the mixing of both air and water at the interface. The motion of the craft is decidedly non-linear and very difficult to scale because most of the governing forces scale by different rules for dimensional similarity. This last fact alone makes model testing particularly difficult to interpret into full scale behavior. This is discussed in more detail in Chapter 8 on Technical Considerations but within certain limits, "rules of thumb" gained from many years of full scale operations of various types of air cushion craft have produced reasonable guidelines for first "scoping" of a high speed SES stability features. These have been summarized in what follows.

Figure 6.14 shows the general sizing parameters of the sidehull width measured with respect to the cushion beam (B_{SH}/B_C), and craft (ship) length to beam ratio of the cushion $(L/B)_C$. Such plots can only be indicative of the stability property because of the variation of the sidehull width (B_{SH}) with draft on any craft. A more correct value would

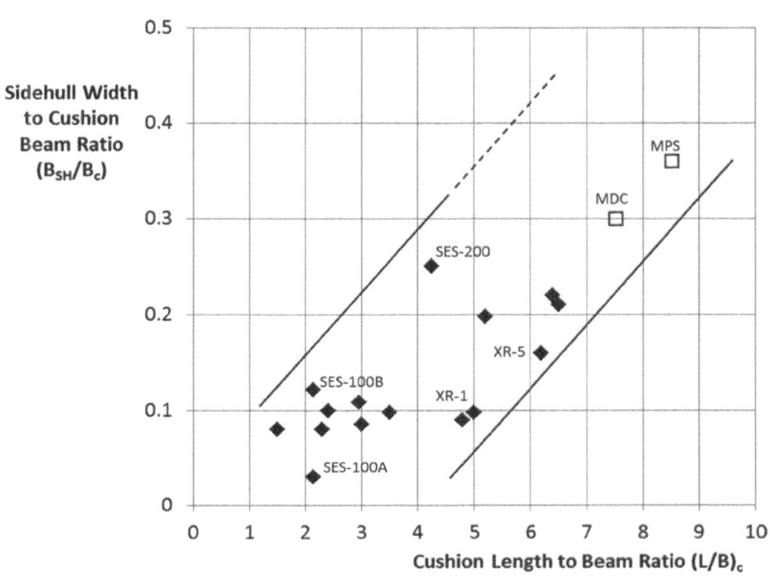

Figure 6.14: Sidehull Width of Sidehull Air Cushion Craft

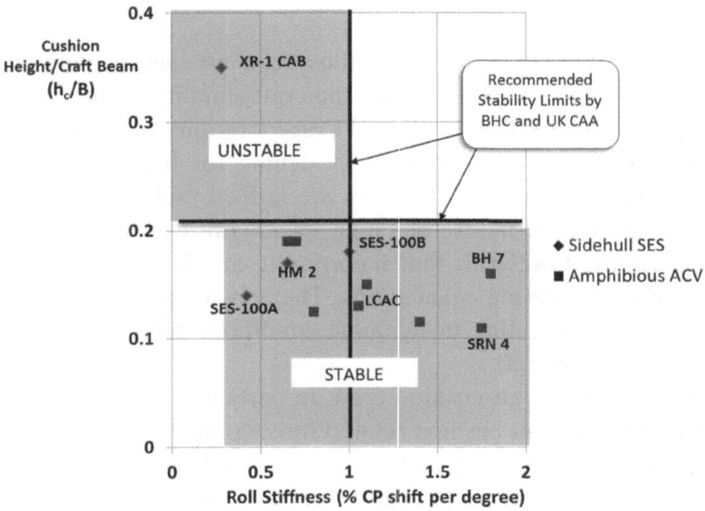

Figure 6.15: Roll Stability Limits for Air Cushion Craft

be the value of B_{SH} at the waterline but for quick indications the widest thickness value has been used in this plot.

From Figure 6.14 can be seen the relative thickness of the sidehulls for the CAB XR-1 series craft and the CAB test craft SES-100A when compared to the SES-100B and other sidehull hovercraft. Also shown in Figure 6.14 are the values of the high length-to-beam ratio (L/B), slower speed SES designs (MDC and MPS) put together by the SES Project Office (PMS 304) (see Chapter 5) just prior to cancelation of the SES Program in 1980.

Figure 6.15 shows the stability regions expressed by the roll "stiffness" (measured as the per cent shift in the center of pressure per degree of roll) as a function of the height of the cushion (h_C/B_C). These are general guidance rules established over a period of time by hovercraft designers (such as the British Hovercraft Corporation, BHC) and incorporated into their designs and in the regulatory agencies in the UK, specifically the Civil Aviation Authority (CAA) who have jurisdiction over such craft in the UK.

The equations behind these charts have been summarized by Mantle (1980)[6] based on the original work by Gordon Elsley and Tony Devereux in 1968[7] and Bob Trillo[8] in 1971 that covered a wide range of amphibious and sidewall hovercraft in the UK, France and the US. This work resulted in the regulations issued by the Air Registration Board (ARB) in 1962[9] and by the Civil Aviation Authority (CAA) in 1972[10] with various updates since that time. It is seen that the values for the MPS and MDC designs (see Chapter 5, Figure 5.10 and Figure 5.11 for more details on these designs) that the design philosophy had reverted to the *thin sidehull* approach. No model tests or stability analyses were performed for these designs and the roll stiffness features need to be checked as discussed below.

In Figure 6.15, the additional key stability parameters of stiffness in roll and the height on cushion are displayed. Based on the criteria established by "best design practice" and by safety regulations it is seen that the XR-1 as built is unstable (proven by the 1964 capsize) and the SES-100A has marginal stability as exhibited by in its early "plow-ins" in 1973.

The design practices of most air cushion craft (amphibious and sidehull types) have held to these lateral stability recommendations. It is noted from the historical data shown in Figure 6.14 and Figure 6.15 that the experience base of operational craft is mainly in the cushion height to beam ratio (h_c/B = 0.10–0.20) and wider sidehull width to craft beam (B_{SH}/B_C = 0.05–0.30) depending on the length to beam ratio (L/B) of the ship. Only the projected and unsubstantiated designs approach higher values of h_c/B_C and lower values of B_{SH}/B_C.

Shown in Figure 6.16 are some projected designs for cushion depth (or skirt height) and in 1980, the designs by the SES Program Office for their MDC and MPS large tonnage surface effect ship designs are included (see Chapter 5) that show very high cushion to beam ratios of h_s/B_C = 0.40–0.50 which appears high, based on the stability criteria developed for such craft and shown above in Figure 6.14 and Figure 6.15. Such high values are not recommended (without extensive model testing in a variety of operating conditions) based on the available theory and design practice and that keeping the ship designs within the bounds shown is probably a prudent design practice, especially for those designs intended to operate at high speeds. Because of the cancelation of the program there was no opportunity for such investigative model tests.

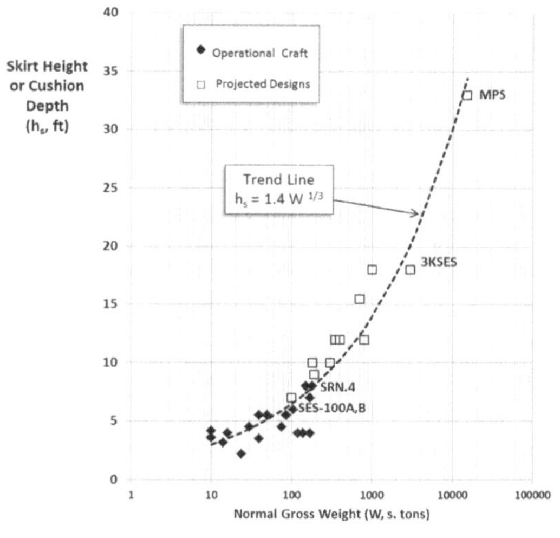

Figure 6.16: Trends in Cushion Height (1960–1980)

The SES-100B followed the more conservative design practice demonstrated by the SRN.4 and BH-7 and other amphibious hovercraft and other sidehull air cushion craft (e.g. the HM.2) craft by ensuring a low value of (h_C/B_C). The roll stability of the SES-100B was then assured through both low (h_C/B_C) and sufficient restoring forces from the thick sidehulls and low deadrise angle to give adequate "roll stiffness" (% C.P. shift per degree).

While it is intuitively obvious that sitting on a bubble of (captured) air is unstable (without some external restoring mechanism, such as buoyant sidehulls) it is informative to see the governing parameters for such conditions. The later elegant treatment by Odd Faltinsen (2005)[11] shows how the depression of the air cushion inside the sidehulls contributes to the changes in the metacentric height and the destabilizing forces in roll for any craft using an air cushion for support contained by sidehulls. The geometric factors of p_C, B_{SH}/B_C, h_C/B and sidehull draft all enter considerations of roll stability.

These findings have their roots in the cautions issued by Cockerell, Warner and others in 1959 at the Princeton Symposium, based on their original designs and operational experience of the first *captured air bubble (CAB)* craft and *sidewall hovercraft* (see Chapter 2).

In summary, the technology of the *thin sidehulls* and *high deadrise angle* design of the SES-100A, coupled with the 100% sidehull length (generating large wetted areas ahead of the CG) caused a compromise in design that the addition of external (and vulnerable) stability appendages did not correct. The *thicker sidehulls* and *lower deadrise angle* of the *partial sidehull* SES-100B showed a more stable design that would later be incorporated into the 3KSES design.

Chapter 8 provides guidance for selecting the sidehull and craft geometry parameters based on work by the Civil Aviation Authority (CAA) in the UK and other regulatory agencies as well as "rules of thumb" generated from hovercraft experience that expands upon the above snapshots of considerations and features.

6.4 Seal System Differences

A second major difference in technologies between the SES-100A and the SES-100B was in the choice of technology in the seal design. The SES-100B, very specifically, was designed and built upon the many years experience base of the British hovercraft work of developing *flexible skirt systems.*

The SES-100A, was designed to explore the novel features of the CAB *rigid planer seals.* The intent of the planer seal was to stop all air leakage from the cushion and to "capture" the air bubble between the sidehulls and the planing seals. The intent being to minimize air leakage (i.e. an "*airtight-watertight airspace*") and thus minimize lift power. The trick was to devise a "ski" or "planing seal" that would maintain contact with the water (even in rough seas) yet not impart any major drag or destabilizing forces to the craft or air leakage. This turned out to be an unsolved problem of the SES Program.

6.4.1 SES-100A Rigid Planer Seal Design

The SES-100A went through three main phases in seal design. These phases are

(a) the initial CAB Planer seal on original structure; then
(b) a retrofitted Bag & Finger Bow Seal on original structure; and finally reverting to
(c) a CAB Planer seal on modified ship structure.

The CAB rigid planer seal was difficult to fit onto the "pointed bow". The cables and downstops of the planing segments were difficult to articulate with the bow geometry shown in Figure 6.17.

It was found that the configuration of the CAB planer seals contributed to the "plow-in" of the SES-100A early in its test program in 1973. The following (paraphrased) description from the 1975 report[12] provides the sequence of events:

"There were two contributing causes for the plow-in events as a result of this particular (CAB) design. The first was due to the large unbroken area that the seal presented to the water surface, such that any wetting rapidly increased due to surface adhesion. The second contributing cause appeared to be due to a drop in cushion pressure in the forward cushion area. This effect was suspected to be caused by momentum exchange between the water and the air incurred by spray generation inside the cushion from the trailing edge of the bow planer seal.

Figure 6.18 shows the diagrammatic bow seal and spray generation and the record of the pressure drop during one of the events.

The inception of the plow-in occurred while the craft was traveling at approximately 45 knots in calm water. The craft was trimmed level with only a slight pitch oscillation of 0.10cps. Instrumentation showed a slight negative longitudinal pressure gradient (lower pressure in the forward area) of 4 percent. All stability appendages (SAD) were clear of the water, which was relatively calm. The plow-in occurred after the craft was being throttled back from full cushion

Figure 6.17: SES-100A Original Bow Shape

operation to partial cushion operation in preparation to stop. The bow and stern seals were being raised gradually to settle into the water.

As can be seen from the traces shown in Figure 6.18, the cushion pressure in the aft area (Frame 3) remained relatively constant throughout the event. The cushion pressure in the forward part of the cushion (Frame 20) rapidly fell off from the average value of 98psf to approximately 50 psf, generating a pressure gradient illustrated by the shaded area in Figure 6.18. Corrective action was taken by lowering the bow stabilizers after the

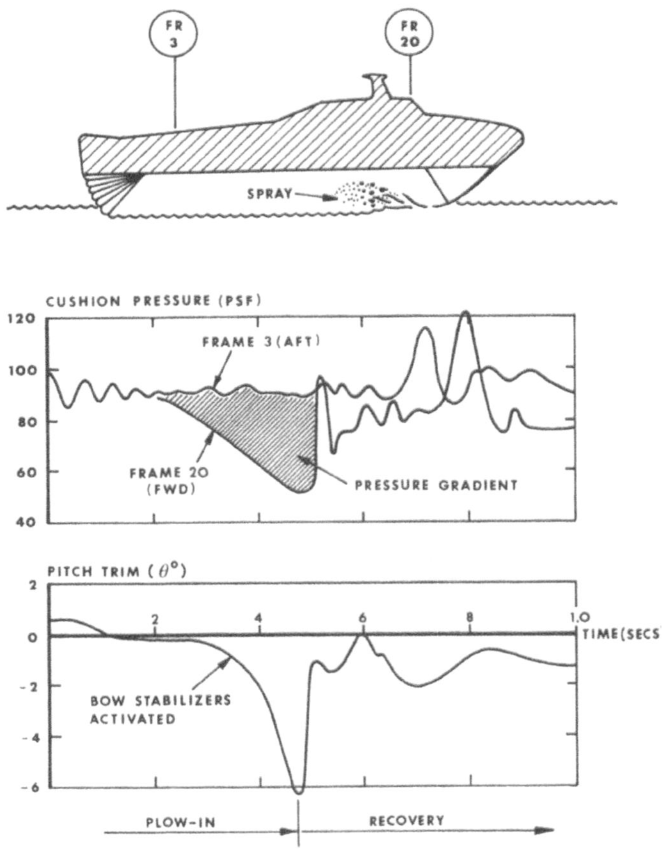

Figure 6.18: SES-100A Plow-in with CAB Planer Seal

> *craft had pitched down 2°. The craft stabilized momentarily and then continued to pitch down (plow-in) to –6° over a period of 4 to 5 seconds. The craft then recovered from the "green water" over the deckhouse and the engines were shut down. The cushion flow remained constant until the craft was at the –2° level and then increased to 5900 cu ft per sec throughout the plow-in. The pressure gradient started well before the pitch down occurred and overpowered the stabilizers (SAD). The craft was not damaged"*

In 1973, after the plow-in of the SES-100A, the bow CAB rigid planer seal was removed during the contractor test program and a flexible "bag and finger" skirt system was installed, so as to minimize the contact between the seal and the water. Actually, because of the difficult geometry posed by the original shaped bow of the SES-100A, the "fingers" of this skirt system were "pericells" or "cones" in the center of the skirt, and fingers on the outer edges to contact tightly with the high deadrise sidehulls of the SES-100A configuration. The experience base of the pericells derived from Aerojet's work for the Navy's JEFF (A) amphibious air cushion craft program. The solution to the plow-in was then as shown in Figure 6.19. This allowed the SES-100A to continue with the contractor test program throughout 1973 and 1974.

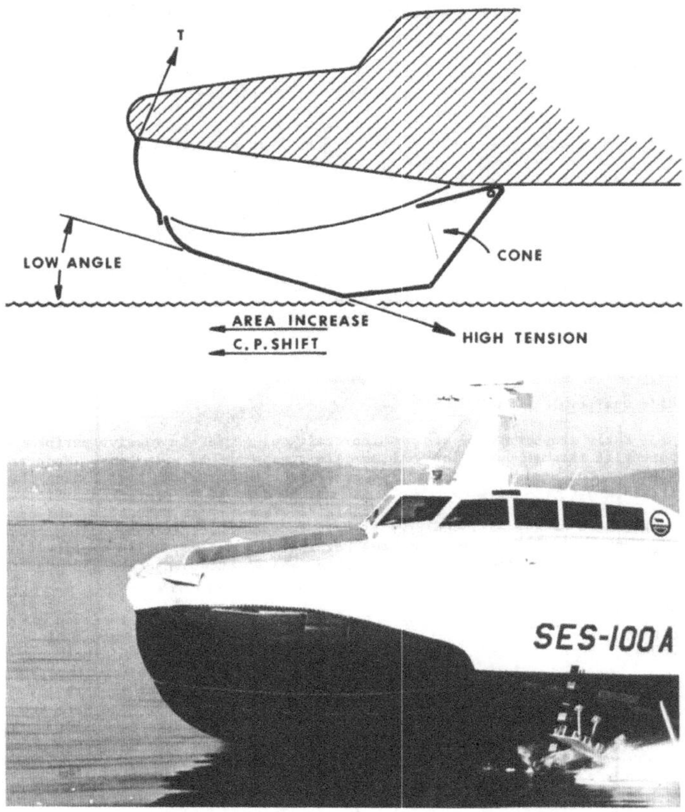

Figure 6.19: SES-100A Retrofitted with Bag & Finger Skirt (1973)

In late 1974 and into 1975 several other major changes were made to the SES-100A. These changes included (1) changing the entire structure forward of the deckhouse (removing the pointed bow) and (2) to now incorporate a wider sidehull similar to the SES-100B with lower deadrise angles. The third change (3) involved the propulsion system. The fully submerged waterjet was replaced with a flush inlet waterjet (see later) and (4) a return was made to the CAB rigid planer seal. The new configuration, that was then tested by the US Navy personnel at the Surface Effect Ship Test Facility (SESTF) in Patuxent River, starting in March 1975, incorporated all these features. The revised configuration of the SES-100A with the CAB planer seal is shown in Figure 6.20.

It is difficult to see the new CAB planer seal in the photograph of the SES-100A but it incorporated the fifteen (15) separate planing segments of the CAB concept, each controlled by cables and down stops. Included in Figure 6.20, is a photo of one of the proposed designs for the 3KSES, that would have the SES-100A design CAB planer seal as shown. The CAB planer seal with its many separate *planer segments* (approx 30 separate elements) can be seen more clearly in the photo of the 3KSES model in Figure 6.20.

SES-100A with CAB Planer Seal Modification

US Navy photo

3KSES Model with CAB Planer Seal

Photo by author

Figure 6.20: CAB Planer Seal on SES-100A and 3KSES Model

The planer seal as tested on the SES-100A (with many variants tried on the XR-1 series of test craft) went through many design changes from rod-stiffened planer elements to unstiffened membrane seals, all controlled by hydraulic reel mechanisms and downstops, and finally a bag and membrane solution. Operational experience of the planer seals on the SES-100A revealed significant structural problems with the planing elements interacting with each other with self-induced damage. Additions of structural elements to minimize interaction only added more weight and higher forces that exacerbated the problem. Because of this experience, there is minimal data on the life of such designs in any practical manner. As the SES program came to a close, changes were being made to the pitch stiffness of the planer seals. But a satisfactory solution for a 3KSES using the CAB planer seal design had not been found.

6.4.2 Pitch Stiffness of CAB Planer Seals on SES-100A

The *"planer seal"* embodied in the CAB concept went through several design iterations that were tested on both the SES-100A and the supporting test craft of the "XR-1 series" of craft. The basic concept of a seal planing on the water surface to minimize air cushion leakage brings with it the associated design problem of *wave following* and *loads* imparted to the seal and to the main ship structure. Such loads have both structural and ride quality ramifications.

The problem of planer seal design is analogous to the *"sprung mass – unsprung mass"* problem of automobile suspension system design. The issue (for ships) is to maintain contact with the water, yet not impart any undue loads to the ship itself. Figure 6.21 shows diagrammatically the main features embodied in the CAB SES-100A bow seal design. In the upper sketch (taken from Warner's patent) shows a planer seal with its restraining "hard" spring. The planing element then generates a lift and drag component as shown. If the drag force is high and the lift force is low, then it is possible to produce a destabilizing pitch moment to the craft. Warner in his notes to the US Navy (in 1930) commented on such destabilizing effects of his design that the US Navy needed to evaluate.

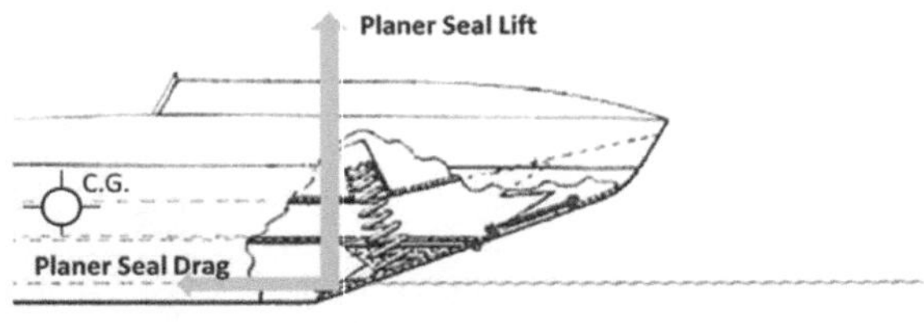

Warner CAB Rigid Planer Seal (Hard Spring) (1929)

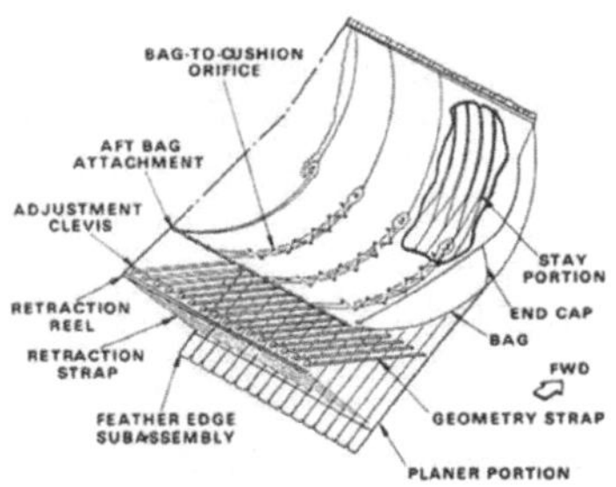

Modified *SES-100A* (Flexible Bag & Planer Seal) (1978)

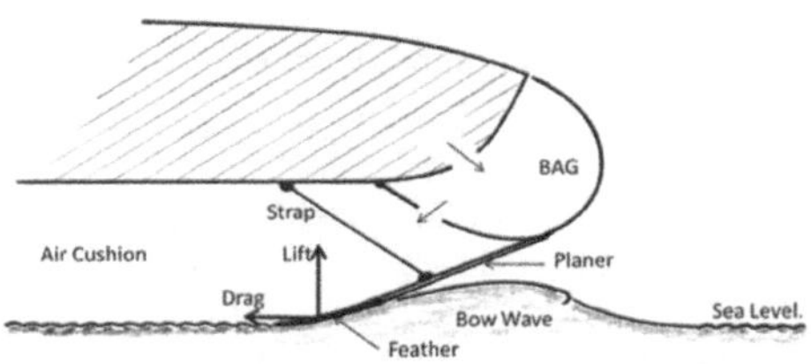

Semi-Soft Spring Planer Seal

Figure 6.21: Hard and Soft Spring CAB Planer Seals

In the lower two sketches of Figure 6.21 are the features of the retrofitted SES-100A "planer seal" design that could be classed as a "semi-soft" spring design. The "stiffness" of the spring in the original Warner CAB design was provided by the mechanical spring shown. The "stiffness" in the SES-100A CAB design was supplied (in later modifications) by the bag pressure (and in some earlier modified versions by a "mid-seal connecting hinge"). The lower left hand drawing shows the last version of the bow seal as used on the SES-100A in its final US Navy tests on Patuxent River, Maryland in 1978. The lower right hand sketch is a simple diagrammatic representation of the bow seal design concept of a "soft spring" bag and "feather" planer seal design. The actual design (on the modified SES-100A) was a mechanically complex arrangement of inflated bow bag and multiple planing elements (15) bolted to each other laterally for a continuous planing surface; plus positioning geometry straps, retraction reels and cables, and down stops. The "planer portion" identified in Figure 6.21 is made up of a series of rigid planing elements (planers) which are positioned with restraint straps. Intraplaner joints constructed of multiple laminations of seal material are bolted to adjacent planer edges to form the bow seal.

Attached to the ends of each planer element is the "feather edge subassembly" as shown in Figure 6.21. Each planer element ends with a "feather" which is actually a thin fiberglass shingle to provide a flexible, light weight planing element at the water interface. A series of straps, stays and restraint cables shown completes the design. The intent of the "feathers" was to maintain contact with the water (and "capture the air bubble" to eliminate or minimize air leakage).

As discussed later, the high accelerations incurred when moving at high speed over water frequently negated this feature and the "feathers" suffered delamination and poor contact. During tests, the (normal) irregular wave action also caused mechanical problems with the intra-planer bolted connections, and tangling with the various stays and restraint cables.

These feather attachments (made from fiberglass) were also added to the bottom of the SES-100A stern seal lobes. The addition of the "feathers" to the SES-100A planer seal was an attempt to introduce a light weight component that would not destroy itself through flagellation (see later). In the final days of the CAB SES-100A, a new design of the seal was being explored to increase the pitch stiffness of the bow seal. The CAB SES-100A designers were struggling, even as the program was being canceled, with how to determine the correct amount of "hard" and "soft" spring features and still maintain acceptable lift and drag characteristics of the seal and provide adequate ride quality. The final tested configuration of the SES-100A incorporated many changes to the seal design (and other features discussed earlier) and it was reported by the SES program Office (PM 304)[13] that a new top speed of 65 knots had been obtained. After the program had been terminated, the final top speed of the CAB SES-100A was officially quoted as "*exceeded 75 knots*"[14] but did not meet the contractually required 80 knots. The difference in the two approaches of seal design for the SES-100A and the SES-100B also changed the nature of flagellation that had a significant impact on maintenance between the two designs. This is covered later under the subject of flagellation. The many variations of the rigid planer seal designs seeking a workable solution in the SES-100A program can be found in the summary report by Ed Butler (SES Program Office) in 1985[15] but none received sufficient test time to determine a final satisfactory solution.

6.4.3 Sidehull SES-100B Flexible Seal Design

The SES-100B, was designed very deliberately, to capitalize on the well-developed design and operational experience in the UK (and elsewhere) of existing air cushion craft (hovercraft), to incorporate the well tested flexible "bag and finger" system from this experience. The flexible "bag and finger" design had received extensive development in the UK over a wide range of craft including such *amphibious air cushion craft* as the 200 ton, 70 knot SRN.4 (see Figure 6.22) and the *sidewall hovercraft* 20 ton, 38 knot HM-2 (see Figure 6.23).

The SRN.4 experience provided data for large tonnage (up to 300 ton displacement in its stretched version, "*the Super 4*") and high speed (up to 70 knots); and the HM-2 experience provided data for application of such flexible seal systems to *sidehull SES* as planned for the sidehull SES-100B design.

The descriptions of these two craft may be found in various issues of Jane's[16] and will not be elaborated upon here. The British Hovercraft Corporation SRN.4 series

Figure 6.22: British Hovercraft Corporation SRN-4 *("The Princess Anne")* Amphibious Hovercraft (200 ton displacement; 70 knots max speed)
Photo permission by GKNAerospace©

Figure 6.23: Hovermarine "HM-2" Sidewall Hovercraft (20 ton displacement; 38 knots max speed)
Reprinted by permission of SNAME

operated from 1969 until 2000. The Hovermarine *"HM-2"* evolved into many variants that were produced to a total number greater than 100 craft that operated in 31 countries around the world with the first craft launched in 1967. These craft provided invaluable data and information on both design and operational experience in a wide range of sea state conditions. The experience from the SRN series of hovercraft and the HM.2 sidewall hovercraft was used extensively in the design of the SES-100B flexible seal system.

Based on this experience and on an extensive model test program including both tow tank and free model tests, the seal system for the SES-100B is as shown in the sketches in Figure 6.24. The similarities with the SRN.4 and the HM-2 skirt systems are apparent.

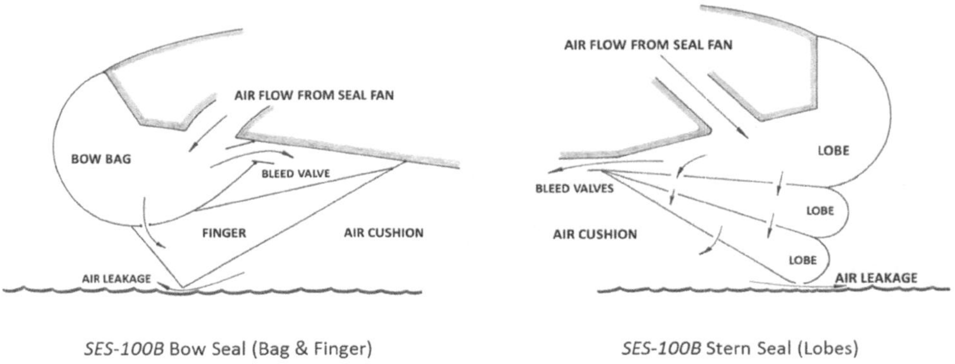

Figure 6.24: SES-100B Flexible Skirt System Diagrammatic

A detailed treatment of the design characteristics of various skirt systems; their geometries, types of materials used and operational experience as applied to both *amphibious* and *sidehull air cushion craft* may be found in the earlier compendiums by this author[17] (1975, 1980).

Because of the high speed requirement of the sidehull SES-100B (80–100 knots) there were concerns about the load alleviation needs from high speed impact of the flexible "bags" with waves. The design requirements based on model test, analysis and operational data from the UK hovercraft experience was to design the bag pressures to "spikes" to be about 3:1 over the operating cushion pressure. Such high speed data from the UK experience with the SRN.4 and HM-2 was somewhat limited however and so an additional conservative safety feature was designed into the SES-100B seal system.

As can be seen from Figure 6.24, this resulted in the incorporation of pressure relief (or bleed) valves in both the hard structure (affectionately called the "colander" during the design phase) and the flexible skirts themselves for the SES-100B.

Figure 6.25 shows these valves (colander) in the hard structure of the bow of the SES-100B that, together with the valves throughout the entire skirt system (see sketches in Figure 6.24) provided the means to relieve loads on the flexible skirt system and to maintain the shape of the inflated seals as the craft traverses the waves at high speed.

One key feature of the SES-100B design is to maintain an even pressure and air flow (with minimal power fluctuations) beneath the "fingers" in the case of the bow seal, and beneath the "lobes", in the case of the stern seal, to avoid unnecessary

Figure 6.25: SES-100B Bow Bag Air Pressure Relief Valves
Photo courtesy of Textron Marine & Land Systems and Textron Inc.

engine loads and to enhance ride quality of the craft. The end result of the design and test process was a bow seal system as shown in Figure 6.26 during an early static test at the Michoud Facility, New Orleans, Louisiana in 1971. A close up of the bow seal system is shown in Figure 6.27.

In any design process involving flexible structures, there is the usual phase of "tailoring" as can be seen from the early static tests in 1971. The interface between the skirt and the hard structure can be seen in Figure 6.25 and Figure 6.26. The extra air leakage in the bow shown in the close up view of the skirt-hull intersection (Figure 6.27) was eliminated after the initial launch by adjustments in the finger attachments. Also, included in Figure 6.27 can be seen how the bow seal meshes with the low deadrise sidehulls discussed earlier on the stability considerations for high speed SES. During the same static tests in 1971 adjustments were made to the stern seal lobes for improved sealing as shown in Figure 6.28.

The advantage of the flexible seal system as used on the sidehull SES-100B (and in all British hovercraft designs) was the deliberate feature of allowing the air to "burble free" from the cushion during extreme wave motion or loading on the seals from wave impact (see Figure 6.22 and Figure 6.23). This leakage of the air flow from the cushion is an essential element in the damping of the heave and pitch motion of an SES, especially as it traverses rough water.

This feature is a significant difference in design philosophy between designing a planer seal to maintain *contact* with the water and (attempt) to maintain an "*airtight-watertight airspace*" as sought in the CAB design of the SES-100A, and an alternative seal design approach that deliberately does *not contact* the water and

Figure 6.26: SES-100B Bow Seal (Bag and Finger) (Static Test 13 August 1971)
Photo courtesy of Textron Marine & Land Systems and Textron Inc.

Figure 6.27: Close Up of SES-100B Bag and Finger and Side Hull interface
Photo courtesy of Textron Marine & Land Systems and Textron Inc.

Figure 6.28: SES-100B Stern Seal (Lobes) (Static Test 13 August 1971)
Photo courtesy of Textron Marine & Land Systems and Textron Inc.

allows the pressurized air to escape as part of the damping mechanism as used in the design of the SES-100B.

The polluted waters in the Intracoastal waterway around New Orleans caused quick discoloration on the side of the craft when sitting for any length of time while hullborne. Such discoloration, however, does allow the observer to see where the waterline of the craft is when both off and on-cushion as seen in Figure 6.26, Figure 6.27 and Figure 6.28. The draft marks are measured from the craft keel and not from the bottom of the rudders and stabilizers that hang beneath the keel.

6.4.4 The Problem of Flagellation

The design of any air cushion craft must contend with the conflicting requirements embodied in the seal design. Both an amphibious air cushion craft and a non-amphibious air cushion craft using some form of sidehulls will incorporate some form of seal that will increase the hard bottom clearance to some required height. The contained air cushion must now serve at least two functions: it first reduces the propulsion power requirements, and secondly serves as a "suspension system" for the ship that contributes greatly to the ride quality as well as to the stability of the ship (discussed earlier).

It is the "suspension system" feature that has plagued most designers of such craft. It is convenient to discuss the problem as an analogue of the automobile suspension system usually treated as a "*sprung-unsprung mass*" problem familiar to automobile designers and transportation engineers concerned with designing roads with acceptable ride characteristics for the general public. Figure 6.29 shows the generic case of an SES traveling over the sea surface with various characteristics of the air cushion and seal system as shown. The stiffness (pitch and heave) is represented by the "spring"; and the air cushion leakage providing the damping for the ship motion is represented by the illustrative "damper" in Figure 6.29.

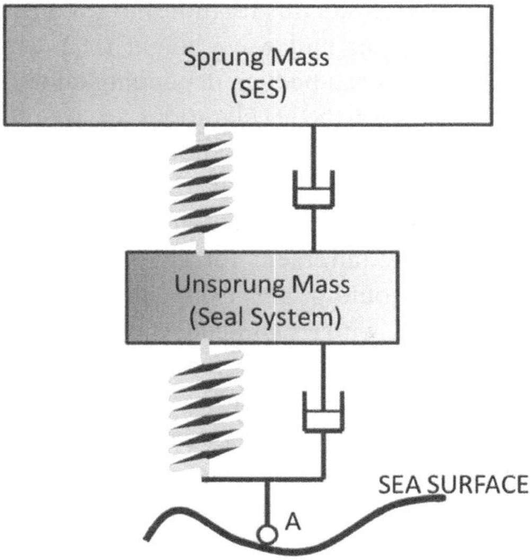

Figure 6.29: Sprung-Unsprung Mass Analogue of SES Seal System

Depending on the particular design of the cushion and the seals, one can show different arrangements of spring and damper but the essential elements are as shown in Figure 6.29. The "Point A" shown in Figure 6.29 deserves discussion as it represents the essential difference between a CAB planer seal that is intended to maintain contact with the water to "plane" to provide some lift force and pitch stiffness to the ship, and also by maintaining contact with the water to ensure the "airtight-watertight airspace" sought by the CAB design to (capture) minimize leakage and lift power; and by a seal system that does NOTrequire continuous contact with the water as designed into the SES-100B. These are the essential differences between the SES-100A various seal designs and the SES-100B seal (skirt) design.

More discussion of the design requirements for such seal "suspension system" features is provided in Chapter 8: Technical Considerations. This section will simply report the results from the SES-100A various planer seal designs and of the SES-100B flexible seal design and how that has manifested itself in the flagellation of the lower edges of the seals as "Point A" moves in relation to the waves during actual test operation. It is necessary first to see what are the accelerations and motions imparted to "Point A" or the lower edges of the seals if indeed contact with the water is maintained.

Both the SES-100A and the SES-100B had to contend with the high accelerations imparted to the seal system as the craft traversed the waves. A simple (first) approach to the understanding of the high accelerations experienced by the seals (or skirt fingers) can be seen from the analysis conducted by Hogben[18] that showed the vertical accelerations experienced by any particle (Point A) following a waveform (assumed to be sinusoidal in the analysis) is given by:

$$n = 2\pi^2 \frac{h_w}{\lambda^2} \left[V + \sqrt{\frac{g\lambda}{2\pi}} \right]^2 \qquad (6.3)$$

Where n = vertical acceleration measured in units of "g's"

h_w = wave height

λ = wave length

V = ship forward speed

Figure 6.30 shows this relationship for a range of ship speeds for the vertical acceleration (n) per unit wave height (h_w). This simple relationship shows that these accelerations can be high depending on the operating conditions. In tests, with non-sinusoidal seas, the accelerations are frequently higher. Based on this simple analysis, a ship operating at 80 knots in Sea State 3 (with a significant wave height of 4.75 feet and a wave length of 62 feet), which was the US Navy requirement, should experience a vertical acceleration of the seal of approximately 15–16g's. While this is a difficult requirement for any seal system (planer or flexible) a more stringent condition would be for a ship speed of 100 knots operating in relatively calm seas (Sea State 1, with a significant wave height of just 1 foot and a short wavelength of 15 feet). This is sometimes referred to as the "cobblestone effect". In that case, as can be seen from Figure 6.30, the vertical acceleration would rise to approximately 100 g's. For the SES-100A, this posed a problem of keeping the rigid planer seal with a weighted planer segment to maintain continuous contact with the water to perfect the "captured air bubble" and to ensure that the mass of the planer segments and cables functioned correctly.

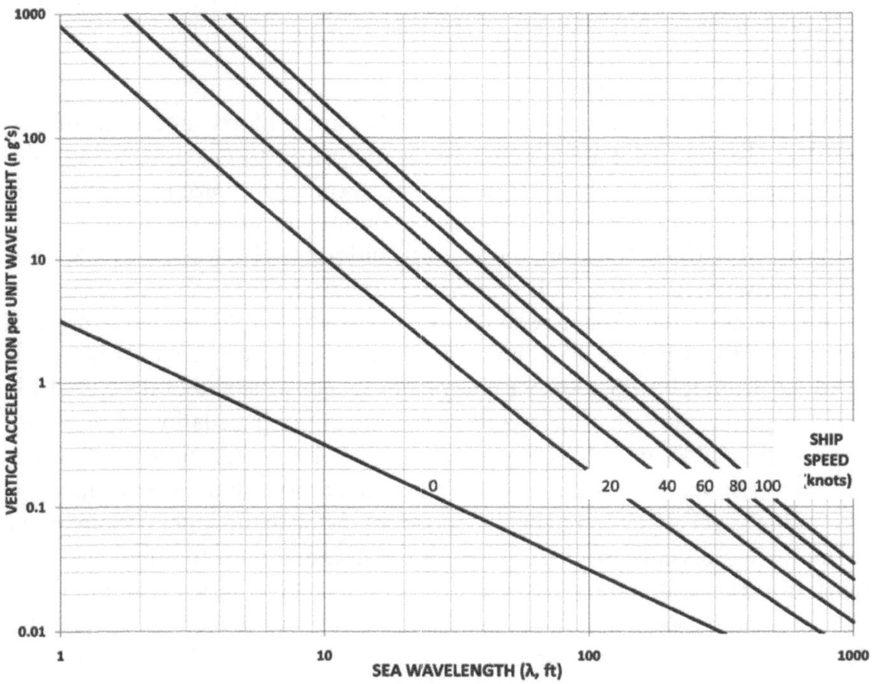

Figure 6.30: Wave Following Vertical Acceleration

Unfortunately, the actual experience with both planer seals and with flexible fingers, the reality was that this simple analysis under-predicted the acceleration levels by at least an order of magnitude indicating that other forces were at work. Small accelerometers were installed on the SES-100B (and on models) and acceleration levels greater than 1000g's were experienced (in some tests accelerations as high as 4000g's were measured) and that "Point A" was definitely NOT maintaining contact with the waves. This indicated other motions like "snap back" responses by the seal material were at work that indicated a complex interaction between cushion pressure, material stiffness of the seal edge (planer or finger) that required more detailed treatments in analyses and model tests. Measurements taken on the SES-100A planer seal showed equally high accelerations (1000–4000 g's). The development of the SES-100A planer seal went through many configurations in an attempt to alleviate this problem but no long term satisfactory solution was found. The "feathers" made of GRP laminates on the lower edges of the planer seals of the SES-100A developed significant de-lamination (bond failure) from flagellation. Various combinations of weave and resin content were tried but none solved the problem. During static tests of the "feathers" the failure mode was due to water seeping (wicking) into the material through the high acceleration motion producing a similar failure through flagellation.

For the SES-100B, that would also have to deal with the same high acceleration at high frequencies, the problem is easier because of the lighter weight of the flexible fabric. In the case of the SES-100B, the high acceleration resulted in the light weight fabric snapping back and forth as it responded to the alternate action of the wave impact and the restoring cushion pressure. This became known as "flagellation" from the UK hovercraft extensive operational experience and contributed to the shortening of the life of the skirt (particularly finger) material. Extensive testing in both the UK and the US was done in the 1960s as designers sought improved wear features of the fingers that operate in the high acceleration environment of high speed hovercraft and SES over waves. Figure 6.31 shows an example of flagellation wear on a test sample of light weight flexible finger material.

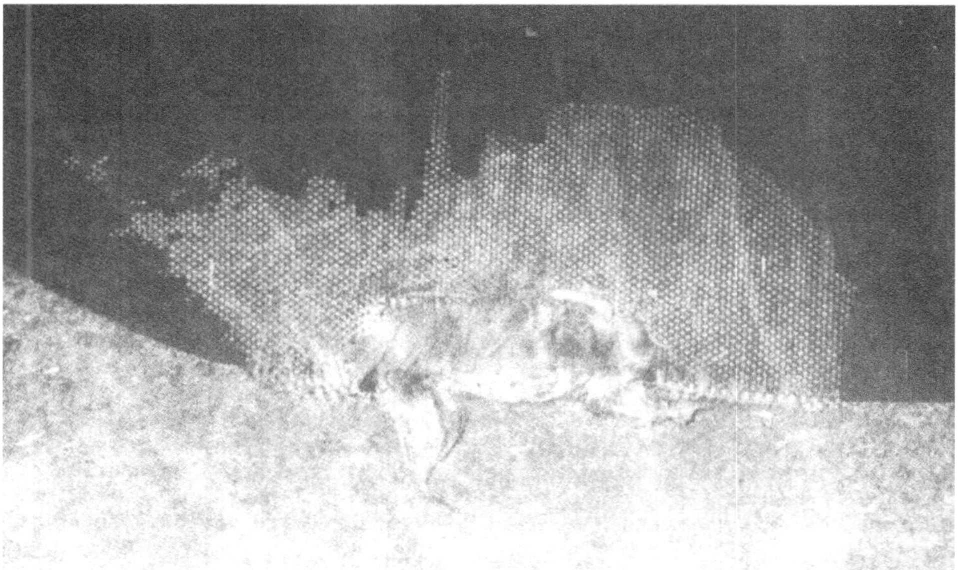

Figure 6.31: Example of Flagellation Wear on Finger Test Sample

Significant research and development on both sides of the Atlantic was conducted in the 1960–1980 period seeking better, longer wear material. Figure 6.32 shows some of this early work conducted in the US material manufacture companies

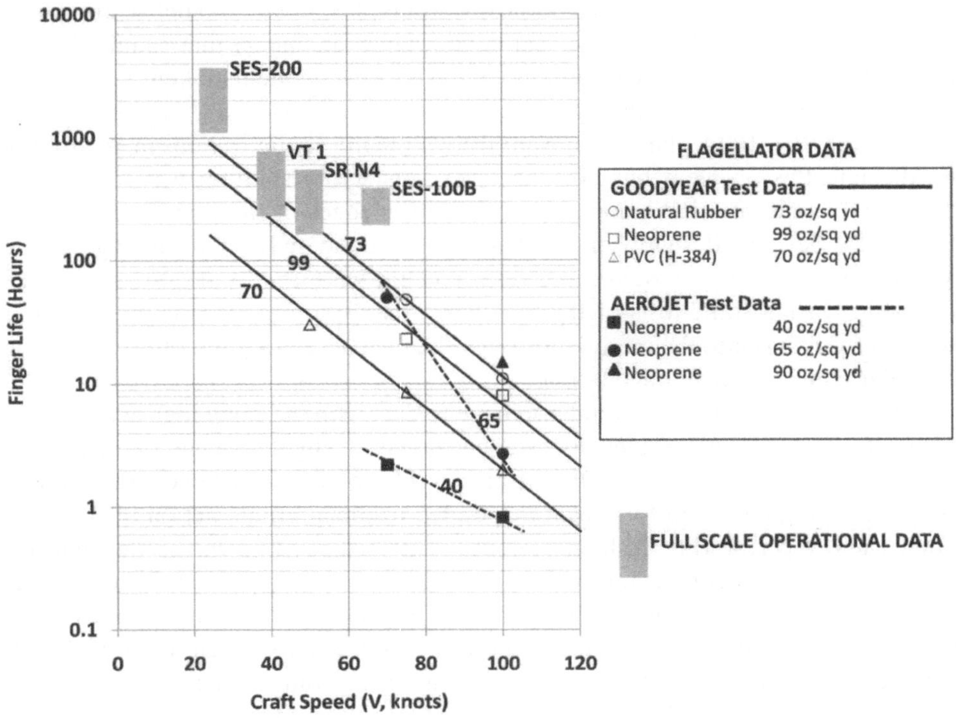

Figure 6.32: Experience on Flexible Skirt (Finger) Life (1968–1983)

compiled by Mantle[19]. As can be seen from Figure 6.32, the life of the flexible fingers degraded significantly with operational speed. Both the test examples in the lab and the operational wear results of actual hovercraft in the UK and air cushion craft in the US show a similar marked degradation of finger life with operational speed of the ship. The life of the SES-200 flexible fingers, obtained in 1982–1983, has been added to Figure 6.32 to the original data compiled in 1975 and shows a similar pattern. The SES-200 data reported a life of 1500–3000 hours but this "improvement" over the life obtained on earlier hovercraft such as the SRN.4 with a reported 400+ hours at an operational speed of 50–70 knots is more due to the slower operational speed of the craft SES-200 (25–30 knots) rather than any major improvement of material.

There is no reliable data on the life of the CAB SES-100A planer seals or "feathers" as the operational time above 70 knots can be measured only in minutes. On the other hand, the SES-100B did experience significant operational life (measured in hours, see later trials results) at speeds up to and including the maximum speed recorded of 91.9 knots. The material engineers projected from the test data that it would be reasonable to document the finger life of the SES-100B in a normal operational profile to be about the same magnitude as the SRN.4 and this data has been added to Figure 6.32 also.

The life of the flexible fingers taken from Figure 6.32 indicates that the life degrades logarithmically, losing about an order of magnitude of operational hours for every 40 knot increase in ship speed.

Above 40 knots, the finger life of such designs can be represented by the approximate formula,

$$\text{Flexible Finger Life (hours)} \approx \frac{3x10^9}{V^4} \tag{6.4}$$

where the ship speed (V) is measured in knots. The decreasing finger life being inversely proportional to the *ship operational speed to the fourth power* is an indication of how severe flagellation is on seal life.

As provided in Chapter 10, where the results of the ANVCE Project are reported, the life of the fingers of the UK hovercraft had increased significantly with improvements in design and materials. The *SRN.4* operating regularly at 70 knots with a reported a life of 400 hours was a significant improvement over earlier slower speed hovercraft and within the realm of a normal maintenance program for a high speed SES. Figure 6.33 shows how the material development programs were making improvements based on operational experience in a variety of operational scenarios. The slow-down of the development of high speed SES after the 3KSES program cancelation in 1980 has left a question mark over where material suppliers might have taken the characteristics of flexible seal material and finger life.

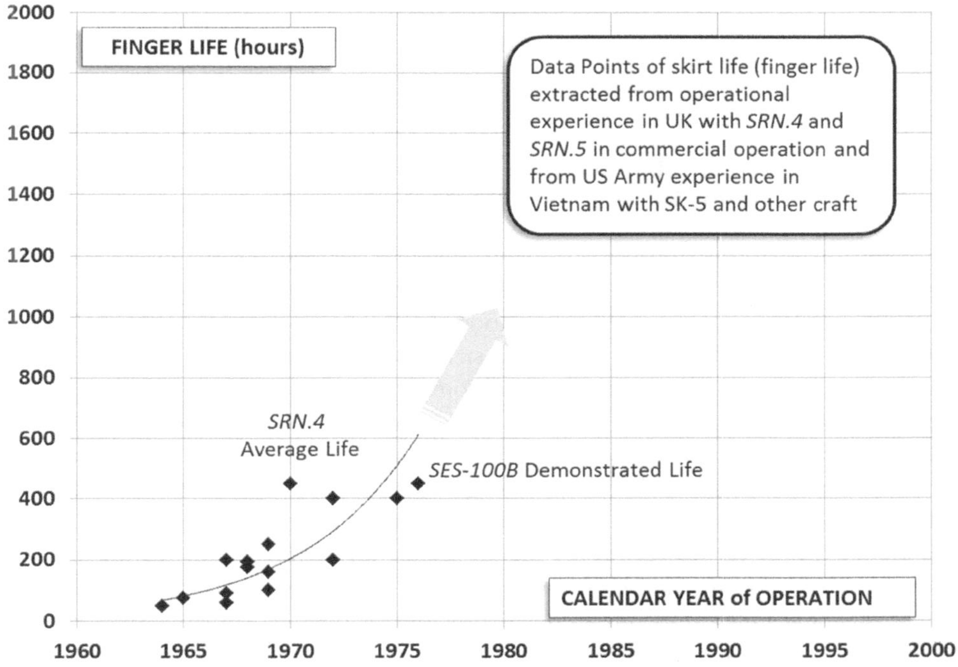

Figure 6.33: Potential For Improved Finger Life

But even with the operational life achieved on the SES-100B and other craft it can be said that a reasonable first step had been achieved. Very roughly, without factoring in operational profiles, this would indicate that an SES operating at an average speed of 50 knots could cross the Atlantic Ocean about 4 or 5 times before replacing the fingers because of wear. Because of this experience, a conscious decision was made in the design approach for the SES-100B and for larger scale

SES to follow a maintenance program for finger wear while the research and development for rigid planer seals continued. Figure 6.34 shows the SES-100B operating successfully with such flexible bag and finger arrangements where the fingers are allowed to oscillate and let the air leak out of the cushion with minimum power loss. It was argued that it was more cost-effective to design into the craft a replacement flexible finger maintenance program rather than to attempt to design a rigid planer element (unsuccessful to date) that did not wear out.

Figure 6.34: SES-100B with Bag and Finger Bow Seal
Photo courtesy of Textron Marine & Land Systems and Textron Inc.

There is a serious unsolved problem, however, in that the scaling of the data from the SES-100B data (and from the other operational hovercraft in the UK) to much larger SES is not very well understood. The material characteristics of the seal material do not scale easily to the larger sizes and much more research and development would need to be done before comfortable predictions could be made. Chapter 8 on Technical Considerations summarize the main concerns and the results achieved to date on scaling material on both strength and flexibility considerations.

In summary, for the different technologies used for the seal systems on the SES-100A and the SES-100B, it was found that a successful design for a *rigid planer seal* as proposed in the CAB design was not accomplished during the testing over the decade of 1969–1979 before the program was canceled in early 1980. On the other hand, the *flexible bag and finger skirt* system developed and used by all UK hovercraft proved successful when applied to high speed SES up to speeds in excess of 90 knots as proven by the SES-100B. If the skirt material manufacturers had continued their development it is suspected that even further improvements would have been accomplished much like the experience gained in the tire manufacturer's world from the solid rubber tires of the early days of low speed automobiles to the highly sophisticated tire material used today in the extreme high speed race cars. See later and Figure 6.40 how the final seal design configuration appeared in the 3KSES design.

After the 3KSES program was canceled in 1980, the SES Program Office (PMS 304) turned its attention to the lower speed 50 knot class SES (see Chapter 5) where their designs now incorporated the flexible *bag and finger* seal design developed successfully on the SES-100B and based on the earlier air cushion craft developments in both the UK and the US. The CAB rigid planer seal concept had been discarded. The issue of scaling of seal systems to larger size SES was still unresolved as the program came to a close in 1980 and a discussion of the subject of scaling is provided in Chapter 8 on Technical Considerations.

6.5 Structural Design Approach

The third main difference in the technologies pursued by the SES-100A and the SES-100B was the basic structure itself. Early in the Preliminary Design phase, the choice to use aluminum was surprisingly resisted by the aluminum manufacturers and suppliers. The concern was that at 100 knots over water, the cavitation bubbles generated on the bottom keels of the sidehulls would seriously erode the material. While designing the SES-100B, the author received a letter from the aluminum supplier at the time absolving themselves from any liability from the use of such materials for the purpose intended because of the suspected damaging effects of cavitation!!

Some comments on the concern of the choice of material for the SES test craft are summarized here from the early work during the design and test phases of the SES-100B[20].

> *"Because of the design high speed of the SES (80 knots), concern was expressed in 1969 as to the ability of aluminum to withstand cavitation erosion. Studies by the aluminum companies and cavitation tests conducted on high speed rotating water tunnels at Chance Vought and at the Naval Applied Sciences Laboratory showed severe degradation at high speeds. Experience on hydrofoils operating for prolonged periods at high speeds had also shown severe pitting in crevices and on surfaces prone to low pressure from the hydrodynamic flow. The aluminum is particularly prone to cavitation damage, due to the high speed flow constantly washing away the natural protective oxide layer of aluminum and thus accelerating the erosion. Consideration was given to the use of titanium "shoes" to the sidehulls because of titanium's particularly good cavitation erosion resistance, but it was finally decided to apply Tarset, the protective paint coating, to the entire underhull and sidehull. This coating was found to be highly successful."*
>
> *"To avoid direct cavitation erosion on the aluminum hull and also to provide the biofouling so prevalent in the southern waters of the Louisiana test area, a Glidden paint system was applied. Because this vinyl based paint system requires four coats of primer, a technique explored on the SES-100B hull was to have each coat a different color such that any cavitation erosion could be detected by color photography. In the test program completed in 1973 (before moving to Panama City Florida for rough water testing), only minor indications of cavitation erosion were found."*

Unrelated to the concern of cavitation damage was an incident that occurred during the manufacturing phase of the SES-100B in 1970 when a technician, who was pressure testing the fuel tanks using a mercury filled manometer, knocked over the manometer causing mercury droplets to scurry across the deck and sink into various crevices. There was some water also present near the spill. The panic that ensued was quite sobering because of the recognition of the extreme corrosive effect of a "mercury, water and aluminum combination".

Quality inspection stress tests conducted within a few hours after the incident confirmed that the strength level of the material had already started to deteriorate. The end result was a cutting out of a 10 foot long mid section of the entire ship before manufacturing could continue! This was a bad day and raises legitimate questions about material choices. The construction of the SES-100B was being accomplished in the NASA Michoud Facility in New Orleans and it was only after the incident was it learned that NASA had a formal restriction against any use of mercury lamps or similar devices in any facility associated with the Space Program because of such dangers!

The two test craft used two different design philosophies in the design of the basic hull structure. They explored different ways to obtain a sufficiently strong structure but sought to minimize as much as possible the structural weight fraction of the craft so as to maximize the payload carrying capability.

The CAB SES-100A used the aluminum common to small boat construction, namely the 5086 alloy which has good weldability and maintains its strength after welding and has a good record for corrosion resistance. It does have a tendency for stress corrosion cracking which is why it is not recommended for aircraft construction. In addition to the use of welded 5086 aluminum for the main structure, the CAB SES-100A used glass reinforced plastic (GRP) for the construction of the deckhouse, the cargo or payload deck and also for portions of the weatherdeck. On launch on 1 September 1971, the SES-100A had a structural weight of 75,746lb for a structural weight fraction of 31% which is a good value for a weight limited craft designed to withstand high loads at high speeds.

The sidehull SES-100B used the higher strength aluminum alloy, namely the 5456 alloy known for its high strength in a marine environment. The SES-100B structure was built, under subcontract from Bell Aerospace, at Levingston Shipbuilding Company near Orange, Texas and delivered (by barge) to the Bell Aerospace Operations at the NASA Michoud Facility in New Orleans, Louisiana on 10 February 1971. Figure 6.35 shows the basic structure of the SES-100B which is shown installed on its construction table in the New Orleans Michoud Facility shortly after being delivered from Levingston Shipyard.

This view is useful in that it shows several key features of the design. The ship is an all welded aluminum structure with simple lines. The shape of the partial sidehull discussed earlier is clearly seen, including the low deadrise shaping that proceeds as high as the wet deck of the ship. The "colander" valves in the bow structure for bow seal load alleviation from wave impact discussed earlier is also seen in the photo.

During the early design phases, it was determined that the motion of an SES in waves tended to be about a rotation point toward the back of the ship. The motions and accelerations were made up of a vertical motion at the CG plus a rotation as it pitched. This meant that the minimum acceleration motion for crew members would be aft and hence the deckhouse is placed toward the rear of the craft. The deckhouse is thus placed high for visibility and toward the rear for ride quality. This is in contradistinction to the SES-100A which has its deckhouse low and toward the front of the ship.

On the day of launch on 22 July 1971, the structural weight of the SES-100B was 69,300lb for a structural weight fraction of 34% of the normal gross weight of the ship, which like the test craft, the SES-100A, was a good value for such a ship. More

Figure 6.35: SES-100B Structure

structural design details of the SES-100A and the SES-100B such as plating design, welding techniques and various structural indices, etc are provided in the earlier cited reference.[21]

A discussion of the design techniques, load factors and other considerations involved in the structural design of the two SES 100 Test Craft and other forms of air cushion craft may be found in Mantle.[22]

6.6 Engines and Their arrangement

The fourth area of difference between the SES-100A and the SES-100B is not a "technology" area of difference *per se* but it did represent a different design philosophy that affected the ease of obtaining data crucial to the design of the full scale ships contemplated for the program. The SES-100A designers chose to use an *integrated* lift and propulsion scheme while the SES-100B was designed to have *separate* lift and propulsion engines.

The engines on the SES-100A required a clutch mechanism to separately control the lift and propulsion systems which proved to be a difficult arrangement in controlling the desired pressure and flow conditions for the (axial fan) lift system as the craft negotiated waves at the same time as adjusting the (waterjet) propulsion system needs.

The lift and propulsion engines of the SES-100B were separate, such that refined control over the pressure and flow requirements for the seals and cushion of the lift system could be operated (and monitored for design purposes) separate from the needs of the propulsion system. One of the key purposes of the SES test craft program was to explore the boundaries of high speed operation in rough water and the ability to adjust the lift system parameters separately from those of the propulsion system elements was significant in achieving this objective.

While not related to the engine system design choices, the design philosophy used in the design of the SES-100B, to make it a useful "test craft" for future SES design, also included the use of thrust meters that were built into the propeller housings (see later on differences between the SES-100A and the SES-100B propulsion systems). These thrust meters on the SES-100B allowed for direct measurement of the thrust of the craft while at speed at the same time as the power measurements were recorded from the propulsion engines. This proved invaluable in the calculation of the drag (= thrust at steady speed) of the SES-100B throughout its entire speed spectrum. The SES-100A did not incorporate such features into its design and thus had to rely on the standard bollard tests to "calibrate" power settings with thrust which is only true at zero speed and so thrust (and thus drag) is only an estimate at speed.

The testing and analysis experiences from the test program of the two design philosophies used for the SES-100A and the SES-100B were significant enough to influence the SES Program Office to direct that the design of the 3KSES would follow the design philosophy of the SES-100B design of separate lift and propulsion engine systems.

6.7 Lift System and Ride Control

The fifth area of difference in key technologies between the SES-100A and the SES-100B design was in the choice of lift system. While there are several different types of fan systems that have been used in air cushion craft designs over the years, the most common types used are variants of the axial fan and the centrifugal fan shown diagrammatically in Figure 6.36.

There are many descriptions and treatments that the reader can read on the many different types of fan systems in use for all forms of air cushion craft together with their characteristics and they have been summarized elsewhere[23] in detail. Generally, one can say that the axial fan offers the greater efficiency but at a price of having sensitive design points in its pressure-flow curves and also being less rugged. The axial flow fan is more vulnerable to foreign object damage (FOD) for example than the centrifugal fan. The centrifugal fan offers a more stable pressure-flow curve and a more rugged design that lends itself well to rough treatment and lower maintenance.

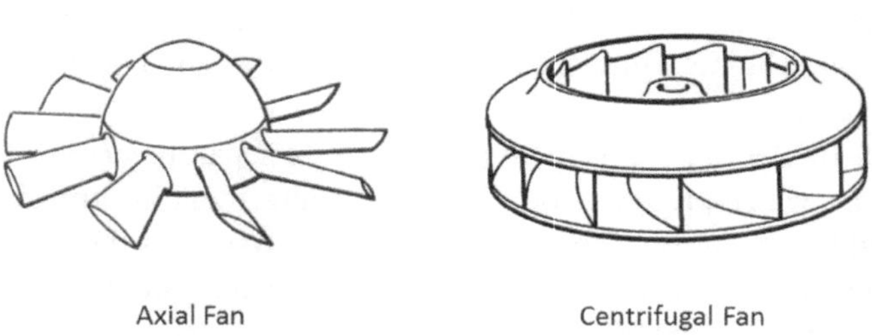

Figure 6.36: Most Common Types of Fans Used

The SES-100A design incorporated

the axial flow fan system and the SES-100B design incorporated the centrifugal fan system that has seen much use throughout the amphibious and sidewall hovercraft in the UK, in the US on the US Navy's LCAC, and throughout the world's hovercraft experience.

Because the SES-100A design was to investigate the features of a unique ride control system where the pitch of the fan blades and of the inlet guide vanes would be varied to accommodate the pressure and flow fluctuations as the craft traversed the waves, it was easier to accommodate these features in an axial fan design. The variations in the pitch of fan and inlet guide vanes were done initially by a transducer that sensed acceleration and then in a later design sensed cushion pressure changes. A similar sensing feature was incorporated later (in 1974) in the SES-100B where the venting vanes in the sidehulls were activated by similar sensors for ride control but with the centrifugal fan system.

The incorporation of such ride control features was found to be quite dramatic rendering an otherwise uncomfortable and unacceptable ride over the waves into a smooth and acceptable ride. Figure 6.37 taken from Mantle (1975)[24] shows the results of the tests on the SES-100A taken during trials off Port Townsend, near Seattle, in the winter of 1973.

The upper diagram shows the results plotted against the parameter (h_w/h_c) which is the ratio of the wave height to the depth of the cushion which was found to be a convenient parameter in the equations for ride control.[25]

The lower diagram shows the result of the tests as a function of ship heading. Similar results to those shown in Figure 6.37 were obtained in the testing of the SES-100A at the SES Test Facility (SESTF) in Patuxent River, Maryland in 1975. As seen from Figure 6.37, reductions of 50% or more in the vertical acceleration of the ship were readily obtained with this system.

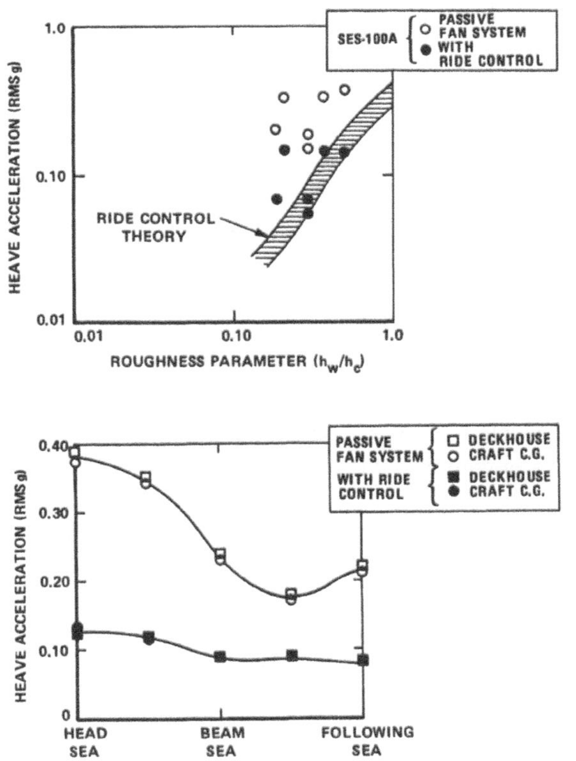

Figure 6.37: SES-100A Ride Control System

During the early Preliminary Design period for the SES program, considerable research was embarked upon by both the SES Program Office and industry into the ride quality criteria to be used in the final designs. There exists voluminous literature on the subject of human tolerance to craft motion and that data and information was integrated into the SES Program as it got underway. Two references

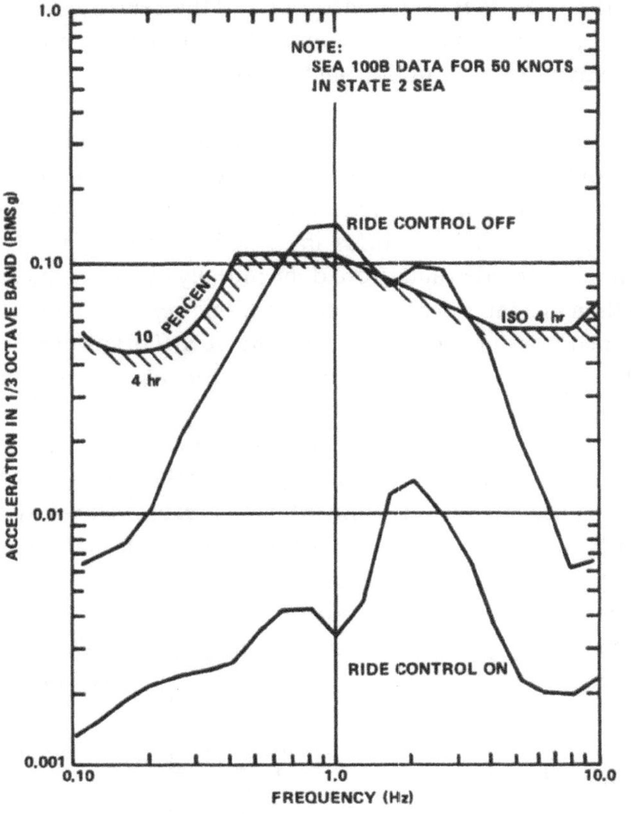

Figure 6.38: SES-100B Broadband Response and Narrow Band Criteria

capture the essence of the criteria development and they are criteria developed by the International Standard Organization (ISO)[26] in 1972 and the work by O'Hanlon and McCauley[27] of Human factors Research in 1973. As can be seen by the dates on these references, the criteria being developed were being done in parallel to the design development of the two SES 100 Test Craft. A complete description of all the criteria developed together with experimental data and theory may be found in Mantle (1980).[28] Figure 6.38 taken from that reference shows the data obtained for the SES-100B with its ride quality system both inactive and active. The particular data shown is for the SES-100B traveling at 50 knots in Sea State 2, which indicates an acceptable ride quality with the simple venting system used in the SES-100B design. The subject of ride quality is not a well defined subject because of many factors that affect how humans react to motion. Much of the data has been collected in controlled experiments on land and the realities of motion sickness over water complete with its psychological implications makes hard and fast rules difficult to establish but a sufficient data base had been established with the SES 100 Test Program to allow more complete experiments and analysis to be performed.

The reader should acquaint himself with the cited references for a complete understanding of each of the regions of motion sickness criteria developed but essentially the data shows that the Test Craft SES-100B exhibited acceptable ride quality over the entire operating spectrum from the low frequencies encounters (0.1–0.6 Hz) where motion sickness is prevalent and the higher frequencies (0.6–4 Hz) where fatigue and work efficiency become factors. Without the ride control system activated, it is seen from Figure 6.38 that at 50 knots in Sea State 2 for a period of 4 hours, crew members would experience some level of fatigue. With the ride control system activated the motion limits were well within the developed criteria by the ISO and others.

At the lower frequencies, it is seen that the motions of the SES-100B at the cited operational speeds and conditions did not induce any form of motion sickness. This was borne out during the test program of the craft in the New Orleans tests, Panama City tests and those by the US Navy in Patuxent River.

The analysis of the motion of an SES and how it is affected by the design of the lift system is covered in Chapter 8 on Technical Considerations. A key factor in the motion of the craft is the ability of the lift system to provide a stable pressure-flow forcing function in the craft dynamics. The centrifugal fan has an inherent stable pressure-flow relationship which was a governing criterion in the selection of such a system for the SES-100B lift system design.

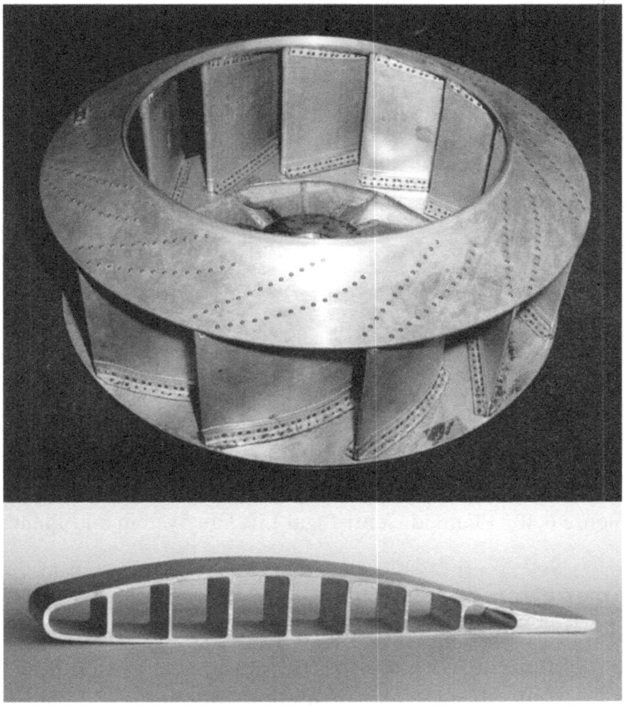

Figure 6.39: SES-100B Centrifugal Fan and Blade Section
Photos by author

Figure 6.39 shows the SES-100B centrifugal fan together with a cross section of the blade. The basic design was taken from the well developed series of fans used in the British hovercraft. A refinement in the design of the blades for the SES-100B fans can be seen in Figure 6.39 where the upper surface trailing edge includes a reflex shape which was found during the developmental period of the design to provide a more aerodynamically stable pressure-flow characteristic around the zero flow condition. This was found beneficial during any possible back flow conditions during any extreme wave pumping action in rough sea operation.

In summary, the differences between the SES-100A lift system and the SES-100B lift system were quite marked. The SES-100A used an axial flow fan system with the pitch of the blades and guide vanes (see Figure 6.2) controlled by fan RPM automatically as the craft air cushion pressure varies as the craft moves through waves. The SES-100B used a simple venting system (see Figure 1.11, Chapter 1) that allowed the cushion to vent as the craft moved through the waves. The ride quality was found to be acceptable for both craft using the available (developed) criteria. The mechanical complexity of the SES-100A axial flow fan system led the SES Program Office (PMS 304) to use a variant of the SES-100B centrifugal fan system in the design of the three thousand to SES (3KSES).

The final version of the use of centrifugal fans into the 3KSES design together with the final version of the planer seal design can be seen in the RMI drawing for the

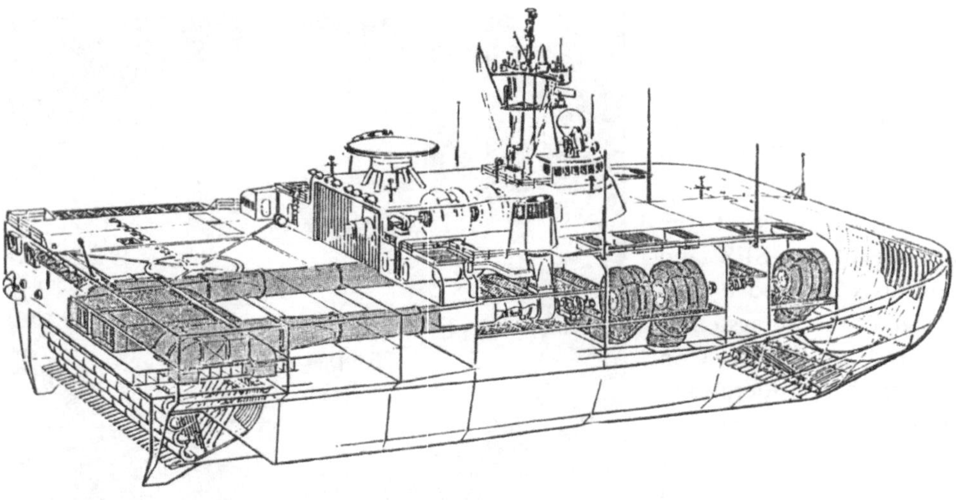

Figure 6.40: Planned Centrifugal Lift Fan System and Planer Seals for 3KSES

3KSES (Figure 6.40). The centrifugal fan system seen in Figure 6.40 incorporated other geometry features to improve the pressure-flow characteristics.[29] As the development work continued on the seal design it is seen that incorporation of the SES-100B "bow bag" for the bow and "lobes" for the stern seal was also integrated into the 3KSES design for a combination of the *flexible bag and finger seal* and the *CAB rigid planer seal*. More details on the 3KSES final design are provided at the end of this Chapter.

6.8 Propulsion System Differences

The sixth and final key technology difference listed in Table 6.1 between the SES-100A and the SES-100B was in the choice of propulsion system. This was a significant difference that needed to be explored for the choice of high speed marine propulsion for such craft as the (planned) large tonnage surface effect ships required to operate at high speed (80–100 knots) in rough seas.

When considering such designs, the designer has two opportunities open to him. They are embodied in the choice of how to produce the thrust (T) and how to lower the drag (D). These two forces are linked in ways that are not obvious as the analysis proceeds. The choice was between *waterjet* propulsion and *marine propeller* propulsion. Other texts provide the background on likely efficiencies and configurations. In both cases, the highest propulsive efficiency occurs when the propulsor is fully submerged but that means that any extension beneath the ship incurs some appendage of some kind that incurs high drag at speed. The trick is to obtain the highest thrust while maintaining low drag.

An additional factor in the propulsion system design decision is the impact of such appendages on the stability and control of the ship. If the same appendages can be used for both propulsion and the stability and control needs, one can foresee an acceptable design. Such variations appeared in the design of the two 100 ton Test Craft as will be discussed.

6.8.1 SES-100A Waterjet Propulsion System

The impact of the *submerged pod* waterjet design of the SES-100A in its original configuration can be seen in Figure 6.41 where the upper sketch is a close up of the strut and inlet taken from Figure 6.2. The contribution of the inlet strut and appendages is a large part of the total ship resistance. At 65 knots (which is at a Froude Number of about 3.0) the drag of the propulsion system is about 40% of the total ship resistance for the SES-100A. The data points shown are for the measured ship resistance of the SES-100A under full scale trials (both accelerating and in steady cruise). The component breakdown is from the original design calculations. Such configurations are an inhibitor to achieving high speeds without expending an exorbitant amount of propulsion power.

In addition to the concerns on normal hydrodynamic drag magnitude shown in Figure 6.41, there is the problem of cavitation that is particularly troublesome at the high speeds contemplated for SES. Based on predictions and full scale measurement of the pod inlet shown it is found that the operational boundaries of the waterjet inlet become more restrictive as the ship speeds are increased. Cavitation forms on both the outer edge and the inner contours of the pod inlet. The ship must be operated within very narrow limits to ensure cavitation free flow as shown in Figure 6.42. Even

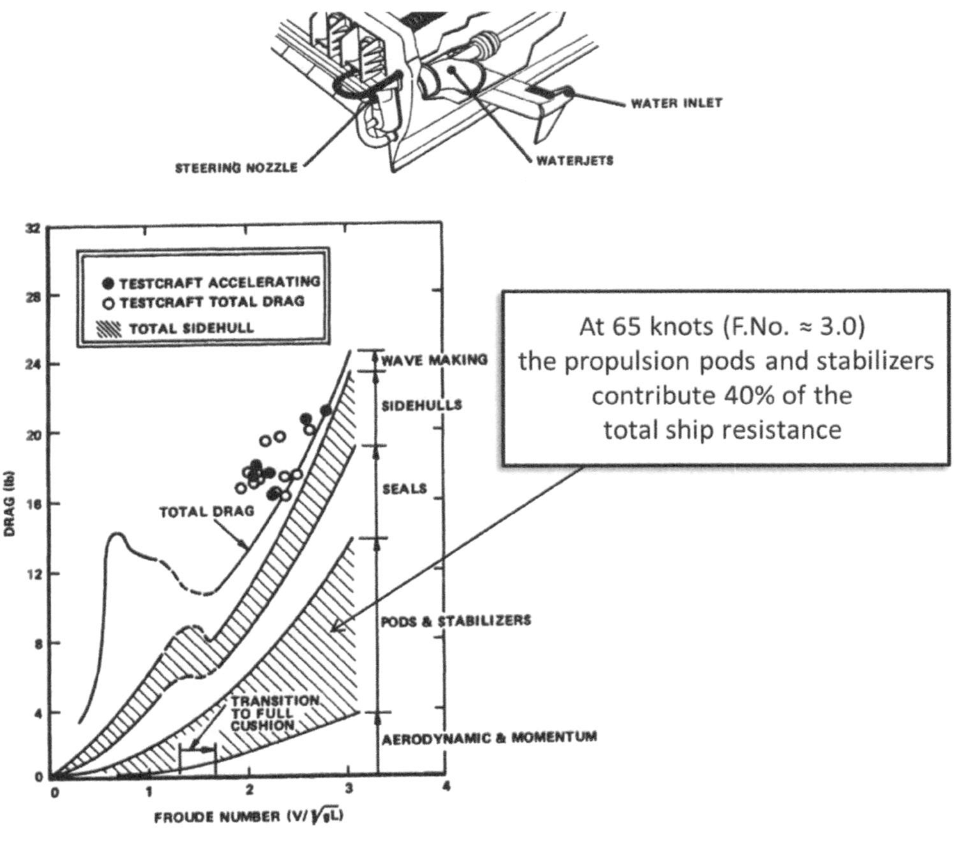

Figure 6.41: CAB SES-100A Drag

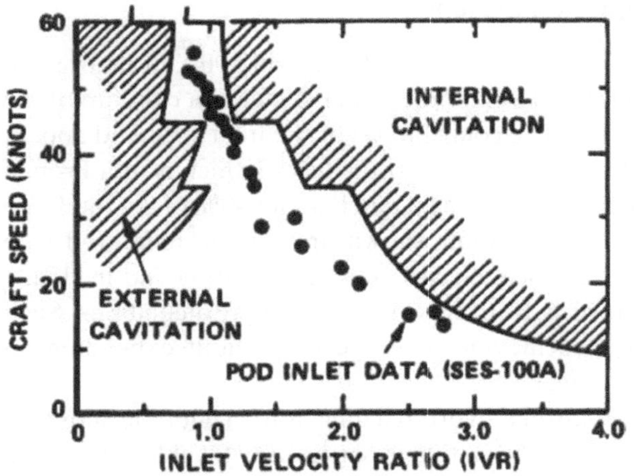

Figure 6.42: Cavitation Boundaries on SES-100A Pod Inlet

at speeds of 50–60 knots the probability of avoiding cavitation either on or in the inlet is not high.

The governing factor is the Inlet Velocity Ratio (IVR), which is given by:

$$IVR = \frac{V_i}{V} \qquad (6.5)$$

where V_i is the stream velocity entering the inlet and V is the ship speed. During the test program a significant amount of data was obtained pertaining to the SES-100A waterjet design and test results.[30] The propulsive efficiency of the SES-100A waterjet system in this configuration was about 45%. The concern about the various features of such a design led the SES-100A designers to change to a flush waterjet inlet and most of the trials conducted by the US Navy crews at Patuxent River were for the craft in that configuration. While changing to a flush inlet certainly eliminated the pod drag, it did mean that an extra fin now had to be added to the craft to ensure adequate directional stability. This added more drag to the craft. Also, because the flush inlet was even with the keel of the ship there was a tendency for air ingestion from the adjacent pressurized air cushion. The problem of cavitation was now exacerbated by the problem of air ingestion which caused an additional "keel" or skeg to be added to the craft to block the air ingestion, again at the cost of drag. Very little documentation of these changes to the craft is available but one can surmise that the lowered maximum speed to 76 knots of the waterjet propelled CAB SES-100A can be attributed in part to these changes. The issue on drag and stability brought on by the choice of seal system has already been discussed.

6.8.2 SES-100B Propulsion System

Chapter 2 discusses some of the design philosophy used in the design of the SES-100B and considerable research and development was undertaken in the early design days on the design of various combinations of *high speed marine propeller* designs for the SES-100B. The development of marine propellers for high speed craft proceeded almost hand-in-hand in development of semi-submerged propellers. It was long recognized that the hydrodynamic resistance of propeller pods was excessive as craft speeds increased (as similarly discussed above on the SES-100A waterjet pod inlet). Accordingly, hydroplanes proceeded in their development (see Chapter 2) with shallow draft and surface piercing propellers operating in the fixed waterplane emanating from the transom of the craft. Much of this work has already been documented[31] and effort here will be concentrated on the specific design used for the SES-100B. There have been several variants of supercavitating propeller designs pursued including the work by Marshall Tulin (1955)[32], Tachmindji and Morgan (1958)[33], Newton-Rader (1960)[34] and others. A more recent survey of such propellers may be found in Faltinsen (2005)[35]. After considerable research and model testing, the author selected the propeller design developed at Hydronautics[36] by Vern Johnson and Marshall Tulin as a result of their original work in hydrofoil development. Figure 6.43 shows the basic profile of the supercavitating propeller together with a photograph of the actual propeller blade of the SES-100B manufactured by Hamilton Standard.

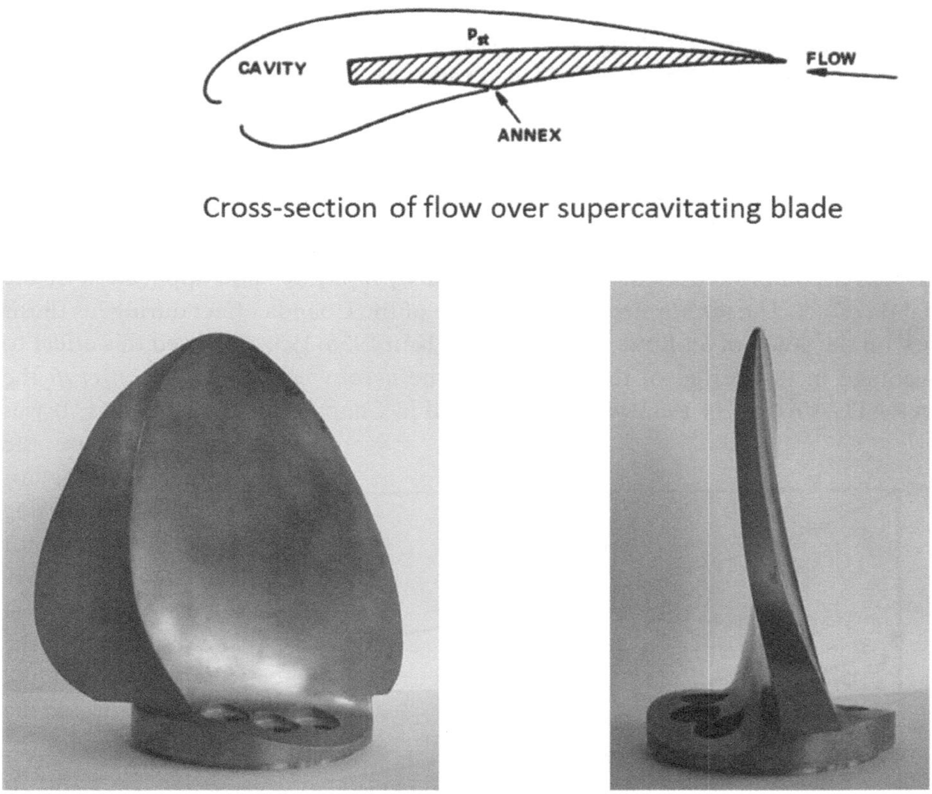

Side View Stern View

Figure 6.43: SES-100B Supercavitating Propeller Blade
Photos by author

The characteristic "scallop" or annex on the pressure face of the Hydronautics supercavitating propeller section can be seen in the upper sketch and again below in the photograph of the actual blade. The propeller blade was manufactured from high grade titanium. The propeller was tested extensively in model scale in water tunnels under proper cavitation conditions and in underway testing on the Test Craft SES-100B for both *fully submerged* and in *surface piercing* modes to gain knowledge of the propeller in terms of its efficiency. Figure 6.44 summarizes the key results and shows how the SES-100B propeller efficiency compared with other supercavitating propellers[37].

From standard texts on propeller design one can show that the ideal propulsive efficiency (η_P) of an unshrouded propeller can be expressed by the equation:

$$\eta_P = \frac{2}{1 + \sqrt{1 + C_T}} \tag{6.6}$$

where the thrust coefficient is given by:

$$C_T = \frac{T}{{}^1\!/_2 \rho_W V^2 A_P} \tag{6.7}$$

where the thrust (T) is non-dimensionalized by the water density (ρ_W); the propeller disc area (A_P) and craft speed (V). The determination of the "best efficiency" varies depending on whether the design is being optimized for the thrust coefficient (C_T) or the advance ratio ($J = \frac{V}{nD}$). The advance ratio (J), expresses the ratio of the craft forward speed (V) to the rotational speed measured by the propeller RPM (n) multiplied by the propeller diameter (D). The references cited in Figure 6.44 are the references used in the original cited reference. The design challenge was how to take advantage of the efficiency of the SES-100B propeller in both the *fully submerged* mode and in the *semi-submerged* mode without the penalty of any pod drag. This was accomplished by a "*variable geometry propulsor*" and application of the Coanda effect. The author first became aware of the Coanda effect during his thesis work on the study of jet flows at Cranfield, UK in 1956–1958. He used this effect to advantage in the design of the jet flap for the *aerodynamic air cushion craft* the MARAD "*VRC-1*" in 1962 which is discussed in Chapter 3 and Chapter 11. It was decided to use the same technique for the surface piercing propeller system for the SES-100B. Figure 6.9 earlier, in the discussion of the SES-100B sidehull shape, showed the opening where the water flowed upward into the propeller. Figure 6.45 shows the basic principle of the design taken from the author's 1974 patent[38] plus a photograph of the actual configuration in the "fully submerged mode".

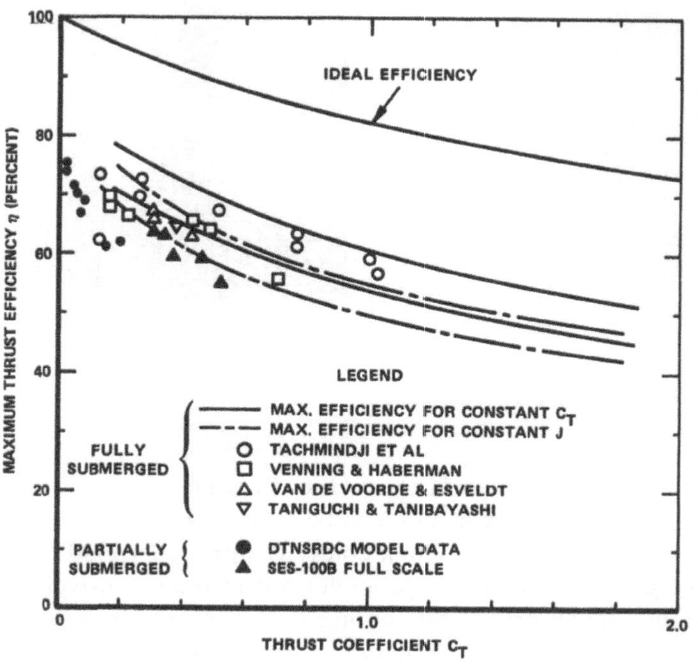

Figure 6.44: Propulsive Efficiency of Supercavitating Marine Propellers

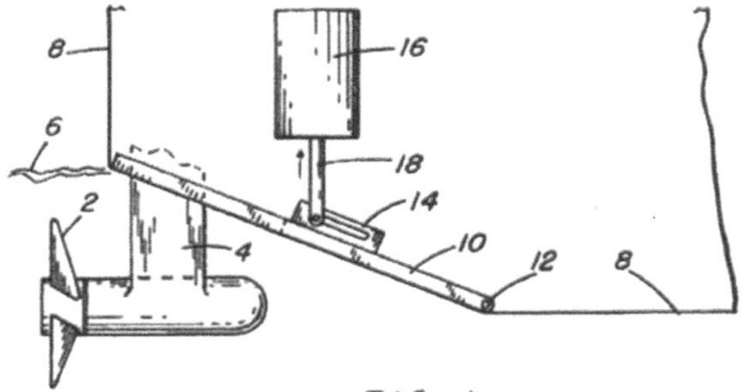

FIG. 1

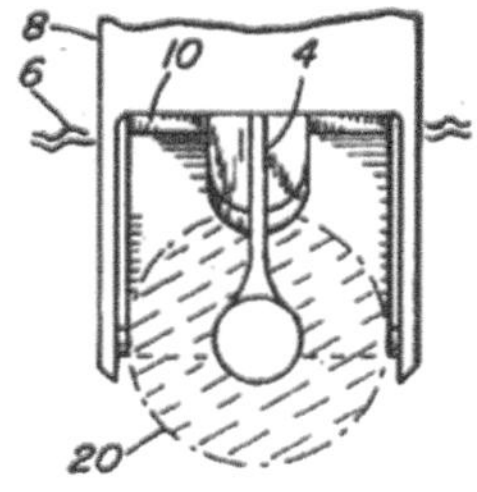

FIG. 2

Figure 6.45: SES-100B Propeller System (Fully Submerged Mode)
Photo courtesy of Textron Marine & Land Systems and Textron Inc.

The essential feature displayed in Figure 6.45 is the shaping that allows the water stream flowing past the ship to curve upwards in accordance with the Coanda effect and allow the propeller to operate in a fully submerged "green water" mode. The upper drawings show the patent drawings where "Fig 1" shows the ramp in the up position and "Fig 2" from the patent shows, by shading, that the entire propeller disc is operating in a fully submerged mode. The lower photograph shows the starboard propeller installation during the outfitting phase at the Michoud Facility, New Orleans in 1971.

Figure 6.46 shows the configuration with the ramps in the down position for operating in the semi-submerged, high speed mode for the SES-100B. Again, from

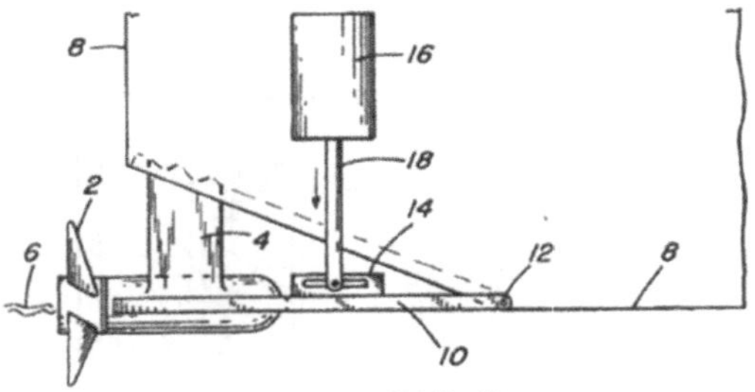

FIG. 3

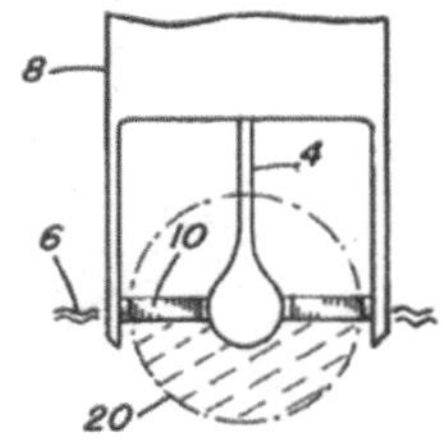

FIG. 4

Figure 6.46: SES-100B Propeller System (Semi Submerged Mode)
Photo courtesy of Textron Marine & Land Systems and Textron Inc.

the patent, "Fig 3" shows the ramp down and "Fig 4" shows that the propeller is operating with half the propeller submerged. An additional feature of the design, as can be seen from Figure 6.45 and Figure 6.46, is that the gearbox of the propulsion system is housed in the wide lower sidehull of the SES-100B making maintenance of that assembly readily accessible, an important consideration in the design of any ship.

The net result of the SES-100B variable geometry propulsor system was a thrust curve as shown in Figure 6.47. The lower speed thrust curve labeled "Fully Submerged" corresponds to the propeller in the configuration shown in Figure 6.45 with the ramp in the up position and the propeller is fully submerged and wetted. As speed is increased, the thrust gradually falls off in the fully submerged mode and a gradual lowering of the ramps will "convert" the propeller to operating as a surface piercing propeller with the ramp configuration as shown in Figure 6.46 in the down position. The thrust in this mode is generated at a higher level as shown in Figure 6.47 by the curve labeled "Partially Submerged".

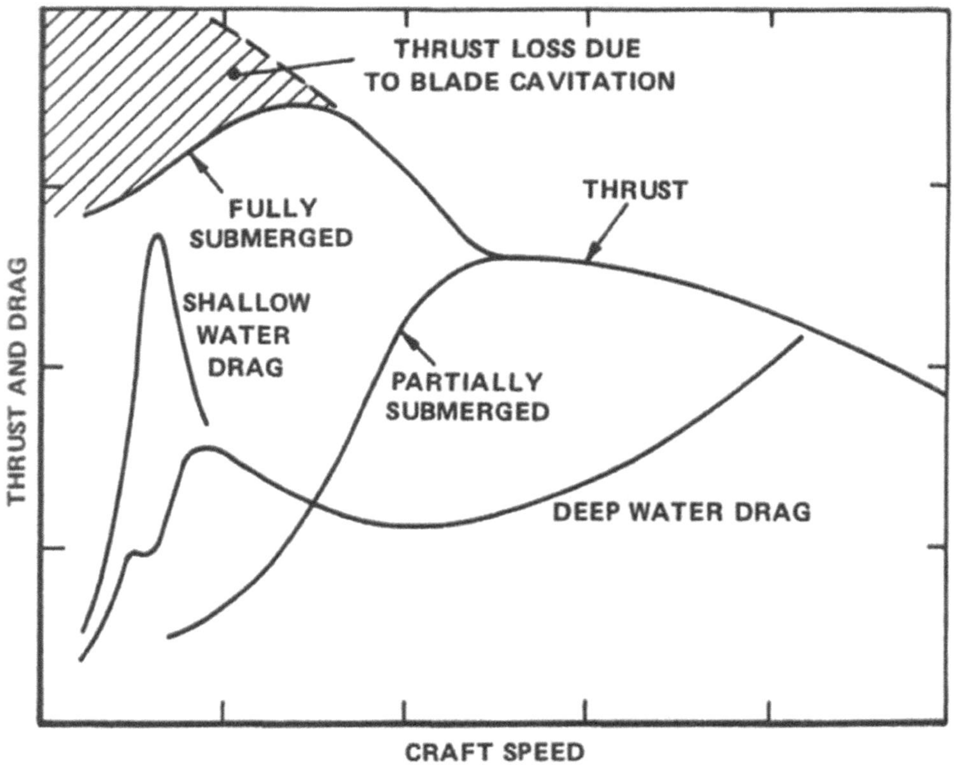

Figure 6.47: Variable Geometry Propulsor Thrust Curve of SES-100B

There will always be some speed range at which a propeller will cavitate but by proper choice of design and operation (by RPM and pitch setting) such cavitation areas can be moved to non-operation areas as shown by the shaded area in Figure 6.47. For the surface piercing mode, while not pursued with the SES-100B propeller system design, it is also possible to vary the pitch setting in a cyclic manner as the blade enters and leaves the water to (a) relieve the loading on the propeller blade and also (b) to lessen the spray generated by the propeller. This feature (if used) would lessen the shaded area of thrust loss in Figure 6.47 and the loss of energy in spray generation of the characteristic "rooster tail" from surface piercing propellers. Such a feature would also lower the radar signature of the wake. But that is left to future designs! The SES-100B did not introduce this unnecessary complication into the propulsion system.

An additional benefit from the variable geometry feature as integrated into the SES-100B is that in the event of any operation in shallow water (where the hump drag is increased over that in deep water) the propeller still provides an adequate "hump thrust margin", which is the difference between the thrust and drag at hump speed. This was of particular interest in the design of the SES-100B because the initial trials were conducted in Lake Pontchartrain which is a constant 15 foot deep lake and was definitely a "shallow water" issue for the tests. The tests were conducted successfully with an adequate thrust margin as can be seen from Figure 6.47 where the shallow water drag estimates have been included in the drag curve set shown. See later discussion on negotiating shallow water hump drag.

6.9 Comment on the Key Technology Differences between the SES-100A and the SES-100B

While not exhaustive descriptions, in the interests of brevity, the above sections have summarized the key differences that were to be explored in the two Test Craft as called for by the initial directive from DDR&E in 1969 (see the beginning of this Chapter). These differences in encapsulated form were:

1. Sidehull shaping
2. Seal System design
3. Structural Arrangement
4. Engine configuration
5. Lift System design
6. Propulsion System design

As discussed in the preceding six sections, the differences between to the two 100 ton Test Craft in these key technology areas were quite marked and had a significant impact on both the projected design for a 3,000 ton class surface effect ship and also on the ultimate decision by both the DDR&E and the US Navy to cancel the program until the risks identified could be eliminated or reduced to acceptable values. As summarized in Chapter 5, the US Navy elected instead to abandon the goal of "100 knots" and turned its attention to more modest speeds of about "50 knots".

In succeeding Chapters, the discussion returns to how the goal of "100 knots" could indeed be achieved.

6.10 Performance of the One Hundred Ton Test Craft

Of the two test craft, it was the SES-100B that met all the objectives set by DDR&E, including the contractual goal of "80 knots or more", and it is worth summarizing the key items that constitute the performance of the craft as determined from extensive testing in Lake Pontchartrain, Louisiana; Panama City, Florida, and Patuxent River, Maryland that sets the stage for design of any larger, high speed ships.

There are various measures of "performance", but usually the "user" of any ship is interested in such items as:

a. Can the ship carry a payload efficiently over a given distance?
b. What are the Cruise Speed and Maximum Speed?
c. What is its turning performance?
d. What are the acceleration and deceleration characteristics?
e. What are the Power Requirements?
f. What is its Fuel Efficiency?
g. Can it operate safely in rough seas?

To avoid the 100 ton Test Craft being designed for speed alone, the US Navy incorporated into the ship specifications various performance requirements. Some indication of these performance requirements of the SES-100A and SES-100B and how they were realized during test is provided in the following sections.

6.10.1 Maximum Speeds of the SES-100A and SES-100B

The main purpose of the 100 ton test craft program was to explore the capability of surface effect ships "*to achieve very high water speeds of 80 knots or more*". Because this was the goal, it is surprising that official documentation of the speeds attained has not been very rigorous with many erroneous speed claims being made for the craft.

The speeds attained by the CAB SES-100A are particularly vague in the Navy reports. Ed Butler, who was the Navy liaison officer throughout the SES-100B program and the "caretaker manager" of the SES Program Office in its waning days in 1979–1980 provided a summary of the Surface Effect Ship Program in his 1985 report[39] and describes the SES-100A as having achieved maximum speeds that are described in different parts of the report as "70 knots", "70+ knots" and "80 knots approx". There were several authors in Ed's report, hence the different speeds quoted. Capt Carl J. Boyd, who was the SES Program Office Program Manager throughout the construction and Navy trials of both 100 ton SES Test Craft, in a News Conference[40] with the Secretary of the Navy J. William Middendorf in 1976, gave the maximum speed of the SES-100A as 76 knots. Raley and Wilson[41], who were responsible for most of the analyses of the two test craft during Navy trials at Patuxent River, also give the maximum speed of the SES-100A as 76 knots with the new flush inlet for its waterjet propulsion. In 1977, the SES-100A was removed from service for major modifications that included:

i. Installation of bow and stern seal retraction mechanisms
ii. Dedication of fans to seal and cushion (similar to SES-100B approach)
iii. Change in sidehull deadrise angle to 45° (from original 60°)
iv. Removal of stability augmentation devices (SAD)
v. Installation of new flexible bag and planer seal system (see Figure 6.21)

With these modifications, the CAB SES-100A was returned to service at Patuxent River in March 1978. Capt Thomas M. Barry, the new Commanding Officer of the SES Program Office (PMS-304) reported that the modified SES-100A soon achieved speeds of 65 knots[42]. No other entries appear in the 1979 issue of Jane's, the last year of operation of the SES Program. Despite the confusion, the author has elected to "log in" a maximum speed of 76 knots for the SES-100A to match the most reliable information quoted above! Note that the reporting of these speeds by the cognizant US Navy program Offices appeared in 1979 and 1985, which is certainly long enough for any late data entry from the Test Program to have been reported and documented.

On the other hand, the speeds attained by the SES-100B were well documented, complete with operating sea conditions, dates and photographs for each speed attained. These photographs are included in Chapter 5. The speeds obtained by the SES-100B are as provided in Table 6.5. There have been different values of speeds quoted in the public literature for the SES-100B also. The on-board "speed-ometer" on the SES-100B, during the high speed in June 1976 run, broke at 89 knots and other test instrumentation recorded 88.88 knots. An independent and probably more reliable speed check was available by an Air Force radar at the nearby Air Base that gave the speed as 91.9 knots. It was this value that was entered into the *Guinness*

Table 6.5: *Achieved Speeds of SES-100B*

Date	Speed and Conditions
22 July 1971	Launched
4 February 1972	First underway run, 20 knots
4 June 1972	40 knots, choppy water
8 September 1972	50 knots, choppy water
1 March 1973	70+ knots, slight chop
8 April 1976	60.8 knots, Sea State II; launched SM-1 missile
May 1976	55 knots, Sea State III (6–8 ft seas)
30 June 1976	91.9 knots, Sea State II

Book of Records! Raley and Wilson give the maximum speed as 89.5 knots during that same record breaking run based on electronic range instrumentation at the nearby Naval Coastal Systems Laboratory. To add to the confusion, another run of 90.3 knots on 2 April 1977 on St Andrews Bay near Panama City, Florida was quoted as "record breaking" in the 1979 Jane's issue. This author is staying with the cross-checked *Guinness Book of Records* value of 91.9 knots!

6.10.2 SES-100B Performance in Rough Water

There was also some confusion as to the speed achieved by the SES-100B during its high speed run in rough seas in May 1976 in US Navy Trials. Capt Carl Boyd, in the same briefing on 30 June 1976 referred to above, gave the speed achieved by the SES-100B in 6 ft waves (Sea State 3) as 55 knots. Ed Butler in his 1985 report above gave the speed during that same test as 35 knots. Raley and Wilson give the speed for that same test as 58 knots! This author settled on 55 knots!

There is a reason why it is difficult to get consistency on speeds in rough water. Figure 6.48 taken from the author's findings in 1980[43] from a wide variety of air cushion craft, shows that the speed in rough water for the SES-100B and similar craft could be represented by the equation:

$$\frac{V}{V_{calm}} = 1 - 0.50\left[\frac{h_W}{h_C}\right]^{3/4} \tag{6.8}$$

where (V) is the speed in rough water and V_{calm} is the speed in calm water *at the same power setting*. The wave height (h_W) and the cushion depth (h_C) are the other parameters in the relationship.

The data for earlier air cushion craft (circa 1965–1970)[44] has been omitted in this version of the chart as it represented the performance of craft using earlier versions of skirt technology that gave much lower speeds in rough water than those craft shown, which include data for the SRN.4, SRN.6, BH.7 and other sophisticated amphibious air cushion craft. The data points for the SES-100B in Figure 6.48 came from trials on Lake Pontchartrain. It is difficult to assure consistency in the data because of several factors such as: the nature of the sea roughness of wind speeds, steepness of the waves and the heading that is not always head on or following seas. It also depends on whether the craft has been designed to contour

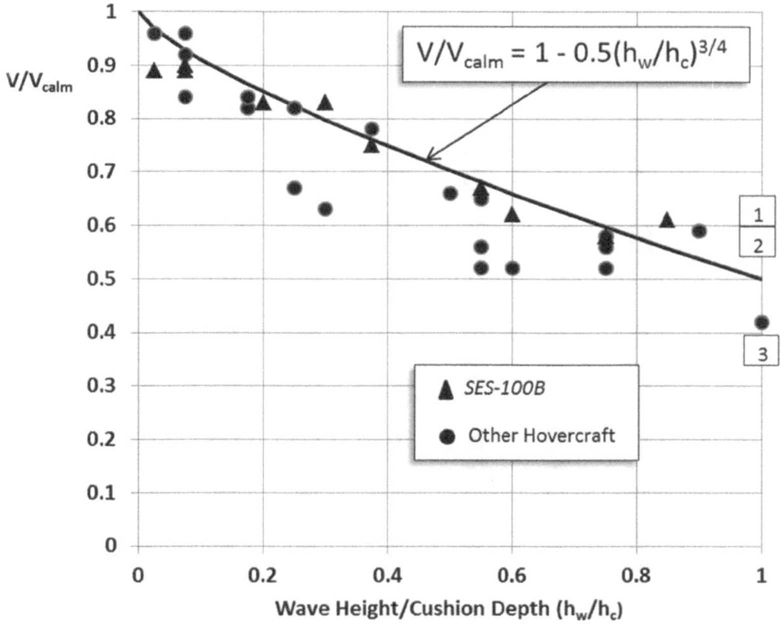

Figure 6.48: SES Speeds in Rough Water

or platform the waves. Oftentimes the operator of the craft has a natural tendency to lower the speed to meet ride quality requirements while heading into rough water, although during test operations all effort is made to keep the power settings the same as for calm sea operation (called the "white knuckle" factor). There is also the difficulty in estimating wave heights and other sea conditions to ensure consistency.

Shown on Figure 6.48 are the three values quoted earlier from the Patuxent River tests and Program Reports mentioned above, which were specifically: 58 knots, 55 knots and 35 knots (labeled as 1, 2, 3 respectively in Figure 6.48). It is seen that all three speeds are plausible, but the higher values appear more consistent with the other data shown, hence the author's preference for quoting 55 knots for the speed of the SES-100B when operating in 6 foot waves which equals the cushion depth (that is for the value $h_W/h_C \approx 1.0$).

The reader should be cautioned about the various speeds that have been reported for these two test craft in the popular press and elsewhere, that can best be described as misleading and not to be believed. The values quoted here, although quite varied, are from reputable sources and the differences are understandable and explainable as outlined above. The most reliable values of speeds have been used in Table 6.2 and Table 6.3 on the demonstrated performance of both the SES-100A and the SES-100B.

As further indication of the speeds attained by the two test craft during Navy Trials on Patuxent River, it is informative to see the test time spent at each of the speed regimes explored to determine characteristics of Surface Effect Ships. Figure 6.49 taken from the report by Raley and Wilson[45] shows the data. The Navy data shows that considerable test time was spent by both the SES-100A and the SES-100B at speeds below and around 50 knots. Above 60 knots, the SES-100A spent about 3½ hours in the testing mode and the SES-100B spent more than 11 hours. The

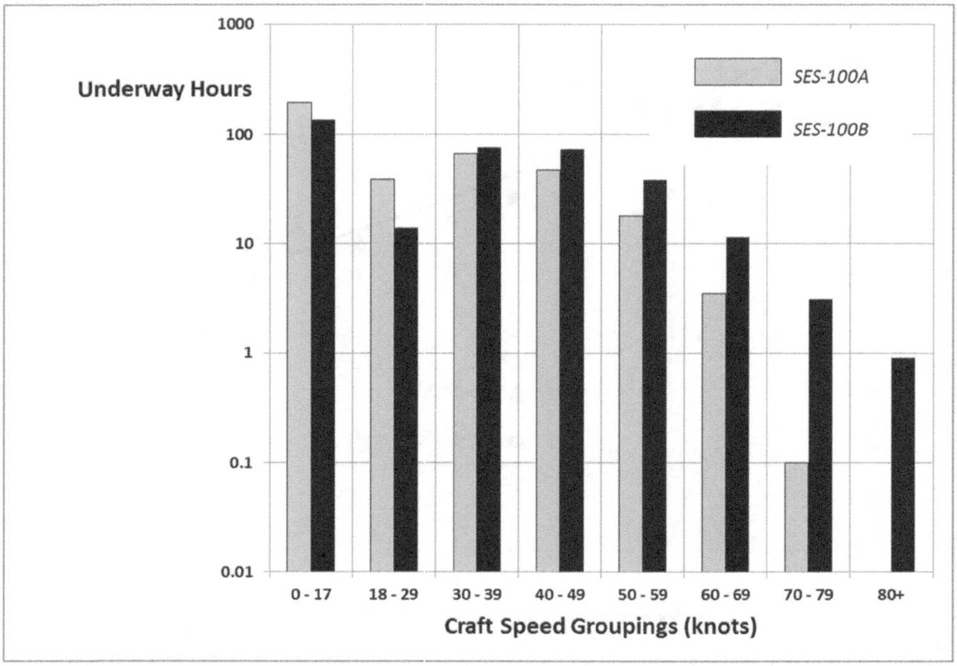

Figure 6.49: Underway Time in Navy Tests of SES-100A and SES-100B

SES-100A spent about 6 minutes at speeds above 70 knots, compared to the 3 hours spent by the SES-100B. Above 80 knots, the Raley and Wilson report shows no time for the SES-100A and about one hour (including the record breaking run of 91.9 knots) for the SES-100B. Documentation of test times after these 1976 tests is not readily available.

6.10.3 SES-100B Habitability Envelope

An important aspect about performance in rough seas is the "habitability" of the craft through use of its ride control system. Figure 6.50 shows the rough sea performance of the SES-100B in terms of speed over the waves and the ride quality as determined using the measurements of heave acceleration in the various conditions of speed and encountered waves. The envelope shows the predicted maximum speeds in Calm Water and in various Sea States up to the specified Sea State III where the Average 1/10th Highest Wave is 6.04 ft (using the Pierson-Moskowitz Sea Spectrum). The predicted maximum speed envelope is shown for both three (3) and two (2) propulsion engines operating at Maximum Continuous Power. The US Navy Requirement was 80 knots in Calm Seas and 30 knots in Sea Sate III. The prediction for maximum speed was 91 knots in Calm Water and 60 knots in Sea State III. The actual demonstrated speeds of 91.9 knots (Calm Water) and 55 (58?) knots (see earlier discussion) in Sea State III are also included in Figure 6.50. While not a requirement, the predicted speeds when operating on two (2) propulsion engines for sustained high speed cruise and endurance are also shown, which are 72 knots in Calm Water and 49 knots in Sea State III as shown.

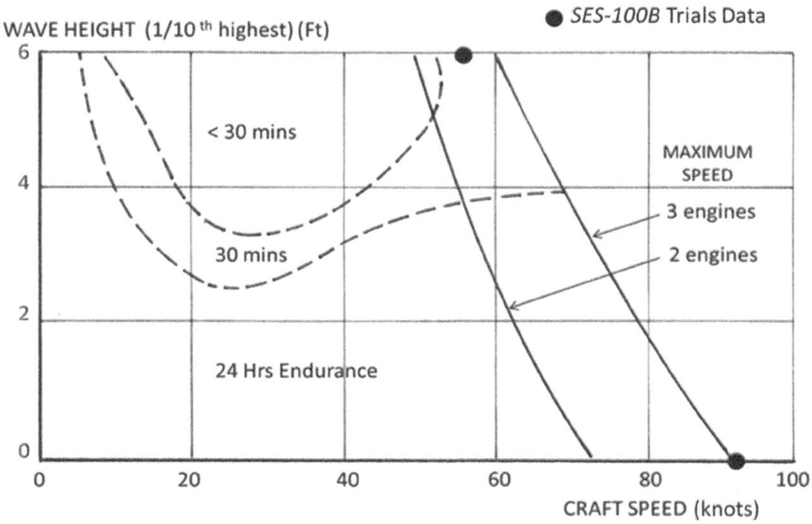

Figure 6.50: SES-100B Habitability Envelope

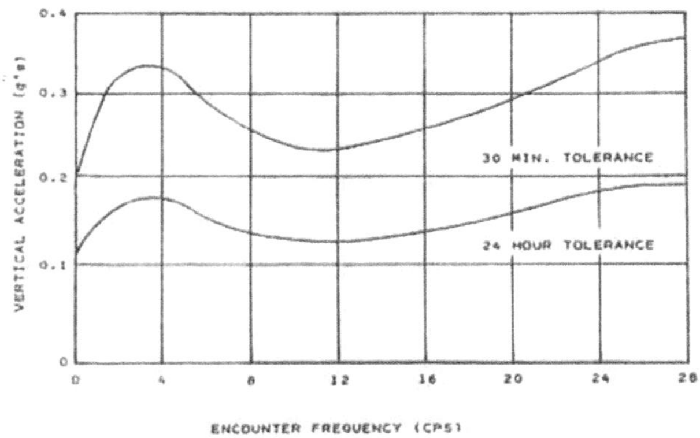

Figure 6.51: SES Program Developed Habitability Criteria Envelopes

In the design of the ride control system for the SES-100B, care was taken to maximize the envelope of personnel endurance limits using the criteria developed in consort with the SES Program Office and the International Standard Organization. The criteria was developed further during the SES Test Craft Program to set endurance limits of different durations from "30 minutes" up to "24 hour endurance" in order to gain experience on setting crew endurance requirements in Navy missions.

Figure 6.38, shown earlier in the discussion of the design of the SES-100B ride control system, showed the heave acceleration results for one particular condition for the SES-100B (50 knots in Sea State II) when compared to the criteria. One set of criteria developed from that work is shown in Figure 6.51. The curves show the boundaries of "30 minute tolerance" and "24 hour tolerance" in terms of heave acceleration which is the predominant factor in determining fatigue and motion sickness limits for crew members. When these habitability criteria envelopes are applied together with a knowledge of the ship motions, it is possible to construct the SES-100B Habitability Envelopes as shown on Figure 6.50. It is seen, using this developed criteria that the ride acceptability for the SES-100B is very agreeable with the ability to operate at high speed in rough seas at the "24 hour tolerance" limits, with only a "pocket" in the operating envelope where the craft would be required to slow down because of habitability needs. These habitability criteria envelopes were developed further during the US Navy ANVCE Project (see Chapter 10).

6.10.4 SES-100B Lift Drag Ratio and Transport Efficiency

If the SES is to be a useful platform it must provide more than just a high speed capability to the US Navy or for any commercial operations as originally sought by the US Maritime Administration (MARAD). Chapter 8 provides more detailed treatment of the three key parameters that influences the design of *high speed dynamic lift craft*, to provide a "useful" platform, but the following provides a snapshot of the values achieved by the SES-100B and how they might apply to the three thousand ton SES (3KSES). They are presented here without further derivation explanation (see Chapter 8 for derivation) for use in the understanding of the data in what follows. These parameters appear in various forms in the literature, not always in a consistent manner and this will be addressed in the interests of clarification.

(Aero) Hydrodynamic Lift Drag Ratio, $$\frac{L}{D} = \frac{W}{D} \tag{6.9}$$

Effective Lift Drag Ratio, $$\left(\frac{L}{D}\right)_{eff} = \left[\frac{W}{D + \frac{\eta_P}{\eta_L}\frac{p_C Q}{V}}\right] \tag{6.10}$$

Transport Efficiency, $$\frac{WV}{P} = \eta_P \left(\frac{L}{D}\right)_{eff} \tag{6.11}$$

In the above equations, Lift (L) equals ship displacement (W) in steady state conditions. The drag (D) is the total drag of the entire ship at speed (V). The *lift drag ratio (L/D)* is the inverse of the usual normalized *specific resistance* or drag-weight ratio (D/W) of the ship.

For those craft that use some form of air cushion support, such as the SES, the lift power enters the equations through the *effective lift drag ratio* $(L/D)_{eff}$ given by Equation 6.10. It is seen from this term that the effect of the air cushion pressure (p_C) and the air flow (Q) generated by the lift system appears in this term, as does the ship speed (V). The *effective lift-drag ratio* also includes the effect of the lift system efficiency (η_L) and the propulsive efficiency(η_P). The lift system term, $\frac{\eta_P}{\eta_L}\frac{p_C Q}{V}$, has the dimensions of drag. In any detail treatment, adjustments have to be made as to how much of the weight (W) is supported by the air cushion (= p_C x cushion area) and how much by the buoyant sidehulls.

The *Transport Efficiency (WV/P)* is the *Effective Lift Drag Ratio* multiplied by the propulsive efficiency (η_P) of the ship's propulsion system. It is by these terms that all the key parameters of the ship including its lift and propulsion systems can be brought to bear on the ship's performance. It is the *Transport Efficiency (WV/P)*, together with the payload carrying capability (see Chapter 7 and Chapter 8), that has a direct correlation with the ability to carry a payload economically and efficiently over a given range.

Over the years, some of these terms have become distorted in the literature causing confusion especially when comparing data from different sources. For example, in the UK in the 1960s it was usual to refer to the *transport efficiency* as the *equivalent lift drag ratio.* In later years, especially in the US, the *effective lift drag ratio* became to be called the *equivalent lift drag ratio.* Hence, it is not always clear when comparing data from different sources as to which form is being used. All effort will be made in this book to retain consistency based on the above "descriptors" of the key terms. In the design of any ship it is necessary to relate these key parameters to those speeds that provide the maximum benefit. The speed at which the *lift drag ratio (L/D)* is a maximum is important in that it provides an indication of the most efficient speed from a hydrodynamic viewpoint. In the aircraft industry, this speed is directly related (in most cases) to the speed for maximum range.

For dynamic lift craft, however, such as the surface effect ship (and its sister the amphibious air cushion craft) the speeds for maximum range occurs at a higher speed due to the complications of the effects of the lift system and propulsion system efficiencies. Because of these lift system effects, the *effective lift drag ratio* $(L/D)_{EFF}$ and the *Transport Efficiency (WV/P)* and *propulsion efficiency* (η_P) play a more dominant role in the design of such dynamic lift craft.

Figure 6.52 collects data from several sources related to the testing of the SES-100B in Lake Pontchartrain, Louisiana; in Panama City, Florida and at the US Navy Test Facility in Patuxent River, Maryland. The solid curves in Figure 6.52 apply to the theoretical predictions and the full scale trials data of

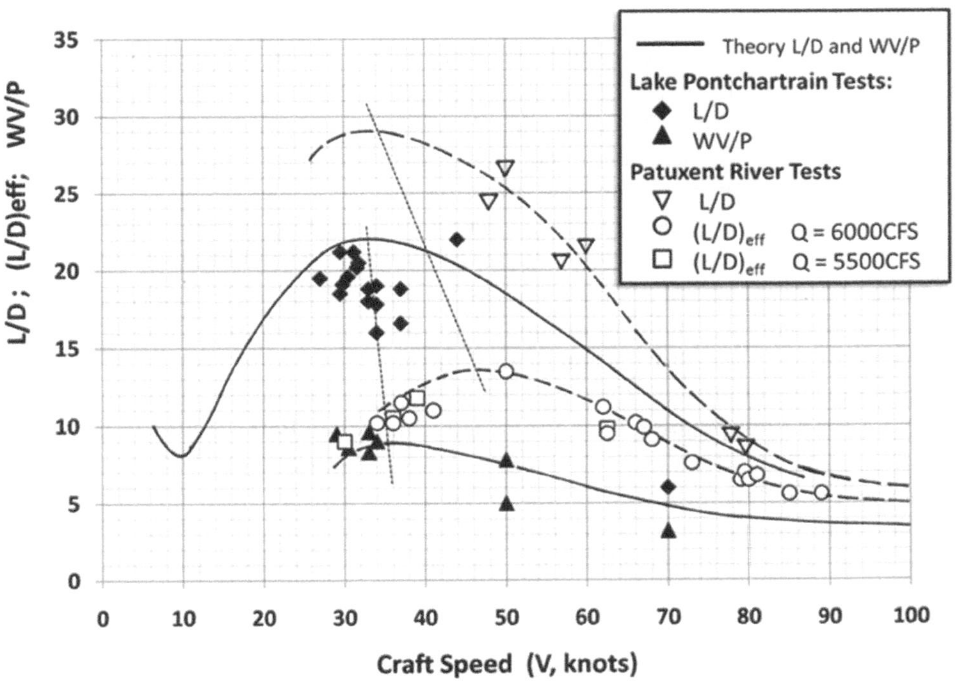

Figure 6.52: SES-100B Lift Drag Ratio and Other Efficiencies

the *lift drag ratio (L/D)* and the *transport efficiency (WV/P)* when obtained while operating in shallow water on Lake Pontchartrain. The dotted curves in Figure 6.52 apply to the *effective lift drag ratio* $(L/D)_{EFF}$ for the full scale trials data obtained while operating the SES-100B in deep water in Patuxent River. While the test data was obtained at different times under different conditions, it is possible to extract important design information applicable to the design of high speed SES.

While it cannot be confirmed with absolute assurance, it is believed that all the data was obtained with the craft operating at its optimum trim angle (lowest drag) and at its maximum air cushion flow conditions (Q = 5,900–6000 CFS). These are critical parameters and the performance can change dramatically if the craft is not operated under common conditions.

For the Lake Pontchartrain tests In shallow water, it is seen from Figure 6.52 that the *maximum lift drag ratio* $(L/D)_{max} = 22$ and occurs at a speed in calm water of about 35 knots. The *maximum Transport Efficiency* ≈ *10* and that occurs at a speed in calm water of about 35–40 knots. Again, it is to be emphasized that these values are for operation in shallow water. As a sanity check on the numbers, it is seen from Figure 6.52 that the value of the *Transport Efficiency (WV/P)* at maximum speed (of 91.9 knots) is about 3.5 and this appears reasonable from the following calculation using the installed powerplant values.

Transport Efficiency at maximum speed:

$$\left|\frac{WV}{P}\right|_{V\max} = \frac{210,000 * 91.9 * 1.689}{550 * (13,500 + 1,860)} = 3.8 \tag{6.12}$$

That is, assuming all the available power can be converted into the propulsion and lift system through its transmission system, one could not expect a *Transport Efficiency (WV/P)* at maximum speed (corresponding to the maximum power) to be greater than the numerical value "3.8". The trials data showed that a value of 3.5 at V_{max} was achieved, which is an acceptable value anticipating transmission losses and other variations not readily apparent from the ship's log.

The trials data obtained later in the Patuxent River tests were obtained in deep water testing where it is seen that a higher *lift drag ratio (L/D)* by about 30% at the lower speeds near "cruise" (~35 knots) was achieved which is not readily explainable from the theory but could be attributed to different trim conditions or a more efficient lift fan flow distribution. The Patuxent River tests also show that the maximum value of *effective lift-drag ratio* $(L/D)_{EFF} = 13.5$ occurs at a higher speed of around 45 knots. Figure 6.53 shows results obtained in both shallow water (Lake Pontchartrain) and deep water (Panama City) for the SES-100B[46] where both ship displacement and craft trim were varied. As might be expected the trim effect is quite marked. The effect of ship displacement and shallow water depth is less clear. These were early tests where the SES-100B was operated "bow up", as a prudent test procedure, and it is possible that a lower trim angle could have been used which would lend credence to the 30% gain in *lift drag ratio (L/D)* experienced in the later Patuxent River tests as experience was gained in operation.

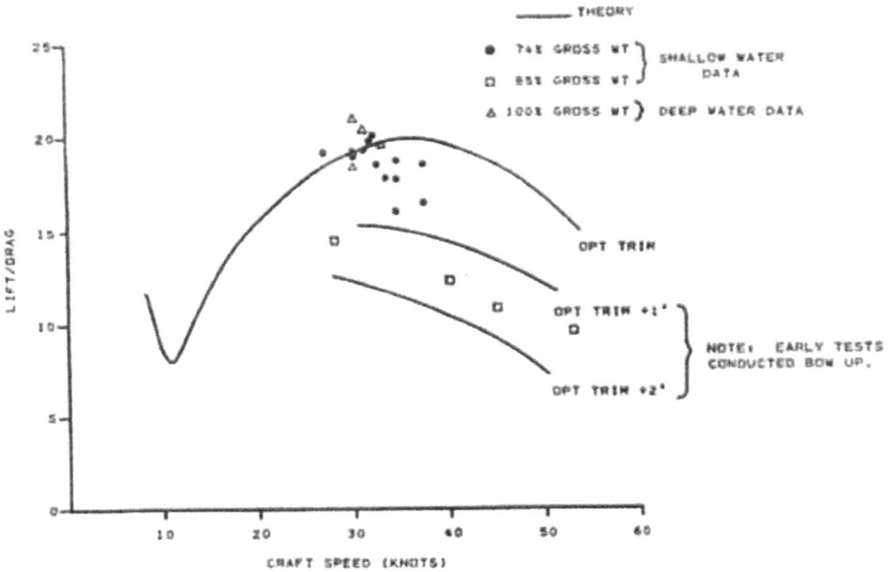

Figure 6.53: SES-100B Lift Drag Ratio at Trim Conditions

Based on the above trials data, it is possible to construct a set of values achieved for the SES-100B for the various key parameters as summarized in Table 6.6

Table 6.6: *SES-100B Efficiencies at Cruise and Maximum Speeds*

	Maximum Value @ Cruise Speeds	At Maximum Speed (91.90 knots)
Lift Drag Ratio (L/D)	22–29 @ 35 knots	6.5
Effective Lift Drag Ratio $(L/D)_{EFF}$	13.8 @ 50 knots	5
Transport Efficiency (WV/P)	10 @ 40 knots	3.5

It is noticed from Figure 6.52, that the *effective lift drag ratio* $(L/D)_{EFF}$ from the Patuxent River Tests was a maximum when operating at a speed of about 50 knots but when the effects of the *propulsive efficiency* (η_P) are included, this reduces the speed for *transport efficiency (WV/P)* to be a maximum at a lower "cruise" speed of approximately 40 knots. The design of the SES-100B propulsion system was geared to achieving propulsive efficiencies at the higher speeds. In any other design approach it would be possible to select different criteria as to where to set the speed for maximum propulsive efficiency. Later in the work associated with the ANVCE Project (see Chapter 10), the designs were geared to a different set of criteria to maximize the range characteristics along with maximizing the speed characteristics of the 3KSES. These are a matter of design preferences and not a limiting design.

A numerical calculation helps to show the effect of the lift system on the "efficiency" of the design when considering it in comparison to other craft. The lift system contributes in two direct ways on (a) the resistance and (b) the power requirements. First, the lift system, through the cushion pressure (p_C) and the air flow (Q) in the creation of the air cushion, decreases the draft of the sidehulls which in turn reduces the hydrodynamic drag and the propulsion power requirement of the ship. This shows up as a reduced value of overall resistance expressed as a D/W value. The second effect is in the computation of the power requirements which appear in the *effective lift drag ratio* $(L/D)_{EFF}$ where the lift system efficiency (η_L), cushion pressure (p_C) and air flow (Q) are included. This can best be illustrated with a numerical calculation using the trials data shown in Figure 6.52.

From Figure 6.52, the maximum value of the *effective lift drag ratio* $(L/D)_{EFF}$ occurs at 50 knots. For this speed and operating condition, the numerical value of $(L/D)_{EFF}$ can be calculated from:

$$\text{Effective Lift Drag Ratio } (\mathrm{L/D})_{\mathrm{EFF}} = \frac{W}{D + \left(\frac{\eta_P}{\eta_L}\right)\frac{p_C Q}{V}} \tag{6.13}$$

$$\left|\left(\frac{L}{D}\right)_{EFF}\right|_{50knots} = \frac{210,000}{8,000 + \left(\frac{0.62}{0.60}\right)\left(\frac{(105)(5,900)}{(50^*1.689)}\right)} \tag{6.14}$$

$$= \frac{210,000}{8,000 + 7,555} \tag{6.15}$$

$$= 13.5 \tag{6.16}$$

It is seen from Equation 6.15 that the effect of the lift system gives a "drag" term that is of the same order as the net resistance (D) of the ship that effectively reduces the hydrodynamic lift drag ratio from a value of 26 (see Figure 6.52) to a value of about half that or $(L/D)_{EFF} = 13.5$.

The propulsive efficiency (η_P) now reduces that term to give a Transport Efficiency as follows:

$$\left|\frac{WV}{P}\right|_{50knots} = \left|\eta_P\left(\frac{L}{D}\right)_{EFF}\right|_{50knots} \tag{6.17}$$

$$= (0.62)(13.5) \tag{6.18}$$

$$= 8.37 \tag{6.19}$$

It is this value that is used to compute the range capability of the ship at that speed. A higher range would be achieved at the lower speed of 40 knots. The reader is cautioned that the literature is replete with false claims by designers who plot the "lift-drag ratio (L/D)" of their designs on charts of "transport efficiency (WV/P)" of competitors' designs. This needs to be carefully watched!

The purpose of the test craft program was to achieve information that would guide the US Navy in its design of high speed ships in the *"80 knots or more"* range of speeds. There are two speeds of interest in the design of ships (as in most vehicles). They are (a) Cruise Speed and (b) Maximum Speed. One must be careful about interpreting the speeds attained by the two test craft and their meaning when applied to the larger scale ships planned. In this case, the aim by the US Navy was to produce a frigate size ship of about 3,000 ton displacement capable of operating in a safe and useful manner at *"80 knots or more"*.

If the surface effect ship was a conventional displacement ship operating at a speed of about 30 knots then it would be correct to simply scale the SES-100B speeds by Froude scaling to obtain the "full scale" speed for the 3KSES, but this would be an incorrect procedure for *dynamic lift* craft that operate at high speeds where other scaling factors (e.g. Cavitation Number scaling) must be considered.

One of the reasons why the two test craft were required to operate at the "full scale" speed of 80 knots or more was because that various factors, such as cavitation, that do not scale by Froude Number. Also, the dynamics of the air cushion also do not scale by Froude Number as do some of the viscous and inertia forces acting on such unique systems as the seals and the compressibility effects in the air cushion itself that become more predominant as the ship size increases. Chapter 8 discusses these scaling factors but with the above reservations, for "quick analysis" a simple Froude scaling can be used – provided the above considerations are understood!

With the above provisos about scaling, the 40 knots *maximum transport efficiency (WV/P) speed* of the SES-100B scales by simple Froude scaling for a 3000 ton size SES to a cruise speed of about 70 knots. If the SES-100B had exhibited a *maximum value of WV/P* at the slightly higher speed of 45 knots, then this would scale to 80 knots and higher for a 3000 ton SES, so it can be safely said that the SES-100B has shown that such a design could be pursued with confidence. One last proviso is that the maximum value of the *Transport Efficiency (WV/P)* achieved which is about 10 for the Test Craft SES-100B will not be the same value exactly for the larger ship either, because of scaling effects and geometric differences but for quick analysis, it is accurate enough to assume a similar value of WV/P for the larger ship.

6.10.5 Comment on Scaling

It is frequently taken for quick calculations to use Froude scaling when projecting to larger craft. This can be a dangerous practice to follow and can lead to erroneous results for the larger SES designs. Some comments on scaling are provided here with more detail given in Chapter 8 on Technical Considerations. The scale factor (λ) between the SES-100 and the three thousand ton SES is about $\lambda = 3.1$ and one would expect, based on Froude scaling, that various characteristics of the larger ship could be scaled using this factor. It is cautioned that the SES-100A and SES-100B were NOT scale models of a 3,000 ton SES, they were designed to explore different technologies that would be used in the *design* of a 3,000 ton SES, which is quite a different thing.

For example, the maximum speed achieved by the SES-100B was 91.9 knots. If the 3,000 ton SES was a geometrically scaled version of the SES-100B, then by Froude scaling one would expect to see a maximum speed for such a ship to be:

$$V_{max} \text{ of 3KSES} = 91.9 \times \sqrt{3.1} = 161.8 \text{ knots!!!}$$

which is clearly <u>*not*</u> the case.

The reason being is that the two 100 ton Test Craft were designed to deliberately explore the <u>full scale speed</u> of 80 knots or more and <u>*not*</u> to operate at a (Froude scaled) speed because of the effects of Cavitation Number scaling, among several other important scaling considerations. These various other characteristics of the various SES systems that also would not scale by Froude Number are discussed in Chapter 8.

If the larger ship was indeed a geometrically scaled up version of the SES-100 craft, one would expect the *lift drag ratio* to be approximately the same (with note taken of the above non-Froude effects) but that value would be achieved at a higher speed. The best that could be said of such a technique is that based on the SES-100B test results, the *lift drag ratio (L/D)* of the 3KSES would be roughly about the same value as that shown in Table 6.6. The differences would be because the drag of the system would be different. One obvious difference is that while the ship may be scaled up in size, the sea conditions over which it must operate do not. The wave heights of Sea State 3 for example, would be a much lower percentage of the cushion depth of the 3KSES than they would be for the SES-100 craft. Also, as discussed previously in this chapter the hull lines of the 3KSES did not match those of the SES-100 Test Craft because of the modifications that had been incorporated as the program progressed. But for coffee table discussions, stating that *lift drag ratios* (L/D) of "about 22 or more" are achievable for high speed 3KSES is a reasonable number to use.

Once the effects of the lift system and the propulsion are incorporated in the performance parameters however, then the "non-scalability" features begin to affect the comparison between SES-100 results and predictions for the 3KSES or any other size ships in a significant manner. For example, the *effective lift drag ratio* $(L/D)_{eff}$ shown in Equation 6.10, will not scale directly by Froude Number scaling because of the influence of the design features of the air cushion through the cushion pressure (p_C) and air cushion flow (Q). The lift and propulsion system efficiencies (η_L, η_P) that are also included in the *effective lift drag ratio* $(L/D)_{eff}$ also do not scale. The term $\frac{\eta_P}{\eta_L}\frac{p_C Q}{V}$ in the *effective lift drag ratio* $(L/D)_{eff}$ has a strong impact with numerical values that can be of the same order as the hydrodynamic drag (D) that varies with both speed and design of the ride control system. Issues of compressibility in the air cushion (that do not scale by Froude Number) also affect the values of these key performance parameters. The design of the air flow system is particularly important. The SES-100B was designed with a design air flow (Q) of 5,900 CFS. This was the result of analyses and model tests that included considerations of wave pumping for improved ride control as well as design considerations of negotiating hump during the acceleration mode from static hover to "super hump" speeds.

If one used a conventional scaling of $\lambda^{5/2}$ one would expect an air flow (Q) design value for the larger 3KSES design of about 5,900 x $3.1^{5/2}$ = 99,800 CFS approx. In the design of the 3KSES, however, because of the relative dimensions of the sea waves to the ship dimensions, and other considerations an actual design value of about 70,000CFS was used to accommodate operation in Sea State 6, with even lower values of air flow (Q) required for operation in Sea State 3. Such "non-Froude scale" considerations have strong influences on sizing of the lift system engines etc. More discussion on such effects is provided in Chapter 8.

Finally, in the scaling of the *transport efficiency (WV/P)*, in addition to the non-scaling effects mentioned above, there is the impact of the propulsion system efficiency (η_P) that will have its own variations with ship speed that can be different from the "scale model". The testing of the SES-100B at 80 knots and above added the additional confidence that the concerns of cavitation at such high speeds have been adequately treated. This does not provide, of course, the necessary scaling of such other key items as the seals (that would require special scaling of the mass and modulus of elasticity and damping characteristics of the heavier seal materials of the larger SES) or the scaling of the propulsion system and other issues. These were the concerns expressed by reviewers of the program (see Chapter 5) that led to cancelation of the program. These cautionary remarks serve as a backdrop to the discussions in Chapter 8 when the larger scale ships are considered.

6.10.6 SES-100B Range

The US Navy requirement was that the SES-100 test craft be designed to have a 1500nm range capability. Figure 6.54 shows the data as achieved by the SES-100B during test in the Panama City, Florida and the Patuxent River, Maryland trials.

The data is shown for operating in either of two modes, first with (a) three propulsion engines and second, (b) a single propulsion engine for more efficient use of power and specific fuel consumption (sfc) settings. The range values shown are computed from the measured values from the tests of propulsion efficiency (η_P); specific fuel consumption (sfc); transport efficiency (WV/P) measured for total power settings (P_L and P_P) and the fuel consumed (W_f). It is seen that the range requirement (of 1500nm) was exceeded with a range of 1600nm at a speed of 60 knots (which is the maximum speed of the ship on one engine and also the best speed for maximum range). On three engines, it is seen from Figure 6.54, that the range was also a respectable 1400$^+$nm at a speed of ~ 52 knots.

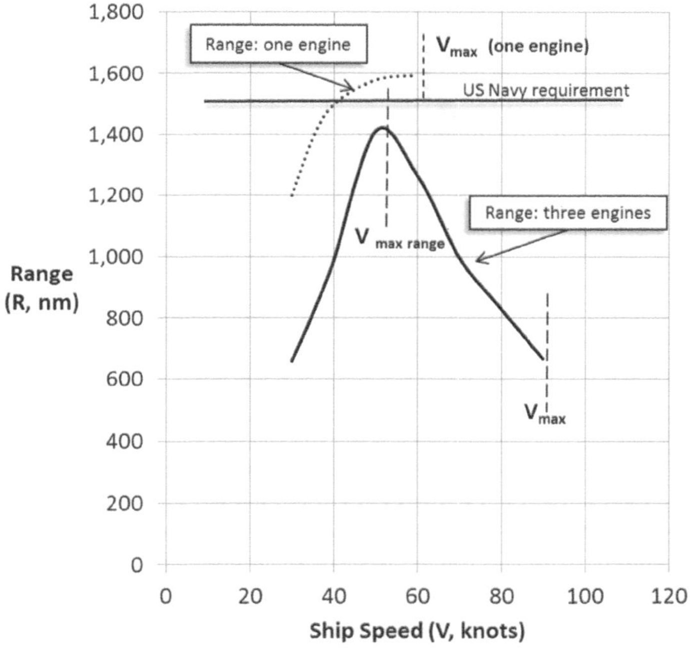

Figure 6.54: SES-100B Range (One and Three Engines)

6.10.7 SES-100B Turning Performance

Because of the high speed nature of the "*80 knots or more*" SES, the US Navy examined such performance characteristics as the turning capability at speed. A series of simulations and model tests were conducted to develop the desired characteristics such as the capability to "bank into the turn", which would contribute to good habitability and ride quality of the SES-100B.

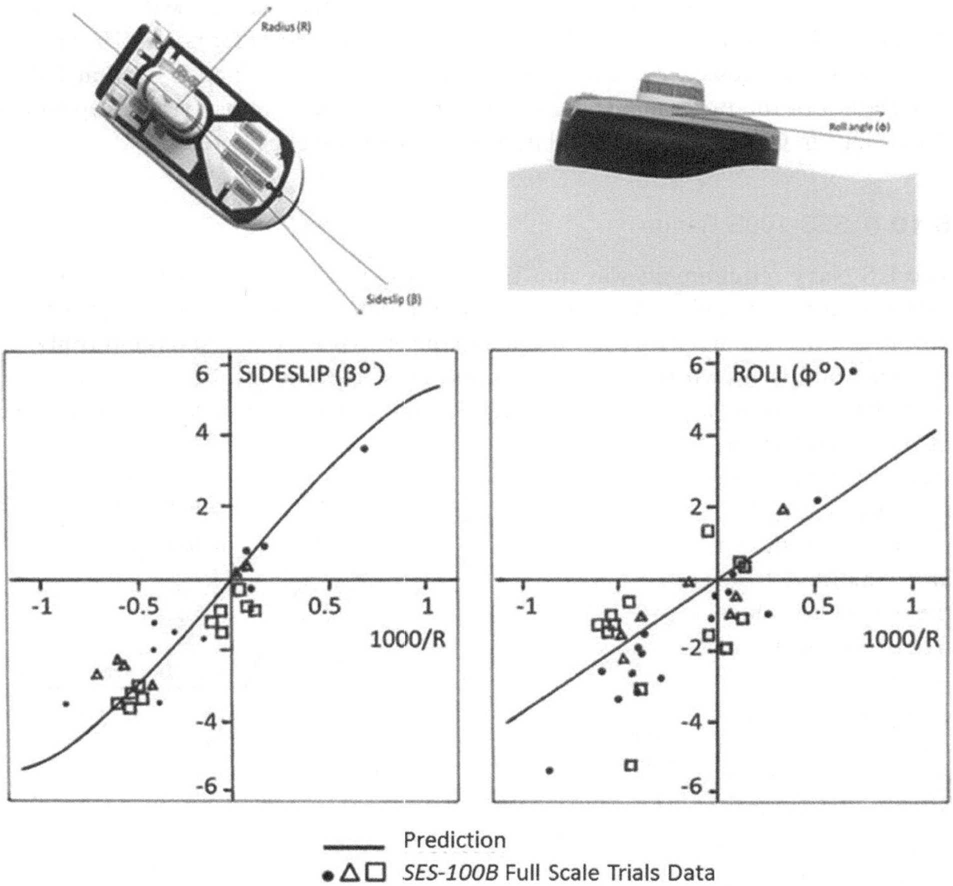

Figure 6.55: SES-100B Turning Performance (Sideslip and Roll)

Figure 6.55 shows the predictions for the SES-100B Test Craft together with results[47] from the full scale trials on Lake Pontchartrain in 1971. Figure 6.55 shows both the sideslip angle (β) and roll angle (Ø) generated in the turn as rudders are activated. The scatter of the full scale data is mainly due to operation in choppy sea conditions and difficulties in measuring small angles. The turns are characteristically flat. For example a turn radius of 1000 ft can be generated by a "banking in" roll angle, Ø = 4° and a sideslip angle, β = 5°.

In 1968–1969, when the US Navy was scoping out the requirements for the SES Test Craft, not much information was available on turn radius of high speed ships and assumptions were made based on the available data from operational hovercraft in the UK and extensive model testing in the US. Based on this background, the requirement established at that time for turning performance was:

(a) Less than 1000 ft radius at speeds below and up to hump speed
(b) No more than 8000 ft radius at all speeds up to maximum speed (specified as 80 knots)

One of the concerns was whether or not the craft could generate sufficient sideslip in cavitating and separated flow conditions to provide the necessary turn radius. As can be seen from both Figure 6.55 and Figure 6.56, that turns could be achieved easily and with small angles of roll, sideslip and rudder. The requirement of the 1000 ft and 8000 ft turn radius was satisfied with a wide margin with modest rudder angles (5°–10°) for the SES-100B. The later test results (1976) obtained from the US Navy Patuxent River trials[48] are included in Figure 6.56.

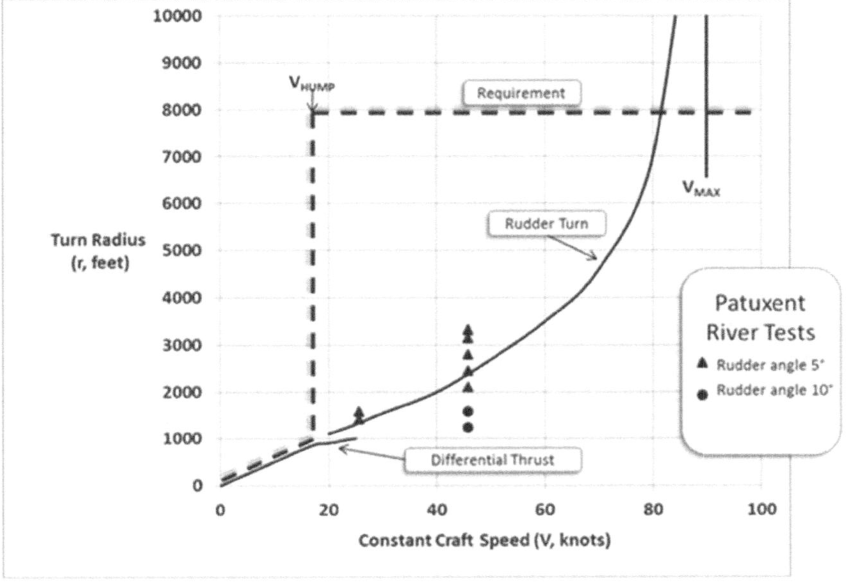

Figure 6.56: SES-100B Turning Performance

Other rudder angles were included in the trials but the data shown provides a good indication of this particular performance feature of the SES-100B. Clearly, the turning radius at the actual maximum speed would be "infinite" so the US Navy requirement was set with the assumed maximum speed to be 80 knots. As shown in Figure 6.56 the turn radius of the SES-100B was indeed less than 8000 ft at 80 knots (measured to be about 7,000 ft). The variations of rudder angle were investigated up to about 15° but the data is not displayed here.

At a craft speed of 65 knots, the yaw rates generated by the rudder turns were measured to be about 0.35–0.45 degrees per second for every degree of rudder angle applied. The turn radius of the SES-100B at 65 knots was as predicted at about 3,600 ft. Turn radius data for the SES-100A is not available.

6.10.8 SES-100B Acceleration Performance

There are two aspects of interest in the acceleration requirements which are related. The first is: *how long will it take to accelerate to cruise speed once the ship is "on cushion"?* The second and related requirement is that the ship not be sluggish as it negotiates hump speed, a critical need for any air cushion craft.

The first requirement, again based on "best practice", was that once on cushion, the ship be capable of achieving 55 knots in one minute. Figure 6.57

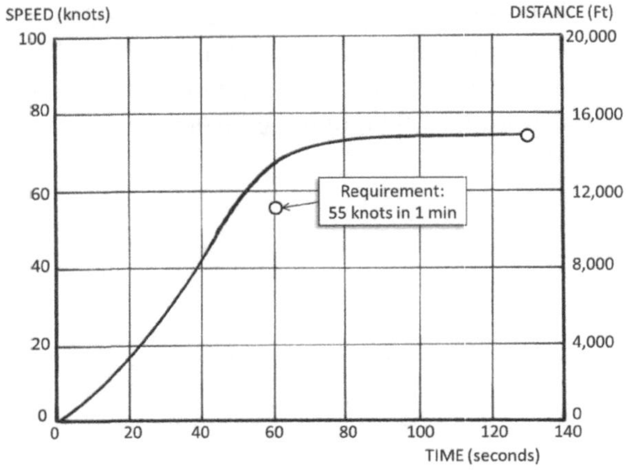

Figure 6.57: SES-100B Acceleration Performance

shows that the SES-100B actually achieved 65 knots in one minute with the three propulsion engines operating at max. continuous power.

The issue of the ship acceleration as it approaches the first (or primary) hump in the drag curve has been a concern for all dynamic lift craft and an extensive review[49] was made of many operational hovercraft (the JEFF air cushion craft, the SRN.4, Vosper Thornycroft VT 100, LCAC and others) as well as hydrofoil craft (TUCUMCARI and PCH-1) to gain some insight into "best practice". From these craft, it was generally accepted that any such dynamic lift craft should have sufficient thrust margin of 0.025 g's. That is,

$$\left|\frac{T-D}{W}\right|_{hump} = 0.025 \tag{6.20}$$

In deep water, as can be seen from Figure 6.47 in the earlier discussion of the SES-100B *variable geometry propulsor* system, this requirement was amply met with a thrust margin over the ship's drag value of 107% (which gives a "sporty" 2.5 times the acceleration suggested from "best practice").

The thrust margin in shallow water, as shown in Figure 6.58, however, proved to be an interesting phenomenon that was not completely resolved during the 100 ton Test Craft Program. Statically, if the theory of shallow water drag was correct, very high hump drag values were predicted, that for the SES-100B would only give a 5% thrust margin – if measured in steady, non-accelerating conditions. To complicate setting the requirements further was the unique feature in operating in shallow water in that if the craft is *accelerating* the predicted high values of shallow water hump drag does NOT occur.

Everest and Hogben[50] earlier, in 1967, had determined from model tests that if the model was accelerating while traversing hump, the predicted hump drag was reduced. Later, Larry Doctors[51] provided an analytical explanation of the parameters involved in such a possibility that depended on both the craft characteristics and the water depth. Doctors predicted in his analysis that drag reductions of 10–15% could be expected if the craft was accelerating. One of the possibilities is that if the craft is moving at high speed, the induced waves do not have time to fully form and thus present less of a resistance.

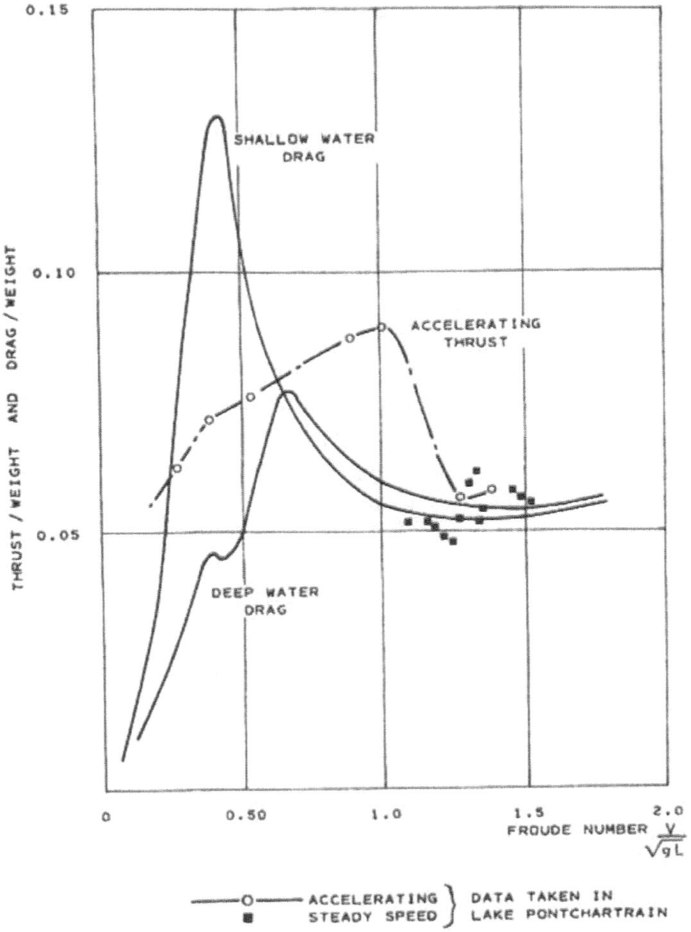

Figure 6.58: SES-100B Acceleration in Shallow Water

The actual experience from the Lake Pontchartrain tests of the SES-100B accelerating through hump in this 15 foot deep lake showed a much higher drag reduction than that predicted, suggesting that other factors were at work in the dynamics between craft and water and the nature of the bottom of the lake. The following is extracted from the reporting[52] of the data at the time. Figure 6.58 shows from the tests in Lake Pontchartrain that the thrust measurements taken while accelerating through hump allowed the craft to move on through hump to the desired cruise speeds. Several reasons for this were explored at the time (1971). It could be that the water (in the lake) was not the critical depth for the SES-100B, or it could be that the wave reflections off the muddy bottom of Lake Pontchartrain were attenuated or it could be that the edge effects of the cushion pressure distribution assumed in the theoretical treatments were sufficient to attenuate the wave drag hump.

In a private communication with Peter Crewe[b] of British Hovercraft Corporation it was learned that any comparison between theory and test of wave drag must take into account the limiting steepness at which the waves break. Crewe referred to the extensive tests on this done for the SRN.4 design and operations. Hence, the thrust margin during shallow water operation of the SES-100B also met the "best practice" value of 0.025g although not quantified directly as Figure 6.58 shows. Because it was not expected that the larger SES designs contemplated were to be operating in such critical depth waters this phenomenon was not explored further in the 100 ton Test Program.

[b] More discussion about Peter Crewe's contributions is provided in Chapter 10 where the author brought him on board as a consultant to the US Navy Advanced Vehicles Concepts Evaluation (ANVCE) Project.

6.10.9 SES-100B Deceleration Performance

For any high speed ship there is an obvious need to know how long and how far it will take the ship to stop. The US Navy placed a requirement in the design specifications that the ship must be able to stop from any speed within a distance of no more than 3000 ft.

Figure 6.59 shows the stopping characteristics (distance and time) of the SES-100B from a predicted speed of 91 knots – the predicted top speed of the Test Craft at the time of the design (1969). It is seen that this requirement was met with a stopping distance of about 2750 feet and a stopping time of about 60–70 seconds.

No specific testing was done with the SES-100A on deceleration characteristics although data is available on the automatic shut-down of the craft from engine shut-off. The SES-100A is fitted with sensors that will shut down the engines if air ingestion is sensed in the waterjet inlets. The problem of cavitation and air ingestion into the flush inlets (from the air cushion) is discussed earlier in the discussion on the propulsion systems for the two Test Craft. The SES-100A experienced such shut downs frequently and it was recorded that from 65 knots, the SES-100A would come to a complete stop in "less than 900 feet". This would not be a normal operating procedure as the strain on the seal system dragging in the water would not be conducive to long seal life but it does provide another data point on the ability of such high speed craft to come to a stop from speed. The SES-100B stopping performance is within a normal operating procedure and satisfied the US Navy requirement as shown in Figure 6.59.

The above summary of performance characteristics for the two Test Craft provided a valuable data set of needed data before proceeding with the design of a 3000 ton SES.

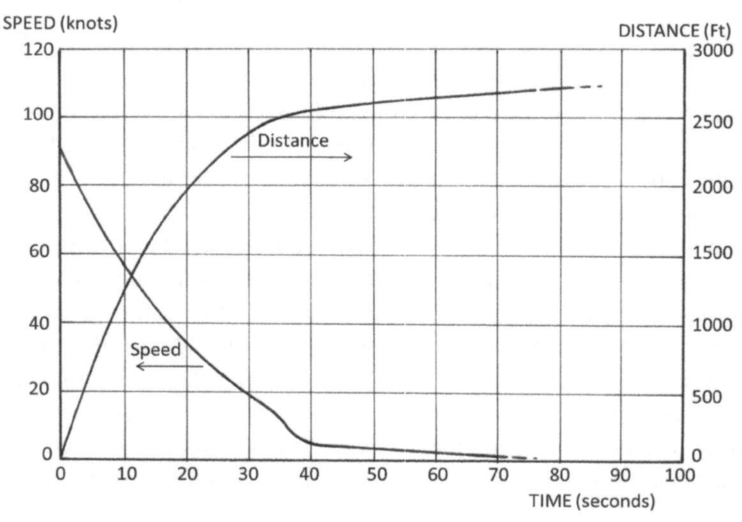

Figure 6.59: SES-100B Deceleration Performance

6.10.10 SES-100B Successfully Meets All Program Objectives

As described above, the SES-100B had accomplished all the objectives of the program as laid down in the original contract release in January 1969 (see beginning paragraphs of this Chapter). As discussed in detail in Chapter 7 on Economic Considerations and in the deliberations of the MARAD Surface Effect Ships for Ocean Commerce (SESOC) Committee in 1965 (see Chapter 4) there are fundamental requirements that must be met in any design. This is summarized well in the SESOC Committee Final Report[53] that says:

> *"The commercial feasibility..... is critically dependent upon the achievement of optimum performance in the following areas: stability and control, effective lift-drag ratio, structure, propulsive efficiency, and fuel economy."*

These features are as equally important in a military ship as they are in a commercial ship. Cost is paramount in both acquisition cost and direct operating cost. The demise over the years of many high speed ships including hydrofoils, air cushion ships and other types are a sobering testament to this real world fact.

The SES-100B was successful in each of the areas of concern. The successes can be grouped under first, technical performance and second, under economic characteristics as identified in the SESOC Committee Final Report[54] discussed in detail in Chapter 4 where the first reviews on large surface effect ships (SES) were initiated.

In addition to the above technical and economic characteristics, the SES-100B demonstrated that the life of the flexible skirt system (a key ingredient in any economic analysis) even at the high operating speeds was compatible with the best experience from the commercial hovercraft operating in England. A value of 400 hours was achieved with the actual operating speed profile of the SES-100B Test Program in the three test areas on Lake Pontchartrain, Louisiana; Panama City, Florida; and Patuxent River, Maryland (see earlier discussion in this Chapter on skirt life and operational hours in various sea conditions for the SES-100B as summarized in Figure 6.32 and Figure 6.49).

The SES-100B met all the required characteristics from the 1969 contract requirements and these are summarized in Table 6.7. In Table 6.7, the acceleration and deceleration numbers come from a combination of calculation and measurements from trials in Lake Pontchartrain (see Figure 6.59). The turning performance comes from trials data in both Lake Pontchartrain and Patuxent River tests (see Figure 6.55 and Figure 6.56).

The values of range quoted are not from a demonstrated continuous cruise but a Breguet Range calculation using measured trials values of propulsion efficiency (η_P), specific fuel consumption (sfc), transport efficiency at constant measured power settings (WV/P), and measured fuel consumed (W_f).

It is difficult to understand that despite these successes with the SES-100B, the US Navy elected to proceed into the 3KSES Phase of the program with many of the failed features of the SES-100A with its problematic waterjet propulsion system and its flawed CAB rigid planer seal. Because of these problems, the SES-100A did not achieve the basic contract speed of 80 knots (actual speed not well recorded but probably around 76 knots) and

Table 6.7: *SES-100B Achieved Successes*

SES-100B Technical Characteristics	**Achieved**
Calm water Speed:	
Contract speed	80 knots
Predicted speed (1969)	91 knots
Actual speed (1976)	91.9 knots in Sea State 2
Rough water performance:	
30 knots in Sea State 3	55 knots in Sea State 3
Stability & Control and Ride Quality	Stable in all axes in calm and rough seas Met all US Navy Habitability Criteria in ride quality
Turn radius: (no more than 8000 ft)	Less than 7000 ft
Acceleration: 55 knots in 60 secs	65 knots in 60 secs
Deceleration: Stop within 3000 ft	Stopped in 2750 ft in 60–70 secs
SES-100B Economic Characteristics	**Achieved**
Lift-drag Ratio (L/D)	High efficiency of 22-29 at 35 knots
Effective Lift-Drag Ratio $(L/D)_{EFF}$	High efficiency of 13.8 at 50 knots
Transport Efficiency (WV/P)	High efficiency of 10 at 40 knots
Structure	Structural Weight (W_S/W) of 34%, well within SOA
Propulsive efficiency (η_P)	High efficiency at 62% at design speed
Fuel Economy (Range of 1500nm)	At design payload (10 tons) and 5 crew on board: 1600nm at 60 knots (single engine operation) 1400nm at 52 knots (three engine operation)

had not resolved the performance and stability issues of the *rigid planer seal* even after an intensive development program with both the contractor and the US Navy.

The original designer of the SES-100A (Aerojet Corporation) had already corrected the stability problem after the "plow-in" of the SES-100A and its original *CAB rigid planer seal* (see earlier discussion and Figure 6.19) by replacing the rigid planer seal with a *flexible skirt system*. Aerojet General then went on and declined to bid on the US Navy 3KSES Program because of these and other (identified) technical problem areas stating that more research and development was essential. This view was supported by others such as the GAO and the Office of Secretary of Defense.

In a surprising move, the US Navy turned the SES-100A over to the new contractor (RMI) who then removed the successful flexible skirt system and installed again a CAB rigid planer seal (see Figure 6.20). This persistence in continuing with a flawed design led directly to the cancelation of the program by the Secretary of Defense in 1980.

If the successful SES-100B design had been pursued into the 3KSES Phase, the program might have been announcing the next step in the program in 1980 rather than its cancelation. The SES-100B had met all the objectives and would scale to a frigate size SES with a speed of 100 knots with minimum risk.

6.11 The Three Thousand Ton Surface Effect Ship

Based on the results of the two 100 ton test craft program, the US Navy issued (after competition) a contract to Rohr Marine Inc (a subsidiary of Rohr Industries) (RMI) in late 1976 for the detail design and other contractual needs of a 3,000 ton class surface effect ship. The history of this program is provided in Chapter 5. Essentially, the ship was to be capable of performing an approved US Navy mission with a payload that included two SH-3H helicopters or one AV-8B Harrier attack aircraft, Harpoon missiles, Mk 32 torpedoes and a TACTAS towed array sonar as part of its ASW mission payload. An artist's rendering of the design is shown in Figure 6.60 and its leading particulars are provided in Table 6.8.

6.11.1 Relationship of 3KSES to SES-100A and SES-100B

In relating the experience base gained from the two 100 ton test craft, the following Table 6.8 characterizes the 3KSES design shown in Figure 6.60, as developed in 1976.

As can be seen, both test craft contributed significantly to the design layout proposed for the 3KSES design. Much of the detailed design features were classified making it difficult to present all the "lessons learned" from the 100 ton test craft program that were integrated into the 3KSES, but the summary in Table 6.9 captures the main features.

What is surprising is that, while the basic features of the CAB rigid seal design were unresolved in the design and testing phase of the SES-100A program, the US Navy made the decision that the 3KSES would use a variant of the CAB rigid planer seal. This would mean that the planned large, ocean going, mission capable ship would be using an undeveloped seal design (at the 100 ton stage) with known performance and reliability problems. Similarly, while the air ingestion and cavitation problems of the SES-100A waterjet propulsion system were not resolved, the

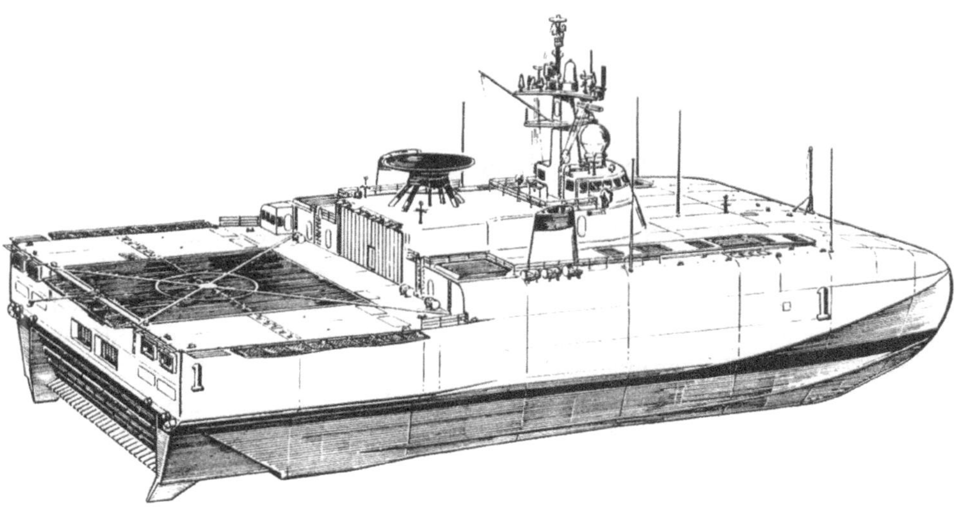

Figure 6.60: Three Thousand Ton SES (3KSES)

Table 6.8: *3KSES Leading Particulars*

Dimensions	
Length Overall	266 ft 3 in
Beam Overall	106 ft
Height (Keel to Deck)	40 ft
Cushion Depth (Keel to Wet Deck)	18 ft
Cushion Area	19,050 sq ft
Cushion Pressure (p_c)	324 psf
Cushion Density (p_c/L)	1.5 psf/ft
Speeds	
Maximum speed (Calm Water)	80 knots
Speed in Sea State 3 (6 ft waves)	TBD knots
Weights	
Normal Gross Weight	6,096,000 lb (3,048 s. tons)
Payload	204 s.tons
Fuel	1,203 s. tons
Weight Empty	1,595 s. tons
Lift System	
Two General Electric LM 2500 marine gas turbines	2 x 22,500 bhp (45,000 bhp)
Three directly coupled centrifugal fans each side	6 fans total
Flexible Bags and Rigid planer seals	Bow and Stern
Propulsion System	
Four General Electric FT 9A-2 marine gas turbines	4 x 36,500bhp max (146,000 bhp)
Two Aerojet two stage waterjets, flush inlets	

Table 6.9: *Features of SES-100A and SES-100B integrated into 3KSES Design*

3KSES Features	SES-100A	SES-100B
Thicker sidehulls and lower deadrise angle similar to		SES-100B design
All welded aluminum structure similar to		SES-100B design
Rigid Planer seals similar to	SES-100A design	
Separate lift and propulsion engine system similar to		SES-100B design
Lift fan system using centrifugal fans similar to		SES-100B design
Ride control system an upgrade from	SES-100A design	
Waterjet propulsion system similar to (modified)	SES-100A design	

US Navy directed that this form of propulsion would be integrated into the 3,000 ton SES design.

Subsequently, when DDR&E directed that a study of alternative SES designs to the US Navy design be pursued, to meet the US Navy mission requirements (see Chapter 10), it was found that when common mission payloads were integrated into the ship design that the ship displacement of the "three thousand ton" SES would have to be increased to the much larger size of 3,600 tons or more. This is discussed in more detail in Chapter 10. This again complicated what the next phase of the "large SES" program should be after the "100 ton" phase of the program had been completed.

A clear driving concern of OSD was that any ship would have to serve *a particular US Navy mission need and not be simply a continued R&D program.*

Chapter 10 discusses all the various candidates including the modification of the above leading particulars for such a ship designed for a US Navy mission. As the US Navy sought answers on whether it was possible to develop a practical 100 knot capability, all of the possible candidates were subjected to a rigorous analysis by the best designers in each area and subjected to peer review in the US Navy ANVCE Project. Chapter 10 shows some of the key conclusions and how the original design for the 3000 ton Surface Effect Ship illustrated in Figure 6.60 compared to these alternatives.

6.12 Conclusion of the US Navy High Speed Large SES Program

The end of the SES program came on 9 January 1980 when the entire program was canceled by the Office of Secretary of Defense (DDR&E) (see Chapter 5). The *CAB SES-100A* was scrapped in 1980 when the program was canceled. The successful *Sidehull SES-100B* was towed to Annapolis and installed on the seawall at the DTNSRDC laboratory as a permanent display in 1983 (see Chapter 5, Figure 5.8) where it stayed until 1999 when, under an unrelated Congressional decreed Base Realignment and Closure (BRAC) process, the US Navy Annapolis laboratory was closed and all assets of the facility disposed of, including the scrapping of the SES-100B in 1999 (see Chapter 5).

References

1. "*Symposium on Ground Effect Phenomena*", Princeton University, Department of Aeronautical Engineering and the US Army TRECOM, 21–23 October 1959.
2. *"A Review of the Sidewall Hovercraft",* by E. G. Tattersall, Jane's Skimmer Systems, 1979, Twelfth Edition. (originally presented to the UK Hovercraft Society 20 September 1978).
3. Mantle, P. J., *"Some Design Aspects of Air Cushion Craft",* Annals of the New York Academy of Sciences, Vol 154, Article 2, pp 924–952, 22 November 1968. Paper presented at the International Congress of Subsonic Aeronautics, New York, N.Y., 3–6 April 1967
4. Mantle, P.J and Hall, A.M., *"Multiple Hull Vehicle with Air Cushion Support"*, US Patent 3,476,069, Filed 24 January 1967, Patented 4 November 1969
5. "Jane's High Speed Marine Transportation, 2009–2010", Ed. Stephen J. Phillips
6. Mantle, P. J., "*Air Cushion Craft Development*", published by GPO (Stock Number 008-047-00291-2) in January 1980
7. Elsley, G.H. and Devereux, A.J., *"Hovercraft Design and Construction",* Cornell Maritime Press, Inc, 1968
8. Trillo, R.L., *"Marine Hovercraft Technology"*, Leonard Hill Press, London, 1971
9. *British Civil Air Cushion Vehicle Safety Requirements (Provisional)* published by Air Registration Board, UK, 1962
10. *British Hovercraft Safety Requirements*, published by Civil Aviation Authority, April 1972
11. Faltinsen, O. M., *"Hydrodynamics of High Speed Marine Vehicles",* Cambridge University Press, 2005
12. Mantle, P.J., "*A Technical Summary of Air Cushion Craft Development*", Report DTNSRDC 4727, October 1975 (ISBN 008-047-00196-7)
13. *"Jane's Surface Skimmers: Hovercraft and Hydrofoils"*, Twelfth Edition, Ed. Roy McLeavy, 1979
14. *"Jane's High Speed Marine Craft and Air Cushion Vehicles"*, Ed. Robert L. Trillo, 1986
15. Butler, Edward A., *"The Surface Effect Ship",* Chapter VI, Modern Ships and Craft, Special Edition of Naval Engineers Journal, American Society Of Naval Engineers, Volume 97, No. 2, February 1985

16. *"Jane's Surface Skimmers: Hovercraft and Hydrofoils"*, Ed. Roy McLeavy, 1979 and *"Jane's High Speed Marine Craft and Air Cushion Vehicles"*, Ed. Robert L. Trillo, 1986
17. Mantle, P. J., *"Air Cushion Craft Development"*, published by GPO (Stock Number 008-047-00291-2) January 1980
18. Hogben, N. *" Hovering Craft Over Water",* Advanced Hydroscience, No. 4, Academic Press, New York, 1967
19. Mantle, P. J., *"Air Cushion Craft Development"*, published by GPO (Stock Number 008–047-00291–2) January 1980
20. *ibid*, page 278
21. *ibid*, page 275–282
22. *ibid*, Chapter VII: Structural Design, pages 253–294
23. *ibid*, Chapter VIII: Lift Fan System, pages 295–329
24. Mantle, P.J., *"A Technical Summary of Air Cushion Craft Development"*, Report DTNSRDC 4727, October 1975 (ISBN 008-047-00196-7)
25. *ibid*, pages 153–180
26. *"Guide for Evaluation of Human Exposure to Whole Body Vibration",* International Standard Organization (ISO) Document ISO/DIS 2631, (1972)
27. O'Hanlon, J.F. and M.E. McCauley, *"Motion Sickness Incidence as a Function of the Frequency and Acceleration of Vertical Sinusoidal Motion"*, Human Factors Research, Inc. Technical Report 1733–1 (Sept 1973)
28. Mantle, P. J., *"Air Cushion Craft Development"*, published by GPO (Stock Number 008–047-00291–2) in January 1980
29. *ibid*, Chapter VIII: Lift Fan System, pages 295–329
30. *ibid*, Chapter IX: Propulsion, pages 331 – 405
31. *ibid*, Chapter IX: Propulsion
32. Tulin, M.P. and M.P. Burkart, *"Linearized Theory for Flows about Lifting Foils and Zero Cavitation Number"*, DTMB Report C-638, February 1955
33. Tachmindji, A.L. and W.B. Morgan, *"The Design and Estimated Performance of a Series of Supercavitating Propellers"*, Proceedings of the Second Symposium on Naval Hydrodynamics, August 1958
34. Newton, R.N. and H.A. Rader, *"Performance Data of Propellers for High Speed Craft"*, Trans. RINA, 103, 2, pp 93–129, 1961
35. Faltinsen, O. M., *"Hydrodynamics of High Speed Marine Vehicles",* Cambridge University Press, 2005
36. Johnson, V. E. and M. P. Tulin, *"The Hydrodynamic Characteristics of High Speed Hydrofoils",* paper at Institute of Aeronautical Sciences (IAS), 29th Annual Meeting, IAS Paper 61–41, New York, N.Y., 23–25 January 1961.
37. Mantle, P. J., *"Air Cushion Craft Development"*, published by GPO (Stock Number 008–047-00291–2) January 1980
38. Mantle, P. J., *"Variable Geometry Propulsor"*, US Patent 3,830,190, Filed 24 May 1973, Patented 20 August 1974
39. Butler, Edward A., *"The Surface Effect Ship",* Chapter VI, Modern Ships and Craft, Special Edition of Naval Engineers Journal, Volume 97, No. 2, February 1985
40. Capt Carl J. Boyd, News Conference with Secretary of Navy J. William Middendorf III in Panama City, Florida, 30 June 1976 after SES-100B made its world speed record run.
41. Raley, F. Carl and Wilson, Fred W.; *"Test and Evaluation of US Navy 100-Ton Surface Effect Ships",* AIAA/SNAME Advanced Marine Vehicles Conference, Arlington, Virginia, AIAA Paper No. 76–860, 20–22 September 1976
42. SES Program Office submittal to Janes' Surface Skimmer Systems for the 1979 issue.
43. Mantle, P. J., *"Air Cushion Craft Development"*, published by GPO (Stock Number 008-047-00291-2) January 1980
44. ibid
45. Raley, F. Carl and Wilson, Fred W.; *"Test and Evaluation of US Navy 100-Ton Surface Effect Ships",* AIAA/SNAME Advanced Marine Vehicles Conference, Arlington, Virginia, AIAA Paper No. 76–860, 20–22 September 1976

46. Mantle, P.J., *"Development of the USN Surface Effect Ship, SES-100B"*, Naval Engineer's Journal, October 1973, pp 65–78)
47. *ibid*
48. *ibid*
49. Mantle, P. J., "*Air Cushion Craft Development*", published by GPO (Stock Number 008-047-00291-2) January 1980. See Appendix C, pages 465–479
50. Everest, J.T. and N. Hogben, "*Research on Hovercraft Over Calm Water*", Transactions Royal Institute of Naval Architects, Vol 109, pp 311–326, 1967
51. Doctors, L. J., "*Wave resistance of an Air Cushion Vehicle in Accelerated Motion",* Naval Architecture and Marine Engineering Report 099, December 1970
52. Mantle, P. J., "*Air Cushion Craft Development*", published by GPO (Stock Number 008-047-00291-2) in January 1980. See Chapter III, Performance Aspects, pp 90–92
53. *ibid*
54. *"Surface Effect Ships for Ocean Commerce (SESOC), Final Report on a Study of the Technological Problems*", by the SESOC Advisory Committee, issued by US Department of Commerce and published by US Government Printing Office, February 1966

7 Economic Considerations

"What Price Speed?"

Giuseppe Gabrielli and Theodore von Kármán, 1950

7.1 Declining American Marine Industry

In the late 1950s and early 1960s, the American marine industry was in trouble. Nicolas Johnson[1], US Maritime Administrator (1964–1965) discussed the problems of the high cost of shipping exports overseas which also meant an increase in the cost of American goods to the foreign consumer. In 1960, the US Share of world shipping was less than 10% despite a major shipbuilding program that had been underway in the US during World War II. Unfortunately, most of the shipbuilding in the US during WW II was embodied in the American Liberty Ship, a standard dry cargo vessel (approx 10,902 dwt) and the T2 tanker (16,543 dwt). The Liberty ship was a wide bodied, bulbous ship with a disappointing top speed of 11 knots. Most of the foreign ships at that time were designed for higher speeds (14 knots and more). This placed the US at a disadvantage in capturing any significant share of the world shipping trade.

This decline in the US share of world shipping, that continues to this day, is shown in Figure 7.1. In addition, there was the problem of high labor and high capital costs, which caused the US government to subsidize the shipbuilding industry at rates at different times through this period of 30% and 50%. The US Maritime Administration needed to find another solution. At that time, the only alternatives available to the shipper was "By Sea" or "By Air" with vastly different costs. The two choices available to the shipper, whether it be a manufacturer of say automobile parts in Detroit or a grain supplier in the mid-West; were to ship by air at 20 cents per ton mile[a] and have his goods arrive within 24 hours, or ship by sea at 5 cents per ton mile and have his goods arrive say, 5 days later. Issues of weight and bulk (space) and needs of roll-on, roll-off capability also entered into the decision of which shipping method to use, but the cost was a significant driver.

[a] The values of 20 cents per ton mile (by air) and 5 cents per ton mile (by sea) are the 1964 values as used at the time by the US Maritime Administration in its planning to examine new forms of economical high speed transoceanic transportation.

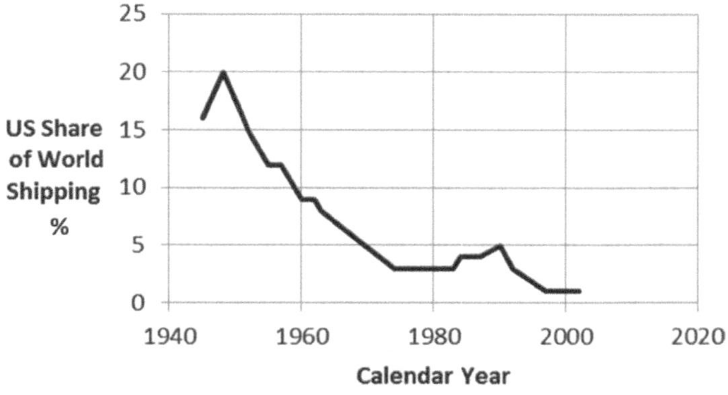

Figure 7.1: US Share of World Shipping

As a further indication of the US Maritime Administration concern was the significant drop in the US shipbuilding construction. During WW II, the US was essentially building 100% of the ships in the world. By 1960 that had dropped precipitously to less than 10 percent as shown in Table 7.1, taken from the data in *Lloyds Register of Shipping*. The Table shows the numbers for the past four decades or more (1960–2005) expressed as percentage of merchant ships built by nations. The numbers are rounded for visibility and may not add to 100%.

Table 7.1: *Merchant Ship Construction (Percentage of World Gross Tonnage)*

	1960	1977	1980	1985	1990	1995	2000	2005
Asia	22%	46%	52%	69%	70%	75%	84%	85%
Europe	66%	41%	25%	18%	18%	17%	11%	10%
Eastern Europe	5%	6%	10%	7%	8%	6%	4%	3%
Others	7%	3%	8%	5%	3%	2%	2%	2%
USA	5%	4%	4%	1%	<1%	<1%	<1%	<1%

All of these concerns led the US Maritime Administration in 1960 to pursue several possibilities of improving the American Merchant Marine. The new possibilities considered and explored by the US Maritime Administration included (a) the nuclear ship N.S. Savannah, (b) the hydrofoil, and (c) the ground effect machine (the name change to hovercraft and surface effect ship came later). These have been discussed briefly in Chapter 4 and will be returned to later.

The merchant marine is particularly vulnerable to certain costs that make it non-competitive to foreign shippers. The higher standard of living of the US labor force and cost of materials had already placed the US merchant marine in a declining situation. Based on internal studies at the US Maritime Administration it was determined that there was a need to improve such things as loading and unloading time, increase ship utilization as well as the basic ship design and also, most importantly, to increase ship speed. The non-competitive low speed of the Liberty ship design was not good. How much the speed should be increased was a subject of further research and led the US Maritime Administration to initiate development of first, the hydrofoil and secondly, the surface effect ship (SES).

Since conventional (displacement) ships typically traveled at speeds of 20–40 knots, with most of the ships traveling at speeds closer to 25–30 knots, and aircraft traveled at speeds of 200–300 knots or more, MARAD reasoned that speeds around 100 knots was

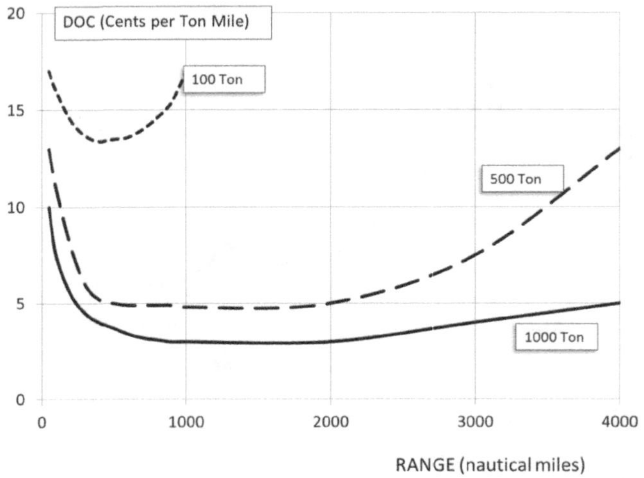

Figure 7.2: MARAD Estimates of Direct Operating Costs (DOC) of "*Columbia*" Class Surface Effect Ships

a good "goal" for economic advantage provided the technologies involved could support such a goal. In 1962, MARAD contracted for the design of the first surface effect ship (SES), the "*Columbia*". See Chapter 3 for descriptions of the surface effect ship "*Columbia*". Figure 7.2, taken from the initial MARAD studies (1960) shows the typical DOC expected. Such estimates were incorporated into the broader surface effect ship studies embarked upon by MARAD in 1965 when it created the Surface Effect Ships for Ocean Commerce (SESOC) Committee in September 1965 upon direction of the Assistant Secretary for Science and Technology, J Herbert Holloman. Analyses of the direct operating costs (DOC) of "*Columbia*"-based ship designs from 100 ton displacement to 1000 ton displacement were encouraging.

Values of about 13 cents per ton mile of the 100 ton class operating over routes of 500 nautical miles and down to less than 5 cents per ton mile for a 1000 ton class SES operating over routes of 1000 to 4000 nautical miles were deemed possible and would meet MARAD's goal of *economic transport at 100 knots over the sea* at costs much less than for air transport.

7.2 Direct Operating Costs

Computing the Direct Operating Costs (DOC) of any vehicle in commercial operation involves many factors making it difficult to make generalized conclusions but some insight can be obtained by concentrating on the main contributors to the cost.

Using the well-established DOC method for transport aircraft known as the Air Transportation Association (ATA) Method[2], the author developed in the early 1960s with his work on short haul aircraft and published later in 1968[3], an equation that captured all the key elements, that gave the direct operating cost in \$/mile as:

$$DOC(\$/mile) = C_1 + C_2\left(\frac{W_f}{R}\right) + \frac{C_3}{V_b} \tag{7.1}$$

This simple equation provided a unique method to compare the DOC of different vehicles without providing the detailed values of any particular case because of all the varying inputs needed for any particular route calculation. The terms in this simplified direct operating cost equation are shown in Table 7.2. The units in each of the terms can be adjusted to whatever system the reader chooses of course, but English units have been chosen here for illustrative purposes. The total operating

Table 7.2: *Direct Operating Cost Comparative Formula*

$$DOC(\$/mile) = C_1 + C_2\left(\frac{W_f}{R}\right) + \frac{C_3}{V_b}$$

C_1 = Fixed crew costs and P.L. & P.D. insurance	C_3 = Variable cost factor (= K_c+K_m+K_a/U)
C_2 = Fuel and oil cost ($/gal)	K_c = Operating crew cost ($/block hour)
W_f = Fuel consumed (gal)	K_m = Maintenance cost (including burden), ($/block hour)
R = Range or stage length (miles)	K_a = Amortization and craft insurance ($/annum)
V_b = Ship block speed (mph)	U = Annual utilization (block hours/annum)

cost (TOC), including indirect costs such as cost of docking facilities, terminal employees, and G&A expense, then becomes:

$$TOC(\$/mile) = DOC(1 + K_i) \tag{7.2}$$

where K_i is the indirect to direct cost ratio. Depending on the type of ship being analyzed, the economic analyses then continue to compute *cost per seat mile* (for passenger vessels) or *cost per ton mile* (for transport ships such as cargo container ships and the like). In the MARAD SESOC work, interest centered on the *cost per ton mile* as shown in Figure 7.2 for one of the candidates considered.

It is not the purpose of this book to explore and document the economic value of a range of ships and concepts. That has been done elsewhere in great detail (and discussed later in appropriate chapters). It is the purpose however to isolate the key parameters that will influence the design choice, especially in pursuit of the goal of high speed ships that can operate economically and efficiently over transoceanic routes.

In determining Direct Operating Cost (DOC), it is seen that three key "technical characteristics" enter the DOC equation (Equation 7.1). They are the ship block speed (V_b); the ship fuel consumed (W_f) and the stage length or range (R). The concept of "block speed" is an accepted term in the aircraft industry that is the actual cruise speed of the vehicle over a given stage length modified to account for the time loss in "takeoff, climb-out, let-down, landing, taxiing" for aircraft. Similar "end game" effects would be included for ships leaving and returning to the dock.

7.3 Bréguet Range Equation

These three terms can be effectively grouped in the Bréguet Range equation, which was developed for the aircraft industry by Louis-Charles Bréguet (Figure 7.3), the famed French aviation pioneer during World War I, based on his work on his own designed and built aeroplanes. Bréguet presented his equation originally to the Royal Aeronautical Society, on 6 April 1922, in a presentation entitled *"Aerodynamical Efficiency and the Reduction of Air Transport Costs"*. It is an amazing but simple piece of work that captures all the main elements for efficient vehicle design.

The original work by Bréguet was for propeller driven aircraft and the form of the equation changes slightly for other propulsion schemes (turbojet, turbofan, etc)

Figure 7.3: Louis-Charles Bréguet (1880–1955)

but the essential elements can be viewed in the original form (see Chapter 8 for other forms). There are also several assumptions in the equation that determine how close the predicted range of an aircraft is when compared to the actual range. The Bréguet Range equation assumes that the aircraft is operated at constant speed, that the fuel consumption is kept to a minimum and constant throughout the cruise and that the ratio of the initial weight (take-off weight) to the final weight (with fuel expended) is as large as possible. Sometimes, for more accuracy, the Bréguet Range equation is computed in increments along the route as the characteristics of power, fuel consumption and other features change as the weight of the vehicle decreases as fuel is consumed. These are refinements however and do not change the basic features so ably captured by Bréguet's elegant equation. As before, English units have been chosen for this particular presentation.

As Bréguet pointed out in his lecture, the three terms in the Range equation (in each of the three brackets in Table 7.3) represent the *design efficiency* of the vehicle as the product of (in Bréguet's words) the *propulsion, aerodynamic* and *structural design efficiencies* respectively. The first of these is the propulsion system efficiency measured by the *propulsive efficiency* (η_p) and the *specific fuel consumption* (*sfc*) of the vehicle (with all of its ramifications in choice of powerplant, mode of operation, etc). The second is the *aerodynamic efficiency* measured by the lift-drag ratio (L/D), of which more will be said later. The third is the *structural design efficiency* measured by the ability of the designer to accomplish a design to withstand all the loads imparted to it and to maximize the amount of fuel (W_f) that can be accommodated while still carrying some designated payload. Given these assumptions, the Bréguet Range Equation is, for the case of propeller driven aircraft, given in Table 7.3:

It makes the following discussion easier, if it is now assumed that the ratio of the fuel weight to the initial weight is small, such that (W_f/W) is much less than unity, so that the natural logarithm can be expanded to be,

$$\log_e\left(\frac{W}{W - W_f}\right) = \frac{W_f}{W} + {}^1\!/_2\left(\frac{W_f}{W}\right)^2 + \ldots\ldots\ldots \quad (7.3)$$

Then, to first order, the expanded natural logarithm can be written,

$$\log_e\left(\frac{W}{W - W_f}\right) = \frac{W_f}{W} \quad (7.4)$$

For a fuel load W_f/W=20%, this approximation underestimates the range by 10%. Within the assumption quoted, the Bréguet Range equation can be approximated by,

$$R = 325\left(\frac{\eta_P}{sfc}\right)\left(\frac{L}{D}\right)\left(\frac{W_f}{W}\right) \quad (7.5)$$

This form of the Bréguet Range equation is a little easier to handle and still displays the three basic *design efficiencies* referred to by Bréguet of the vehicle to transport goods

Table 7.3: *Bréguet Range Equation*

$$R = 325\left(\frac{\eta_P}{sfc}\right)\left(\frac{L}{D}\right)\left(log_e \frac{W}{W - W_f}\right)$$

R	Range (nautical miles)
Sfc	specific fuel consumption (lb per HP hour)
η_P	Propulsive efficiency (non-dimensional)
L/D	Aerodynamic efficiency (Lift/Drag, non-dimensional)
W	Weight at start of cruise (tons)
W_f	Weight of fuel consumed at end of cruise (tons)
log_e	Natural logarithm

over a given range. The relationship to the direct operating cost (DOC) Equation 7.1, given earlier, also becomes a little more recognizable. For any given vehicle, it can be seen that for maximum range, it is normal to operate the vehicle at a cruise speed for maximum value of (L/D) or (WV/P) which would occur at a reduced power setting. The reader is referred to the standard texts of aeronautical engineering as to how to compute that speed for which the (L/D) or (WV/P) is a maximum (see Chapter 8).

In the aircraft industry, there are other operating conditions to that given by the Bréguet range equation that are sometimes used by pilots depending on the need of the mission. These include operating for Maximum Endurance (time), an important characteristic for air loiter aircraft on surveillance missions, rather than Maximum Range (distance) for example, but such variations are not considered here further as the intention in this book is to concentrate on the economical and technical aspects of operating *efficiently at high speed* over the sea.

One more step is required before the economic equation for DOC can be addressed in terms of seeking the best platform and form of *surface effect ship* under consideration. For equilibrium flight, the lift (L) of the aircraft must equal the weight (W). Also, the power expended to propel the aircraft (P) must overcome the drag (D) of the aircraft when flying at speed (V), such that

$$\eta_P \mathrm{P} = \mathrm{DV} \tag{7.6}$$

This then allows the *aerodynamic efficiency* term (L/D) to be written,

$$\eta_P \frac{L}{D} = \frac{WV}{P} \tag{7.7}$$

where (η_P) is the propulsive efficiency of the vehicle.

The approximate form of the Bréguet Range equation can now be written,

$$R = 325\left(\frac{1}{sfc}\right)\left(\frac{WV}{P}\right)\left(\frac{W_f}{W}\right) \tag{7.8}$$

In this form, one can see how the designer must work with the three Breguet efficiency terms to maximize the range potential of his vehicle. This form of the equation needs to be interpreted differently, however when considering vehicles other than aircraft and conventional ships or more correctly, any vehicle that requires two distinct power sources to function. This important difference will be returned to later when discussing the technical considerations of surface effect ship designs.

It is seen that the propulsion system efficiency and the aerodynamic efficiency combine when considering the power requirements for an efficient design. This will become more evident when the effective lift-drag ratio is considered (see later with Equation 7.23). In summary, in terms of economical considerations, the two important equations are the DOC equation and the Range equation, as shown here:

$$DOC(\$/mile) = C_1 + C_2\left(\frac{W_f}{R}\right) + \frac{C_3}{V_b} \tag{7.9}$$

$$R(miles) = 325\left(\frac{1}{sfc}\right)\left(\frac{WV}{P}\right)\left(\frac{W_f}{W}\right) \tag{7.10}$$

With these simple equations one can see how the technical characteristics of the vehicle influence the direct operating cost (DOC) of the vehicle. Other economic terms considered in any economic evaluation are (a) the productivity, measured in *payload ton miles* and (b) the *payload ton miles per pound of fuel.* Using the above equation for the range, these economic measures become:

$$\text{Productivity (payload ton-miles)} = 325\left(\frac{1}{sfc}\right)\left(\frac{WV}{P}\right)\left(\frac{W_f}{W}\right)W_p \tag{7.11}$$

$$\text{Payload ton-miles per pound of fuel} = 0.163\left(\frac{1}{sfc}\right)\left(\frac{WV}{P}\right)\left(\frac{W_p}{W}\right) \tag{7.12}$$

Where the payload (W_p) is measured in short tons and the range (R) is measured in nautical miles (with these particular "constants" in the equation). This derivation explains why some analysts plot curves of (W_pV/P) instead of (WV/P) because the weight (W) or ship displacement "cancels out" in the above Equation (7.12) but this is a corruption of two different ship design efficiencies in the comparison and combining them can mask important differences between vehicles and is not recommended.

7.3.1 The Three Bréguet Efficiencies for Economic Transport

As summarized above, Bréguet in 1922, highlighted three efficiencies that were important to maximize the range of any transport vehicle. Using his words, they were the *propulsion, structural design* and *aerodynamic efficiencies* to obtain what he called the *design efficiency* of the vehicle which equated directly to range. These have been shown in one of its various forms in Equation 7.10. An outline is given on each of these efficiencies to set the stage for the technological development that followed in the 1950s until today to achieve high speed, efficient transport at sea over transoceanic routes.

A comment first, is worth making on Bréguet's choice of the word "efficiency" in his work that has been echoed here in this author's work and used throughout. "Efficiency" is not just the ratio of some parameter to the ideal, which gives rise to, say, efficiency is "80% or some other number of some ideal value". Efficiency is used here in the well understood context of "well engineered" or "efficient design" meaning that all systems, sub-systems and components have been engineered and integrated into the total design so as to give the "best design possible". When one is buying an automobile, the car salesman has no problem in understanding the buyer's request for the most "fuel efficient" car on the lot! The salesman would not understand, in this context, some "fuel efficiency" expressed as some number between zero and 100% of some ideal value! What Bréguet was striving for, in his term "*design efficiency*" was what combination of design features of *propulsion, structural design* and *aerodynamic efficiencies* would provide the best vehicle to meet some predetermined requirement, such as "maximum range" or "minimum direct operating cost". The phrase "*structural design efficiency*" as used by Bréguet has a broader context than just "structure" as will be discussed in what follows. It is the design of all systems within the vehicle structure to give the best overall design to carry a given payload efficiently and economically over some given distance. It is in this context that the discussions in this book examine these features as they apply to the problem of obtaining "efficient", economic transportation for high speed over the ocean.

Propulsion Efficiency Bréguet was referring here specifically to the ability of the designer to include powerplants with good fuel efficiency (*sfc*) and the *propulsive efficiency* (η_P) of the propulsor (e.g. propeller). The propulsive efficiency is more correctly included in the aerodynamic efficiency as shown in Equation 7.10 and is discussed later. The fuel efficiency is best measured by the specific fuel consumption as it appears in Bréguet's equation,

$$\text{Propulsion Efficiency} = \frac{\eta_P}{sfc} \tag{7.13}$$

Historically, the specific fuel consumption has improved over time and varies with the type of powerplant and power rating of the engine as shown in Figure 7.4 taken from Mantle (1966)[4] and updated in 1975 and 1980 with improved powerplant data.

New power plant developments since that time have been added to Figure 7.4 and this indicates that while major changes in this parameter have occurred in engine performance, all to the advantage of the vehicle designer (with lower values of *sfc*), it is usually outside the ability of the vehicle designer to modify the powerplant and its fuel consumption characteristics other than to simply integrate it into the vehicle design.

The gas turbine usually chosen in high speed hovercraft and surface effect ship design because of its low specific weight (from 0.20 lb/HP to 1.0 lb/HP over an engine power range of 1,000 to 20,000 HP) has a high specific fuel consumption (*sfc*) varying from 0.40 to 0.80 lb/HP.hr over an engine range of 1,000 to 20,000 HP with *sfc* improving with increased power. Over this same HP range, the diesel engine, due to its high thermal efficiency cycle, has lower *sfc* values of 0.30 to 0.40 lb/HP.hr. In the last ten years (since 2000), further improvements on fuel efficiency have occurred.

The characteristics of gas turbines (and similarly for aircraft piston engines which tend to have *sfc* values between the gas turbine and the diesel engine) are fairly standard and usually accepted by vehicle designers in their vehicle designs. Except in special cases, the *propulsion efficiency* ($=\eta_p/sfc$) referred to by Bréguet, is normally treated as an input to the vehicle design rather than anything that can be controlled by the vehicle designer as distinct from the remaining two "efficiencies" (*structural design efficiency* and *aerodynamic efficiency*) in the Bréguet Range equation but is included here for completeness. Improvements in fuel types, emission standards, materials development have all served to improve the efficiency in terms of fuel consumption, so important to increasing range shown in Bréguet's equation. Engine manufacturer handbooks provide the necessary data for any particular vehicle design.

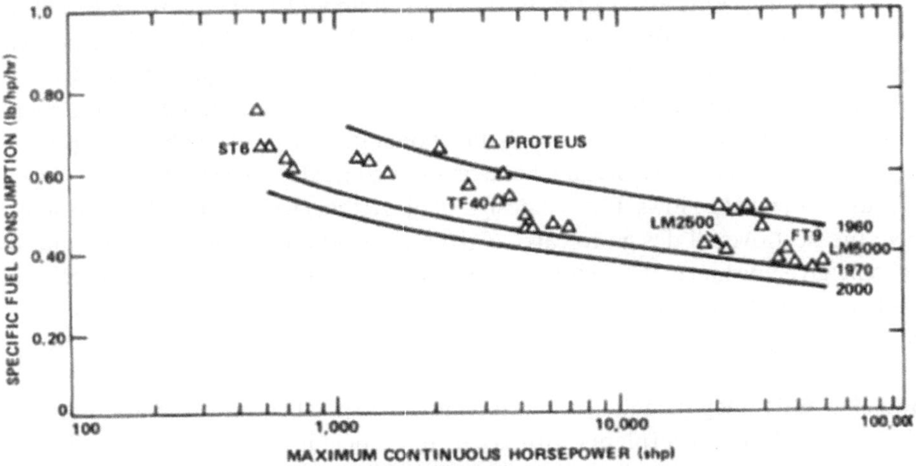

Marine Gas Turbines

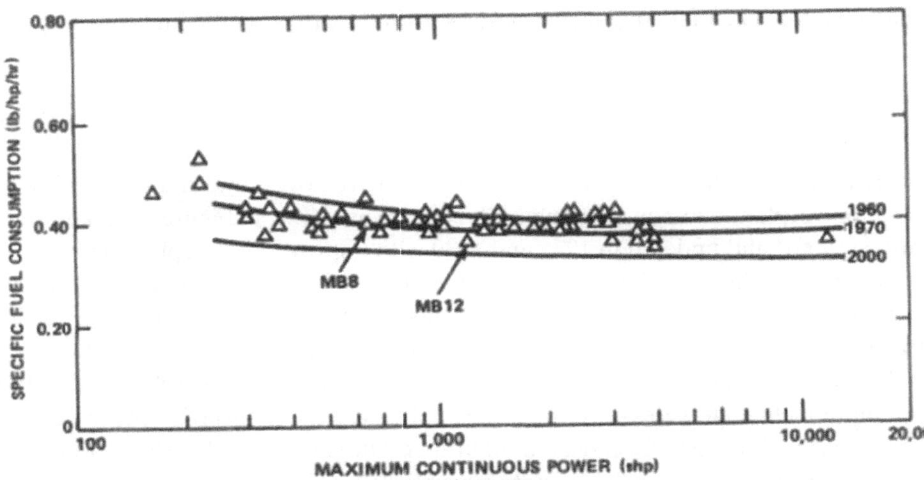

Diesel Engines

Figure 7.4: Trends in Specific Fuel Consumption with Power Setting over Time

Aerodynamic Efficiency and Transport Efficiency The second term in Bréguet's Range equation, that has received considerable attention in the development of the high speed surface effect ship, is the *aerodynamic efficiency*, defined by Bréguet as,

$$\text{Aerodynamic Efficiency} = \frac{L}{D} \tag{7.14}$$

which is related to the term (WV/P) in his Range equation by,

$$\frac{L}{D} = \frac{1}{\eta_P} \frac{WV}{P} \tag{7.15}$$

where (η_P) is the *propulsive efficiency* of the propulsor, not to be confused with Bréguet's use of the term *propulsion efficiency* measured by (η_P/sfc), which is the first term in his Range equation. This second efficiency term has received considerable attention over the years in the development of the surface effect ship and is considered in the following section where the terms of *aerodynamic efficiency* and *transport efficiency* are explored in detail.

Effective Lift-Drag Ratio In all the original work of comparing vehicles prior to the late 1950s, the above use of the aircraft derived Bréguet range equation into the economic considerations for ships was a valid evaluation approach. This is because whether "aircraft" or "ships" were being propelled, there was only one power source (P) installed to propel the vehicle forward. The number of engines does not effect the total power installed in the vehicle. Ships were essentially supported by *buoyant forces* from the water and in some cases *dynamic forces* on planing hulls. Aircraft were supported (once they were airborne) by the *aerodynamic forces.* In both cases, there was only the need for propulsion power (P) to propel the vehicle (aircraft or ship) forward.

With the advent of the air cushion supported vehicles such as hovercraft and the surface effect ship in all its forms, the situation changed and additional power sources were required. This extra power was required to lift (or lubricate) the craft and was quite distinct from the power required to propel the vehicle forward.

This then meant that the total power installed in the vehicle (hovercraft, surface effect ship) would now be the sum of the propulsion power (P_P) and the lift power (P_L), that is:

$$P = P_P + P_L \tag{7.16}$$

The propulsion power is related to the drag (D) and speed (V) of the vehicle through,

$$\eta_P P_P = DV \tag{7.17}$$

The lift power (P_L) would depend on the air cushion pressure (p_c) and the air cushion flow (Q), such that,

$$\eta_L P_L = p_c Q \tag{7.18}$$

The total installed power in the vehicle would then be expressed as,

$$P = \frac{DV}{\eta_P} + \frac{p_c Q}{\eta_L} \tag{7.19}$$

By substitution, it is easy to show that the Lift-Drag ratio (L/D) used in the earlier work on aircraft and ships would now be replaced in the Bréguet range equation for hovercraft and surface effect ships by a new term called the *Effective Lift Drag Ratio* $(L/D)_{EFF}$ as shown:

$$\left(\frac{L}{D}\right)_{EFF} = \frac{W}{D + \frac{\eta_P}{\eta_L}\left(\frac{p_c Q}{V}\right)} \tag{7.20}$$

This new term came into being when considering "air cushion assisted" vehicles. It is the familiar aerodynamic lift-drag ratio (L/D) modified by the air cushion effect, $\left(\frac{\eta_P}{\eta_L}\right)\left(\frac{p_c Q}{V}\right)$, which has the dimensions of drag. Note that as the air cushion flow (Q) goes to zero, the effective lift-drag ratio $(L/D)_{EFF}$ returns to the aerodynamic lift-drag ratio (L/D) The complete derivation of the Bréguet Range equation as it applies to hovercraft and surface effect ships, incorporating the lift system effects embodied in the Effective Lift Drag ratio $(L/D)_{EFF}$ above may be found in Mantle[5] (1980) and earlier cited references. The Transport Efficiency is then defined as

$$\text{Transport Efficiency (WV/P)} = \eta_P \left(\frac{L}{D}\right)_{EFF} \tag{7.21}$$

The reduction of the effective lift-drag ratio by the system propulsive efficiency relates the efficiency to the installed power in the vehicle.

When comparing vehicles whether on an economic basis or on a technical basis it is essential to include this "modifier" of the standard aircraft derived Bréguet Range equation or any other calculations involving power. Frequently from test data of actual vehicles, one must use the installed power for propulsion (P_P) and the installed lift power (P_L) for assessments because of the difficulty of determining the actual system efficiencies (η_P, η_L) and the air cushion flow (Q) in any actual operations.

With these provisos in mind, it is now worth delving a little deeper into how the vehicle characteristics seen in both the Bréguet Range equation and in the Direct Operating Cost (DOC) equation manifest themselves in the operating economics. This was the subject of a seminal Thurston lecture and paper in 1950 put together by two pillars of engineering, Professor Theodore von Kármán and Giuseppe Gabrielli, in their paper *"What Price Speed?"*[6]

Structural Design Efficiency The third efficiency addressed by Bréguet is that related to the ability of the designer to design his craft in such a manner as to transport a given Payload (passengers or cargo) over the distance or range required. This parameter has occupied ship designers as well as aircraft designers since the beginning of transport as designers have put pencil to paper (and quill to papyrus!). There is first the problem of designing the structure itself which is especially an issue to high speed ship designers who must strive for light weight yet retain sufficient ruggedness to withstand the pounding of the waves.

The author in 1966[7] showed the trends of all vehicles (ships, aircraft and the existing air cushion craft) structural weights as shown in Figure 7.5. Since the original data was shown, the projected data for "large SES" from the design work of the 1980s has been included as the technology has advanced.

The trends shown in Figure 7.5 demonstrate the advantage of conventional ships with their inherent shape and buoyant support from the water that results in lower structural weights; and the disadvantage of aerodynamically supported aircraft that must operate at very high speed with dynamic loading. Since this original collection of data in 1966, it is informative to see that after half a century of intensive development, the structural weights of the best designed hovercraft and surface effect ships are still significantly higher than conventional "ships" and "aircraft". The State-of-the-Art (SOA) curve shown is the envelope of the state-of-the-art at that time (1966) but it is important to note, has not improved significantly since then. It is given by,

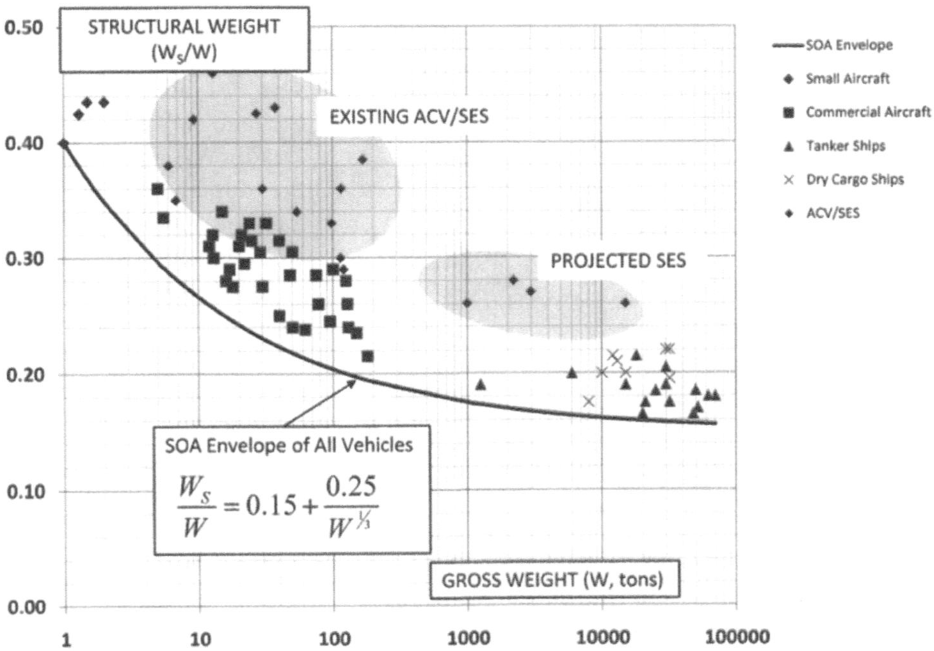

Figure 7.5: Historical Trends in Structural Weight

$$\left|\frac{W_S}{W}\right|_{env} = 0.15 + \frac{0.25}{W^{1/3}} \tag{7.22}$$

More consideration on actual structural weights for consideration in cost estimates appears later in this Chapter. An additional 0.10 to Equation 7.22 would give a SOA line through the air cushion craft data.

In the 1960s, as "hovercraft" and "SES" were being designed there was considerable concern on how to design a lightweight structure (usually of high grade marine aluminum) that could or should be designed to be light enough to be supported by an air cushion yet at the same time be rugged enough to withstand the poundings of wave action at high speed. In the next chapter on Technical Considerations, the various techniques that have been tried and incorporated into the existing designs such as the 100 ton class SES built and tested, which had structural weight fractions of greater than 30%, is presented.

A further manifestation of the Bréguet *structural design efficiency* is the ability of the designer to design his vehicle to "make room" in his vehicle for the rest of the essential sub-systems and components such as engines, controls, electrical systems, and of course the payload and fuel. This is best measured by the Empty Weight (W_E) of the vehicle that includes everything except the Payload (W_p) and the Fuel (W_f). The combination of the Payload and the Fuel Load is called the Disposable Load (W_D), such that,

$$W = W_E + (W_p + W_f) \tag{7.23}$$

where Disposable Load (W_D) is given by,

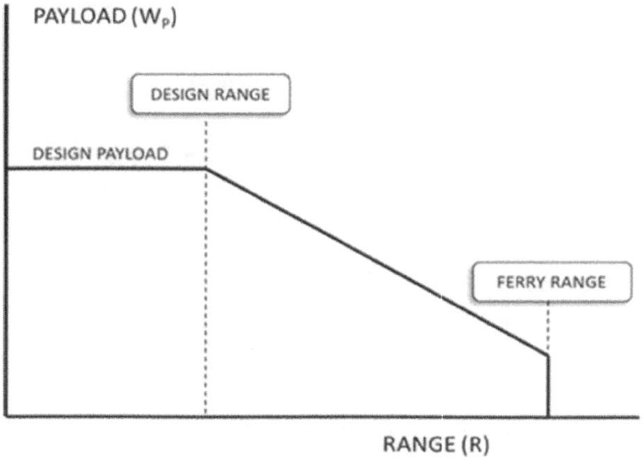

Figure 7.6: Typical Payload –Range Characteristic for Vehicles

$$W_D = W_p + W_f \quad (7.24)$$

The trade between Payload (W_p) and Fuel Load (W_f) is now more the domain of the commercial operator and less so of the ship (or aircraft) designer. Once the vehicle is built, it may choose to carry large payloads over short distances or smaller payloads over longer distances, all within the same vehicle. Each vehicle typically has in its design specifications a Payload-Range chart similar to that sketched in Figure 7.6

The vehicle typically has a "design payload" and corresponding "design range". It also has a longer "ferry range" that at the end of its cruise has just sufficient fuel (reserve fuel) for final maneuvering, emergency fuel and other contingencies. Commercial aircraft for example, by FAA regulations must always have sufficient fuel on board to fly to an alternate airport in times of bad weather or emergencies. Similar rules are applied to commercial shipping. "Rules" for dynamic vehicles such as hovercraft and in particular winged surface effect ships are less defined.

Because of this variation in any vehicle on payload (W_P) and fuel load (W_f), it is more informative and characteristic of the vehicle types and ability (as Bréguet states) to compare his *structural design efficiency* through examination of the Disposable Load (W_D) capability of different vehicle types. Such a comparison shows how the designer was efficient (or not) in his design to accommodate the load carrying capacity of the vehicle.

Figure 7.7, taken from Mantle[8] shows the trends of a range of vehicles from ships to aircraft for the Disposable Load (W_D/W) capability. Included in Figure 7.7, is some data for hovercraft that predicted at that time (1967) that one could expect a Disposable Load (W_D/W) capability of such craft to be about 40%.

Since that original collection of data, there have been actual high speed surface effect ships built plus several new projections of designs. That new information is provided in Figure 7.8. One can see in comparing the two sets of data (two and a half decades later) that the prediction of 40% disposable load capability is still valid – even though there are some new ambitious projections in the literature!

On examination of Figure 7.8 and comparing it with the broader set of data for all classes of marine vehicles that the hovercraft and the surface effect ship typically have less "room" or capability for Disposable Load (W_D) when compared to conventional aircraft and ships (see Figure 7.7) because of the need for extra systems such as lift fan systems, skirts and seals, etc, but it is still an acceptable value.

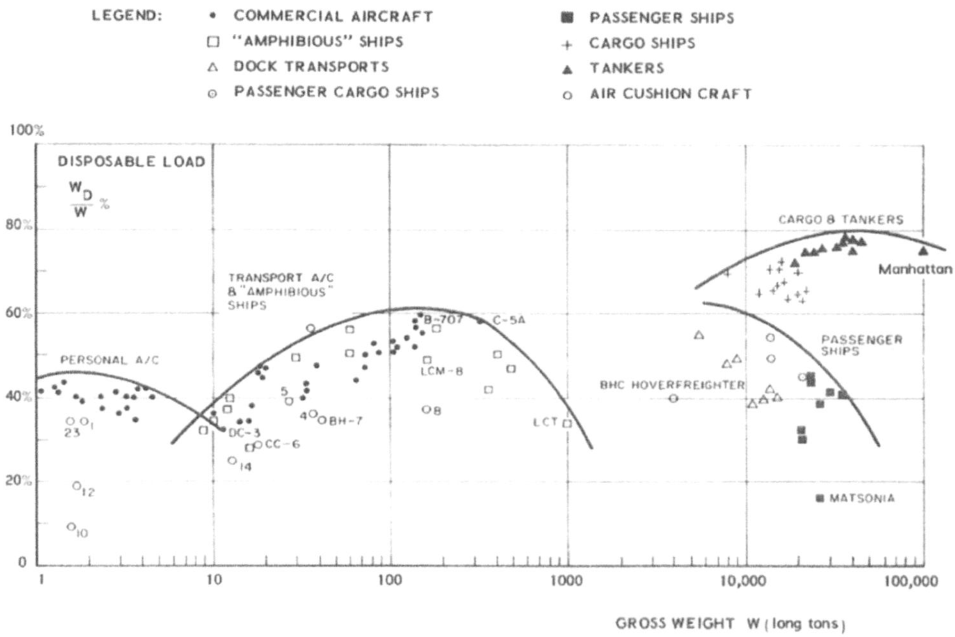

Figure 7.7: Historical Trends in Disposable Load of Transport Vehicles (1967 data)

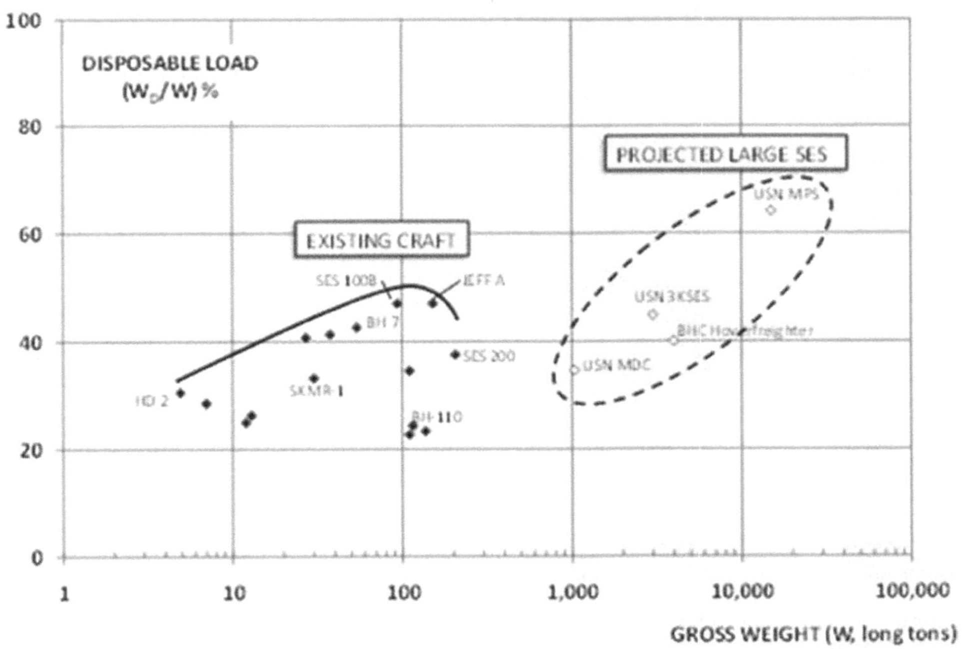

Figure 7.8: Trends in Hovercraft and Surface Effect Ship Disposable Load (1963–1980 data)

The (1980–1985) US Navy projections for large surface effect ships in the 15,000 ton displacement class are in the same groupings as conventional displacement ships that do not have to contend with the extra "lift system" and the associated weight penalty, hence these projections appear to be optimistic but are included here without further comment. In later Chapters, as each class of high speed marine craft are discussed in more detail additional plots of Payload (W_P) and Disposable Load ($W_D = W_P + W_f$) for the various candidates for high speed are examined. Especial interest will be given to those craft seeking the "sustainable one hundred knots at sea" for either economic or military advantage. The designs of Sidehull Surface Effect Ships (seeking designs speeds of around 100 knots) and Winged Surface Effect Ships (seeking design speeds of around 400 knots) will receive a detailed examination. See Chapter 8 and Chapter 11.

7.4 What Price Speed?

Professor Theodore von Kármán (Figure 7.9) was a revered figure and a *"tour de force"* in the aeronautical world. Born in Budapest, Hungary in 1881, Theodore von Kármán had an illustrious career over many decades.

In 1930 he assumed direction of the newly established California Institute of Technology (CALTECH). In 1944 he co-founded the NASA Jet propulsion Laboratory. In 1951, he received approval to establish one of his personal goals, the NATO Advisory Group for Aeronautical Research and Development (AGARD)[b] in Paris and remained Chairman of that illustrious group until his death in 1963. Dr Theodore von Kármán's achievements, awards and accomplishments are too numerous to list here but the libraries will provide ample coverage of this brilliant engineer.

Figure 7.9 Dr Theodore von Kármán (1881–1963)
Courtesy of the Archives, California Institute of Technology

Dr Theodore von Kármán believed strongly in "simple and direct truths" and did not tolerate half hearted treatment of aeronautical matters from his students. In Professor von Kármán's last few years, he was in declining health and tended to spend most of his time in Paris at his creation AGARD and only returned to CALTECH occasionally. When he did, as Professor Emeritus of CALTECH and specifically the Guggenheim Aeronautical Laboratory

[b] On 1 January 1998, AGARD was merged with the NATO Defense Research Group (DRG) to form the current Research & Technology Organization (RTO) at NATO, a group now of over 3000 engineers and scientists involved in defense technologies and military operational analyses.

at the California Institute of Technology (GALCIT), Pasadena, Professor von Kármán would make a habit of dropping by the lecture hall unannounced. It was his habit to sit in the front row on the aisle, so that the speaker at the podium could not fail but to see him. Professor von Kármán was quite hard of hearing and wore a (what would now be called an old fashioned and bulky) hearing aid conspicuously around his neck with the rotary volume control in full view. If the speaker became too verbose or too obtuse or off subject, Dr von Kármán, keeping his gaze steadily on the speaker, would slowly reach across his chest and without fanfare deliberately turn off his hearing aid, much to the consternation of the speaker. It tended to keep most speakers, especially students, explicit and on top of their subjects if the "great man" was listening to them talk. It became a "badge of honor" at CALTECH and encouragement to students to give a talk without Dr von Kármán turning off his hearing aid!

One of the quotes of Dr Kármán, showed his insistence of "practicality" for engineers' work, it is:

> *"scientific results cannot be used efficiently by soldiers who have no understanding of them and scientists cannot produce results useful for warfare without an understanding of the operations".*

This emphasized his pursuit of simple understandable ideas that work, are based on sound principles, and have a sound base of practicality.

Dr Giuseppe Gabrielli (Figure 7.10) was a renowned Italian aircraft designer. Born in Caltanissetta, Sicily he rose to prominence with his designs of many Italian military aircraft, most notably the FIAT G.50 and G.55 World War II fighters, and the G.80, one of the first jet aircraft in Italy.

He had studied at the Technical University in Aachen, Germany where he earned his doctorate under Dr. Theodore von Kármán and where the two became firm friends and collaborated on many projects together since that time in 1926.

Figure 7.10 Dr. Giuseppe Gabrielli (1903–1987)

When Dr. von Kármán formed AGARD (see above) in 1951, Dr Gabrielli served as the Italian delegate. Dr Gabrielli's achievements are also too numerous to list here but he rose through FIAT from Chief Designer to President and Chairman and remained so until his death in 1987.

His work with Dr. Theodore von Kármán on the Thurston Lecture paper, *"What Price Speed?"*, was a labor of love as the two renowned engineers sought to understand the underlying principles of locomotion.

Quoting from the paper:

> *"Means of locomotion on the ground, on the surface of, and within water, through the air, and perhaps through empty space, compete in an ever-growing effort toward higher velocities".*

Gabrielli's and von Kármán's study was intended to seek the underlying principles without too much concern on gathering exhaustive data on different vehicles. It is believed that this goal was achieved to a remarkable degree considering that this work was done leisurely together sitting in either Gabrielli's living room in Rome or von Kármán's living room in Altadena in California.

There are two charts from their paper, *"What Price Speed?"* which have attracted most attention in the high speed naval community. They are collected envelopes of the current state-of-the-art (in 1950) for many means of locomotion for what the authors call, *"specific power"* and *"specific resistance"*. This data has direct impact on the economical equations discussed earlier in this Chapter. The Gabrielli and von Kármán interest was much broader than that of concern here with improvements in just "sea" and "air" transport, for their paper covered all forms of locomotion, including pedestrians, horses, cyclists, motorcycles as well as various types of ships and aircraft.

In the interests of historical accuracy, the original charts are shown here as Figure 7.11 for *"specific power"* and for *"specific resistance"*. Keeping the original material on display is important to preserve here because of the many later distortions and misinterpretations that have unfortunately appeared in the literature on these important seminal findings by Gabrielli and von Kármán.

In the Gabrielli and von Kármán paper the terms *specific power* and *specific resistance* are defined as:

$$\text{Specific Power} = \text{HP/ton} \tag{7.25}$$

$$\text{Specific Resistance} = P/WV \tag{7.26}$$

These terms and their significance in the design of efficient vehicles are discussed later in this Chapter. The collected data is shown on Figure 7.11.

In the context of discussing achieving high speed at sea, it is helpful to extract just the "sea vehicles" and the "air vehicles" from the total set of transport means examined by Gabrielli and von Karman. This has been done in the following. Figure 7.12 shows the *specific power* (HP/ton) for just the sea and air vehicles. It will be seen that there are two general groupings as might be expected from knowledge of the medium used in supporting the vehicles. In the lower left of the diagram are the boundaries of sea vehicles such as submarines, battleships and destroyers. In the upper right hand of the diagram are the air vehicles such as fighter aircraft, commercial aircraft and helicopters. The sea vehicles are essentially supported by buoyancy forces of the water. The air vehicles are essentially supported by dynamic forces resulting from passage through the air. The lone class of air vehicles, the airship, supported by buoyancy forces in the air falls between the two main groupings. More will be said about the unique class of airships in later Chapters.

The dotted line in Figure 7.12 is Gabrielli and von Karman's "achievement level" limit line that represents the state of the art (in 1950) that the vehicles exhibit in terms of *specific power*. This limit line has moved to the right in later years as such new forms of efficient aircraft design (and advent of the supersonic airplane not part of the "set" in

1950) have appeared on the scene. Future "achievement levels" in technology have moved this "limit line" to the right as advanced material, power plant, light weight structures, etc have been incorporated in the designs. In Chapter 8 on Technical Considerations more is said about such limits. In any event, in the intervening years there has been no fundamental shift in the basic characteristics of the two groupings of sea and air vehicles. The general upward trend in *specific power* as speed is increased, is by amounts roughly proportional to maximum speed squared.

From an economic considerations viewpoint, the major research and design development has centered on the melding of the two main groupings to achieve some form of efficient high speed transport over the sea. To help focus the discussion on seeking a unique form of high speed transport over the sea, Figure 7.13 has been constructed from the data shown in Figure 7.12, which shows the envelopes of the two main classes of vehicles: sea vehicles and air vehicles. From such an examination it is seen that there is an area of *specific power* between the two classes that has become known as "The Gap". As sea vehicles seek to go faster they face a steep

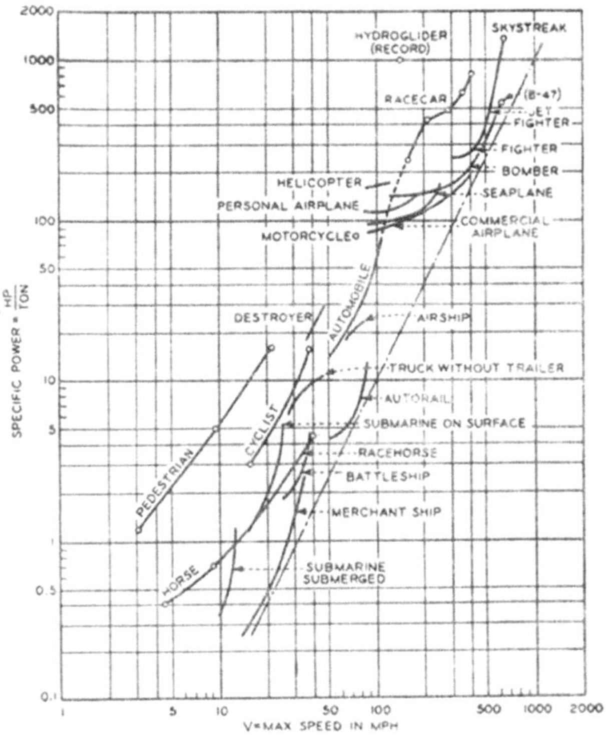

(a) SPECIFIC POWER

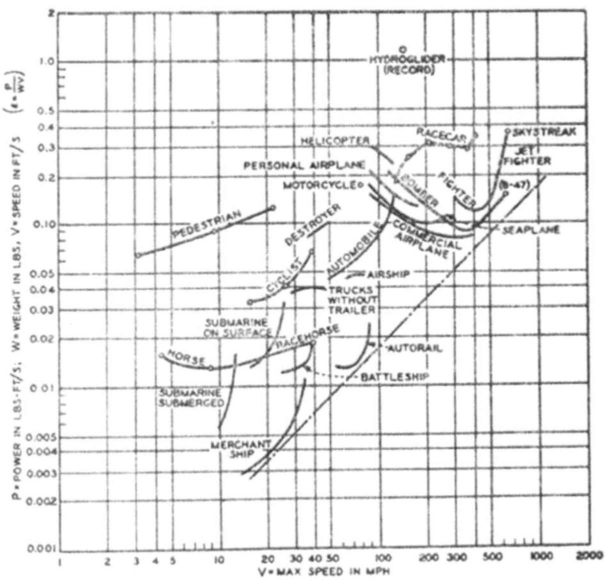

(b) SPECIFIC RESISTANCE

Figure 7.11: Original Specific Power and Specific Resistance Charts by Gabrielli and von Karman

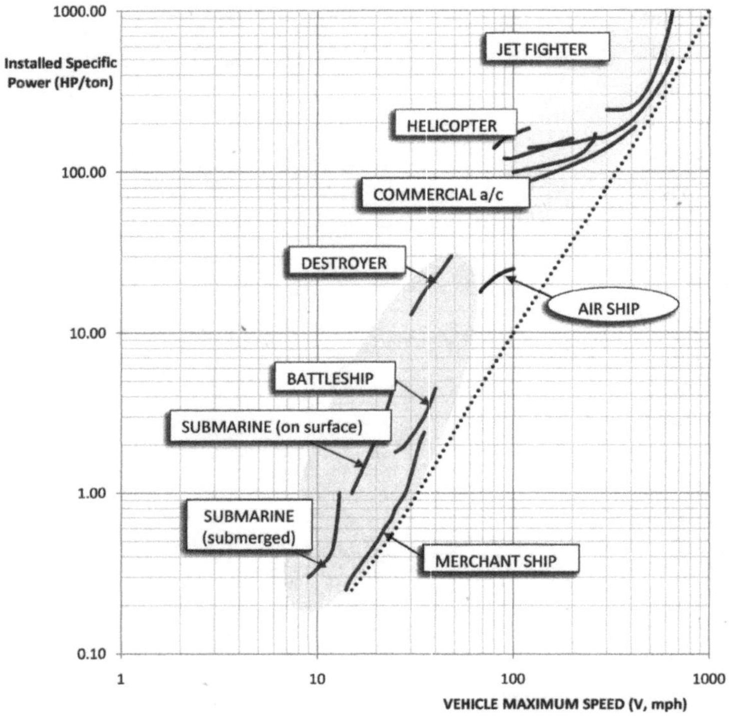

Figure 7.12: Specific Power of Sea and Air Vehicles
(from Gabrielli and von Karman, 1950)

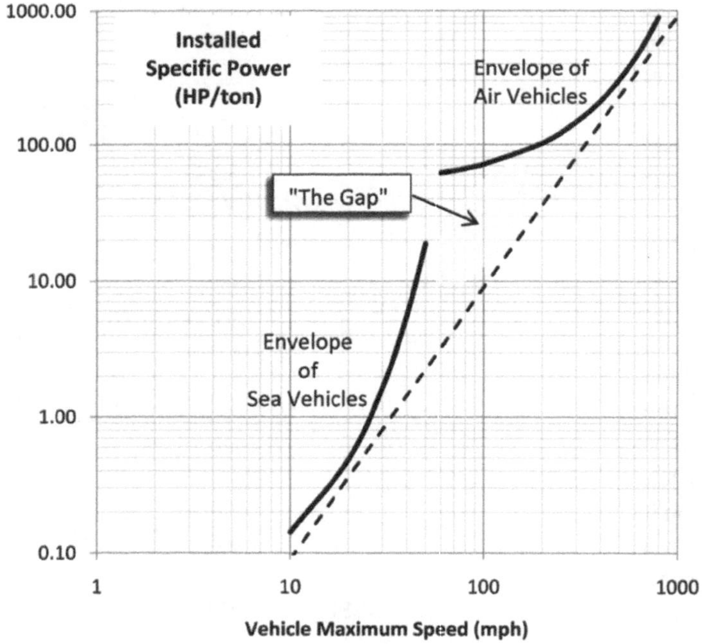

Figure 7.13: Envelopes of Specific Power for Sea and Air Vehicles
(Gabrielli-von Kármán data, 1950)

increase in power required as speeds approach anything much above 40–50 mph. A similar rise in power is required in air vehicles as they seek to travel faster than about 500 mph as they approach the speed of sound (about 660 mph at 35,000ft altitude). This has led designers of high speed ships to seek speeds of around 100 knots but with much lower *specific power* values shown in "The Gap", say about 20–30HP/ton in round numbers. Chapter 8 on Technical Considerations expands on the physical principles that cause this increase in *specific power*.

To bring the discussion more in line with the economic considerations in high speed ship design attention is now given to the second set of data prepared by Gabrielli and von Kármán and that is the *specific resistance* data shown in Figure 7.11. As before, selecting just the sea and air vehicle data from the total set, Figure 7.14 can be constructed. As for the *specific power* groupings, a similar set of groupings exist for the *specific resistance* data of sea and air vehicles. Here, "The Gap" is even more pronounced around the area of interest. The second chart of interest taken from the Gabrielli-von Kármán paper is that called by the authors as "*specific resistance*" defined by them as

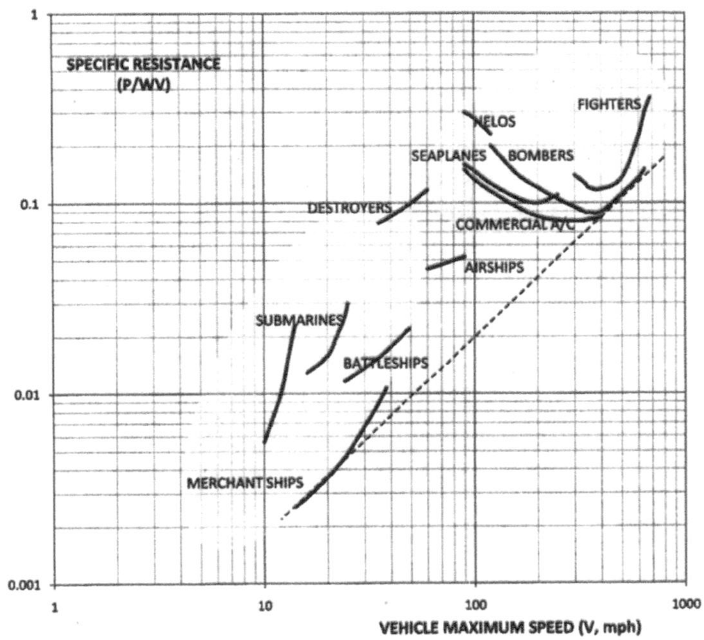

Figure 7.14: *Specific Resistance* (P/WV) for Air and Sea Vehicles (Gabrielli-von Kármán data, 1950)

$$\text{Specific Resistance, } (\varepsilon) = \frac{P}{WV} \tag{7.26}$$

Gabrielli and von Kármán define specific resistance as the ratio between the *tractive force* (P/V) and the weight of the vehicle. As Gabrielli and von Kármán say in their paper,

> *"Evidently, if one compares two vehicles and considers the work necessary for a given transport performance, i.e., the transportation of the same gross weight over the same distance, this work is directly proportional to ε".*

That is to say, since the work done by the vehicle is the weight times the distance traveled and that power is the work per unit time, then *specific resistance* can be written:

$$\text{Specific resistance } (\varepsilon) = \frac{(\text{Power})(\text{Time})}{(\text{Weight})(\text{Distance})} \quad (7.27)$$

and, so

$$\text{Specific resistance } (\varepsilon) = \frac{\text{Power}}{(\text{Weight})(\text{Speed})} = \frac{P}{WV} \quad (7.28)$$

as used by Gabrielli and von Karman as a true measure of the economical use of power to achieve some desired speed. Figure 7.14 shows the data for *specific resistance* (ε = P/WV), again with the extraneous vehicles (cyclists, race cars, etc) removed, leaving just "ships" and "aircraft" so as to focus on the issue under discussion. As before, for the *specific power*, it is seen from Figure 7.14, that there are two distinct groupings of the *specific resistance* (ε = P/WV) that collect "sea" and "air" vehicles with again the example of the lone vehicle of the air ship (that is in "the gap" between the groupings) that is supported differently from either ships or aircraft. Gabrielli and von Kármán make the point that the parameter of *specific resistance* (ε) provides a useful tool for the comparison of different types of transportation[c] since,

> *"it gives an indication of the price one has to pay in power for speed, and it helps to find out whether a certain system of transportation is suitable for further speed increase without penalty of higher specific power requirement or whether one is near the economical limit".*

In similar vein to that done for the *specific power* above (Figure 7.13), envelopes of the *specific resistance* (ε) for the two groupings of sea and air vehicles from Gabrielli and von Karman's data is shown in Figure 7.15. In this format, the difference between the two groupings is even more pronounced with the rapid increase in *specific resistance* (ε) as speed is increased. These envelopes are general in nature and should not be treated as being precise boundaries but more in the light of groupings of general characteristics in each case. Gabrielli and von Kármán make the point very clearly in their paper that their curves are general in nature and are not intended to be too prescriptive for obvious reasons. The major difference between sea and air vehicles in terms of *specific power* and *specific resistance* are quite marked as shown in Figure 7.13 and Figure 7.15. The reasons behind the rapid increase in resistance is discussed more fully in Chapter 8 on Technical Considerations as the effects of compressibility in both air and water are examined. This is a fundamental issue in the search for a new form of vehicle that can operate as a new class of vehicles that operate in the interface between water and air. For *specific resistance*, it is found that the general upward trend in (P/WV) is directly proportional to the vehicle speed.

There are the same two distinct groupings of "air" and "sea" vehicles for the reasons cited. It can be seen immediately from this work, the driving force behind the US Maritime Administration effort in the early 1960s to seek that form of transportation that would "bridge the gap" between the displacement ship with its 5 cents per ton mile economics and that of the conventional aircraft at its 20 cents per ton mile characteristic direct operating cost.

In honour of Bréguet's original terminology for *aerodynamic efficiency* term WV/P (see earlier and Equation 7.7 and Equation 7.8) and in recognition of

[c] Note that in the context of total vehicle design, even naval speedboats and fighter aircraft are involved in "transportation" as they must still carry a payload efficiently into the mission.

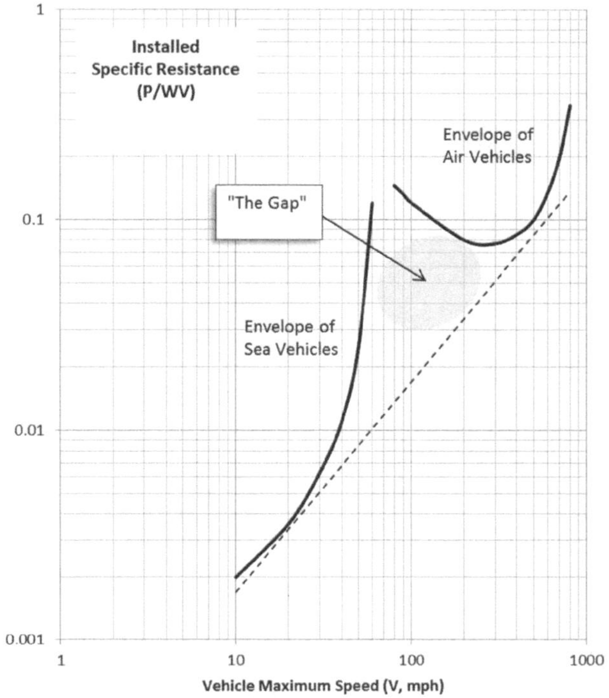

Figure 7.15: Envelopes of *Specific Resistance* for Sea and Air Vehicles (Gabrielli-von Karman data, 1950)

Gabrielli and von Kármán's method for comparing vehicles for suitability for efficient and economical transportation, the author in 1966[9], in 1967[10] and later, coined the term,

$$\text{Transport Efficiency (TE)} = \frac{WV}{P} = \frac{1}{\varepsilon} \tag{7.29}$$

which is the reciprocal of Gabrielli & von Kármán's *specific resistance* (ε) with especial recognition given to including the power consumption for the air cushion in air assisted vehicles such as surface effect ships (through the *effective lift drag ratio* $(L/D)_{EFF}$ discussed earlier) of interest to both the Maritime Administration and the US Navy. Again, it is emphasized that in the context of "design efficiency" as used in the Bréguet Range Equation, it is not appropriate or meaningful that the numerical values of the *Transport Efficiency* (TE) be between zero and 100%!

As the aeronautical and marine industry pursued high speed marine vehicles in the 1960s (after Gabrielli and von Kármán had produced their treatment of transportation vehicles) it was of interest to examine how such new vehicles fared in terms of Transport Efficiency (WV/P). In 1966 and again in 1967, as a measure of the promise of air supported ships, Mantle[11] summarized the existing data at that time on the *Transport Efficiency* (TE) of various high speed marine vehicles including hydrofoils, amphibious hovercraft and various forms of sidehull surface effect ships and displayed them for comparison with existing conventional ships and aircraft. This was done to see if the "promise" of these new vehicles was at least possible in

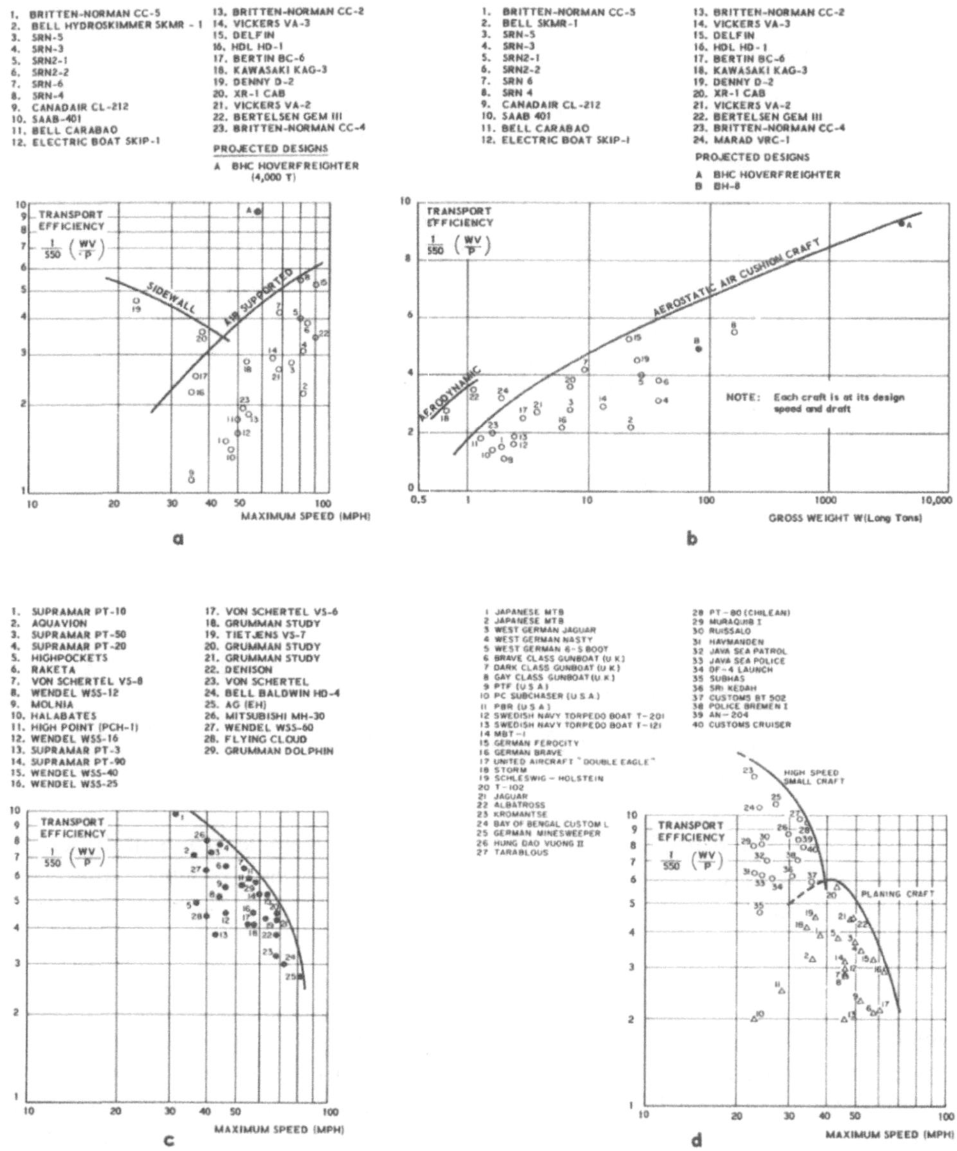

Figure 7.16: Collected *Transport Efficiencies* (TE) of High Speed Marine Vehicles (1967) Originally appeared in New York Academy of Sciences, International Congress on Subsonic Aeronautics, Paper "Some Design Aspects of Air Cushion Craft" by author (see references)

bridging the gap. These summaries are included here as Figure 7.16[d]. In Figure 7.16, the first grouping in the upper left hand corner of the chart, labeled (a) is for all existing (at that time) amphibious and sidehull surface effect ships. There are two

[d] The constant of "550" in these 1967 original charts was the result of keeping "P" in units of "horsepower" instead of power (ft lb/sec) but the numbers remain the same with the recognition that the *Transport Efficiency* (WV/P) is non-dimensional. That unnecessary clarification has been dropped here.

distinct groupings as might be expected; the sidewall (and sidehull) craft to the left at lower speeds and the amphibious hovercraft to the right at higher speeds. It was reasonable to expect that the efficiency of amphibious hovercraft would be more efficient at higher speeds than sidewall (sidehull) craft all else being equal and this is borne out by the data. The upper right hand chart, labeled (b) in Figure 7.16, was a look at the possible scaling effects on the *transport efficiency* as size or displacement is increased for amphibious hovercraft and sidehull surface effect ships and other air cushion supported vehicles. This observation was borne out by the available data.

In the lower left hand corner of Figure 7.16 the collected data (c) of existing hydrofoils at that time (both surface-piercing and fully submerged hydrofoils) is displayed. In the lower right hand corner, the existing data (d) on high speed planing boats was also collected. State-of-the-art envelopes were drawn around each of these classes of high speed marine vehicles as can be seen in Figure 7.16.

7.5 Transport Efficiency of Vehicles since 1967

The envelopes of the high speed marine vehicles (hydrofoils, sidewall hovercraft, hovercraft and planing craft) from Figure 7.16 have been superimposed on the general chart originally prepared by Gabrielli-von Kármán (except their data of *specific resistance* (ε) has been inverted to compare directly with *transport efficiency, WV/P*) for use in the economic equations on cost, in Figure 7.17. There are several important changes made to the original data by Gabrielli-von Kármán. First, since their seminal work in 1950, several important developments have occurred. These include the advent of the jet aircraft, the nuclear submarine, and monohull design improvements. These additions have been included in Figure 7.17 to "modernize" the data and to indicate the nature of the various improvements that have occurred in both sea and air vehicles.

The submarine is an interesting case in that during WW II and before, submarines were all diesel powered requiring them to spend large amounts of time on the surface replenishing batteries and engine operation. This led to designing their hulls to be more seaworthy on the surface of the sea. Such shapes were not the most efficient when submerged. When the nuclear powered submarine came on the scene (USS Nautilus commissioned 30 September 1954) the shape of the hull was designed for more efficient operation underwater giving rise to the increased value of *transport efficiency* as shown on Figure 7.17 compared to the original data supplied by Gabrielli-von Kármán on diesel submarines.

Other additions to the envelopes of *transport efficiency* embodied in improved hydrodynamic shaping of destroyer hulls, labeled "New DD" in Figure 7.17 to replace the old destroyer designs used in the original charts. Improved tanker hull designs have also been added in Figure 7.17.

The envelopes around the sea and air vehicles have been drawn loosely because these envelopes were never precise in the beginning (Gabrielli-von Kármán make this point clearly in their paper) to merely indicate the major regions dominated by the classes of vehicles shown. This is done deliberately to emphasize that the focus here is on "the gap" between the two groupings of "air" and "sea" vehicles to set the stage for any possible development to achieve new ship designs that would meet the goals laid down by MARAD in 1960 and endorsed later by the US Navy as their program got underway in 1969. Many later writers have added more detail to these

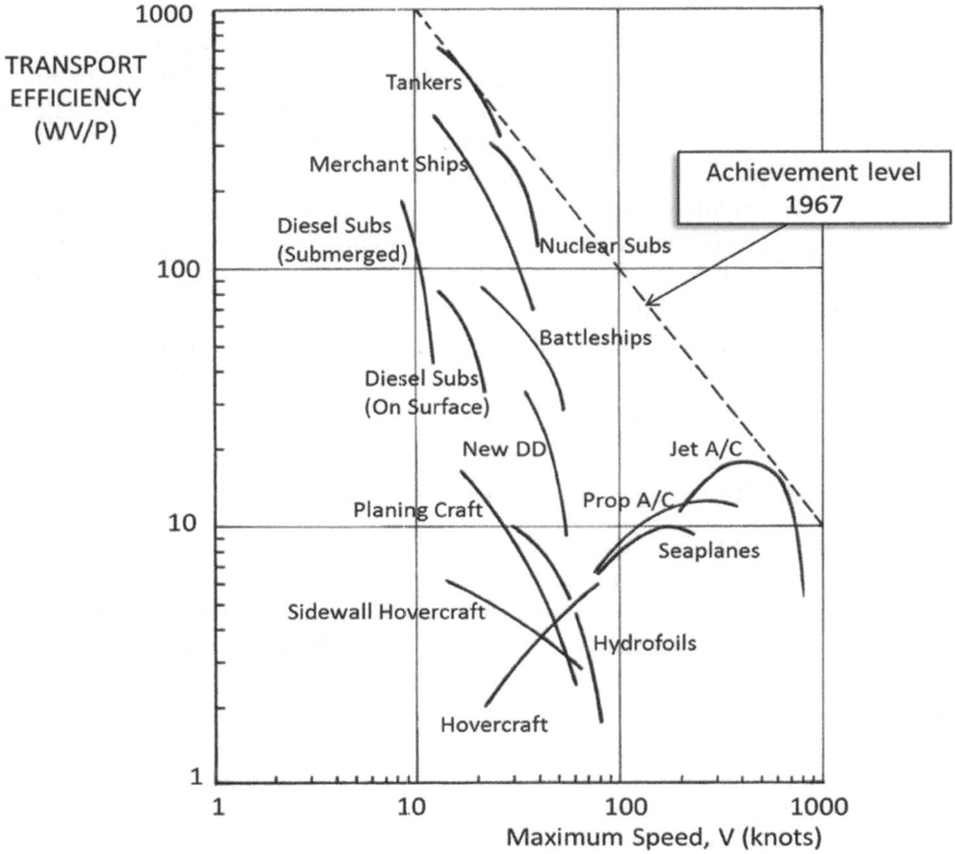

Figure 7.17: Envelopes of State-of-the-art of Transport Efficiency (TE) of Vehicles (1967)

charts as conventional ships and aircraft improve in efficiency with improved power plants, more efficient structure and higher strength-weight materials, etc. This is indicated by the "achievement level" dotted line that is used to show improvement as time and technology improves each class of vehicles.

The 1967 "achievement level" can be expressed by the simple relationship:

$$\left|\frac{WV}{P}\right|_{achievement} = \frac{10,000}{V} \tag{7.30}$$

Where the maximum speed of the vehicle, V is expressed in knots. There is no real significance of this limit line that merely tells the reader where vehicles operating in two entirely different mediums have reached their highest values as of 1967. It is true, nevertheless, that only in very special circumstances have any new improvements in sea or air vehicles since 1967 "crossed" this "achievement level" line.

The issue before the Maritime Administration and the US Navy, however, is whether or not it was possible to bridge "the gap" between sea and air vehicles with *efficient* and *economical* types of ships of sufficient size for transoceanic transport. *The decade from 1969 to 1979 was dedicated to achieving that goal with the large surface effect ship.*

From the data shown over a wide range of sea and air vehicles especially those designed for high speed over the sea that it has been extremely difficult to reach into that most oft quoted "gap". It is safe to say that designers have been sorely tested to achieve such a goal. As will be discussed in Chapter 8 on technical considerations, that very few designs have demonstrated a transport efficiency "goal" that could be expressed by:

$$\text{Transport Efficiency Goal}: \quad \text{"WV/P Greater than 10 at 100 knots"} \qquad (7.31)$$

It is important to note that in the charts on *specific resistance* in the Gabrielli-von Kármán work and in the *transport efficiency* charts in Figure 7.16 and Figure 7.17 the data is for vehicles operating at their maximum speed (V_{max}) and total installed power (P). It is recognized that when the vehicle is operating for maximum range, both the speed, usually called cruise speed (V_{cruise}) and power consumption will be at lower values.

As pointed out in the original work, however, it is the maximum speed and the total installed power that "sizes" the vehicle. Once the vehicle has been sized, then one can determine the performance for cruise conditions. By analogy, one can visualize a personal automobile sized on cruise power with no power available to accelerate or maneuver, would yield what is commonly called a "gutless wonder"!! The design of SES in terms of maximum and cruise speeds and other characteristics that define a "useful" ship are discussed in more detail in Chapter 8 on technical considerations.

The intent here is to show the State-of-the-Art as both the US Maritime Administration and the US Navy entered the era 1969–1980 seeking a design that met or exceeded the "goal" to operate economically and usefully in "the gap". The optimistic claims by many designers in the early 1960s failed to materialize as realistic designs took hold. As discussed in Chapter 4, both MARAD and the US Navy were predicting in 1965 a 10,000 ton CAB design with almost double the above *transport efficiency* with a value of WV/P = 18 at a maximum speed V_{max} = 129 knots! The actual experience after an intensive two decades (1960–1980) of development could not achieve this ambitious design claim.

7.6 The Problem Facing the High Speed Ship Designer

An even more dramatic illustration of the problem facing designers of high speed marine craft is to show the envelopes of the various types of vehicles displayed in Figure 7.17 but to display them in collected groupings of "sea" and "air" vehicles as shown on Figure 7.18. In Figure 7.18 all the envelopes of the various sea vehicles have been enclosed in a "sea vehicle" boundary. A similar boundary has been wrapped around all the "air vehicles". That is, except for the special case of the airship, to be discussed separately.

It is seen with this display, the envelope of the "sea vehicles" suffers a precipitous drop of the transport efficiency (WV/P) as the speed is increased much past about 30 knots. A similar precipitous drop occurs for "air vehicles" past about 600 knots. There are definite physical reasons for these "barriers" that are analyzed in detail in

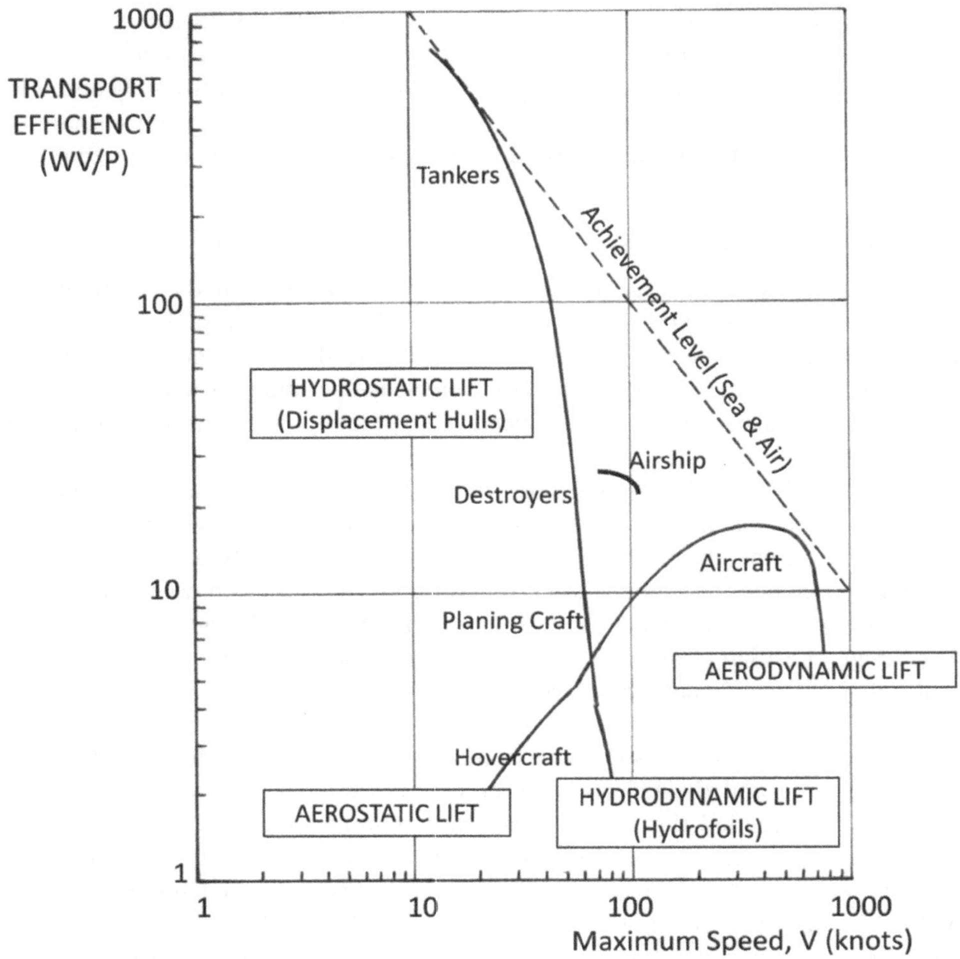

Figure 7.18: "Envelope of Envelopes" of Sea and Air Vehicles Transport Efficiency

Chapter 8 based on Froude Number (water interface) and Mach Number (air compressibility) phenomena. For this Chapter on economical considerations, it is sufficient to note that as ships move from slow speed hydrostatic support (for displacement hulls) to hydrodynamic lift support (such as in planing craft and hydrofoils) large expenditure in propulsion power is incurred that relates directly to cost in both acquisition and operation of such craft. A similar set of phenomena occurs as those craft supported by some form of aerodynamic lift (hovercraft and aircraft) move to higher and higher speeds. The air cushion craft or "hovercraft" in Figure 7.18 are labeled "aerostatic" as their lift is generated by on-board means and not through movement through the air. Such craft then blend into the aerodynamically supported aircraft set, which also suffer the increased expenditure in power to attain their high speeds. Each of these mechanisms lead to increased costs of the vehicles. The technical considerations on these mechanisms at work in vehicle design and operation are discussed in Chapter 8. The aerodynamic air cushion craft (WIG, channel flow, etc) are treated separately in Chapter 11.

The lone class of air vehicles that operate as an aerostatic supported vehicle, namely the air ship certainly enjoys a high transport efficiency (WV/P) with values greater than about WV/P = 20 at speeds around 100 knots. Because it is fully supported by buoyancy in the air it suffers a different problem from the other "high speed" vehicles in that it must have a large enclosed vehicle volume to provide the supporting lift mechanism. A re-examination of the airship was included in the US Navy Advanced Naval Vehicles Concept Evaluation (ANVCE) Project in 1976 and this is discussed further in Chapter 10.

As the discussions are expanded to discuss in more detail how best to achieve a "100 knots at sea" capability, Figure 7.18 will be useful as a "template" to assess how various proposed designs have fared in seeking an economical, efficient means of operating at "100 knots at sea".

7.7 Acquisition Cost of High Speed Marine Craft

The earlier sections of this Chapter have concentrated on the design characteristics that influence the Direct Operating Cost (DOC) of the ship. These were the propulsion efficiency (η_p); the critical transport efficiency (WV/P); and the available fuel load (W_f/W). These same characteristics also directly influence the Acquisition Cost of the craft. The Acquisition Cost is the price (before adding in profit) to be paid to acquire the ship after the necessary Research & Development (R&D) Costs have been expended. It is common practice to refer to the Life Cycle Cost (LCC) by the equation:

$$\text{Life Cycle Cost (LCC)} = \text{RD Cost} + \text{Acquisition Cost} + \text{Operating Cost} \quad (7.32)$$

Some writers include the reduction in LCC by inclusion of the scrap value at the end of the vehicle's life but this refinement is not included in what follows.

The Acquisition Cost will vary depending on the number of ships purchased. Typically, there is a Lead Ship followed by ships of lower cost (affected by learning curves) for additional Follow-On craft or ships. One can see here the potential for a wide range of costs for any particular craft type depending on many factors. The learning curves vary greatly among manufacturers; the time over which additional craft are procured; change orders on follow-on craft improvements are integrated into the follow-on craft; and many other intangibles. Hence, it is not possible to be precise in presenting Acquisition Costs any more than it is to be precise about Direct Operating Costs (DOC) that depend in turn on variables such as fuel costs, facilities arrangements, insurance costs, etc. But it is possible to establish general trends that can directly influence decisions on designs and especially in the all important "cost-effectiveness" calculations for craft intended to serve in a variety of military missions.

At the outset, it can be said that "high speed ships" are expensive. The question is: *"Is the increased cost offset by an improved commensurate mission effectiveness?"*

With the above provisos in mind, Figure 7.19 collects a set of costs of operational craft including surface piercing hydrofoils, fully submerged hydrofoils, sidewall hovercraft and amphibious air cushion craft (hovercraft) and operational

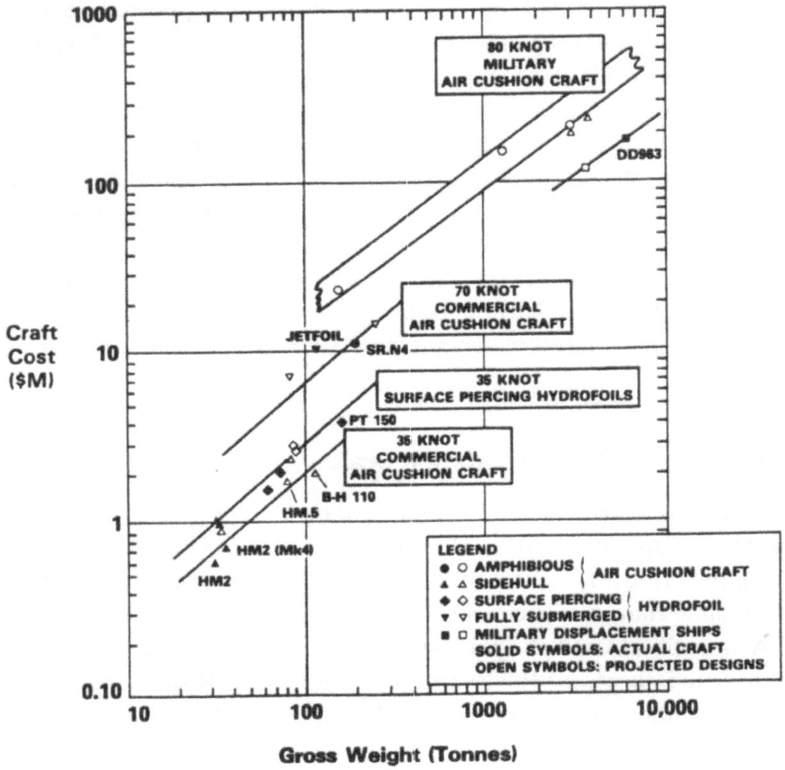

Figure 7.19: Cost of High Speed Marine Craft (1976 $)

displacement hull navy ships such as the destroyer DD 963. Superimposed on Figure 7.19 are some collected designs from the ANVCE Project (see Chapter 10) of air cushion craft (both amphibious and sidehull designs).

These costs were compiled in 1976[12] and 1977[13] and would need to be updated by the appropriate inflation indices to provide an indication of "today's costs", but the main issue here is the relative costs between the various types of craft and ships which are as applicable today as they were in the 1970s when the original data was compiled. Later in this Chapter, a comment is given on some related issues with simple application of "inflation indices" and "consumer price indices" to earlier data.

As can be seen from Figure 7.19, there is a definite impact of both "size" and "speed" in the determination of the acquisition cost of the craft as well as the type of craft under consideration. There have been countless authors who have presented cost estimating relationships (CER) but an interesting and broad treatment that is useful in determining cost trends is the work by Dix and Riddell in 1976[14]. In their work, Dix and Riddell found that the cost of any vehicle was determined by the empty weight (W_E) and the installed power (P). They found that from a wide range of collected data of various vehicle types, that the cost could be "estimated" from the equation:

$$\$_{acq} = 1000 W_E + 1200 \frac{P}{Q^{0.33}} \tag{7.33}$$

In this simple equation, the Acquisition Cost of the vehicle ($\$_{acq}$) can be estimated from knowing the empty weight (W_E); the total installed power (P); and the quantity built (Q). In the usual treatment of costs and learning curves, the cost here is the average cost over a batch of craft built. Dix and Riddell remind the reader that their method is general in nature and the cost "estimated" by this method could be off by a factor of two, but what is important here is the relative cost of one type of vehicle when compared to another.

The effect of power (and thus speed) is very pronounced in all vehicles. In the context of this book, it is expected that some form of powered lift vehicle, such as the *air cushion craft* in its many forms would be most likely to be the candidate of choice for a large transoceanic, high speed ship. Accordingly, it is interesting to see how the Dix and Riddell cost relationship would predict costs for such craft.

The total installed power ($P=P_L+P_P$) for air cushion craft from the smallest craft to the largest projected air cushion supported ship design (e.g. 10,000 ton SES) show a remarkable consistency in the power-weight relationship. The collected data from 1975–1980[15] has been included in the data shown in Figure 7.20. Generally, it was expected that the sidewall hovercraft would show a lesser total power than the later higher speed SES. This can be seen in Figure 7.20 where the sidewall hovercraft used for commercial operation in England (the Denny *D.1* and *D.2* and the Hovermarine *HM.2*) do indeed show a lower total power requirement. However, once the US Maritime Administration and the US Navy sought much higher speeds in their designs, such advantages of the *sidewall hovercraft* were lost in the total power requirement as can be seen in Figure 7.20 where the *sidehull SES* designs exhibit

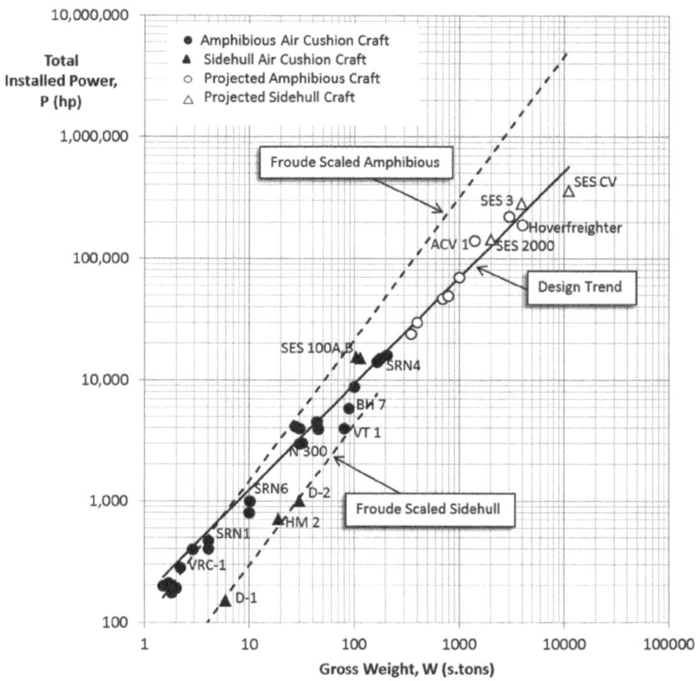

Figure 7.20: Total Installed Power of Air Cushion Craft

similar high power requirements as the amphibious forms of air cushion craft. The design for the British Hovercraft Corporation (BHC) 4000 ton *Hoverfreighter* (see Chapter 4, Figure 4.16) and the large SES designs from the US Navy (see ANVCE Project in Chapter 10) all follow this same general trend. The difference between the design trend and Froude scaling is discussed in more detail in Chapter 10.

The data in Figure 7.20 was found to follow the simple relationship for total power (P):

$$P = 165 W^{7/8} \tag{7.34}$$

In this relationship, the total power (P) is expressed in units of horsepower and the full gross weight (W) is expressed in short tons. Returning to the Dix and Riddell formulation, if the above relationship for air cushion craft is substituted in Equation 7.33 and the assumption is made that the empty weight (W_E) is 60% of the gross weight (W); and that a buy of Q = 25 craft is also assumed, then the acquisition cost of air cushion craft could be given as:

$$\$_{acq} = 600 W + 67,715 W^{7/8} \tag{7.35}$$

If this form of the Dix and Riddell acquisition cost equation (Equation 7.33 and Equation 7.35) is superimposed on the data in Figure 7.19, then one sees a display like Figure 7.21 where the dashed line represents Equation 7.35 with the demonstrated power levels of high speed air cushion craft included.

While the above equations for acquisition cost ($\$_{acq}$) certainly cannot be used to predict specific craft costs, it does provide an indication on the trends in vehicle costs

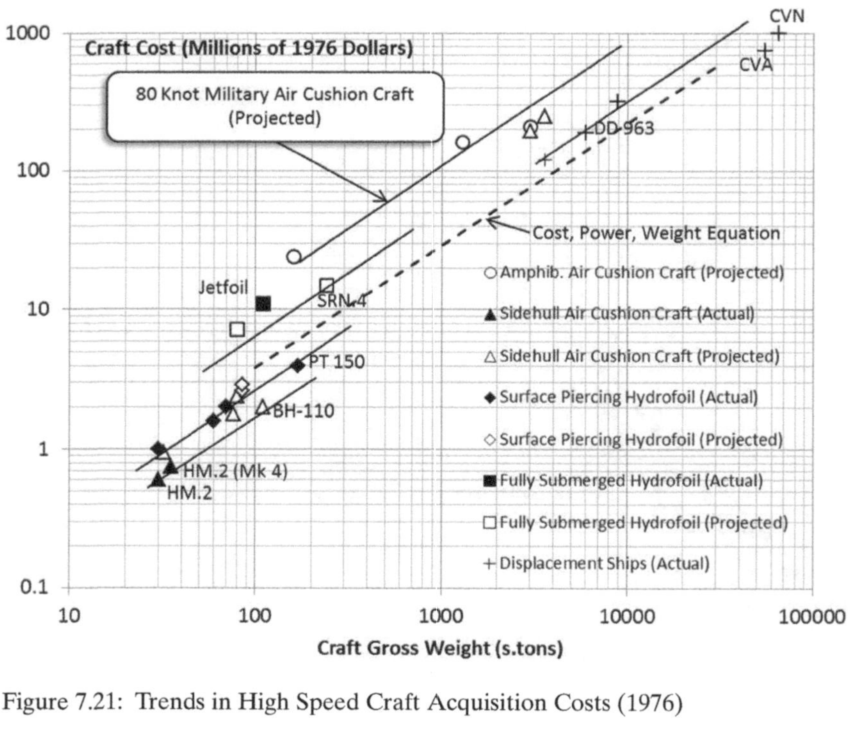

Figure 7.21: Trends in High Speed Craft Acquisition Costs (1976)

and especially the strong influence of the total power installed (P) on the final outcome of acquisition cost. All the costs listed in the band labeled "This Study" in Figure 7.21 refer to the collected costs from the ANVCE Project discussed in Chapter 10.

It is seen that the dashed lines represents the acquisition costs trends of high speed air cushion craft to a reasonable degree. The ANVCE Project Costs were approximately double the Dix and Riddell estimates and probably more representative of the expected costs for military high speed ships with specific military payloads. The second term in Equation 7.35 (the power term) dominates the acquisition cost by a factor of about 65:1 for the smaller sizes (around 100 tons displacement) and about 35:1 for the larger sizes (around 10,000 tons displacement). Even while Dix and Riddell warn us that the actual costs could be off by about 2:1, because of the spread of their data base, it still provides a guide as to where to influence the design to reduce costs. Figure 7.21 also shows that the costs of a 2000–3000 ton class SES tends to be about the same cost as a 10,000 ton displacement class conventional ship. Thus the smaller, high speed SES would have to demonstrate a significantly better military effectiveness to offset that cost difference. This was one of the contributing problems to the high speed SES program as the US Navy tried to demonstrate the military effectiveness or usefulness of the 3KSES program. This is in addition to the concerns expressed about the technical maturity of the various proposed unproven design concepts (see Chapter 5).

A further illustration of the cost impact on the military effectiveness can be appreciated from plots such as that shown in Figure 7.22. In that chart, the ANVCE Costs for several "3000 ton class" high speed concept designs have been compared to a similar size frigate, the FFG 7 (at 3648 tons displacement). If a conventional displacement hull is "pushed" to higher speeds, the power expenditure becomes

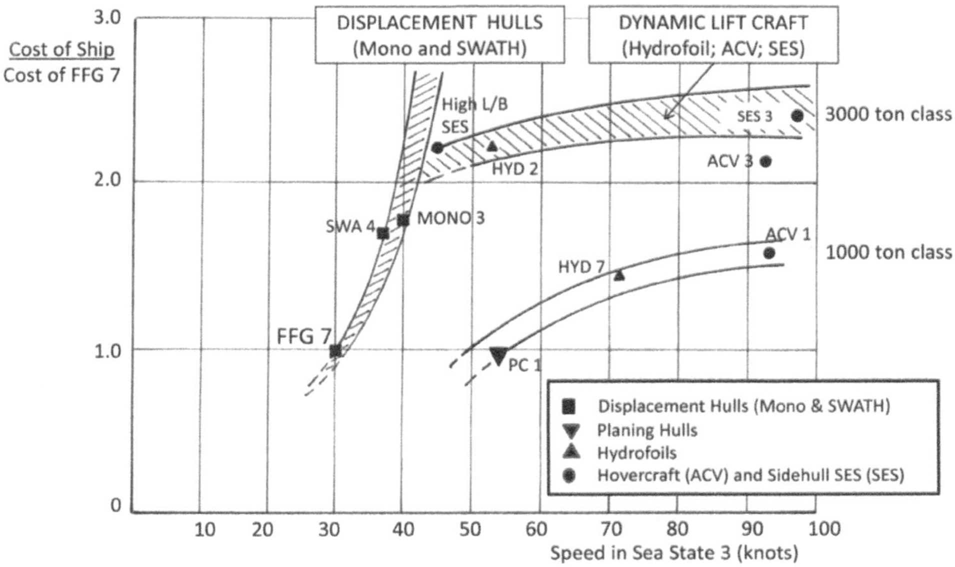

Figure 7.22: Relative Cost of Speed

exorbitant, causing the costs to rise exponentially as shown as high speed versions of displacement ships were pursued in the ANVCE Project. These high speed displacement designs labeled "MONO 3" and "SWA 4" are the monohull and SWATH designs respectively, designed for 40 knots speed rather than the 30 knot speed of the FFG 7. One can see that any further increases in speed (and power) would produce a design more than 3:1 in cost over the conventional FFG 7 ship.

Once the decision has been made to switch to some form of dynamic lift (hydrofoil, air cushion craft,..) then the increase in ship speed comes at a more modest increase in cost. The issue then becomes: *does the 2 to 3 times in cost incurred by such concepts come with a commensurate increase in military effectiveness?* Unfortunately, the various designers were unable to show such a marked increase in military effectiveness for high speed SES (around 100 knots operational speeds). In the ANVCE Project, it was found that if the design speed goal of 100 knots was lowered to say 45–50 knots then some cost advantages appeared. This can be seen by the design point labeled "High L/B SES" in Figure 7.22. For lower speeds (than 100 knots) the geometry of the ship changes to be more slender (higher L/B) with the slenderness ratio increasing for lower speeds. In this case, from the ANVCE Cost Analysis, it was found that the exorbitant cost increase incurred with a conventional displacement hull was less for this form of sidehull SES which was "only" 2:1 over the FFG 7 cost compared to the >3:1 for displacement hulls. The issue remains as whether even the 2:1 cost increase can be offset by commensurate increases in military effectiveness.

For interest, the cost data for the "1000 ton class" high speed designs from the ANVCE Project are included in Figure 7.22. The 1090 ton planing craft (PC 1); the 983 ton hydrofoil (HYD 7) and the 1264 ton hovercraft (ACV 1) are shown. The 1090 ton planing craft design (PC 1) costs about the same as the 3648 ton FFG 7 (about three times the size of the PC 1). Was going from 30 knots to 55 knots worth the 3:1 reduction in ship size and loss of military payload and military effectiveness?

7.8 Inflation Indices and Cost Trends Over Time

Much of the design work on high speed craft was done for the US Navy in the late 1970s with a specific concentrated cost analysis done on a wide range of different designs all done to a common standard in the ANVCE Project. It would be tempting to simply adjust those costs to today by applying an Inflation Index calculation. Figure 7.23 displays the inflation indices from 1976 to 2011 using the data from the Bureau of Labor Statistics[16]. These have been plotted in Figure 7.23 which shows that the buying power of the dollar has declined to where it would require $3.89 in 2011 to acquire the same object that cost $1 in 1976. The annual fluctuations in the inflation index over that same time period shows that after a major adjustment in 1980 (incidentally the same year the 3KSES program was canceled) the inflation increase each year has been steadily declining from about 4% per year to today's value of about 2% per year.

Over such a long time period however, many other factors that would influence the Acquisition Cost of the ship would come into play. In the above trends in costs no information has been provided on how the costs are divided between material and labor. The costs have been lumped together as a single entity. Any detailed cost

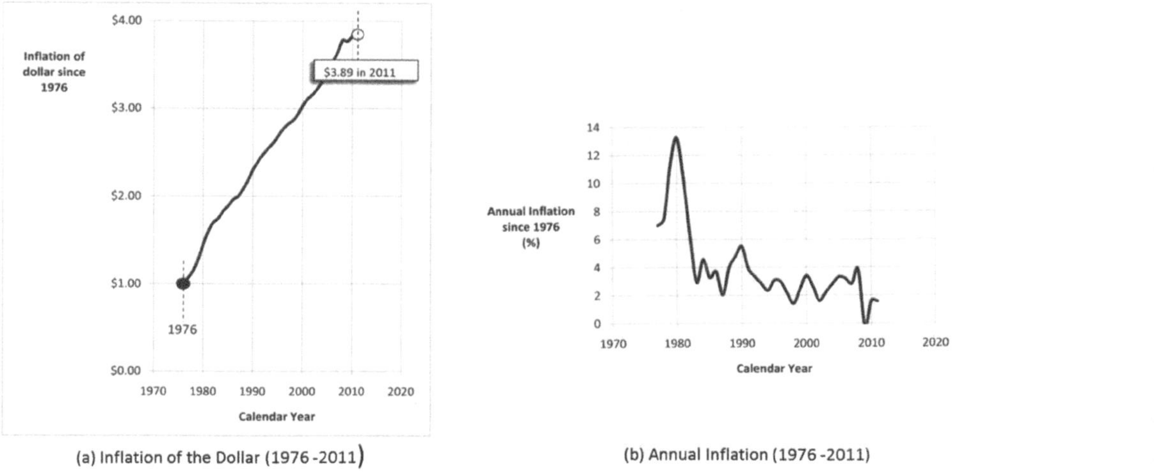

Figure 7.23: Inflation (1976–2011)

estimation would realistically go through a stringent cost analysis of material and labor. These subdivisions of the cost are significantly affected by changes in labor rates that vary widely among manufacturers. The decisions to "make or buy" in the acquisition process is a constant debate in any Program Manager's Office. The cost of materials is subject to international trade conditions and availability in the market place. As a small indication of how the cost is built up on any system between labor and material Figure 7.24 shows the collected costs from the ANVCE Project on the structure of steel ships and on the structure of aluminum hulled high speed marine craft.

In Figure 7.24, on the left hand side the curve is one of the ANVCE Cost Estimating Relationships (CER) for steel hulls. Superimposed on that same chart is some data collected by the author that shows how the material costs remained fairly constant and the driver in the cost reduction as size increase was due mainly

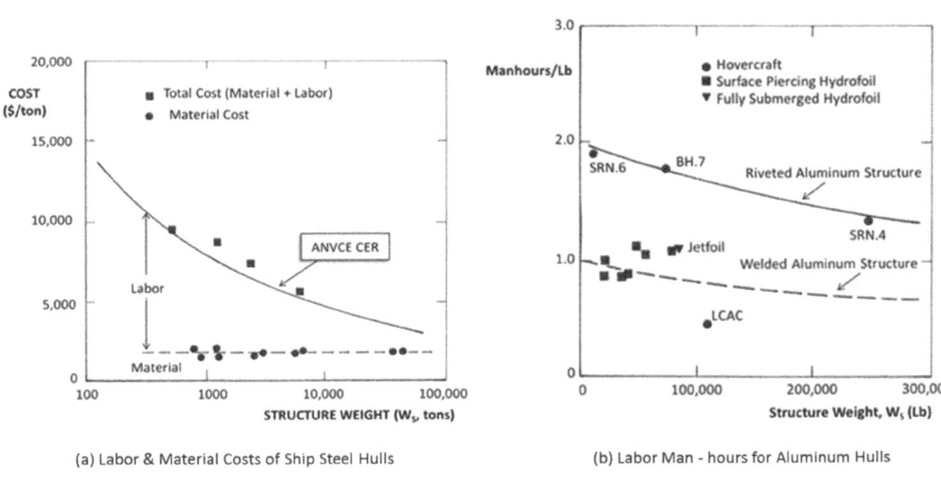

Figure 7.24: Examples of Labor and Material Costs for Ship Hulls (Group 100)

to the reduction in manhours required for construction. Such a division between Material and Labor Costs was not included in the ANVCE Project or in the Dix and Riddell analyses. How the labor manhours would change over time from say 1976 to 2011 with such improvements as computer aided manufacture and other improvements would not be reflected by the CERs used here.

Also, while the data is somewhat sparse, Figure 7.24 also shows some collected data on the difference between welded aluminum structures and riveted aluminum structures as used in such craft as hydrofoils and hovercraft. Again, such differences that could change over time were not included in the ANVCE Project or in other broad CER treatments. Because of this, it is not considered prudent to simply inflate all the cost data gathered at the time of design and construction of such craft in the late 1970s period to gather an estimate of "costs" for today's construction. It would be sufficient to say that the costs would be significantly higher than those quoted and the effort here has been more on comparison between ship types in the same time period rather than extrapolation to later time periods.

7.9 Weight and Cost Algorithms

The previous sections have shown the general trends in the Acquisition Costs of craft designed to operate at "100 knots at sea". The emphasis has been on trends rather than any prescriptive cost estimation. Cost estimation is usually a jealously guarded capability by individual manufacturers and it is difficult to obtain costs, especially those set to some common standard for fair comparison between competing concepts. It is possible to approach detail costs, however, through the use of standardized cost estimating relationships (CER). Most CERs are based on the weights of the subsystems and their build-up to total craft weights. As discussed above, the weight is a strong parameter to estimate the cost. The US Navy has produced at different times, standardized procedures for ensuring consistency in the weight and cost estimation. These have been issued in various documents such as the Military Handbook (MIL-HBDK-881) in its various updates and issues.

The US Navy has also issued the Ship Work Breakdown Structure (SWBS) for standardized treatment of weights and the work (manhours, material costs, etc). The introduction of new craft forms such as hydrofoils and air cushion craft also introduced changes to the SWBS as new features like "struts and foils" and "lift systems including skirts and seals" were introduced into ship inventories. These changes were starting to be introduced in 1973[17] and updated in 1977 to include hydrofoils and air cushion craft.

Figure 7.25 shows the SWBS as it applies to such advanced ship forms in the context of this book. It shows how the Weight Groups add up to the Empty Weight (W_E) discussed earlier in the general trend analyses of costs and then how the Useful Load or Disposable Load is added to the Empty Weight to give the final Gross Weight (W) or as sometimes called, the All Up Weight (AUW) of the vehicle. As can be seen from Figure 7.25, the all-important lift system is treated by the US Navy SWBS as part of the Auxiliary Systems for the ship. For the hydrofoil, this would include the struts and foil systems. For the air cushion craft (hovercraft, SES) this would include the lift engines, fans, flexible structure such as the skirts and related components.

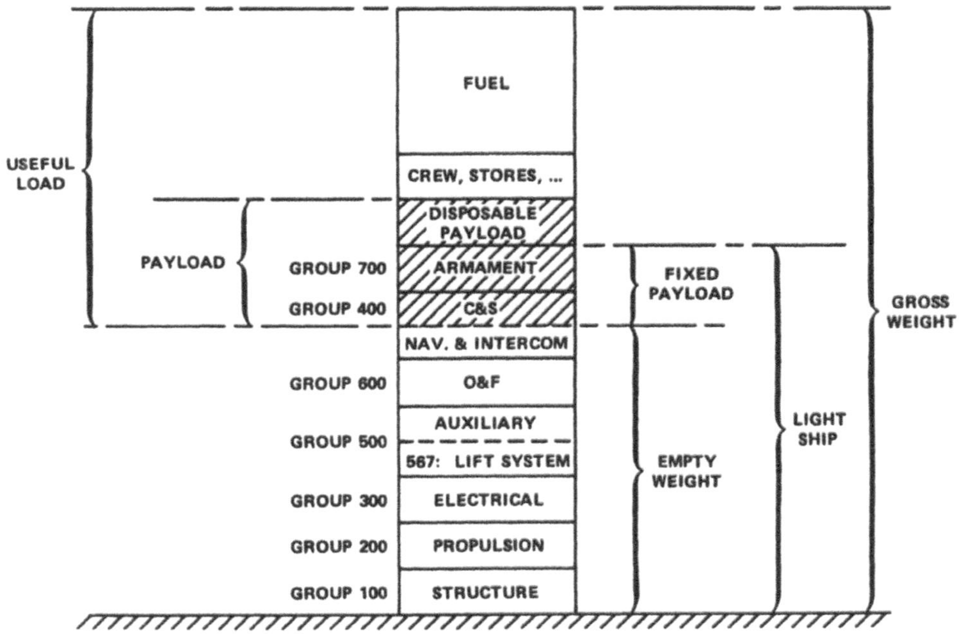

Figure 7.25: Ship Work Breakdown Structure (SWBS)

The Cost Estimating Relationships (CER) are then based on these seven Weight Groups as shown. The Empty Weight (W_E) is made up of five of the Weight Groups. The Fixed Payload is then made up of Group 400 and Group 700, which when added to the Empty Weight gives the Light Ship Weight as shown. The addition of the Disposable part of the Payload (e.g. ammunition, missiles, ordnance) then adds to the Fixed Payload for the Total Payload of the ship. The other term of "Useful Load" is shown in Figure 7.25 and easily identifiable.

In researching data of various ships, it is found that not all weight breakdowns and the associated cost estimates follow this format consistently but a sufficient data base has been built up over the years to form some trends. Later issues of the Military Handbook such as MIL-HDBK-881A issued on 6 April 2005 on the *New WBS* and the NAVSEA Instruction 4790.18 issued on 10 April 2007 on the *Expanded SWBS* have made minor changes in the SWBS that includes reassigning the dynamic craft weights to other subgroups. This has tended to introduce confusion into the data collection and distorts attempts to provide reliable trends. Fortunately, most of the work on dynamic lift craft such as hydrofoils and air cushion craft was done prior to these changes and thus any distortion in the data trends has been minimized.

In 1980[18], a detailed data base was compiled for high speed air cushion craft in their many forms that has been used successfully in the Cost Estimating Relationships (CER) and will not be repeated here in its entirety. Three Weight Groups are of particular interest in the light of the earlier discussion on general trends in total cost estimating for Acquisition Cost. These are the Group 100: Structure; Group 200: Propulsion and a subgroup of Group 500 Auxiliary, namely Group 567 on Lift System. These are discussed here as major drivers in the cost. Some of the key weight group algorithms are summarized here as are the Cost Estimating Relationships (CER).

SWBS Group 100: Structure

The structural weight (W_S or W_1) of amphibious air cushion craft (hovercraft, ACV) and non-amphibious air cushion craft (sidewall hovercraft, sidehull SES) is a key item in that it typically makes up about 60% of the Empty Weight (W_E) of the ship and thus plays directly into economics as discussed earlier in connection with the Direct Operating Cost (DOC). Figure 7.26 shows the collected data from the operational British Hovercraft Corporation craft such as *SRN.5, SRN.6* and *SRN.4* (see Figure 4.13, Chapter 4); the US Navy test craft, *SES-100A* and *SES-100B* (see Chapter 6) and other craft including design projections for craft up to 10,000 ton displacement Gross Weight air cushion craft.

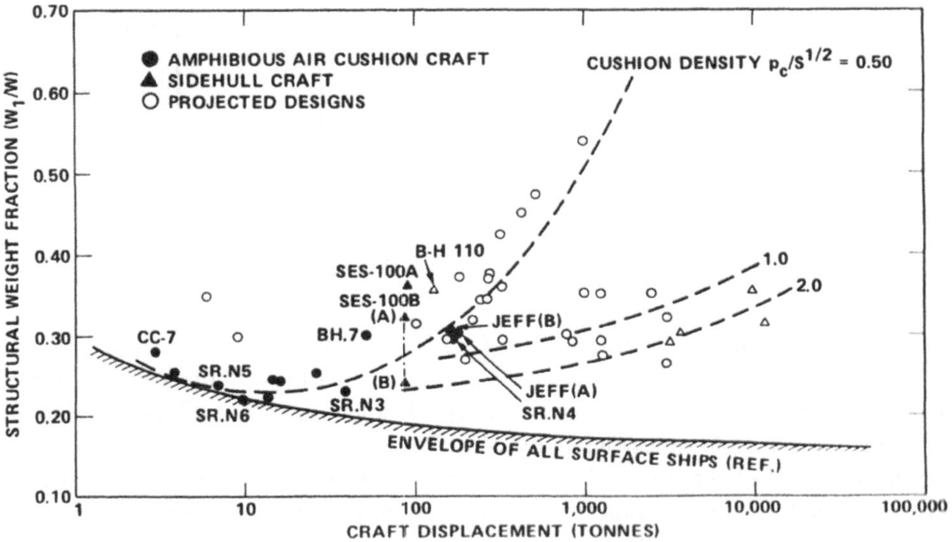

Figure 7.26: Structural Weight of Air Cushion Craft

The envelope of all ships types structural weight labeled (REF) in Figure 7.26 was shown earlier and the equation for the envelope was shown to be:

$$\left|\frac{W_S}{W}\right|_{env} = 0.15 + \frac{0.25}{W^{1/3}} \tag{7.36}$$

All the projected designs of air cushion craft have structural weight fractions significantly higher than this envelope value as seen in Figure 7.26. The various designers of high speed ships have struggled to lower the structural weight in their designs with the aim to achieve economical values.

A design technique used in the designs to lower the structure weight is to make the ship "more dense" by designing the craft for higher values of the Cushion Density, defined as:

$$\text{Cushion Density} = \frac{Cushion\ Pressure}{\sqrt{CushionArea}} = \frac{p_c}{\sqrt{S}} \tag{7.37}$$

This parameter is an important parameter in the design of several systems including the structural weight W_S/W, or W_1/W to use the nomenclature compatible with the US Navy SWBS and Group 100 for structure. It is also a key parameter in the design of the lift system where any requirement for a high cushion pressure (p_c) may place the design beyond the state of the art (SOA) of single stage fan systems normally used in existing air cushion craft, which would be an additional reason for higher costs of the ship. This was found to be the case in the design of the large SES in the 10,000 ton or more displacement category. For such large ship designs, it was found that high values of Cushion Density of around $p_c/\sqrt{S} = 2.0$ are required. For such large ships, of the type sought by both the US Maritime Administration and the US Navy, the structural weight could be estimated by the equation:

$$\frac{W_1}{W} = 0.21 + \frac{W^{1/3}}{\left(\frac{p_c}{\sqrt{S}}\right)^{1/3}} \tag{7.38}$$

More discussion on this key aspect of designing high speed SES is provided in Chapter 8 on Technical Considerations. It is included here to emphasize that all design treatments to achieve acceptable values of structural weight tend to push the designer into expensive systems to the detriment of the economics of the design.

SWBS Group 200: Propulsion

As has been shown, the Propulsion System is a large driver in the cost of the vehicle of any type. The data for a wide range of propulsions systems and vehicle types is shown in Figure 7.27 taken from the 1980[19] data set. As might be expected there is a wide range in weights for craft propelled by marine gas turbines and

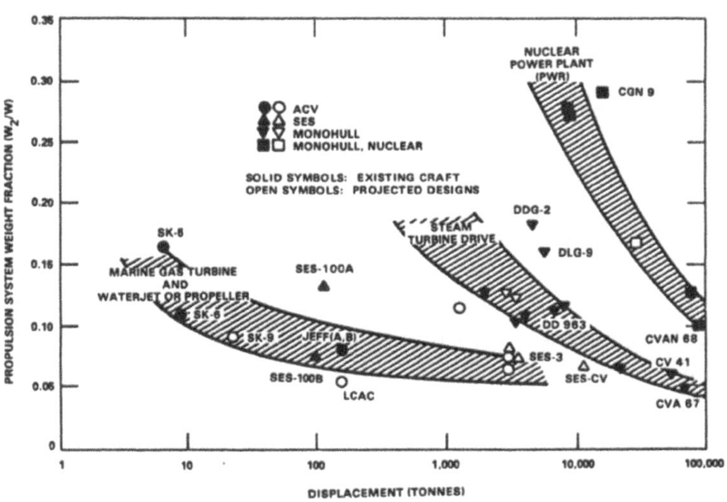

Figure 7.27: SWBS 200: Propulsion Systems Weight

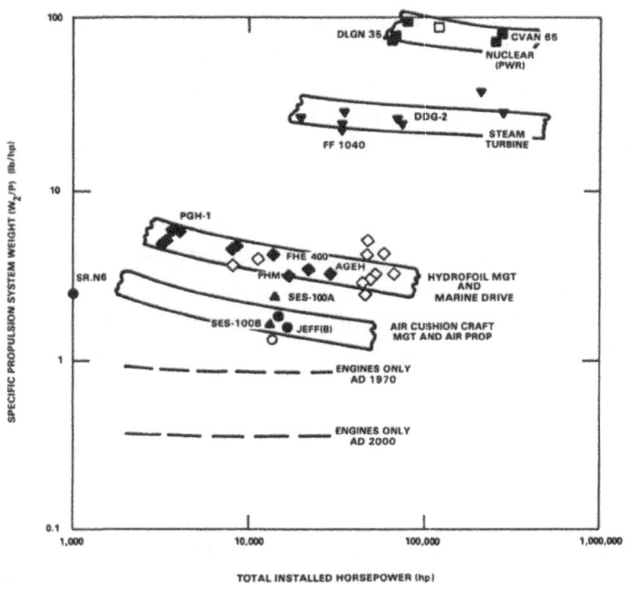

Figure 7.28: Specific Propulsion Power of Ships

propellers or waterjets to those craft propelled by steam turbine engines and further for those propelled by nuclear powerplants.

In the size of ships of concern to the US Maritime Administration (10,000 tons to 15,000 tons displacement) and even at the US Navy interest of frigate and destroyer size ships (3,000 to 5,000 ton displacement) it can be seen that the propulsion system weights of the high speed marine craft tend to converge to values around ~7% of the ship displacement.

More insight into the effect of power on these weights can be seen from the collection of data on the specific power weights (Lb/HP) taken from the same data set and shown in Figure 7.28. In this case it can be seen that the installed propulsion power measured as *specific power* as weight of the propulsion system per installed horsepower (W_2/HP) contributes strongly to the weight and may be expressed by the relationship:

$$\frac{W_2}{P} = 1.25 + \frac{74}{\sqrt{P}} \tag{7.39}$$

This specific power can then be related to the SWBS Group 200 weight by the relationship:

$$\frac{W_2}{W} = \left(\frac{W_2}{P}\right)\left(\frac{P}{WV}\right)(V) \tag{7.40}$$

This relationship shows how the propulsion power system weight (W_2/W) is related to the *specific power* (W_2/P_P) of the system being used together with the transport efficiency (WV/P) and ship speed (V) discussed earlier in the economic equations.

It is important to note that this data and the associated equations are indirectly associated with the lift system power (P_L) that is included in the SWBS Weight Group 500 (Auxiliary Systems). It is part of the design process to use the lift power (P_L) to reduce the needed propulsion power (P_P) so that these equations can be interpreted correctly.

SWBS Group 500: Auxiliary Systems

The weight of the lift systems in air cushion craft under the US Navy SWBS is a little more complicated in that this important system is treated as part of the auxiliary systems that are unrelated to the lift system. If the lift systems are separated from the "other auxiliary systems" the data looks like the data shown in Figure 7.29. There is no similarity between the auxiliary systems of air cushion craft and those in conventional displacement ships.

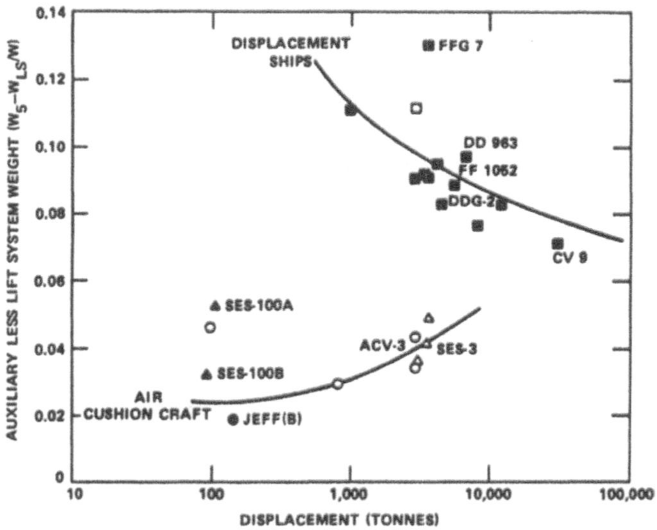

Figure 7.29: SWBS 500 Less Lift Systems

For the air cushion craft the weight can be expressed by:

$$\frac{W_5 - W_{LS}}{W} = 0.0024W^{1/3} + \frac{0.06}{W^{1/3}} \tag{7.41}$$

In this expression, (W_5) is the total weight of SWBS Group 500 and (W_{LS}) is the weight of the lift system in Group 500. The data set is somewhat sparse and the trend curve shown is only approximate tending more specific data, but for the purposes of this book, this part of the system (about 2–4% of the craft weight) is not a driving parameter in the economic design of the ship.

The lift system (W_{LS}) is made up of the lift engines, fans and associated equipment and the flexible skirt system (W_{SK}) which is included in (W_{LS}).

Data for both these subsystems within Group 500 is shown in Figure 7.30 for a wide range of actual operational craft and projected designs of air cushion craft.

There is a sufficient data base for these important subsystems of air cushion craft to provide reliable trends. These two important subsystems (part of SWBS 567) can be represented by the expressions:

$$\frac{W_{LS}}{W} = 0.044 + \frac{0.08}{W^{1/3}} \tag{7.42}$$

$$\frac{W_{SK}}{W} = 0.014 + \frac{0.08}{W^{1/3}} \tag{7.43}$$

As can be seen from Figure 7.30, the lift systems of air cushion craft make up about 5% of the ship displacement in the frigate size ship, of which the skirt systems contribute about 2% for the same size ships.

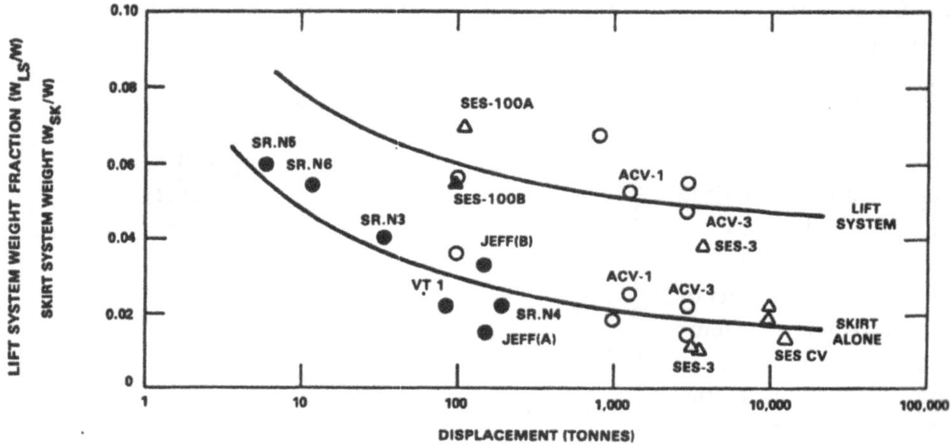

Figure 7.30: Lift System and Skirt System Weights of Air Cushion Craft

Cost Algorithms

Estimating the costs of any vehicle can be a complex endeavor and developing simple formulae to give reliable costs is not a simple matter. Many approaches have been used by many researchers. Some have treated the cost as based on "parts count". Such an approach is used in many industries including electronics, armament suppliers and others. Other approaches use the weight groups and power levels as discussed in the above sections. Others include complexity factors and other means of separating vehicle types.

Some interesting reading of the various approaches may be found in the References. The RCA PRICE[20] model has been used to success in many programs. The PRICE Method uses subsystem weights, their enclosed volumes and the manufacturing complexity features. In 1982, Salmon and Locke[21] used the PRICE Method to conventional surface ships with some success. In 1984, Jones, Mayo and Clark[22] applied the PRICE Method to hovercraft and found good agreement.

The method found most useful by this author[e] is based on the work by James Moy (US Navy) for the ANVCE Project in 1976 and published in 1977[23]. In a similar vein to the other cost estimators, Jimmy Moy's work is based on subsystem weights (using the US Navy SWBS); the installed power and a set of complexity levels. Moy expresses the cost of a ship by the expression:

$$\$_{ship} = \sum_{i=1}^{n=9} (CER)_i + profit \tag{7.44}$$

That is, there are nine groups (i = 1–9) in the US Navy SWBS for which seven CERs are directly related to the construction of the ship as discussed above in the section on weight algorithms and two CERs directly related to the design, engineering and

[e] In what follows, the reader should recognize that the absolute costs (and thus the constants in the equations) will change for today's costs but that the relative values will probably remain the same, which is the driving concern here in design choice. It will require another detailed cost study to verify these assumptions.

construction services (Groups 800 and 900). To complete the "cost" estimate a profit is added to represent the current accepted values in the ship construction industry to generate the "price".

The $(CER)_i$ set in Equation 7.44 is then expressed by:

$$(CER)_i = K_i K_{i,j} X_i^{n_i} \tag{7.45}$$

In this formulation:

K_i = a constant associated with SWBS Group *(i)*
$K_{i,j}$ = a constant which reflects the scaling of Group *(i)* according to the unique features of that Group
X_i = the independent variable reflecting the weight or power of the group being estimated

In what follows, the weights and power are used from the weight algorithms shown earlier in this Chapter. The complete listing of all nine CERs may be found in Moy's report[24] that provides the explanations of the various "scaling" (to use Moy's words) that describes whether the system is a "conventional aluminum hull" or a "high tensile strength aluminum hull" and other determinants in the design. Such variations provide some indication of the complexity factors in the design and some attempt to sort out manufacturing manhours involved in each of the different designs. The data base contained both well established and some less well defined data points. It represented the most complete data base and CER method available at the time (1976) as the various high speed marine craft were being designed and constructed.

It should be pointed out that the method was primarily designed for the estimating the cost of military ships. This brings with it the additional complication that in US Navy ships there is a considerable amount of Government Furnished Equipment (GFE) that is not included in the cost as it is accounted for in a different accounting "slot" in the US Navy Budget System. These additional items are usually for the Payload or Weapons Systems and for associated Government Costs, such as Change Orders and other items. Typically, such GFE can add 50% to the cost of the basic ship. This was true for the particular set of hovercraft as well as for conventional ships. Provided some care is taken in the cost estimating method one can draw reasonable conclusions about the cost of commercial ships but the main emphasis here has been on military ships designed for some (identified) mission with its appropriate cost of the military payload.

Cost of Follow-On Vehicles

The next level of complication is the recognition of the "learning curve" in manufacture where the cost of any follow-on craft after the first one built is usually less by some industry standard amount. This was referred to earlier in the formulation by Dix and Riddell (see Equation 7.33) where the effect of the quantity built (Q) was included in the Acquisition Cost estimate. The learning curve in construction was first noted by T.F.Wright in 1936[25] when the mass production of aircraft and other military hardware first occurred. Wright found that the cost of the airplane tended to decrease by a constant factor for each doubling of the number built. For example, Wright found an *80% learning curve* in the production of aircraft which meant that craft number 4 cost 80% of craft number 2; craft number 8 cost 80% of craft number 4, and so on. It was found by later researchers that this "learning curve" characteristic occurred in other industries but with different learning curve values for each industry or product. In the shipbuilding industry, which is more manpower intensive the learning curve is not as pronounced as in the highly automated airplane (and automobile or electronics) industry. A typical value for ship construction is closer to 90–95% learning curve. Wright formulated this learning curve with the expression:

$$\$_Q = \$_1 Q^{[\ln(lc)/\ln 2]} \tag{7.46}$$

The terms in the cost equation are:

$\$_Q$ = cost of the Qth craft
$\$_1$ = cost of the first craft
lc = learning curve "slope"
ln = natural logarithm

The slope of the curve of craft cost ($\$_Q$) versus quantity built (Q) is not the value of the learning curve, of course, because of the factor 2 in the doubling of the quantity (Q). If Equation 7.46 is written as:

$$\$_Q = \$_1 Q^m \tag{7.47}$$

Then if the learning curve is "80%" then the slope of the cost curve (m) would be (m = –0.319). If the learning curve was "95%", which is more appropriate for shipbuilding, then the slope of the cost curve would be (m = –0.074). There are variations of the learning curve used in different industries. These variations include "cumulative average cost" of succeeding batches of units made but these subtleties get swamped by the many other vagaries in costs of ships and are pursued here.

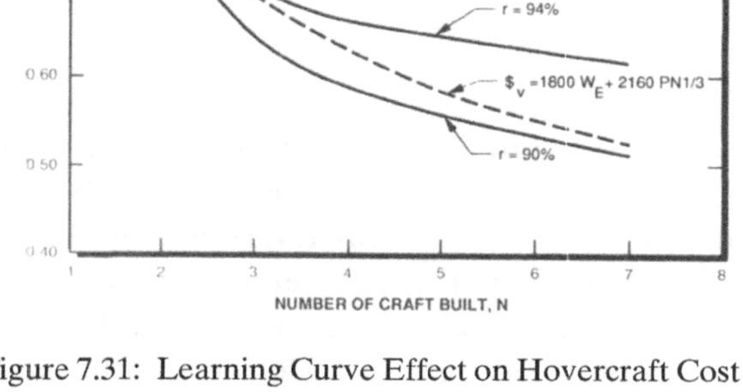

Figure 7.31: Learning Curve Effect on Hovercraft Cost Example

Figure 7.31 compares the cost of follow-on craft when estimated by the detailed Moy method used in the ANVCE Project together with the approximate formula given by Dix and Riddell shown earlier. This particular example was for a hypothetical 100 ton displacement air cushion craft with an installed power of 15,000 HP. The example indicates that the approaches used here are in general agreement at least for small quantity buys. The two approaches give different results for large quantity buys. The data base for large quantity buys of high speed air cushion craft does not exist so firm conclusions cannot be given at this time. These differences will be returned to once the detailed treatment of the SWBS Weight & Power Group method has been discussed.

Detailed Cost Estimating Relationships

The complete set of the ANVCE CERs is provided in the Moy Report. In similar vein to that discussed for the SWBS weight groups, the examples of the Group 100, Group 200 and Group 500 will be shown in this Chapter as an indication of the cost drivers in high speed air cushion craft.

Group 100 Structure Cost Estimating Relationship

From Moy[26], the CER for the structure weight cost of the lead ship is given by:

$$\$_{100} = 22,543 K_{1,j} W_1^{0.772} \tag{7.48}$$

The data base covered conventional steel hulls to high grade aluminum such that the cost difference between these types was expressed in the $K_{1,j}$ set of factors:

Table 7.4: *SWBS 100 Structure Cost Factors*

$K_{1,j}$	Description of data set
1.000	Conventional steel hull
1.570	HT steel hull
2.192	Conventional aluminum hull
5.587	HT aluminum hull

This particular format of the CERs provides a further insight in cost drivers in that in this case, it is found that the cost of high strength aluminum hull structures of the type contemplated for large, high speed SES costs approximately 5–6 times as much as conventional steel hulls, all else being equal. This formulation does not differentiate between material and labor costs as discussed earlier but certainly provides a clue as to where the costs are in such designs. In the ANVCE Project, the actual costs of particular craft used in the data base are not made public as they reveal internal costing methods of manufacturers and the intent here is to determine trends in costs of high speed ships and not to delve into individual manufacturer's competitive data.

Group 200 Propulsion Cost Estimating Relationship

From the same data set, the CER for the (non-nuclear) propulsion system cost of the lead ship Moy gives:

$$\$_{200} = 1,236 K_{2,j} P_P^{0.808} \tag{7.49}$$

In this case, a wide range of data base was available and the cost difference between the various choices is expressed by the factor $K_{2,j}$ as shown in Table 7.5. In this case, there is a significant difference and complexity in the choices of propulsion systems and it is difficult to put the systems into "convenient" boxes but the data base was rich enough to at least set up six separate and distinct types of propulsion schemes that allow reasonable choices to be made as to the likely costs of any new design. Table 7.5 provides that data set. For high speed marine craft of the type under consideration, the most likely set of data applicable for high speed air cushion craft (SES) is that with the cost factor $K_{2,j}$ = 1.502, i.e. the system with gas turbine propulsion and high speed propulsors. It is seen from the data base that many of the other choices or propulsion systems are actually more costly.

Table 7.5: *SWBS 200 Propulsion System Cost Factors*

$K_{2,j}$	Description of data set
1.000	CODOG/CODAG powerplant, high speed marine propulsors and straight drive train
1.502	Gas turbines (GT/CODAG) powerplant, high speed marine propulsors and right-angled drive train
1.979	Steam turbine powerplant, low speed, controllable pitch (CP) propellers and long shafts
2.346	GT powerplant, low speed CP propellers and reduction gears
3.281	GT powerplant with complex drives
3.437	Nuclear, steam pressurized water reactor powerplant, low speed fixed pitch propellers, straight drive train

Group 500 Auxiliary System Cost Estimating Relationship

Here the data collection was not as easy to separate as for the other types of systems. As discussed earlier for the Auxiliary Systems Weight Group (see Equations 7.41, 7.42 and 7.43) the lift systems are included as SWBS Group 567 in the larger Group 500 for Auxiliary Systems. This unfortunate grouping has complicated the understanding of the costs of the "auxiliary systems" but the data base[27] was collected under ship type as shown in Table 7.6. The cost estimating relationship for the Group 500, that included both the lift system and unrelated equipment was grouped into one cost grouping by:

$$\$_{500} = 62,983 K_{5,j} W_5{}^{0.782} \tag{7.50}$$

Table 7.6: *SWBS 500 Auxiliary System Cost Factors*

$K_{5,j}$	Description of data set
1.000	Steam propelled displacement ship
1.528	GT propelled displacement ship
4.161	Fully Submerged hydrofoils
5.370	Air Cushion Craft

In the weight group data collection (see Figure 7.29) there is very little correlation between the auxiliary systems in conventional displacement ships and those in the high speed marine craft that include unique systems such as struts and foils in the hydrofoil class of ship and fans, skirts and related items in air cushion craft. There is also a distinct difference between surface piercing hydrofoils systems and those of fully submerged hydrofoil systems. Because the weight of these dynamic craft systems is low (approximately 2–4% of the ship's weight) hopefully this grouping of systems does not distort the cost trends too much. This is an area that definitely needs work. The grouped data is as shown in Table 7.6.

Based on the data collected on several existing operational air cushion craft, it is seen from Table 7.6 that the weight of the Auxiliary Systems (that include the Lift System) the tendency is for the weight of these additional systems to cost >5:1 over conventional displacement ship auxiliary systems. This is a weight penalty for any dynamic lift craft and especially for the air cushion craft (hovercraft, SES).

The remaining Group Cost Estimating Relationships (CER) including the additional CERs that estimate the Design and Engineering Costs (Group 800) and the Construction Services Costs (Group 900) must be included for the total cost estimates. All these items must be included in any particular design to put together the true total cost of any given Lead Ship. The costs of First Follow-On Ship where the design costs have been "sunk" and then the Follow-On Ships in a construction "Buy Package" where Learning Curves can be introduced to capture the lower costs for large quantities of ships built.

Frigate Sized SES Cost Example

Using the earlier cited weight and power algorithms together with the ANVCE cost algorithms, it is possible to construct the acquisition cost of the lead ship of a frigate sized SES. If this example case is an SES of 3600 tons, with an installed total power of 215,000 HP and a top speed of 100 knots, the estimated weight breakdown and the cost would be as shown in Table 7.7 using the algorithms referred to above. While these are not detailed estimates but values using the trend lines expressed by the algorithms, there are certain characteristics that appear. It is seen that the three SWBS Groups 100, 200 and 500, contribute about 75% of the cost of the empty ship (or 39% of the cost of the complete ship if design, engineering and construction services are included). This is a significant cost driver in the design.

If the same algorithm approach is used to estimate the cost of a frigate sized ship of the same type as the FFG 7[f] then the cost estimates shown in Table 7.8 are obtained. This generic "FFG-7" has a displacement of W=3648 tons; total power installed of P=41,000HP and has a top speed of 30 knots. Because of the lower speed (V_{max} = 30 knots), the power is not such a cost driver and the same three SWBS groups now contribute about 58% of the empty ship cost or 19% of the total ship cost.

It should be mentioned (again) that cost estimating is a dangerous pursuit in the absolute because of the many variables and intangibles that come into play. While the effort in this book is on the cost and performance of high speed ships that will

[f] Note that these are generic "FFG-7 like" costs and not the actual costs of the FFG-7.

Table 7.7: *Example 3600 ton High Speed SES*

SWBS Group	Weight (tons)	%W_E	%W	Cost ($M) (1976)	%Total Ship Cost
100: Structure	1105	51.1	30.7	28.2	12.5
200: Propulsion	238	11.0	6.6	29.0	12.8
300: Electric Plant	80	3.7	2.2	9.9	4.4
400: Command & Control	116	5.4	3.2	9.4	4.2
500: Auxiliary (inc. Lift Sys.)	323	14.9	9.0	31.0	13.7
600: Outfit & Furnishings	182	8.4	5.1	8.4	3.7
700: Armament	118	5.5	3.3	2.1	0.9
Empty Weight (W_E)	2162	100	60.0	118.0	52.1
800: Design & Engineering				69.1	30.5
900: Construction Services				9.7	4.3
Total Lead Ship Cost				196.8	87.0
Profit (15%)				29.5	13.0
Total Lead Ship Price				$226.3M	100%

Table 7.8: *Generic Cost Estimate for 3648 ton frigate (FFG-7 like)*

SWBS Group	Weight (tons)	%W_E	%W	Cost ($M) (1976)	%Total Ship Cost
100: Structure	1316	47.0	36.1	8.7	7.3
200: Propulsion	277	9.9	7.6	9.9	8.5
300: Electric Plant	200	7.1	5.5	12.6	10.7
400: Command & Control	104	3.7	2.9	4.0	3.3
500: Auxiliary (inc. Lift Sys.)	476	17.0	13.1	19.6	16.6
600: Outfit & Furnishings	328	11.7	9.0	9.8	8.3
700: Armament	100	3.6	2.7	1.8	1.6
Empty Weight (W_E)	2801	100	76.8	66.3	56.3
800: Design & Engineering				19.1	18.2
900: Construction Services				6.0	5.7
Total Lead Ship Cost				91.4	87.0
Profit (15%)				13.7	13.0
Total Lead Ship Price				$105.1M	100%

complicate the cost estimating of untried technologies, the conventional world of ship design is not without its wide variations in cost estimates. In Table 7.8, using the algorithm approach the cost of a "FFG-7 like" ship is expected to be about $105M in 1976 dollars. The actual FFG-7 was an innovation in design at the time with such novel features as a steel hull plus aluminum superstructure and other innovations. It also had some complications in that it was restricted in growth size with small compartments not amenable to upgrade to newer equipment. The Lead Ship contract for $94.4 million was awarded to Bath Iron Works in October 1973. That price is not far off from the estimate given in Table 7.8 of $105M. The price increased in 1978 when a combination of factors came into play. Several items had not been included in the original contract and there were some design problems (e.g. new fin stabilizers and associated electronics were required). A towed sonar array was added that was intended to be in the original package. These changes raised the average cost of the ship in 1978 to be $194M. But the "generic" cost estimate of $105M is at least in between to the actual costs as the FFG-7 program got underway.

The cost of the equivalent sized SES with a projected speed of 100 knots is seen to be of the order of between 2:1 and 2.5:1 over the cost of the generic FFG-7 using the same algorithms. These were the values used in Figure 7.22 under the discussion of the relative cost of speed.

7.10 Conclusions of Economic Considerations

This chapter has sought to bring out the many considerations when designing a ship or craft to achieve some specific economic or useful goal. The tendency to concentrate on some focused technical feature in a design program has frequently caused a downfall in the acceptance of the design for a larger purpose. The designer should always keep in mind not only the technical risks in the proposed design but also how the design is treated as a whole to meet some final objective. Concentration on just one technical feature is not the best way to get an acceptable design.

In the Acquisition Cost of the craft under consideration, this chapter has shown the many characteristics and design choices that must be made to achieve the final design for a given cost goal. Trades between ship systems and subsystems should always be considered if an economical design is the final planned outcome. The impact of weight of the different subsystems (W_i); the installed power (P); and the design maximum speed sought (V_{max}) have been shown in the above discussions in the impact on the final cost. In the case of dynamic lift craft such as the SES, the trade between lift power (P_L) that influences the Auxiliary Systems (Group 500) weight and the propulsion power (P_P) that influences the Propulsion System weight (Group 200) is an important trade as to the impact on the craft cost.

Once the craft has been built, this chapter has shown the impact of the performance characteristics of propulsion efficiency; the transport efficiency (WV/P); and the empty weight (W_E) that have a direct effect on the operating economics through the Direct Operating Cost (DOC) of the ship.

Chapter 8 delves into more technical detail on each of these technical considerations in the design of the high speed ship.

References

1. Nicolas Johnson, *"The American Partnership: The US Merchant Marine"*, speech given at the Mid-Gulf Conference on Transportation & Industrial Modernization for Profit and Growth, New Orleans, La., 9 February 1965
2. *"Standard Method of Estimating Comparative Direct Operating Costs of Turbine Powered Transport Airplanes"*, Revised version dated October 1967, Air Transportation Association, Washington, DC.
3. Mantle, Peter J. and Lavis, David R., *"Domain of the Air Cushion Craft"*, paper given at the Society of Automotive Engineers (SAE), New York, New York, 29 April 1968
4. Mantle, Peter J. *"Interface Craft for Future Transportation"*, ASME, 1966 National Transportation Symposium, San Francisco, California, 1–6 May 1966
5. Mantle, Peter J., *"Air Cushion Craft Development"*, published by US Navy in 1980 (ISBN 008-047-00291-2)

6. Gabrielli, Giuseppe and Theodore von Karman, *"What price Speed?"*, Mechanical Engineering, October 1950 Issue.
7. Mantle, Peter J. *"Interface Craft for Future Transportation"*, ASME, 1966 National Transportation Symposium, San Francisco, California, 1–6 May 1966
8. Mantle, Peter J. *"Some Design Aspects of Air Cushion Craft"*, International Congress of Subsonic Aeronautics, The New York Academy of Sciences, New York, N.Y., 3–6 April 1967
9. Mantle, Peter J. *"Interface Craft for Future Transportation"*, ASME, 1966 National Transportation Symposium, San Francisco, California, 1–6 May 1966
10. Mantle, Peter J. *"Some Design Aspects of Air Cushion Craft"*, International Congress of Subsonic Aeronautics, The New York Academy of Sciences, New York, N.Y., 3–6 April 1967
11. *ibid*
12. Capt Tom Meeks and Peter J. Mantle, *"The Advanced Naval Vehicle Concepts Evaluation"*, AIAA/SNAME Advanced Marine Vehicles Conference, Arlington, Virginia, Paper No. 76–846, 20–22 September 1976
13. Mantle, Peter J., *"Advanced Concepts for Sea Control"*, Society of Automotive Engineers Aerospace Meeting, Paper No. 770966, Los Angeles, CA, 14–17 November 1977
14. Dix, D.M. and Riddell, F.R., *"Projecting Cost-Performance Trade-Offs for Military Vehicles"*, Astronautics and Aeronautics magazine, September 1976
15. Mantle, Peter J., *"Air Cushion Craft Development"*, published by US Navy in 1980 (ISBN 008-047-00291-2)
16. US Bureau of Labor Statistics, http://www.bls.gov/data
17. *"Ship Work Breakdown Structure"*, Naval Ship Systems Command, NAVSHIPS 0900–039-9010, March 1973
18. Mantle, Peter J., *"Air Cushion Craft Development"*, published by US Navy in 1980 (ISBN 008-047-00291-2)
19. *ibid*
20. *"PRICE"*, Manual issued by RCA PRICE Systems, Edition 3, 1984
21. Salmon, C.L. and A.J. Locke, *"The Application of PRICE to Ships and Other Large Structures"*, paper presented at the PRICE Users Symposium, Munich, GE, 14–16 September 1982
22. Jones, R., Mayo, D.L. and Clark, D.J., *"The Application of the RCA PRICE Parametric Cost Model to Air Cushion Vehicles"*, paper presented to International Society of Parametric Analysts, Sixth Annual Conference, San Francisco, CA, 16 May 1984
23. Moy, James, *"Ship Construction Cost Estimating Relationships for Advanced Naval Vehicle Concept Evaluation (ANVCE) Study"*, Naval Ship Engineering Center, Report 6112DE-045–77, August 1977.
24. *ibid*
25. Wright, T. F., *"Factors Affecting the Cost of Airplanes"*, Journal of the Aeronautical Sciences, Vol 3, February 1936, pp 122–128
26. Moy, James, *"Ship Construction Cost Estimating Relationships for Advanced Naval Vehicle Concept Evaluation (ANVCE) Study"*, Naval Ship Engineering Center, Report 6112DE-045–77, August 1977.
27. *ibid*

8 Technical Considerations

> *"When you can measure what you are speaking about, and express it in numbers, you know something about it. When you cannot express it in numbers, your knowledge is of a meager and unsatisfactory kind; it may be the beginning of knowledge, but you have scarcely, in your thoughts, advanced to the stage of science."*
>
> Lord Kelvin (1824–1907)

To determine the "performance" of a vehicle in the broadest sense of the term, one must consider many design characteristics and how they must be applied to achieve some "mission need". The words "mission need" are specifically chosen instead of "performance requirements" to emphasize that if the vehicle or ship is not designed to meet some (customer defined) mission need, the design will quickly fail and programs canceled. This is true whether one is considering the performance of the family automobile or the military naval warship.

As discussed in Chapter 6 in relation to the SES-100A and SES-100B Program, the stability and control characteristics of any (high speed) ship must also be considered in conjunction with the specific performance needs related to speed, power and weight. Shaping of the ship for performance optimization must also include the necessary shaping for stability and control. Shaping for performance may not always be the best shape for stability. Indeed, they are frequently in conflict in vehicle design. Designing simply for speed alone is insufficient. In Chapter 6, the various measures of "performance" were listed as:

a) Can the ship carry a Payload efficiently over a given distance or Range?
b) What is the Cruise Speed and the Maximum Speed?
c) What is the turning performance?
d) What are the acceleration and deceleration characteristics?
e) What are the Power Requirements?
f) What is the Fuel Efficiency?
g) Can it operate safely in rough seas?
h) Is it stable and controllable in all normal modes of operation?

Chapter 7 provided a driving force behind most of these "requirements" by use of the Direct Operating Cost (DOC) Formula, given as:

$$DOC(\$/mile) = C_1 + C_2\left(\frac{W_f}{R}\right) + \frac{C_3}{V_b} \tag{8.1}$$

The Operating Cost can be expressed in units of choice, of course, and use of English units and metric units will be used where appropriate based on data points from the literature. The constants C_1, C_2, C_3 are expanded upon in detail in Chapter 7 on Economic Considerations (see Chapter 7, Table 7.2) but depend on individual parameters for the ship in question (such as crew costs, insurance, maintenance, etc). While Equation 8.1 was developed for commercial vehicles, the designer of military vehicles would be sorely remiss to ignore economically acceptable operating costs in any design. *The cancelation of most of the high speed vehicle programs in the US Navy (hydrofoils, hovercraft and SES) is testament to this reality.*

What is expanded upon in this Chapter are the "technical" features embodied in the design of the high speed ship. The governing technical parameters in the Economic Equation 8.1 are the range (R), fuel expended (W_f), and speed (V_b). Chapter 7 also introduced the Range Equation that shows how specific vehicle characteristics determine the range (R). For aircraft and conventional ships, it was shown that the Breguet Range equation (*for propeller driven craft*) could be written as:

$$R = 325\left(\frac{\eta_P}{sfc}\right)\left(\frac{L}{D}\right)\left(\log_e \frac{W}{W - W_f}\right) \tag{8.2}$$

Which, in its approximate form to first order, can be written as:

$$R = 325\left(\frac{\eta_P}{sfc}\right)\left(\frac{L}{D}\right)\left(\frac{W_f}{W}\right) \tag{8.3}$$

Additional technical characteristics now enter the performance considerations through such craft characteristics as propulsion efficiency (η_p/sfc); the aerodynamic efficiency or lift-drag ratio (L/D); and the fuel load (W_f/W).

The *"aerodynamic (or hydrodynamic) efficiency" (L/D)*, recognized by all aircraft and ship designers, can be directly related to the ability to transport a given weight (W) at a given speed (V) for a given expenditure of power (P), which is a measure of work efficiency, through the *"transport efficiency (TE)"*, defined as:

$$\text{Transport Efficiency (TE)} = \frac{WV}{P} = \eta_P\left(\frac{L}{D}\right) \tag{8.4}$$

Chapter 7 also showed how for the air cushion craft, that needed to expend an additional power, the lift power (P_L), to achieve the performance of the craft, these Breguet Range Equations had to be modified such that the Transport Efficiency (WV/P) would now include the lift power expenditure, such that the aerodynamic L/D ratio would be modified to become the Effective Lift Drag ratio $(L/D)_{EFF}$, where:

$$\left(\frac{L}{D}\right)_{EFF} = \frac{W}{D + \dfrac{\eta_P}{\eta_L}\left(\dfrac{p_c Q}{V}\right)} \tag{8.5}$$

This form of the *effective lift drag ratio* shows that, in equilibrium conditions (L = W), the $(L/D)_{EFF}$ is shown to be a function of the propulsive efficiency (η_P) and the lift system parameters of cushion pressure (p_c), air cushion flow (Q), and lift system efficiency (η_L). It is seen also that the craft speed (V) now enters the L/D ratio in an additional term. This additional term $\frac{\eta_P}{\eta_L}\left(\frac{p_c Q}{V}\right)$ has the dimensions of drag. For the case of zero air flow (Q = 0) the equation for $(L/D)_{EFF}$ reduces to the conventional lift drag ratio (L/D). The Transport Efficiency (TE) now becomes modified also, to be:

$$\frac{WV}{P} = \eta_P \left(\frac{L}{D}\right)_{EFF} \tag{8.6}$$

As shown in Chapter 7 on economic considerations, the Transport Efficiency (TE) was a driving force in the design of most vehicles. The power consumed in pushing the ship to higher speeds was a significant characteristic of the various candidates for high speed and was the main driver in the cost of many such craft. This leads directly to understanding of the separate components that add up to the total drag (D_{total}) or resistance of the vehicle, and this is discussed first in what follows. See later for more discussion of Transport Efficiency (WV/P) and its significance in vehicle design.

8.1 Drag of High Speed Air Cushion Craft

The form of the drag of high speed air cushion craft will change depending on the type being considered (amphibious and non-amphibious; aerostatic or aerodynamic) and a "universal" law cannot be put down that will cover all types. Fortunately, the breakdown of the components that build up to the total resistance in each type is readily understood and can be treated separately for both design and evaluation. There is sufficient overlap in the many types that allows a basic formulation to be written as:

$$D_{total} = D_{wave} + D_{aero} + D_{mom} + D_{sk} + D_{SH} \tag{8.7}$$

The total drag (D_{total}) is seen to be made up of the following main components:

D_{wave} One of the largest drag components is the drag of moving the depression of the pressurized air cushion through the water. This is certainly true for both the amphibious and non-amphibious aerostatic type of air cushion craft but less so for the aerodynamic form of air cushion craft in its cruise mode.

D_{aero} Above the waterline, air resistance of the craft is a strong consideration that increases with speed. It is caused by the frontal resistance, called Profile Drag (D_o) for all craft. For the aerodynamic form, an additional component of D_{aero} comes into play called the Induced Drag (D_i). The consideration of the total aerodynamic drag ($D_{aero} = D_o + D_i$) will be treated in this Chapter and again in Chapter 11 when the aerodynamic form is treated in detail.

D_{mom} The so-called Momentum Drag is caused by the energy transfer as the pressurized air flow (Q) is generated onboard the craft and the momentum is exchanged into the lifting mechanism. The form of this drag will be different depending on the type of air cushion craft (aerostatic or aerodynamic).

D_{SK} The sealing mechanism around the craft whether amphibious or non-amphibious will contribute some form of skirt or seal drag.

D_{SH} In the non-amphibious form of craft there will be some form of sidehulls to provide hydrodynamic lift for stability, sealing of the air cushion (in combination with the skirt system) and to a lesser extent a contribution to the lift of the craft. In most cases, this component of drag will apply to the aerodynamic form of craft only in specific circumstances as will be discussed. For convenience of discussion, the drag of the sidehulls will include the resistance of any appendages such as stability and control devices (fins, rudders, etc). It will also include any resistance from propulsion devices such as pods, inlets and the like. Such items can be strong contributors to the total drag and this is discussed where appropriate in context. In any detailed treatment of such terms, the designer will find himself working between "drag" and "thrust" estimation as the propulsion system interacts with the drag of the hull through such items as "thrust deduction" and issues of cavitating flow in propulsion inlets.

There are other contributors to craft resistance such as spray drag (D_{spray}) and drag over various land surfaces (ice, sand, etc) for the amphibious type. Such drag contributions have been treated elsewhere[1] and will not be discussed in detail here as they are usually not "drivers" in the basic design decisions for high speed air cushion craft. Each of these drag components take on different forms in both calm water and in rough seas. A complete derivation of all the components (for both calm and rough water) was treated complete with available test data in 1980[2] but there are some features in terms of selecting "best candidates" for high speed design (say, 100 knots) that will be expanded upon in the text.

8.1.1 Cushion Induced Wave Drag (D_{wave})

This component of the total drag has received considerable analytical and model test treatment since before the air cushion craft was invented. The hydrodynamics of air depressions in the water was treated early on by such notables as Sir Horace Lamb (1849–1934) and Sir Thomas Havelock (1877–1968). Sir Havelock[3] gave the solution for a point source pressure distribution, while Sir Horace Lamb[4] developed an approximate two-dimensional theory for the induced wave drag of a moving depression generated by a cushion pressure (p_C) and length (L). Since that initial work, there have been several adaptations and extensions as specifically applied to the air cushion craft. Peter Crewe and Wilf Eggington[5] developed a result for the induced wave drag of air cushions based on Lamb's work specifically for air cushion craft. In those referenced RINA Transactions, Crewe & Eggington of British Hovercraft Corporation (BHC) compared their theory with model data done in connection with the first hovercraft SRN.1 and found reasonable agreement with the two-dimensional theory for the case of deep water. More complete treatments for the induced wave drag of air cushion craft then followed by Newman & Poole[6] in 1962; Barratt[7] in 1965; and Doctors[8] in 1970. These were complete treatments including three dimensional effects such as the effect of cushion planform, water depth, and edge (shape) effects. Everest & Hogben[9] conducted extensive test model tests in both deep and shallow water and compared the results with the theoretical predictions by Newman & Poole and by Larry Doctors. Everest & Hogben also compared their model data with the theoretical work by Barratt. Everest & Hogben found fairly good agreement in deep water, especially around the primary hump. Progressively poorer agreement occurs in the secondary and tertiary humps[a]. Indeed the character and even existence of the secondary and tertiary humps resulted in much discussion among researchers. The theoretical treatments by both Newman & Poole for unshaped depressions and Doctors for shaped depressions provide a good basis for use in comparative analysis. Figure 8.1 (a) and (b) show the results expressed as a non-dimensional wave drag against Froude Number for a range of cushion length to beam ratios (L/B). Both Newman & Poole and Doctors express their results in a form for wave drag expressed as an induced wave resistance coefficient (R_C) defined as:

$$R_C = \frac{D_W/_W}{\left(\frac{4}{\rho_W g}\right)\left(\frac{p_C}{L}\right)} = f(F, L/B) \tag{8.8}$$

In this formulation:

D_W = Induced Wave drag of the moving air cushion
W = Weight of craft
ρ_W = Density of water (1.99 slugs per ft^3 for salt water)
g = Acceleration due to gravity (32.2 ft/sec/sec)
p_C/L = Cushion density (psf/ft)
F = Froude Number $\left(\frac{V}{\sqrt{gL}}\right)$
L/B = Length to Beam ratio of air cushion

Except in the region around the primary hump, the differences between the shaped and unshaped cushion wave drag are minor, especially in the high speed region ($F > 2$) which is of interest to the high speed air cushion craft designer. A quick look at Figure 8.1 shows why, for high speed air cushion craft, the design choice would be low values of (L/B) say around L/B = 2; and for slower air cushion craft, the design choice leans toward higher values of L/B. The primary hump is the larger increase in wave drag around Froude Numbers $0.5 < F > 1.0$. As can be seen from Figure 8.1, the secondary hump occurs at lower values of Froude Number $0.3 < F > 0.4$. As has been mentioned earlier this "secondary" hump does not always appear in practice. A cautionary note based on Figure 8.1 is that while it is clearly advantageous to use higher L/B (more slender) craft for the lower speeds one has to be careful that while the design may avoid the primary hump, the secondary hump becomes more dominant and occurs at a lower Froude Number. Using Figure 8.1 as an example, the secondary hump at $F \approx 0.3$ is actually higher than the primary hump at $F \approx 1.2$ for a slender craft with L/B = 6.5 and occurs at a lower speed in the region where higher (L/B) craft are intended to operate.

Some comparisons between the various results by all the researchers mentioned above show some additional insights. Figure 8.2 shows the comparison for the maximum value of the wave drag expressed in the non-dimensional resistance (R_C) form at their respective "hump speeds, F_H".

While the actual analytical formulations by Newman & Poole and by Doctors are complex it is possible to collect the results from Figure 8.1 and plot them on

[a] See, for example, Lamb's "*Hydrodynamics*" for detailed discussion on "primary", "secondary" and "tertiary" humps.

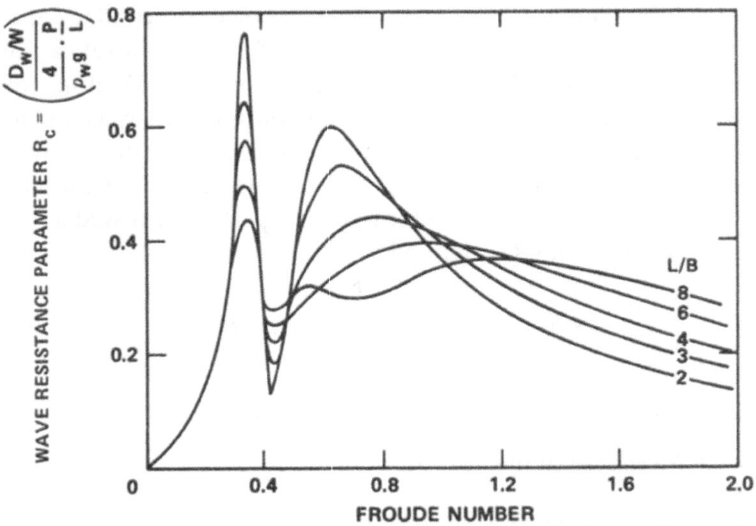

(a) Newman & Poole Theory for Air Cushion Wave Drag (Unshaped Depressions)

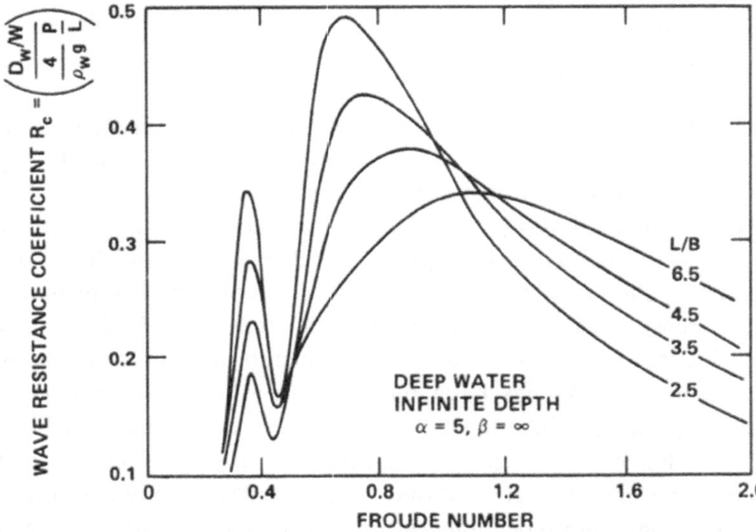

(b) Doctors' Theory for Air Cushion Wave Drag (Shaped Depressions)

Figure 8.1: Two Wave Drag Theories for Traveling Air Cushion Depressions

Figure 8.2. This shows that while the theories by Barratt, Newman & Poole and Doctors are similar in functional form, they give different levels of magnitude for the hump drag. For the lower values of L/B, more suitable for high speed air cushion craft, it is seen that the theory of Barratt more closely matches the available data by Crewe & Eggington and by Everest & Hogben. While not shown here, the trials data for the SES-100B (see Chapter 6) also match these values.

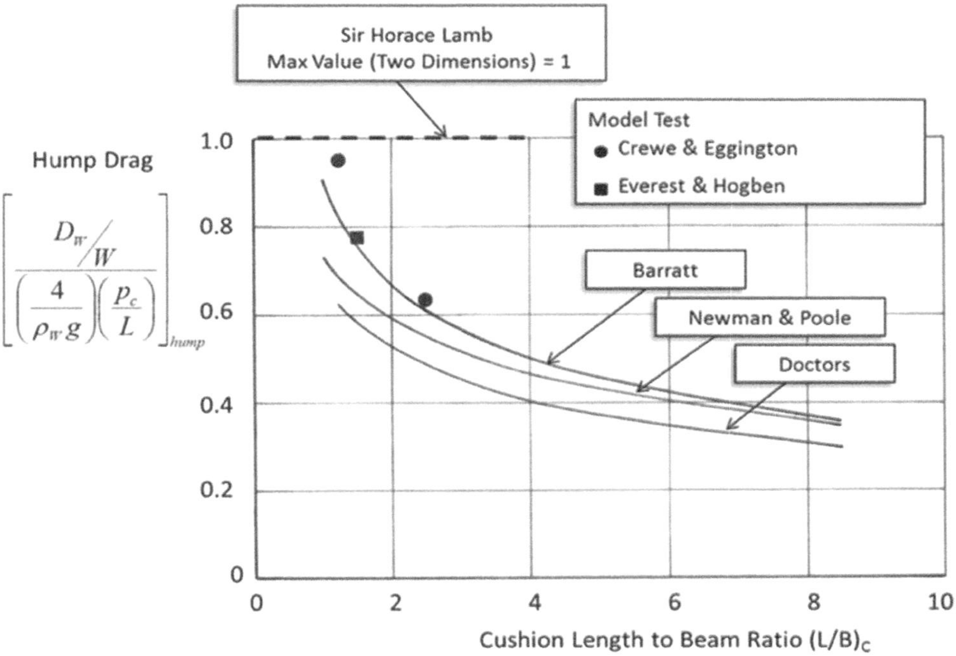

Figure 8.2: Hump Drag of Air Cushion Craft

The trend line shown is through the maximum values of the induced wave drag coefficient (R_C) for the Barratt theory and can be represented by the simple relationship:

$$\left|\frac{D_W/W}{(4/\rho_W g)(p_C/L)}\right|_{hump} = \frac{1}{\sqrt{L/B}} \tag{8.9}$$

This prediction also matches the model test results shown in Figure 8.2 from Crewe & Eggington and also from Everest & Hogben. The theoretical results from Doctors would be the same functional form but about 20% less magnitude. There are insufficient controlled test results available to resolve that 20% difference, but since the full scale test results from the SES-100B also show similar values to those shown on Figure 8.2, it is probably prudent to use Equation 8.9 as a first iteration in any design process.

For comparison, the two dimensional (2D) result for the maximum value of the induced wave drag as given by Sir Horace Lamb has been included in Figure 8.2 and is expressed as:

$$\left[\left|\frac{D_W/W}{(4/\rho_W g)(p_C/L)}\right|_{hump}\right]_{2D} = 1 \tag{8.10}$$

While there are some small differences between the magnitude of the primary hump drag between the various treatments and test, there is much less difference between the speed at which this rise in drag occurs. Figure 8.3 collects the hump speeds (F_H) by the different researchers.

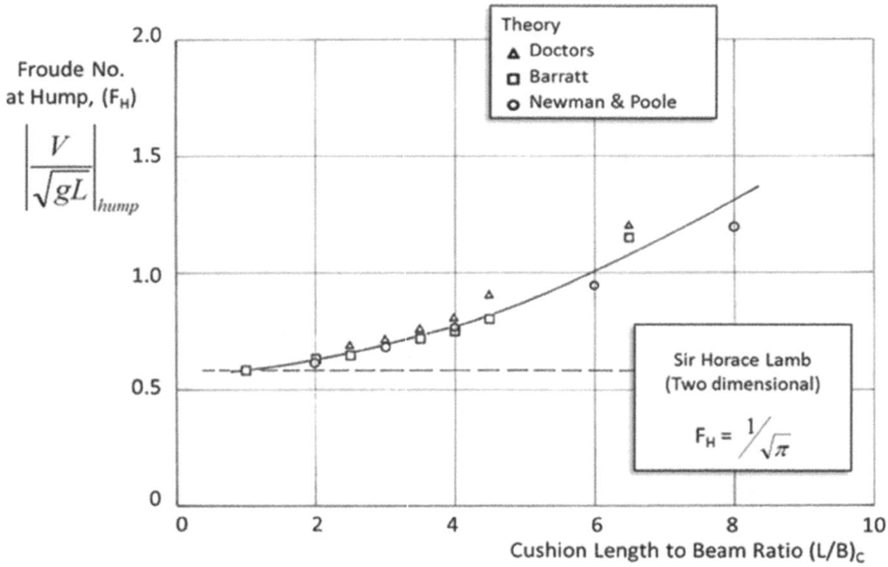

Figure 8.3: Hump Speed of a Moving Air Cushion

Again, in the analytical expressions for the speed at which the wave resistance is a maximum, it is possible to extract those speeds from charts such as Figure 8.1 and plot them on Figure 8.3. It is interesting to note that the prediction by Crewe & Eggington for two-dimensional air cushion depressions, using the work by Lamb, for the hump speed was:

$$F_{hump} = F_H = \frac{1}{\sqrt{\pi}} \tag{8.11}$$

It is seen that all the above theories show that as the L/B is reduced the speed for maximum wave drag asymptotically approaches the two dimensional value, $1/\sqrt{\pi} = 0.564$.

A "rule of thumb" for estimating the hump speed (F_H) can then be constructed from Figure 8.3 as:

$$F_{hump} == \frac{1}{\sqrt{\pi}} + 0.03\left(\frac{L}{B}\right)^{3/2} \tag{8.12}$$

Of particular interest to the designer of high speed air cushion craft is how the wave drag of the air cushion falls away as speed is increased. The particular "superhump speeds" of interest are $1.2 < F > 2.0$ which would cover the speeds of interest of a 3,000 ton frigate size SES with a cruise speed of around 70 knots and a maximum speed of about 100 knots. It is seen from Figure 8.1 that the induced wave drag is falling off rapidly with increased speed. Unfortunately, the theoretical expressions for the induced wave drag in both the Newman & Poole's analysis and Doctors' analysis do not lend themselves well to "easy-to-use" closed form expressions. However, if one takes a graphical representation of Doctors theory from Figure 8.1 and applies a simple curve fit, one gets a result such as Figure 8.4.

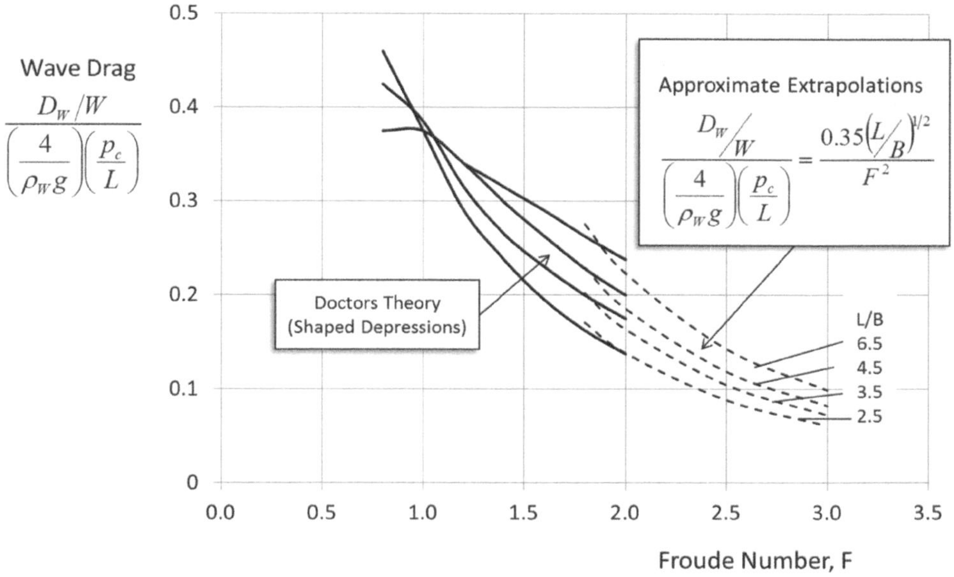

Figure 8.4: Induced Wave Drag at Superhump Speeds

While the agreement between the approximate expression and the exact equations from Doctors becomes less attractive as L/B is increased this is not seen as a problem because at high speed the interest is directed at the lower values of L/B as shown (around L/B = 2.5 and less). In which case, the approximate expression for the "superhump" speeds becomes[b]:

$$\left|\frac{D_W/W}{(4/\rho_W g)(p_C/L)}\right|_{superhump} = \frac{0.35(L/B)^{1/2}}{F^2} \tag{8.13}$$

This expression is reasonably accurate for speeds of $F > 1.5$ and $L/B < 4.5$. The expression may be used for other values of speed (F_H) and slenderness ratio (L/B) but with decreasing accuracy that should be treated with caution.

In summary, it is seen that sufficient corroborative evidence exists to characterize the major drag component (D_{wave}) to be as given in Table 8.1.

With these expressions as given in Table 8.1, it becomes possible to include the induced wave drag (D_{wave}) of the air cushion craft in a consistent and closed form manner that lends itself to analytical treatment and the seeking of design points such as cruise speed (V_{cruise}) and speed for maximum L/D that will influence the final design selections. As can be seen and expanded on later, the induced wave drag is a major component of the total drag of the craft that greatly controls the choice of such design points and other technical and economic considerations in any final design.

By way of an example, Figure 8.5 shows the predicted induced wave drag (D_{wave}) for the SES-100B using the above equations, as part of the total drag of the craft.

[b] It is possible to find a better curve fit formula but only at the expense of losing an analytic equation for the drag useful for drag analysis of the total ship (see later development of lift drag ratio (L/D) equation).

Table 8.1: *Characteristics of Air Cushion Craft Induced Wave Drag*

Induced Wave Drag Feature	Equation
1. Maximum value (at Hump Speed)	$\left.\left\lvert\dfrac{D_W/W}{(4/\rho_W g)(p_C/L)}\right\rvert\right._{hump} = \dfrac{1}{\sqrt{L/B}}$
2. Hump Speed (Froude Number)	$F_{hump} = \dfrac{1}{\sqrt{\pi}} + 0.03\left(L/B\right)^{3/2}$
3. Superhump Induced Wave Drag	$\dfrac{D_W/W}{(4/\rho_W g)(p_C/L)} = \dfrac{0.35(L/B)^{1/2}}{F^2}$ Valid for F > 1.5; L/B < 4.5

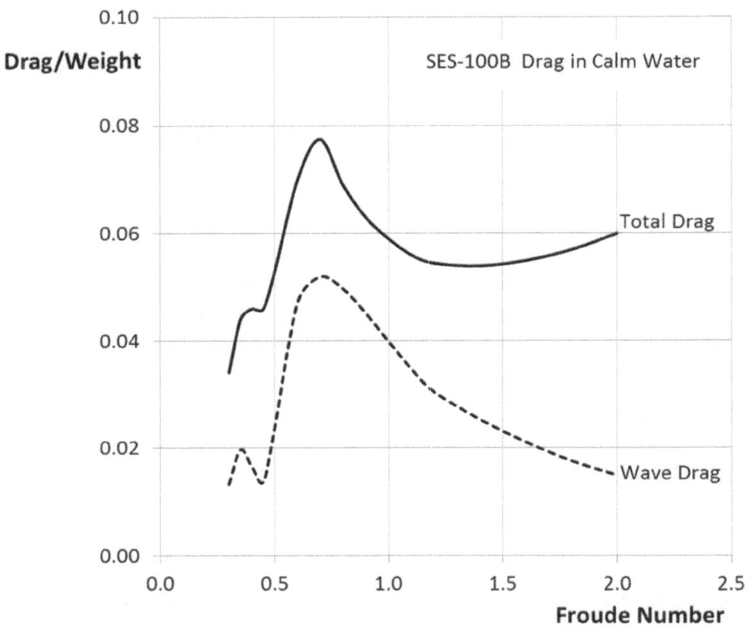

Figure 8.5: Total Drag and Induced Wave Drag of SES-100B (Calm Water)

The remaining components over and above the induced wave drag (D_{wave}) shown in Figure 8.5 are the major components of aero drag (D_{aero}), momentum drag (D_{mom}), skirt or seal drag (D_{sk}), and sidehull drag (D_{SH}) to comprise the total drag (D_{total}).

8.1.2 Aerodynamic Drag (D_{aero})

The aerodynamic drag of an air cushion craft can generally be written in the form:

$$D_{aero} = C_{D_o}\, 1/2\, \rho V^2 S_r \tag{8.14}$$

In this formulation, the drag coefficient (C_{Do}) is based on some reference area (S_r). The air density (ρ) would be the sea level value ($\rho = 0.00119$ slugs /ft^3) and the forward speed (V) measured in ft/sec in English units. Normally, for vehicles such as air cushion craft, the aerodynamic drag is that referred to as the profile drag and is referenced on the frontal area (S_F). Any induced drag from the aerodynamic lift would be minimal to negligible except in the case of the aerodynamic air cushion craft where an additional term that varies inversely with the speed squared called induced drag. The treatment of this special case is referred to Chapter 11 on aerodynamic air cushion craft. So restricting the treatment to aerostatic type air cushion craft where aerodynamic lift is negligible, the aerodynamic drag can be written as:

$$D_{aero} = C_{D_o}\, 1/2\, \rho V^2 S_F \tag{8.15}$$

Because it is common practice to refer all the drag components to the planform area (or cushion area) of the craft (S), then the equation for the aerodynamic drag becomes

$$D_{aero} = C_{D_o}\left(\frac{S_F}{S}\right) 1/2 \rho V^2 S \tag{8.16}$$

Historically, the profile drag coefficient (C_{Do}) based on frontal area (S_F) ranged from values of $C_{Do} = 0.25$ for very streamlined shapes, to $C_{Do} = 0.38$ for the more blunt or boxy shapes. Based on wind tunnel tests, the SES-100B was measured to have a profile drag coefficient, $C_{Do} = 0.32$. This value would be representative of most of the SES configurations considered to date.

In terms of the drag-to-weight ratio, the aerodynamic drag can then be written;

$$\frac{D_{aero}}{W} = C_{D_o}\left(\frac{S_F}{S}\right)\frac{1/2 \rho V^2}{p_C} \tag{8.17}$$

This will be shown later to be a useful expression when the speed becomes high enough to where the aerodynamic forces become significant. Alternatively, the expression for the aerodynamic drag for most air cushion craft without aerodynamic lift can be written as a function of Froude Number, in the form:

$$\frac{D_{aero}}{W} = C_{D_o}\left(\frac{\rho g}{2}\right)\left(\frac{S_F}{S}\right)\frac{F^2}{p_C/L} \tag{8.18}$$

8.1.3 Momentum Drag (D_{mom})

The momentum drag (or ram drag) is that force due to the rate of change of momentum in accelerating the cushion air to craft speed. The momentum drag can then be written as:

$$D_{mom} = \rho Q V \tag{8.19}$$

Within this deceptively simple formulation is the problem of correctly determining the correct amount of air cushion flow (Q). The correct amount of air flow depends on the geometry and function of the lift system. In the early designs of hovercraft, the air flow (Q) was channeled through the "loops" or "cells" of peripheral jet flexible skirts. By the turning of the air jets, the curvature of the flow generated the lift pressure (p_c) with an associated air flow (Q). As the designs of the flexible skirts evolved, the peripheral jet gave way to the plenum design where the air flow entered the cushion area directly and the flow was controlled simply by the leakage beneath the skirts. The analysis of both these types of lift systems together with test results was done in detail in the 1980 report[10] and will not be repeated here. The reader is referred to these earlier reports for a more detailed treatment of all the various skirt systems and the effect on the pressure, flow and power requirements for the lift system design of air cushion craft. Because most craft designs have now settled (more or less) into the modified plenum lift systems, the equations here will be restricted to those types. There is an additional complication in that the leakage beneath the skirts or seals will be quite different as might be expected between the calm water case and the leakage in rough water operation. These two cases will be treated separately.

Air Flow in Calm Water The air pressure in the plenum beneath the craft is:

$$p_C = \frac{W}{S} \tag{8.20}$$

In which case, the air flow exiting beneath the craft, can be written:

$$Q = \sqrt{\frac{2p_C}{\rho}}(Ch)D_C \tag{8.21}$$

In this expression for the air flow (Q), the exit area under the craft is the hemline length of the leakage area (C) times the air gap (h). Because of the flow conditions, it is normal to recognize the *vena contracta* in the flow with a discharge coefficient (D_C). Typically, for sharp edges such as underneath the skirts a value of $D_C = 0.64$ would be a likely value. In the case of the fully amphibious air cushion craft, the length of the leakage area would be $C = 2(L + B)$ for rectangular shaped craft[c]. For sidehull type craft, the leakage length (C) would depend on the configuration of the partial length sidehull such that:

$$\frac{C}{2B} = 1 + \frac{L}{B}\left(1 - \frac{L_{SH}}{L}\right) \tag{8.22}$$

Included in this simplified expression for the "leakage length" is the partial sidehull length (L_{SH}).

In calm water, the expression for the momentum drag can be written (for amphibious air cushion craft where $L_{SH} = 0$):

$$\left|\frac{D_{mom}}{W}\right|_{calm} = \frac{2\sqrt{2\rho g}(1 + L/B)(h/L)D_C}{(p_C/L)^{1/2}}F \tag{8.23}$$

In the calm water case, it is seen that the momentum drag (D_{mom}/W) is a direct function of the craft length-to-beam ratio (L/B), the cushion density (p_C/L), the air gap beneath the skirt (h/L), and the speed of the craft expressed in terms of Froude Number (F). Most of the data on air flow beneath the skirts was generated for the amphibious type of craft and concentration was on the "air gap" beneath the skirt. This air gap was important not only in the generation of the air flow (Q), it was also

[c] In these and following expressions, it is recognized that the dimensions L, B, C etc refer to the geometry of the air cushion and not the overall ship dimensions

driven by the recognition that skirt flagellation and wear was influenced strongly by the air gap (h/L). For a more general treatment, especially for other air cushion craft forms such as the sidehull air cushion craft and partial sidehull types, the geometric relationships for the air gap become less tractable and concentration on the air flow (Q) itself is more useful. This is also true for test programs where it is extremely difficult to measure air gaps but it is relatively easy to determine the air flow directly. In this case, it is convenient to express the momentum drag directly as a function of the air flow (Q). A flow coefficient is defined as:

$$C_Q = \frac{Q}{VS} \tag{8.24}$$

Where the actual air flow (Q) through the system is non-dimensionalized by the craft's forward speed (V) and the cushion area (S). The momentum drag can be written in the more measurable form as

$$\frac{D_{mom}}{W} = \frac{\rho g C_Q}{p_C/L} F^2 \tag{8.25}$$

It is shown later in the treatment of the drag polar for such craft that the flow coefficient (C_Q) plays an important part in contributing to the determination of the optimum value of $(L/D)_{EFF}$ and in the determination of the cruise speed (V_{cruise}). It is seen from Equation 8.25, that if the flow coefficient (C_Q) is held constant then the momentum drag would increase in proportion to the square of the speed. This would require however that the air flow (Q) be increased in direct proportion to the craft speed (V) (see Equation 8.24). This would be somewhat overly cautious in an operational context and in a practical case the flow would be increased at a slower rate with speed depending on the craft design and operational conditions. To provide some insight to the key parameters, the momentum drag can be expressed more directly in terms of the actual air flow (Q) in the form:

$$\frac{D_{mom}}{W} = \frac{\rho\sqrt{g}\,{}^{L}\!/_{B}}{p_C/L}\left(\frac{Q}{L^{5/2}}\right)F \tag{8.26}$$

In this form, the relationship between actual air flow (Q) and craft speed (V or F) and craft dimensions can be seen more directly. Other factors, specifically the rough water requirement for air flow will help determine how much air flow is required and how the demands of air replenishment during operation in the waves (wave pumping) determine the generation of lift system air flow.

Air Flow in Rough Water In rough water operation, the calculation of the air flow (Q) is more complex. In rough water operation, the lift system that is generating the cushion pressure (p_C) and the air flow (Q) for the desired air gap (h/L); must, in addition, provide sufficient air flow for the desired ride quality; and it must provide sufficient air flow to instantaneously replenish the cushion swept away by the wave action as the craft traverses the waves. This latter requirement is significant in that if, after waves have "pumped" out the cushion, there is insufficient lift power (P_L) to replenish the cushion, the craft will rapidly lose altitude and pitch into the waves. Wave pumping then, is a general problem associated with the replenishment of the cushion air during rough water operation over the complete operational range of speed and sea states for the particular craft in question. A particular troublesome case is when the craft responds at the encounter frequency. The complexities of the craft dynamics and the irregular or random nature of sea wave motion have made the predictions of such phenomena difficult. Also, the (lift) power to provide a desired air gap, adequate ride quality, and wave pumping are not additive and depends on the craft conditions, seal or skirt compliance and operator technique.

It is common practice to calculate the power (and air flow) required in each planned operating condition and then make a selection for the final design. Prudence would suggest selecting the maximum air flow case. Mel Beardsley (a contemporary of Christopher Cockerell and one of the originators of air cushion craft in the US) prepared an analysis[11] specifically for the sidehull type of SES for the selection of the air flow (Q) required. While done specifically for the sidehull type of craft, the results are equally applicable to the amphibious type of air cushion craft.

Beardsley considered two extreme cases in his analysis: (1) the case of maximum wave pumping flow requirement for zero deck motion and (2) the case of constant cushion volume (zero wave pumping). For the case of maximum wave pumping (and thus zero vertical acceleration) the air flow is given by:

$$Q = Bh_W V \sin(\pi L/\lambda) \tag{8.27}$$

In this expression, h_W and λ are the wave height and wavelength respectively of the oncoming waves. Secondly, for the case of zero wave pumping (and thus maximum vertical acceleration):

$$\frac{d^2z}{dt^2} = -\frac{\pi}{10}\frac{V^2}{L} \tag{8.28}$$

Beardsley concludes that the practical amount of air flow (Q) is somewhere between the extreme case of maximum flow (Equation 8.27) to remove all vertical acceleration, and zero flow (Equation 8.28), other than the calm water hover power requirement, to result in maximum acceleration levels.

To help determine the best amount of air flow (Q) the question becomes: will the wave pumping requirement be more or less than the basic requirement to satisfy the momentum drag requirement? After some algebraic manipulation, the air flow for wave pumping (Q_{WP}) (zero vertical acceleration) can be written:

$$Q_{WP} = \pi \frac{h_W}{\lambda} VS \left| \frac{\sin(\pi L/\lambda)}{\pi L/\lambda} \right| \tag{8.29}$$

Similarly, the air flow required to satisfy the momentum requirement was shown to be:

$$Q_{mom} = \sqrt{2/\rho}\,(p_C)^{1/2} ChD_C \tag{8.30}$$

These two flow requirements can now be compared through the rearranged inequality:

$$\pi\left(\frac{h_W}{\lambda}\right)\left|\left(\frac{\sin \pi L/\lambda}{\pi L/\lambda}\right)\right| F < \mathrm{Q} > \sqrt{\frac{2}{\rho g}}(p_C/L)^{1/2}(Ch/S)D_C \tag{8.31}$$

It is not obvious which will give the greater or lesser air flow (Q) from Equation 8.31 and it must be determined in individual cases as to which will give the required air flow for the design. Beardsley[12] ran several calculations for the *SRN.5* hovercraft (an operational craft at the time in 1967) and found that the wave pumping power for the *SRN*.5 was 235HP for the 15,000 lb craft operating at 70 mph (~61 knots) in 2.5 ft waves. It was assumed that the worst case wave pumping condition would occur in such waves when $L/\lambda = 1/2$ with L = 28 ft. For the same *SRN.5* craft operating with a 4 inch air gap (h/L = 0.019) and a 50% lift system efficiency the computed lift power was 280HP. Thus, in that case the basic calculation of power using the momentum flow exchange calculation was the dominant value. This is not a general conclusion because for larger craft operating at lower values of air gap and higher sea states, the wave pumping requirement dominates. This is more likely for large SES traveling at high speed over rough seas.

8.1.4 Skirt or Seal Drag (D_{SK})

The 1980[13] report provided a detailed treatment of the drag of skirt systems in both calm water and in rough water for a wide range of operational air cushion craft. The majority of that data is for the well established and researched amphibious air cushion craft. Very little data is available in a systematic manner for the drag of skirts from sidehull craft and most sidehull designs have had to rely on empirical means and model tests for drag estimation. It was found during the research on air cushion craft at both British Hovercraft Corporation in the UK and Bell Aerosystems in the US that the skirt drag of air cushion craft was also dependent on whether the craft is operating in calm water or in rough seas.

Calm Water Skirt Drag (D_{SK}) The skirt drag in calm water was also found to be treated as the sum of two components, as follows:

Wetting Drag (D_{SKw}):	A drag component related to craft speed, hemline length and air flow leakage
Wavemaking Related (D_{SKwm}):	A drag component related to cushion wavemaking generated by the cushion pressure of the craft

The skirt wetting drag (D_{SKw}) was found to be best represented by (based on model tests):

$$D_{SKw} = K\left(\frac{h}{C}\right)^{-0.34} C\sqrt{S}.q_W \tag{8.32}$$

In this formulation, C is the hemline length, S is the cushion area and q_W is the dynamic pressure of the water ($1/2\rho_w V^2$). The constant K is strongly dependent on the geometry and cushion density of the particular craft and skirt.[14] Variations in K range from 2.13×10^{-6} for the low cushion density *SRN.4* to 3.71×10^{-6} for the high cushion density US Navy JEFF(B) air cushion landing craft. The author found that an average value of $K = 3.46 \times 10^{-6}$ gave a reasonable fit over a wide range of cushion densities (p_C/L) such that the non-dimensional form of the skirt wetting drag in calm water can be written as:

$$\frac{D_{SKw}}{W} = 0.221 \times 10^{-3} \frac{1 + {}^{L}/_{B}}{\left({}^{L}/_{B}\right)^{1/2}} \left(\frac{h}{C}\right)^{-0.34} \frac{F^2}{p_C/L} \tag{8.33}$$

For most types of air cushion craft, this drag is quite small, of the order of about 2–5%. The larger skirt drag component is that due to the effect on the skirt from the induced wave making of the cushion.

Wavemaking Related Calm Water Skirt Drag (D_{SKwm}) From the Arctic study cited above, the data from both the BHC and Bell model tests showed that the wavemaking related skirt drag (D_{SKwm}) was additive to the induced wavemaking drag discussed earlier (see Table 8.1) and could be represented by:

$$\frac{D_{SKwm}}{D_{wave}} = \frac{1.374}{\left({}^{p_C}/_{L}\right)^{0.259}} - 1 \tag{8.34}$$

This empirical relationship and the model test data is shown in Figure 8.6.

As can be seen from Figure 8.6, the wavemaking related skirt drag can add an additional 25–30% of the induced wave drag of the pressurized air cushion to the calm water drag prediction for the craft.

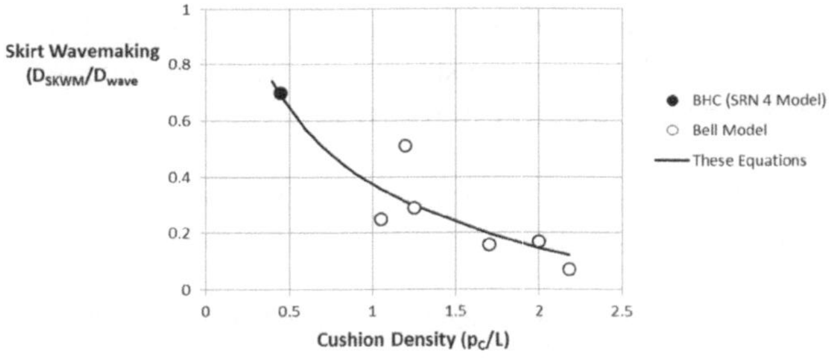

Figure 8.6: Calm Water Skirt Drag (Wave making related)

Rough Water Skirt Drag Precise methods of predicting skirt drag in rough sea operation are not easily come by. It is also generally accepted by air cushion craft designers that any method used to predict such drag components are only generally acceptable for sea conditions where the wave height (h_W) is no more than about 80% of the cushion height (h_C). Above this value there is almost certain to be hard bottom contact and the drag of the craft (and skirts) is more dependent on craft response than on skirt geometries.

Model tests conducted at Bell Aerospace[15] show a tendency for the skirt drag in rough water to be some function of wave height (h_W), cushion height (h_C) and the finger height (h_f). How much of the skirt is made up of the bag and how much is made up of the finger has received a considerable amount of research and development by BHC and others since the early days of hovercraft design. Figure 8.7 shows the history of the skirt development in this regard. The history shows the tendency to gradually reduce the size of the pressurized bag and to increase the length of the fingers (h_f) to become a higher percentage of the cushion height (h_C).

The functional relationship was found to be:

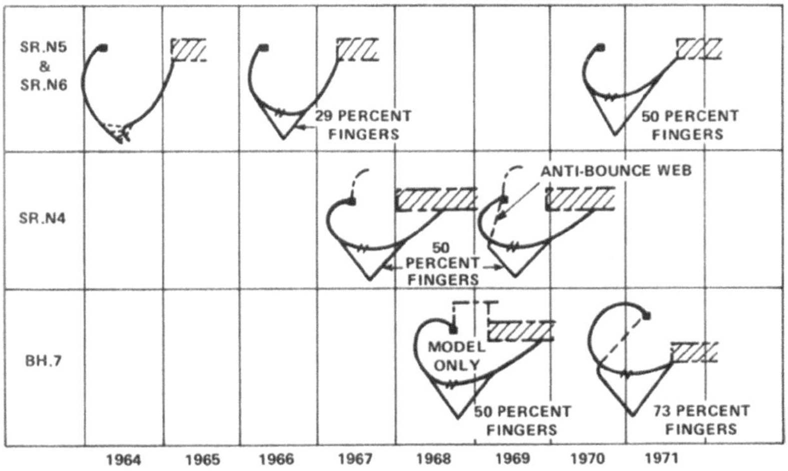

Figure 8.7: BHC Early Bag and Finger Skirt Development

$$\frac{D_{SKrough}}{q_W(C\sqrt{S})} = f\left(\frac{2h_W}{h_C + h_f}\right) \tag{8.35}$$

The author explored this dependence on a variety of air cushion craft from different sources (both the US and the UK) and found that the skirt drag in rough water ($D_{SKrough}$) collapsed to a simple form, as shown:

$$\frac{D_{SKrough}}{q_W(C\sqrt{S})} = 20 \times 10^{-5}\left[\frac{2h_W}{h_C + h_f}\right]^{5/3} \tag{8.36}$$

This relationship is shown on Figure 8.8 together with the measured data from the various US and UK craft identified in the chart. It can be seen that an encouraging agreement exists between this relationship and the actual data from a wide range of air cushion craft. The key geometric term that relates wave height (h_W), cushion height (h_C) and finger length (h_f) can be written:

$$\left[\frac{2h_W}{h_C + h_f}\right]^{5/3} = 2^{5/3}\left[\frac{h_W/h_C}{1 + h_f/h_C}\right]^{5/3} \tag{8.37}$$

As the finger length (h_f) approaches the height or depth of the cushion (h_C), which has been the development trend, it is seen that the geometric term in Equation 8.35 approaches the value (h_W/h_C). If one plots the available data of the performance of air cushion craft in rough seas against this "roughness parameter" (h_W/h_C), the effect of skirt design becomes readily apparent. This is shown in Figure 8.9.

The effect of skirt geometry on the drag of the air cushion craft can be seen quite dramatically in the collected data in Figure 8.9 where the early skirt designs were much stiffer and presented more resistance to the craft causing a significant fall-off in craft speed in rough waters. As the skirt designs were modified (see Figure 8.7) with longer and more flexible fingers the resistance from the skirts was much less and the speed in rough seas increased. In the post 1975 period, the skirts were much improved and the fall-off in craft speed had been minimized and the speed could be expressed by:

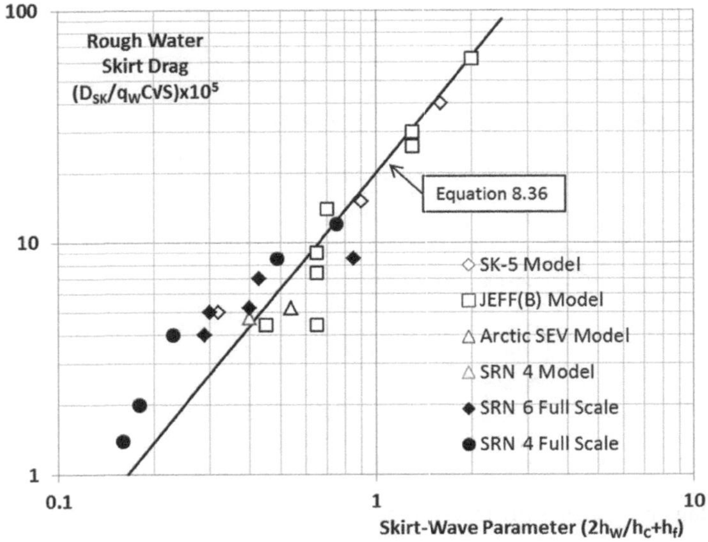

Figure 8.8: Rough Water Skirt Drag

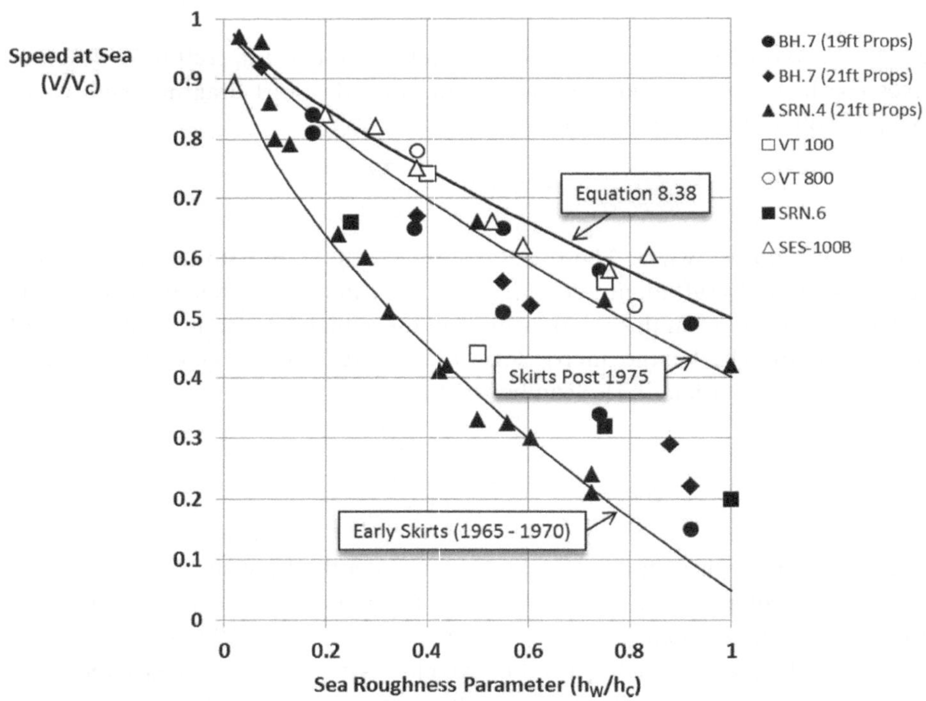

Figure 8.9: Speed of Air Cushion Craft in Rough Seas

$$\frac{V}{V_{calm}} = 1 - 0.50\left(\frac{h_W}{h_C}\right)^{3/4} \tag{8.38}$$

During the Test Program on the SES-100B (see Chapter 6) it was found that this simple relationship gave a good representation of the sidehull craft as well as for the amphibious craft. The influence of the "sea roughness parameter" (h_W/h_C) in both the drag estimate of the flexible skirts and of the speed in rough seas of the craft is clearly seen. Expressing this skirt drag in terms of the drag weight ratio as for the other components, the expression becomes:

$$\frac{D_{SKrough}}{W} = 1.28 \times 10^{-3} \frac{1 + {}^{L}\!/_{B}}{\left({}^{L}\!/_{B}\right)^{1/2}} \left[\frac{2h_W}{h_C + h_f}\right]^{5/3} \left(\frac{F^2}{{}^{p_C}\!/_{L}}\right) \tag{8.39}$$

8.1.5 Sidehull Drag (D_{SH})

The shape of the hydrodynamic hulls along the sides of the craft has been a contentious issue in the development of high speed SES. There were essentially two schools of thought in the US SES program. The first school envisioned the hulls being nothing more than non-buoyant "sideboards" as written up in the 1964 Captured Air Bubble (CAB) patent. The intention was that such sideboards would have minimal drag (skin friction only) and that stability of the craft would come from some other (and undefined) "artificial means". The second school that evolved from the UK operational sidewall hovercraft in the UK, required that the hulls be shaped to not only provide minimal drag but also to provide the needed roll, pitch and yaw stability to the craft. This meant that such *sidehulls* would have higher drag than the *sideboards* of the CAB concept but would incorporate the necessary stability features.

The essential features of the two design concepts:

- Captured Air Bubble Surface Effect Ship (CAB SES)
- Sidehull Surface Effect Ship (Sidehull SES)

are shown in Figure 8.10 and Figure 8.11. It can be seen from these two generic sketches showing the principal differences that the intention of the *CAB SES* (see Figure 8.10) was to provide minimal hydrodynamic resistance by (a) use of "thin sideboards" with no capability for either buoyant lift or dynamic lift and thus no (or minimal) pressure drag, and (b) use of "rigid planer seals" to maintain constant contact with the water and thus "capture the air bubble" with no air leakage and thus minimal lift power required. Since such a concept is fundamentally unstable in the transverse plane, the *CAB SES* required (undefined and not shown) additional artificial stability means.

The principle behind the *Sidehull SES* (see Figure 8.11) is to integrate the required stability into the sidehulls directly and design the hydrodynamic sidehull shape so as to optimize the design by trading off stability and performance characteristics into the basic sidehull shape. Additionally, the *Sidehull SES* concept recognized the significant difficulty in designing any rigid device that would "plane" on the water with no air leakage or avoidance of the tremendous accelerations and hydrodynamic loads from such hard contact. This resulted in the incorporation of the well-developed flexible skirt system from the amphibious hovercraft world for the bow and stern seals. Also, the trade in the *Sidehull SES* concept was to integrate the added power (for air leakage) with the reduced power of flexible seal drag.

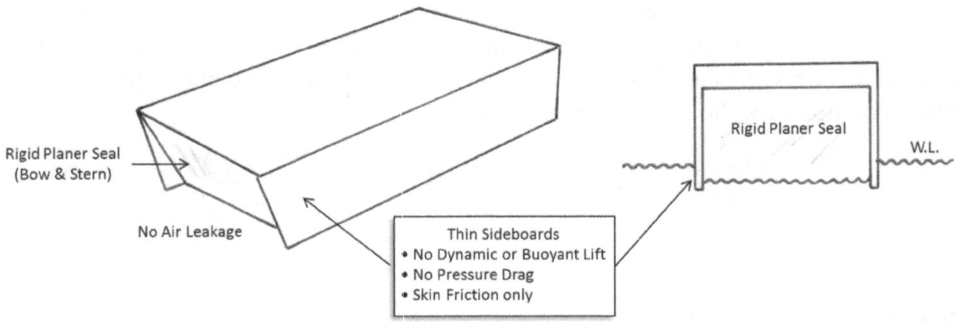

Figure 8.10: Captured Air Bubble Surface Effect Ship (CAB SES) Basic Features

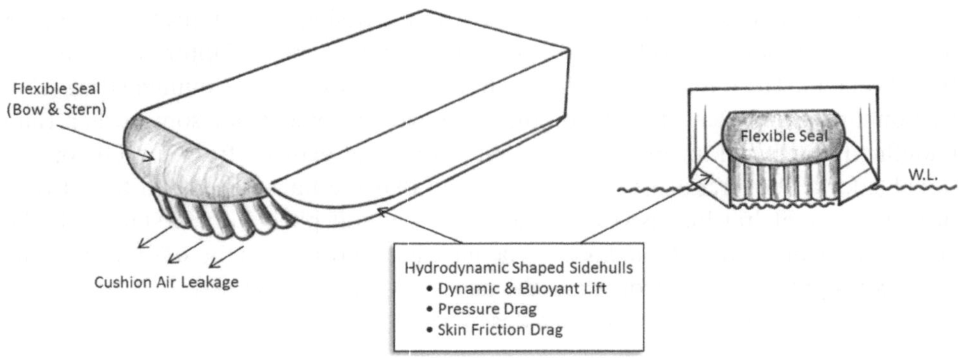

Figure 8.11: Sidehull Surface Effect Ship (Sidehull SES) Basic Features

The difference between these two extremes was most apparent in the competing designs of the two craft in the 1969–1980 US SES Program: (a) the SES-100A test craft based on the CAB concept and (b) the SES-100B test craft based on the UK sidewall hovercraft design. The description of these two craft and their detailed differences is explored in Chapter 6. The difference in the two hull designs is seen in Figure 8.12 for the SES-100A design and Figure 8.13 for the SES-100B design. The SES-100A was designed to have essentially a "thin sideboard" sealing the cushion

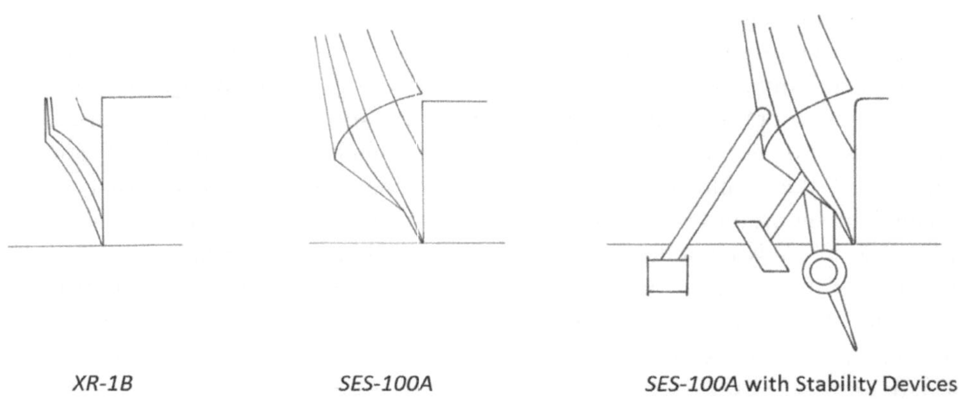

Figure 8.12: Comparison of XR-1B and SES-100A Lines (to same scale)

while the SES-100B had a more conventional high speed hull shaping. It was realized by the SES-100A designers that such a high deadrise angle hull (60° bow to stern)[d] similar to the original CAB XR-1 would not have sufficient stability and accordingly added the "artificial stability" features called for in the CAB patent as shown in Figure 8.12. In contrast, the lines for the SES-100B with its wider sidehull, lower deadrise angle and built-in stability are shown in Figure 8.13.

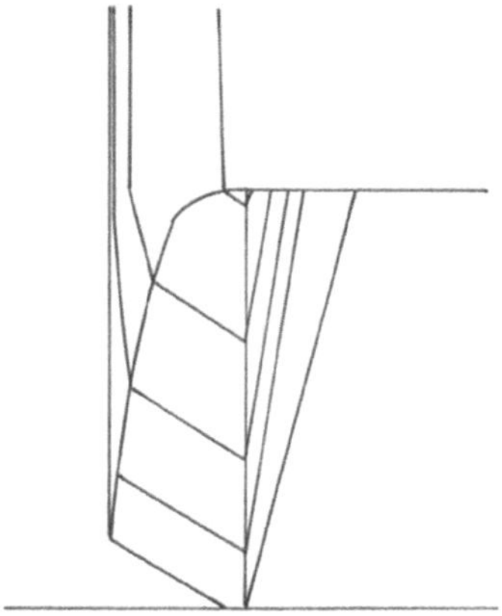

SES-100B Bow View (Starboard)

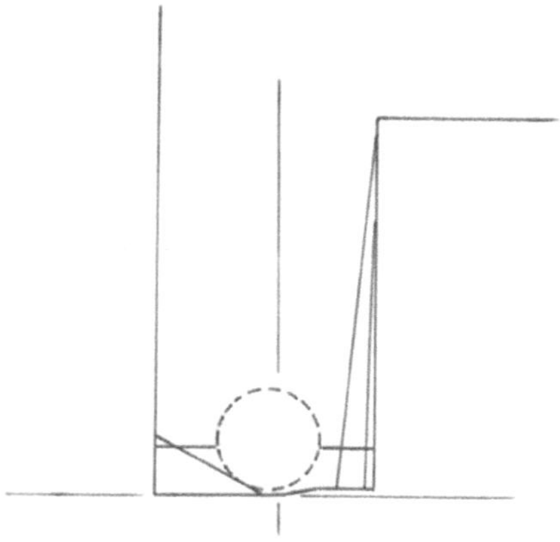

SES-100B Stern View (Port)

Figure 8.13: SES-100B Bow and Stern Views of Sidehull Lines

In Figure 8.12 for the SES-100A, it can be seen that the steep deadrise angle (60°) is modeled after the XR-1 CAB hull. The XR-1 went through many changes after the craft's capsize and fatal accident in 1964 and one of the modifications was the XR-1B for which a lines drawing is available and shown in the left hand drawing of Figure 8.12. The design philosophy was that a minimum thickness would still be used (for low drag) and that extra thickness for stability would be added above the keel and water line. This meant, of course, that roll stability still would not be achieved except at high angles of roll which would not be a good feature for any operational craft (military or commercial). The SES-100A design followed the same CAB design philosophy but added a greater thickness at a lower height above the keel line. Roll stability would still not be achieved except at high roll angles. In the case of the SES-100A, the quoted "artificial stability" devices were then added as shown (Figure 8.12) to ensure that the required stability would be achieved. Chapter 6 describes the various features in more detail, including the artificial stability devices and the instability experienced during initial trials of the SES-100A in 1972. Figure 8.13 shows the competing design of the SES-100B where stability was built into the design at the outset as shown. Sufficient sidehull thickness and low deadrise angle (30°)was incorporated into the sidehull design at the outset. There are several other requirements that need to be integrated into the design that include need for stability in all modes and the need to integrate the propulsion system into the hull design.

[d] The lesson learned on the XR-1 program (see later) of not using such a high deadrise angle was not passed on to the SES-100A program.

The effect of the lower deadrise angle for the SES-100B is readily apparent from comparing that shown in Figure 8.13 with the high deadrise angle design of the SES-100A shown in Figure 8.12 as it brings with it an automatic increase in thickness and buoyancy. The SES-100B incorporated a partially submerged propeller in its design (see Chapter 6 for complete description) and the lines of the SES-100B sidehull flare out toward the stern of the craft to accommodate this propulsion system. This flaring is able to be accommodated because the design is to operate in supercavitating flow ($F > 2$) and there is no need for any "streamlining" or "boat tailing" as might be used for slower speed vessels. Additionally, because of this flaring toward the stern, extra buoyancy and extra lift is available behind the craft center of gravity (C.G.) that contributes significantly to the yaw stability of the craft.

It is this integration of performance, stability and propulsion features into a single coherent design that makes the SES-100B an efficient design of lower drag; higher lift-drag ratio (L/D) and a higher transport efficiency (WV/P) that contributed to its world record breaking top speed of 91.9 knots in 1976 while still retaining an efficient and stable design.

From Sideboard to Sidehull The development of the SES-100A hull lines started with the (unstable) XR-1 design and progressed as analysis and test dictated into improved hull form. After the initial failure of the XR-1 CAB with its thin sideboard lines and the compromised stability of the SES-100A, the US Navy embarked on a systematic program to move away from the "thin sideboards" of the CAB concept and to investigate new "shaped sidehull" forms. Robert Waldo at Aerojet General Corporation in 1968, summarized much of the US Navy research in his paper on *Special Problems in Surface Effect Ships*[16]. The US Navy explored several different sidehull shapings to achieve all the objectives in performance, stability and integration with the propulsion system. The trials and tribulations of the basic CAB design as embodied in the XR-1 series of manned craft are ably summarized in the US Navy SES Project Office paper[17], an extract of which is given here:

> *"The XR-1 was the Navy's first SES "manned model". The water jet propelled craft was characterized by a rather high cushion length/beam ratio = 3.5. The rigid "barn door" seals were fixed. Numerous improvements were made during the development programme: a larger J-85 engine was installed, enabling the XR-1 to reach 60 knots; sidewall deadrise was increased; the J-85 was replaced by a T-53 engine and air propeller, seals were articulated (of rigid elements) and made moveable. On December 8, 1964, while making a turn at moderate speed, the craft heeled, pitched down and flipped over. One man was killed.*
>
> *Following this tragic event the craft was extensively modified to increase its dynamic stability: cushion width was increased from 10 to 15 ft; sidewall width from 12 to 18 in and deadrise reduced to 45 degrees....."*

The report continues, describing more modifications needed to improve the craft stability and performance characteristics, as follows:

> *"The "A" configuration was converted to the "B" configuration by removal of the air propeller and installation of two Pratt and Whitney Seajet G-1A waterjet pumps..... The XR-1C design.. began in late 1972. Conversion was begun in Spring 1973. The rigid articulating seals were replaced by flexible, rubberized fabric seals, each consisting of a rubberized-fabric "ski" stiffened longitudinally by steel ribs... Ride qualities were*

greatly improved over those of the XR-1B.... The XR-1C showed definite roll-in tendencies during turns. In contrast, the rigid seal XR-1B maintained slightly rolled-out attitudes during turns".

It is unfortunate that all of these modifications were made on a man carrying craft weighing 17 tons or more and with speeds up to 60 knots and not on models in controlled tests in hydrodynamic facilities where all of these predictable instabilities could have been easily resolved in a safe manner. Later craft, such as the XR-3 were used to explore various modifications seeking a new hull form for the CAB SES.

One of the research tools used was the small test craft the XR-3. It was built to explore several features of SES but in particular was used to explore different sidehull shapes designed to operate in supercavitating flows. Figure 8.14 (a) shows the XR-3 lines in the "as built" configuration. As can be seen, the lines still retained the high deadrise angle in the forward sections but the lines flared out toward the stern similar to the SES-100B design to accommodate the propulsion system. The intention in SES design was to remove fully submerged propulsion schemes (with their associated high drag) and to include either flush inlet water jets or semi-submerged propeller designs. Model testing of several configurations were done in this same time period and earlier, and basic results are summarized in several US Navy Naval Ship Research & Development Center (NSRDC) reports, of which just one is referenced here to capture the main points.[18] All of the model testing confirmed that with the geometry of the XR-1B sidehull lines, the restoring roll moment was still of insufficient magnitude to provide the necessary stability to an otherwise unstable CAB air cushion. All indications were that the deadrise angle should be lowered to at least 45°. As described in Chapter 6, this increases the likelihood that the restoring force will pass above the craft C.G. and thus would be stabilizing.

From the XR-3 model tests, it was found that by adding chines to the design, improvements in both performance and stability were attained. While not incorporated into the craft, the tests showed that incorporating a low chine (see Figure 8.14 (b)) the roll stability was improved due to an effective decrease in the deadrise angle. The single chine version (not shown) had lower drag than the original XR-3

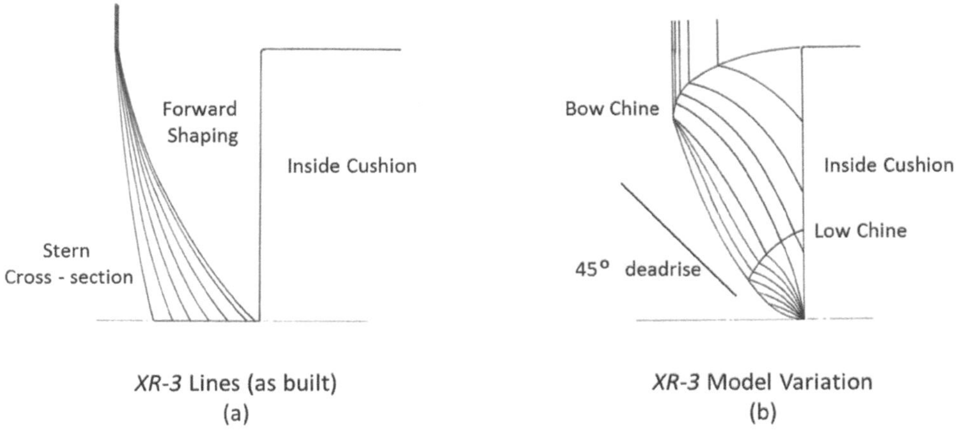

Figure 8.14: Sidehull Lines of XR-3 (as built and one proposed modification)

configuration by about 34%. The difference between the single and the double chine designs (shown) was minimal in terms of drag, but as mentioned above, the double chine version had improved stability. While not dwelt on here because of the complexities of individual designs, improvements in both yaw and pitch stability is achieved with these changes over the basic CAB "thin sideboard" hull design as well as the improved roll stability.

The stability of the SES-100B sidehull configuration was achieved with the basic integration of performance needs and stability needs into one hull form (see Chapter 6) as shown on Figure 8.13. In addition to the basic shaping of the sidehull itself is the effect of the two sidehulls in the overall configuration and stability of the ship. The basic features of the SES-100B pulled heavily on the ongoing experience in the UK with the operational experience of the sidewall hovercraft Hovermarine HM.2 and other work at Christopher Cockerell's HDL company.

From the drag calculations, model tests and from the full scale trials data, the hull lines of the SES-100B resulted in lower drag values (and higher speeds) than the SES-100A lines and retained the necessary stability in all three modes of roll, pitch and yaw. The setting of the stability features for the SES-100B was based on a comprehensive model test program and recognition of some overall basic characteristics retained in the design because of recognition of the ongoing (at the time) research and development in sidewall hovercraft in the UK. The comprehensive model test program included a radio controlled free "flight" model that allowed study of the non-linear cross coupling between roll, pitch and yaw that is not possible from linear tow tanks or analysis. This approach avoided the instabilities from such interaction. Figure 8.15 shows the SES-100B radio controlled model in test.

Figure 8.15: SES-100B Radio Controlled Model in Test
Photo courtesy of Textron Marine & Land Systems and Textron Inc.

Sidehull Design for Lower Speeds

The lines of the SES-100B were designed to give performance and stability at high speed (at approximately 100 knots). The results from the SES-100B program were used later to advantage for the design of *sidehull SES* intended to operate at lower speeds (approx 40 knots).

Figure 8.16 shows the lines developed on the stable SES-100B designed for the expected high speed of 100 knots and the later developed Bell-Halter SES (see Chapter 5) designed for 40 knots. In such designs the stability is built into the hull shaping in similar manner to that designed for high speed and low drag. As can be seen from Figure 8.16, the same basic shape is used except a spray rail has been added along the chine. This feature contributes to improved stability at the lower speeds and to reduced spray (and the resultant spray drag). In addition, it is possible to increase the sidehull width inside the cushion to provide additional buoyancy and to ensure that the wet deck of the craft is clear of the water during hullborne operations. This has the advantage of avoiding the normal suction forces when the craft transitions from the hullborne mode to the air cushion mode.

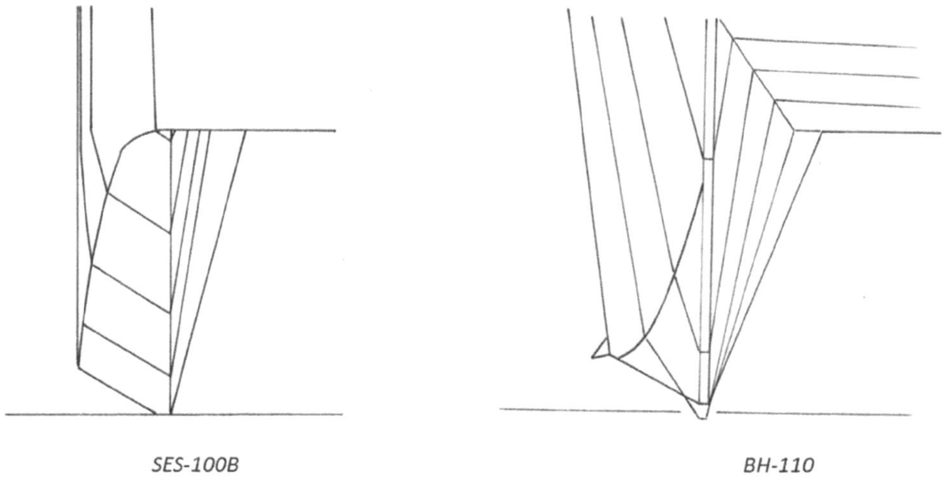

Figure 8.16: High Speed and Lower Speed Sidehull SES Lines Comparison

Over the period of development of SES (1960–1980) there has not been a systematic treatment of such hulls in the same manner as that done for conventional planing hulls or seaplane hulls, where very definite characteristics have been exhaustively studied complete with basic theories and multiple model tests to generate widely applicable design tools. On the other hand, most of the work on SES design has been that tied to specific configurations and not to generating general rules. Because of the failed CAB design using "thin sideboards" much of the developmental activities had been directed at producing "corrective" actions to the *thin sideboard* CAB configuration rather than embark on a more comprehensive set of generic sidehull/air cushion designs for general future design guidance.

Figure 8.17 shows, for example, modifications made for model testing to the original XR-1 CAB hull design to correct for the roll instability of the craft. As can be seen from Figure 8.17, the original XR-1 shaping had insufficient shaping to provide the necessary restoring moment in roll and the corrective action was to add thickness in the stern sections of the hull with the intent to provide a restoring roll moment as the craft rolled (albeit to large angles). Unfortunately, this correction also meant that the yaw stability was compromised when the craft yawed and the increased lift aft caused the craft to pitch down resulting in an unfavorable pitch-yaw stability interaction.

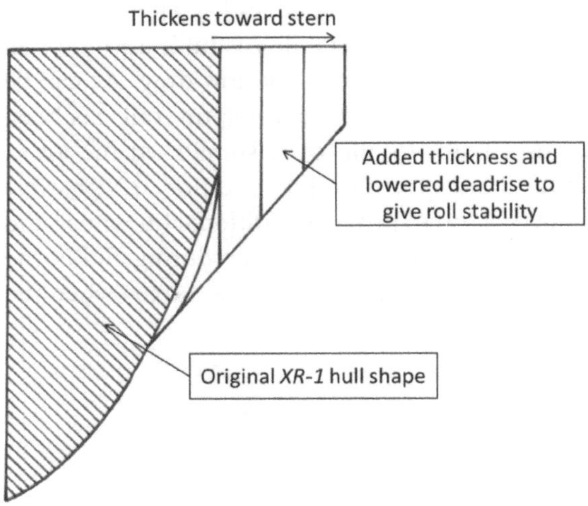

Figure 8.17: Modifications to XR-1 CAB hull to give Roll Stability

Such techniques serve to illustrate that the entire design of the hull from both performance (drag) and stability (in all modes) viewpoints must be considered as a "package" before any final design is pursued.

During the period 1964–1967, the US Navy pursued various solutions to give stability to the "thin sideboards" of the CAB design. One approach pursued by the US Navy is shown in Figure 8.18. In 1967, Hydronautics[19] under contract with the US Navy proposed combining the "thin sideboards" of the CAB concept with "pontoons" to provide the necessary stability as shown in Figure 8.18. The thesis was that the "thin sideboards" could be retained in order to achieve low drag but that the stability would be added through "pontoons" as shown. According to the designers, the shaping of the pontoons was such that the combined effect of the downwash and the sidewash behind the forward (asymmetric) pontoons and the crossflow beneath the thin sideboards resulted in complete elimination of the wetting of the outside of the thin sideboards at high Froude Number operation.

With this intriguing concept, the calculations by Hydronautics showed that for a 4,000 ton displacement ship with an overall L/B = 2 it was predicted that the lift drag ratio (L/D) would be L/D = 36 at 100 knots!! Such a highly optimistic value was not verified in practice as there were no model tests done for this intriguing configuration to substantiate such a concept. All later follow-on work in SES design and test

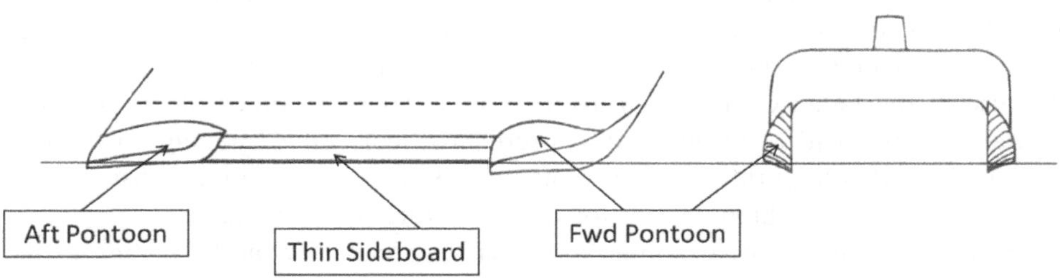

Figure 8.18: Hydronautics SES Design (1967)

indicated that such high values were optimistic by a factor of at least 3:1 but the concept is documented here awaiting some possible future confirmation.

In 1969, this same principle (of retaining thin sidehulls and augmenting the stability by some other means) was essentially integrated into the SES-100A design where the thin *sideboard-sidehulls* were augmented in stability with the addition of the "stability augmentation devices" (SAD) as shown in Figure 8.19 and discussed earlier. These SAD features were removed later as they were found to be ineffective and dangerous and were replaced by thicker sidehulls with lowered deadrise angles more in line with the design of the SES-100B, which were subsequently incorporated in the modified SES-100A design (see Chapter 6).

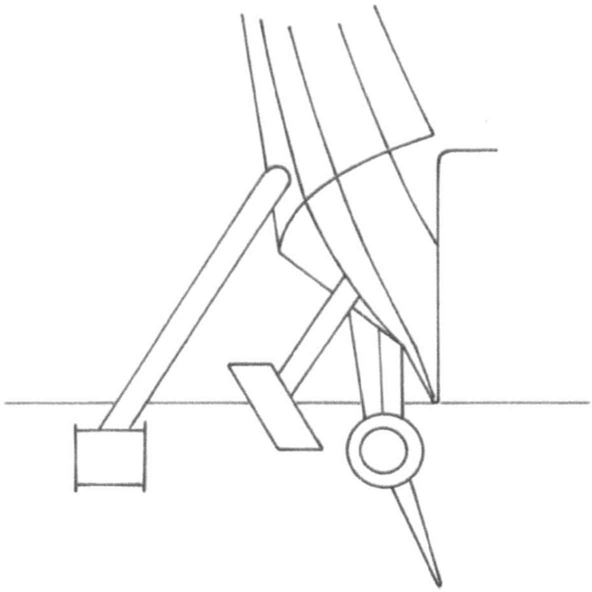

Figure 8.19: SES-100A with Stability Augmentation Devices (SAD)

The US Navy continued to pursue combining the thin sideboards of the CAB concepts with other means to provide the necessary stability. This included conducting research on augmenting the CAB lines with various additional skegs beneath the keels. Extensive model testing conducted in 1967[20] showed that it was possible to achieve adequate stability by such means but not in all modes.

It was also hoped that adding skegs and ventral fins would also help alleviate air ingestion from the pressurized cushion air into the flush inlets for waterjet propulsion (also extensively pursued by the US Navy for high speed SES design) but this proved not to be adequate in the SES-100A full scale tests and limited the SES-100A high speed operation to much less than the specified 80 knots.

These same 1967 model tests on CAB stability also showed that simply lowering the deadrise angle (with the commensurate hull thickening) achieved the same stabilizing features but without the detrimental side effects of the thin sideboards and stability augmentation devices.

As can be seen from the above discussion, the design of the sidehulls in SES design is a complex subject not readily collapsible into any general form as for the other components contributing to the drag of the craft has been treated in the earlier sections. Hopefully, some general guidelines have been summarized here for future use in anticipation of some novel (but practical!) hullforms yet to be devised.

The above brief summary has outlined some of the many geometric configurations used in sidehull design for high speed SES, with some small excursions into lower speed SES. It is seen that only general guidelines can be given for the drag of the sidehull (D_{SH}) in the same vein as the other drag components but there are some general features that can be identified.

Sidehull Shaping for Performance and Stability It should be clear to the reader from the above discussion that SES sidehull design would have benefited from a systematic program of careful variation of hull design parameters complete with an extensive model test program to produce clear and usable design tools for final designs. The sidehulls needed to be designed to produce the sought after minimum drag but at the same time provide the necessary stability (and control) over the full range of speeds and sea conditions in its planned operational envelope.

Only an incomplete data base has been built up in the two decades of SES development (1960–1980) and later, especially in the crucial area of designing for very high Froude Numbers ($F \approx 2$–3) and slender sidehulls ($(L/B)_{SH} > 10$) with asymmetric waterlines. Some indications of design trends and possible areas to pursue are summarized in what follows. The drag of SES sidehulls will be made up, in varying degrees depending on the application, of the following components:

(a) Sidehull wavemaking
(b) Base Drag
(c) Form and Pressure Drag
(d) Spray Drag
(e) Ski Friction Drag
(f) Planing Drag

Because of the nature of the sidehull SES configuration it is difficult to separate the contributions to drag from the main support of the pressurized air cushion and that of the sidehulls themselves. Figure 8.20 shows the generic hull form of the SES. There is a hydrodynamic interaction (wave action, crossflow, etc) between the cushion and the sidehull that make it difficult to separate the effects. In contradistinction to its

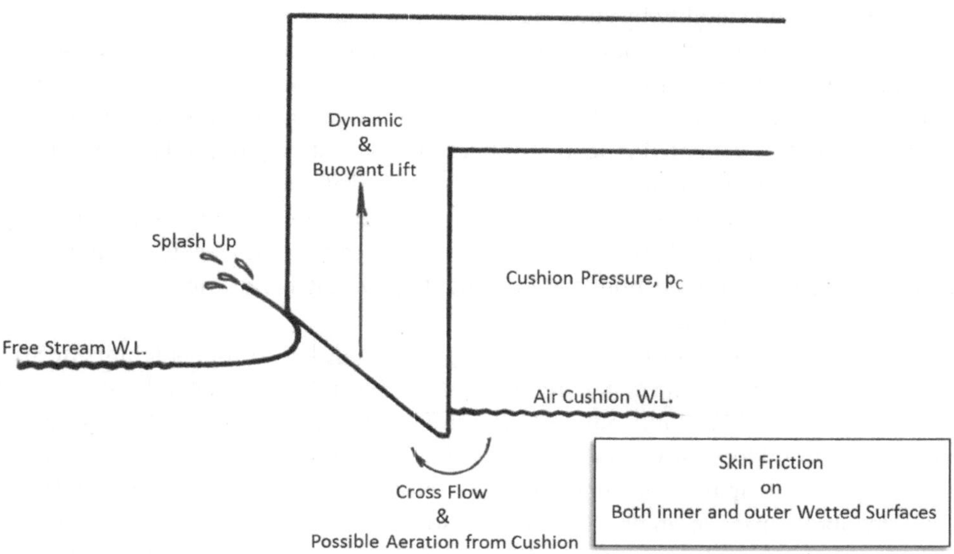

Figure 8.20: Generic Air Cushion Craft Sidehull

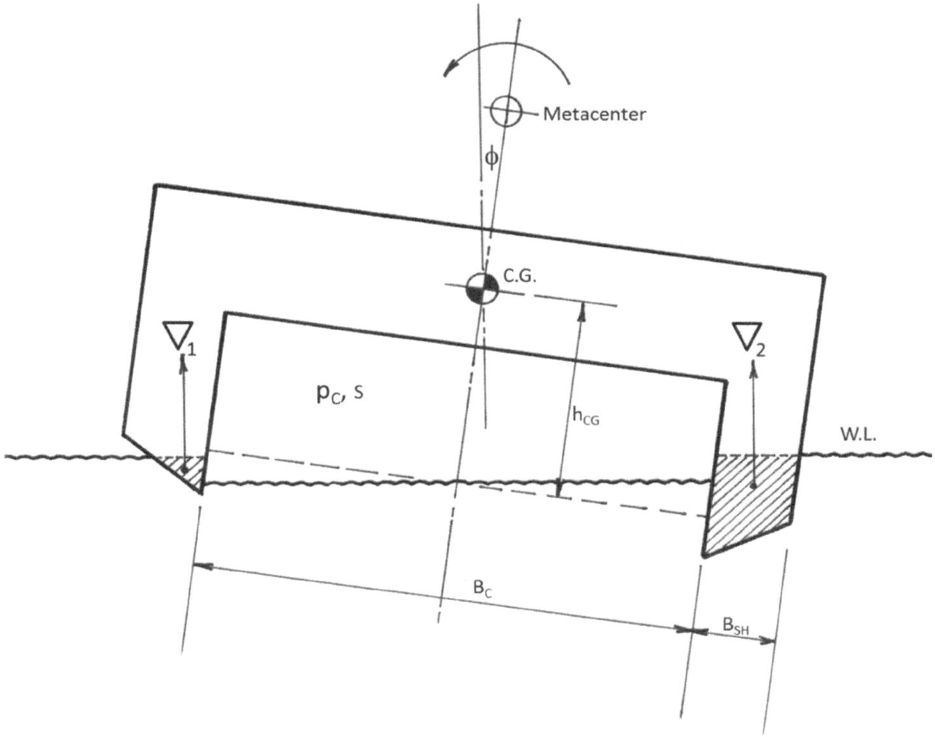

Figure 8.21: Sidehull SES Transverse Stability at Zero Forward Speed

cousins; *planing hulls* and *seaplane hulls*; for which there is a solid data base, the SES sidehull must operate in asymmetric water levels (W.L.) between the lowered inside level caused by the air cushion pressure (p_C) and the outside free stream water level. In addition to all the contributors to sidehull drag listed above is the complication in all cases from operation in rough water.

These complications have affected both the theory of such hulls and in particular in devising suitable model test procedures that will separate the effects. In many cases, designers have used the data base from that available from the rich data base on planing hulls and seaplanes operating with symmetrical waterlines and adapted them wherever possible to the SES sidehull design, not always with success.

An indication of the geometric properties of sidehulls on the stability and performance in the design of the sidehulls in any Sidehull SES can be seen from examination of the cross-section of the craft in a roll position at zero forward (ship) speed (Figure 8.21).

In the illustration of Figure 8.21, the craft is at rest and has rolled to the right an angle of ϕ degrees. This has decreased the volume of displaced water on the left hand sidehull (∇_1) and increased the displaced volume on the right hand sidehull to (∇_2).

In equilibrium conditions, the weight of the craft can be written:

$$W = p_C S + \rho_W g(\nabla_1 + \nabla_2) \tag{8.40}$$

To simplify the algebra, it will be assumed that $W \approx p_C S$ in what follows.

The restoring moment in roll (M_ϕ) can then be written (using conventional sign notation):

$$M_\phi = -[-p_C S(h_{CG} - 0.5h) + \rho_W g h_{CB}(\nabla_2 - \nabla_1)]\sin\phi \qquad (8.41)$$

In this formulation several assumptions have been made. The height of the craft center of gravity (CG) is taken to be measured from the mid-point of the inside depressed water level (shown by the dotted line intersecting with the depressed water level). In equilibrium static conditions the water level would be depressed by an amount:

$$h = \frac{p_C}{\rho_W g} \qquad (8.42)$$

Also, it has been assumed that Archimedes principle for objects with asymmetric water levels would be satisfied by treating the two "submerged" volumes as ∇_1 and ∇_2 as shown. The two buoyancy forces, given as $\rho_W g\nabla_1$ and $\rho_W g\nabla_2$ act through their centers of buoyancy measured as height h_{CB} from that same water level. For more complex geometries it would be necessary to integrate the forces along the hull but that precision does not add to the understanding of the main principles discussed in this context.

It is common practice to express the static stability of any hydrodynamic vessel in terms of the per cent center of pressure shift per degree of roll, which in this case would be expressed as:

$$\text{\% c.p. shift per degree} = \frac{M_\phi}{WB_C} \times 100 \qquad (8.43)$$

In which case, the static roll stability can be shown to be, by combining equations 8.41 and 8.43:

$$\text{\% c.p. shift per degree} = -\sin\phi\left[-\left(\frac{h_{CG} - 0.5h}{B_C}\right) + \left(\frac{\rho_W g}{p_C/L}\right)\left(\frac{1}{L/B}\right)_C\left(\frac{h_{CB}}{B_C}\right)\left(\frac{\nabla_b}{SB_C}\right)\right] \qquad (8.44)$$

As can be seen from Equation 8.44, the roll stability of the sidehull SES is a complex function of several basic measurable craft dimensions which can be expressed as:

$$\text{\% c.p. shift per degree} = f\left(\frac{h_{CG}}{B_C}, \frac{h_{CB}}{B_C}, \frac{L_C}{B_C}, \frac{p_C}{L}, \frac{\nabla_b}{SB_C}\right) \qquad (8.45)$$

Each of these characteristics is easily determined in any particular design. While general rules cannot be specifically written down because of the many geometric variations possible, it can be taken from the experience of many designs that certain "boundaries" can be established and dominant characteristics identified. In Chapter 6, charts used during the SES Program showed where the various craft "sit" in regard to these stability parameters. They are repeated here for completeness. Figure 8.22 shows the computed values of% c.p. shift per degree, expressed as $\frac{M_\phi}{WB_C} \times 100$. This shows the strong influence of the basic geometric parameter of the height (depth) of the craft cushion (h_C/B), which is related to the stability parameter (h_{CG}/B_C) in Equation 8.44 by a constant.

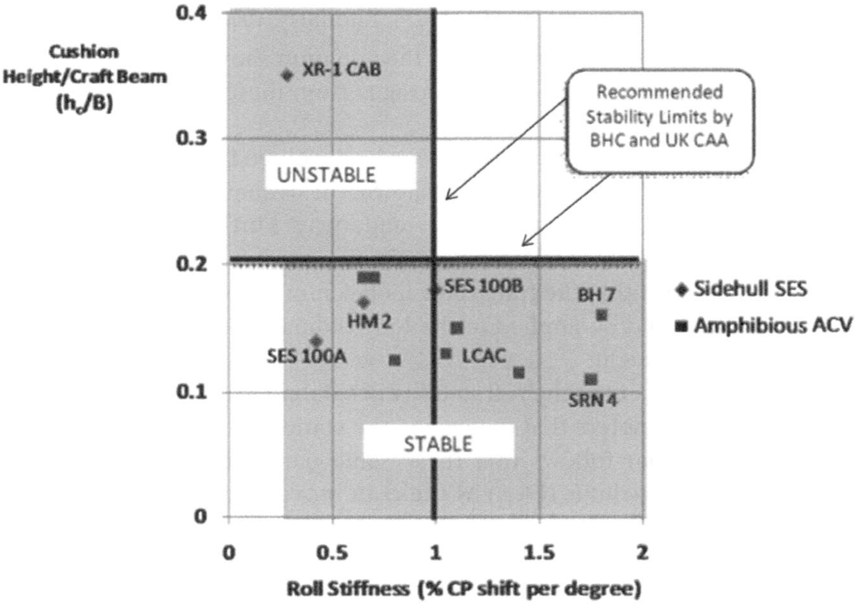

Figure 8.22: Roll Stability Characteristics of Various Air Cushion Craft

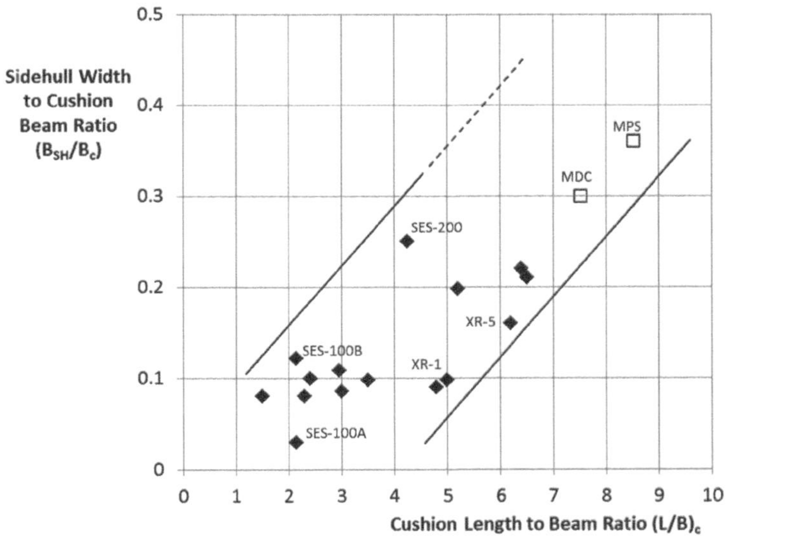

Figure 8.23: Sidehull Width and Slenderness Ratio Effect on Roll Stability

From successful operational craft, Roll Stiffness (expressed as % CP shift per degree) between 0.2 and 2% is a satisfactory design. Figure 8.23 shows the statistics of various sidehull SES that demonstrates two other key parameters in Equation 8.44 which are the ratio of sidehull beam to craft beam (B_{SH}/B_C) (see Figure 8.21) and the slenderness ratio of the craft (L_C/B_C).

In the above "rules of thumb" used some of the geometric parameters would need refinement in any particular case to extract them for final use. For example, the parameter $({}^{B_{SH}}/_{B_C})$ is included in the term $({}^{\nabla_b}/_{SB_C})$. Similarly, the lengths (L, L_C, L_{SH}) need to be used appropriately in each term in the equation shown. One can see the "more stable" designs and the "less stable" designs from these charts (Figure 8.22 and 8.23).

While not derived here, similar equations can be developed for the SES "underway" where, in addition to the buoyant forces shown, the dynamic planing forces are included to calculate the static stability while underway. Further, in any detailed treatment for specific designs, the criteria would be to ensure for static stability that the buoyancy center is such that the transverse metacenter is above the CG. Similar geometric constraints would be applied in the longitudinal direction for a positive longitudinal metacentric height.

The above basic treatment of the roll stability of sidehull SES has shown the key geometric and shape parameters that determine the static stability of such craft. It will now be shown in what follows that these same parameters have a marked influence on the drag of sidehulls (D_{SH}) as the craft moves through the water and dynamic forces enter the equation as well as the basic buoyancy forces to determine the lift and drag characteristics of *sidehull SES.*

Sidehull Lift and Drag It is generally accepted that the main drag components of hulls (D_{SH}) can be represented by the equation:

$$D_{SH} = \left(\Delta \tan\tau + \frac{C_f 1/2 \rho_W V^2 S_{SH}}{\cos\tau}\right) \tag{8.46}$$

In this formulation, Δ is the vertical force acting on the sidehull; τ is the trim angle of the sidehull (which may or may not be the wave slope of the air cushion); C_f is the skin friction coefficient; ρ_W is the density of water, and S_{SH} is the reference wetted area of the sidehull. The vertical force, Δ, is made up of two components, namely the dynamic lift force (Δ_d) and buoyant force (Δ_b), where:

$$\Delta_d = C_{LSH} {}^1\!/_2 \rho_W V^2 S_{SH} \tag{8.47}$$

$$\Delta_b = \rho_W g \nabla_b \tag{8.48}$$

The three key parameters that determine these forces (dynamic and buoyant) are the sidehull lift coefficient (C_{LSH}); the displaced volume of the sidehulls (∇_b) and the sidehull wetted surface area (S_{SH}).

Unfortunately, despite the almost \$1 billion expended on SES development in the major development period 1960–1980, there is very little available data on the treatment of lift and drag characteristics of sidehulls so crucial to good SES design (with different deadrise angles inside (β_i) and outside (β) of the air cushion) and operating in asymmetric water levels in both low or high speed flow conditions (cavitating or otherwise). The high speeds in supercavitating flows and the presence of pressurized air in the cushion with its associated cross flow (especially with high slenderness ratio sidehulls), further complicates any generalized treatment.

Fortunately, during the period 1930–1950, when interest in seaplanes and planing hulls was at its peak, driven by wartime development needs, a significant data base of both theoretical treatments and model test results was established that can be used to advantage to understanding the hydrodynamic force characteristics (lift and drag) of sidehulls so crucial to SES development. This extensive data base can be drawn upon to aid in the understanding of sidehull SES lift and drag (and stability) characteristics.

Lift and Drag Characteristics of Planing Hulls and Seaplanes Both conventional planing hulls and seaplane hulls have many different configurations, shapes and features. Figure 8.24 shows just two examples out of the thousands of designs that exist. Some hulls have hard chines (as shown) or have rounded bilge forms (usually for slower speeds). The deadrise angle usually starts high at the bow giving a sharp entry into the water then rapidly flattens out toward the stern to give stability to the craft. Seaplanes follow very similar reasoning with steps incorporated to raise the stern sections out of the water as the design is optimized for both aerodynamic flight and take-off and landing characteristics.

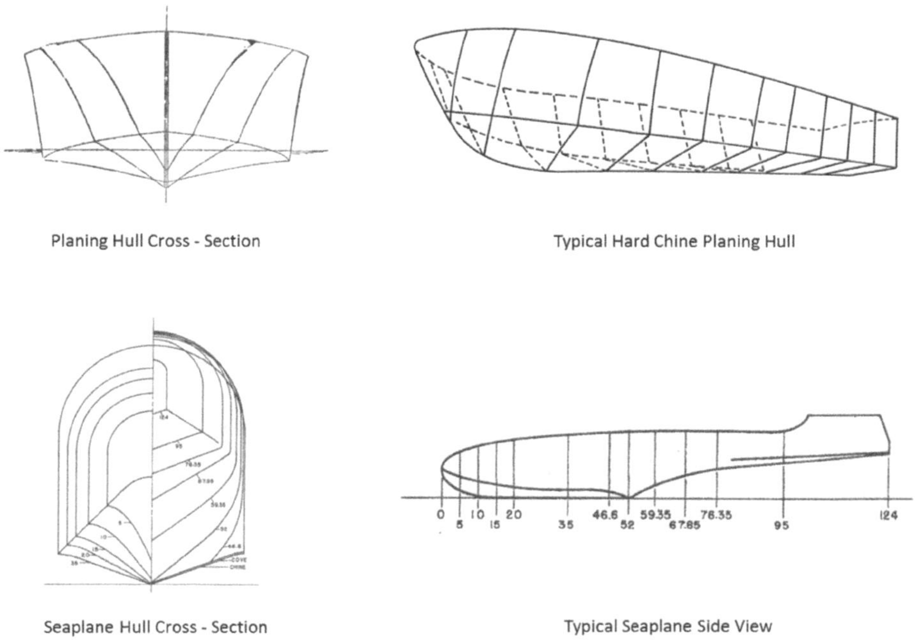

Figure 8.24: Typical Planing Hull and Seaplane Geometries

Because of the complex geometries involved, the vast majority of researchers into the lift, drag and stability characteristics of such hull forms have had to resort to much more simplified geometries in order to generate theories and design guidelines. The majority of the work by the researchers has concentrated on using (i) flat plates and (ii) simple prismatic forms to generate the performance equations. Figure 8.25 shows the usual approach taken in the research and development of the equations used in theory and test.

In the upper sketch of Figure 8.25, the flat plate is shown with a *constant* width of (b) and a wetted length of (l) and planing at an angle of attack or trim angle(τ). These are the fundamental geometric parameters used in the theories, test and empirical relationships. Clearly, the edge of the flat plate is representative of the (hard) chine and the results would be different if the water is above or level with these edges.

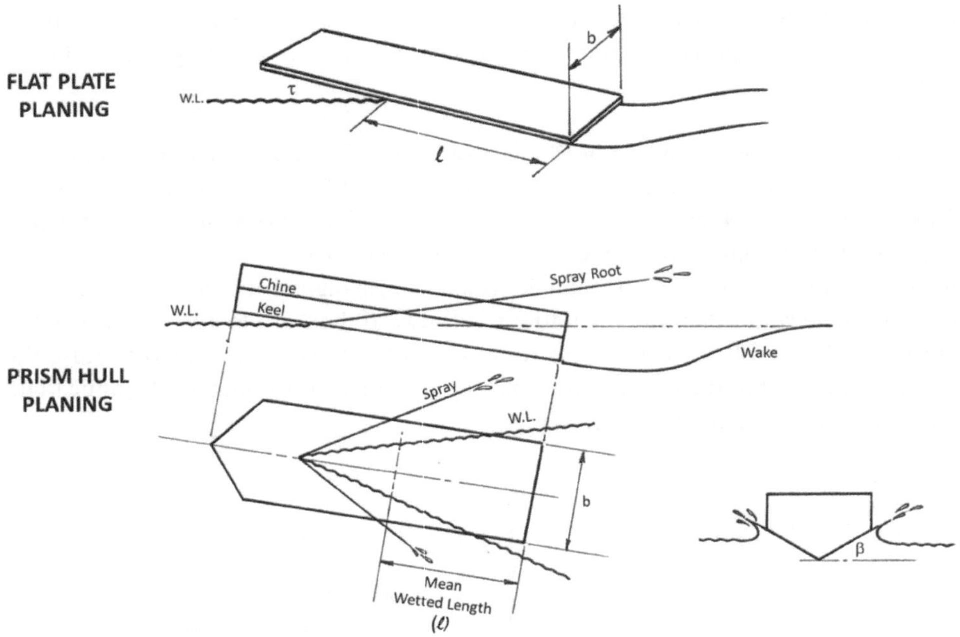

Figure 8.25: Flat Plate and Prism Hull Forms Planing

The all-important deadrise angle (β) is treated through assumptions of simple prismatic forms of *constant* deadrise angle (β) along the length of the hull as shown in the lower sketch in Figure 8.25. The geometry even in the simple prismatic form requires approximations to handle waterlines and splash-up from the spray roots on the hulls. A large data base exists over a wide range of parameters, test conditions, and research establishments to provide a high confidence level in the theory and test of planing forms.

Flat Plate Planing Theory and Test It is helpful to first define some key parameters that are used in describing the lift and drag characteristics of such shapes. They are:

$$\text{Aspect Ratio, } A = \frac{b}{\ell} \tag{8.49}$$

$$\text{Slenderness Ratio, } \lambda = \frac{\ell}{b} \tag{8.50}$$

$$\text{Speed Coefficient, } C_V = \frac{V}{\sqrt{gb}} \tag{8.51}$$

$$\text{Froude Number, } F = \frac{V}{\sqrt{g\ell}} \tag{8.52}$$

These parameters are related to each other by:

$$A = \frac{1}{\lambda} \tag{8.53}$$

$$C_V = \sqrt{\lambda} F \tag{8.54}$$

Further, it will be found in the literature that different researchers use different ways of describing the lift and drag in various coefficient forms, expressing lift and drag coefficients based on either the beam (b) or the reference wetted area (S), which again can be related to each other from the above relationships.

$$\text{Lift Coefficient (based on beam)},\ C_{Lb} = \frac{\Delta}{{}^1\!/_2\,\rho_W V^2 b^2} \tag{8.55}$$

$$\text{Lift Coefficient (based on area)},\ C_{LSH} = \frac{\Delta}{{}^1\!/_2\,\rho_W V^2 S_{SH}} \tag{8.56}$$

The subscript "SH" has been added to emphasize that the area used in the flat plate work relates to the sidehull area ($B_C L_C$) used here. These two forms describing the lift coefficient of the hull are directly related, using the above equations, as:

$$C_{Lb} = \lambda C_{LSH} \tag{8.57}$$

In the context of this book, these different forms for the coefficients will be translated from the researchers work as necessary to provide consistent treatment of lift and drag for sidehull SES drag analyses.

The theoretical work on lifting flat plates can be traced to the original work by Lanchester (1907)[21] and Prandtl (1913)[22] that became known as the Lanchester–Prandtl Lifting Line Theory[e] and later, the Lifting Wing or Surface Theory. This theory gives the lift coefficient (based on area) as:

$$C_L = \left(\frac{2\pi A}{2 + A}\right)\tau \tag{8.58}$$

In this formulation, the two dimensional lift curve slope is given as (2π) and the three dimensional correction assuming an elliptical pressure distribution is given by the inclusion of the aspect ratio (A). Such a formulation was developed for flat plate airfoils. The theory showed that the lift increased linearly with the angle of attack or trim (τ). This lifting line theory also gave the induced drag from this lift that can be expressed as:

$$C_{Di} = \frac{{C_L}^2}{\pi A e} \tag{8.59}$$

This expression for the induced drag includes a *planform efficiency factor* (e) to accommodate other circulation distributions on the lifting plate. An elliptical circulation distribution assumed $e = 1$.

In 1941, Robert Jones[23] at the Langley Memorial Aeronautical Laboratory, known for much of the work on low aspect ratio wings in the US, proposed "correcting" the lifting line theory to include the effect of chord with the equation:

$$C_L = \left(\frac{2\pi A}{2 + \xi A}\right)\tau \tag{8.60}$$

[e] See Chapter 11 for discussion and application of the Lanchester-Prandtl lifting line theory to wing-in-ground effect (WIG) craft.

This correction for the effect of chord was given as:

$$\xi = \frac{b+l}{b} = 1 + \frac{1}{A} \tag{8.61}$$

In which case, the revised form for the lift coefficient would be given as:

$$C_L = \left(\frac{2\pi A}{A+3}\right)\tau \tag{8.62}$$

Researchers, after the initial work by Lanchester and Prandtl, explored these simple theoretical results applied to planing surfaces on the water by adjusting the lift curve slope and how it depended on aspect ratio (A) or slenderness ratio (λ). As might be expected, for high values of aspect ratio (A) or low values of slenderness ratio (λ) the various theories and test data approach the lifting line theory values but depart from them for low aspect ratio (A) or high slenderness ratio (λ) shapes.

Prof Korvin-Kroukovsky[24] at Stevens Institute of Technology (1950) based on the work by Wagner[25] (1932) applied the lifting line theory to planing flat plates on water. Korvin-Kroukovsky analyzed the case of planing flat plates taking into account the vertical deflection of the water and the effects of transverse flow to give the result:

$$C_L = \left(\frac{0.73\pi A}{A+2}\right)\tau + 0.88\tau^2 \tag{8.63}$$

In Equation 8.63 the trim angle(τ) is given in radians. In terms of the slenderness ratio (λ) this relationship can be written as:

$$C_L = \left(\frac{0.04}{1+2\lambda}\right)\tau + 0.00027\tau^2 \tag{8.64}$$

In this formulation, Korvin-Kroukovsky found that for better correlation with model test results, the theoretical constant of ($^{2\pi}/_{57.296} = 0.110$) from the lifting line theory had to be reduced to 0.04. In Equation 8.64, the trim angle (τ) is quoted in degrees. The cross flow term for the wave flow shows up in this derivation as a function of the trim angle squared. Earlier (1949), Korvin-Kroukovsky[26] joined by fellow researchers at Stevens Institute of Technology went further on developing an empirical relationship for planing flat plates to include the buoyancy term.

This resulted in the empirical equation:

$$C_{Lb} = \tau^{1.1}\left[0.0120\lambda^{1/2} + 0.0095\left(^{\lambda}/_{C_V}\right)^2\right] \tag{8.65^{f}}$$

If this relationship is rewritten in terms of lift coefficient based on area, and the speed coefficient (C_V) is replaced by Froude Number (F), the relationship for lift coefficient can be expressed as:

[f] In 1954, Dan Savitsky and Joseph Neidinger at Stevens Institute of Technology proposed a different value of the buoyancy term to be $0.0055^{\lambda^{5/2}}/_{C_v^2}$ for "very low values of Speed Coefficient (C_V)" for a better fit with model data. However, because the sidehulls for SES operate at much higher speed coefficients, the original equation is retained here which has the advantage of being a more tractable form in terms of Froude Number.

$$C_L = \tau^{1.1}\left[\frac{0.0120}{\sqrt{\lambda}} + \frac{0.0095}{F^2}\right] \quad (8.66)$$

In this simpler and more recognizable form, the trim angle (τ) is in degrees. The second term is the buoyancy force which rapidly decreases as speed is increased. Figure 8.26 shows this relationship for the lift coefficient of flat plates. In the development of this relationship to match the experimental results, Korvin-Kroukovsky et al. replaced the cross flow term that depended on the square of the trim angle and modified the first term instead by raising its exponent to n = 1.1. This will be returned to later when other researchers' data are reviewed and the range of interest for the slenderness ratio to those more compatible with sidehull SES are discussed.

Figure 8.26 is a useful chart to show the effect of the buoyancy forces. For the SES-100B, the sidehull slenderness ratio is approximately $\lambda = 8.5$ and at 80 knots the Froude Number, $F = 3.0$. With these values, the Korvin-Kroukovsky formula (Equation 8.66) would show that the buoyancy force is approximately 20% of the total sidehull lift at that speed.

A "rule of thumb" used by planing hull designers is that while "planing" is an imprecise term it is frequently taken that planing is said to begin when the buoyant forces reduce to about 20% of the dynamic forces. From the Korvin-Kroukovsky empirical formula (Equation 8.66), this would occur at a speed proportional to the fourth root of the slenderness ratio:

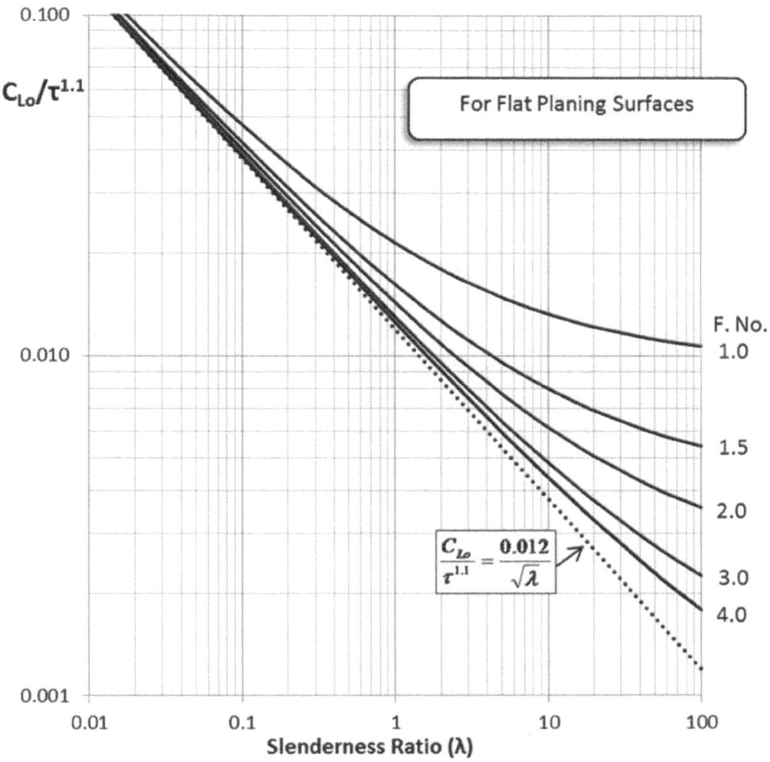

Figure 8.26: Lift of Flat Plates (Korvin-Kroukovsky, Savitsky & Lehman) (1949)

$$|F|_{planing} \cong 2\sqrt[4]{\lambda} \tag{8.67}$$

As a "sanity check" on the above empirical relationship at least for the buoyancy term, consider the buoyancy generated by the volume of displaced water for the flat plate shown in Figure 8.25. The displaced volume of water would be:

$$Volume = {}^1\!/_2 bl^2 \sin\tau\cos\tau \tag{8.68}$$

The buoyancy force would then be given by:

$$\Delta_b = {}^1\!/_4 \rho_W g l^2 b \sin 2\tau \tag{8.69}$$

In coefficient form, this would be:

$$C_{\Delta b} = \frac{\Delta_b}{{}^1\!/_2 \rho_W V^2 l b} = \frac{\sin 2\tau}{2F^2} \tag{8.70}$$

For small angles (with τ expressed in degrees), this can be given to a good approximation as:

$$\frac{C_{\Delta b}}{\tau} = \frac{0.017}{F^2} \tag{8.71}$$

If this result is compared to the buoyancy term in Equation 8.66, it suggests that there is some other factor (distorted water boundaries?) to cause an almost 2:1 difference in the buoyancy term from the model test data even allowing for the difference in the exponent of the trim angle. It is important to note in the Korvin-Kroukovsky empirical relationship (Equation 8.66) that it was based on model test data that only covered flat plates with slenderness ratios, $\lambda \leq 5.5$. While this was adequate for conventional planing craft, it was out of the range of geometries used for sidehull SES designs. Typical values of SES sidehull slenderness ratios are in the range of $\lambda \to 10 \to 20$ or more.

The model test data used by Korvin-Kroukovsky et al. for their empirical relationship was collected by Sottorf (1932, 1933)[27] and Sambraus (1936)[28] in Germany and Shoemaker (1934)[29] in the US for low slenderness ratio flat plates. Hence, there is a reasonable question as to the general applicability of the empirical relationship to sidehull SES use. Fortunately, there are other data sets and theoretical treatments that extend the data base for higher slenderness ratios more applicable to SES design which can be examined and compared to the original extensive treatment by Korvin-Kroukovsky, Savitsky and Lehman.

In 1949, Fred Locke[30] at the Bureau of Aeronautics, US Navy Department conducted an empirical study covering a wide range of hull types with slenderness ratios up to $\lambda \approx 30$. Based on the early work by Bollay[31] and others already referenced, Locke reasoned that the lift of such shapes be represented by:

$$C_L = \eta\zeta K\tau^n \tag{8.72}$$

In this formulation, the "constants" (η, ζ) are shape parameters to interpret data of other than flat plates (e.g. shaped airfoils). The factors (K, n) were found to depend strongly on the aspect ratio (A) of the flat plates. Using a data base of airfoils, planing flat plate surfaces and yacht hulls in yaw, Locke found consistent agreement in the data. Expressing Locke's data in terms of slenderness ratio (λ) instead of aspect ratio (A), Figure 8.27 and Figure 8.28 shows his results. Applying a curve fit to Locke's data gives:

$$K = \frac{0.0203}{\lambda^{0.6554}} \tag{8.73}$$

$$n = 1.144\lambda^{0.09} \tag{8.74}$$

Two immediate observations from Locke's data and Figure 8.27 and Figure 8.28 are that the lift coefficient (C_L) decreases with increase in slenderness ratio (λ), similar to the Korvin-Kroukovsky results; and secondly the exponent (n) varies between 1.0 and 1.6 depending on the slenderness ratio. This is to compare with the "average value" of n = 1.1 used by Korvin-Kroukovsky for geometries more appropriate to planing craft. As the slenderness ratio (λ) is increased to SES sidehull values it is seen that the value of the exponent "n" increases to values of n > 1.4. A large data base became available in 1952 by Weinstein and Kapryan[32] that extended the slenderness ratios out to values of $\lambda \approx 8.5$ which are more applicable to SES sidehull geometries. Figure 8.29 shows the test results.

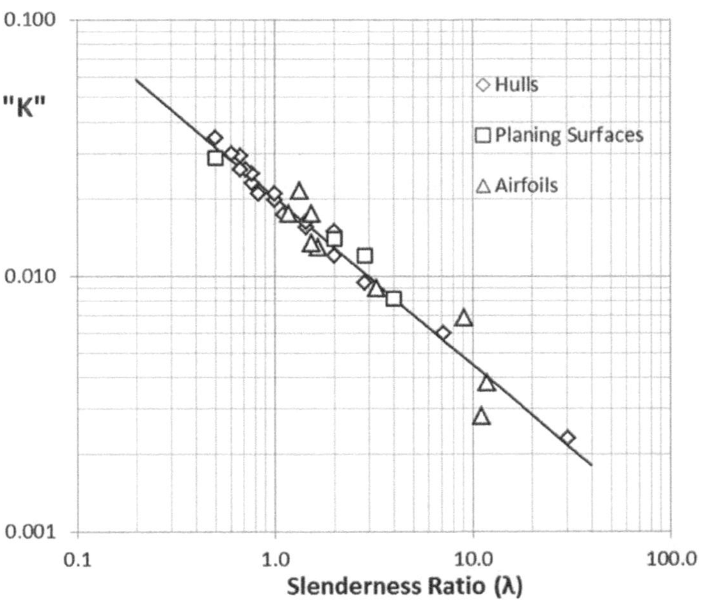

Figure 8.27: Locke's "K" Factor for Lift Coefficient

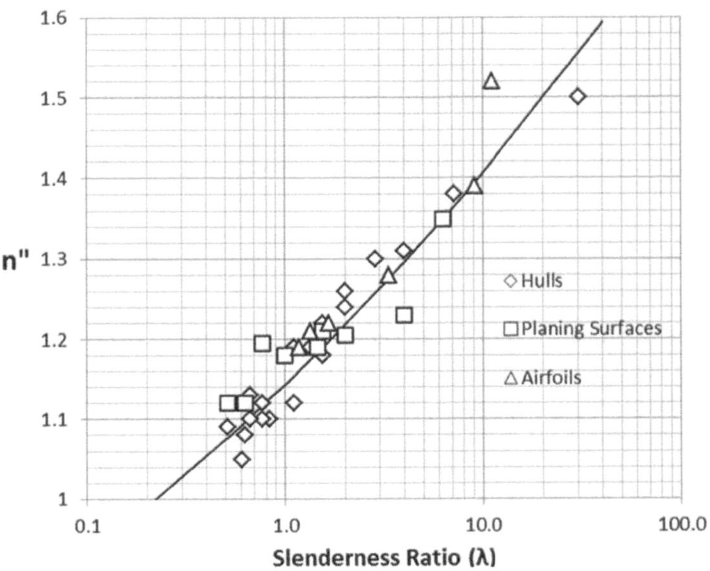

Figure 8.28: Locke's Exponent "n" for Lift Coefficient

Superimposed on the Weinstein and Kapryan test results are the

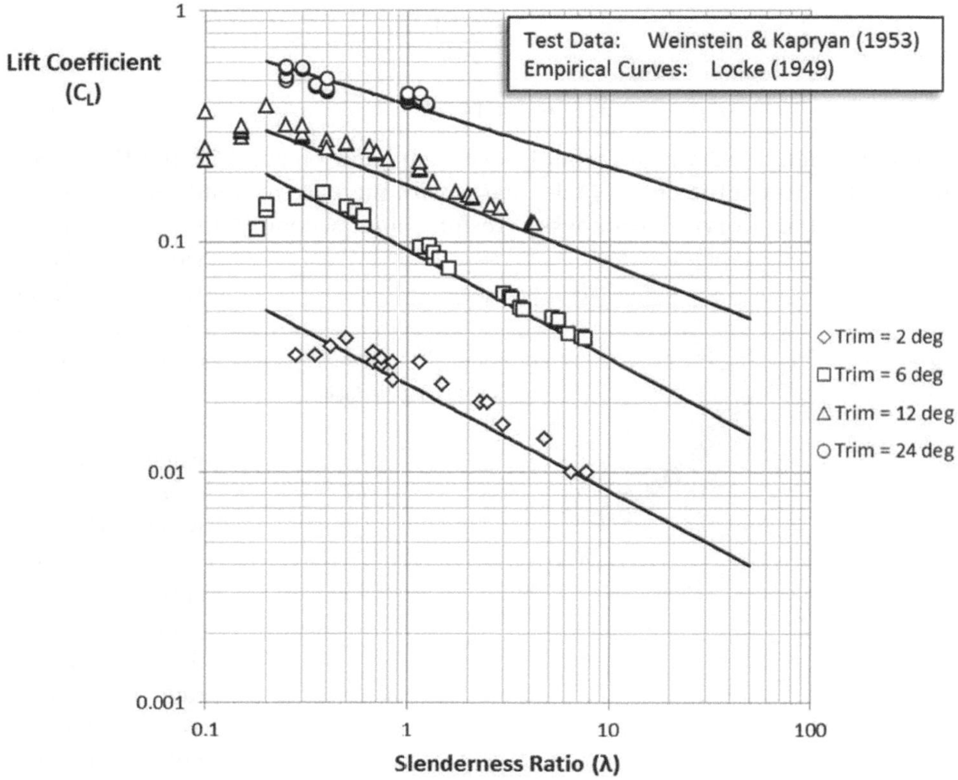

Figure 8.29: Weinstein and Kapryan Test Results for Rectangular Flat Plates

empirical relationships developed by Locke using Equations 8.72, 8.73 and 8.74. It is seen that there is good agreement over a wide range of trim angles (τ) and slenderness ratios (λ). There is less of an agreement at the low values of slenderness ratio ($\lambda \leq 0.5$) where the aspect ratio is large ($A \geq 2.0$) and the lift curve slope approaches the lifting line theory results. This change in the functional form can be seen from the work by Charles Shuford[33] in 1954.

Working from the lifting line theory Shuford developed the theoretical relationship:

$$C_L = \left(\frac{0.5\pi A}{1+A}\right)\tau(1 - \sin^2\tau) + \sin^2\tau\cos\tau \tag{8.75}$$

In Equation 8.75, the term "$\sin^2\tau$" is due to the suction force from the large negative pressures at the leading edge of the plate. The term "$\sin^2\tau\cos\tau$" is to account for the cross flow on the plate. Expressing Shuford's theory in terms of the slenderness ratio (λ) and expressing the trim angle (τ) in degrees gives the equation:

$$C_L = \left(\frac{0.0274}{1+\lambda}\right)\tau(1 - \sin^2\tau) + \sin^2\tau\cos\tau \tag{8.76}$$

Shuford's theory is plotted against Weinstein & Kapryan's data for the case $\tau = 2\text{deg}$ in Figure 8.30.

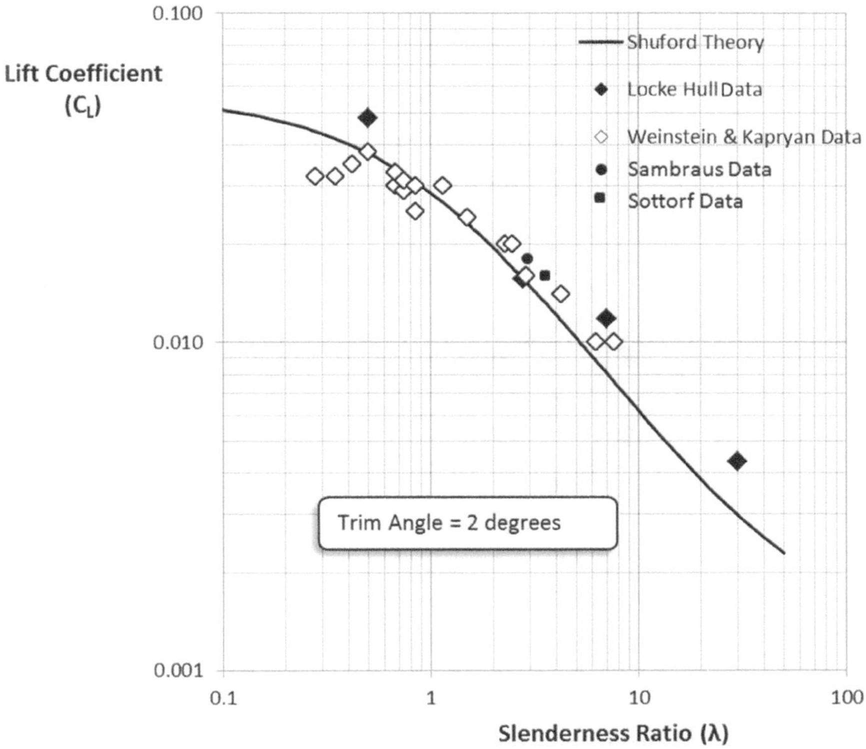

Figure 8.30: Shuford's Theory and Weinstein & Kapryan's Data for Lift Coefficient

Also included in Figure 8.30 is some of Locke's data for hulls (taken from his yawed yacht hull data). Just one data point has been taken from the Sottorf and Sambraus data because of the difficulty in extracting their data from the many cross-plots that were necessary to find a common data point for a specific trim angle. This is because their test technique set the lift coefficient and let the trim angle vary. This can be seen from Figure 8.31 that shows the Sottorf and Sambraus (two independent test series) of trim angle versus slenderness ratio for constant lift coefficient. It is seen that despite the necessary cross plotting from the various researchers' data there is a good agreement in the final value of lift coefficient versus slenderness ratio for constant trim angles.

Some appreciation of the difficulty of interpreting the various theories, empirical relationships and documented test data for the lift coefficient can be made by examining Figure 8.32 where various contributors to planing understanding have been collected. Figure 8.32 is Figure 8.30 with the empirical relationships of Locke and Korvin-Kroukovsky et al. added. The empirical relationship of Locke shown is that given by Equations 8.72, 8.73 and 8.74. The empirical relationship shown for Korvin-Kroukovsky et al. is that given by Equation 8.66 except that the buoyancy term has been dropped. Probably if a common value of speed (Froude No.) could be found for the empirical relationships, the Korvin-Kroukovsky formulation in Figure 8.32 would be in closer agreement with the Locke relationship.

It can be seen that at low values of slenderness ratio (λ) the Shuford theory more closely reflects the form of the Weinstein & Kapryan test data. At high values of

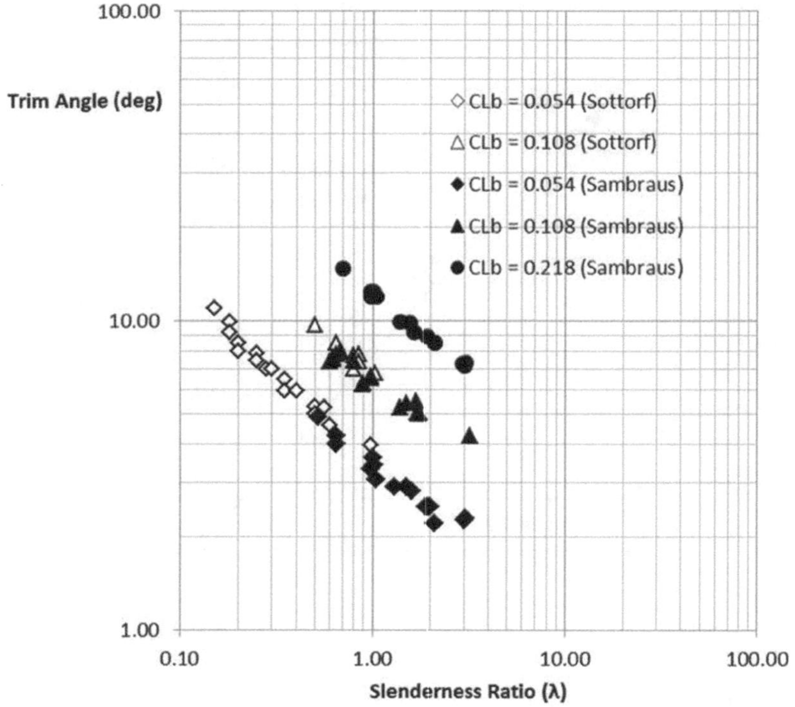

Figure 8.31: Sottorf and Sambraus Test Data for Trim Angle at Constant Lift Coefficient

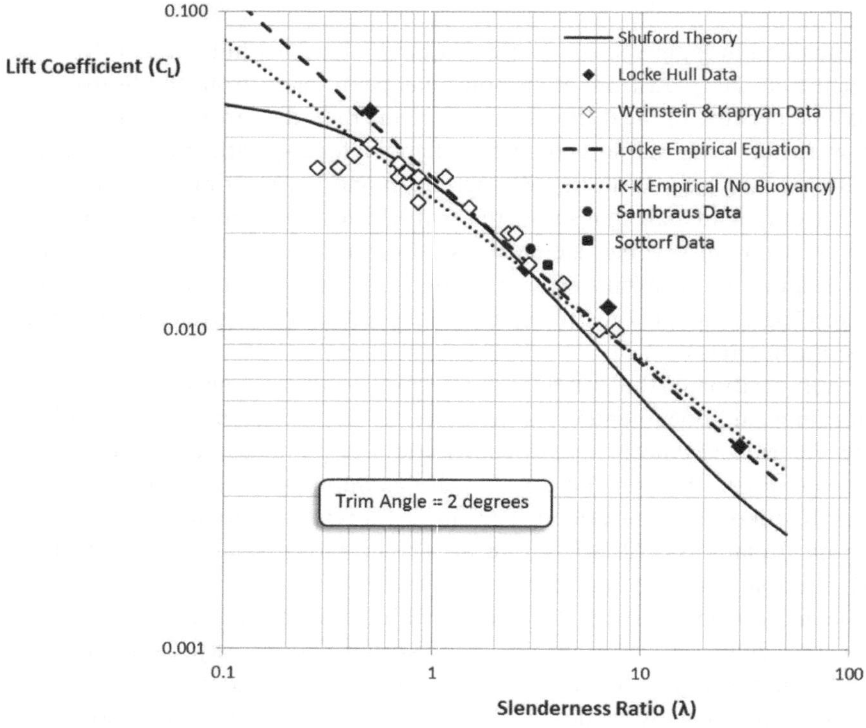

Figure 8.32: Comparison of Theory, Empirical Relationships and Test Data (For Flat Plates)

slenderness ratio, more applicable to SES sidehull geometries, the Shuford theory is less representative of the data probably because it does not include the effect of buoyancy. Both the Locke and the Korvin-Kroukovsky empirical relationships do match the test data at the high values of slenderness ratio (including buoyancy) as well as for the mid range of slenderness ratio more applicable to planing craft for which the equations were derived.

From these plots it is believed that a reasonable estimate can be made for determining the lift coefficient of planing flat shapes for sidehull geometries ($5 < \lambda < 50$) for SES hull forms absent actual data from sidehulls operating in asymmetric water levels and the presence of a pressurized air cushion.

It now remains to determine the effect of the all-important deadrise angle (β) that will impact both the stability and the performance of the SES.

Effect of Hull Deadrise on Lift Coefficient Many of the same researchers who examined the lift characteristics of flat plates also examined the reduced lift caused by adding deadrise to the hull. Figure 8.33 shows some of the results. The additional researchers included in Figure 8.33 are Springston & Sayre[34] in 1955 at the US Navy Towing Basin who studied the particular case of a prismatic hull with deadrise $\beta = 50$ degrees.

While the data base is limited to values of slenderness ratio $2 < \lambda < 4$, it does provide an indication of the severe fall-off in lift as the deadrise angle is increased. Many different algebraic formulations may be found in the literature describing this fall off but none adequately matches the available experimental data listed. One simple expression put forward by Locke is the linear relationship shown, which is:

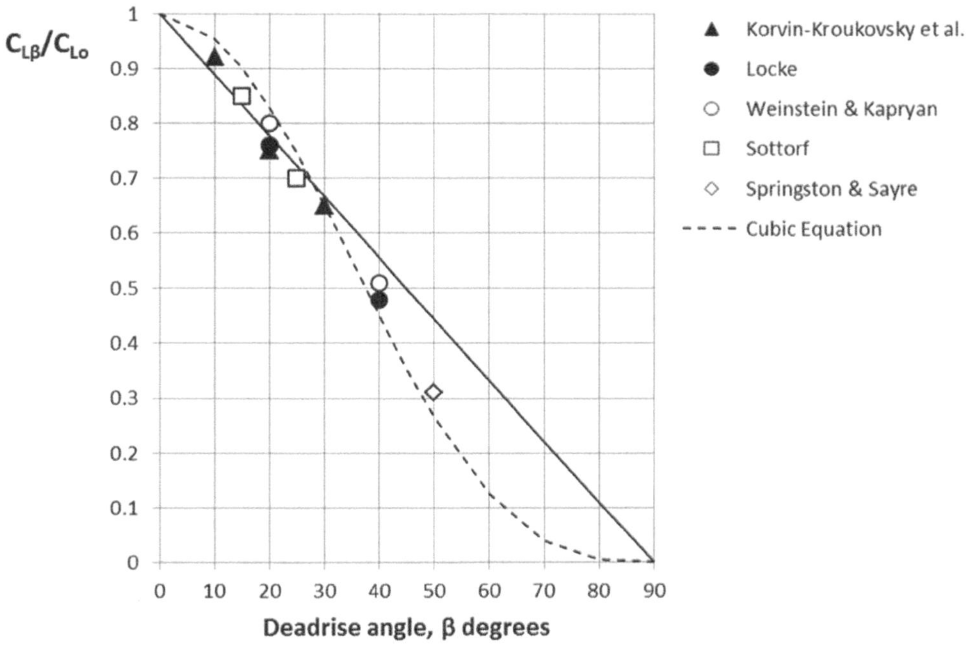

Figure 8.33: Effect of Deadrise Angle on Planing Lift

$$C_{L\beta} = C_{Lo}\left(1 - \frac{\beta}{90}\right) \tag{8.77}$$

In this expression, for clarity of the different terms, C_{Lo} is the lift coefficient of the flat plate geometries previously discussed and $C_{L\beta}$ is the lift coefficient of the prismatic hull. As can be seen, this linear relationship gives a good approximation up to about $\beta \approx 30$ degrees but overestimates the lift for the higher deadrise angles. An expression that the author finds useful for its wider applicability of deadrise angles is a simple curve fit cubic relationship (see Figure 8.33), given by:

$$C_{L\beta} = C_{Lo}\cos^3\beta \tag{8.78}$$

This relationship describes the experienced significant loss of lift at the higher deadrise angles. As an example, it is seen that the lift of a 30 degree deadrise prismatic hull is 65% of the flat plate value but a hull with a 60 degree deadrise has only 12% of the lift of an equivalent flat plate. It also correctly predicts no lift for a thin sideboard craft (β = 90 degrees). Such tests illustrate the problem of achieving sufficient sidehull lift for adequate transverse stability discussed earlier for the different SES hull forms. This "correction" for deadrise can be applied to all the flat plate results shown earlier. The Locke empirical results fit over a wide range of slenderness ratios more applicable to the SES sidehull geometry. If the factor is applied to Locke's equation to give for the lift of prismatic hull forms it becomes:

$$C_{L\beta} = \eta\zeta K\tau^n\cos^3\beta \tag{8.78}$$

In this expression, the shape factors (η, ζ) are as before and the (K, n) factors that depend on the hull slenderness ratio (λ) are as given by Equations 8.73 and 8.74.

However, the empirical relationship developed by Korvin-Kroukovsky, Savitsky and Lehman shown earlier (see Equation 8.66), although developed over a narrow slenderness ratio data base, Figure 8.32 shows that it matches the broader based Locke data. It has the advantage that the dynamic and buoyancy terms are separated, which as will be shown later, provides better insight into the contributing factors to both lift and drag.

If the "cubic" deadrise factor is applied, then the lift coefficient given by Equation 8.66 for the flat plate can be expressed for the prismatic form as:

$$C_{L\beta} = \tau^{1.1}\left[\frac{0.0120}{\sqrt{\lambda}} + \frac{0.0095}{F^2}\right]\cos^3\beta \tag{8.79}$$

It should be pointed out that Korvin-Kroukovsky et al. in their 1949 report gave an expression to account for deadrise angle (β) but it required a two-step process and did not give a good match with the available experimental evidence especially at the high values of deadrise angle of interest to SES sidehulls.

The form shown in Equation 8.79 was found to be more generally applicable for use in the predictions for sidehull SES hull forms. Again, the proviso is that none of the experimental data by the researchers was done for asymmetric hull forms operating in asymmetric water levels and pressurized air cushions.

From this relationship, the lift (dynamic + buoyant), (Δ_{d+b}) from the sidehulls can be expressed as:

$$\Delta_{d+b} = C_{L\beta}\tfrac{1}{2}\rho_W V^2 S_{SH} \tag{8.80}$$

Where $C_{L\beta}$ is given by Equation 8.79 for both dynamic and buoyant forces. Note that the skin friction drag is not included in this relationship.

The associated drag term from this lift is given by:

$$D_{d+b} = \Delta_{d+b} \tan\tau \tag{8.81}$$

Application of Planing Hull Results to SES Hullforms It has already been discussed that the applicability of conventional planing surfaces data to SES hull forms is somewhat circumspect in the absence of actual experimental data but it is believed that a reasonable approximation can be made if the two asymmetrical sidehulls are treated as one symmetrical hull for the purposes of determining the dynamic lift for the ship. Figure 8.34 illustrates this approach using the SES-100B hull form. This will need verification by model test of course, but is a first estimate for hull lift of craft with asymmetric water levels and hull geometry.

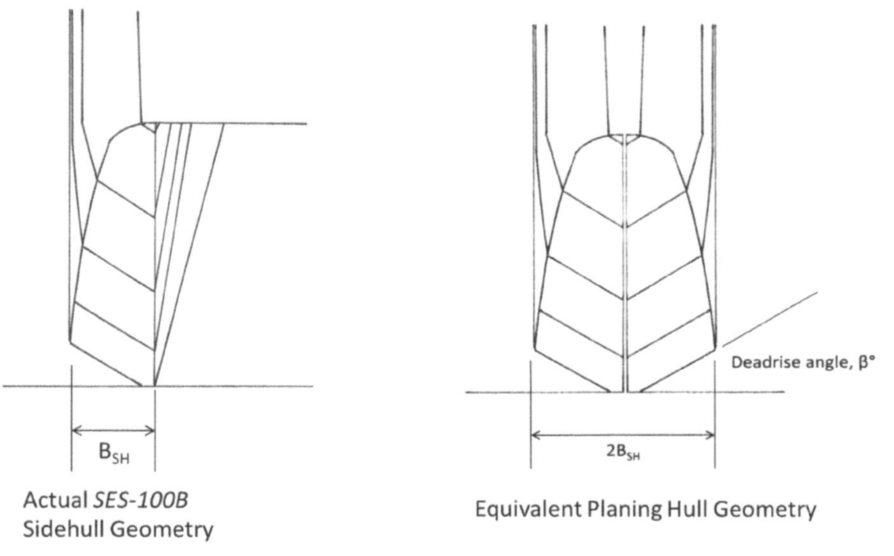

Figure 8.34: SES-100B Actual Sidehull and Equivalent Planing Hull

This completes the available data for the dynamic and buoyant lift of SES sidehulls using the available data base of planing flat plates and prismatic hulls. The remaining drag term is the skin friction drag (D_{SF}) which, as will be discussed, is strongly geometry dependent.

Sidehull Skin Friction Drag The skin friction (second term in Equation 8.46) depends on the wetted area (S_{SH}); the skin friction coefficient (C_f), that depends on surface roughness and speed, and the trim angle (τ). Because the skin friction acts in the plane of the flow, the force in the horizontal or drag direction, requires resolving by "cos τ".

To aid the discussion, Figure 8.35 shows the main geometry under discussion. Later variations in the geometry will be introduced as needed. At the lower speeds and at high trim angles, the dominant term is the pressure drag ($\Delta \tan \tau$) but at higher speeds and with lowered trim angles more applicable to high speed air cushion craft, the skin friction drag dominates and this drag term will be presented here. It is important to note that while the pressure drag term is minimized at speed by design, the sidehull must still retain sufficient vertical force (Δ) to provide the necessary restoring roll moment M_Φ to the craft (see Equation 8.41), otherwise the same instabilities experienced by early craft designs would be incurred. This is especially important in high speed air cushion craft designs where the response times in the complex interactions in roll, pitch and yaw could cause catastrophic accidents.

The skin friction drag of the sidehulls (D_{SKsf}) is dependent upon the wetted area (both inside and outside) of the sidehulls. This wetted area (S_{SH}) is a complex function of the geometry of the sidehulls and the operating conditions. There are three main components to this skin friction drag. The first is due to the immersion depth (d_i) of the sidehull relative to the mean wave slope of the cushion; the second is due to the wetted area above the mean wave slope on the outside of the sidehull; and third, the drag generated by the spray pattern generated by the sidehull called "splash-up".

From Figure 8.35 (upper sketch), the wetted area on the inside of the craft (per sidehull) is given by:

$$S_i = K_i d_i L_{SH} \tag{8.82}$$

In this simplified equation, a shape factor (K_i) has been included, to account for any increased area inside the cushion area such as that shown in Figure 8.16 for the

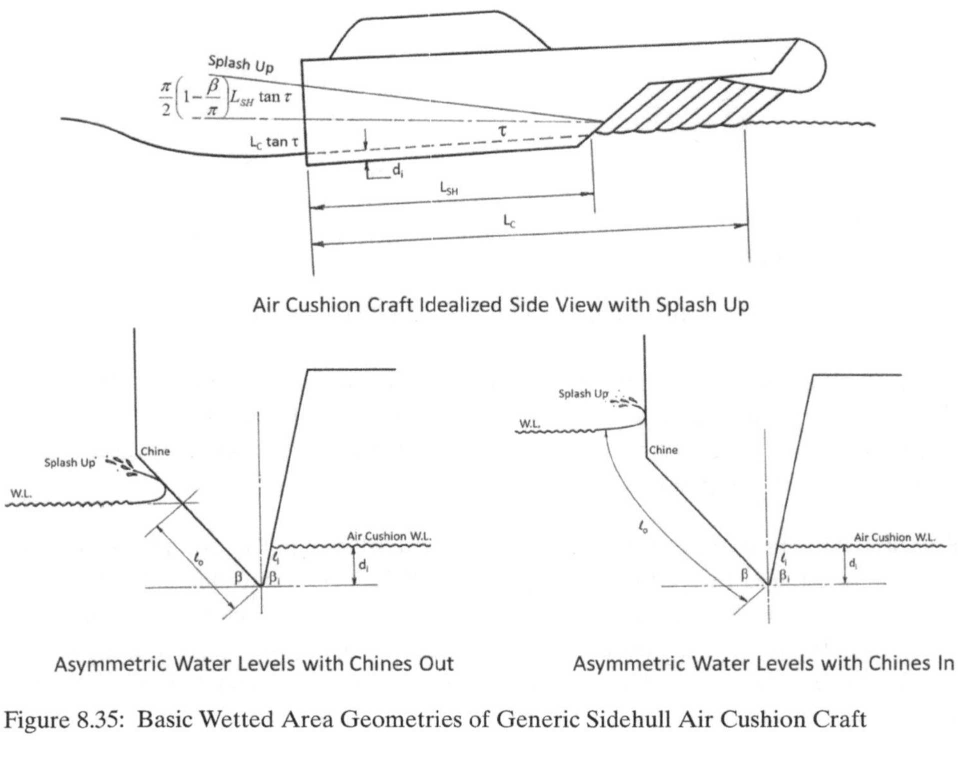

Figure 8.35: Basic Wetted Area Geometries of Generic Sidehull Air Cushion Craft

BH-110. If the inside of the sidehull is sloped inward at some angle (β_i) such as shown in Figure 8.16, then the shape factor would be:

$$K_i = \frac{1}{\sin\beta_i} \tag{8.83}$$

The wetted area on the outside of the sidehull (S_o) would have a similar wetted area below the mean wave slope plus the two wetted areas shown in Figure 8.35. Without the splash-up area, the outside wetted area (per sidehull) would be given by (approximately):

$$S_o = K_o\left(d_i L_{SH} + \left[\tfrac{1}{2} L_C{}^2\tan\tau - \tfrac{1}{2}(L_C - L_{SH})^2\tan\tau\right]\right) \tag{8.84}$$

The shape factor on the outside of the sidehull would depend on whether the craft is operating with "chines in" or "chines out". For the "chines out" case (see Figure 8.35, left hand sketch), the wetted area would be vertical depth increased by the shape factor:

$$K_o = \frac{1}{\sin\beta} \tag{8.85}$$

For the case of "chines in" operation (see Figure 8.35, right hand sketch), the determination of the wetted width (ℓ_o) would best be estimated from the particular shape under consideration and aided by model test. In the above formulation of the sidehull geometry presented in Figure 8.35, it is recognized that there are many variations possible on sidehull geometry. Figure 8.36, taken from the author's 1967 patent[35] on the Partial Length Sidehull (PLS) shows just three possible variations, all of which would require a different mathematical formulation of the wetted areas.

As seen In Figure 8.36, the sidehull could be shaped with "steps" familiar to planing hull and seaplane designers alike (see middle sketch) or the sidehull could be

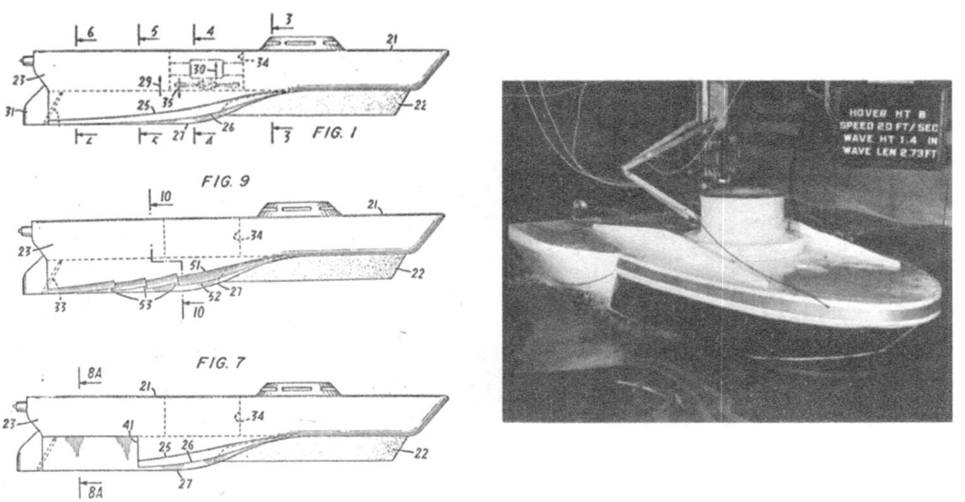

Figure 8.36: Partial Length Sidehull SES and Model Test (1965–1966)
Photo permission by New York Academy of Sciences

positioned anywhere along the full length of the ship to control the location of the hydrodynamic center of pressure (CP) for stability and control purposes (see lower sketch).

While it is recognized that such geometric variations are possible and would have to be considered in any particular design, it is useful to stay with the simplified (and more usual) shaping of the sidehull as shown in Figure 8.35 and Equation 8.83 such that some general guidelines of sidehull drag can be determined. Again, even in this simplified representation of the sidehull, the geometry is assumed to be such that the deadrise angle (β) is constant along the hull. Adjustments can be made for any particular shaping.

The third area of wetting is that caused by "splash-up" as shown in Figure 8.35. As might be expected the determination of splash up is not easily made and many researchers over many decades have formulated different methodologies for producing an easy to use formula for splash-up. The late Peter Payne, a renowned engineer in many areas and disciplines, tackled this problem and the reader is referred to his work on the subject where he has challenged many of the earlier formulations. His 1994 paper[36] summarizes that earlier work and puts forward new formulations.

The subject is complex and so only a general approximation used with success in seaplane (flying boat) design in the UK will be used here for easy use.

Some indication of the energy loss due to this spray pattern can be seen in Figure 8.37 which shows the sidewall hovercraft, HM 2 operating in relatively calm water at about 40 knots. The complex shape of the wetted area from this spray pattern is clearly visible in Figure 8.37.

The splash-up factor is given as a factor related to the deadrise angle and the depth of the depression, in which case the increased wetted area on the outside of the sidehull would be given by (per sidehull):

Figure 8.37: Splash-Up Wetting of Sidehulls on HM.2 Sidewall Hovercraft

$$\text{Splash-Up wetted area} = \frac{\pi}{4}\left(1 - \frac{\beta}{\pi}\right) L_{SH}^2 \tan\tau \tag{8.86}$$

The total wetted area on the outside of the sidehull (per sidehull) would then be given by:

$$S_o = K_o\left(d_i L_{SH} + \left[\tfrac{1}{2} L_C^2 \tan\tau - \tfrac{1}{2}(L_C - L_{SH})^2 \tan\tau\right] + \frac{\pi}{4}\left(1 - \frac{\beta}{\pi}\right) L_{SH}^2 \tan\tau\right) \tag{8.87}$$

The total wetted area (both inside and outside) would then become, after some algebraic simplification, (per sidehull):

$$S_{SH} = K_i d_i L_{SH} + K_o\left[d_i L_{SH} + \tfrac{1}{2}\tan\tau\left(2L_C L_{SH} - L_{SH}^2\left\{1 - \frac{\pi}{2}\left(1 - \frac{\beta}{\pi}\right)\right\}\right)\right] \tag{8.88}$$

The skin friction drag (D_{SHsf}) is given by:

$$D_{SHsf} = C_f \tfrac{1}{2} \rho_W V^2 S_{SH} \tag{8.89}$$

In this formulation, the skin friction drag (D_{SHsf}) is seen to be a function of the coefficient of friction (C_f) that is a function of Reynold's Number (R.No.) and surface roughness and the wetted area (S_{SH}). The wetted area for the craft in this case would be twice the value given by Equation 8.87, which is for one sidehull. Because the drag components here have been expressed in terms of the drag-weight ratio, then the skin friction drag for a sidehull air cushion craft can be written (per ship):

$$\frac{D_{SHsf}}{W} = C_f \rho_W g \frac{F^2}{(p_C/L_C)} (L/B)_C (L_{SH}/L_C)^2 \times\left[\left(\frac{d_i}{L_{SH}}(K_i + K_o) + \tfrac{1}{2}\tan\tau\left\{2\frac{L_{SH}}{L_C} - \left(1 - \frac{\pi}{2}\left[1 - \frac{\beta}{\pi}\right]\right)\right\}\right)\right] \tag{8.90}$$

While somewhat cumbersome, Equation 8.89 collects all the geometric factors of the (partial) sidehull air cushion craft into one equation and is easily calculable. Of significance, is the fact that the skin friction drag of the sidehulls is intricately tied to the air cushion parameters through the trim angle (τ), where:

$$\tan\tau = \frac{D_{wave}}{W} \tag{8.91}$$

The equation for the induced wave drag (D_{wave}) of the pressurized air cushion has been provided earlier (see Equation 8.8 and Equation 8.13). This emphasizes the difficulty of analyzing the drag of the sidehull separately from the induced wave drag of the cushion. Standard texts on fluid dynamic flow will provide the various equations developed by many researchers for the skin friction coefficient.

The form found useful is that developed by Schultz-Grunow[37] (for flow along planer surfaces) which works well for air cushion craft. The skin friction coefficient is given (for turbulent flow) by Schultz-Grunow as:

$$C_f = \frac{0.427}{\left(\log(R.No.) - 0.407\right)^{2.64}} \tag{8.92}$$

The Reynold's Number (R.No.) is defined as:

$$R.No. = \frac{VL_{SH}}{\nu} \qquad (8.93)$$

This Reynold's Number is based on the wetted length (L_{SH}) and the kinematic viscosity of the water, which can be taken to be $\nu = 2.359 \times 10^{-5}$ ft^2/sec. The normal operating conditions for such craft imply that the flow will not be laminar such that Equation 8.91 gives an accurate estimate of the skin friction coefficient (C_f). In some applications, an additional skin friction coefficient is added for surface roughness of craft in normal operating marine conditions to account for deterioration of the hull surface and marine growth. The additional skin friction coefficient varies between 0.0001 and 0.0005 over and above the value given by Equation 8.92. Sighard Hoerner[38] provides an excellent treatment of the subject providing the various equations for skin friction put forward by Schoenherr, Prandtl, von Karman and others.

Because Equation 8.92 is somewhat difficult to visualize, Figure 8.38 shows the skin friction coefficient (as provided by Schultz-Grunow) for those values of Reynold's Number above the transition regime from laminar to turbulent flow.

By way of a numerical example, compare the skin friction coefficients between a 100 ton air cushion craft and a 3000 ton air cushion craft. By Froude scaling, these two craft are scaled by approximately 3:1. If the 100 ton craft was operating at 58 knots then the 3000 ton craft would scale to 100 knots by Froude scaling. This would increase the Reynold's Number between the two craft by about $3^{3/2}$ or slightly greater than 5:1. From Figure 8.38 this would give about 25% drop in skin friction

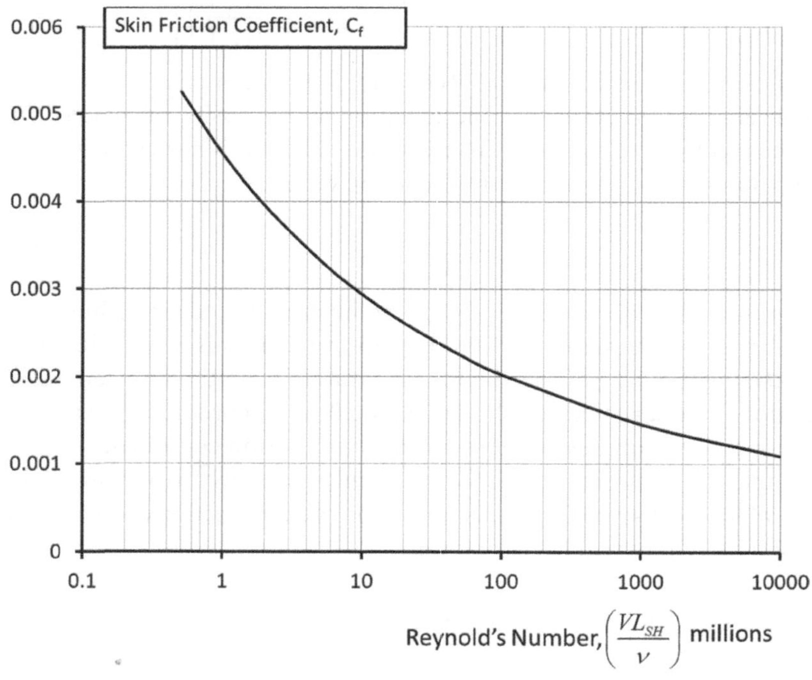

Figure 8.38: Skin Friction Coefficient for Turbulent Flow (Schultz-Grunow)

coefficient for the larger craft. The different scaling laws for each of the drag components in air cushion craft make it difficult to give blanket scaling of the performance. This is an important consideration when comparing performance numbers between different vehicles.

Sidehull Wavemaking Drag There has been no model testing of isolated sidehulls operating with asymmetric water levels (representing inside and outside of the air cushion) but some work on thin hulls and their wavemaking characteristics was conducted by the US Navy in 1952[39] (unrelated to hovercraft and SES development) that provides some indication of the magnitude of this drag component. This work by Weinblum, Kendrick and Todd showed that the wavemaking drag of thin slender hulls could be represented by the expression,

$$D_{SHwave} = \frac{8\rho_W g}{\pi}\left(\frac{B_{SH}{}^2 d_{SH}{}^2}{L_{SH}}\right)C_{WM} \tag{8.94}$$

This expression (for a single sidehull) shows that the drag depends on the geometric characteristics of sidehull beam (B_{SH}); sidehull length (L_{SH}), sidehull immersion depth (d_{SH}) and a drag coefficient (C_{WM}) that depends on Froude Number (based on sidehull length) and shape.

Figure 8.39 shows how this drag coefficient (C_{WM}) varies with both Froude Number and shape. If the sidehull wavemaking drag is expressed in the more familiar non-dimensional form of drag-weight ratio, then the drag can be written:

$$\frac{D_{SHwave}}{W} = \frac{8\rho_W g}{\pi}\left(\frac{(d_{SH}/L_C)^2(L_C/B_C)}{(L_{SH}/B_{SH})^2(p_C/L_C)}\right)C_{WM} \tag{8.95}$$

In this form, it can be seen how the sidehull drag will depend on the cushion parameters; the relative dimensions of sidehull to cushion and the trim angle, as well as the shape and speed dependence through C_{WM}. The two shapes tested were the thin parabolic hull shape and a thin hull with a transom. From earlier discussion of both stability needs and propulsion system accommodation, this latter hull form is more likely to be applicable to SES design. The sidehull immersion (d_{SH}) requires some discussion. As seen in Figure 8.35, the sideull immersion will vary along the length of the sidehull (L_{SH}), the trim angle (τ) of the craft and the shape of the waterline. Again, it is cautioned that one can only interpret these results approximately for thin sidehulls operating in asymmetric waterlines from these symmetric waterline results.

It is assumed that the trim angle of the sidehull(s) is the same as the trim angle of the wavemaking air cushion, such that:

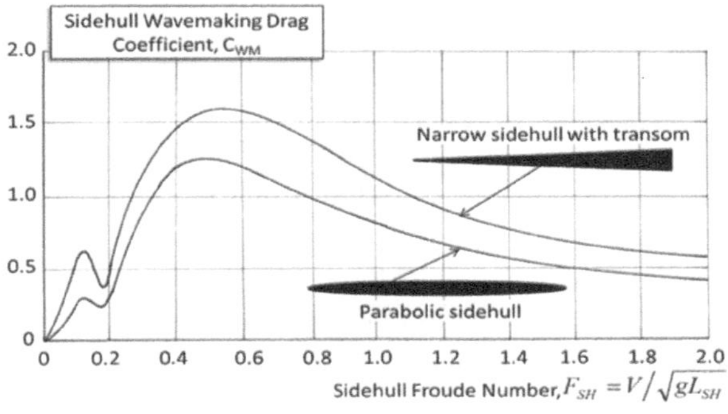

Figure 8.39: Sidehull Wavemaking Drag Coefficient

$$\tau = \tan^{-1}\left(\frac{d}{L_C}\right) = \tan^{-1}\left(\frac{D_{wave}}{W}\right) = \tan^{-1}\left(\frac{D_{wave}}{p_C S_C}\right) \tag{8.96}$$

This shows that the wavemaking drag of the sidehulls is intricately tied to the wave drag of the cushion (D_{wave}) discussed earlier through the trim angle (τ). In any particular case, any offset of the sidehull keel line from the static waterline would have to be included in the angle of sidehull trim.

The immersion depth for the sidehull (d_{SH}) can be approximated by the relationship:

$$d_{SH} = d_i + {}^1\!/_2 d \tag{8.97}$$

The two depth terms are as shown in Figure 8.35 and defined as:

d_i immersion depth relative to the mean wave slope (8.98)

d depth of cushion at stern = $L_c \tan\tau$ (8.99)

As a further complication, the immersion depth (d_i) is not an easily determined parameter. In practical designs, the immersion depth (d_i) is frequently increased over that needed in calm water by an amount sufficient to avoid leakage of the air flow (Q) needed for wave pumping.

The wave pumping requirement was discussed earlier in the treatment of the momentum drag (ρQV). A good "rule of thumb", that seems to work in most cases, is to assume that the immersion depth (d_i) be taken to be one half the average wave height for the design requirement of the craft.

8.2 Total Drag Estimation

Armed with the above equations for each of the drag components it is possible to predict the total drag of the sidehull SES to a reasonable accuracy recognizing the many empirical relationships that have to be employed in the drag estimation process.

The total drag of the ship, expressed as a drag/weight ratio would then be:

$$\frac{D_{total}}{W} = \sum\left(\frac{D_{wave}}{W}, \frac{D_{aero}}{W}, \frac{D_{mom}}{W}, \frac{D_{skirt}}{W}, \frac{D_{SH}}{W}\right) \tag{8.100}$$

The inverse of the drag/weight ratio is the lift-drag ratio for the craft:

$$\frac{L}{D} = \frac{1}{D_{total}/W} \tag{8.101}$$

These equations were applied to the SES-100B and compared to the two sets of full scale trials data collected on Lake Pontchartrain by the contractor in 1972 and by the US Navy trials on Patuxent River in 1976. This comparison is shown in Figure 8.40.

The difference between the two sets of trials data is discussed in more detail in Chapter 6 on the SES Program. A rigorous comparison between the different configurations was not conducted but the following points may be made. The early trials of the SES-100B were conducted on Lake Pontchartrain in New

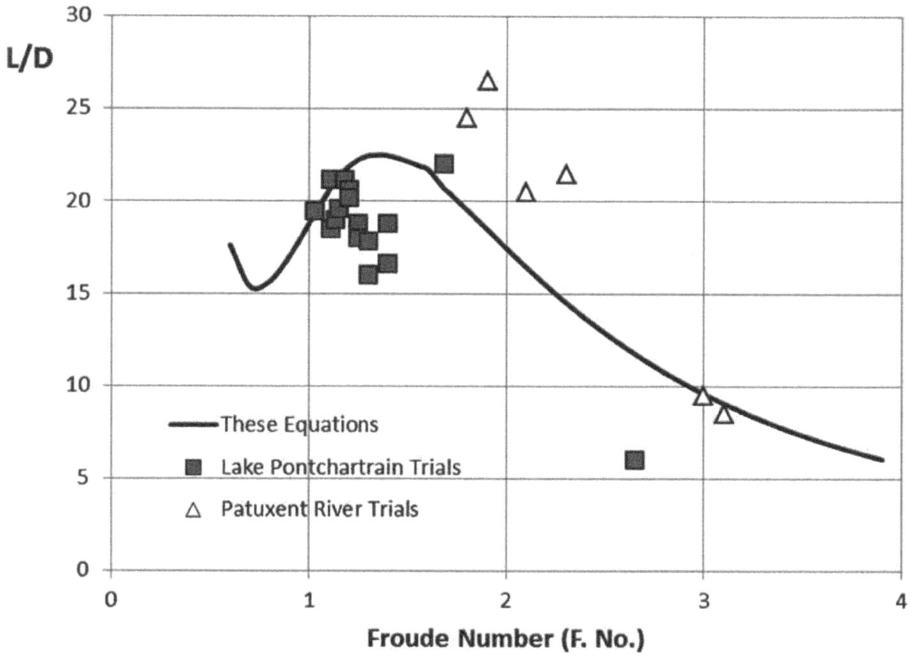

Figure 8.40: SES-100B Lift Drag Ratio (Prediction and Full Scale Trials Data)

Orleans where the water was shallow (approx 15 feet deep) and it was recognized that the wave drag would be higher than in deep water. The trials were also conducted in choppy seas (see photos in Chapter 6) and would contribute to higher drag. Also, for safety reasons the tests were conducted with a higher than optimum trim angle. The US Navy tests were conducted in the deep waters of the Patuxent River and tests were run in calm water. The experience gained over the intervening four years also allowed the trim angle to be more tightly controlled and lowered.

The scatter of the full scale trials data illustrates the problem of obtaining data from a dynamic craft at high speed where the drag is highly dependent on critical parameters that are not easily controlled. For example, a considerable effort was spent during the contractor trials on Lake Pontchartrain to determine the value of the maximum lift-drag ratio (L/D) and the speed at which it occurs. As can be seen from Figure 8.40, the drag variations were considerable not only because of the operational issues already cited but the vagaries of predicting the air cushion wave drag (D_{wave}) as the cushion pressure (p_c) and air flow (Q) varied due to the choppy waters. The extreme sensitivity of the dynamic craft drag to the sidehull immersion (d_i) also contributed to significant variations in the drag values as seen.

Enough confidence in the trials data is obtained from the data collected however (see Chapter 6 for more complete SES-100B trials data, specifically Figure 6.52) to say that a maximum value lift-drag ratio $(L/D)_{max}$ of between 22 and 29 is not an unreasonable number to assume. The equations developed above give a conservative value suitable for design purposes. As will be developed later, the speed for this maximum value of lift-drag ratio was found to be about 35 knots (F ~ 1.5) for the SES-100B. How these values vary with scale is discussed later. With these provisos, the complex interaction of all the drag components predicted by this method above and their sensitivity to geometric parameters has resulted in a reasonable estimate of both the functional form and the numerical value of the expected lift-drag ratio for SES.

8.3 Design Speeds for SES

The characteristic shape of the lift-drag ratio curve for dynamic vehicles like the SES-100B shown in Figure 8.40 highlights the fact that certain "optimum speeds" are possible that are not available to the conventional displacement hulled ship. There are three speeds of interest to the ship designer and they are displayed on Figure 8.41.

These three speeds can be characterized by the following definitions:

1. Maximum Speed	Determined by "thrust = drag" at maximum power setting from the propulsive system
2. Speed for Minimum Drag	Speed where drag is a minimum and where L/D is a maximum
3. Speed for Maximum Transport Efficiency	Speed where product of Lift-Drag Ratio (L/D) and propulsive efficiency (η_P) is a maximum

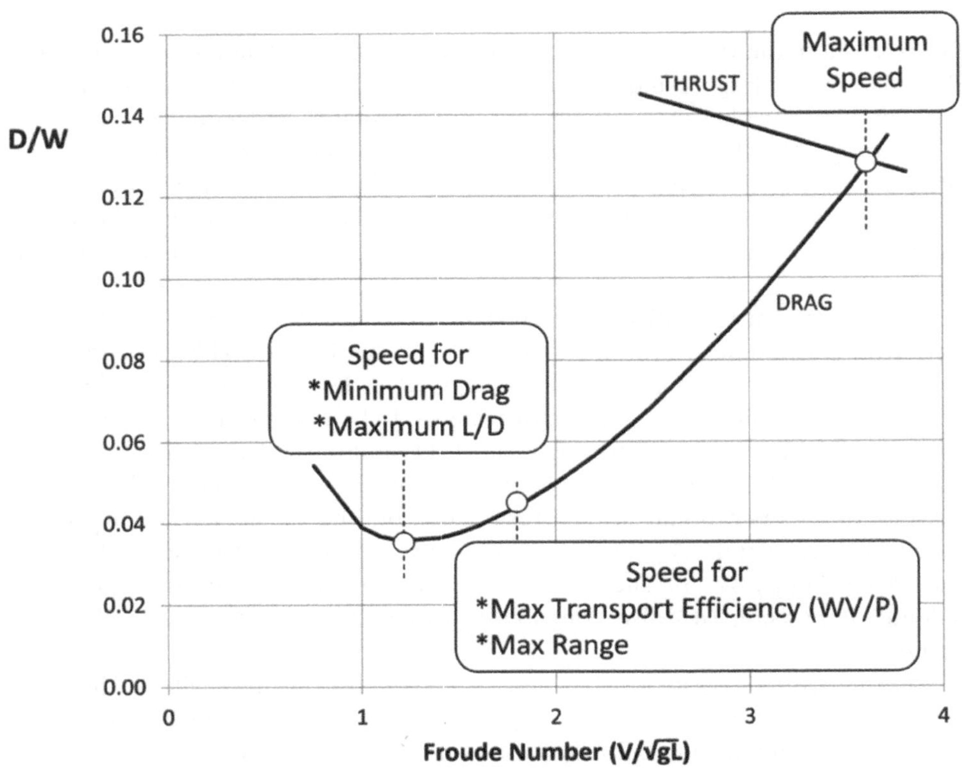

Figure 8.41: Design Speeds for Surface Effect Ships

The speed for Maximum Transport Efficiency (WV/P) will normally be higher than the speed for Maximum (L/D) because of the usual design practice to maximize the propulsive efficiency (η_P) at as high a speed as possible closer to the maximum speed of the craft. Two clarifying points on these definitions are necessary. The first relates to the "Maximum Speed" definition. In actual ship design (especially for military ships) it is necessary to clarify if the power setting of the engines is at "*maximum continuous power*" or "*military power*" or similar designations from the engine manufacturer as this will determine which thrust level is being discussed. The second clarification is that in the case of air cushion supported ships, the "lift-drag ratio" (L/D) must include the additional effect of the lift power to give the Effective Lift Drag Ratio $(L/D)_{EFF}$ as given earlier in this Chapter.

The design terms of significance are:

$$\text{Lift Drag Ratio}: \frac{L}{D} = \frac{W}{D} \tag{8.102}$$

$$\text{Effective Lift} - \text{Drag Ratio}: \left(\frac{L}{D}\right)_{EFF} = \frac{W}{D + \frac{\eta_P}{\eta_L}\left(\frac{p_C Q}{V}\right)} \tag{8.103}$$

$$\text{Transport Efficiency}: \frac{WV}{P} = \eta_P\left(\frac{L}{D}\right)_{EFF} \tag{8.104}$$

A final clarifying point is that the Maximum Range of the SES can then be determined with the additional inclusion of the fuel load term, "$\log_e \frac{W}{W-W_f}$" or its approximate term, "W_f/W". Each of these terms was outlined at the beginning of this Chapter as design determinants. In each case however, the underlying determining characteristic is the drag curve as shown on Figure 8.41. For convenience, the Range Equation (*for propeller driven craft*) is repeated here to show how each of the terms mentioned influence the range of the ship. The constants have been selected to give the range in nautical miles.

$$R = 325\left(\frac{1}{sfc}\right)\left(\frac{WV}{P}\right)\log_e\left[\frac{W}{W - W_f}\right] \tag{8.105}$$

The governing parameter is the lift drag ratio, (L/D), modified by lift system parameters (η_L, p_C, Q), which can be determined given the drag equations developed above and summarized in Equation 8.100.

Speed for Maximum Lift-Drag Ratio $(L/D)_{max}$

The speed for which the lift-drag ratio is a maximum is a key design parameter. The maximum value of the lift-drag ratio (L/D) will occur at some speed that will depend on the characteristics of the craft and its operating conditions. Each of the drag components described vary with speed in different ways. This can be seen readily in Table 8.2 using the equations already developed in this Chapter.

In what follows, attention will be restricted to the calm water case, although the same approach can be used for rough water operation.

Table 8.2: *Calm Water Drag Components Dependence on Speed (Froude Number)*

Drag Component as Function of Speed	"Constants" in Equations
$\frac{D_{wave}}{W} = \frac{K_1}{F^2}$	$K_1 = 0.35\left(\frac{4}{\rho_W g}\right)\left(\frac{p_C}{L}\right)\left(\frac{L}{B}\right)^{1/2}$
$\frac{D_{aero}}{W} = K_2 F^2$	$K_2 = C_{Do}\left(\frac{\rho g}{2}\right)\left(\frac{S_F}{S}\right)\left(\frac{1}{p_C/L}\right)$
$\frac{D_{mom}}{W} = K_3 F$	$K_3 = \frac{\rho\sqrt{g}\,L/B}{p_C/L}\left(\frac{Q}{L^{5/2}}\right)$
$\frac{D_{SK\,wetted}}{W} = K_4 F^2$	$K_4 = 0.221x10^{-3}\left[\frac{1+L/B}{(L/B)^{1/2}}\right]\left(\frac{h}{C}\right)^{-0.34}\frac{1}{p_C/L}$
$\frac{D_{SK\,wavemaking}}{W} = K_5\left(\frac{D_{wave}}{W}\right)$	$K_5 = \frac{1.374}{(p_C/L)^{0.259}} - 1$
$\frac{D_{SH\,dynamic}}{W} = K_6\left(\frac{D_{wave}}{W}\right)^2 F^2$	$K_6 = \frac{0.012}{\sqrt{\lambda}}\left(\frac{\rho_W g}{2}\right)\left(\frac{S_{SH}}{S}\right)\cos^3\beta$
$\frac{D_{SH\,buoyancy}}{W} = K_7\left(\frac{D_{wave}}{W}\right)^2$	$K_7 = 0.0095\left(\frac{\rho_W g}{2}\right)\left(\frac{S_{SH}}{S}\right)\cos^3\beta$
$\frac{D_{SH\,friction}}{W} = \left[K_8 + K_9\left(\frac{D_{wave}}{W}\right)\right]F^2$	$K_8 = C_f\frac{\rho_W g}{p_C/L}\left(\frac{L}{B}\right)\left(L_{SH}/L_C\right)^2\left(\frac{d_i}{L_{SH}}\right)(K_i + K_o)$
	$K_9 = C_f\left(\frac{\rho_W g}{2}\right)\frac{(L/B)\left(L_{SH}/L_C\right)^2}{p_C/L}\left[2\frac{L_{SH}}{L_C} - \left(1 - \frac{\pi}{2}\left\{1 - \frac{\beta}{\pi}\right\}\right)\right]$

From Table 8.2, it can be seen how each of the drag components vary with speed. Some vary directly with the speed (Froude Number, $F = V/\sqrt{gL}$); some are proportional to F^2: some are proportional to the inverse of F^2 and so on. The "constants" of proportionality depend on craft geometric characteristics and other known properties (e.g. C_{Do}, C_f). In order to make the following analysis of the lift-drag ratio (L/D) tractable, some simplifying but important assumptions were made. The first is that the trim angle (τ) is the same as the slope of the air cushion, i.e.:

$$\tan\tau = \frac{D_{wave}}{W} \tag{8.106}$$

This introduces a cumbersome equality into the drag equations where:

$$\tau^{1.1}\tan\tau = 0.0116\left[\tan^{-1}\left(\frac{D_{wave}}{W}\right)\right]^{1.1}\left(\frac{D_{wave}}{W}\right) \tag{8.107}$$

While a cumbersome expression, as a result of the empirical relationship introduced by Equation 8.66, it can be readily handled in any computerized spreadsheet and shows the continued importance of the trim angle in the drag calculations.

It is also assumed that for the skin friction term, the alignment of the flow along the sidehulls can be treated as if in the plane of the ship such that Cos τ =1 to keep the algebraic complexities to a minimum. Because of the low trim angles around the design speeds under discussion (approximately 1–2 degrees) these simplifying assumptions do not mask the essential features which will determine the desired design speeds for SES (*speed for maximum L/D; speed for maximum range*, etc).

With these assumptions, the total drag of the ship can be written as:

$$\frac{D_{total}}{W} = \frac{a_1}{F^2} + a_2 F + a_3 F^2 + a_4 + \frac{a_5}{F^4} \tag{8.108}$$

The constants a_1, a_2, a_3, a_4,a_5 are easily determined from the values given in Table 8.2 and are:

$$a_1 = K_1(1 + K_5 + K_1 K_6) \tag{8.109}$$

$$a_2 = K_3 \tag{8.110}$$

$$a_3 = K_2 + K_4 + K_8 \tag{8.111}$$

$$a_4 = K_1 K_9 \tag{8.112}$$

$$a_5 = K_7 K_1^2 \tag{8.113}$$

The speed at which the lift-drag is a maximum is also the speed for minimum drag, hence will occur when the derivative of the drag curve is zero. That is:

$$\frac{\partial\left(^{D_{total}}/_{W}\right)}{\partial F} = 0 = -\frac{2a_1}{F^3} + a_2 + 2a_3 F - 4\frac{a_5}{F^5} \tag{8.114}$$

This becomes an intractable equation to solve for the speed (Froude Number) but certain practical assumptions can produce some easy-to-use design formulae.

While it was shown earlier that the dynamic lift and buoyancy forces from the sidehulls were essential to achieve static stability (especially in the transverse plane), the drag from these lift forces are small compared to the other terms in the drag equation. This allows the terms (a_4, a_5) to be dropped, which greatly simplifies the drag equation. The derivative in that instance then becomes:

$$\frac{\partial\left(^{D_{total}}/_{W}\right)}{\partial W} = 0 = -\frac{2a_1}{F^3} + a_2 + 2a_3 F \tag{8.115}$$

The speed for minimum drag (and maximum lift-drag ratio) can then be determined from the solution to the quartic equation:

$$2a_3 F^4 + a_2 F^3 - 2a_1 = 0 \tag{8.116}$$

While Equation 8.116 can be solved by normal means of solving quartic or bi-quadratic equations it is still a cumbersome process with little insight gained as to the driving design parameters in the equations. By making some practical assumptions as to the relative magnitude of the drag components, it is possible to derive some "rules of thumb" with some approximate solutions to the speed equation.

Because the momentum drag (D_{mom}) is a small fraction of the total drag (typically around 10% or less) and tends to be a slow function with speed, then it is possible to drop that term recognizing that any estimate without the momentum drag would tend to give a slightly higher speed than if included. In which case, the term (a_2) can be dropped and the simplified equation for the total drag is approximated by:

$$\frac{D_{total}}{W} = \frac{a_1}{F^2} + a_3 F^2 \tag{8.117}$$

Which, upon inversion, gives the lift drag ratio:

$$\frac{L}{D} = \frac{F^2}{a_1 + a_3 F^4} \tag{8.118}$$

This simplified equation is useful to determine the main design parameters in determining the lift drag ratio for the craft. It also provides an estimate for the numerical value, recognizing that the terms involving a_2, a_4 and a_5 have been dropped for ease of algebraic solutions. These dropped terms are related to the buoyancy forces, effect of deadrise on sidehull drag and the momentum drag for the craft. These must be included in any final drag analysis but can be dropped for ease of analysis for early design purposes. This formulation tends to *underestimate* the value of the actual lift drag ratio.

By equating the derivative with respect to Froude Number of Equation 8.117 to zero yields the solution for the speed for maximum lift-drag ratio as:

$$|F|_{(L/D)\max} = \sqrt[4]{\frac{a_1}{a_3}} \tag{8.119}$$

After some algebraic manipulation, the speed where the lift drag ratio is a maximum can then be written:

$$|F|_{(L/D)_{\max}} = 0.136\sqrt[4]{\frac{\left(p_C/L\right)^2\left(L/B\right)^{1/2}}{\tfrac{1}{2}C_{Do}\left(\rho/\rho_W\right)\left(S_F/S\right) + C_f\left(L/B\right)\left(L_{SH}/L_C\right)^2\left(d_i/L_{SH}\right)\left(K_i + K_o\right)}} \tag{8.120}$$

In Equation 8.120, it has been taken that the skin friction coefficient (C_f) to be an average value over the speed range for any particular craft or ship. By way of illustration, when the parameters of the ship SES-100B are inserted into Equation 8.120 it predicts the speed (Froude Number) for maximum L/D to be at F = 1.6. The full equations where the momentum drag is included gives a speed of F = 1.4 as seen on Figure 8.40. This is a general rule in that Equation 8.119 will slightly *overestimate* the speed for maximum lift-drag ratio because of the simplifying assumptions made.

Earlier, in the determination of the correct flow (Q) for the SES a special case of constant flow coefficient ($C_Q = Q/VS$) was considered which assumes a different variation of momentum drag (D_{mom}) with speed which could be integrated into the above equation but it is not a general rule and the approximate form developed here is preferred as a guide in design.

Maximum Lift Drag Ratio

Once the speed at which the lift-drag ratio is a maximum is determined, it becomes a simple matter to determine the value of $(L/D)_{max}$ by substitution of Equation 8.119 into Equation 8.118 to give:

$$\left(\frac{L}{D}\right)_{max} = \frac{1}{2\sqrt{a_1 a_3}} \tag{8.121}$$

As before, this result has dropped the (a_2, a_4, a_5) for algebraic ease only. This results tends to slightly *underestimate* the actual value of $(L/D)_{max}$ but is sufficiently accurate for early design work. These "quick design" rules can be displayed for the lift-drag ratio curves of SES in Figure 8.42

The actual predicted curve for the SES-100B taken from Figure 8.40 has been used in Figure 8.42 merely to indicate the types of numbers that can be expected for a high speed SES. The equations for maximum lift-drag ratio $(L/D)_{max}$ and the speed at which it occurs $(F)_{L/Dmax}$ are the general forms developed from the drag components. The maximum speed for the craft is determined from the thrust levels available. The other design speeds of interest, namely speed for maximum Effective Lift Drag Ratio $(L/D)_{EFF}$ and speed for Maximum Range $(V_{max\ range})$ cannot be displayed on Figure 8.42 as they are vehicle specific and depend on the lift system (η_L, p_C, Q) and propulsion system (η_P, P) characteristics of the particular craft being analyzed. But the equations developed for $(L/D)_{max}$ and the

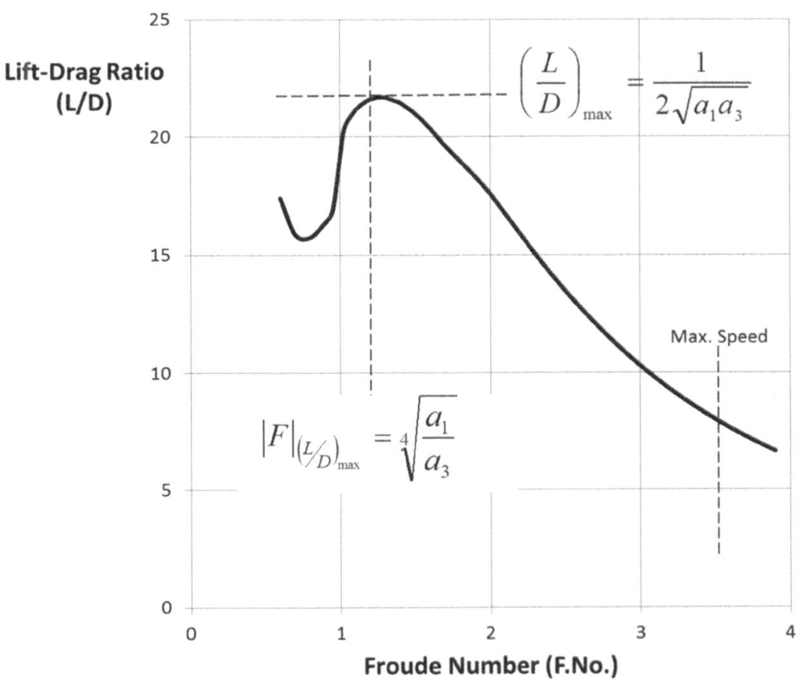

Figure 8.42: Lift Drag Ratio and Design Speeds

speed at which it occurs can be determined readily from the above results as displayed.

IMPORTANT NOTE. It is the term that contains the all-important flow (Q) term that incurs the need to solve a quartic equation and it is emphasized that dropping that term to get a closed solution as provided in Equations 8.119 and 8.121, is a device to get preliminary results. It must be left for other writings to include the flow term in a closed form solution for more detailed design rules.

8.4 Analysis of the Gabrielli-von Kármán Specific Resistance Curves

Now that an appreciation of how the lift drag ratio characteristics of dynamic craft such as the surface effect ship (SES) has been treated in some detail, it is informative to see where such dynamic craft "fit" in the general spectrum of vehicles of all types and to assess some of the claims made by designers of such craft seeking a solution to the "100 knots at sea" dream in the 1960–1980 era of heavy investment and activity. The 1950 seminal work by Prof Gabrielli and Prof Theodore von Karman[40] provides an insightful treatment of the problem. A brief history of these stalwarts and brilliant engineers in the field of engineering is given in Chapter 7. Almost as a labor of love (to use Prof von Karman's words) these two gentlemen prepared a very complete compilation of the drag and powering characteristics of a wide range of vehicle types. Their original data is shown in Chapter 7. This data covered ships, aircraft, automobiles, trains as well as horses, cyclists and other transportation vehicles. Because Gabrielli and von Karman's work was published in 1950, there is no data for air cushion supported craft (Sir Christopher's hovercraft did not appear until 1959). In 1967, the author prepared a compilation[41] of the available data on air cushion craft and other high speed marine craft (planing boats, hydrofoils and other types) and superimposed that data on the plots by Gabrielli and von Karman. Chapter 7 gives that data complete with identification of the specific craft involved. The reader is referred to Chapter 7 for the more detailed description of all the compilations. Figure 8.43, taken from Chapter 7 (Figure 7.17) shows the various groupings of interest to the subject at hand.

The original data by Gabrielli-von Karman was expressed, using their terminology as *specific resistance*, defined as:

$$\text{Specific Resistance}\,(\varepsilon) = \frac{P}{WV} \tag{8.122}$$

In the context of this book, the data has been inverted to give the Transport Efficiency (TE):

$$\text{Transport Efficiency}\,(\text{TE}) = \frac{WV}{P} = \frac{1}{\varepsilon} \tag{8.123}$$

In this work, it became clear that there were at least two distinct groupings of Transport Efficiencies; one related to sea vehicles and the other related to air vehicles. This can be seen from Figure 8.44.

What is being measured by the Transport Efficiency (WV/P) is the ability of the craft to transport its weight over a given distance ("work done") for a given power expenditure (work done per unit time). The higher the value of WV/P, the more efficient the craft in using its power to move the craft at speed.

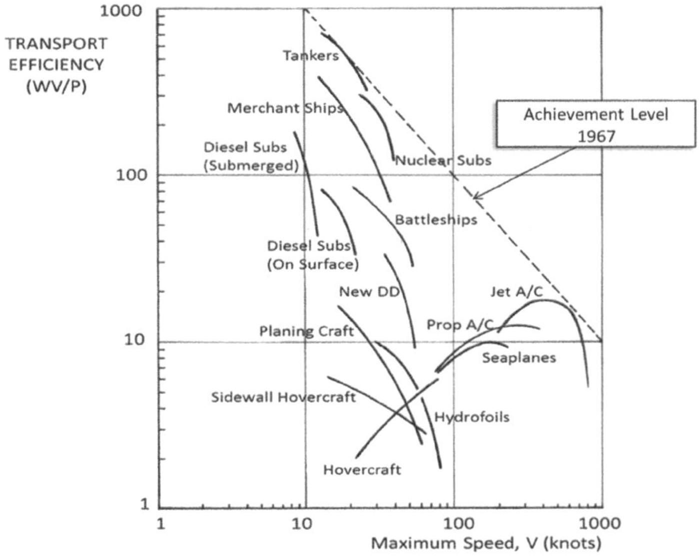

Figure 8.43: Groupings of Sea and Air Vehicles Transport Efficiencies

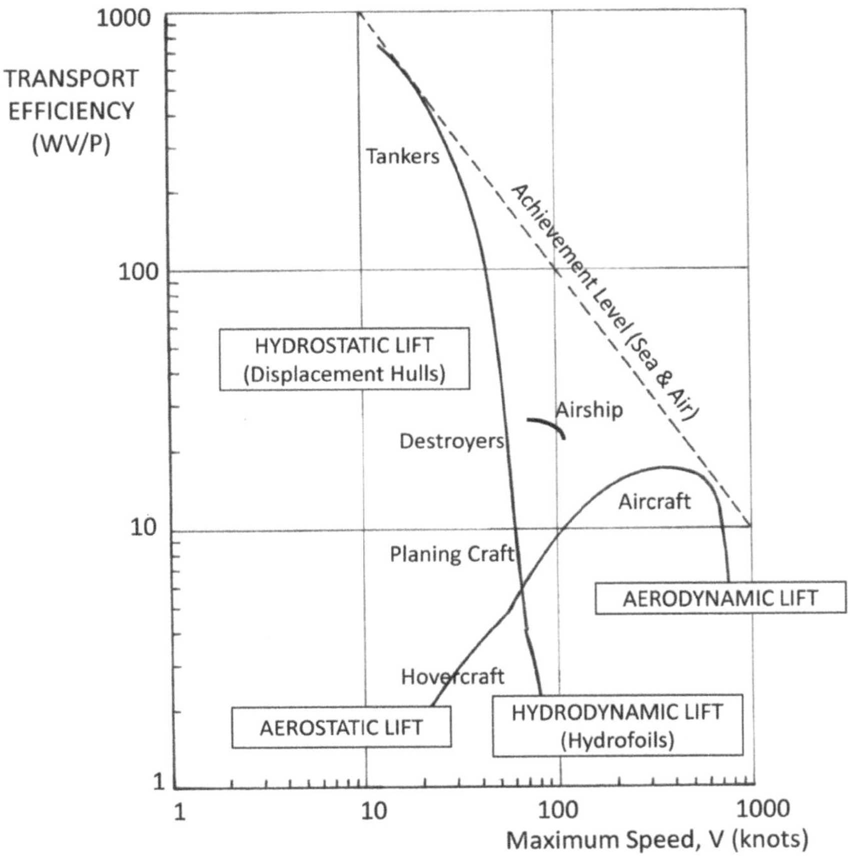

Figure 8.44: "Envelopes of Envelopes" of Sea and Air Vehicles Transport Efficiency

Aerodynamic and Hydrodynamic "Barriers"

In Figure 8.44, it is seen that the groupings of vehicle Transport Efficiencies are determined by whether the vehicle under discussion fits into one of four groupings determined by the dominant forces at work in their lifting mechanisms, viz: *hydrostatic, hydrodynamic, aerostatic* or *aerodynamic*. It is also noticed that there are essentially two "barriers" to increased efficiency as speed is increased. These "barriers" are not absolute as vehicles have been designed to push past these "barriers" but only at the expense of large amounts of power and corresponding loss in Transport Efficiency. These two "barriers" are related to "*wave drag*" but the term "*wave drag*" has completely different meanings with very different physical mechanisms at work in both air and water. Figure 8.45 shows these general boundaries of both the "sea" and the "air" vehicles and the "barriers" to increasing efficiency.

An explanation of these "barriers" clarifies the limits to vehicle performance.

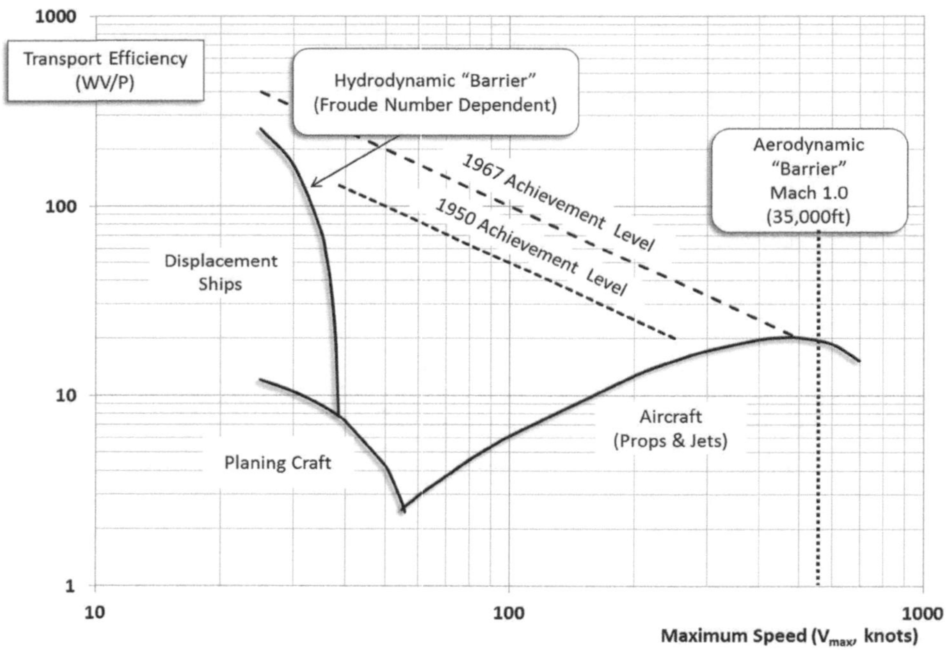

Figure 8.45: Hydrodynamic and Aerodynamic "Barriers"

Aerodynamic "Barrier" to Transport Efficiency

Aircraft are fully immersed in their medium (air) and as the speed is increased the air is compressed as the aircraft moves through the air and a pressure wave moves outward from the aircraft by molecular collision[g]. When the energy level reaches a

[g] The actual mechanism is described by Bernoulli's equations for compressible fluids and these are found in any standard text on Aerodynamics.

certain threshold the air molecules collapse and "*shock waves*" are formed around the aircraft. Because the pressure waves move at the speed of sound these shock waves form around the aircraft emanating from the nose of the aircraft at an angle given by $\mu = \sin^{-1}\left(\frac{1}{M}\right)$. There are other points of the aircraft that generate other shock waves but this simplified treatment explains the essential points. In this relationship for the angle of the shock waves, "M" is the Mach Number given by:

$$M = \frac{V}{a} \tag{8.124}$$

The speed of sound (a) is determined by the thermodynamic relationship

$$a = \sqrt{kgRT} \tag{8.125}$$

The constant (k) is the ratio of specific heats for air; (g) is the acceleration due to gravity; (R) is the gas constant and (T) is the absolute temperature. The temperature (T) varies with altitude and hence the speed of sound varies also (a ~ $\sqrt{T}$) such that at sea level, the speed of sound is approximately 660 knots on a Standard Day and about 575 knots at 35,000 ft. Aircraft tend to fly at such altitudes over long distances due to these variations in atmospheric properties for more efficient flight.

The *wave drag* of aircraft, referred to above, will increase rapidly as the speed of the aircraft approaches the speed of sound which at altitude will be about 575 knots. There is a "Divergence Mach Number" before M = 1 where the compressibility effects cause a rise in drag as shown in Figure 8.46 for a typical profile drag coefficient C_{Do} versus Mach Number for a set of swept wings[42]. The drag for the complete aircraft would follow a similar pattern. At lower speeds, the drag coefficient is that of the profile drag of the aircraft.

As speed is increased the compressibility effects begin to be exhibited and the (wave) drag starts to rise at some Divergence Mach Number (M < 1) that depends on the geometry and other factors of the aircraft. In the example shown, the Divergence Mach Number occurred at 0.70 to 0.90 depending on the sweep angle of the wing. The drag coefficient continues to rise with the addition of the wave drag component until the aircraft becomes supersonic (M = 1) after which the drag coefficient begins to fall (Figure 8.46). As aircraft designers push their designs to such high speeds it is seen that this increased drag causes a loss of Transport Efficiency (WV/P) as speeds approach the Divergence Mach Number which in this example would be at speeds of about 400 to 520 knots at 35,000 ft as shown in Figure 8.44 and Figure 8.45 that shows the significant fall-off in the boundary of Transport Efficiency of aircraft at the higher speeds (M → 1 and beyond).

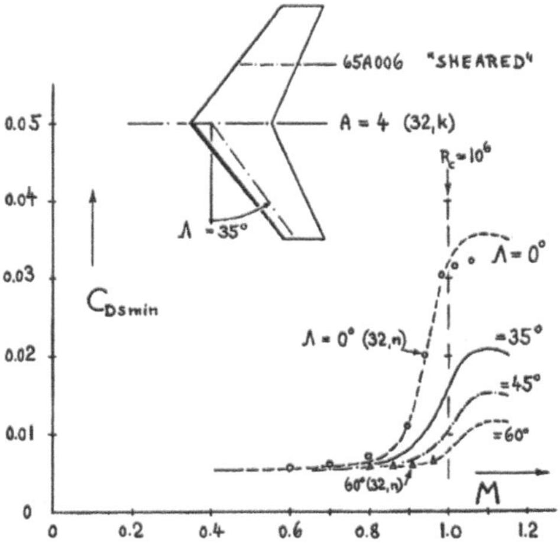

Figure 8.46: Typical Profile Drag Coefficient of Swept Wings in Air

Hydrodynamic "Barrier" to Transport Efficiency

A very different set of physical mechanisms are at work for ships that operate at the interface of the two mediums (air and water). A very different form of *wave drag* determines the boundaries of displacement hull forms and there is no similar speed where drag will rise and then start to fall. In the case of displacement hulls (ships) pushing through water the drag continues to rise with the continued fall off in Transport Efficiency as shown in Figure 8.44 (and in the original charts by Gabrielli and von Karman). In the case of water vehicles, there is no equivalent condition to the aerodynamic "barrier" of ($\frac{V}{a} = 1$). In the marine case, the "barriers" are the boundaries themselves that depend on more than speed. These boundaries depend on both Froude Number and the relationship between the ship and the waves it generates as it moves through the water. This resistance of the water against the ship is called "wave drag" and should not be confused with the very different "wave drag" of aircraft discussed above.

Much of the original work on the wave drag of displacement hulls as they move through water was done by Sir Thomas Havelock in 1908[43] who was instrumental in the work on moving pressure depressions discussed earlier in relation to the wave drag of air cushions. Sir Havelock showed that there was a complex interaction between both transverse waves and longitudinal generated waves as the hull moves through the water and the drag is caused by the addition and cancelation of the forces by these waves depending on whether the ship length and the wavelength were in phase or not. Such action causes "humps" and "hollows" in the drag curve of any particular ship. If a Froude Number based on wavelength is defined as ($F_\lambda = V/\sqrt{g\lambda}$) Sir Havelock showed that the "humps" in the drag curve formed at values:

$$F_\lambda = 1/\sqrt{2\pi n} \quad \text{where n} = 2, 4, 6, \ldots\ldots \tag{8.126}$$

In this relationship, λ is the wavelength of the induced waves (measured crest to crest). Using this formula, this gives humps in the drag curve at Froude Numbers ($F = V/\sqrt{gL}$):

$$F = 0.163,\ 0.199,\ 0.282,\ 0.399, \ldots\ldots \tag{8.127}$$

In actual practice, these humps do not extend into the higher speeds. Wigley[44] in 1942 conducted tests on particular hull forms and showed that these humps in the drag curve do indeed occur in the lower Froude Number range. Figure 8.47 shows one of Wigley's results for a particular hull form that matched Havelock's predictions to an excellent degree. As might be expected, while the functional form will be as Havelock and Wigley have shown, the specific numerical values of the humps in the drag curve and the speed at which they occur will depend on the actual hull shape. It has been a difficult task to analytically define hull shapes and it has become normal practice to characterize hull shapes by means of shape parameters such as block coefficients, prismatic coefficients, midship coefficients, waterplane coefficients and other ways of defining ship shapes. Two of these coefficients used by Wigley are:

$$\text{Prismatic Coefficient, } C_p = \frac{\nabla}{L_{pp}A_m} \tag{8.128}$$

$$\text{Midship Coefficient, } C_m = \frac{A_m}{Bd} \tag{8.129}$$

With such coefficients, it becomes possible to at least generally classify the types of hulls under consideration. In the definition of the Prismatic Coefficient, (∇) is the volume of the water displaced by the ship; (L_{PP}) is the length of the ship measured between the perpendiculars, and (A_m) is the cross-sectional area of the midship. Hence, this coefficient allows for the distribution of the volume of the ship to be categorized. A low value of (C_P) would mean a ship with full midship with fine ends. For the Midship Coefficient, (C_m), this coefficient expresses the ratio of the largest cross-section of the hull (midships) to a rectangle of the same beam (B) and depth (d) of the underwater hull. Hence, a low value of (C_m) would indicate a slender ship while a high value would indicate a boxy or bulky ship shape. The other coefficients provide further characterizations of the hull shapes, all of which would determine the magnitude of the drag values in Wigley's charts. The particular hull used by Wigley as shown is for a ship hull shape with ($C_P = 0.64$); ($C_m = 0.667$) and a beam to depth ratio ($^B/_d = 3$)

While clearly the above brief discussion is a simplified overview of the various characteristics of displacement hulls, the boundaries of the Transport Efficiencies shown in Figure 8.44 and Figure 8.45 show that ship designers have followed these "rules" in getting the most efficient hull that they can in their designs. Specifically, ship designers have sought to pick design speeds that avoid the hump drag values predicted by Havelock and shown by Wigley and selected design speeds more in line with the "hollows" in the drag curve.

As a further illustration, consider Figure 8.48 that shows the ship speeds of a wide range of ships as a function of their ship length. Curves of constant Froude Number corresponding to the humps in the drag curves predicted by Sir Havelock (as per Equation 8.127) have been included as the solid lines. A dotted line not related to the Sir Havelock "humps" corresponding to a Froude Number of F = 0.10 has been added to capture the cargo and container ships that are designed to stay below these hump drag speeds.

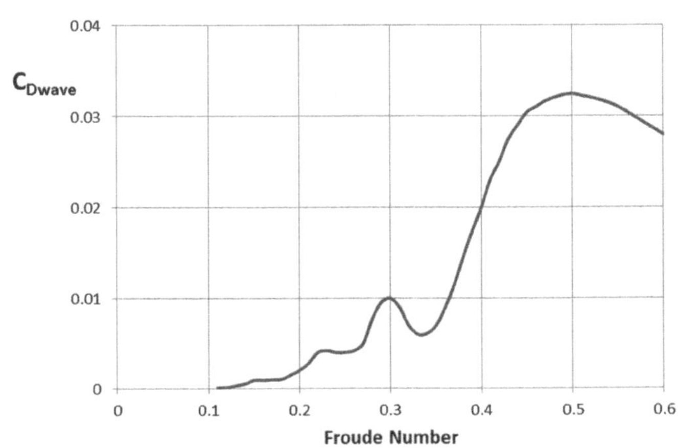

Figure 8.47: Wigley's Results on Wave Drag of a Displacement Hull

Some immediate observations are that most ships have followed these "rules" except in specific cases where deliberate departures were used in the

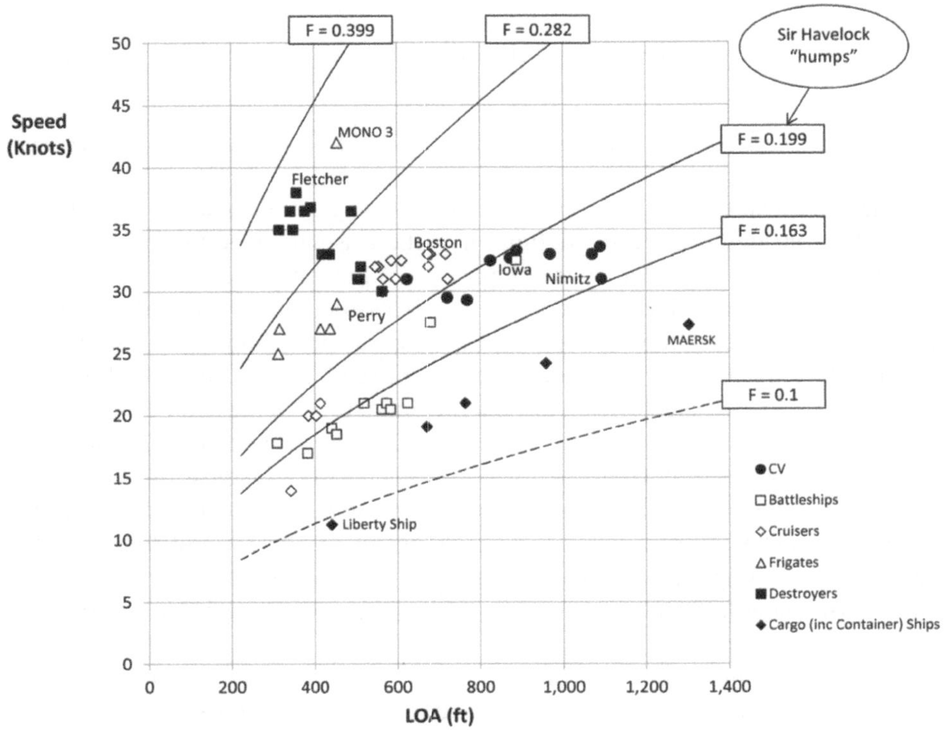

Figure 8.48: Ship Speeds and Sir Havelock "Humps"

design for some specific design purpose. The destroyers of WW II pushed "the envelope" to achieve speed but at a cost of power (as shown later). The cargo ships stayed away from the hump drag speeds for economical reasons. The Liberty Ship shows up dramatically, in that the ship was clearly underpowered for its ship length with a speed of only 11 knots that contributed to the loss of "market share" on the world's shipping trade that left the US in a declining competitive position after WW II. Chapter 7 discusses this major issue that led to the US Maritime Administration's seeking a high speed solution for the US merchant fleet. The apparent departure from the Sir Havelock curves for the aircraft carriers (CV) is probably due more to the policy of the US Navy not to be specific in quoting exact ship speeds in the Public Press for security reasons rather than any hydrodynamic reason. Carrier speeds are usually quoted as "30 knots or more", hence the leveling out of the quoted speeds for these ships. The lone data point labeled "MONO 3" is the baseline displacement frigate prepared for the ANVCE Project discussed in detail in Chapter 10. It is included here to illustrate that as for the WW II destroyers, the high speed of 42 knots was more the effect of overpowering of the ship for its length than by any improvement in hull form. Some license has been taken in this discussion in that the actual speeds for the hump drags will vary depending on hull shape (block coefficient, prismatic coefficient, etc) but the general trends are as indicated per the Wigley example.

Given these results by Havelock (1908) and verified by test by Wigley (1942) it becomes possible to understand the boundaries of the Transport Efficiency of displacement hulls and why there is an almost precipitous fall-off in Transport

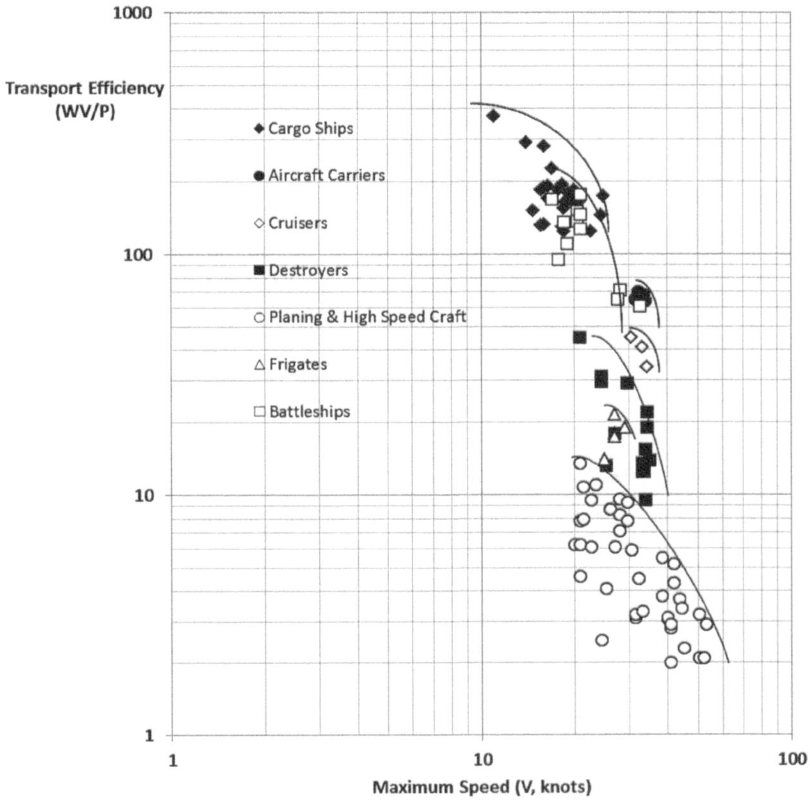

Figure 8.49: Transport Efficiency of Displacement Ships and High Speed Craft at Maximum Speed (knots)

Efficiency (WV/P) as the speed (Froude number) is increased to values much above $F \approx 0.60$–0.70. This can be seen by examining the data from a wide range of ships based on the work by Gabrielli and von Karman which showed that the various classes of ships (aircraft carriers, battleships, cruisers, etc) all tended to coalesce into specific groupings of Specific Resistance against speed (in mph). As would be expected, Figure 8.43 showed these same groupings and boundaries for the Transport Efficiency (the inverse of Specific Resistance) against speed (in knots). While the original paper by Gabrielli and von Karman did not show the specific ship data, a similar compilation using published data on ships has been gathered and shown in Figure 8.49. As might be expected the same groupings of the ship data appear, all showing a significant drop in efficiency as speed is increased.

The work by Havelock and Wigley and others would suggest that the ship designers would be influenced by the characteristic drag curve shown in Figure 8.47 in selecting their speeds for the ships. If the same data used in Figure 8.43 and Figure 8.49 is now plotted against Froude Number for these same ships the data shows a more consistent collapsing of the Transport Efficiency against the hydrodynamic parameter Froude Number. This is shown in Figure 8.50. It is seen immediately that the data of the several different types of ships collapses to a common "barrier" as the design stays generally in the "hollows" of the drag curves predicted by Havelock. If

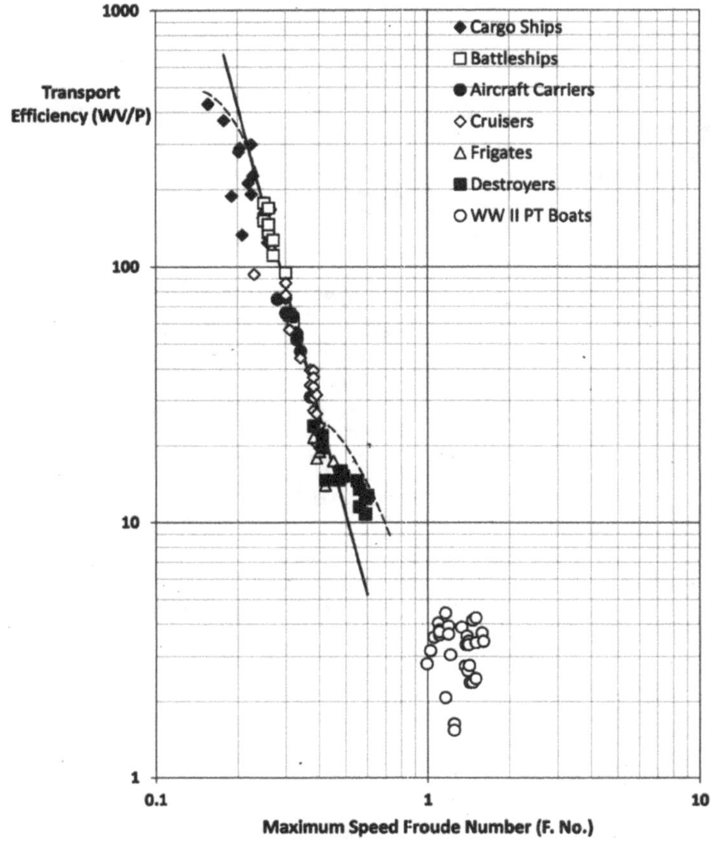

Figure 8.50: Transport Efficiency of Displacement Ships & Planing Craft at Maximum Speed Froude No.

a more detailed analysis was made of these ship designs it would be found that the many variations in hull shaping (through block coefficients, prismatic coefficients, etc) would cause the variations that are blurred in a common overview such as that shown from the available published data. Clearly the variations are small enough for the common trend line shown for displacement ships in Figure 8.50. There are two notable exceptions seen in Figure 8.50.

The first is in the "low speed" Froude Number range that is mostly associated with the cargo ship category (that includes tankers, container ships and bulk carriers). It is seen that the Transport Efficiency is lower than the trend line (shown by the upper dotted line). This is probably due to the designer's choice to use higher "bulkier shaped hulls" (e.g. higher block coefficients) to carry more cargo at the expense of "hydrodynamic efficiency" to improve the economics of the ship. The second "departure" from the trend line is in the "higher speed" group of destroyers (DD Class) where a conscious decision was made for designed destroyers of WW I and WW II to achieve high speed for the ship with sleek lines for hydrodynamic efficiency at the expense of ship volume (shown by the lower dotted line).

Finally, in those cases where speed was paramount it was necessary to forgo hydrodynamic efficiency and the specific lower group shown on Figure 8.50 is the planing craft groupings of WW II PT Boats. It is seen that the downward trend in hydrodynamic efficiency (Transport Efficiency) continues as speed is increased (and size or displacement is reduced). While the PT Boats are mainly planing craft the group includes some semi-planing craft which causes the spread in the data as the "draggier" hulls are included. The data on "high speed small craft" has been omitted from Figure 8.50 for clarity leaving the specific planing craft designed for speed in military applications (PT boats). The complete set of such high speed craft is shown

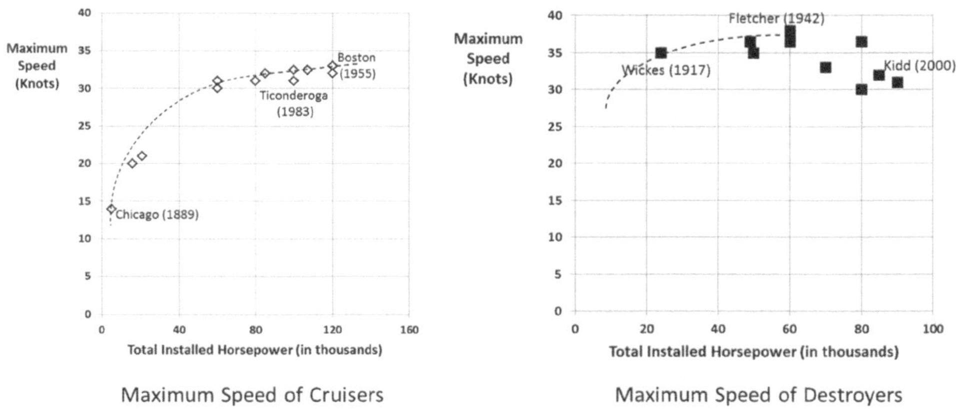

Figure 8.51: Maximum Speeds of Cruisers and Destroyers

in Figure 8.49 and the identification of the specific craft is given in Chapter 7, Figure 7.16 where the economic considerations are discussed. While the emphasis in this book is how to "efficiently achieve 100 knots at sea", it is informative to see where the base data lies after the best designers in the land have produced their ship designs to accomplish the mission whether for military craft or commercial transport. To indicate the significant drop in Transport Efficiency as a function of speed can be seen in the trend line shown in Figure 8.50 which can be represented by:

$$\left|\frac{WV}{P}\right|_{boundary} = \frac{0.675}{F^4} \tag{8.130}$$

Two examples, taken from the data set shown illustrate the problem facing the displacement ship designer in obtaining high speed. The maximum speeds attained by the Class of "Cruisers" and the Class of "Destroyers" are as shown in Figure 8.51.

In all cases it is seen that tremendous amounts of power are expended to achieve diminishing increase in speed as the design pushes against the hydrodynamic "barrier" shown in Figure 8.50 and Equation 8.130. In the case of the cruisers, the installed horsepower doubles and triples as the maximum speed design point is pushed for an extra 2–3 knots above 30 knots for the cruisers. The 1955 *Boston* (one of the *Baltimore* Class of heavy cruisers) at slightly less than 18,000 tons displacement required 120,000 HP to achieve its top speed of 33 knots. The later guided missile cruiser (CG) the 1983 Ticonderoga at over 9,500 tons displacement required 100,000 HP to achieve its top speed of 31 knots. The combination of weight (displacement), length and speed produced ships that pushed against the "barrier" shown in Figure 8.50 to achieve the efficiency sought by the ship designers. Larger expenditures of power would not have generated any measurable improvement in speed.

A similar characteristic in asymptotic speed improvement can be seen for the destroyer class of ships in Figure 8.51. During WW II emphasis was placed on speed and the DDs were capable of speeds up to 35 knots or more but only with extreme expenditure of power. The 1942 *Fletcher* Class required 60,000 HP to achieve its top speed of 38 knots for a ship displacement of slightly less than 3,000 tons. It is the

Fletcher Class that shows ships "pushing" past the trend line in Figure 8.50 at a sacrifice in Transport Efficiency ($\frac{WV}{P}$ ~ 13). Later designs of DDs, specifically the *Spruance* and *Kidd* Class (in 2000) accepted lower speeds (approx 30–31 knots) for a larger size (approximately 9,000 tons) to achieve higher Transport Efficiency ($\frac{WV}{P}$ ~ 24) that "fit" on the trend line – an almost 2:1 improvement in efficiency for a modest sacrifice in speed (approx 7 knots). "Pushing through" water at high speed proved prohibitive in power expenditure using displacement hulls and attention turned to planing on top of the water to achieve speeds above 30 knots. The Class of ships (or craft) shown in Figure 8.49 that achieved speeds of approximately 50 knots achieved this at the cost of two important design parameters: (i) size and (ii) Transport Efficiency. While the size of displacement ships displayed in Figure 8.50 and Figure 8.51 can be measured in "1000's" of tons, the planing craft class has had to restrict itself to "100's" of tons in order to keep the power expenditure to practical values. At the same time, despite the size reduction, the Transport Efficiencies of such planing craft are still very low. The drop in Transport Efficiency is at least 3:1 in most cases. Most of the WW II PT craft shown in Figure 8.50 had displacements of 35–120 tons.

In 1976, as part of the Advanced Navy Vehicles Concepts Evaluation (ANVCE) Program, the US Navy sought a planing hull ship design of large displacement to conduct military missions. The details of that design are provided in Chapter 10. The design, that was not produced, representing the best technology and design available at that time, was of a craft of 1109 long tons with a projected maximum speed of 55.8 knots. The projected Transport Efficiency of the design was $\frac{WV}{P}$ = 3.9 at that maximum speed. While that ANVCE design was not built, it is interesting to compare this design with the *Destriero*, a high speed private yacht that won the Blue Ribbon award for its non-stop run across the Atlantic without refueling at an average speed of 53.1 knots in 1992. Figure 8.52 shows the two craft. There is a danger, of course, in comparing two craft designed to different standards but a general appreciation of the problem facing the designer in achieving high speed can be seen. The PC-1 was a design study as part of the US Navy ANVCE Program

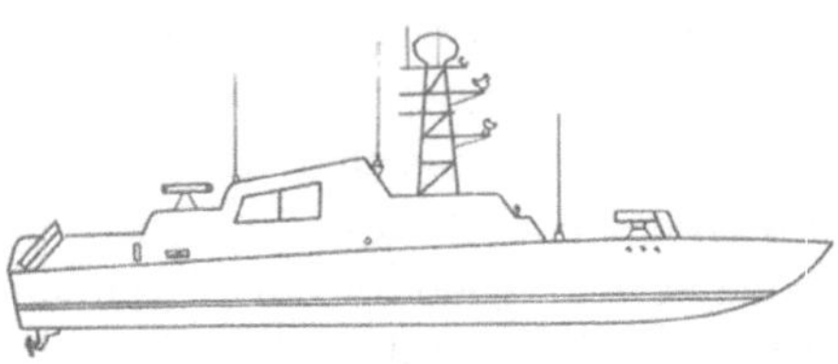

ANVCE US Navy Planing Ship Concept (PC-1)

Private Yacht *Destriero*
Photo permission by Fincantieri SpA

Displacement:	1,109 tons	Displacement:	900 tons
LOA:	250 ft	LOA:	220 ft
Max Speed:	55.8 knots	Max Speed:	59.4 knots
Installed Horsepower:	108,570HP	Installed Horsepower:	60,000HP

Figure 8.52: Comparison of Two Large High Speed Planing Craft

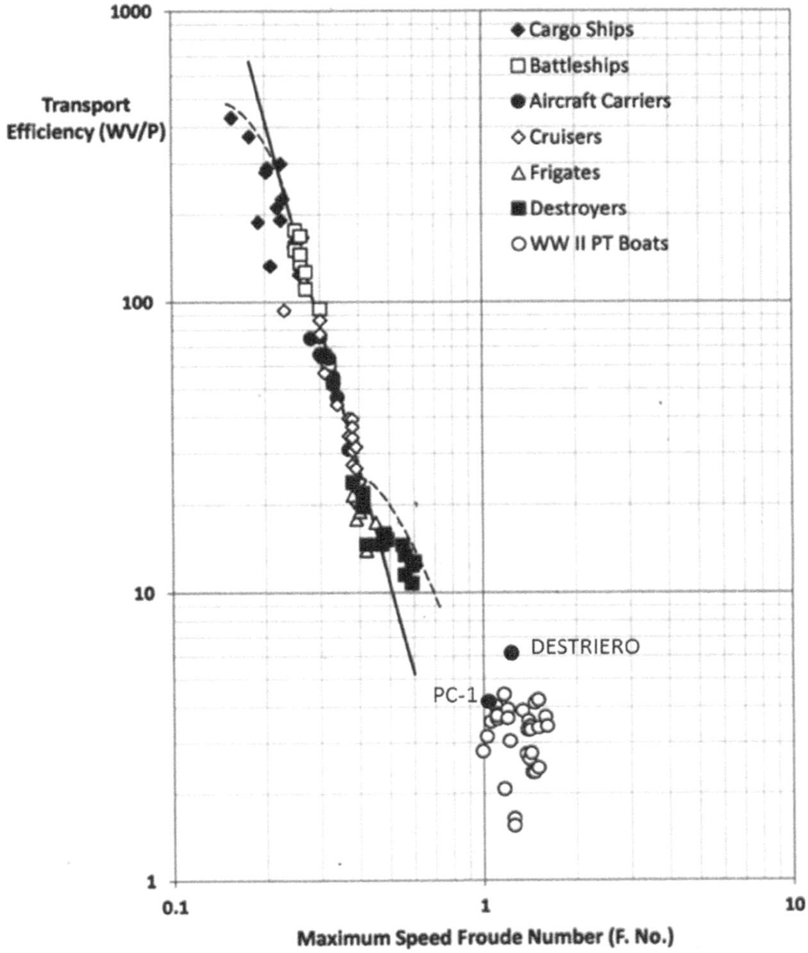

Figure 8.53: Large High Speed Planing Craft Transport Efficiency

where the craft was designed to Navy (military) standards and required to carry a specified military payload. The *Destriero*[h] is a private yacht and the payload is not known. The *Destriero* won the Blue Ribbon award for crossing of the Atlantic non-stop without refueling on 9 August 1992 at an average (service) speed of 53.1 knots. The ship records show that the maximum speed was closer to 60 knots at times during the crossing from Ambrose Light, New York, US to Bishop Rock lightship in the Scilly Isles, England, a distance of 3,106 nautical miles. It was, at that time, the largest ship built in light aluminum alloy. The current ship owner is Bravo Romeo (Ireland).

These two craft (the design concept, PC-1 and the operational yacht *Destriero*) represent the best of designs of the large planing craft. Their Transport Efficiencies are displayed on Figure 8.53 to demonstrate the difficulty, even with advanced technology that has appeared on the scene after the earlier ship designs displayed on Figure 8.50, to make the necessary significant improvements in hydrodynamic

[h] See Fincantieri website http://www.fincantieri.com for more data on this high speed yacht.

efficiency needed. Later yacht designs have been built and demonstrated speeds up to 60 and 70 knots (again without military standards or military payload). The key message remains is that the designing of high speed ships using either *displacement hull* technology or *planing hull* technology will probably not be attaining the high efficiency required at the high speeds sought by the US Maritime Administration in its early goal for economic advantage or by the US Navy for its "100 knot Navy" and that other technological avenues would have to be explored.

More discussion of the relative "boundaries" of efficiency for other types of vehicles (hovercraft, hydrofoils, wing-in-ground effect vehicles, etc) is provided in Chapter 10 that describes the US Navy Advanced Navy Vehicles Concept Evaluation (ANVCE) Program where the Navy sought a systematic evaluation of ship types to achieve *efficient, high speed, practical ships designed for a military mission*.

8.5 Comparing Vehicles on Design Speed or Maximum Speed

A movement started around 1998 to "adjust" the Gabrielli-von Karman approach to use *design speed* instead of *maximum speed* as a design tool. This was an unfortunate misconception of how vehicles are designed. As clearly stated by Gabrielli and von Karman in their work, the specific resistance (ε) (see Equation 8.122) varies with speed and the way the vehicle is operated. To quote from their original work:

> *". . . . . it may occur that a vehicle built for a higher velocity can be more economical at a lower speed, not using its full power, than a vehicle designed for the same lower speed at its maximum velocity. The exact meaning of the curves (see Chapter 7, Figure 7.11 in this book) is the statement that, according to present experience, in order to design a vehicle of a certain type for a given maximum speed, at least as much power has to be installed as is shown. . . ."*

The *design speed* will vary according to the intent of the designer and will vary from one class of vehicles to another (for reasons that may not be readily apparent to the analyst examining the final results). This has been demonstrated in the charts shown earlier. The *design speeds* of some destroyers in WW II were set to be higher than the "Froude barrier" for military advantage. Some tankers selected their *design speeds* to be lower than the "Froude barrier" with bulkier shapes for economical advantage. For most of the displacement ships the *maximum speeds* and the *design speeds* were selected to avoid the "humps" in the drag curves that are Froude Number dependent which depend greatly on the choice of hull form. The planing boats used for military purposes were designed for *maximum speed* for military advantage and any lower "design speeds" were incidental.

The SES-100B, as a Research & Development Test Craft, had three[i] *design speeds*, viz: approx 35 knots for maximum lift drag ratio $(^{L}/_{D})_{\max}$; approx 40 knots for maximum Transport Efficiency $(^{WV}/_{P})_{\max}$ for maximum range, and approx 92 knots to demonstrate ship characteristics at (full scale) supercavitating speeds, but it had to be *sized* to accommodate the power needed for its *maximum speed*.

[i] Actually four design speeds if one includes the speed to achieve the contract specified range, which was 60 knots on ONE engine to achieve 1600nm range. The specified required range for the SES 100 test craft was 1500nm.

In addition, a key parameter in determining the economical viability of a vehicle is its Useful or Disposable load (W_D) capability (sum of payload and fuel). How much Useful Load can be carried will be determined by its Empty Weight (W_E) of which the structural weight (W_S) is a large percentage (approximately 60% or more). The structural weight is determined by the loads on the structure and the pressure loading from impact with the waves at its *maximum speed* in the *operational sea states.*

It is the *maximum speed* that sizes the vehicle not the lower *design speed(s).*

8.6 Assessment of SES-100B Drag Theory and Test

A key step along the way in attaining a high speed ship was the 1969–1980 program of the "100 ton testcraft". The complete program is described in Chapter 6. The SES-100B met all the objectives of the program and much was learned on the technology and engineering aspects of such craft. The goal was to provide a frigate size ship (approx 3000 ton displacement) capable of speeds of 100 knots. If Froude scaling was the only important parameter in the design, then designing a 100 ton ship of about 56 knots ($\lambda \sim 3{:}1$) would have been sufficient but this would not have yielded the key information of the cavitation effects and other considerations that do not scale according to Froude Number. Hence, it was also necessary to conduct trials at the "full scale speed" of 100 knots. The 100 ton SES were designed to have this extra speed capability built into the design so as to explore the hydrodynamic characteristics in (non–Froude scaled) supercavitating flow conditions as well as to explore the basic characteristics of a Froude scaled ship.

The data on the SES craft is provided in Chapter 6. What is provided in this chapter is an analysis of the key parameters, using the drag equations developed here (see Table 8.2), to understand the test results obtained in the Builder's Trials Program in 1971–1975 on Lake Pontchartrain (Louisiana) and off Panama City (Florida); and the US Navy Trials in 1976–1980 on Patuxent River (Maryland). Figure 8.54 shows the predictions for the drag of the SES 100B Test Craft. The curve labeled "SS 0" applies to the drag prediction using the equations developed earlier in this chapter for *calm sea* operation.

As was noted in the development of the equations, it was assumed for analytical simplicity, that the wave drag (D_{wave}) could be approximated by an equation ($D_{wave} \propto {}^{1}/_{F^2}$). Equation 8.13 gives the complete expression for high Froude Numbers. Such an approximate equation is only valid for Froude Numbers ($F > 1.5$). Accordingly, in Figure 8.54, the dotted line is the value of the wave drag using the full expressions from Doctors' prediction method for calm water (SS 0) operation (see Figure 8.1(b)) for speeds around the hump speed. These two curves together then provides an estimation of calm water (SS 0) drag over the full range of Froude Numbers from low speed up to the maximum speed attained of the SES-100B ($F \approx 3.48$).

There are no reliable analytical methods for predicting the drag of SES type craft in rough water operation and resort must be made to empirical relationships that the user must develop as necessary to reflect the various geometries and operating conditions deemed most likely. The curves labeled SS I, SS II and SS

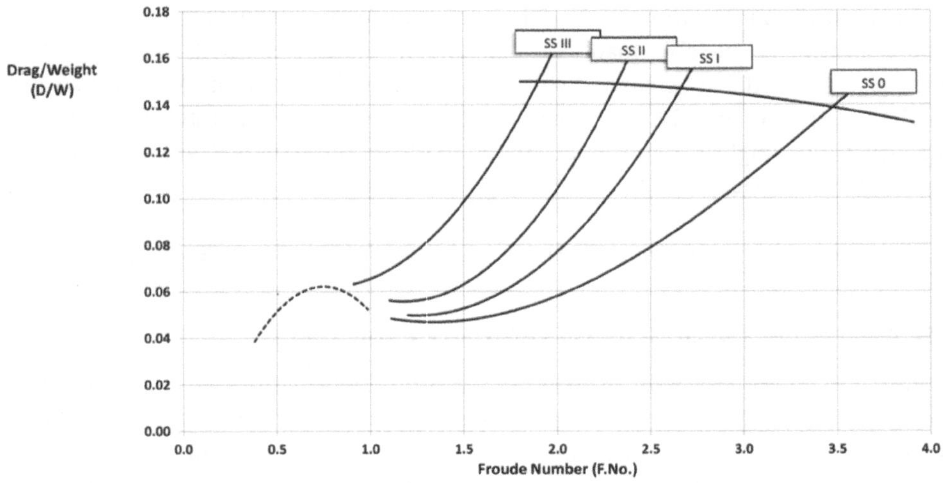

Figure 8.54: SES-100B Drag Prediction

III (for Sea State I, Sea State II and Sea State III) respectively in Figure 8.54 are such derived curves making assumptions on wave heights, wave lengths and other characteristics that are discussed here as the Trials data is included. The thrust curve has been added also in Figure 8.54 to show (by intersection with the drag curves) the estimated maximum speeds (V_{max}) in the various Sea States. These intersections correspond roughly with the speeds given by the formula developed earlier:

$$V_{SS} = V_{SS=0}\left[1 - 0.5\left(\frac{h_W}{h_C}\right)^{3/4}\right] \tag{8.131}$$

The correlation of this formula with the available data is shown in Chapter 6, Figure 6.48 for a wide range of air cushion craft. In Equation 8.31, (V_{SS}) is the ship speed in the Sea State; ($V_{SS} = 0$) is the ship speed in Calm Seas; (h_W) is the wave height in the associated Sea State, and (h_C) is the depth of the air cushion (wet deck height above the mean inside water level). These predicted speeds match the stated speeds by the US Navy[45] from the Trials, of 91.9 knots in Sea State 0, "in excess of 60 knots" in Sea State II, and "up to 58 knots" in Sea State III. Obtaining correlation of full scale trials data with prediction is a formidable task because of the extreme difficulty of controlling test parameters.

8.6.1 Sea States, Wave Heights and Wave Lengths

It is extremely difficult to define Sea States in a manner that will make their use in analytical treatments consistent from one set of data to another. The term "Sea States" originated as a mariners' imprecise term on how "rough weather at sea" was to be described. Descriptions of Sea States and their component parts has received much analysis in the maritime literature providing ranges of key parameters such as wave height, wave length, cyclic frequency, wave celerity, wave energy and many other terms. While treatments of the energy in sea waves is paramount to their

Table 8.3: *Pierson-Moskowitz Wind and Sea Scale*[46]

Sea State	Average Wave Height (ft)	Significant Wave Height (ft)	Average 1/10th Highest Wave Height (ft)	Average Wave Length (ft)	Wind Speed (knots)
Sea State I (SS I)	0.29–0.84	0.46–1.34	0.59–1.70	6.1–17.62	5–8.5
Sea State II (SS II)	1.16–2.12	1.86–3.38	2.36–4.30	24.39–44.44	10–12
Sea State III (SS III)	2.28–2.97	3.64–4.6	4.62–6.04	47.8–62.43	13.5–16

understanding for means of describing the waves, analytical treatments suitable for ship drag estimation has proven illusive. Some general rules have been derived over centuries of development. Using the Pierson-Moskowitz Wind and Sea Scale (there are many others), the Sea States under discussion are as shown in Table 8.3. Other key characteristics of sea descriptions are available in the literature.

An overview of the various theories for Sea State descriptions may be found in the author's earlier 1980 report. As can be seen from Table 8.3, the definition of Sea States is not precise. There are gaps in values between the Sea States. There is a wide range of values in each category, with numbers varying between by 2:1 to 4:1. The wind speeds associated with the Sea States are large percentages of the ship speeds under discussion. All these characteristics cause problems when predicting or comparing performance in rough seas. This was apparent when examining the Trials Data of the SES-100B.

All the Full Scale Trials Data used here is that by the US Navy (except for Lake Pontchartrain data from Builder's Trials). The US Navy Trials data is from the 1976 report by Raley and Wilson[47] and the later 1978 report by Wilson, Wells and Heber[48]. There is a large data base in the references that allows some confidence in the prediction methods but not sufficient data to be conclusive.

The US Navy Trials reports leave much of the key information unreported. The Sea States quoted do not indicate which values in Table 8.3 were used in labels of Sea State I, II and III for the tests. In the actual tests various methods were used for wave heights and other characteristics of the sea.

In Panama City, Florida use was made of oceanographic instrumentation on platforms in the Gulf of Mexico to record local wave conditions. Each Test Craft was outfitted with a boom with an on-board radar altimeter to measure the wave heights (Figure 8.55). None of this data was reported in the published reports, hence, it is unknown if the wave conditions were in agreement with Table 8.3 or which mid-range values were selected for the labeling of the data. The wide spread in the recorded data shown suggests that the measurements were not consistent. Also, the reported test data is for varying settings of air cushion flow (Q) that hampers comparison of the data in each of the Sea States. The variations obtained in Sea State III showed drag values that were *less* than for lower Sea States. Insufficient data is presented in the reports cited to resolve these anomalies but an assessment of the drag of surface effect ships can be ascertained by reviewing the data in Figure 8.56.

Figure 8.55: SES-100B Operating in Gulf of Mexico
US Navy photo

8.6.2 Drag of SES-100B in Calm Seas

Figure 8.56 (a) shows the drag data in Calm Seas (SS 0) and in SS I. Only two sets of data are available, each at different values of air cushion flow Q = 5900CFS for the Calm Seas test and Q = 4100 CFS for the tests in SS I, so a direct comparison is not possible. The wide scatter of data also hampers the comparison. The data for Sea State I indicate a lower drag than in Calm Water around the hump speed but in line with the prediction at higher speeds. Without data for the same air flow rates (Q), this difference is difficult to resolve. The prediction for Calm Seas using the earlier developed equations appears to be in general agreement except that it *overpredicts* the drag at low speed and *underpredicts* the drag at high speed. Despite this difference, the measured top speed of the SES-100B recorded at 91.9 knots in calm water matches the prediction in Figure 8.56 (a). The major departure from predictions appears in the measured drag values (not surprisingly) in the higher Sea States.

8.6.3 Drag of SES-100B in High Sea States

The wide spread of the data in Figure 8.56 (b) and (c) requires some discussion. Ignoring for the moment any difference between prediction and measurement, it is noticed that some of the measured data shows *lower* drag in Sea State III than that measured in Sea State II which does not seem logical. The US Navy reports cited present the data without comment providing no guidance as to these differences (seen in both the 1976 and the 1978 reports). Assuming that the thrust meters in the SES-100B were working correctly and allowing for some inaccuracies in measuring wave heights, wavelengths and wind speeds it is possible to conjecture why the results are as shown. There is also the uncomfortable statement in the 1978 US Navy Report that the 1976 report included data at different weights because of fuel

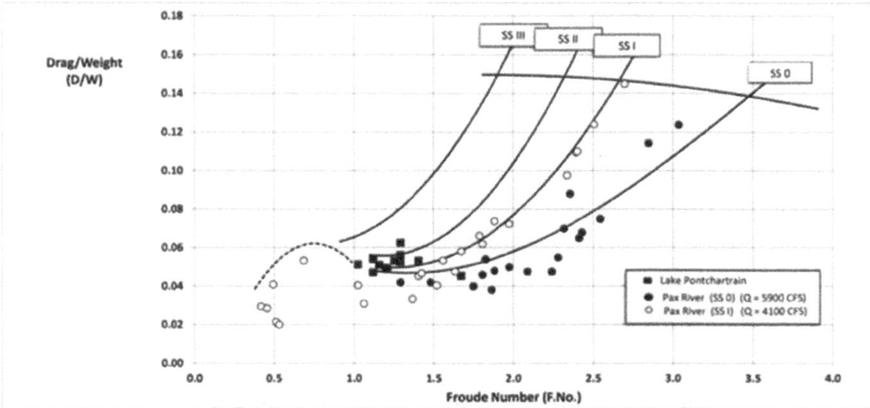

(a) *SES-100B* Drag in Calm Seas and Sea State I

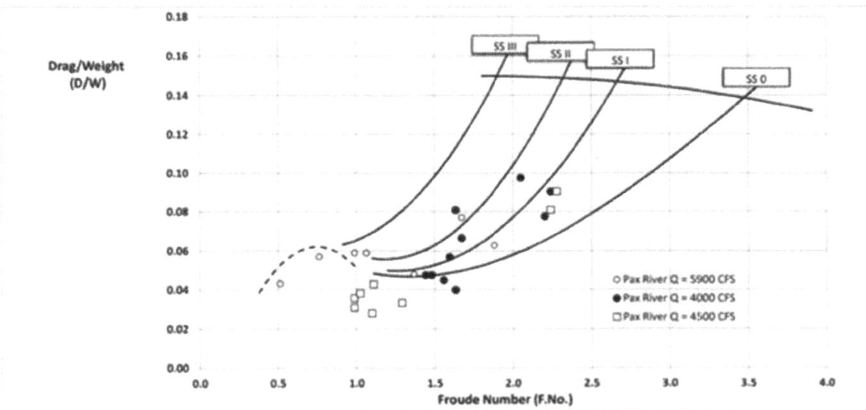

(b) *SES-100B* Drag in Sea State II

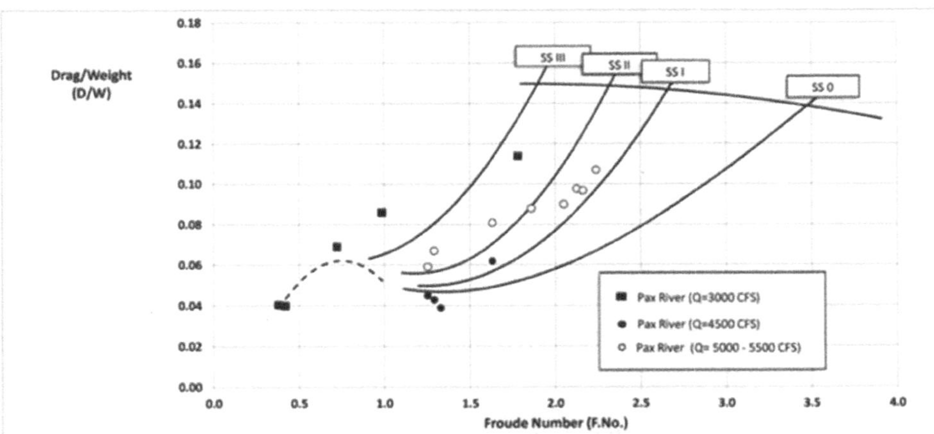

(c) *SES-100B* Drag in Sea State III

Figure 8.56: SES-100B Drag Data from Full Scale Trials

burn-off and so the 1978 report normalized its data on a "drag/weight" basis but no information was given as to the applicable weights for the drag (in pounds) used in the 1976 report. Hence, Figure 8.56 is a combination of drag data at different weights, some normalized and some not!

There are three major contributors to the drag of a sidehull SES operating in rough seas. These are (i) the wetted areas of the sidehulls giving rise to the skin friction drag; (ii) the drag of the skirts; and (iii) the dynamic effects from the fan and cushion system. Figure 8.57 illustrates these features.

8.6.4 Drag of the Sidehulls in Rough Seas

The drag of the sidehulls in Calm Water has been treated earlier (see Table 8.2 and supporting text). In rough seas the treatment is more complex. In calm seas, the depth of immersion of the sidehulls (d_i) is kept to a minimum for low drag. In rough sea operation, it is normal to increase the immersion to minimize the loss of air flow underneath the keel as shown in Figure 8.57. This is not a hard and fast rule because of the trade between sidehull drag and wave pumping requirements and it would have been useful if this trade had been explored in the SES-100B trials. The significant change in drag shown in Figure 8.56 (c) showed high drag for Q = 3000 CFS and low drag for Q = 4500 CFS but then high drag again for Q = 5000–5500 CFS. The increased wetted area (shaded area in Figure 8.57) will be strongly dependent on the ratio of the sidehull length (L_{SH}) to the wavelength of the oncoming waves (λ). That increased wetted area (ΔS_{SH}) can be expressed as a function of the ship and wave parameters as:

$$\Delta S_{SH} = f\left(h_W, L_{SH}, \frac{2\pi V}{\lambda}, t\right) \tag{8.132}$$

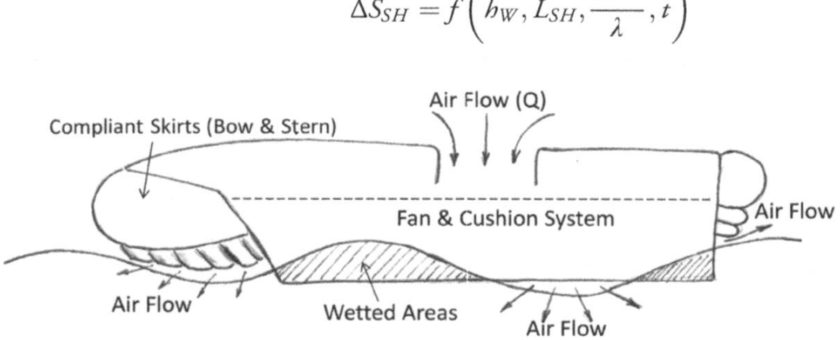

Figure 8.57: Mechanisms for Drag in Rough Seas

By assuming certain wave shapes (e.g. sinusoidal) it is possible to generate approximate expressions for this wetted area but they would be imprecise at best. This was done as a first approximation for the curves labeled (SS I, SS II, SS III) in Figure 8.56.

This imprecision coupled with the lack of documented data on measured wave heights (h_W) and wavelengths (λ) in the US Navy Trials Program make it an exercise in futility to seek correlation between prediction and measurement of this critical component. The 1978 US Navy Report also suggested in their analysis that the wetted area in rough seas on the outside of the sidehulls would be the same as the wetted area in calm seas. As the photo of the SES-100B shows in Figure 8.58, this is an unlikely situation. Other photos of the SES-100B (see for example Chapter 5 photos of the SES-110B operating in rough seas) also show this not to be the case.

Figure 8.58: SES-100B in Rough Seas
Reprinted by permission of SNAME

The amount of wetted area will also change with time (t) as the wave passes along the sidehull depending on wave celerity, ship speed and ship length, wavelength and wind speed.

It would have been useful if the trials had been run on a statistical basis keeping basic parameters (such as air flow, Q) constant for a *time averaged* value of drag. This would have reduced the scatter. Unfortunately, since the SES craft have now been scrapped this cannot be done and one is left with conjecture from Figure 8.56 (c) as to which is the best representation of the drag in Sea State III and what should the design setting be for the air flow (Q). In Figure 8.56 (a) the full scale trials data suggests that a high flow value (Q = 5900 CFS) would give the lowest drag in Calm Water and Sea State I operation. In Figure 8.56 (b), for operation in Sea State II, the trials data suggests that there is no clear setting of air flow for the lowest drag as the data points for Q = 4000 CFS to Q = 5900 CFS show a wide scatter with no clear trend in air flow setting. Finally, in Figure 8.56 (c), for operation in Sea State III, the trials data suggests that a mid-flow setting of Q = 4500 CFS provides the lowest drag. One would not have been surprised if the Trials results had been reversed with the higher Sea States requiring a higher flow, not less, to account for wave pumping needs.

The Trials drag data in Figure 8.56 (b) for Sea State II is for both head seas and following seas with no discernible difference detected. From Table 8.3, the wind speed for Sea State II is about 10 knots. At a mid-range ship speed of 50 knots the air speed in those tests would be 50±10 knots between head and following seas. The aerodynamic drag at 50 knots is about 10% of the total predicted drag in SS II, so it would be expected that the drag variance would be about $\left({}^{60}/_{40}\right)^2$ times the ship speed value or about 2:1. That is, one would expect about another 10% difference (~2,000lb) in the data between head seas and following seas. Such a difference does not show up in the Trials data. The US Navy Trials report states that their measurements are accurate to ± 1,300lb, so it is possible this variance is lost in the data spread. This apparent inconsistency in the Trials results (especially the effect of flow rate, Q) suggests that the interaction between sidehull drag, skirt drag and wave pumping effects is not fully understood. There is no comment in the published Trials Reports as to these differences and scatter or what might be their cause. Some possible conjecture follows in the discussion of the skirt system, the fan system and the cushion system.

8.6.5 SES-100B Skirt System Design for Rough Water Operation

Unlike the rigid planer seal of the CAB concept that requires the seal to maintain contact with the water (planing) to capture the air bubble to ensure little or no flow ($Q = 0$); the SES-100B flexible skirt system is designed following the hovercraft skirt designs that require some flow ($Q \neq 0$) with small air gaps and no contact with the water to minimize drag (see Figure 8.57). There have been a few instances in the development of hovercraft of the skirts "grabbing" the water through surface friction and contributing to instabilities and pitch down of the craft. In the light of that hovercraft history, a unique feature was built into the SES-100B skirt design with the incorporation of a "soft spring/hard spring" accumulator system. This feature is discussed in Chapter 6 on the design of the SES-100B and a photo of the "colander" bow that provides this capability is shown in Figure 8.59.

(a) Quick Inflate/Slow Deflate Accumulator
Photo courtesy of Textron Marine & Land Systems and Textron Inc.

(b) Quick Inflation of Bow Seal upon Launch
(No Engines Running)
Photo courtesy of Textron Marine & Land Systems and Textron Inc

(c) Slow Deflation of Bow Seal
During Operation in Rough Seas
US Navy photo

Figure 8.59: SES-100B Bow Seal Design for Proper Response in Sea States

The design concept was to provide a capability to inflate the bag and fingers fairly quickly such that on the forward "stroke" the skirt acted like a "soft spring". During rough sea operation when the water forces pushed against the skirt system at high speed the resistance of the air flowing back into the plenum (on the backward stroke) through the "colander", slowed the deflation rate and the skirt system acted like a "hard spring". By this technique the bow skirt system would always try to maintain its geometry and thus avoid any suction forces from deflated skirts dragging in the water.

This feature can be seen on the SES-100B during its launch on 22 July 1971 in Figure 8.59 (b). The launch was performed by a lowering of the ship into water by crane (the boom is visible above the craft). The skirts immediately inflated as seen in the photo (with no fan engines running and the inlet vanes closed). After rapidly inflating, the bow skirt remained inflated for some time as the craft sat in the water with the cushion air trapped by the sidehulls and the skirts. The bow skirt retained its configuration even during rough water operation as shown in Figure 8.59 (c) and Figure 8.58, suggesting that the design concept was working as designed.

It is possible, of course, during rough water operation, occasionally the fingers would drag on the water and cause an increase of drag under certain conditions. It would have been useful to have had a photographic record of SES-100B bow skirt movements while underway in rough seas that could be correlated with recorded drag measurements. There was no reporting of such incidents in the US Navy Trials of "skirt grabbing" so it is left as conjecture as to whether unplanned skirt drag (due to deflation and extra wetting) contributed to the wide scatter of drag data in the Trials results shown in Figure 8.56.

8.7 Fan and Cushion System Dynamics

The third possibility of drag variations can be tied to the interaction between the fan system and the air cushion of the craft (together with the effects of seal compliance). The drag characteristics of a dynamic craft such as the surface effect ship will depend on the dynamics in pitch and heave motion. The effects in the other axes come into play as the craft maneuvers. The Trials data did not report on the motion of the SES-100B in the various Sea States, that is, no reporting of heave motion or pitch angle about the mean wave slope was provided. A small change in pitch angle would give rise to more or less wetted area of the sidehulls with the corresponding change in drag. As the craft heaved, there would be a change in immersion depth with a corresponding change in drag (of both sidehulls and skirts). While no specific data was made available, it is possible to illustrate the effects. The characteristic variation of cushion pressure (p_C) and air flow (Q) is as shown in Figure 8.60 for a cushion system where multiple lift fans act in parallel operation. The SES-100B had eight (8) directly coupled lift fans in parallel operation. The design point for the cushion pressure would then be given by the expression:

$$p_C = p_O + \frac{\partial p_C}{\partial Q}(Q - Q_O) \tag{8.133}$$

The term (p_o) is the equilibrium value of the cushion pressure normally taken as (W/S) ignoring the lift from the sidehulls. The selection of the fan design is made such that the cushion slope ($\partial p_C/\partial Q$) is negative for static stability in heave.

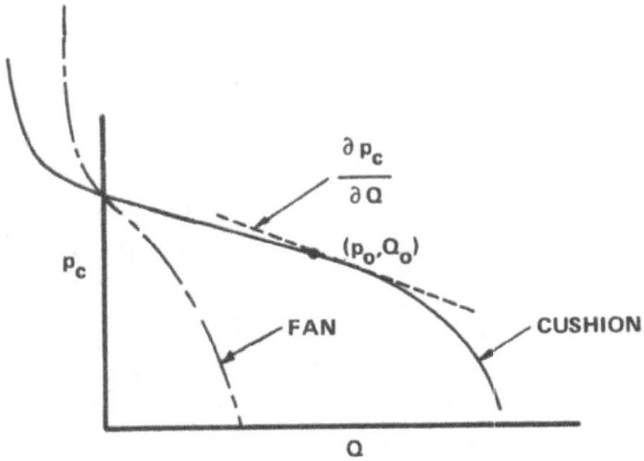

Figure 8.60: Fan and Cushion Pressure-Flow Characteristic

As the craft traverses the waves the pressure (p_C) and flow variations (Q) would follow the "cushion curve" shown in Figure 8.60.

The heave response of any air cushion craft operating in waves was treated by Mantle in 1962[49] and a brief outline of those results follows. That original work was done for peripheral jet hovercraft but the results are generally applicable for plenum type craft such as the SES-100B and others.

In the analysis, it is assumed that the craft is moving over a waveform given by:

$$h_W = A\cos(Kx - \omega t) \tag{8.134}$$

This wave form would correspond to that shown in Figure 8.57 where (A) is the wave height, (K) is the wave frequency in distance ($K = 2\pi/\lambda$) and ω is the wave frequency in time ($\omega = 2\pi V/\lambda$). As discussed above, x, is the distance along the sidehull. To reduce the problem to manageable levels, the complex motion of wave pumping and escape area variations in actual craft motion is treated instead as a simple piston heave motion where the forcing function is an "effective wave height, h_{We}" constant along the craft length. This effective wave height is defined as:

$$h_{W_e} = \frac{1}{L_C}\int_0^{L_C} A\cos(Kx - \omega t) \tag{8.135}$$

Upon integration, the effective waveform is then given as:

$$h_{W_e} = A\left|\frac{\sin\frac{\pi L_C}{\lambda}}{\frac{\pi L_C}{\lambda}}\right|\cos(\phi - \omega t) \tag{8.136}$$

Hence, the simple piston analogy gives the actual wave height multiplied by a function:

$$F\left(\frac{L_C}{\lambda}\right) = \left|\frac{\sin\frac{\pi L_C}{\lambda}}{\frac{\pi L_C}{\lambda}}\right| \tag{8.137}$$

This *effective wave amplitude function*, $F(\frac{L}{\lambda})$ is shown in Figure 8.61. The length (L) has been left generic in that it may be that some intermediate length between the cushion length (L_C) and sidehull length (L_{SH}) might be more appropriate depending on the magnitude of the different lengths and the water interaction. There is also a phase shift with this artifice of the simple piston analogy, given by: $\phi = {}^{\pi L}/_{\lambda}$. It is suggested that damping and coupling effects between pitch and heave motion will "round out" the discontinuities at the integer values of ${}^{L}/_{\lambda}$ in Figure 8.61. To see if this result matched actual data, the author collected data from two different sources from Tinajero & Fresh (1960)[50] and Uyeda (1961)[51] from hovercraft models tested in different waves. Applying the Effective Wave Amplitude Function, $F(L/\lambda)$ to the data from these sources gave the result shown in Figure 8.62. It can be seen that the data did generally collapse to a common non-dimensional base. The data covered a range of craft length to wavelength ratios, $0.417 \leq L/\lambda \leq 1.668$ and a range of wave height to wavelength ratios, $0.005 \leq A/\lambda \leq 0.020$. This range of values approximates the Sea State ranges shown in Table 8.3.

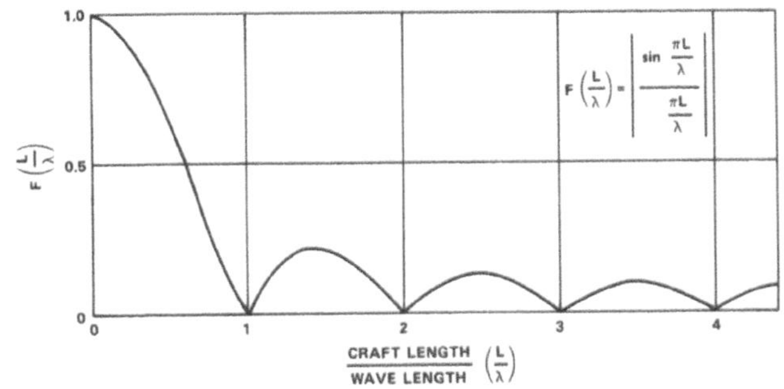

Figure 8.61: Effective Wave Amplitude Function

In the 1962 analysis, the heave motion of the craft, for small heave perturbations (H) about an equilibrium skirt air gap (h) was described by the second order differential equation:

$$\frac{d^2H}{dt^2} + 2\zeta\omega_n \frac{d}{dt}(H - h_W) + \omega_n^2(H - h_W) = 0 \tag{8.138}$$

In terms of the magnification ratio shown in Figure 8.62, the solution to the heave equation can be written:

$$H(t) = MF(L/\lambda)A\cos(\phi - \omega t) \tag{8.139}$$

The amplitude of the wave height (A) would be the wave height shown in Table 8.3. The magnification ratio (M) in Figure 8.62 is expressed in terms of the damping ratio (ζ) and the natural frequency (ω_n) by the equation:

$$M = \frac{1}{\sqrt{\left[1 - \left(\frac{\omega}{\omega_n}\right)\right]^2 + \left[2\zeta\frac{\omega}{\omega_n}\right]^2}} \tag{8.140}$$

In the 1962 analysis, the author gave approximate expressions for the heave stiffness ($S\,\partial p_C/\partial Q$) and damping ratios ($\zeta$) applicable for peripheral jet hovercraft which were the most prevalent type at the time. The natural frequency (ω_n) was determined to be:

$$\omega_n \approx \sqrt{\frac{g}{h}} \tag{8.141}$$

This simple result, where (h) is the average air gap could be used in a quick estimate of the cushion stiffness. The damping ratio was found to be a little more complicated and was found to be a function of the air flow (Q) and cushion pressure, natural frequency and thickness parameters of the peripheral jet. Later in 1969, Wheatley[52] developed the expressions specifically for the plenum air cushion craft and for conditions of *constant lift power* (P_L), the natural frequency and damping ratios were:

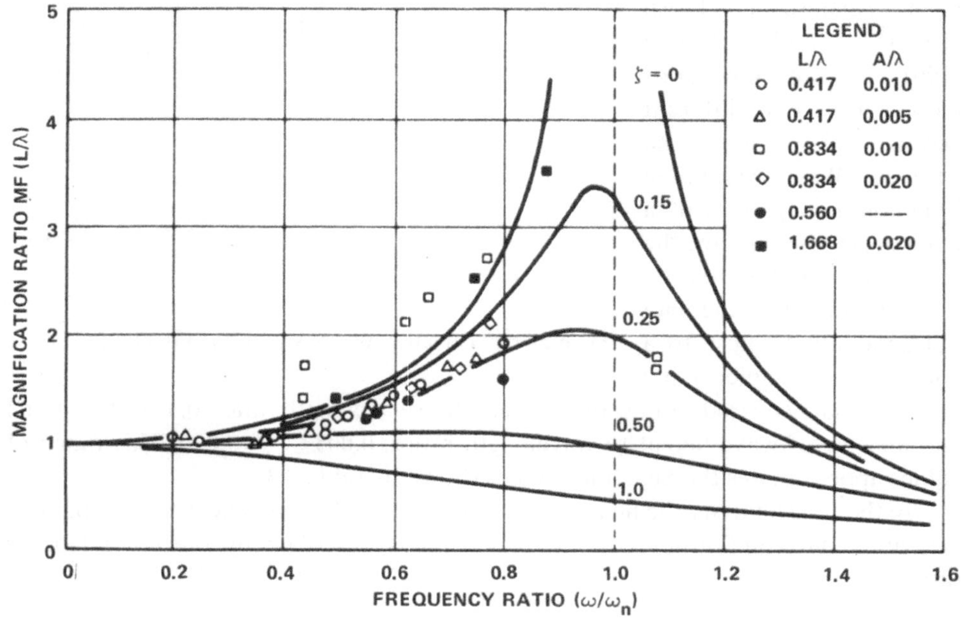

Figure 8.62: Normalized Heave Response Over Waves

$$\omega_n = \sqrt{\frac{2g}{h_a}\left(1 - \frac{2}{3}\frac{p_C}{p_H}\right)} \tag{8.142}$$

$$\zeta = \frac{gS}{Q\omega_n}\left(1 - \frac{2}{3}\frac{p_C}{p_H}\right) \tag{8.143}$$

In the above expressions:

h_a average height of the air gap beneath the skirt
p_H fan total head
S cushion area
Q total cushion flow

For the case of ($p_C/p_H = 0.75$), a not atypical value for air cushion craft, these results reduce to the values given earlier for the natural frequency and damping ratio.

$$\omega_n = \sqrt{\frac{g}{h_a}} \tag{8.144}$$

$$\zeta = \frac{gS}{2Q\omega_n} \tag{8.145}$$

Other researchers also examined the dynamics of the "*skirt-fan-cushion system*" and showed different effects but all with a common theme. Breslin[53] in 1968 analyzed the specific case of surface effect ships and included such factors as the compliance of the skirt; free surface deflection and cushion flow. Breslin developed an expression for the natural frequency that was a complicated relationship that showed that the natural frequency could be represented by a functional form:

$$\omega_n = f\left(\frac{Q}{p_C}, \frac{\partial p_C}{\partial Q}, K_S\right) \tag{8.146}$$

If the skirt compliance term (K_S) was dropped from Breslin's more complete equations, the author found that the expression for natural frequency could be simplified to:

$$\omega_n = \sqrt{\frac{g}{h}}\sqrt{\frac{\frac{Q}{p_C}\frac{\partial p_C}{\partial Q}}{1+\frac{Q}{p_C}\frac{\partial p_C}{\partial Q}}} \tag{8.147}$$

This shows how the characteristics of the cushion (and fan) pressure-flow system affect the natural frequency in heave. The similarity between this result and Equation 8.141 and Equation 8.142 can be readily seen albeit with different parameters being considered. The values of cushion pressure (p_C), cushion slope $\partial p_C/\partial Q$ and air gap (h) are all recognizable terms that can be measured during test.

It will be noticed that the expression for the damping ratio (ζ) includes the air flow (Q). Earlier, in development of the equations for the drag components (see Table 8.2) the wave pumping requirement in the determination of the air flow depended on the same *effective wave amplitude function*, $F\left({}^{L}/_{\lambda}\right)$. For zero heave acceleration of the deck, it was found (see Equation 8.29) that the air flow for adequate wave pumping was given by:

$$Q_{WP} = \pi\frac{h_W}{\lambda}VSF\left(\frac{L}{\lambda}\right) \tag{8.148}$$

These results, derived from examining the effects of ship geometry, wave geometry and air flow, show that both the motion and the operating conditions can have significant impact on the drag of the craft. While not expanded upon here, similar treatments for the pitch motion of the craft also show significant impact on the drag. Pitch variations around the mean wave slope of ±2° cause significant changes of the wetted area of the sidehulls with the associated changes in craft drag. Notice from Figure 8.58, that the pitch angle is considerably more than the steady state trim angle calculated from the wave drag of the air cushion at that speed (which is $\tau < 1°$).

In the light of these predictions, it is useful to re-examine the drag data from the US Navy full scale trials of the SES-100B to see if it is possible to explain the wide scatter in the reported data. Upon examining these relationships and the test conditions in the US Navy Trials, several observations can be made. The size of the ship (SES-100B) in relation to the wavelengths encountered would indicate that the tests included operations where the ship length to wavelength ratio $\left({}^{L}/_{\lambda}\right)$ varied sufficiently to cause major changes in the effective wave amplitude function $F\left({}^{L}/_{\lambda}\right)$ and the magnification ratio (M) that would suggest that the heave motion of the ship changed significantly. Also, the craft speeds were such that the ratio $\left({}^{\omega}/_{\omega_n}\right)$ approached the resonance value $\left({}^{\omega}/_{\omega_n} = 1\right)$ in several instances. Because these conditions were heavily influenced by the flow settings (Q) (see for example Equation 8.145 and Equation 8.148) the question remains as to how the heave

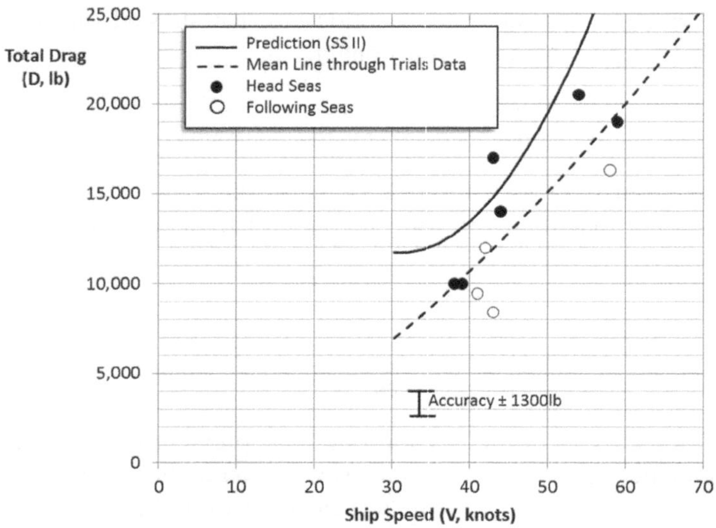

Figure 8.63: SES-100B US Navy Trials Data (Sea State II)

response of the SES-100B affected the immersion levels of the side-hulls as the craft passed through the waves[j].

Unfortunately, these can only be conjectures as the data of cushion pressure (p_C), fan total head (p_H), cushion pressure-flow relationship ($\partial p_C/\partial Q$); trim angle (τ) variations and heave acceleration levels were not made available as part of the reported data. Although somewhat conjectural in the light of the missing information, it is useful to review the SES-100B data from the US Navy Trials for the particular case of tests in Sea State II. Figure 8.63 shows the data taken from the Trials reports used earlier and included in Figure 8.56. In Figure 8.63, the difference between head seas and following seas is now identified. While the following seas drag data is generally less than for the head seas case (although not in all runs) the difference cannot be explained by the change in aerodynamic drag alone. It is seen that departure from the mean line is far greater than the stated accuracy of the test measurements (+5,000 to –4,000 lb variation compared to the ± 1,300 lb accuracy stated in the US Navy report). Some other mechanism must be at work to cause these significant differences in drag.

Table 8.3 shows a wide spread in data to describe SS II, but if the average values of average wavelength and wave height are used i.e. $h_W = 1.6$ feet and $\lambda = 34$ feet, then ${}^{h_W}/_{\lambda} \approx 0.05$. With a cushion length $L_C = 61.5$ feet for the SES-100B, then ${}^{L_C}/_{\lambda} \approx 1.8$. If it is assumed that the average air gap (h_a = 2ins) (highly speculative), then the natural frequency would be about $\omega_n \approx 13.9$ rads/sec. For a craft speed of 40 knots, the encounter frequency would be $\omega = 2\pi V/\lambda = 12.5$ rads/sec. That is, the craft would be traveling at a speed where (${}^{\omega}/_{\omega_n} = 0.89$) which is close to resonance (see Figure 8.62). The effective wave amplitude function $F\left({}^{L_C}/_{\lambda}\right) \approx 0.10$ under these conditions suggests a large dampening of the wave amplitude. The damping ratio would be $\zeta = 0.55$ for an air flow value of 4,000 CFS (the stated air flow from the US Navy Trials report) which from Figure 8.62 corresponds to a highly damped motion. The wave pumping flow required under these conditions would be (from Equation 8.146), $Q_{WP} \approx 2000$ CFS which would suggest that the Trials value of 4000 CFS was too high for that particular test possibly giving rise to the excessive drag value shown in Figure 8.63.

[j] Recall that according to the work by Korvin-Kroukovsky et al. (see earlier discussion on planing hull data), the sidehulls of the SES-100B are not planing in these Trials.

These are of course, highly speculative calculations based on the very imprecise data made available. It emphasizes the need to carefully control the test parameters and record all information.

Despite these concerns, the US Navy Trials data provided valuable information sorely needed to understand the characteristics of high speed air cushion supported ships, but not sufficient to draw meaningful conclusions for future designs. It is indeed unfortunate that the program was canceled approximately one year after the Trials Reports were made available foreclosing any possibility of assessing the key parameters in determining the drag of such craft in high seas.

8.8 Dynamic Similitude and Scaling Laws

The air cushion craft is unique among the candidates for high speed at the interface between air and water. It is subject to both hydrodynamic and aerodynamic forces. Even at low speed, the air cushion craft must contend with the dynamics of the air flow moving through its system. Scaling of any results from models or Test Craft to larger sea going ships must ensure dynamic similitude if the expected performance is to be achieved.

Systems are said to be dynamically similar if (a) they are geometrically similar and (b) if the forces acting on the "model" are in the same ratio to each other as similar forces act on the "full scale" ship. Because aerodynamic and hydrodynamic forces scale differently, this gives a unique problem to air cushion craft designers.

The forces acting on air cushion craft (SES, hovercraft, WIG, ...) can be expressed as:

$$\text{Force} = f(L, V, \upsilon, \rho, \rho_W, g, p_{amb}, p_C, Q, h_W, \lambda, \ldots) \qquad (8.149)$$

In this functional form, the key parameters that influence the design of the craft are:

- L characteristic length of the craft. This length would be chosen depending on many considerations but would usually be the length of the cushion (L_C) or the length of the sidehull (L_{SH}) or, in the case of the aerodynamic form of air cushion craft, the chord (see Chapter 11)
- V craft speed
- υ kinematic viscosity of water
- ρ density of the air
- ρ_W density of water
- g gravitational constant
- p_{amb} ambient pressure (or more directly, the vapor pressure, p_v)
- p_c air cushion pressure
- Q air cushion flow
- h_w wave height of the sea
- λ wave length of the sea

Using the Buckingham π theorem of dimensional analysis, independent dimensionless groups from the above list can be derived to describe the forces acting on the craft. Many of these dimensionless numbers will be recognized by aerodynamicists and hydrodynamicists alike. Each of the dimensionless numbers represent ratios of different types of forces acting on the craft, as shown. Some immediately recognizable dimensionless numbers are:

$$Reynold's\,Number\,(R.No.) = \frac{Inertia\,Forces}{Viscous\,Forces} = \frac{VL}{\upsilon} \tag{8.150}$$

$$Froude\,Number\,(F.No.) = \frac{Inertia\,Forces}{Gravitational\,Forces} = \frac{V}{\sqrt{gL}} \tag{8.151}$$

$$Cavitation\,Number\,(\sigma) = \frac{Water\,Static\,Pressure\,above\,Vapor\,Pressure}{Kinetic\,Energy\,per\,Unit\,Volume} = \frac{p_C - p_V}{\frac{1}{2}\rho_W V^2} \tag{8.152}$$

Because of the general familiarity of these non-dimensional numbers some "short cuts" have been taken in their description in Equations 8.150, 8.151 and 8.152. Some pertinent comments are:

Reynold's Number (R.No.)

The Reynold's Number (R.No.) expresses the ratio of the inertia forces acting on the craft to the viscous forces. At low R.No., the viscous forces dominate and the flow would be smooth and *laminar*. For high values of R.No., the flow becomes *turbulent*. As might be expected, the skin friction drag is higher for turbulent flow. For dimensional similitude, the full scale ship and the "model" craft would be operated at the same R.No. If the subscript "m" denotes model scale and "FS" denotes full scale ship, then operating at the same R.No. means that the "model" craft would have to operate at a speed:

$$V_m = \left(\frac{L_{FS}}{L_m}\right) V_{FS} \tag{8.153}$$

Froude Number (F.No.)

Strictly speaking, the ratio of the forces in Equation 8.151, is the square of the dimensionless number shown but common practice has given rise to the form shown and used in the majority of work in hydrodynamics. The Froude Number (F.No.) expresses the ratio of the inertia forces acting on the craft to what William Froude referred to as the *wave making* forces that are proportional to (g_L). For dimensional similitude, the full scale ship would be operated at the same F.No. as the "model" craft. This means that the "model" craft would have to operate at a speed:

$$V_m = \sqrt{\frac{L_m}{L_{FS}}} V_{FS} \tag{8.154}$$

Cavitation Number (σ)

The cavitation number expresses the potential for the water to cavitate. Cavitation is the formation and then implosion of cavities in the water when subjected to rapid changes in pressure. These rapid changes in pressure will cause vapor filled cavities that will collapse as the ambient pressure approaches the vapor pressure of water ($p_v = 2.3kPa$ or approx 48.73psf at 68°F). The formation and collapse of these cavities cause forces large enough to damage the exposed surfaces of the ship. This can happen at high speed through the water from either the craft itself or from propulsive devices. Because the water is the same for both model and full scale, then the vapor pressure (p_v) is the same and so for dynamic similitude of super-cavitating flows, the "model" craft would have to operate at the same speed as the full scale ship:

$$V_m = V_{FS} \tag{8.155}$$

From these brief descriptions, it is clear that it is not possible to simulate all characteristics of viscous forces (skin friction); wave making forces, and cavitation characteristics in the same "model". Various "tricks" are used in tow tanks, etc to change surface roughness to simulate full scale, but it is not possible to simulate Reynold's Number, Froude Number and Cavitation Number in the same way on the same ship. Because of the concern about operating at high speeds where damaging cavitation was known to occur it was decided to have the 100 ton craft operate at the full scale speed of 100 knots, even though a Froude Scale speed of 56 knots would have been sufficient for normal hydrodynamic Froude Number scaling. It is important to note that the 100 ton SES is not a direct "scale model" of the 3000 ton frigate envisioned by the US Navy. It is the technologies that have been scaled, not the ship.

Not discussed here in detail, but an example of the difficulty in scaling is that the skirt compliance (K_S), referred to earlier, in understanding the craft motion and drag characteristics, would also have to be scaled by different scaling laws to capture the true skirt motion and drag, and to properly simulate structural characteristics (stress levels, wear patterns, etc). Features of the material such as the modulus of elasticity, unit weight, etc would have to be modeled in order to fully understand how a much larger ship would behave at sea. Apart from minimal exploratory studies, both analytical and lab tests, that were accomplished during the period 1960–1980, no major effort was built into the SES program to gain an engineering level of design expertise of these critical design issues of skirt behavior.

Cushion Density ($^{p_c}/_L$)

Another scaling factor determined from the dimensional analysis is the cushion density which appears as;

$$\text{Cushion Density} = \frac{p_C}{\rho g L} \tag{8.156}$$

The non-dimensional cushion density expresses the ratio of the aerostatic forces to the gravity forces. Because the air density (ρ) and the acceleration of gravity (g) are constants in this context, it is common practice to refer to the cushion density as:

$$\text{Cushion Density} = \frac{p_C}{L} \tag{8.157}$$

In this dimensional form it has the dimensions of *psf per foot* (in English units). The cushion density is an important parameter in both the hydrodynamic design of the craft and the structural design (see later). In the drag equations for air cushion craft it was found to be a dominant factor in the determination of the wave drag (D_{wave}) and in other drag components such as the skirt drag (D_{SK}) and aerodynamic momentum drag (D_{mom}) (see Table 8.2).

As shown later, the cushion density (p_C/L) significantly affects the proportions of the craft and the loads acting upon it, which in turn significantly affect the structural weight of the ship. The cushion density is a key parameter in the design.

Pressure Number (*k*)

The original 1967 dimensional analysis[54] also produced a non-dimensional number that expresses the ratio of the *aerodynamic forces* to the *aerostatic forces* acting on the ship, such that:

$$\text{Pressure Number(k)} = \frac{\textit{Aerodynamic Forces}}{\textit{Aerostatic Forces}} = \frac{\frac{1}{2}\rho V^2}{p_c} \tag{8.158}$$

A *low* value of the Pressure Number (k) means that the aerostatic (cushion pressure) forces dominate, which is the case for conventional air cushion craft and SES. A *high* value of the Pressure Number (k) means that the aerodynamic forces dominate, which is the case for aerodynamic air cushion craft such as WIG and Channel Flow craft and *ekranoplan* which are discussed more fully in Chapter 11.

The Pressure Number (k)[k] can be related to two other numbers; the Froude Number (F.No.) and the Cushion Density ($^{p_C}/_L$), as shown in Equation 8.159:

$$k = \frac{\rho g}{2} \frac{F^2}{^{p_C}/_L} \tag{8.159}$$

Hence, from a scaling point of view, it is possible to conduct experiments at the same Pressure Number provided the Froude Number and Cushion Density are also scaled by Equation 8.159.

Most air cushion craft and SES designs operate at low values of Pressure Number ($0.20 < k < 0.60$) with the majority operating in the lower range. At these speeds, the aerodynamic forces are about 20% of the aerostatic forces and the shaping of the craft for aerodynamic purposes is less important. In Chapter 11, it is found that the craft are designed for high values of Pressure Number ($k > 1$) and the external shaping of the craft is appropriately different and behaves differently to its aerostatic cousins.

[k] The *Pressure Number* defined here should not be confused with the *Euler Number* ($\mathrm{Eu} = \frac{p_1 - p_2}{\rho V^2}$) which compares the local static pressure drop ($p_1 - p_2$) in a particular flow, such as over a restriction (dam, weir, etc) to the dynamic pressure in that *same* flow stream. The *Euler Number* is used to characterize the losses in the flow in standard problems of hydraulics. It is also used in characterizing cavitation phenomena in that same flow. On the other hand, the *Pressure Number (k)* compares *two* flows, i.e., (a) the *aerodynamic lift* (pressure) of the air cushion craft, and (b) the *static lift* (pressure) of the underneath air cushion generated by a separate fan system.

Flexible Structure Scaling

A discussion of the structural weight of air cushion craft is provided later, but it is appropriate at this juncture to provide some insight on the scaling of *flexible skirt structures*, which are part of the total craft structure, because of the direct relationship of skirt compliance (K_S) and other considerations in the drag of air cushion craft. The scaling from model results to full scale craft or from Test Craft to larger ocean going ships of the skirts is a key area not fully resolved during the 1960–1980 period or later.

The correct scaling for air cushion craft skirts or seals is not well understood or systematically tested. A more in-depth treatment of the subject complete with test results from actual craft is provided in the 1980 work by this author. Some indications from that work are included here to illustrate the problem and provide guidance as to where any future research should be guided. It has already been shown that the acceleration levels of any part of the ship (skirt) directly following the waves was given by:

$$n = 2\pi^2 \frac{h_W}{\lambda^2} \left[V + \sqrt{\frac{g\lambda}{2\pi}} \right]^2 \tag{8.160}$$

In this equation, (h_W) is the wave height; (λ) is the wave length and (V) is the craft speed. As can be seen from Equation 8.160, the scaling of the forces from such accelerations does not fit well with the scaling laws outlined above. It is also known that the actual accelerations on SES seals are significantly higher than those expressed by this equation (see discussion of Trials Data on the 100 ton SES craft in Chapter 6) and is probably due to complex additional flexing of the seal edge induced by the wave acceleration.

The weight of skirt materials will heavily influence the stiffness and compliance, with a direct impact on the skirt drag. Figure 8.64 shows skirt weights of past air cushion craft and projected designs for larger craft in the 1,000 ton to 5,000 ton displacement class. During the early years of hovercraft development several schools of design philosophy were followed as "best design practice" was sought. Some believed that heavy material was best (the motion was less) and would not wear as much. Another school believed that light weight designs were better (the forces were less). The data from the various craft from these "schools" is shown in Figure 8.64, with a considerable difference shown in the weight of the skirt systems. During the US Navy ANVCE Program, some effort was made to explore these different effects.[55] Data from this work has been included in Figure 8.64 shown by the open symbols.

The correct scaling for skirt material sufficient to gain confidence for the design of large SES was not determined. Some idea of the problem can be seen from Figure 8.64. For example, the scaling for the bag material would be different than for the finger material. The points labeled "B" in Figure 8.64 assume that the material weight would be Froude scaled. Also, the points labeled "A" in Figure 8.64 assume that the *stress levels* in the load carrying material remain constant. In reality, one would expect that specific weights in between these values might be more appropriate.

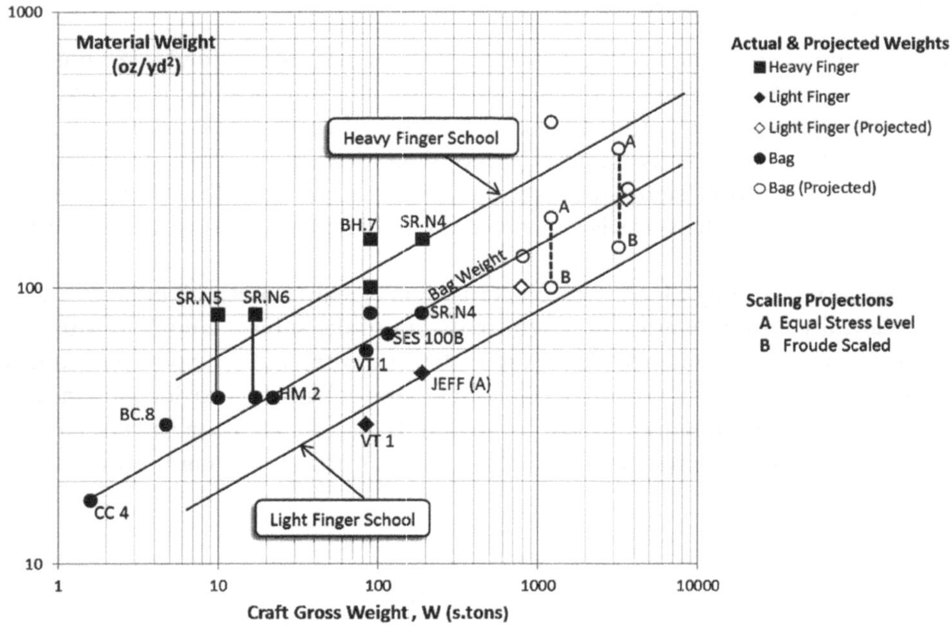

Figure 8.64: Skirt Material Specific Weight

The scaling laws for the finger material are even less clear. For example, the scaling of the finger material would be expected to depend on the basic specific weight which would scale[1] by $\lambda = {}^{L_{FS}}/_{L_m}$, but due to the vibration characteristics it would be expected that the bending stiffness (EI) should be considered, which by Froude scaling would scale by λ^4 – a considerable difference!! This latter Froude scaling does not seem plausible and it is left to future research to determine which scaling laws would apply to larger, faster air cushion craft and SES using flexible skirt seals. No reliable information is available for the scaling of rigid planer seals of CAB concepts.

8.9 Some Statistical Scaling Relationships

Because of the many variables in the design features of air cushion craft and the many assumptions that have to be made in the scaling "laws", it is useful to see how the many craft that have been built follow any scaling trends and compare them to the scaling laws discussed above.

Cushion Length and Cushion Pressure Design Trends

The size of air cushion craft was heavily influenced by the design philosophy prevalent in the UK where it was believed that lower cushion pressures would

[1] It is hoped that the reader can navigate between λ, being used as both a scale factor $({}^{L_{FS}}/_{L_m})$ and also as the sea wavelength in this discussion!

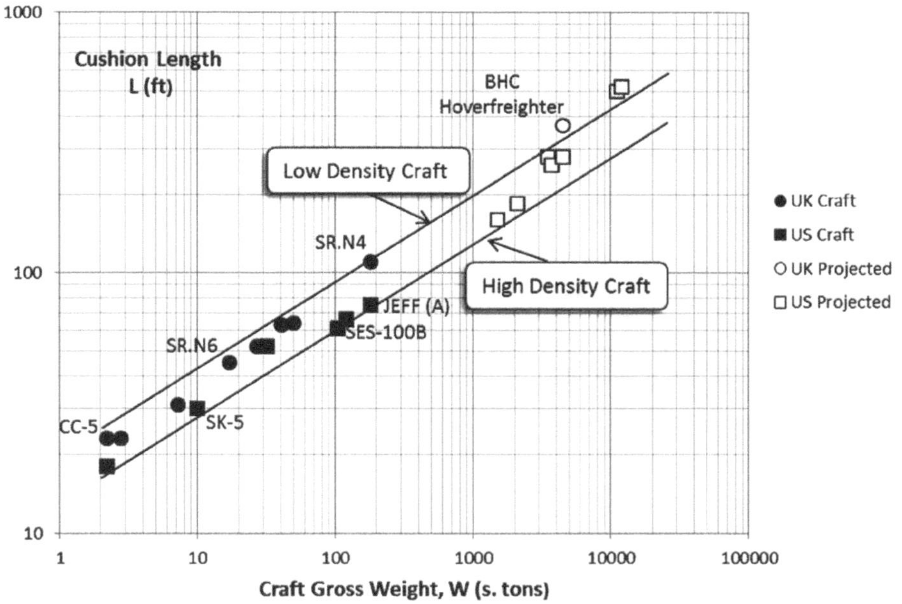

Figure 8.65: Size Trend of Aerostatic Air Cushion Craft

mean less power to overcome the wave drag. In the US, the design philosophy was to make the craft more compact which resulted in higher cushion pressures. These two "classes" of craft designs was evident in the size of the craft measured in terms of the length of the cushion (L_C). While not prescriptive, most British hovercraft exhibited low cushion density values ($^{p_C}/_{L_C} < 1.0$ psf per foot) and the US designs followed high cushion density values ($^{p_C}/_{L_C} > 1.0$ psf per foot). Figure 8.65 shows the size trends measured as the craft cushion length (L_C) as a function of craft displacement (weight) (W) for these two design philosophies.

The trends can be represented by the following equations:

$$\text{Low Cushion Density } \left(^{p_C}/_L < 1\right): \; L_C = 20W^{1/3} \tag{8.161}$$

$$\text{High Cushion Density } \left(^{p_C}/_L > 1\right): \; L_C = 12.5W^{1/3} \tag{8.162}$$

In these trend relationships, the cushion length (L_C) is measured in feet and the weight (W) is measured in short tons. It is seen that generally, not surprisingly, the aerostatic air cushion craft data base follows a Froude scaling for the overall dimensions of the craft. As shown in Chapter 11, where aerodynamic air cushion craft are discussed, these size trend "scaling laws" do not apply. The two design trends for the cushion pressure (P_C) of air cushion craft as shown on Figure 8.66 are:

$$\text{Low Cushion Density } \left(^{p_C}/_L < 1\right): \; p_C = 10W^{1/3} \tag{8.163}$$

$$\text{High Cushion Density } \left(^{p_C}/_L > 1\right): \; p_C = 19W^{1/3} \tag{8.164}$$

As before, English units of psf for cushion pressure and short tons for weight are used.

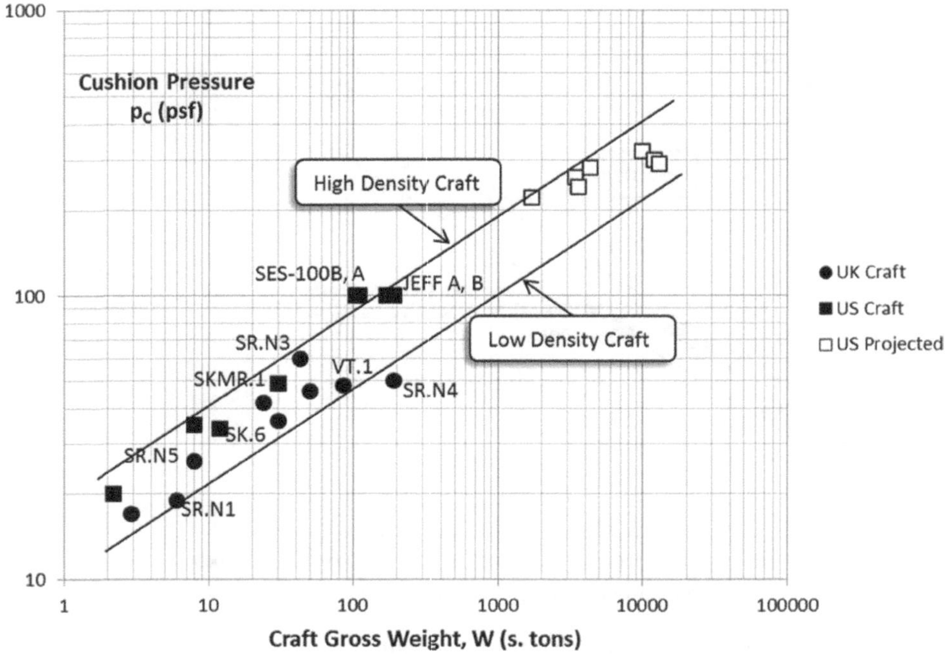

Figure 8.66: Cushion Pressure of Aerostatic Air Cushion Craft

From Equations 8.161 to 8.164, it can be seen that the design trend has been to follow Froude scaling such that the cushion density ($^{p_c}/_L$) is independent of scale once the basic decision has been made to follow either low cushion density design philosophy or high cushion density philosophy. The reader is reminded that the statistical trends shown here are the result of design trends only and not laws of physics and are provided for guidance and insight on various designers' lessons learned from actual craft production and operation. Again, these scaling trends apply to aerostatic type craft and not the aerodynamic type which are discussed in Chapter 11. A similar trend can be seen in the available data base for the cushion pressure as seen in Figure 8.66.

Total Installed Power and Lift Power Design Trends

Some of the problems facing the high speed air cushion craft designer when projecting to the larger sizes start to become apparent when the power requirements are computed. In the 1960s, the amphibious air cushion craft and the sidewall air cushion craft had reached an engineering level of sophistication sufficient to use as a starting point for the following discussion on scaling.

Figure 8.67 shows a wide range of craft both operational and projected from a modest 1 ton displacement experimental hovercraft to a projected 10,000 ton displacement sidehull SES aircraft carrier (SES CV). These craft show an installed power from 100hp to greater than 300,000 hp. The 10 ton displacement SRN 6 amphibious air cushion craft with a maximum speed of 60 knots together with the 20 ton displacement HM 2 sidewall hovercraft with a maximum speed of 40 knots

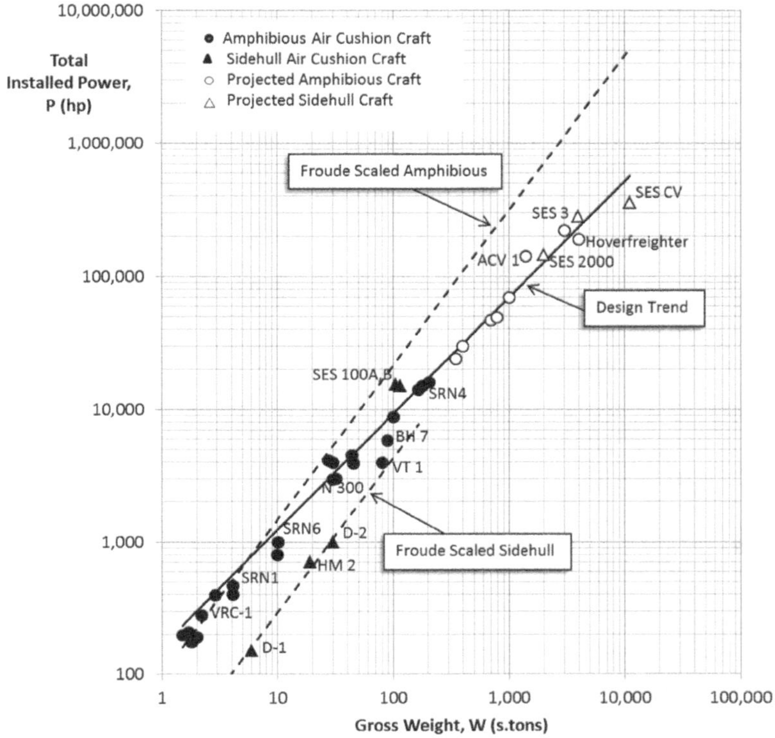

Figure 8.67: Total Installed Power for Aerostatic Air Cushion Craft

represent the state of the art in the late 1960s. By Froude scaling, the power would increase with size according to:

$$P \propto \lambda^{7/2} \propto W^{7/6} \tag{8.165}$$

This Froude scaling law has been superimposed on Figure 8.67 to illustrate how the power *would* have scaled to the larger craft if the designs had followed such a law. Commensurate with this Froude scaling would be an increase in speed that would have scaled the 60 knot SR N6 to a speed of 190 knots for a 10,000 ton displacement amphibious air cushion craft! A similar hypothetical Froude scaling *would* have scaled the sidewall hovercraft HM 2 to about 115 knots for the same displacement!

The realities of the drag phenomena already discussed precludes such scaling, of course, and both the actual built craft and projected ship designs scaled by a more modest scaling as shown by the trend curve through the data points. This trend curve is given by:

$$P_{total} = 165 W^{7/8} \tag{8.166}$$

Both the amphibious and the non-amphibious air cushion craft have settled into this trend line.

Included in the total installed power is the lift power (P_L) and it was found that a very similar design trend has appeared in all the operational and projected craft as shown in Figure 8.68. Again, the early designs have followed Froude scaling because

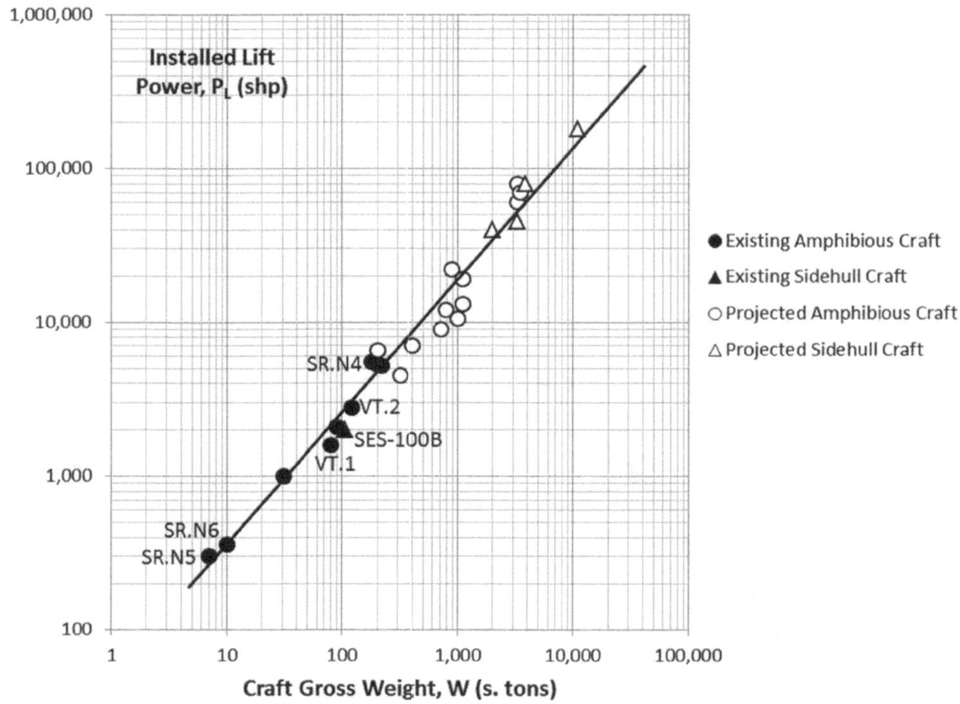

Figure 8.68: Installed Lift Power of Aerostatic Air Cushion Craft

of the lower speeds but as the speeds and size increased the design trends moved away from Froude scaling to give the trend line:

$$P_{lift} = 45W^{7/8} \qquad (8.167)$$

From the earlier discussion on wave pumping requirements for operation in different sea states it is interesting to see that the scaling is such that the relationship between lift power (P_{lift}) and total power (P_{total}) has remained relatively constant at about 27%. The old "rule of thumb" that the lift power of air cushion craft is about one third of the total power appears to be holding, despite the different sea states, speeds and sizes for the different ships.

Transport Efficiency Design Trends

Charts such as Figure 8.67 and Figure 8.68 provide quick overviews of the propulsion and lift power installed in aerostatic air cushion craft but they provide no information as to either speed capability or efficiency of the design. One must combine power, speed and displacement in the manner discussed earlier to see how each design compares with its peers in accomplishing some design objective. All the craft either built or projected to be built were designed with different objectives in mind. Figure 8.69 shows how the various craft group in achieving their Transport Efficiency values.

As stated above, it is the maximum installed power (to accomplish its maximum speed) that sizes the craft and Figure 8.69 shows these values as a function of the maximum speed for the two groups or "classes" of craft: (a) the amphibious air cushion craft and (b) the sidehull air cushion craft. The steady relationship between power and size has now been replaced by a wide scatter of data covering craft speeds from 20 knots to 100 knots (in calm water). The scatter is reduced (slightly) if the speed is expressed non-dimensionally as Froude Number as shown in Figure 8.70.

After three decades of development and operational experience from the data base shown, some immediate observations can be made:

- There is no definitive separation of *sidehull* air cushion craft and *amphibious* air cushion craft except that sidehull craft are generally more efficient at the lower speeds and amphibious air cushion craft are generally more efficient at the higher speeds. This is as might be expected from considerations of the drag of sidehulls in the water versus the extra lift air leakage of amphibious craft. This observation immediately becomes distorted however when the propulsion system is considered with its domains of efficiency for water propulsion and air propulsion
- As speed is increased, the craft tend to crowd against the "Froude barrier" in similar vein to displacement hulls. A precipitous fall-off in Transport Efficiency is seen as the Froude Number increases much above F.No. = 2 to 3.
- After intense development and technological exploration, resulting in the projected designs from the US Navy Advanced Concepts Evaluation (ANVCE) program (see Chapter 10) it is seen that the efficiencies of the advanced concepts (sidehull or amphibious) have not increased the level of Transport Efficiency much beyond the historical levels already achieved. The level achieved has not broken through the "ceiling" of WV/P = 10. This result has had a dampening effect on the many claims in the 1960s and later where advocates of different designs offered values of Transport Efficiency of 20 or more as they sought economical advantage for high speed marine transportation. Chapter 4 outlined the impetus for such craft to seek high speed ships that could operate on the open seas at competitive advantage to "expensive" air craft.

Figure 8.71 shows these same results plotted in context with conventional displacement ships and aircraft (both propeller driven and jet aircraft). In this context, the comparison is even more dramatic. The displacement ship data has been taken from Figure 8.48 and the planing craft data has been omitted for clarity. The envelope of the displacement ship data shown is a generous "envelope of envelopes" from Figure 8.48 to show the general trend for comparison with the air cushion craft data. The air cushion craft names have been omitted for clarity In Figure 8.71, but are clearly identified in Figure 8.69 and Figure 8.70.

Some license has been taken in the comparison in that the transport efficiencies for the air cushion craft are applicable for calm (or near calm water) operation while the displacement ship data (partly due to their lower speeds) are applicable to a wider range of sea state. In that sense, the data shown for the air cushion craft is the "best performance" values and thus the "ceiling" of WV/P = 10 should really be adjusted downward for practical expected levels in operational sea states.

The factor of 2:1 *less* efficiency than the Transport Efficiency values initially projected by the US Navy and MARAD in the 1965 era to start the major push to develop high speed ships for both commercial and military purposes, is a sobering

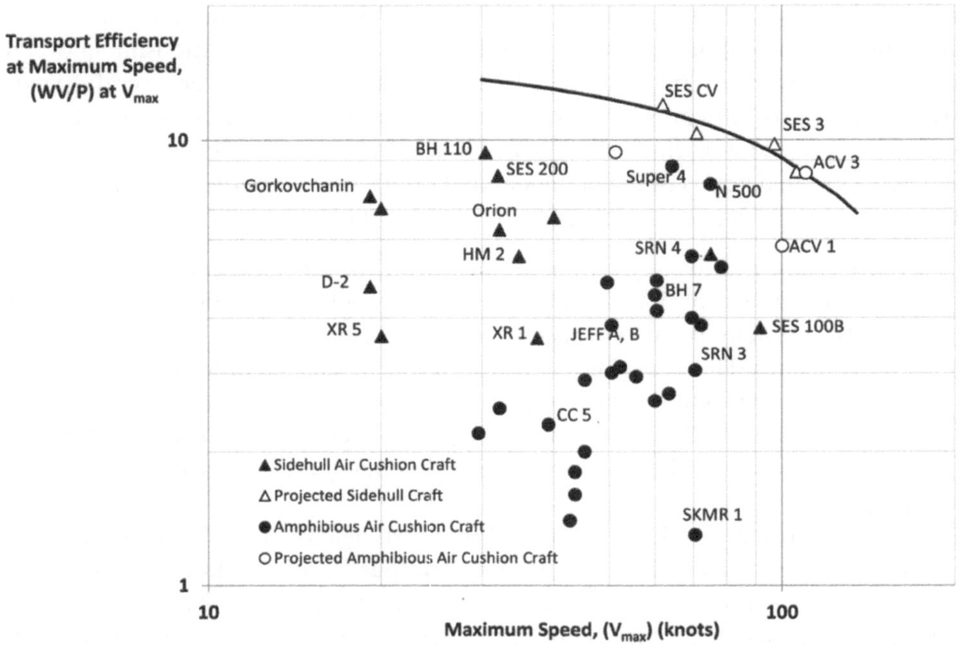

Figure 8.69: Transport Efficiencies of Aerostatic Air Cushion Craft at *Maximum Speed*

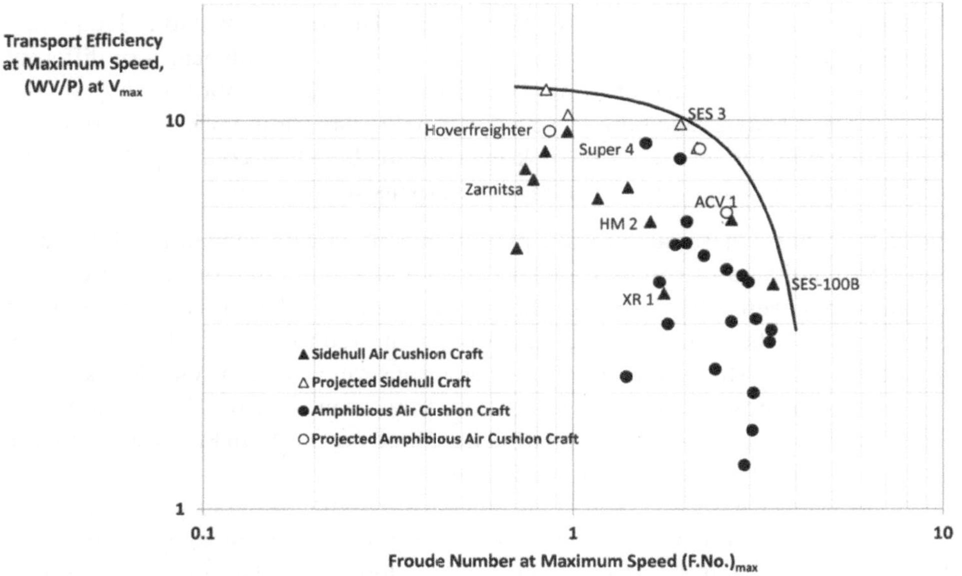

Figure 8.70: Transport Efficiencies of Aerostatic Air Cushion Craft at *Maximum Froude Number*

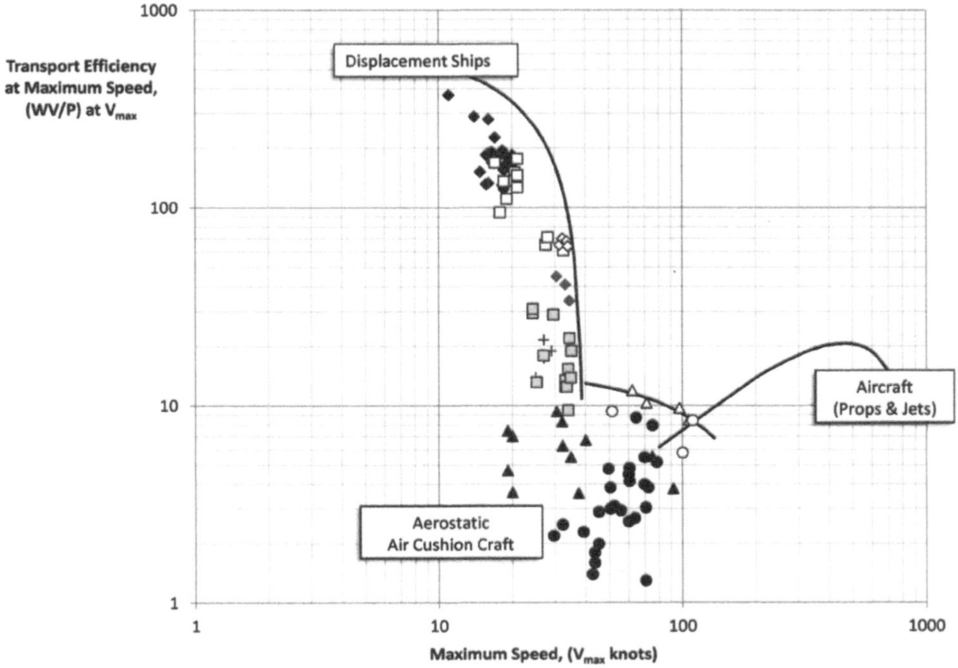

Figure 8.71: Transport Efficiencies of Aerostatic Air Cushion Craft, Ships and Aircraft

realization after two decades of intense development. The envelope of the aircraft data has been included in Figure 8.71 to emphasize that the driving reason behind the development by MARAD was to seek high speed ships with efficiencies somewhere between those of ships and those of aircraft. As the data shows, this was only marginally met for a few of the projected designs, again raising questions of economic advantage.

8.10 Scaling from SES-100B to Frigate Size SES

Some insight can be gained into the possibilities and limitations of aerostatic air cushion craft in meeting “high speed, high efficiency” ships by comparing the data from the SES-100B Test Craft and the projected performance for the frigate size SES. In Figure 8.71 (and in Figure 8.69) the frigate size SES (SES 3) designed as part of the 1976–1979 Advanced Navy Vehicles Concepts Evaluation (ANVCE) Program is shown as the point at *WV/P = 10 at a speed of 100 knots.* It was found during the ANVCE Program that the amphibious air cushion craft (ACV 1 and ACV 3) could achieve higher speeds as shown but at a loss of efficiency.

Early proponents of the high speed SES were projecting transport efficiencies of *twice* these values and at speeds approaching 130 knots (see Chapter 4 on the initial work by the US Maritime Administration). Such unrealistic performance is not likely especially for the aerostatic class of air cushion ships. It is informative to examine the results of the SES-100B program and see how those results can be scaled to a frigate class SES. The performance of the SES-100B has already been summarized earlier in this Chapter and the attained lift-drag ratio (L/D) can be

SES-100B

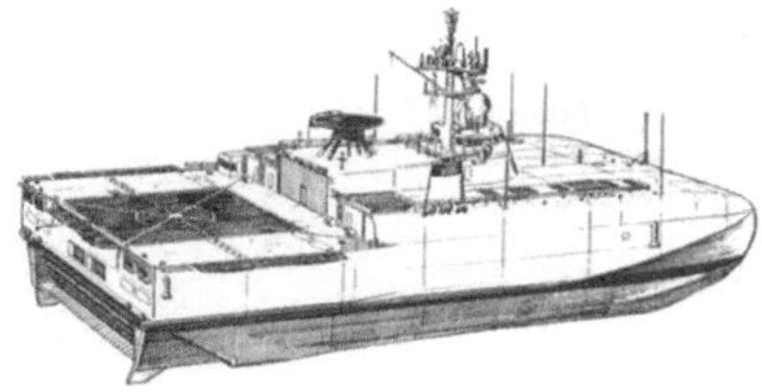

SES 3

Photo permission by Textron Marine & Land Systems and Textron Inc

SES-100B		SES 3	
W	105 s. tons	W	3,450 s. tons
LOA	77.7 ft	LOA	273 ft
BOA	35.0 ft	BOA	107 ft
Cushion Pressure	105 psf	Cushion Pressure	295 psf
Max. Cushion Flow	5,900 cfs	Max. Cushion Flow	55,000 cfs
Propulsion Power	13,500 shp	Propulsion Power	160,000 shp
Lift Power	1,860 shp	Lift Power	50,000 shp
Max speed (Calm Seas)	91.9 knots	Max speed (Calm Seas)	97 knots
Scale Factor	$\lambda = 1$	Scale Factor	$\lambda = 3.2$

Figure 8.72: Comparison of SES-100B and Frigate size SES (ANVCE SES 3)

compared to the predictions made during the US Navy Advanced Vehicles Concept Program for a frigate size SES. Figure 8.72 shows the basic characteristics of both craft. More detailed information on the frigate size SES is provided in Chapter 10 where a major effort was undertaken by the US Navy to determine the nature of such large high speed ships and what development would be necessary.

A true comparison will require study of the military mission, payload and operating conditions (see Chapter 10) but concentration here will be on the hydrodynamic scaling principles. After the design studies were completed as part of the ANVCE Program, one of the designs for the frigate size sidehull SES called the SES 3 is shown in Figure 8.72 which is approximately a 3:1 scale increase over the test craft SES-100B. The lift-drag ratio (L/D) of that design, taken from the study is shown in Figure 8.73. The prediction for the SES-100B using the equations developed earlier is included in Figure 8.73 for an analysis of the scale effect. Clearly, upon examination of Figure 8.72, it is seen that the two craft shown are not geometrically similar (a requirement for dynamic similitude). The shape of the sidehulls is different, the seal systems are of different designs, and the SES-100B was propelled by supercavitating semi-submerged propellers and the SES 3 (in this version shown) was propelled by waterjets. Each of these differences have a marked impact on the final performance but there is enough similarity to demonstrate some key points at least at the *hydrodynamic level* in scaling to large, high speed ships based on the air cushion principle. Figure 8.73 shows the two sets of prediction of the lift drag ratio variation with speed up to the maximum speeds for each craft (91.9 knots for the SES-100B and 97 knots for this version of the SES 3). At first glance, it would be assumed that there is very little similarity. Also, the expected increase in efficiency (L/D) with increase in ship size did not occur[m]. If now, the results are displayed as a

[m] A slight complication in the comparison is that the prediction for the SES 3 is done for operation in SS III which would scale down to SS I for the SES-100B but that difference would be small.

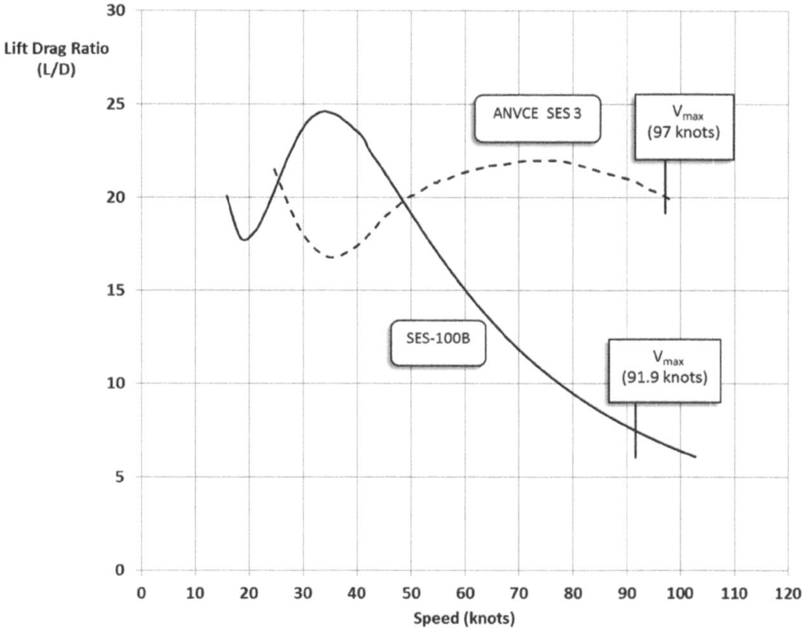

Figure 8.73: Lift Drag Ratio of SES-100B and Frigate Size SES as function of speed

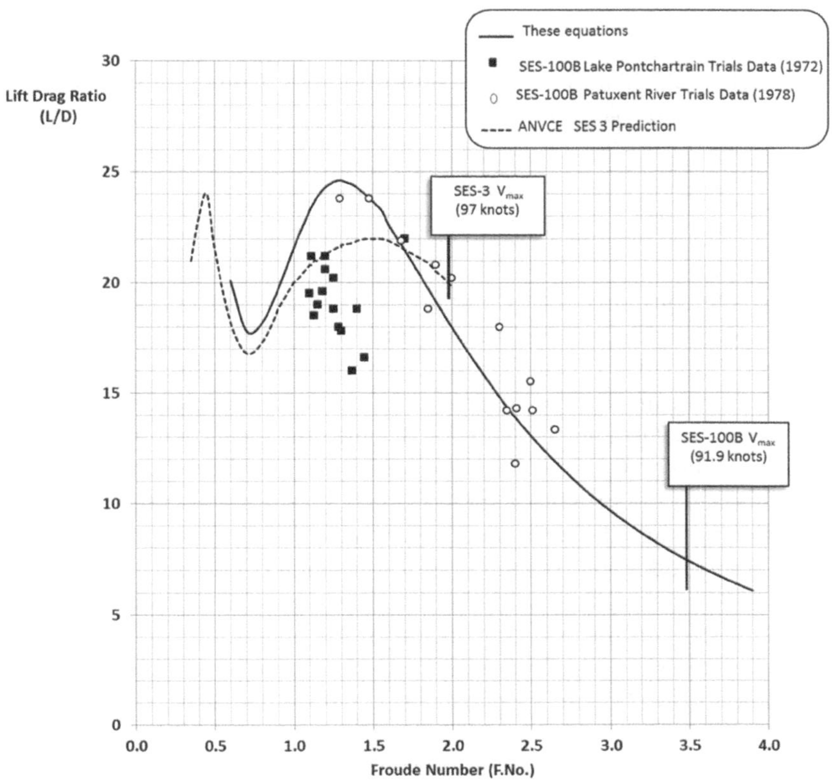

Figure 8.74: Lift Drag Ratio of SES-100B and Frigate Size SES as function of Froude Number

function of the non-dimensional Froude Number ($V/\sqrt{gL}$) instead of the actual speed (V, knots), a different perspective emerges as shown in Figure 8.74. The family resemblance between the two craft now starts to appear. The two sets of Trials data for the SES-100B from Lake Pontchartrain and Patuxent River have been added to Figure 8.74 to keep the discussion on a real basis.

As stated earlier, the SES-100B had several very specific design speeds to achieve. Two are discussed here. The first was to demonstrate efficient performance (high L/D) at a speed that would scale to a frigate size ship. The second design speed was to demonstrate safe and efficient performance at the "full scale" supercavitating flow speeds of 80 knots or more. That is why the frigate size SES "tops out" at a Froude Number of F.No. = 2.0 while the SES-100B is required to operate at a speed of F.No. = 3.5.

8.10.1 Magnitude of Lift Drag Ratio

If all the forces acting on the ship were scaled by Froude Number then the lift-drag ratio would be constant and independent of ship size. This is because both Lift (= weight) and the Drag would scale by the same scale factor of λ^3. This is not the case, however because the skin friction scales by Reynold's Number (R.No.) which would generate a *lower* drag and thus a *higher* lift drag ratio of the frigate size ship compared to the SES-100B. But again, as can be seen from Figure 8.74, this is not the case. An indication of the contributing causes may be found in the expression for the maximum value of the lift drag ratio given by Equation 8.121. If just the dominant drag contributions of *cushion wave drag, skirt or seal drag* and the *sidehull skin friction* are retained, then in terms of ship geometry, Equation 8.121 can be expressed in more recognizable terms as:

$$\left[\frac{L}{D}\right]_{max} = \frac{1}{2\sqrt{A\left(\frac{L}{B}\right)^{1/2}\left[B\frac{f(L/B)}{(h/C)^{1/3}} + C(L/B)C_f\right]}} \quad (8.168)$$

In Equation 8.168, the constants A (wave drag), B (skirt drag) and C (sidehull skin friction drag) are geometric constants that are included in the tabulation of all the drag components in Table 8.2. It is the relative magnitude of the three terms under the square root that determines if the lift-drag ratio will be larger or smaller between the two craft (the SES-100B and the SES 3). Because the cushion density (p_C/L) and the slenderness ratio (L/B) are less for the SES 3, it would be expected that the wave drag would be less and provide a greater lift-drag ratio for the SES 3 but the geometry of the sidehulls and the seals are different and the drag terms in the square brackets overrode this reduction to give a net lower value of the lift drag ratio as shown in Figures 8.73 and 8.74. This is driven primarily by the fact that there was not dynamic similitude between the two designs. Also, the assumption of the air gap leakage for the skirt drag is a significant factor in the two drag components. The impact of air cushion flow (Q) and air gap (h/C) has been discussed earlier in this chapter.

As the speed is increased past about F.No. = 2.0, the reduced skin friction coefficient (C_f) of the larger ship will start to show its effect through the effect of the higher Reynold's Number and the drag reduces and the lift drag ratio (L/D) starts to improve over that of the smaller craft. This is seen to occur as speed increases in Figure 8.74 as it approaches its maximum speed of 97 knots.

8.10.2 Speed for Maximum Lift-Drag Ratio

From Equation 8.119, the speed at which the lift drag ratio was a maximum was approximately given by:

$$|F|_{\left({}^{L}/_{D}\right)_{max}} = \sqrt[4]{\frac{a_1}{a_3}} \tag{8.169}$$

In this approximate expression, the terms a_1 and a_3 captured key characteristics of the craft that again scaled by different scaling laws. Specifically:

$$a_1 = K_1(1 + K_5 + K_1 K_6) \tag{8.170}$$

$$a_3 = K_2 + K_4 + K_8 \tag{8.171}$$

From Table 8.2;

$$K_1 \propto \left(\frac{p_C}{L}\right)\left(\frac{L}{B}\right)^{1/2} \tag{8.172}$$

$$K_2 \propto C_{D_o}\frac{S_F}{S}\frac{1}{{}^{p_C}/_{L}} \tag{8.173}$$

$$K_3 \propto \frac{{}^{L}/_{B}}{\left({}^{p_C}/_{L}\right)}\left(\frac{Q}{L^{5/2}}\right) \tag{8.174}$$

$$K_4 \propto \frac{1 + {}^{L}/_{B}}{\left({}^{L}/_{B}\right)^{1/2}}\left(\frac{h}{C}\right)^{-0.34}\frac{1}{{}^{p_C}/_{L}} \tag{8.175}$$

$$K_5 \propto \frac{1}{\left({}^{p_C}/_{L}\right)^{0.259}} \tag{8.176}$$

$$K_6 \propto \frac{\cos\beta}{\sqrt{{}^{L_{SH}}/{}_{B_{SH}}}} \tag{8.177}$$

$$K_8 \propto C_f\frac{{}^{L}/_{B}}{{}^{p_C}/_{L}}\left(\frac{L_{SH}}{L}\right)^2\left(\frac{d_i}{L_{SH}}\right) \tag{8.178}$$

These expressions (the full equations are given in Table 8.2) show how the overall geometry of the ship $({}^{L}/_{B}, {}^{L_{SH}}/_{L_C}, {}^{h}/_{C})$; the cushion density $({}^{p_C}/_{L})$; the cushion flow $(Q, {}^{h}/_{C})$; the geometry of the sidehull $({}^{S_{SH}}/_{S}, {}^{L_{SH}}/_{B_{SH}}, {}^{d_i}/_{L_{SH}}, \beta)$; aerodynamic shape $(C_{D_o}{}^{S_F}/_{S})$; and skin friction (C_f) all influence the speed at which the lift drag ratio is a maximum. Of these, if geometric similarity (by Froude scaling) is maintained, only the skin friction term will not scale according to Froude and the speed at which the lift drag ratio is a maximum will be at a slightly different scale speed than the "model" (in this case, the SES-100B).

Using the above equations, the Froude Number for which the lift drag ratio is a maximum is F = 1.26 (approx 33 knots) for the SES-100B which matches the full equations solution and the Trials data to a reasonable degree given the scatter of the data points. Because the skin friction coefficient (C_f) will be less for the frigate size ship (SES 3), the Froude Number will be slightly higher than for the SES-100B which is in general agreement with the results shown on Figure 8.74 at a Froude No. of F = 1.5 or approximately 75 knots. A direct Froude scaling from the SES-100B speed would be approx 62 knots showing that the reduced skin friction drag has a significant effect on the speed for maximum lift drag ratio. Note that these calculations can only give an indication in these examples because (a) the SES-100B is not an exact "similitude" model and (b) all the specific geometric characteristics of the craft are not available in this overview.

8.10.3 Scaling of Transport Efficiency

Scaling for the Transport Efficiency (WV/P) becomes more complex than for lift drag ratio ($^L/_D$) because now several non-scalable parameters come into play. Recall that the Transport Efficiency is given by:

$$\frac{WV}{P} = \eta_P \left(\frac{W}{D + \frac{\eta_P}{\eta_L} \left(\frac{p_C Q}{V} \right)} \right) \tag{8.179}$$

As discussed above for the lift drag ratio ($^L/_D$), if Froude scaling was the correct scaling law, then:

$$W \propto \lambda^3;\ V \propto \sqrt{\lambda};\ P \propto \lambda^{7/2} \tag{8.180}$$

Hence, by Froude scaling:

$$\frac{WV}{P} = \text{constant, independent of scale} \tag{8.181}$$

This result is not true, of course, because of the factors already discussed as a result of the many different forces acting on the ship, such as the viscous (or skin friction) forces and other considerations. The additional factors to consider in the Transport Efficiency (WV/P) over and above the hydrodynamic and aerodynamic considerations for the lift drag ratio ($^L/_D$) are the propulsive efficiency (η_P); the lift system efficiency (η_L); and the air flow rate (Q) that, as already discussed, is intimately tied to the Sea State through the wave pumping requirement. Hence, for dynamic similitude, the Sea State would also have to be scaled which, of course, is an unlikely situation in any practical sense.

Despite these differences, it is informative to compare the Trial results for the SES-100B in terms of Transport Efficiency with the prediction for the SES 3. This is shown in Figure 8.75.

The one set of data available from the 1976 Trials at Patuxent River is displayed for two settings of the air cushion flow rate (Q = 5500 cfs and 6000 cfs). As for the lift drag ratio ($^L/_D$) the family resemblance is seen for the

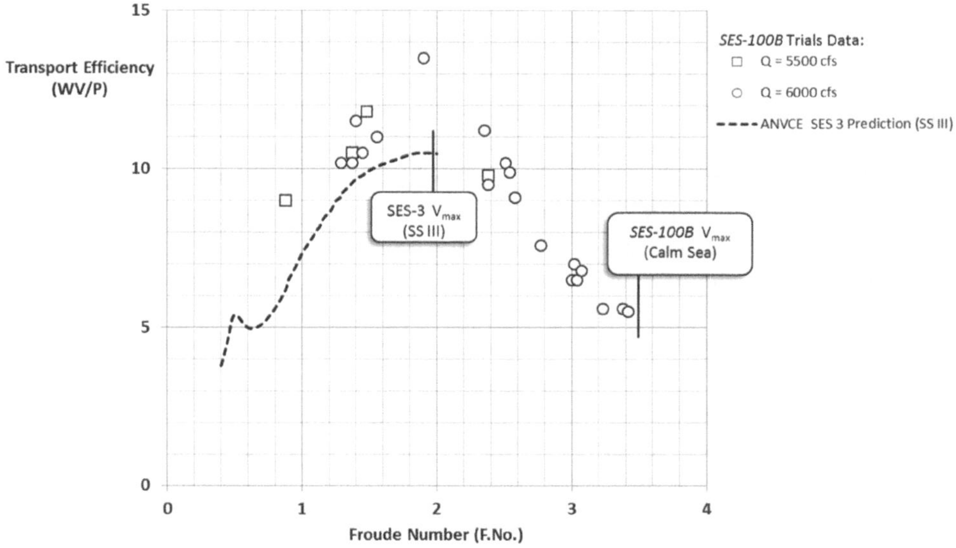

Figure 8.75: Transport Efficiency of SES-100B and SES Frigate

Transport Efficiency (WV/P). The prediction for the SES 3 was made for operation in SS III with a significant wave height of $h_W = 4.6$ feet which scales to SS I for the SES-100B. Unfortunately, the Trials data for the SES-100B in Sea State One is inconclusive (see Figure 8.56) because the tests were done at "off design" air flow rates (Q = 4100 cfs instead of 5900 cfs) but these differences will be ignored in what follows so that some general guidelines can be discussed. In the ANVCE Program (Chapter 10), the decision was made to use SS III as the baseline as this was the most likely low Sea State condition for generating general guidelines for maritime vehicle design.

The air flow rate of Q = 6000cfs Froude scales to the frigate size as:

$$Q_{SES3} = 6000\left(\lambda^{5/2}\right) = 137,506cfs \tag{8.182}$$

Such a value overestimates the air flow required for air cushion craft operation. From Equation 8.27, the air flow needed for ship deck zero vertical acceleration is given by:

$$Q_{wavepumping} = Bh_W V \sin\left(\frac{\pi L}{\lambda}\right) \tag{8.183}$$

At just one speed (50 knots) in Sea State III (with $h_w = 6ft; \lambda = 62ft$), this gives an air flow requirement of Q= 54,147cfs. From Figure 8.72, the designers have selected a lift system capable of generating Q=55,000cfs, which appears reasonable. Common practice based on actual craft as shown in Figure 8.68 showed lift systems with power:

$$P_{lift} = 45W^{7/8} \tag{8.184}$$

For the SES 3, this "trend" relationship gives a lift power of 56,000 shp. The SES 3 designers show a lift power of 50,000 shp, (see Figure 8.72) which although seemingly low is reasonably close given the vagaries of all the available data. This result is below that determined by Froude Scaling (see Figure 8.68) for the reasons already cited. It is unfortunate that the Trials for the SES-100B did not explore a consistent variation of tests on this crucial parameter in determining air cushion flow requirements over the range of likely Sea States for SES operation[n].

While it is disappointing that the US Navy Trials data did not zero in on this key feature of high speed SES (especially in the light of the data scatter in the drag curves), it is believed that sufficient emphasis has been given on extracting scaling laws from both the dimensional analysis and correlation with available test data from actual operational craft. While there was one lone data point in calm sea operation of the SES-100B of $WV/P = 13.5$ at 50 knots (F.No. = 2), the level of Transport Efficiencies around "10" as shown in Figure 8.71 is a reasonable value to be expected for large, high speed SES based on aerostatic design principles.

An important observation from these results is that the values of Transport Efficiency either derived from these equations or obtained by operational craft are about <u>*one half*</u> to <u>*one third*</u> the magnitude claimed by early proponents of such craft. This drop in efficiency was a significant contributor to the cancelation of many of the projects planned by the US Maritime Administration (MARAD) and the US Navy. There were other contributing reasons expanded upon here and in other chapters.

8.11 Useful Load, Disposable Load and Empty Weight

In addition to the *aerodynamic efficiency* (L/D) of the craft there is the efficiency of the design to carry a payload W_p). The total weight of the craft can be expressed as:

$$W = W_E + (W_p + W_f) \tag{8.185}$$

In this expression:

$$\text{Empty Weight} = W_E \tag{8.186}$$

$$\text{Useful Load or Disposable Load } (W_D) = W_p + W_f \tag{8.187}$$

In Equation 8.187, W_P is the craft payload and W_f is the fuel load. In craft design, it is usual to have a *design payload* (W_P) selected for a *design range* (R). In addition, the craft will have a ferry range in which the payload is "converted' to fuel to give the *ferry range.* Figure 8.76 shows this payload-range chart diagrammatically. The *useful* or *disposable load* (W_D) is thus the total load that can be interchanged between payload and fuel load as necessary in the design.

Because the choice of "how much payload (W_p)" and "how much fuel load (W_f)" is a variable in the design process, it is usual to compare vehicles on a disposable load (W_D) or empty weight (W_E) basis. In the US Navy ANVCE Project (see Chapter 10) this was done for a wide variety of vehicles to establish a "state-of-the-art" set of data.

[n] A competing design for the SES 3 in the ANVCE Program showed for a ship displacement of 3658 s. tons and a top speed in Calm Seas of 106 knots, the lift power was computed as $P_{lift} = 80{,}000$ shp and an air cushion flow capability of 75,000 cfs, which is still much less than Froude scaling would predict.

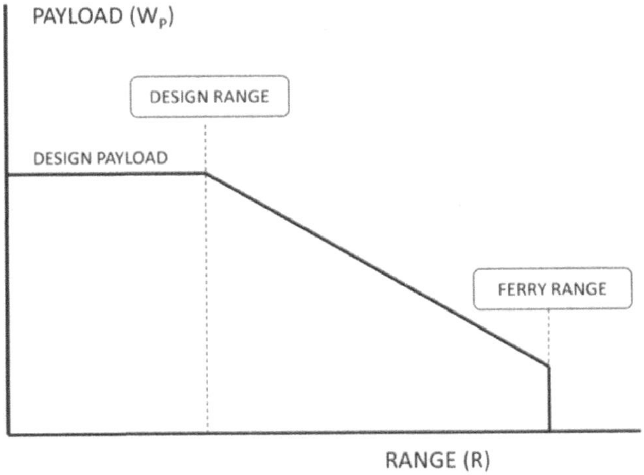

Figure 8.76: Typical Payload-Range chart

Two charts are reproduced here from Chapter 10 and Chapter 11 to summarize the data for the two types of craft of most interest. Figure 8.77 shows the data for the disposable load (W_D) of *air* and *surface craft* and Figure 8.78 shows the data for the empty weight (W_E) for the *aerodynamic form of air cushion craft.* It was a matter of convenience to work between disposable load ($W_D = W - W_E$) for Figure 8.77 and empty weight (W_E) for the aerodynamic air cushion craft in Figure 8.78. The relation between the two is simply:

$$\frac{W_D}{W} = 1 - \frac{W_E}{W} \tag{8.188}$$

In Figure 8.77, it is seen that for the *aerostatic forms of air cushion craft*, a disposable load (W_D) of about 40–50% is not an unreasonable expected value.

As can be seen in Figure 8.78, the data spread for the available aerodynamic form of air cushion craft is quite marked and due primarily to the preponderance of projected design values and limited actual craft. These differences are discussed in detail in Chapter 10 and in more detail in Chapter 11. The more reliable data from actual aerodynamic forms of air cushion craft would tend to show that a disposable load (W_D/W) of around 50% is a probable expected value pending more detailed design treatments. As discussed in Chapter 11, the PAR WIG design data (no actual craft built) of 20–30% for (W_E/W) does not appear practical.

More discussion of the component weights that make up the empty weight is provided in Chapter 7 (Economic Considerations) and Chapter 10 (US Navy ANVCE Project).

8.12 Design Efficiencies and Performance Measures

It was Louis-Charles Bréguet (1880–1955) who introduced the idea of three design efficiencies for aircraft (which have now been extended to all craft in comparison studies). Bréguet's work was expanded upon in Chapter 7 when the economic considerations for vehicle design were discussed. These three "efficiencies" were (in Bréguet's treatment) identified in his Bréguet Range Equation[o]:

[o] As before, in this discussion the use of the propeller driven craft equation will be used. See Chapter 11 for the variations of this equation for jet engine powered craft.

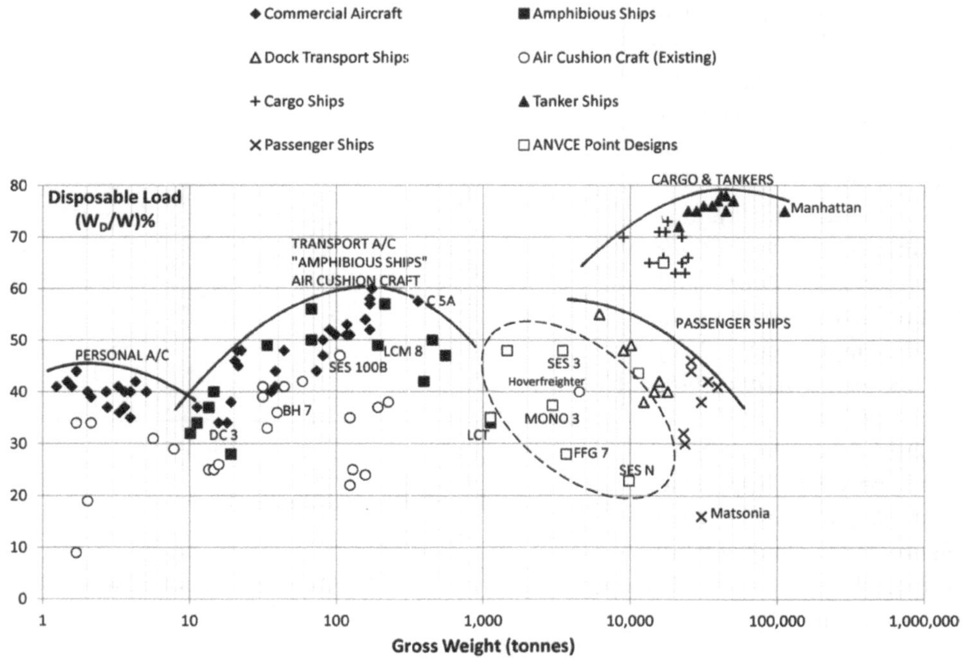

Figure 8.77: Disposable Load (W_D) of Air and Surface Craft

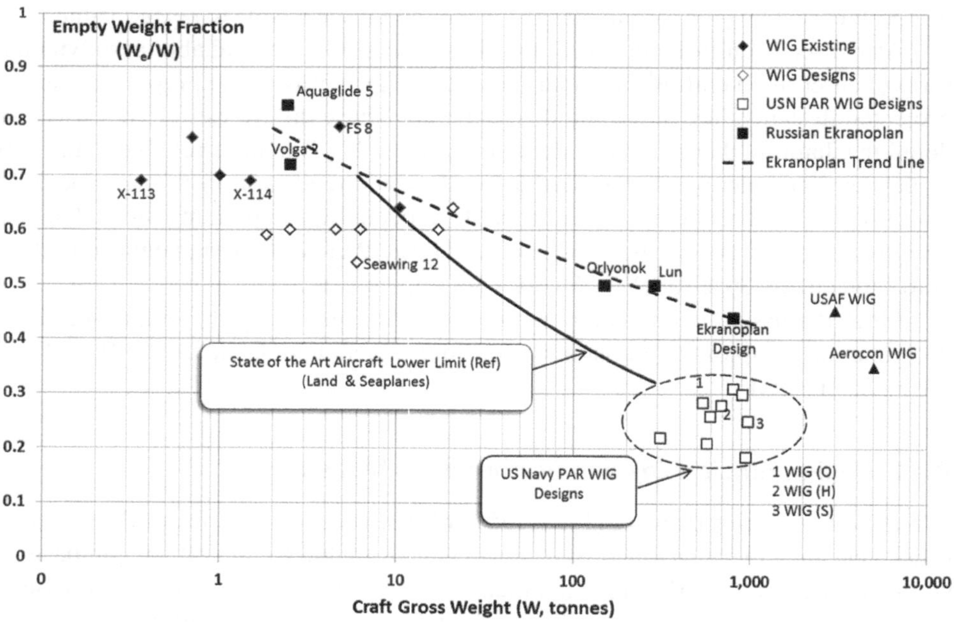

Figure 8.78: Empty Weight (W_E) of Aerodynamic Air Cushion Craft

$$R = 325\left(\frac{\eta_P}{sfc}\right)\left(\frac{L}{D}\right)\left(\log_e \frac{W}{W - W_f}\right) \tag{8.189}$$

Bréguet called the three terms in the parentheses in Equation 8.189 as his three efficiencies:

$$\text{“Propulsion Efficiency”} = \left(\frac{\eta_P}{sfc}\right) \tag{8.190}$$

$$\text{“Aerodynamic Efficiency”} = \left(\frac{L}{D}\right) \tag{8.191}$$

$$\text{“Structural Efficiency”} = \left(\log_e \frac{W}{W - W_f}\right) \tag{8.192}$$

These three efficiencies have been placed in quotes in the above equations to honor the names given in the original work by Bréguet. Even though new forms of craft (including the jet aircraft) have appeared on the scene since his original work in 1922, the essential importance of these three terms remain unchanged. His equation captures the three essential components of designing any vehicle designed to carry a payload (W_p) over a distance (R) in an efficient manner. The first efficiency relates to designing a propulsion system that has a high system efficiency and consumes the minimum amount of fuel. The second efficiency called aerodynamic efficiency by Bréguet which is the familiar lift drag ratio (L/D). As discussed in this and other chapters, this becomes the Effective Lift Drag Ratio $(L/D)_{EFF}$ for air cushion craft given by:

$$\left(\frac{L}{D}\right)_{EFF} = \frac{W}{D + \frac{\eta_P}{\eta_L}\left(\frac{p_c Q}{V}\right)} \tag{8.192}$$

The importance of this "*aerodynamic efficiency*" has been discussed in the earlier sections of this chapter and elsewhere in the book. The relationship to the Transport Efficiency (TE) is given by;

$$\text{Transport Efficiency (TE)} = \frac{WV}{P} = \eta_P\left(\frac{L}{D}\right)_{EFF} \tag{8.193}$$

The third efficiency called the "*structural efficiency*" by Bréguet is not just the efficiency of the structure of the vehicle (although it is invariably the largest component of the empty weight) but the efficiency of the combination of all components making up the empty weight. The statistics of this key component have been collected above in Figure 8.77 and Figure 8.78 for all air and surface vehicles, where:

$$\text{“Structural Efficiency”} = \log_e\left(\frac{1}{1 - W_f/W}\right) \tag{8.194}$$

This last efficiency becomes dependent on each specific design arrangement where a design trade is made between payload (W_P) and fuel load (W_f) in the make-up of the disposable load ($W\text{-}W_E$).

If a first order approximation is made to the natural logarithm in Equation 8.189 and Equation 8.194, then one gets the simpler form:

$$R(miles) = 325\left(\frac{1}{sfc}\right)\left(\frac{WV}{P}\right)\left(\frac{W_f}{W}\right) \quad (8.195)$$

Two economic measures frequently used in evaluating vehicles can then be written:

$$\text{Productivity (payload ton-miles)} = 325\left(\frac{1}{sfc}\right)\left(\frac{WV}{P}\right)\left(\frac{W_f}{W}\right)W_P \quad (8.196)$$

$$\text{Payload ton-miles per pound of fuel} = 0.163\left(\frac{1}{sfc}\right)\left(\frac{WV}{P}\right)\left(\frac{W_P}{W}\right) \quad (8.197)$$

The constants in the above equations have been chosen for English units (miles, tons, pounds, etc).

Given these equations it becomes a simple matter to compare the performance of two vehicles (subscripts 1 and 2) in the following manner:

$$\text{Range}: \frac{R_1}{R_2} = \frac{(sfc)_2}{(sfc)_1}\frac{({}^{WV}/_P)_1}{({}^{W}V/_P)_2}\frac{({}^{W_f}/_W)_1}{({}^{W_f}/_W)_2} \quad (8.198)$$

$$\text{Payload-Range}: \frac{(W_P R)_1}{(W_P R)_2} = \frac{(sfc)_2}{(sfc)_1}\frac{({}^{WV}/_P)_1}{({}^{WV}/_P)_2}\frac{({}^{W_f}/_W)_1}{({}^{W_f}/_W)_2}\frac{(W_P)_1}{(W_P)_2} \quad (8.199)$$

$$\text{Payload ton-miles per pound of fuel}: \frac{(\text{Ton}-\text{miles/pound})_1}{(\text{Ton}-\text{miles/pound})_2} = \frac{(sfc)_2}{(sfc)_1}\frac{({}^{WV}/_P)_1}{({}^{WV}/_P)_2}\frac{({}^{W_P}/_W)_1}{({}^{W_P}/_W)_2} \quad (8.200)$$

The statistical charts given throughout this chapter give some idea of the variations possible of the above key parameters in any particular case and specific calculations would have to be done for detailed and complete analysis.

8.13 Summary

This chapter has provided some general guidelines as to the necessary analyses needed to be done in determining craft performance and stability, especially when evaluating performance in the light of other requirements of design such as cost, military use and other factors. More detailed treatment is covered in the other chapters where specific issues are addressed.

References

1. Mantle, P. J., *"Air Cushion Craft Development"*, published by US Navy in 1980 (GPO Stock No. 008-047-00291-2)
2. *ibid*
3. Havelock, Thomas H., *"The Effect of Shallow Water on Wave Resistance"*, Proceedings Royal Society, Series A, Vol 100, No. A706, 1922
4. Lamb, Horace, *"Hydrodynamics"*, published by Dover Publications, Sixth Edition, 1932. First edition 1879.

5. Crewe, Peter R. and Eggington, Wilfred J., *"The Hovercraft – A New Concept in Maritime Transport,"* RINA Lecture, 19 Nov 1959. See also Transactions of Royal Institute of Naval Architects, 1960
6. Newman, J.N. and Poole, F.A.P., *"The Wave Resistance of a Moving Pressure Distribution in a Canal"*, DTMB Report 1619, March 1962
7. Barratt, M.J., *"The Wave Drag of Hovercraft"*, Journal of Fluid Mechanics, Vol 22, Part 1, pp 39–47, 1965
8. Doctors, L. J., *"The Wave Resistance of An Air Cushion Vehicle"*, University of Michigan, December 1970.
9. Everest, J.T. and Hogben, N., *"Research on Hovercraft Over Calm Water"*, Transactions of Royal Institute of Naval Architects, Vol 109, pp 311–326, 1967
10. Mantle, P. J., *"Air Cushion Craft Development"*, published by US Navy in 1980 (ISBN 008-047-00291-2)
11. Beardsley, Mel W., *"Wave Pumping and Its Effects on the Design and Operation of Air Cushion Ships"*, US Navy Marine Engineering Laboratory, MEL Tech memo 48/67, March 1967
12. *ibid*
13. Mantle, P. J., *"Air Cushion Craft Development"*, published by US Navy in 1980 (ISBN 008-047-00291-2)
14. *"Parametric Design Analysis and Concept Exploration"*, Arctic Surface Effect Vehicle Program, Bell Aerospace Final Report 7416–950001, Vol 1, February 1973
15. Chaplin, John B., *"Amphibious Surface Effect Vehicle Technology – Past, Present and Future"*, AIAA/SNAME Advanced Marine Vehicles Conference, Paper No. 74–318, San Diego, CA, 25–27 February 1974
16. Waldo, Robert D., *"Some Special Problems in Surface Effect Ships"*, Journal of Hydronautics 2, Vol 3, July 1968
17. Handler, Eugene (US Navy Project Office), *"Major Test Facilities Used in Development of the Large Surface Effect Ship"*, Hovering Craft & Hydrofoil, Volume 17, No. 3, December 1977
18. Wilson, Robert A., *"Design Parameters Affecting Turn Stability of Captured Air Bubble Vehicles"*, NSRDC Report No. 2965, 1968
19. Hydronautics, *Final Technical Task Report on Improved SES Performance*, Volume II by V. Blaes, J. Witte, R. Barr, O. Scherer, D. Wiggett, Report No. 834–1, 1967
20. Wilson, R. A., *"Captured Air Bubble Vehicle Stability Tests"*, AIAA/SNAME Conference on Advanced Marine Vehicles, Paper No. 67–349, Norfolk, Virginia, May 1967
21. Lanchester, F. W., *"Aerodynamics"*, Constable and Co., pp 139–178, London, 1907
22. Prandtl, Ludwig, "Flüssigkeitsbewegung", Handworterbuch der Naturwissenschaften, Vol 4, pp 101–140, Gustav Fischer, Jena, Germany, 1913
23. Jones, R.T., *"Correction of the Lifting Line Theory For the Effect of Chord"*, NACA TN No. 817, July 1941
24. Korvin-Kroukovsky, B. V., *"Lift of Planing Surfaces"*, Journal of Aeronautical Sciences, Readers Forum, pp 597–599, September 1950
25. Wagner, von Herbert, *"Hauplaufsätze Uber Stoss– und Gleitvorgünge an der Oberfläche von Flussigkeiten"*, ZAMM, pp 193–215, August 1932
26. Korvin-Kroukovsky, B. V.; Savitsky, Daniel; Lehman, William F., *"Wetted Area and Center of Pressure of Planing Surfaces"*, Experimental Towing Tank, Stevens Institute of Technology Report 360, August 1949
27. Sottorf, W., *"Versuche mit Gleitflächen" (Experiments with Planing Surfaces)*, Werft-Reederei-Hafen, 1932 & 1933. Re-issued under translation by J.Vanier as a NACA Tech Memo No.739, March 1934
28. Sambraus, A., *"Gleitflächenversuche bei grossen Froudeschen Zahlen und Tragflügelvergleich"*, Luftfahrtforschung, Vol 13 No. 8, pp 269–280, 20 August 1936. Translated by J. Vanier and re-issued as *"Planing-Surface Tests at Large Froude Numbers—Airfoil Comparison"*, NACA TM No. 848, February 1938

29. Shoemaker, James M., *"Tank Tests of Flat and V-Bottom Planing Surfaces"*, NACA Tech Note No. 509, November 1934
30. Locke, Fred W. S. Jr., *"An Empirical Study of Low Aspect Ratio Lifting Surfaces with Particular Regard to Planing Craft"*, Journal of the Aeronautical Sciences, March 1949, pp184–188
31. Bollay, W., *"A Contribution to the Theory of Planing Surfaces"*, Proceedings of Fifth International Congress of Applied Mechanics, p 474, 1938
32. Weinstein, Irving, and Kapryan, Walter J., *"The High Speed Planing Characteristics of a Rectangular Flat Plate Over a Wide Range of Trim and Wetted Length"*, NACA Tech Note 2981, 1953
33. Shuford, Charles L., *"A Review of Planing Theory and Experiment with a Theoretical Study of Pure-Planing Lift of Rectangular Flat Plates"*, NACA Tech Note 3233, August 1954
34. Springston, George B. Jr., and Sayre, Clifford L., Jr., *"The Planing Characteristics of a V-Shaped Prismatic Surface with 50 degrees Deadrise"*, US Navy DTMB Report 920, February 1955
35. Mantle, P. J., and Hall, A. M., *"Multiple Hull Vehicle with Air Cushion Support"*, US Patent 3,476,069 filed 24 January 1967, patented 4 November 1969
36. Payne, Peter, *"Contributions to Planing Theory"*, Ocean Engineering, Vol 21, No. 3, pp 257–309, 1994
37. Schultz-Grunow, *"New Skin Friction Drag Law"*, Lufo 1940, p239; Translation NACA Tech memo 986
38. Hoerner, Sighard, *"Fluid Dynamic Drag"*, pub by author, Library of Congress Card No. 64–19666, 1965
39. Weinblum, George P., Janet J. Kendrick and M. Allison Todd, *" Investigation of Wave Effects Produced by a Thin Body, TMB Model 4125"*, DTMB Report 840, November 1952
40. Gabrielli, T.G., and Th. Von Karman, *"What Price Speed?"*, Mechanical Engineering, pp. 775–781, October 1950
41. Mantle, Peter J., *"Some Design Aspects of Air Cushion Craft"*, Annals of the New York Academy of Sciences, Vol 154, Article 2, pp 924–952, November 1968. Previously presented at the International Congress of Subsonic Aeronautics, New York, N.Y., 3–6 April 1967
42. Hoerner, Sighard, *"Fluid Dynamic Drag"*, pub by author, Library of Congress Card No. 64–19666, 1965
43. Havelock, Thomas H., *"Propagation of Waves Produced by a Travelling Disturbance"* (Short form of title), Proceedings of Royal Society Series A, LXXXI, pp 398–430, London 1908
44. Wigley, W.G.S., *"Calculated and Measured Wave Resistance of a Series of Forms"* (Short form of title), Transactions of Royal Institute of Naval Architects (RINA), Vol 84, pp 52–74, 1942
45. Raley, F. Carl and Fred W. Wilson, *"Test and Evaluation of US Navy 100-Ton Surface Effect Ships"*, Paper No. 76–860, AIAA/SNAME Advanced Marine Vehicles Conference, Arlington, Virginia, 20–22 September 1976
46. Pierson, W. J., and L. Moskowitz, *"A Proposed Spectral Form for Fully Developed Seas Based on the Similarity Theory of S. A. Kitaigorodskii"*, Journal of Geophysical Research, Vol 69, No. 24, pp 5181–5190, 1964
47. Raley, F. Carl and Fred W. Wilson, *"Test and Evaluation of US Navy 100-Ton Surface Effect Ships"*, Paper No. 76–860, AIAA/SNAME Advanced Marine Vehicles Conference, Arlington, Virginia, 20–22 September 1976
48. Wilson, R.A., S.M. Wells and C.E. Heber, *"Powering Prediction for Surface Effect Ships Based on Model Results"*, AIAA/SNAME Advanced Marine Vehicles Conference, Paper No. 78–744, San Diego, California, 17–19 April 1978
49. Mantle, P. J., *"On the Dynamic Heave Motion of Single Peripheral Jet GEMs Over an Undulating Surface"*, Vehicle Research Corporation Paper No. 9 (May 1962)

50. Tinajero, A.A. and J.N. Fresh, *"Aerodynamic Response of a Seven Foot Ground Effect Machine Flying Over Uneven Surfaces"*, DTMB Report 1436 (1960)
51. Uyeda, S.T., *"Study of Loads and Motions of Two Types of Ground Effect Machines"*, General Dynamics Convair Report ZH-150 (1961)
52. Wheatley, J.H.W., *"Heave Suspension Characteristics and Power Requirements for a Plenum Air Cushion"*, National Physical Laboratory Hovercraft Report No.9, December 1969
53. Breslin, J., *"The Dependence of Surface Effect Ship Heave Response on Model Scaling Parameters as Deduced from a Simplified Linear Analysis"*, Surface Effect Vehicle Summer Study Report for Institute of Defense Analyses", October 1968
54. Mantle, P.J., " *Some Design Aspects of Air Cushion Craft,* " Annals of the New York Academy of Sciences, Vol 154, Article 2, pp 924–952, November 22, 1968. (Paper also presented at the International Congress of Subsonic Aeronautics, New York, N.Y. 3–6 April 1967
55. Meeks, Capt T.L., and P.J. Mantle, *"The Advanced Naval Vehicle Concepts Evaluation"*, AIAA/SNAME Advance Marine Vehicles Conference, Arlington, Virginia, Paper 76-846, 20 September 1976

9 Navy Military Operations Considerations

To be prepared for war is one of the most effectual means of preserving peace.
President George Washington Speech to Congress, 8 January 1790

In the 1960s, the studies by the US Maritime Administration on the *economics* of high speed ships showed definite advantage to ships that could transport goods at speeds approaching 100 knots. Later, studies by the US Navy recognized the *military* value of "high speed" but could not be specific about such a classification as "a 100 knot Navy" in general and only in specific missions could the advantage of such a speed be quantified. The one navy mission area that could show benefit was the anti-submarine warfare (ASW) mission where it could be shown that a ship that sprinted ahead of the fleet; drop to "listening speeds" with a towed array; and sprint ahead again such that the average speed of such a ship would be comparable to the "fleet speed" of the naval force was of value if the sprint speed was close to 100 knots. In this instance, taking into account the characteristics of the platforms and sonar technology at the time, the speed capability of "100 knots" appeared to be an advantage. Other navy missions did not show a specific need and other ship characteristics played an equal or greater role than just speed.

In earlier centuries, there was no question that being a fast ship "*that goes in harm's way*" was a definite advantage, but as technology advances in weapon systems blossomed, the speed of the *platform* became less a factor than that of the *weapon*. It *may* have been possible, in yesteryear, by ship maneuvering to avoid an enemy's cannon ball – but not from today's anti-ship missile. In the 1970s some tests were done with high speed hydrofoils where, by taking advantage of the hydrofoil's high maneuverability capability, some successes were achieved in "dodging" (break lock) a single anti-ship second generation missile but this was not the general case. Advances in missile technology since the 1970s would most certainly negate any high speed ship's maneuverability defense against today's brand of missiles (cruise or ballistic). The value of "high speed" and in "100 knots in particular" would have to stand the scrutiny of a broad based mission analysis. Only sporadic reports during the development period of high speed ships have tackled this subject in any meaningful and quantifiable way.

In 1979, the US Navy published the results of its two year study; Advanced Naval Vehicles Concepts Evaluation (ANVCE) Project, that tackled this complex problem.[1] See Chapter 10 for more details of that project's findings. One of the challenges faced in that study was the temptation to consider the design of high speed ships in the context of "the last war" and not in preparation of what the next war or conflict might need. This was especially important as it was already becoming clear to Navy planners at the time that technology was advancing on all fronts and that the very nature of the "enemy" was changing. Within a decade of the issuance of the ANVCE Report, the world had dramatically changed and Strategic Planners in the defense departments around the NATO nations were recognizing that the likely wars in the future would not be between two major adversaries with an established battleground ("line in the sand") between them but a series of conflicts with adversaries who were not constrained by national boundaries and "normal" means of fighting. The US learned this hard lesson during the Vietnam conflict when it was learned that high technology weapon systems designed for "the last war against a similar minded enemy" simply did not work against an adversary who fought a different kind of war (jungle warfare in the case of Vietnam) with non-conventional and non-high technology weapons.

In 2001, this bitter lesson was learned again in the US where the (non-national state) adversary chose not to fight on the military battlefield but moved the fight into the middle of the civilian populace; and further, used our own systems (civilian aircraft) against us as the weapon. Today, in 2012, the definition of warfare is still being analyzed with no guarantee that the "right weapons" or the "right operational tactics" are being chosen. The global reach of missiles has forced the US to consider establishing missile silos on its soil equipped with intercontinental range missiles. As of today, there are only a minimum number of "bullets in the holster" with this defense architecture with large portions of the US left unprotected. Any change in the threat direction would place the US into a "Maginot Line" defense.

NATO Europe is still planning how it should defend itself against missile attack and the latest NATO planning suggests it will not have a planned defense capability until at least 2020 and maybe much later. That is, Europe would not have a NATO defense capability until a decade from now. Also, NATO is considering changing from its original 1990 plan of a *land based* "fixed" system into more of a mobile *sea based* system because of the (real) uncertainty of the next threat direction. In this defense equation, the US Navy is developing missile equipped ships that must be capable of "speeding" to designated threat areas on short notice. Would 100 knot ships improve this situation?

Coupled with the uncertain nature of "the next war" scenario and how best to "preserve the peace", is the ever increasing time it is taking to develop new weapon systems. Almost a generation (25 years or more based on collected statistics)[2] is a typical time period required to develop and field new weapon systems of all types. While such an explosion of technology and the new nature of warfare itself were not yet on the horizon in 1976, the ANVCE Project and the US Navy planners of the time sought to understand where high speed ships would fit into this scenario of non-conventional forms of warfare. The ANVCE Project specifically explored the value of speed in the military missions of transit, convoy, search, pursuit, maneuver, avoidance and in the specific mission of amphibious warfare. With such variability in the nature of future warfare and the time taken to develop such capabilities, it would seem that a *greater degree of flexibility* needs to be integrated into the *designs of the future platforms* and concentration on *just* the speed of the platform may not be the correct approach.

Some comments on military operations and how they might benefit from high speed platforms with unique capabilities follows.

9.1 Speed and Amphibious Capability

In the design of air cushion craft, the ability of such craft to operate over both land and sea was a useful ability in many missions. The English dictionary defines amphibious to be *"capable of operating on both land and water"*. That is quite a different meaning than that used by the military in their definition of amphibious warfare; which is *"military operations by both land and naval forces against the same object"*. Craft designed for use in amphibious warfare, even today, are designed to *stop at the shoreline.* This is because of their conventional and simple hull design and use of marine propulsion. Figure 9.1 shows a typical *amphibious* landing during WW II. The particular amphibious landing shown is that by the US infantry and marines landing on Omaha Beach at Normandy on D-Day on 6 June 1944.

Figure 9.1: Amphibious Landing on Omaha Beach 6 June 1944
US Army photo

In the case of the Normandy landings, some 160,000 Allied forces landed along a 50 mile stretch of French coastline of relatively smooth sandy beach. Because the coastline was heavily fortified by the occupying Nazi Germany, the vulnerability of

Figure 9.2: Vulnerability of Infantry during Amphibious Landing (D-Day)
US Army photo

the Allied landing forces, first as they approached the beach from the sea (at about 6–10$^+$ knots depending on sea conditions in the landing craft shown unloading at the shoreline), and second, as the troops and materiel were setting up their beachhead. All done under the strafing gunfire from the enemy fortifications further inland. An even more graphic display of the vulnerability of troops during amphibious landings is shown in Figure 9.2, that shows troops landing during the same invasion at Normandy on 6 June 1944. Because of the slope of the beach it was not always possible for the "amphibious" landing craft to even get to the shore line and the troops were required to wade ashore under fire from the enemy fortifications on the cliffs above the beach, barely visible, in the background. Both Figure 9.1 and Figure 9.2 show the vulnerability of troops and materiel to enemy fire while the beach head is being established. Such amphibious landings suffered heavy losses during World War II from enemy gunfire and further, self-inflicted injuries from dangerous landing craft ramp operations in rough seas.

Figure 9.3 shows another amphibious landing, this time by the US Coast Guard in WW II, unloading equipment, that shows that wide open beaches cannot be assured around the world's coastlines to the landing forces, increasing their vulnerability even more as the enemy can now focus his gunfire onto the only available landing spots for shallow draft landing craft.

Much later, this serious vulnerability to amphibious warfare changed in 1977 with the introduction of the *air cushion craft*. In the period 1965–1970, the US Navy, including its Marine Corps (USMC), conducted studies on how to improve its landing craft by taking advantage of the work ongoing in the UK on hovercraft. The aim was twofold: (a) increase the speed of the landing craft to reduce the "time to shore", and (2) to remove the constraint of having to stop at the shoreline to disembark troops and equipment. Contracts were let to two companies: Aerojet General for a

Figure 9.3: Not All Amphibious Landings have Smooth Unobstructed Beaches
US Army photo

prototype craft, JEFF (A) and Bell Aerospace for a prototype craft, JEFF (B). A description of the program was given by James Schuler, the US Navy Program Manager in 1977.[3] The winner was Bell Aerospace with its first craft being launched and making its first run, off the coast near the US Naval Coastal Systems Laboratory, Panama City, Florida on 17 October 1978. On that particular run, the craft traveled at a speed of 42 knots – some 3 to 4 times faster than the "amphibious" landing craft of World War II. After the prototype program was complete, the first production air cushion landing craft (LCAC-1) built by Bell Aerospace (now Textron Marine & Land Systems) rolled out on 2 May 1984 and delivered to the US Navy unit ACU-5 in Panama City, Florida on 14 December 1984. About 100 craft were built and delivered by the contractor over the 16 year period 1984–2000. This number included 8 craft to Japan.

The introduction of the *air cushion* supported landing craft changed the Amphibious Force capability significantly, improving both the performance of landing craft and a significant improvement in the survivability of the infantry and marines during the assault. Figure 9.4 shows an LCAC crossing the shoreline during the military exercise, *Talisman Sabre*, on 30 July 2011 with the Australian Defence Force. If the terrain will permit, with no obstacles greater than 4 feet in height, the LCAC can deliver its cargo (troops and equipment) even further inland. Figure 9.5 shows such a landing where the LCAC has landed inland well away from the shoreline during the 15th Marine Expeditionary Force humanitarian aid mission in

Figure 9.4: LCAC crossing the shoreline
USMC photo

Figure 9.5: LCAC Unloading Away from the Shoreline
USMC photo

California on 13 January 2005. It is seen that the loading and unloading is done on dry land with the shoreline in the background, which is a marked improvement over the vulnerability of troops in WW II style landing craft under enemy fire.

As a significant improvement in capability given to the Amphibious Forces by air cushion landing craft over that experienced by the World War II landing craft, it is unfortunate that this capability was not embraced to any significant degree as judged by the small numbers of craft ordered by the US Navy. Only about 91 craft (LCAC) were delivered to the US Navy which is a small number compared to the more than 35,000 conventional landing craft used in World War II of all types (LCM, LCVP, LCP, LVTP).

As the LCAC comes to the end of its service life[a], the US Navy is proceeding with follow-on craft to the LCAC and is currently in the early design phase of an

[a] The LCAC will be extended to 2028 through its service life extension program (SLEP)

improved version called the Ship to Shore Connector (SSC). The SSC is an upgraded version of LCAC with different engines, increased payload (from 60 tons to 74 tons) and other features for improved Reliability & Maintainability. The speed is a little slower (at 35 knots^{+}) but a slightly longer over-the-horizon (OTH) range of 25nm instead of 15nm. Only 73 SSC craft have been ordered with the first delivery planned for 2017 and the remainder to be delivered over an eight year period through 2025. Meanwhile, the US Navy Amphibious Forces continue to use conventional landing craft similar to those used in World War II shown in Figure 9.1, Figure 9.2 and Figure 9.3. While the original LCAC were designed and built by Textron Marine & Land Systems; the SSC will be designed by the US Navy. The US Navy believes that this will lower the costs. The initial buy of the SSC is for 8 craft with a price tag of $44 million each, in 2012 dollars.

Despite what would appear to be distinct advantages in the use of air cushion technology to landing craft, the numbers (given above) ordered by the US Navy suggests that the US Navy is not convinced that the advantages outweigh some of the characteristics of such craft. An insight into this can be ascertained by reviewing some of the work by the Naval Studies Board of the National Research Council. Periodically, the US Navy asks the Naval Studies Board to review the Navy's plans and give recommendations on new avenues to pursue. Two recent study reports capture the "mood" from the reviews. In 1988, the Naval Studies Board[4] saw merit in the use of the air cushion principle and also in the inclusion of the evolving wing-in-ground-effect (WIG) technology. A quote from that report is:

> *"Lift for the more conventional forces will be lighter and faster than today's (for example, the panel has visualized a fast, long range assault vehicle that combines many of the advantageous features of both WIGs and LCACs)".*

By 1997, however, the US Navy was becoming less enthusiastic as seen from the Naval Studies Board Report of that year[5]. A quote from that study[b] that was evaluating the Navy and the Marine Corps for the period 2000–2035 had the following quote:

> *"The ship-to-beach transport burden will fall on LCACs, which, with their 60-ton payloads and 25–40 knot speed, are very capable. However, an LCAC is expensive to operate ($3,000 per hour), and its aluminum construction makes it somewhat vulnerable to damage".*

Because of this, the Naval Studies Board recommended instead, a *conventional high speed displacement hull* craft, called "sea sled", capable of carrying a 180 ton payload at 30 knots. This is essentially an upgraded hull form version of the World War II craft. It would be constrained to stop at the shoreline like the World War II craft, which is an unfortunate step back in capability and vulnerability of the troops.

The US Navy has had to deal with all high speed vehicles and ships being proposed and built at different times (air cushion craft, hydrofoils, WIGs, etc); all offering high performance but carrying a high price tag in both acquisition and operating costs. With constrained Defense Budgets, it should not be a surprise that such programs are canceled or only small numbers procured (as in this case with

[b] Navy Studies Board Report, *Volume 8: Logistics* that contains the quote was a separate report issued in 1997 after the original study work was completed in 1995.

LCACs and SSCs). In the context of military operations considerations discussed in this chapter, this remains an unsolved problem for the high speed ship designer.

9.2 Amphibious Warfare: Past, Present and Possible Future

Given the development of amphibious assault and the nature of changing warfare, it is possible to postulate how the amphibious assault mission could evolve into a more flexible capability by using the unique feature of the *air cushion* and further to integrate into the evolution the added value of *aerodynamic lift* (of the type discussed in Chapter 11 and earlier chapters). Figure 9.6 illustrates three assault paths from the sea. The first path, called "yesterday" in the upper left of Figure 9.6, shows the conventional landing craft of World War II. It shows the slow speed of the assault and the constraint of stopping at the shore line to transfer troops and equipment. The middle path in Figure 9.6 labeled "today", shows the value added of the air cushion craft that allows a faster trip to shore plus the ability to go inland without stopping. It is capable of traversing certain obstacles depending on the skirt height of the craft. The new path, that is shown in the lower right-hand of Figure 9.6 labeled "future", could evolve from today's technology and is the use of "wing-in-ground-effect" of one form or another (see Chapter 10 and Chapter 11) to overcome the terrain limitation even more than that offered by the conventional air cushion craft. Figure 9.3 provided a good example of such terrain obstacles in amphibious assault that would be avoided by such an improved capability.

The integration of aerodynamic lift into the craft design brings with it the automatic increased speed (of the order of 100 knots or more). Such a craft would approach at high speed in "ground effect" (IGE) and then, if necessary, climb up out

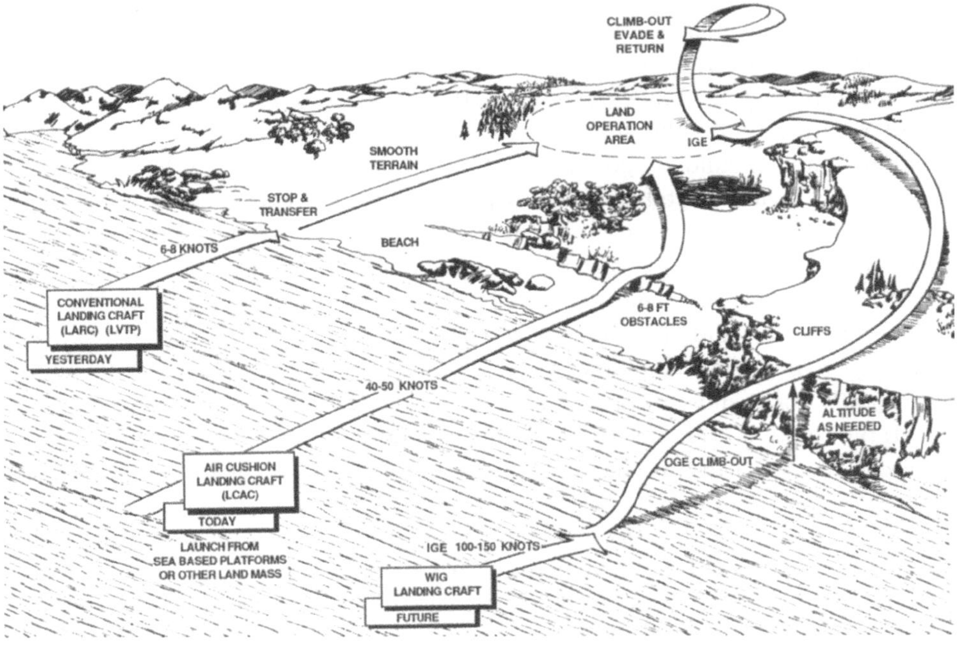

Figure 9.6: Possible Amphibious Assault Evolution: Past, Present and Future

Figure 9.7: Demonstration of Russian *ekranoplan* "*Orlyonok*" on beach of Black Sea (c.1990)

of ground effect (OGE) to "jump" over any high terrain encountered. The actual form of wing-in-ground-effect capability has appeared in different forms since the 1960s.

Chapter 1 summarizes the various forms of WIGs that have been designed and flown in the US, Russia, Germany and elsewhere since that time up to the present day. The Russian efforts that began in the 1960s and first reported by the US Defense Intelligence Agency (DIA) in 1967 were known during the US Navy ANVCE Project and the technical details of much of that work is given in Chapter 10 and expanded upon in Chapter 11. One such craft was the "*Orlyonok*" (Eaglet) and Figure 9.7, taken from Chapter 1, shows the craft parked on the beach after having come ashore from a simulated amphibious assault. The photograph shows the *Orlyonok* being shown to a group of foreign dignitaries sometime in 1990.

While the *Orlyonok* and the other Russian *ekranoplan* prototypes had their shortcomings they did show that it was possible to develop an *aerodynamically supported craft* for an amphibious assault that included the ability to fly (albeit inefficiently for a short distance) over terrain geography not available to World War II type landing craft or even to the more modern air cushion craft LCAC or SSC as discussed earlier in this chapter. A close-up of the offloading ramp is shown in Figure 9.8.

Figure 9.8 shows that the bow section has been hinged open and the ramps lowered such that an armored vehicle and troops can exit.

Other versions of such aerodynamic assault landing craft are included in Chapter 10 and Chapter 11 that show definite promise for further improvements in amphibious assault and other military operations. Some of the problems

Figure 9.8: Close-up of the offloading Ramp of the *Orlyonok*

associated with an *ekranoplan* type of aerodynamic air cushion craft including the problems of significant spray and high powered jet exhaust are discussed in detail in Chapter 11 together with some proposed solutions.

9.3 Speed and Stealth

A major concern for military ships is the ability to avoid detection while conducting their missions at sea. To minimize detection, the ship designer must produce a design that has (a) minimum visual signature (usually accomplished by some form of camouflage painting scheme); (b) a low radar cross section (RCS); and (c) a low heat or infrared (IR) signature. With today's advances in detection technology of radars and IR sensors, it has become a significant concern to ship designers to design ships that can avoid detection. Satellite imagery (mainly IR but also radar) coupled with airborne radar and IR detection capability is of particular concern. The following comments apply to the very difficult problem of lowering the RCS of ships, a technology that has not progressed significantly over the past few decades because of the inherent difficulty of operating at the interface of water and air with its complicated physics. To date, it has not been possible to combine both speed and stealth at the same time into a maritime ship, although there are some possible solutions (see later).

The radar cross section of any target (ship) is a complex measure of many features. In the standard definition of radar cross section, the RCS is a measure of the proportion of the incident energy (of the target) reflected back to the transmitting radar. More specifically, it is the measure of the far field reflected energy. This reflected returned energy is the complex result of many parameters that include the transmitted wavelength; the shape and orientation of the target with respect to the transmitting radar; the distance to the target; the reflectivity of the material (including whether the material of the ship is coated or not with reflective or radar absorbing materials); the inevitable losses in the system and other parameters such as the efficiency (gain) of the antenna. Because of the techniques used in measuring the back scattered energy in terms of steradians and the solid angle of the receiver as seen from the target, the radar cross section (RCS) is measured as an area (square meters).

Standard texts on radar technology (see for example Skolnick[6]) provide the derivation of the various ways of expressing the RCS for different types of radars. For convenience, a pulse radar is assumed here such that the RCS can be written as:

$$\sigma = \frac{P_r(4\pi)^3 R^4 L_T L_W}{P_t G^2 \lambda^2 F^4} \tag{9.1}$$

In this expression for RCS:

P_r	Power transmitted and received by the target (watts)
P_t	Peak power of the pulsed signal from the radar (watts)
R	Distance (range) from the radar to the target (meters)
L_T	Total losses in the system
L_W	Losses introduced by water content along the path of the transmitted signal
F	Propagation factor due to curved earth and Lloyd's mirror effect (see Skolnick)
G	Gain (efficiency) of the antenna ($= {}^{4\pi A_e}/_{\lambda^2}$)
A_e	Effective area of the antenna
λ	Wavelength of the transmitted signal from the radar

It is usually computed that the RCS (σ) is expressed in units of square meters and is quoted as $\sigma(\mathrm{m}^2)$ or σ(sq m) in the literature. The RCS of maritime targets covers a wide range of values from say a small dinghy floating in the water with an RCS of around 1 m^2 to an aircraft carrier of over 1,000,000 m^2. This wide range of values has led to the standard treatment of handling large numbers by using the decibel system and RCS is given as:

$$\sigma(dBsm) = 10\log_{10}\left(\sigma\left(m^2\right)\right) \tag{9.2^c}$$

This expression is given graphically in Figure 9.9

The variation of RCS between ship types varies widely but typically a coastal patrol vessel would have an RCS of about 10–20dBsm, a frigate size ship would have an RCS of about 20–40dBsm, and an aircraft carrier would have in excess of 50–60dBsm depending on radar frequency and many other factors already mentioned.

The main contributors to the total RCS of the ship can usually be identified as having three main components such as:

1. Ship Hull Shape (including any appendages)
2. Ship Mission Systems (including navigational systems, weapon systems & sensors)
3. Ship wake (and wave-hull interaction)

Figure 9.10 shows where the three general areas are for a typical frigate size ship, using the FFG 59 frigate (USS *Kauffman*), one of the "long hull" versions of the Oliver Hazard Perry-class frigates, as an example. The USS *Kauffman* was one of the last US Navy frigates built in the 20th century.

[c] In its correct form, the term in the logarithm bracket should be expressed as the ratio of the target RCS (σ) relative to a 1m^2 target, but this nicety has been omitted here for clarity.

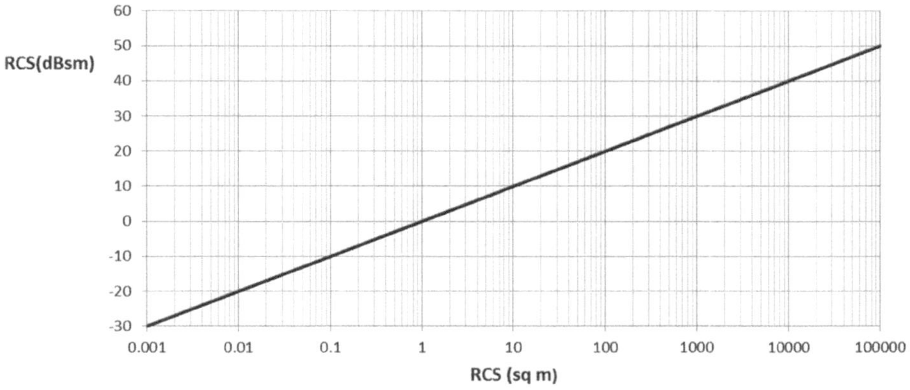

Figure 9.9: Radar Cross Section (RCS) Conversion (m^2 to dBsm)

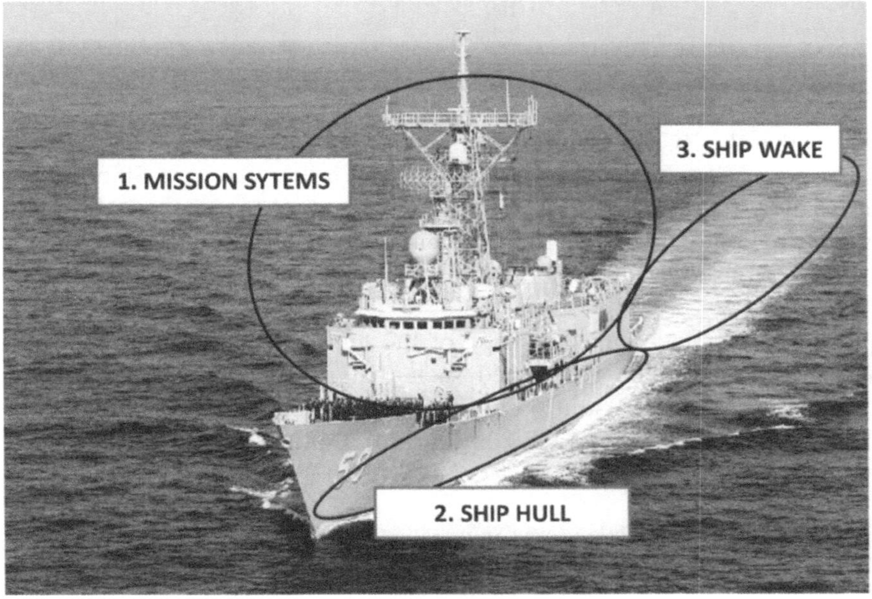

Figure 9.10: Three RCS Signature Groupings of a Ship
US Navy photo

The advances in radar technology over the last half century has given ship designers pause for concern in designing ships to avoid detection on the open seas. Attempts to minimize the total ship signature have not met with great success as may be illustrated by examining the three main RCS signatures highlighted in Figure 9.10 using typical values. In this constructed example of likely values (assuming an X-band detecting radar)[d]:

1. Ship Hull	RCS (hull)	= 30dBsm
2. Mission Systems	RCS (systems)	= 35dBsm
3. Ship Wake	RCS (wake)	= 20dBsm

[d] For illustrative example only and not the actual values for the FFG-59.

The RCS will vary depending on whether the orientation of the ship to the radar is "bow on" (as used here in this example) or "broadside". These values will be used for the illustrative example.

What is the RCS signature of the total ship? To get that answer one must recognize that using the "dBsm" form of quoting an RCS as used in the example is merely an artifice for handling large numbers, and it is necessary to revert to the "area" form of square meters radar cross section as expressed in Equation 9.1 to obtain the actual value of the total ship RCS signature as shown in Table 9.1

Table 9.1: *Example RCS Calculation (Basic Ship)*

Contributor	σ(dBsm)	σ(sq m)
Ship Hull	30	1000
Ship Mission Systems	35	3,160
Ship Wake[e]	20	100
Ship Total RCS	36.29	4,260

It is seen that the RCS of the total ship when expressed in dBsm is very close to the largest value of the contributing RCS, which is, in this case, the Ship Mission Systems. That is, *Total Ship RCS* is 36.29dBsm compared to the topside *Mission Systems* contribution of 35 dBsm. This problem is akin to the everyday problem of listening to the acoustic signature of radios on the beach. One only hears the loudest radio!

Suppose, by clever design, the shape of the hull, with possible use of radar absorbing materials, can be redesigned such that its RCS can be reduced to $0.001 m^2$ (i.e. the RCS of the hull is reduced from +30dBsm to −30dBsm) (see Figure 9.9). What is the new total RCS of this ship? Table 9.2 shows the result.

Table 9.2: *Example RCS Calculation (New Ship)*

Contributor	σ(dBsm)	σ(sq m)
Ship Hull	−30	0.001
Ship Mission Systems	35	3,160
Ship Wake	20	100
Ship Total RCS	35.13	3,260.001

As can be seen, even with magically making the ship hull almost "disappear" to searching radar, the RCS of the total ship is still at a high value of RCS (again close to the largest contributor to the ship signature) at a level of 35.13 dBsm in this example. This simple constructed example illustrates the extreme difficulty of designing a ship

[e] A nominal value of wake RCS is given here for illustration. It is a strong function of the environmental conditions (of wind and wave action, "sea clutter", grazing angle, water-air mixture at the surface among other difficult to predict parameters) as well as the disturbance due to the ship itself causing wide variations in "ship wake" RCS.

to evade detection from today's radars. *All* contributors to the RCS must be reduced – including that of the ship's wake, if one is to reduce the total ship's RCS to low undetected values. While not expanded on here, the contributions to the ship RCS of the mission or weapon systems and all other additional systems above the deck of a ship is significant. From Figure 9.10, one can see the many sharp corners, corner reflectors and many complicated geometries of the shipboard mission systems, all of which add significantly to the ship RCS. The RCS of the deck stanchions alone can equal the RCS of the hull in many ships. In Chapter 2, Figure 2.1, a photograph of an aircraft carrier is shown (CVS-9, USS *Essex*) that shows the additional problem of aircraft on deck. The RCS of each aircraft on deck can have values of 10–30dBsm which is a major source of detection over and above the typical RCS of aircraft carriers as platforms themselves (50–60dBsm). If the ship is to have any military mission capability at all, such weapon system equipment must take on a different form.

Since the early 1970s intense effort has been undertaken by several nations to design ships that are more stealthy so as to avoid detection by enemy radar. The problem has been the extreme difficulty of counteracting the large amount of energy that the ship imparts to the water as it moves at high speed through it. That energy is transferred to its visible wake. Three ships designed to have lower hull RCS signatures are shown in Figure 9.11 that illustrates this problem.

(a) USN Littoral Ship *USS Independence LCS-2*

(b) Swedish Navy Stealth Ship *Visby* Corvette
Photo Credit: Jimmie Adamsson/Swedish Armed Forces

(c) Royal Norwegian Navy Stealth *SES LCS*
Photo Credit: Sjøforsvaret/Forsvarets mediesenter

Figure 9.11: Some Examples of RCS Reduction Techniques
US, Swedish and Norwegian Navy photos

In Figure 9.11 (a) is shown the US Navy Littoral Combat Ship (LCS-2), the *USS Independence*. It is built by Austal USA and was commissioned on 16 January 2010. It has a full load displacement of 3104 tonnes and has a maximum speed of 44 knots.

It can be seen from the photograph that the designers concentrated on a low hydrodynamic drag hull design and incorporated shaping to the upper hull to minimize the RCS. Also, the clean upper decks show that close attention was given to reducing the amount of corner reflectors and other RCS producing features of the ship mission systems. However, from a stealth viewpoint, it is seen that the ship still has to deal with transferring energy from the hull and propulsion system into the wake which will still exhibit a persistently large RCS for tracking radars.

In Figure 9.11 (b) is shown the Swedish Navy Stealth ship "*Visby*" corvette. It has a full load displacement of 640 tonnes and a top speed of 35+ knots. It was commissioned on 16 December 2009. The *Visby* was specifically designed to have stealth features with use of camouflage paint to reduce the visual signature and upper hull and superstructure design with reflecting surfaces and other stealth features including a hull structure built entirely of carbon fiber reinforced plastics. As can be seen from the photograph, however, the choice of a conventional monohull hull for the basic ship shape produces large amounts of spray and a large wake which will negate much of the hull RCS reduction sought in the stealth ship design when operating at high speed.

In Figure 9.11(c) is the Royal Norwegian Navy coastal corvette *Skjold*[f] Class LCS. It was commissioned on 17 April 1999. It has a full load displacement of 274 tonnes and has a classified maximum speed in calm water of 60+ knots. It is of a *sidehull SES* (air cushion craft) design (see Chapter 1) with low drag sidehulls and flexible bow and stern seals to contain the pressurized air cushion between the sidehulls. This fast patrol boat was designed to have extensive stealth features incorporated. The camouflage painting can be seen in the photograph. The use of sloping sides of hull and superstructure to deflect the radar incident energy can be seen. Many other stealth features include use of anechoic coatings of radar absorbing materials (RAM) and a combination of glass and carbon reinforced plastics in the main structure. The windows are flush with the surrounding structure (no coamings) and the glass contains radar reflective screens. Again, however, like the other ships in Figure 9.11, the choice of hull shape and propulsion systems leave the ship with a large detectable wake from enemy radars.

It is a difficult task to add stealth features to a ship and still retain its military mission capabilities. In the 1970s work began to take shape in several nations to devise new hull forms that would eliminate this wake problem (of easily detectable RCS values) and still result in a military mission capable ship. Some of the hull forms considered are shown in Figure 9.12. On the left of Figure 9.12 is the conventional monohull. It can be a displacement hull, planing craft hull or other variant but they all have the common characteristic of a *large waterplane area* that must be pushed through the water creating a disturbance and wake. To minimize the waterplane area (and drag), various new hull forms started to appear.

The first concept was to push the displacement volume *deep* in the water away from the circulatory water surface action and to place the main part of the ship with its mission payload, crew etc, *above* the waves. This left a thin strut-like structure slicing through the water with a *small waterplane area*. Variations of this idea produced a *small waterplane area single hull* (SWASH) and *small waterplane area*

[f] Norwegian for "shield".

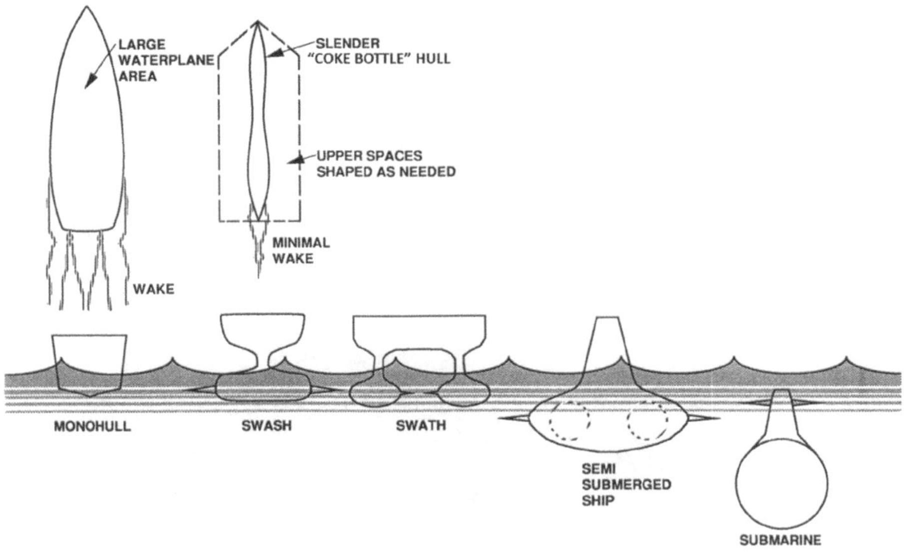

Figure 9.12: Some Possible Hull Forms to Minimize Ship Wake

twin hull designs as shown in Figure 9.12. The design of the strut (or struts) then began to optimize the water-air interaction to determine a waterplane shape that minimized wave drag and drag of the wake. This led to the "coke bottle" concept familiar to supersonic aircraft designers to shape the fuselage-wing combination to present a smooth shape to move through the air. A similar (but not identical) solution is found with the strut – hull combination for SWASH and SWATH operating at the interface between water and air (two mediums), a much more complicated interaction than for aircraft (in one medium).

A further refinement was to design a ship that was called a *semi-submerged ship* as shown in Figure 9.12 where most systems are housed in a submarine-like hull away from wave celerity and only a small part of the ship protrudes through the water-air interface for certain mission purposes. This is only a concept as no operational craft have been pursued or built. Finally, to eliminate the surface wake altogether is to operate as a fully submerged submarine, as shown on the right hand side of Figure 9.12.

While this solves the wake problem of ships as shown in Figure 9.11, there are all the usual issues of using submarines for many naval missions, such as communication while underwater. This led to other approaches for ship design to incorporate stealth.

9.3.1 *Sea Shadow* Stealth Ship

In early 1975 a new ship concept started to take form in the Lockheed Skunk Works in Burbank, California[g]. That concept tackled the problem of minimizing the wake problem for a military useful ship. The *Sea Shadow* project was conducted not only

[g] The Skunk Works relocated to Palmdale, California in 1989 when Lockheed (now Lockheed Martin) reorganized its aircraft operations at the then Burbank Airport.

Figure 9.13: *Sea Shadow* Stealth Ship (At rest)
US Navy photo

under the corporate "Skunk Works rules of secrecy" it also was conducted for the US Navy under the secrecy rules of Special Compartmented Information (SCI) programs or as popularly known, a "black program". Even though the project was subsequently declassified in 1993, the basic features, critical technology, method of construction and planned mission for larger size ships, must remain classified under SCI program rules. In the context of this book that is not a problem as publicly released US Navy photographs of the ship after the project was declassified in 1999 and on public display tell the story in terms of how to reduce hull and wake RCS in a surface ship platform. Figure 9.13 shows the *Sea Shadow* (bow on) as it gets ready to leave the San Francisco area on 18 March 1999 and motor to the San Diego Naval Station and the Suisin Bay Reserve Fleet in Benicia, California where it was housed until sold for scrap in 2012 to the Bay Ship & Yacht Company, California for \$2.5M[h]. Figure 9.14 shows the *Sea Shadow* Stealth Ship underway.

These public release US Navy photos show the basic concept to be that as shown in Figure 9.12, except that the hull shape above the sea surface exhibited the sloped surfaces for RCS reduction as shown in Figure 9.11. The slope angles are chosen to deflect the enemy radar of searching aircraft or satellites. This sloping faceted technique is familiar to that used on other aircraft products of Lockheed Martin such as the stealth fighter F-117 and other aircraft[i]. As demonstrated in other ships designed to reduce their RCS, care has been taken to avoid having any mission

[h] Again, because of SCI classification rules, the buyer had to agree to scrap the ship immediately to avoid any possible "reverse engineering" of its stealth characteristics.

[i] There are other shaping technologies available such as used by Northrop Grumman on the B-2 stealth bomber.

Figure 9.14: *Sea Shadow* Stealth Ship (Underway)
US Navy photo

equipment on the outside of the ship. Clearly, this is a very mission specific design consideration.

While not expanded upon here, other photographs in the open literature show different standing wave patterns along the sloping sides of the *Sea Shadow* Stealth Ship that depend on ship speed than that shown in Figure 9.14 of the *Sea Shadow* underway. This is because the design of the "coke bottle" hull and strut interaction is for a particular design ship speed and that the wake generation (and associated RCS) will vary with this design speed selection.

Apart from the classified mission application of the *Sea Shadow* it was mainly a research tool to help understand such hull-strut interactions with the standing wave to optimize on RCS values, as well as research into above water shaping to elude aircraft and satellite searching radars. In this regard, it did provide valuable data in the design of later US Navy ships seeking low RCS such as the DDG-1000 among others. The *Sea Shadow* was finally scrapped in 2012. One of the conditions when the ship was put up for sale in 2009 was that any buyer must agree to cut up the ship immediately and not reveal the actual underwater shape of the unique hulls that produced these minimum wake signatures. Public photographs taken of the *Sea Shadow* taken inside its "hangar", the Hughes Mining Barge, after the program was declassified, do provide some general idea of the submerged shaping, use of foils for stabilization and control, and type of propulsion used, but the specific key details remain classified. The technology of SWATH-like hulls and their strut design is an unexplored area that must await future research on reducing the RCS of surface ships and their wakes together with the required unique propulsion systems. To date, it has only been possible to incorporate stealth at relatively slow speeds and no techniques have yet been devised to achieve stealth at speeds approaching 100 knots.

9.3.2 The 100 knot Submarine

Over the past half century there have been many embryonic attempts to achieve high speed under water (where the stealthiest designs are most likely). None have progressed much past design studies or model tests. Some effort has been placed in using supercavitation means, in which the craft is "encased" in its own gas bubble caused by operating at supercavitating speeds and thus with minimum hydrodynamic resistance and water disturbance. This technique has been used successfully in torpedo design by different navies. The Russian Navy has the *Shkval* (Squall) torpedo, deployed in 1990+, in which the specially shaped nose cone ejects gas forward (from its liquid fuel rocket engine) that is then deflected outwards and rearward to encase the torpedo in its own supercavitating bubble. Much higher speeds than 100 knots have been claimed for this torpedo (albeit for short range). Other nations, including the US and Germany, have variants of this concept in some of their torpedo designs.

There have been some limited attempts to scale up this technology into a full scale submarine to achieve 100 knots or more. DARPA initiated such an effort in 2005, with work by Electric Boat Company, but no reliable technical data has yet been released or confirmed even in the model scale. In 2006, plans were underway to build a 100 foot long test craft submarine, called *Underwater Express*, built on that principle for DARPA and to conduct trials. An at sea demonstration scaled vehicle was tested without achieving the projected speeds and the program was concluded in 2012. China also claims to have a similar plan for a "100 knot submarine", but again no reliable information has been made available at time of writing.

Accordingly, such a concept must be left in this "space holder" pending availability of reliable data on whether this is even feasible or practical. Whether a ship design can be devised to use some unique form of hull shaping (similar to that used in the submarine-like submerged hulls of the *Sea Shadow*) or a new hull form (and propulsion system) concept like the original DARPA proposed "100 knot submarine", it remains for future research and writings to determine if this is even possible in a practical mission capable ship. Because of the need for submarines to "run silent", the noise generated by such supercavitating designs will require especial attention. Other technical issues would include the probability of reattachment of the flow (with immediate sudden drag increase) during any maneuvering of the submarine.

9.3.3 Stealthy Surface Piercing Submarine

A possible concept that would incorporate the lessons learned from the stealthy *Sea Shadow* ship program with the advantages of the intrinsically stealthy submarine is displayed in Figure 9.15. The basic hull (or double hulls) of the "100 knot submarine" could be integrated with the surface piercing strut of the *Sea Shadow* ship to provide several features that could alleviate many of the problems of submarines. These include solving the communication problem with submarines of communicating through the air-water interface. The stealthy "strut" or sail could house Low Probability of Intercept (LPI) communications, boundary layer control devices, stealth radar, missile launching tubes (SAM and SLCM) and RPV launch tubes. None of these features in themselves pose insurmountable technology challenges but the combination would make a formidable naval fighting vessel, capable of both high speed and stealth. The shaping of the hull (coke bottle or other shaping) together with the technologies discussed above in the DARPA design may provide the sought after speed and stealth military combination feature. The hydrodynamic issues of such a concept will require some new theoretical analyses and experimentation to provide a wide operating speed spectrum, as the submarine would be designed to operate in both the fully submerged mode (for transit?) and either submerged or in the surface piercing mode for mission operation.

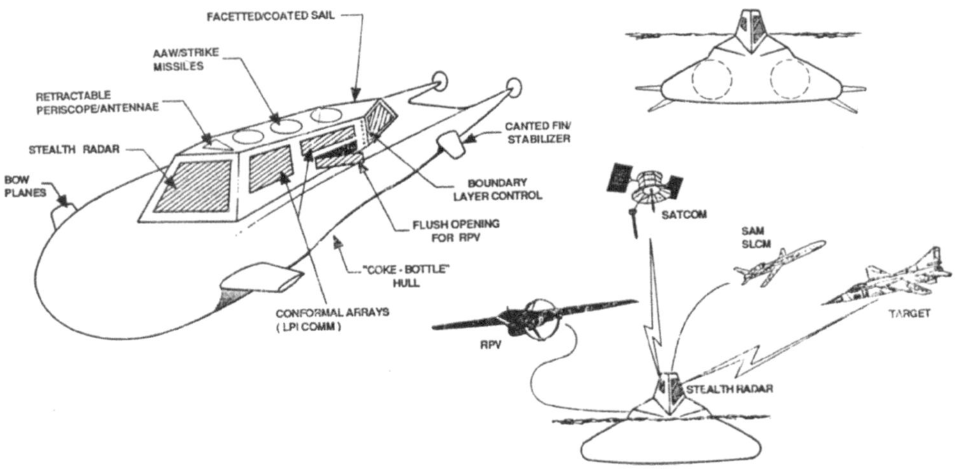

Figure 9.15: Concept for a Stealthy Surface Piercing Submarine

9.4 Summary

There are many military operations considerations in designing high speed naval ships. Some of those considerations have been identified in this chapter. With advances in radar (and other sensor) technology it is becoming increasingly difficult for the (surface) ship designer to design a ship that can be capable of avoiding detection on the surface of the world's oceans. More work needs to be done to combine speed and stealth in a military capable platform and some possible solutions have been identified here for further hydrodynamic research. Unlike in the aeronautical world, the combination of high speed and stealth is an unsolved problem in the marine world.

Also, speed in military missions needs to be worked in a broader context so that special purpose or high speed ships can be integrated with the (slower speed) fleet of conventional displacement ships. The use of a high speed ship like the surface effect ship (SES) that would sprint ahead of the fleet, then slow down with a towed sonar array to listen for attacking submarines, then sprint ahead again as the slower conventional hulled fleet caught up, is one attractive mission. This *ASW mission* was included in the US Navy ANVCE Project discussed in more detail in Chapter 10.

Finally, a better use of the air cushion principle needs to be integrated into *amphibious warfare* to reduce the vulnerability of troops and equipment in assault on theater beaches. Various possible designs are included in the other chapters of this book (see for example, Chapter 11).

References

1. *"Advanced Naval Vehicles Concepts Evaluation (ANVCE) Project",* Prepared by Chief of Naval Operations (OP-96), Report No. CNO LTR 96/C194136, Vols 1,2,3,4; December 1979
2. *"The Missile Defense Equation: Factors for Decision Making"*, by Peter J Mantle, book published by AIAA, January 2004, ISBN 1-56347-609-6
3. *"The Amphibious Assault Landing Craft Program",* by James L. Schuler, Naval Engineers Journal, Vol. 85, No. 2, April 1977
4. *"NAVY-21: Implications of Advancing Technology for Naval Operations in the Twenty-First Century",* Naval Studies Board Report published by National Academy Press, NRC: NSB:046, (Eleven Volumes), 1988
5. *"Technology for the United States Navy and Marine Corps, 2000–2035"*, Naval Studies Board Report published by National Academy Press, Volume 8: Logistics,1997
6. Skolnick, Merrill I., *"Introduction to Radar Systems",* 2nd ed., McGraw-Hill, New York, 1980

10 Advanced Naval Vehicles Concepts Evaluation (ANVCE) Project

It is better to know some of the questions than all of the answers.
James Thurber, 1894–1961

I do not understand; I pause; I examine.
Michel Eyquem de Montaigne, 1533–1592

The two decade period 1960–1980 was the crucial period in the US where first, the US Maritime Administration (MARAD) and second, the US Navy sought large ships capable of operating at 100 knots on the high seas for both economic and military advantage. During that two decade period there were two key review milestones when the then current Administration became nervous about the choice of concept and called for a re-evaluation of the MARAD and US Navy choice. The first occurred in 1965 when the US Department of Commerce directed that a re-evaluation be made of the many high speed ship concepts available. This was the Surface Effect Ships for Ocean Commerce (SESOC) Committee. The results of that committee's work are discussed in detail in Chapter 4.

The second "call for a re-evaluation" was in 1975, exactly one decade later, when the Office of the Secretary of Defense called for a major review because after a decade of development, the choice by the US Navy was not performing as advertised. This resulted in the US Navy's Advanced Naval Vehicles Concepts Evaluation (ANVCE) Project (1976–1979). The ANVCE Project wrapped up its work and published its findings just before the US Navy canceled the high speed SES program on 9 January 1980. This chapter outlines the results from the ANVCE Project.

On 30 June 1976, the *sidehull Surface Effect Ship* SES-100B attained a speed of 91.9 knots in a slight choppy sea off the coast of Panama City, Florida. Despite that achievement many in the technical community and in The Pentagon were becoming concerned that the claims of the "100 knot Navy" advocates simply did not hold water. The technological problems that were surfacing based on analysis and test and operational experience both in the US and in the UK (where hovercraft technology and operation was quite advanced) produced "red flags" about the viability in several key areas of technical maturity in the US Navy choice. At the same time, the issue of the form of the proper Navy mission requirements was also raising questions in the Defense and Congressional community.

These concerns resulted in 1975 with the DDR&E, Dr. Malcolm R. Currie, in a memo dated 13 February 1975 (see Chapter 5 for the chronology of events), directing the Navy to slow the SES program down and to properly analyze the technology in a consistent manner and to examine all the possible "contenders" for a *high speed ship* (especially one designed to operate efficiently at 100 knots) and in addition, to conduct an in-depth *mission analysis* to place the program in operational Navy perspective. Twelve years earlier, the US Navy had already pulled back from such high speeds in the hydrofoil development program when the US Navy hydrofoil FRESH 1 had overturned at 80 knots on 18 July 1963, and caused the hydrofoil program to be re-directed to a relatively slower speed closer to 50 knots rather than the planned 100 knots. The technologies at the lower speeds (even though at 50 knots it was more than *double* the current Navy Fleet speeds) were much more attainable and reduced the risks markedly. Concern was growing in OSD, GAO and elsewhere that a similar "too fast" goal was being cemented into the SES program without prudent development and proper management.

It was about this time (circa 1974) that the US Navy and others were beginning to suggest changes to how the Navy should perform its many missions and the structure of the "new" Maritime Strategy was emerging that suggested that a re-examination be made of the *mission of the SES* and how it fitted in to the overall Navy Maritime Strategy. In 1974, Adm Stansfield Turner, then President of the US Naval War College (and later to serve as Director of Central Intelligence) introduced the idea of "Sea Control" to the new thinking on the US Navy Maritime Strategy based on the post-Vietnam period and the change in the growth and form of naval power around the world. The US Navy responded immediately and in July 1975, the US Navy set up in the Chief of Naval Operations System Branch (OP 96)[a] in The Pentagon, a major effort called the Advanced Naval Vehicles Concepts Evaluation (ANVCE) Project. An ANVCE (pronounced "AN-vee") Project Office was formed in OP-96 led by Project Officer CAPT Thomas L. Meeks (OP-96V), a *military operations specialist*; and Deputy Project Manager & Technical Director, Peter J. Mantle, an *independent advanced marine vehicle expert*, (using the US Navy descriptors). The team of experts in this $10M project encompassed national and international experts in all the designated areas. They came from Government, industry and academia to conduct analyses, model testing and full scale trials wherever possible to provide the necessary confidence level in the findings.

While the main thrust of the guidance from DDR&E was a major reorientation of the FY-76 SES program, the DDR&E went further:

[a] The Systems Analysis Division was formed in the late 1960s. CNO ADM Elmo Zumwalt used the Division primarily to guide him in setting directions for the US Navy (he was one of OP-96 early Directors). In 1970, ADM Zumwalt abolished the Long Range Planning Division (OP-93) and shifted such functions to OP-96. Sometime later, In 1983, CNO ADM James Watkins closed down OP-96 and shifted all its functions to the Program Resource Appraisal Division (OP-91) and to the new Director of Naval Warfare Division (OP-095). In 1992, CNO ADM Frank Kelso made significant changes to OPNAV including changing the "OP" codes to "N" codes – with also several Division number changes! There have been more organizational changes since that time and as recently as 2009 under CNO ADM Gary Roughead (2007) OPNAV organization, such analyses would now be done in the new Integration of Capabilities and Resources Division (N8) or perhaps in the new Information, Plans and Strategy Division (N3/5).

> *"Conduct R&D and studies necessary to evaluate the technical and military potential and costs of other Advanced Navy Vehicles (ANV) concepts which offer possible alternatives to the SES"*

The DDR&E memorandum was expanded upon in the SES Program Element guidance in the budget of that year with the following text:

> *"Where insufficient information exists for evaluation, analysis and testing of models on small scale prototypes will be performed as necessary to provide it"*

The evaluation was to include long-range aircraft, with airborne and/or seaborne loiter capability, wing-in-ground (WIG) effect aircraft, airships, high length-to-beam (L/B) ratio SES and low L/B SES, hydrofoils, and small-waterplane-area, twin-hull (SWATH) ships. In the language of the Project Directive:

> *"The objective of the Advanced Navy Vehicles Concepts Evaluation is to examine the technology now being developed for carrying naval weapons systems and its applicability to military missions. The study will evaluate each potential advanced naval vehicle in the light of technological feasibility, affordability, and applicability to an existing or projected naval mission. The evaluation will include examination of all potential naval vehicles using advanced technology, as well as the related areas of technology which may impact on advancing naval vehicles. It is to consider, but not be restricted to:*
>
> *(1) long-range, air and sea loiter aircraft,*
> *(2) wing-in-ground effect aircraft,*
> *(3) lighter-than-air vehicles,*
> *(4) surface effect ships,*
> *(5) hydrofoil ships, and small water(plane) area ships*
>
> *Critical examination of the technical literature, conduct of point design and model tests where necessary to fill gaps in the engineering data, examination of potential weapon system employment and interface problems, estimates of development and production costs and the impact of each vehicle examined on present or potential Navy mission areas are the principal tasks to be undertaken."*

By direction, to keep the scope bounded, submarines and carrier based aircraft were not included in the evaluation project. As the scope of the ANVCE Project began to take shape, there was recognition that simply designing ships based on some new technology alone was a flawed approach. Historically, when submarines and aircraft were first used in naval applications, earlier in the 20th century, it became necessary to realign the structure of naval missions to make effective use of their unique capabilities. While it is evident that new technology often requires or permits redefinition of missions, the exploitation of new technology sometimes must await the realization of the part it can play in a new or expanded mission. While nothing is currently in view that promises such a radical shift in naval capabilities and missions as that which followed the introduction of aircraft and submarines, new vehicle technology areas must be examined because one area might possibly provide an increase in mission effectiveness or a significant reduction in cost per unit. In a fiscally constrained environment (as true today as it was in 1975), the question of whether a particular vehicle concept should be developed for a particular role is one of considerable complexity; it involves a close interaction among platform technology, weapon and sensor technology, cost and mission needs. The ANVCE Project thus focused on the integration of technology and design, together with cost and mission effectiveness, keeping a watchful eye on possible new missions or new ways of fighting a war or contingency operations. Another guiding force in the ANVCE Project was the 1974 CNO directed "*Project 2000*", which was a guidance document on the political, economic, and strategic world environment through the turn of the century and of a fleet considered appropriate to that environment. One of the guiding statements in that CNO directive was:

> "*. . . future naval vehicles must be capable of handling the diverse threat and, to make sure that they will have the necessary flexibility to meet future requirements over a long service life, they must be capable of performing multiple missions. . .*"

This guidance tended to mitigate against design concepts that were directed toward a single mission purpose that could easily become obsolete if the threat changed. Today, with the increase in a non-standard threat that might be grouped under the umbrella of "terrorism" this is even more important. As the Vietnam War showed that "modern technological weapons" designed for earlier wars were not necessarily the best solution for land and air warfare in the Vietnam jungle scenario, today's new threat of *terrorism* or *asymmetric warfare* also changes how one approaches modern naval ship design.

The ANVCE Project was a two and half year project (July 1975–December 1978)[b] at a funding level of $10 million in 1975 dollars. It involved industry, government laboratories and agencies (NAVSEA, NADC, DTNSRDC, NISC and others) and independent experts from across the world. A Chief of Naval Operations (CNO) oversight Executive Board (CEB) was established together with an independent Technical Review Panel (TRP) made up of experts from Navy offices, MIT, NASA, CNA, APL and others. The Technical Review Panel (TRP) was chaired by Dr Alan Berman, Naval Research Laboratory (NRL). Of the $10 million, approximately $7 million was allocated to technical assessments that included comprehensive state-of-the-art summaries, Point Designs (to focus the evaluation) and model tests and in some cases, full scale trials operation evaluations. The remaining funds were used for Mission Analysis and for Cost Analysis.

Within the ANVCE Project technical assessment, as an additional effort, assessments were contracted with world renowned "outside the US Navy system" independent experts, and in particular with:

Sir Christopher Cockerell	Amphibious Hovercraft and Sidewall Hovercraft
Baron Hanns von Schertel	Surface Piercing Hydrofoils
Dott. Ing. Leopoldo Rodriquez	Surface Piercing Hydrofoils
Peter Crewe	Hovercraft, Seaplane technologies and design
Peter Payne	Supercritical Planing Hull technology and design

At the time of writing this book, the author notes sadly that these friends and associates, all giants in their field have all passed on, but their willingness to participate and provide their invaluable written assessments and advice at the time of the ANVCE Project was personally appreciated by this author in his many

[b] The Final Report was issued by the Office of Chief of Naval Operations (CNO) in December 1979

discussions with them as the ANVCE Project work progressed. If there was a common piece of advice from these stalwart practitioners of the art it was: "*simpler is better*!"

Working with Sir Christopher Cockerell has been a rewarding experience over the years. His insight and ability to zero in on the basics of his inventions was remarkable. The author later, in 1976 and 1977, brought Sir Christopher on-board to help on the US Navy Advanced Vehicles Concepts Evaluation (ANVCE) Project as an independent consultant and expert to help evaluate the *amphibious air cushion vehicle* and *sidehull surface effect ship* concepts that the US Navy were pursuing for the "100 knot" goal.

Baron von Schertel and Leopoldo Rodriquez were remarkable in their ability to describe hydrofoil technology in practical engineering terms and raising legitimate questions about the cost of introducing the high technology fully submerged hydrofoil into a naval environment. Peter Crewe, as head of the Projects Office in Saunders Roe, England, when the author was working there in the early 1950s, was a giant in the field of hovercraft technology, seaplane design and other areas and provided valuable advice again at a practical level not usually seen in the industry. Peter Payne was an amazing engineer capable of delving into intricate technologies and producing new thought and design possibilities. Sitting together on his deck watching the sunset in Annapolis and arguing over where the US Navy was going provided an "outside the establishment" viewpoint that was valued as the ANVCE Project progressed.

The results of the ANVCE technical evaluation were derived from 165 separate analyses, surveys, and experiments consuming 70 percent of the project. Nine generic concepts were explored in depth and, from them, 23 Point Designs were developed (which were necessary for costing and military mission analysis). In the context of this book, much of the work on aircraft design and related matters in the ANVCE Project will not be repeated here[c] and the key results for mainly *"surface bound"* concepts will only be discussed as it relates to the original goal of "100 knots at sea".

It is hoped that the many lessons learned from the work will, in their summary form, provide a backdrop for the reader on the "rise and fall" of the original "100 knot Navy" aims of the US Navy over that crucial decade 1969–1979 after which the approximately $1 billion program to develop a "100 knot SES" was canceled. The SES program was redirected at that point to the current activities on "slower speed" SES missions. Officially, the goal of "100 knots" has not been re-examined since that time which is considered unfortunate because the seeds of such an approach were just beginning to emerge in 1976! To avoid any conflict with any US Navy plans or official policies, the work of the ANVCE Project summarized in this book has been taken from the public domain. Specifically, the papers presented at various professional societies by Capt Tom Meeks, Peter Mantle (author) and Cdr 'Corky' Graham, all of the ANVCE Project Office[1,2,3] and others will be used as the main source data for this Chapter.

10.1 Types of Vehicles Considered

Using the guidance from the DDR&E directive, a set of specific vehicles (marine and air) were analyzed in depth. Figure 10.1 shows the set of nine basic types

[c] Although many interesting results emerged from that part of the ANVCE Project related to aircraft!

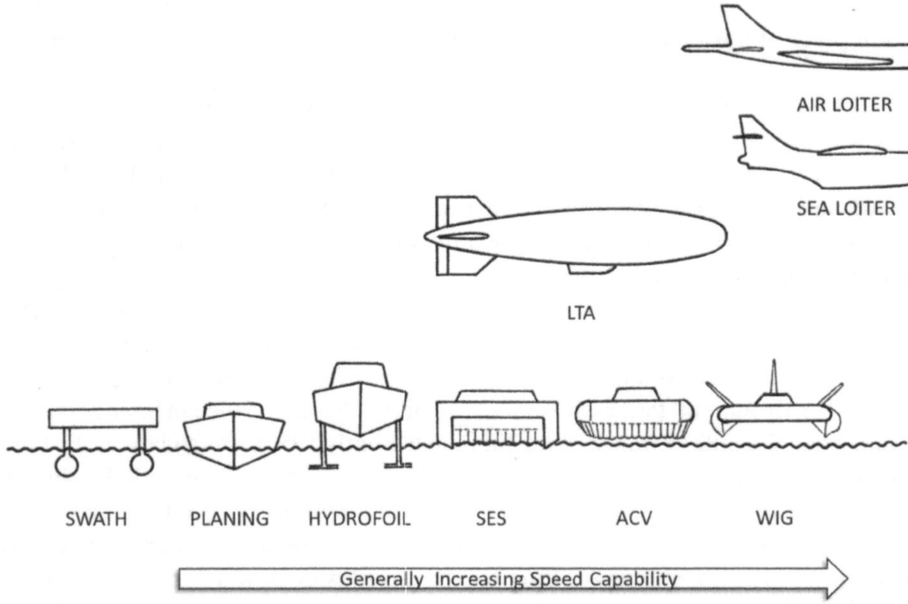

Figure 10.1: Spectrum of ANVCE Project Vehicle Types

displayed in an increasing speed domain (left to right). While not a rigorous positioning of concepts, it is helpful to place the vehicle concepts in some order for discussion and it is generally recognized that, all else being equal, SWATH ships would be the slowest vehicle and aircraft would be the fastest vehicle with the range of speeds covering 30 to 300 knots or more. As Sea States are considered, some slight shift in position occurs. For example, in high Sea States, the SWATH ship is actually faster than conventional displacement mono-hulled ships. Again, as is intuitively obvious and borne out by analysis, the sidehull Surface Effect Ship would be slower than the amphibious air cushion craft because of the hydrodynamic resistance of its sidehulls. This simple categorization changes slightly when considerations of the propulsion systems are included. The "goal" of "100 knots" would appear somewhere in the middle of this set of ANVCE vehicle types.

10.1.1 The ANVCE Point Designs

While, in the ANVCE Project, considerable work on the state-of-the-art technology was undertaken in each of the areas first, before any Point Designs were produced; in the context of this book, the selected Point Designs are summarized here first so as to provide a visual reference of why the particular designs were created and where the technologies limiting the "100 knot goal" can be identified.

Based on preliminary mission assessments and appreciation of the ongoing Maritime Strategy, it was found convenient to group the designs into "common mission groups" such as:

1. Surface Combatants (Near Term)
2. Air Capable Surface Combatants (Near Term)
3. Air Capable Surface Combatants (Far Term)
4. Non-Air Capable Surface Combatants (Far Term)
5. Multi-Purpose Air Combatants
6. Single Purpose Air Combatants
7. Aircraft Carrier Group
8. Logistics Group

In the ANVCE Project, "Near Term" was defined as those concepts where the technology extant in 1970 would be used with a planned Initial Operational Capability (IOC) of 1985. Similarly, "Far Term" applied to technologies that might reasonably be expected to be available in 1985 with a corresponding IOC of 1995–2000. This was based on recognition of how long it takes to develop any ship concept in the US Navy (and elsewhere). It takes approximately 16 years from start of Category 6.3 (Advanced Development) to Initial Operational Capability (IOC)[d], which means that for a vehicle to have an IOC of 1985 it must be based on technologies available around 1970. Similarly for an IOC of 2000, the technologies can be based on new technologies circa 1985 (give or take a few years). Because of the concentration in this book on "100 knots at sea", most of the aircraft related Point Designs developed in ANVCE will not be addressed. The special case of the Wing-in-Ground Effect (WIG) vehicles that operate "between" the surface ship concepts and aircraft concepts will be addressed as the "100 knot at sea" goal is examined as they offer a possible solution. Some versions of the WIG were grouped as surface vehicles and some as air vehicles depending on their design features and mission.

10.1.2 Common Combat Suites

A frequent problem when comparing vehicles is that oftentimes, vehicles are compared that were initially designed for completely different purposes giving what should be obvious: flawed comparisons. This problem was very evident in the original Maritime Administration (MARAD) studies in 1965 with their Surface Effect Ships for Ocean Commerce (SESOC) work (see Chapter 4) that was the genesis for much of the US Navy activities in SES development. To avoid a repeat of that problem, great effort was made to lay down a common requirement for each concept yet at the same time allow for the uniqueness of that concept to be incorporated.

This balancing act is demonstrated in Figure 10.2 where the mission payload was provided to each "point designer" in a common package. If the designer could accommodate the common mission payload and add some extra features that he felt could be incorporated at the same time, without jeopardizing his concept, then that was allowed.

[d] In the intervening three decades since the ANVCE project, the Acquisition Schedules of vehicle development have not improved much despite frequent revisions of the Department of Defense Acquisition policies and directives! Indeed, in many cases, the development times for new vehicles have slowed down and taken much longer.

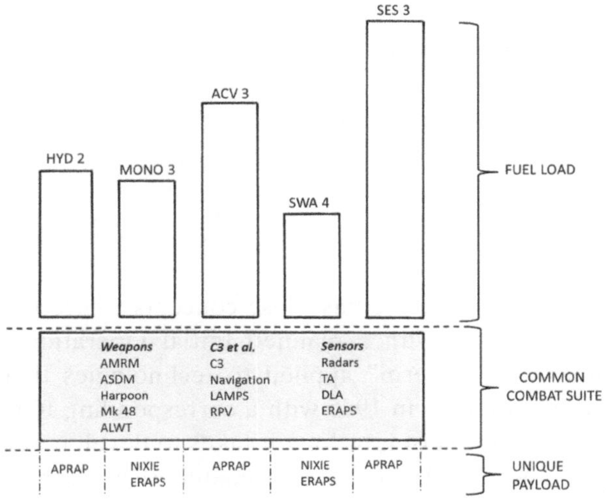

Figure 10.2: Common Mission (ASW) Payload for Point Designs (example for Surface Combatant Group)

The mission payload shown in Figure 10.2 applied to the mission of Anti-Submarine Warfare (ASW) for the Air Capable Surface Combatant Group of vehicle concepts considered to be the most likely mission for the high speed SES in the "3KSES Program" that was ongoing at the time of the direction from DDR&E to the US Navy. The shaded area comprises the common mission payload suite provided to each designer. "Below the line", certain unique additional mission payload was allowed if the concept could show that its unique platform concept could accommodate that payload for improved mission performance.

An example of matching the platform unique characteristic with the weapon or sensor was with the APRAP sensor as shown in Figure 10.2. The active/passive reliable acoustic path (APRAP) sensor requires a platform with a high speed capability to offset the "dead-in-the-water" listening time so as to maintain a high speed of advance (SOA) in its mission and keeping up with the (conventional displacement ship) fleet. Hence, it made sense to allow the high speed concepts of hydrofoil, ACV and SES to "add" this key sensor to the common payload to make a better integration of platform capabilities with the combat suite capabilities. Similar matching of platform and combat suite capabilities were used for the other candidates.

"Above the line" is where each concept would have to include the amount of fuel necessary to accommodate the range requirement specified for all designs. As just one example, the fuel load of the slower speed SWATH design (SWA 4) would be less than for the high speed SES Point Design (SES 3) as shown in Figure 10.2. The Disposable Load of each concept was allowed to vary but the common Payload was retained for all.

It is believed that each ship designer could then go about designing the best and most efficient design he could while accommodating the common mission payload for all "candidates"; include the required fuel to meet the range requirement for all candidates; and also include any unique features that the designer deemed advantageous to his design. In this example, the Useful (or Disposable) Load of each design, made up of payload (W_P) and fuel load (W_f) varied, but the payload was essentially common for the mission.

Comparing vehicles to a "common standard" is not a simple arithmetic comparison from set formulas!! Such a technique summarized here allowed the designers to provide their best design while still satisfying a common mission requirement.

10.2 Point Designs

As the designers went to work, the various concepts settled into categories for Sea Control along the lines provided in the US Navy's "*Project 2000*". Broadly speaking, these Point Designs settled into *Surface Vehicles* and *Air Vehicles.* These are shown in what follows.

10.2.1 Surface Vehicle Point Designs

For the surface bound vehicles, the ship concepts could be grouped into *"1000 tonne class"; "3000 tonne class";* and the *"aircraft carrier"* group[e]. For the main purpose of this book, which is to evaluate the best candidate for a surface ship capable of conducting a naval mission at 100 knots, the discussion will center on both the "1000 tonne class" and the "3000 tonne class" ships. This latter group most closely associates with the "3KSES" comparable to a *frigate class combatant* which was the genesis of the DDR&E directive to evaluate possible alternatives (see Chapter 5). For fair comparison with displacement ships a displacement ship based on the same requirements and advanced technology was added to the mix as "MONO 3" as shown in Figure 10.3 As the designers settled on their best designs to meet the requirements set, the size of the individual "Point Designs" appeared as shown (drawn to scale) in Figure 10.3.

The acronyms used in Figure 10.3 are as shown in the following Table 10.1.

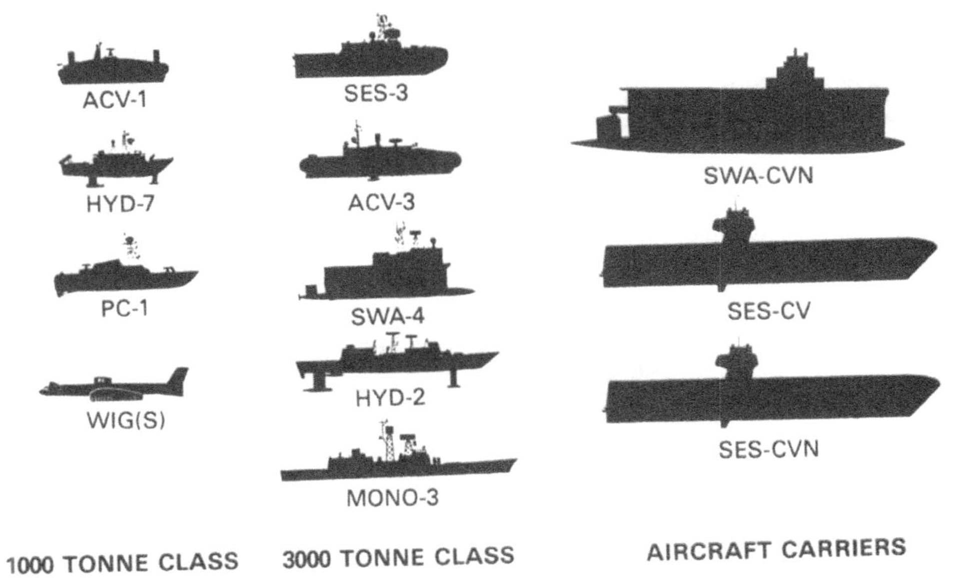

Figure 10.3: Surface Vehicles for Sea Control

[e] In order to bring together all the international data to a common standard for analysis, the decision was made to use the metric system throughout the ANVCE Project.

Table 10.1: *ANVCE Surface Vehicles*

Acronym	Vehicle Type
ACV	Air Cushion Vehicle (amphibious)
HYD	Hydrofoil (fully submerged foils)
PC	Planing Craft
WIG(S)	Wing-in-Ground Effect Craft (Soft endplates)
SES	Surface Effect Ship
SWA	Small Waterplane Area Twin Hull Ship
MONO	Monohull Displacement Ship (for comparison purposes)

10.2.2 Air Vehicle Point Designs

Similarly, for the Air Vehicles, the point designers produced the set of concepts shown in Figure 10.4. The Air Vehicle concepts fell into three basic categories; *"Single Purpose"; "Multi-purpose"* and *"Logistics"*. All of these concepts were significantly higher in speed than the "100 knots" sought by the original aim of the program but they provided a valuable context for the evaluation.

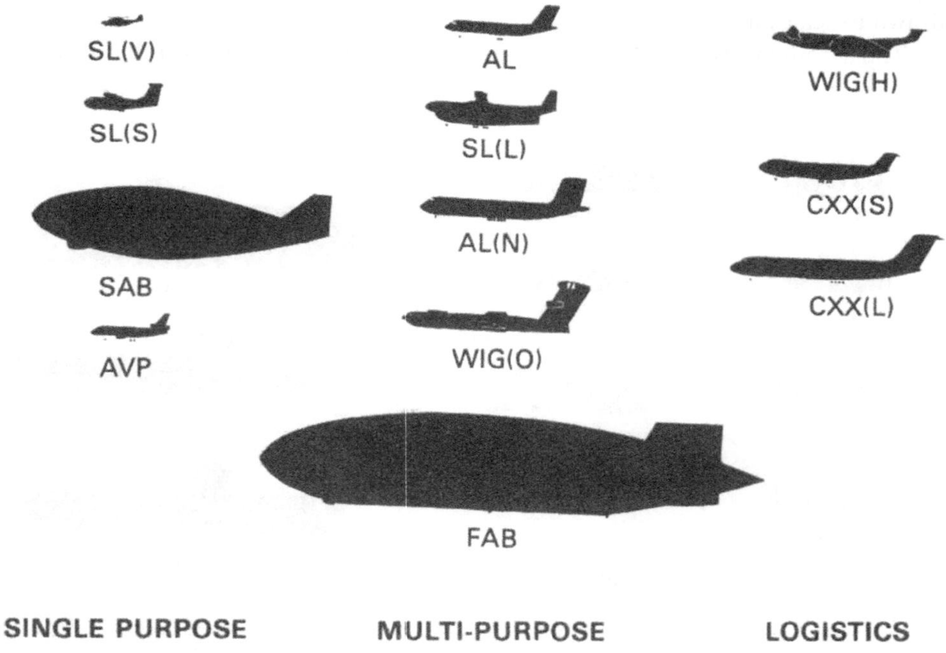

Figure 10.4: Air Vehicles for Sea Control
Copyright ©1977 SAE International. Reprinted with permission from 770966

The acronyms for these concepts are as shown in the following Table 10.2

While the main emphasis for discussion in this book is on "surface vehicles", the unique feature of the wing-in-ground effect craft that can operate in both ground effect (on the sea surface) and out of ground effect in certain design concepts requires that attention be given to such unique forms of "high speed at sea" vehicles.

Table 10.2: *ANVCE Air Vehicles*

Acronym	Vehicle Type
SL(V)	VTOL Sea Loiter Aircraft
SL(S)	Small Sea Loiter Aircraft
SL(L)	Large Sea Loiter Aircraft
SAB	Semi Air Buoyant Airship
FAB	Fully Air Buoyant Airship
AL	Air Loiter Aircraft
AL(N)	Air Loiter Aircraft (Nuclear powered)
WIG(O)	Wing-in-Ground Effect Aircraft (with out of ground effect flight capability)
WIG(H)	Wing-in-Ground Effect Craft (with hard endplates)
CXX	Long range cargo aircraft (USAF) (for comparison purposes)

These will receive detailed review in Chapter 11 on Winged Surface Effect Ships. The main purpose here has been to display the various candidates in context of the spirit of the DDR&E request to examine other possible ways of achieving the US Navy program goals.

Two groups studied (out of many) in the ANVCE Project are shown here for discussion. They are the so-called *"Air Capable Surface Combatants, 3000 tonne Class"* that most closely associates with the frigate class envisaged for the original 3KSES and the *"Non-Air Capable Surface Combatants, 1000 tonne Class"* that appeared to be more in line with the "state-of-the-art" in high speed ship design. These two groupings are shown in Figure 10.5 and Figure 10.6 respectively. Each of these groups, as outlined above was designed to common standards as closely as could be done from the many design choices, designer skills and not-too rigorous mission requirements. It is to be recalled that while "100 knot Navy" was a convenient label for elevator speeches, it was necessary to recognize the reality of the sea surface and set the requirement in more concrete terms. It was decided to specify 100 knots in calm seas; 80 knots in Sea State 3 and 50 knots in Sea State 6.

Once the Point Designs were done it is interesting to see how they compared to this "requirement". Using just the Sea State 3 requirement of 80 knots as a talking point; the maximum speed shown in the SES and ACV concepts were between 90 and 97 knots in Sea State 3. The hydrofoil craft had, under comparable conditions, showed design speeds of 53 to 71 knots depending on ship size and design. The high speed planing craft ship showed a top speed of 55 knots and the advanced monohull displacement ship had a top speed of 40 knots under the same set of requirements. The vehicle concept that exceeded the "speed requirement" was the wing in ground effect vehicle which was quoted by the designers as 200 knots for WIG(S); 200–390 knots for WIG(O) and 280 knots for WIG(H). These speeds are so different for the WIG designs than the other "surface bound" concepts and have less technical maturity, that a more detailed discussion is reserved for Chapter 11 on Aerodynamic Air Cushion Craft. Mission profiles were not placed on these designs but from Chapter 2 it is noted that Sea State 3 occurs about 50% of the time in the North Atlantic and about 30% of the time in South East Asia. So in any practical application of these Point Designs, the top speed would be considerably less than the values shown here.

	**SES-3** Gross Weight (W, tonnes) 3,658 Length x Beam (m) 81.2 x 33 Payload (W_P, tonnes) 391 Range at Cruise Speed in SS 3 3,208 nm @ 92 knots Maximum Speed in SS 3 97 knots
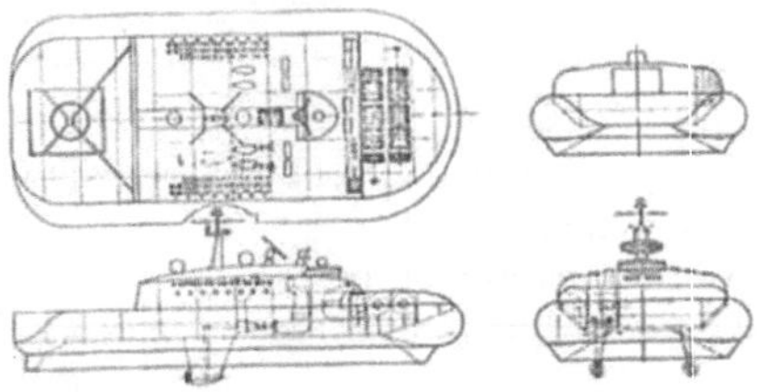	**ACV-3** Gross Weight (W, tonnes) 3,000 Length x Beam (m) 84.4 x 41.5 Payload (W_P, tonnes) 350 Range at Cruise Speed in SS 3 2,600 nm @ 93 knots Maximum Speed in SS 3 93 knots
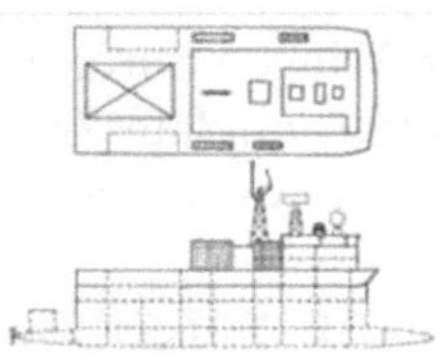	**SWA-4** Gross Weight (W, tonnes) 3,408 Length x Beam (m) 93 x 32 Payload (W_P, tonnes) 364 Range at Cruise Speed in SS 3 3,950 nm @ 19 knots 1,330 nm @ 37 knots Maximum Speed in SS 3 37 knots
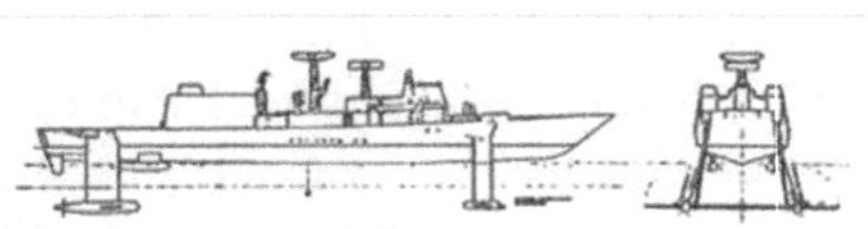	**HYD-2** Gross Weight (W, tonnes) 2,408 Length x Beam (m) 111.2 x 24.7 Payload (W_P, tonnes) 335 Range at Cruise Speed in SS 3 Foilborne 2,600 nm @ 45 knots Hullborne 5,900 nm @ 14 knots Maximum Speed in SS 3 53 knots
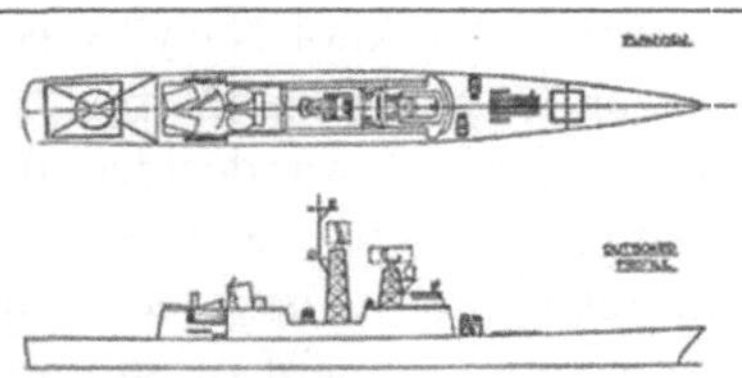	**MONO-3** Gross Weight (W, tonnes) 2,934 Length x Beam (m) 126 x 14.2 Payload (W_P, tonnes) 392 Range in SS 3 Max speed 1,647 nm @ 40 knots Cruise Speed 6,641 nm @ 15 knots Maximum Speed in SS 3 40 knots

Figure 10.5: ANVCE Point Designs for Air Capable Surface Combatants (3,000 tonne Class) (Far Term)

These design "points" (Figure 10.5 and Figure 10.6), were prepared by the best experts from industry and government in the respective areas for these ships. In the case of the "amphibious" ACV, the particular Point Design shown on Figure 10.5 is the water propeller propelled design. In the ANVCE Project, other design variants were examined that used air propeller propulsion (free air propellers and shrouded air propeller variants) that gave speeds that varied depending on the design of about 90–93 knots, again meeting the requirement.

More design analyses would have to be done to determine whether there is any advantage to a frigate size ship coming up on land! While air propulsion certainly

ACV-1

Gross Weight (W, tonnes)	1,264
Length x Beam (m)	58.03 x 29.77
Payload (W_P, tonnes)	129
Range at Cruise Speed in SS 3	1,450 nm @ 90 knots
Maximum Speed in SS 3	90 knots

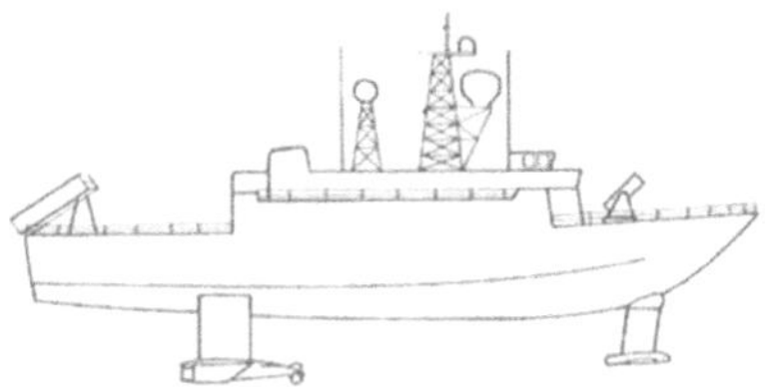

HYD-7

Gross Weight (W, tonnes)	983
Length x Beam (m)	60.4 x 13.7
Payload (W_P, tonnes)	114
Range at Cruise Speed in SS 3	
Foilborne	850 nm @ 71 knots
Hullborne	953 nm @ 25 knots
Maximum Speed in SS 3	
Subcavitating foils	55 knots
Supercavitating foils	71 knots

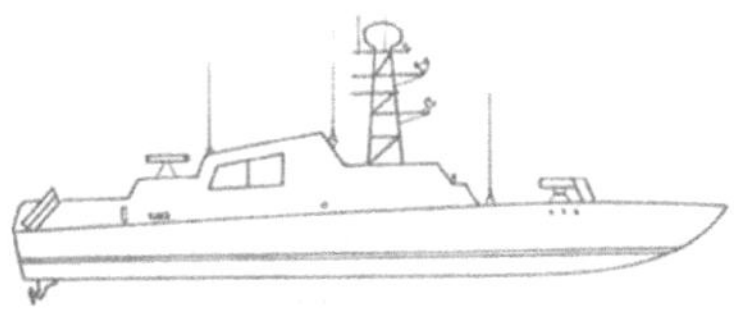

PC-1

Gross Weight (W, tonnes)	1,090
Length x Beam (m)	76.2 x 14
Payload (W_P, tonnes)	122
Range at Cruise Speed in SS 3	7,500 nm @ 15 knots
	1200 nm @ 52 knots
Maximum Speed in SS 3	55 knots

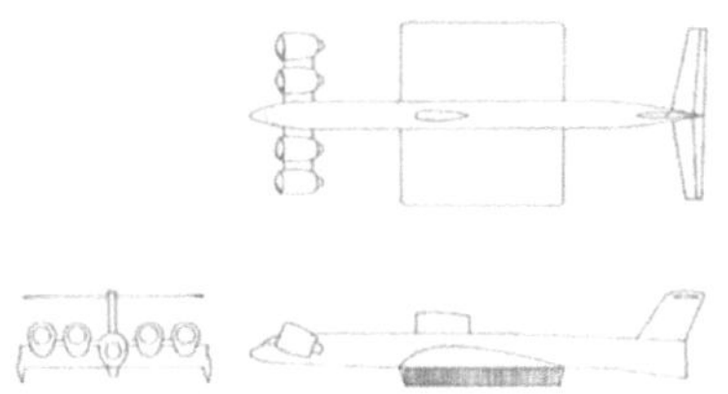

WIG(S)

Gross Weight (W, tonnes)	907
Length x Beam (m)	79.63 x 32.92
Payload (W_P, tonnes)	107
Range at Cruise Speed in SS 3	4,616 nm @ 200 knots
Maximum Speed in SS 3	200 knots

Figure 10.6: ANVCE Point Designs for Non-Air Capable Surface Combatants (1,000 tonne Class) (Far Term)

gives the required speeds and satisfies getting over hump easier, it is at a disadvantage in deck space for an otherwise "air capable ship". Such design variants would be pursued in any particular case.

As part of the "design mix" a baseline of current craft (both operational and ongoing designs) was included in the analysis. This set included US Marine air cushion craft (LCAC); amphibious landing craft (LCPC); mine countermeasures craft (MCM); coastal patrol craft (CPIC); fully submerged hydrofoil (PHM) and other surface craft. A similar set of operational air craft was also included as a baseline in the analyses of the air vehicles not reported on in this book.

In summary, these Point Designs provide much information for cost estimation and for mission analyses. They also provided basic points to include in the technical maturity assessments that would be necessary to answer the questions placed by the DDR&E directive, the GAO studies and Congress.

10.3 Summary of Technological Issues to be Resolved

In the context of seeking best solutions for "100 knots at sea" only a few of the concepts studied in the ANVCE Project will be expanded on in this book. Table 10.3 summarizes which concepts will be discussed.

Table 10.3: *ANVCE Concepts for High Speed at Sea*

Concept evaluated in ANVCE Project	Discussed in this book
Surface Effect Ships (SES)	Yes
(Amphibious) Air Cushion Craft	Yes
Hydrofoil (Fully Submerged & Surface Piercing)	Briefly
Monohull Displacement Ship	Briefly
Planing Craft (Conventional & Supercritical)	Briefly
Small Waterplane Area Twin Hulled (SWATH) Ships	No
Air Loiter Aircraft	No
Sea Loiter Aircraft	Briefly
Lighter Than Air (LTA) Vehicle	No
Wing-in-Ground-Effect (WIG) Craft	Yes

The complete set of analyses for all concepts is provided in the 1979 US Navy ANVCE Report.[4] In what follows, a brief overview is provided for hydrofoils and planing craft because, although they do not provide a 100 knot capability, especially in the larger ocean going sizes, they do represent a valid data base for alternative solutions to some of the needs for high speed in the US Navy (as well as in the private sector). Because the US Navy would be adding a new capability to the Fleet it was considered invaluable to compare with the best conventional monohull displacement that would also be available using advanced technology. Hence, a baseline displacement ship was included in the mix. As will be expanded upon in this Chapter, one of the recommendations by ANVCE was to take the necessary development in smaller steps in ship size to allow the technology to develop in a more reliable manner than the fast paced program of the US Navy at that time. This would allow the distinct advantages available to the US Navy from such concepts to become more apparent and available on a cost-effective basis.

Concentration here will be on discussing the technical issues and recommended solutions in the class of vehicles commonly grouped under the headings of *surface effect ships, air cushion craft*, and *wing –in-ground-effect craft*. Even within these latter classifications there are important sub-classes that have a significant effect on the choice of candidate for high speed at sea. The discussion of the ANVCE findings on technical issues and solutions follows under these last headings.

10.3.1 Selected Key Limiting Technologies in SES and Hovercraft

A significant number of technologies that affect the design of high speed ships was examined in the ANVCE Project and an in depth discussion is provided in the 1979 ANVCE Report. A selected few of those technologies that could cause limitations in the pursuit of large, high speed ships are reproduced here to indicate where the boundaries are in several key areas.

Structural Weight A fundamental issue in high speed ship design is whether a structure can be achieved that is both strong enough to withstand the impact loads from the sea waves at high speed yet light enough such that the ship can carry a useful load. Figure 10.7 shows collected data from a wide range of vehicles from light aircraft to tanker ships of up to 10,000 tons or more ship size. Each type of craft shown operates in a different medium and at different speeds and with very different geometries. Aircraft have complex structures comprising fuselages, wings and appendages and operate at high speeds (100–300+ knots) immersed in air but under dynamic loading conditions. The air cushion craft has a more compact structure and operates in the speed range (35–100 knots) at the interface of air and water. The tanker ships (at 10–25 knots) are fully supported by water buoyancy and have geometries of a more simple shaping but with internal decking. It is seen that the structural weight, expressed as a fraction of the gross weight or displacement ($^{W_S}/_{W}$) varies from about 40% for the lighter, dynamic vehicles down to less than 20% for the larger, simpler, buoyant support vehicles. Some examples of air cushion craft are included in Figure 10.7 to indicate the current state of the art for such craft that tend to have structural weight fractions in the mid range of all vehicles of about 30–35%. A considerable amount of development has gone into exploring if lighter values could be achieved without sacrificing safety or structural integrity.

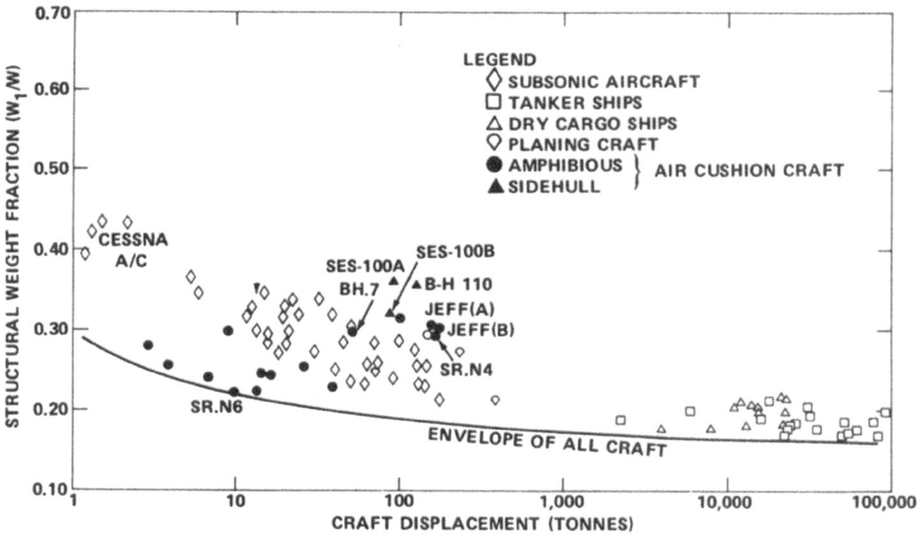

Figure 10.7: Structural Weight of Various Craft

The envelope shown in Figure 10.7 of structural weight of all craft can be expressed by the relationship:

$$\frac{W_S}{W} = 0.15 + \frac{0.15}{W^{1/3}} \tag{10.1}$$

Much of the development of the air cushion craft, especially those designed for high speed was involved with determining whether the structural weight could be reduced from the early high values. In the 1980 report, the author developed an equation for the structural weight of an idealized air cushion craft that related all the major determinants of the loading to give:

$$\frac{W_S}{W} = \frac{F(t, E, ...)}{p_C} + \frac{n}{\sigma} G\left(L/_B, H/_L, ...\right) W^{1/3} \tag{10.2}$$

The first term, governed by $F(t,E,\ldots)$ includes the local properties of the panels and stiffeners and reflects the buckling criteria effect (through the Young's Modulus E) and the minimum gauge thickness (t) of the structural members. The second term, governed by the geometric function, $G(L/_B, H/_L.....)$, reflects some of the overall sizing or geometric effects on the design. It includes the overall shape through slenderness ratio $(L/_B)$ and the craft depth or height to length ratio $(H/_L)$ reflecting the beam strength. This second term also includes the load factor (n) that is dominated by the impact pressures caused by contact with the waves; and the average stress level (σ) in the structure.

This functional form was applied to a set of air cushion craft (both operational and conceptual) as shown in Figure 10.8. The functional form when applied to the data base can be expressed by:

$$\frac{W_S}{W} = \frac{0.25}{W^{1/3}} + \frac{0.04}{\left(p_C/S^{1/2}\right)^{1/3}} W^{1/3} \tag{10.3}$$

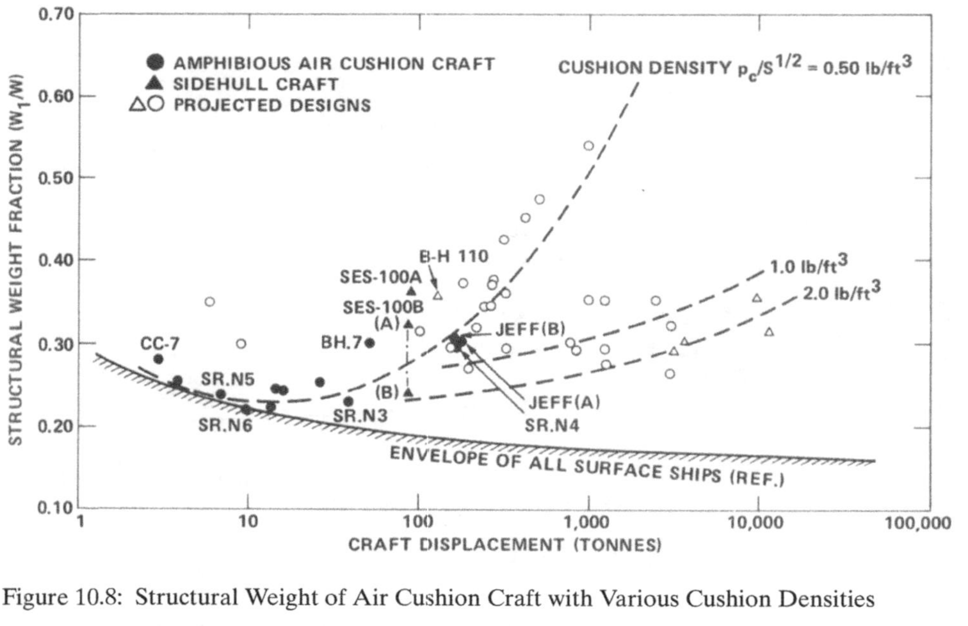

Figure 10.8: Structural Weight of Air Cushion Craft with Various Cushion Densities

It is seen that in this form, the trend is determined by two overall effects, one inversely proportional to the cube root of the gross weight and one directly proportional to the cube root of the weight. That is:

$$\frac{W_S}{W} = \frac{A}{W^{1/3}} + BW^{1/3} \tag{10.4}$$

The "constants", A and B are as shown in Equation 10.3. In the constant B, the cushion density, $\left(p_c / S^{1/2} \right)$ that is a key factor in the determination of the drag of the craft (see Chapter 8, Table 8.2) also appears here as a key structural parameter. It was convenient to express the cushion density in a more general form in terms of cushion area (S) rather than cushion length (L)in this treatment to normalize the effect of (L/B) ratio.

The effect of the cushion density is quite marked as it reflects, among other attributes, the compactness of the craft and its structure. For low values of cushion density $\left(p_c / S^{1/2} = 0.5 lb/ft^3 \right)$ it was found that a good agreement was found with the available craft. For higher values of cushion density approaching 1–2 lb/ft^3 there is less agreement. This is probably due to many other contributions to the structural weight as the size (displacement) is increased. This would include the effect of the number of decks, for example, more likely in the larger sizes and included in the design points of the projected designs. The effect of speed would become more important for the larger sizes as the impact pressures become dominant. Also, the idealized structure used to develop the equation was based on a box-like structure without the effect of sidehulls (more applicable to the amphibious form of air cushion craft). This point is illustrated by the calculations for the SES-100B. Point (A) in Figure 10.8 is for the SES-100B "as built" complete with its sidehulls. This value shows the structural weight fraction $W_S / W = 0.33$. The Point (B) is a calculation for a hypothetical form of the SES-100B without the sidehulls weight.

The various design concept points shown in Figure 10.8 are taken from the ANVCE Project work as part of the evaluation of the many concepts under consideration for a high speed SES capable of operating at high speed over waves.

It is the second term in Equation 10.2 and Equation 10.3 that influences whether the structural weight can be reduced to lower values as the ship size is increased. As Equation 10.2 shows, the load factor (n) is a strong contributor to that structural weight. During the ANVCE Project, the experience base from the UK hovercraft industry was brought to bear on the problem. Peter Crewe[5] (British Hovercraft Corporation) who was brought on as a consultant to the ANVCE Project had developed early in the development of hovercraft an expression for the wave impact load factor (load divided by craft weight) given by:

$$n_W = \frac{0.12 K_1 V_v V}{W^{1/3} \left(1 + r_x^2\right)^{2/3}} \tag{10.5}$$

In this expression:

n_W	=	wave impact load factor (load divided by weight)
K_1	=	empirical factor
V_v	=	craft relative vertical velocity (fps)
V	=	craft speed (fps)
r_x	=	distance from craft C.G. to impact point (measured parallel to the centerline) divided by the radius of gyration
W	=	craft weight (lb)

The craft relative vertical velocity (V_v) is a key determinant in the load and expresses the net vertical velocity of the downward motion of the craft combined with the upward motion of the oncoming wave. Crewe gives this relative velocity as:

$$V_v = v_S + \frac{2.26\pi h_W}{\sqrt{\lambda}} \tag{10.6}$$

The sink velocity of the craft (v_s) is difficult to express in any general form because it depends on the dynamics of the motion of the particular geometry and mass characteristics of the individual craft. In absence of other information, Peter Crewe recommends the use of $v_s = 2fps$ as a general value. It is to be emphasized that this value was determined from test data of the first hovercraft SR.N1 and other values should be investigated for other craft. The sea state is represented in Equation 10.6 through the wave height (h_W) and wave length (λ).

The above equation for the load factor was incorporated into the British Hovercraft Safety Requirements in 1972[6]. Along with that expression for the load factor was the expression for the wave impact pressures acting on the hard structure of the craft during wave impact:

$$p_{\max} = 0.0324 K_2 V_v V \tag{10.7}$$

This peak impact pressure (measured in psi) is that pressure acting on a localized area due to wave impact. It is attenuated toward the outer edges of the surface, such that the average pressure would be somewhat less. The determination of the impact pressures of high speed hull forms has been the subject of much research using flat plates, wedges and other shapes to gain insight on the impact pressures. The linear relationship between the *craft velocity* and the *impact pressure* has been observed in many high speed air cushion craft both with and without the attenuation effects of those craft with an encircling air cushion within a flexible skirt structure.

The linear relationship between bow wave impact load factor and the impact pressure with the craft speed determined from observations of the SR.N1 in the UK was also observed in the US during the testing of the SKMR 1 which was tested before the addition of skirts in 1963. This provided a confirmation of the impact pressures without any load or pressure attenuation effect of an air cushion.

The empirical factors K_1 and K_2 are shown in Figure 10.9. These charts

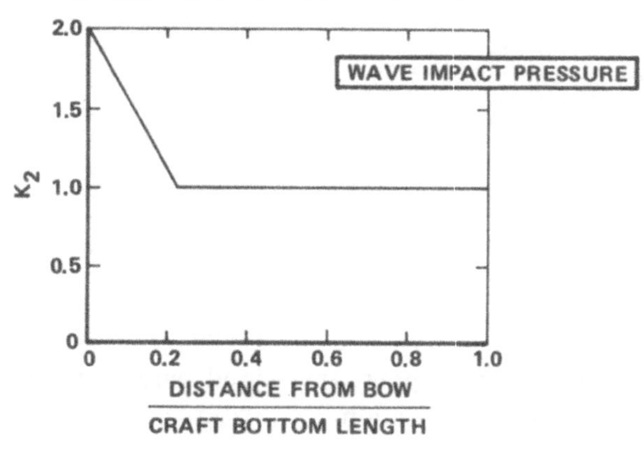

Figure 10.9: Load and Hull Impact Pressure Factors

show how impact with waves is attenuated further aft of the bow. While these results were determined using specific geometries on specific craft, it has been found that they apply over a wide range of both amphibious and sidehull SES alike.

"Rule of Thumb" on Impact Pressures As the early design work on both amphibious and non-amphibious air cushion craft developed, a certain consistency appeared that can be illustrated with an example. The SES-100B demonstrated approximately 55–58 knots in Sea State 3 (see Chapter 6, Figure 6.50). Sea State 3 has a significant wave height (h_W) of about 6 ft and a wavelength of about 62 ft. If these values are fed into Equation 10.6 and Equation 10.7 using the empirical factors from Figure 10.9, one gets the result that $p_{max} \approx 0.83V$, where $V$is in knots. Similarly, if the results for Sea State 4 are used in the same equations, one gets the result, $p_{max} \approx 0.95V$. This consistent result led to the "rule of thumb" (allowing for a safety margin):

$$p_{max}(psi) = V(knots) \tag{10.8}$$

This "rule of thumb" was used in many of the US designed air cushion craft as a starting point in the design process. The impact pressures on the plating of the four US air cushion craft taken from the author's 1980 report are shown in Figure 10.10. The British design philosophy was a little different, in that the designers took into account the load alleviation effect of the air cushion (an assumed 30–40%) and this effect on the impact pressure design values is shown in Figure 10.11. Because the US craft were designed for higher speeds and the technology for such high speed craft structural design was (is) not well defined the higher impact pressure design approach for the US craft was deemed appropriate.

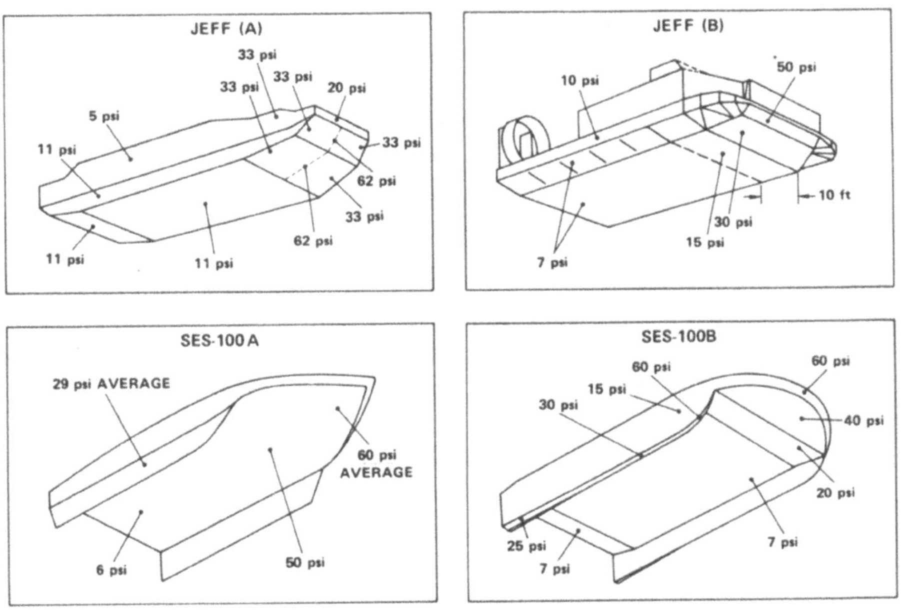

Figure 10.10: Hull Impact Design Limit Pressures (US Air Cushion Craft)

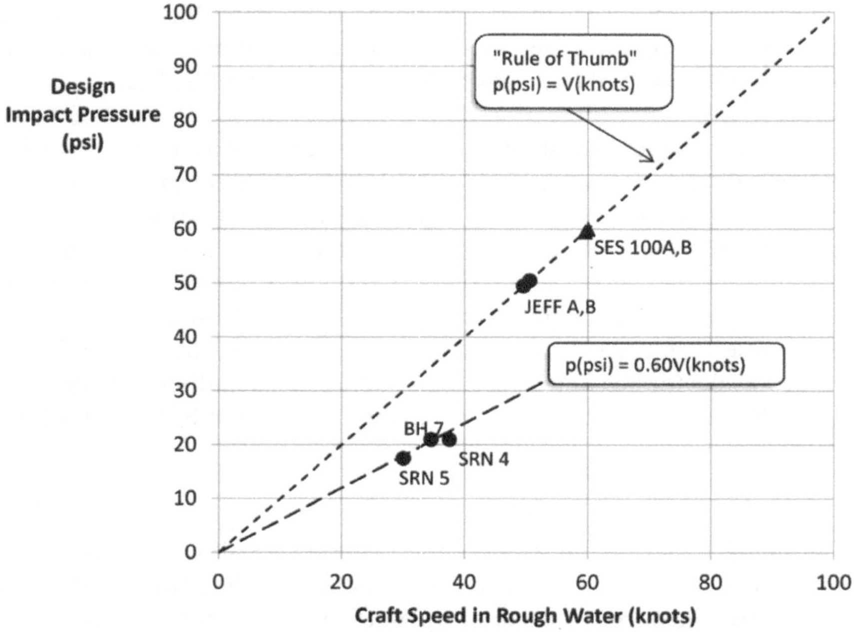

Figure 10.11: Bow Impact Pressure Design Criteria

In 1994, some 15 years after the completion of the ANVCE Project, another review was made of high speed air cushion craft, specifically wing-in-ground-effect craft which was conducted by the Advanced Research Projects Agency (ARPA)[7]. In their Final Report, ARPA learned that the impact pressures on the hull used by the Russian designers were quite high. While the Popular Press would advertise that the Russian *ekranoplan* designs would clear the water at low speeds due to the air ingestion from the efflux from the inclined lift augmentation engines (PAR in US parlance), in reality the craft traveled a long distance in "Take-Off" mode through the waves to avoid large hydrodynamic loads on the wing trailing edge flaps. The *ekranoplan* LUN, for example had a "take-off run" of at least a kilometer or more, which it must do in waves if the mission calls for it. The take-off speed, meaning the speed at which the LUN can sustain aerodynamic lift without hull support occurs at about 65% of the cruise speed. Hence, the LUN will remain in contact with the water surface (waves) for a wide speed range (in excess of 200 knots) despite activation of the PAR system. Indeed, all engines (PAR and cruise) must be operational throughout this take-off mode in high speed contact with the waves.

The advertised cruise speed for the LUN is 243–297 knots (according to the ARPA report). In the same report, it is reported that the maximum impact pressure experienced on the LUN was 294 psi. This is similar to the "psi equals knots" rule of thumb! If this is true, then Figure 10.11 could be re-plotted as Figure 10.12. Clearly, there is a paucity of data on this critical design parameter for high speed air cushion craft but as a "holding pattern" until more reliable data comes along this data is summarized here.

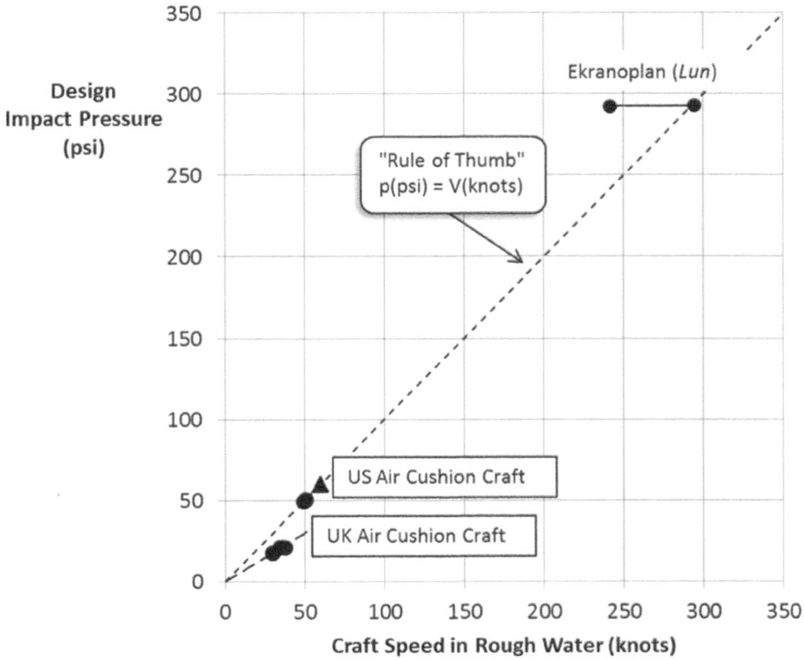

Figure 10.12: Impact Pressures including Russian LUN values

In 1952, Robert Smiley[8], in analysis of impact loads on seaplane hulls, developed the relationship between the craft speed (V), sink speed (v), trim angle (τ) and the impact pressure ($p_{\max}$), as:

$$p_{\max}(psi) = {}^{1}\!/_{2}\,\rho_W f^2 \tag{10.9}$$

The "equivalent planing velocity", (f), is given by:

$$f = V + \frac{v}{\tan \tau} \tag{10.10}$$

By making some reasonable assumptions about sink speed (vertical velocity of the hull with respect to the wave) and average trim angle during motion of the craft, it is found that this formula gives similar results to the "rule of thumb" given by Equation 10.8 found so successful in air cushion craft design. The complication of the load alleviation due to the presence of the air cushion cannot always be relied upon, and for safety or design purposes is not used. See earlier equations by Peter Crewe on these factors.

An indication of the effect of design impact pressures on structural weight can be seen in Figure 10.13 where the plating weights of several high speed craft (air cushion craft and hydrofoils) has been collected from various sources. This graphic also includes the effect of the associated form of construction. To the left of the chart are seen data for air cushion craft, mainly taken from the British Hovercraft Corporation set of craft plus the lone US data point for the US Marine Corps air cushion craft prototype, JEFF A. The chart shows the effect of the different forms of construction from riveted panels to welded panels including those with PVC foam

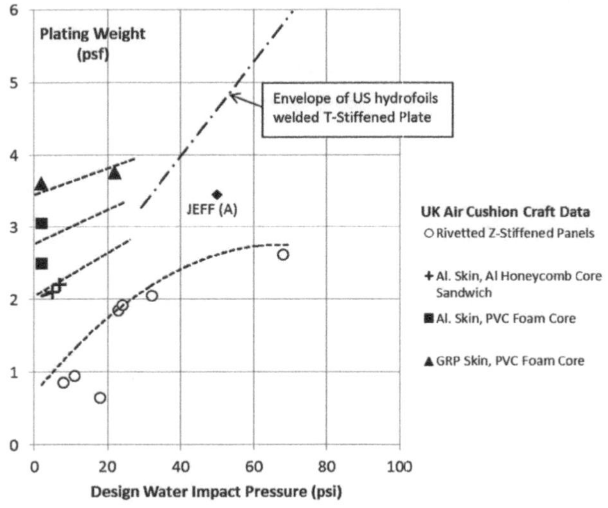

Figure 10.13: Effect of Impact Pressures on Weight

cores. The envelope of welded T-stiffened plate data from US (fully submerged) hydrofoils has been taken from Heller and Clark.[9]

In the data, the weight of the panels includes allowances for the panel edge fittings. It is seen that there is a wide range in the data but all show a significant increase in weight due to the impact pressure from waves. As impact pressures increase from around 20 psi to 80 psi the weight of the panels increase by a factor of 3:1 or more which will have a significant effect on the structural weight. If the impact pressures rise to the levels shown in Figure 10.12 one would expect a major effect on the structural weight of craft designed for such high speeds.

Later in this Chapter, when the ANVCE WIG designs are discussed, this observation is expanded upon where the large impact pressures experienced in the Russian *ekranoplan* designs led to large structural weight fractions.

Structural Weights of ANVCE Point Design Surface Vehicles Once the ANVCE Point Designs were complete, where the designs were to a common standard, the structural weight was displayed against a backdrop of all vehicles in Figure 10.14. The envelope of all vehicles from Figure 10.7 has also been included in Figure 10.14. The two dotted ellipses surround, (1) existing air cushion craft that have seen operational service, and (2) the ANVCE Point Designs. It is seen that the existing craft exhibit structural weight fractions of 45-30% over displacements of 5-160 tonnes. The ANVCE Point Designs structural weight fractions of 36-23% for 1,000–10,000 tonnes displacements. Included in the ANVCE set is the baseline advanced technology displacement hull ship, MONO 3 projected at 25% structural weight fraction. This baseline displacement ship is to be compared with other displacement hull ship values of 30% (FFG 7 shown) and other displacement ships (not shown) of 34% (FF 1033); (35%) DLGN 35 and 44% (DLGN 35). After the ANVCE Project was completed, the SES Program Office set about designing larger displacement SES at a slower design speed of about 50 knots. Those design concepts are described in Chapter 6. They are the MDC and MPS. In those two cases, the US Navy projected an optimistic 26% structural weight fraction.

The original design candidate of a 3,000 tonne sidehull SES (SES 3) was estimated at a 30% structural weight fraction and 32% in the amphibious version (ACV 3). That 2% difference is more due to different estimates by two different contractors rather than any fundamental concept difference. These values were deemed to be reasonable projections based on the large data base available from

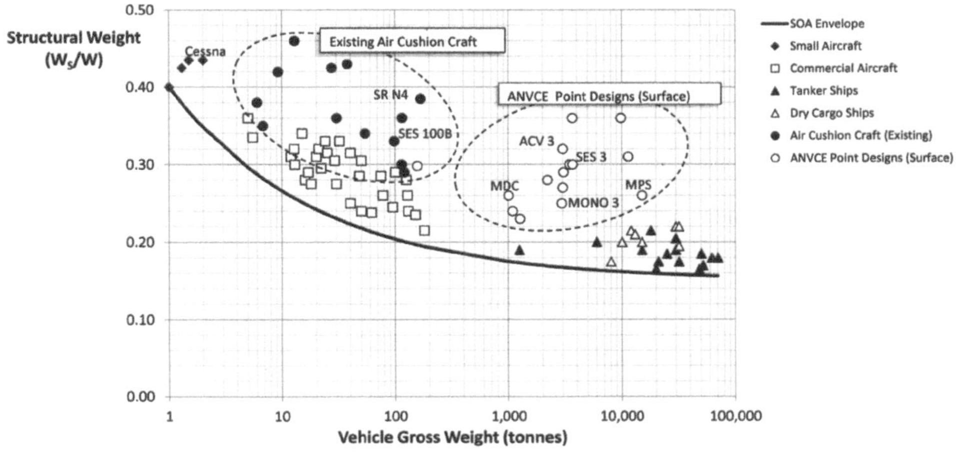

Figure 10.14: Structural Weight of ANVCE Point Designs (Surface Vehicles)

operational craft and estimates including the effect of larger displacement and higher speeds. These design points are also in agreement with the projections predicted using the parametric Equation 10.3 and Figure 10.8.

Disposable Load The weight (or displacement) of craft can be expressed by:

$$W = W_E + (W_p + W_f) \tag{10.11}$$

In this expression;

$$\text{Empty Weight } W_E = \sum_1^n (W_1 + W_2 + \ldots\ldots W_n) \tag{10.12}$$

That is, the Empty Weight (W_E) is the sum of all the subsystem weights (W_n). In the US Navy Ship Work Breakdown Structure (SWBS)[10] and its various updates up to the present day, there are usually seven major weight sub groups in the Empty Weight (or Light Ship Weight in the US Navy parlance). These are as shown in Table 10.4. In the make-up of the Empty Weight (W_E) it is usually the structure that is the largest component. The structure weight is alternatively called W_{100} or W_1 or W_S in the literature. In the 1980 report, the author showed a wide data base of the seven subgroup weights for air cushion craft for all these groups. In Figure 10.14, the wide spread of the structural weight fractions, is from 30% to 45% for existing craft and from 20% to 35% for projected designs. These fractions for structural weight translate into 60% or more of the Empty Weight of such craft.

In Equation 10.11, the difference between the Empty Weight (W_E) and the Gross Weight (W) is the Useful Load or Disposable Load (W_D) (using the different terminologies of the aircraft and marine industries) made up of the Payload (W_P) and the Fuel Load (W_f). That is,

$$\text{Disposable Load (Useful Load) } = W_D = W_P + W_f \tag{10.13}$$

A glance at Table 10.4 will show that there are various adjustments necessary when talking weights depending on the context. For example, there are "disposable load"

Table 10.4: *US Navy Ship Work Breakdown Structure*

Weight Sub Group		Weight
Light Ship		
Group 100	Structure	XXX
Group 200	Propulsion	XXX
Group 300	Electrical	XXX
Group 400	Command & Surveillance (C&S)	XXX
Group 500	Auxiliary Systems (including Lift Systems)	XXX
Group 600	Outfit and Furnishings (O & F)	XXX
Group 700	Armament	XXX
Light Ship Weight (Empty Weight) (W_E)		XXX
Loads		
Crew & Provisions		XXX
Stores and Fresh Water		XXX
Disposable Payload (Ordnance, Cargo,. . .) (W_P)		XXX
Fuel (W_f)		XXX
Total Loads		XXXX
Gross Weight		XXXXX

items defined in Equation 10.13 and "non-disposable" load items such as Crew & Provisions and Stores. This leads to other breakdowns such as "Operational Weight Empty" (OWE) that is the Light Ship Weight plus Crew & Provisions, which is a combination of Light Ship Weight items and Load items in Table 10.4. Hence, when compiling data, it is necessary to be specific. The decision to include "Flexible Structure" such as the weight of skirts in the Group 500 (Auxiliary Systems) rather than in load bearing structure, which is included in Group 100 (Structure) is a natural choice. The propulsion engines are included in Group 200 but the lift engines are included in Group 500 along with the flexible structure skirt systems.

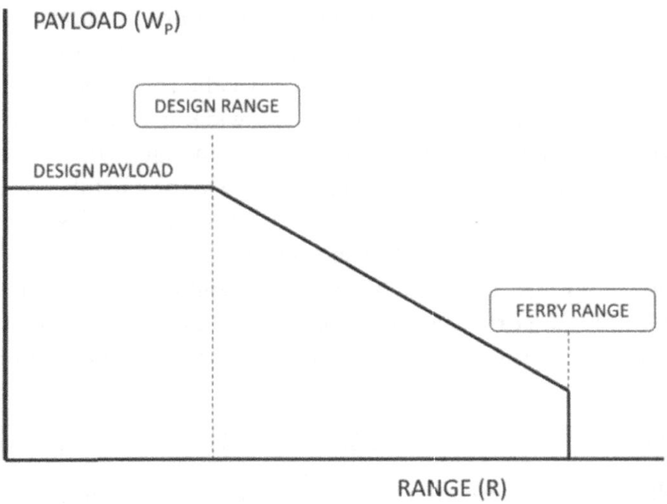

Figure 10.15: Typical Payload-Range Chart

The definition of "payload" (W_P) and "fuel" (W_f) in the Disposable Load definition also requires some clarification. Figure 10.15 shows a typical payload-range chart for any vehicle. The vehicle is designed to carry its payload over some specified distance. This would be called "design payload" and "design range". That design range will normally include some "built-in" requirement whether for commercial or military use. Commercial aircraft are required, for example, to include in its range calculations sufficient fuel to fly the aircraft to an alternate airport (usually within 150 miles from the planned flight path) for safety reasons. Other requirements are applied for military vehicles. Hence, when designing the craft to meet some range requirement these "extra miles" must be included. An additional range definition that is included in the design of any commercial or militarily useful craft is the Ferry Range that is a calculated range assuming the payload was converted to fuel as shown in Figure 10.15. When comparing craft characteristics from the literature these different "payload-range" relationships need to be understood – but not always made clear in the various documents!

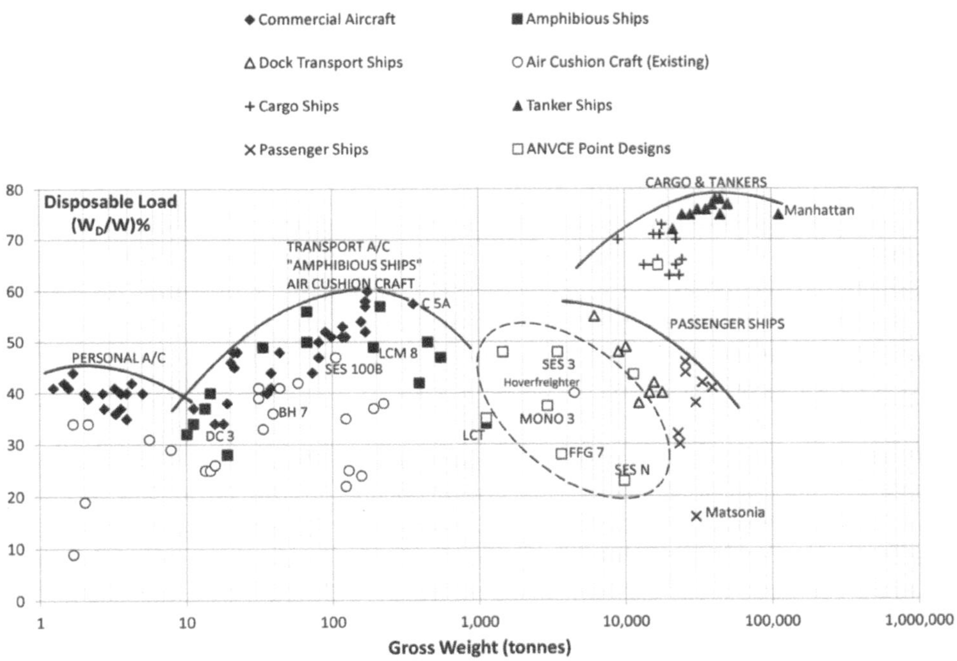

Figure 10.16: Disposable Load of All Craft

With these provisos, it is informative to see where the ANVCE Point Designs of Surface Vehicles in terms of Disposable Load (W_D) expressed as a fraction of the Gross Weight, fit into a backdrop of existing craft. Figure 10.16 shows this comparison. Envelopes of existing operational vehicles both past and present are shown in the four main groupings. The existing air cushion craft fall under the two groupings labeled "*Personal Aircraft*" in the 1–10 tonne weight class and a broader grouping labeled "*Transport Aircraft, "Amphibious Ships"* and *"Air Cushion Craft"* in the 10–1000 tonne weight class. The smaller craft are those more in the sports and demonstration craft groupings.

The main set is in the second grouping. It is seen that air cushion craft have less disposable load capability than their cousins, conventional aircraft and "amphibious" ships because of the penalty of the weight of the lift systems (engines, flexible structure, lift fans, etc). The disposable load varies between 30% and 45% (with a few outliers) of the Gross Weight. The two groupings for the larger displacement ships, in the 10,000 to 100,000 tonne class belong to the slower displacement ships (10–25 knots). Because of the slower speeds and the inherent buoyancy support of the water allows for lower structural weights (see Figure 10.14) and higher disposable loads.

The ANVCE Point Designs are grouped inside the dotted line ellipse. The baseline designs for FFG 7 and MONO 3 displacement ships using the same US Navy SWBS system is included in the ANVCE set shown. Disposable load fractions of 35–48% appear for these advanced craft designs that nestle between today's air cushion craft and the passenger ship groupings. For reference, the British Hovercraft Corporation (BHC) projection for a 4000 tonne "Hoverfreighter" with a disposable load fraction of 40% is included in the middle of the ANVCE Point Designs in Figure 10.16. The expected low value of disposable load for the nuclear powered SES, the 10,000 tonne SES N is also shown.

The impact of these values of Disposable Load (W_D), together with the results for the Transport Efficiency (WV/P), on the economic performance of advanced naval vehicles is discussed later in this chapter.

Lift Systems Technology for Air Cushion Craft The second technological area evaluated by ANVCE pertained to those technologies germane to the lift systems. This included (a) the lift fans to generate the needed cushion pressure (p_c) and (b) the flexible skirt structures. The issue with the skirt systems was the type of material required and the life of that material when subjected to the high accelerations when traversing waves at high speed.

Lift Fan Systems As air cushion craft are projected into the larger sizes such as frigate size ships (approx 3000 tonnes) and larger into aircraft carrier sizes or ocean going commerce ships (> 10,000 tonnes), the cushion pressures required begin to stretch the state of the art in lift fan design. Figure 10.17 shows the projected cushion pressure for both high and low cushion density $({}^{pc}/_{L})$designs (see Chapter 8, Figure 8.66 for discussion on cushion densities). Included in Figure 10.17 are the ANVCE projected designs. The single stage fan state of the art pressure limit is about 400 lb/sq ft as shown. If the normal pressure losses are accounted for, this puts the 250–300 psf values expected for large SES cushion pressures very close or above the SOA single stage fan limit.

Single Stage Fan Design Fan design is a well-developed art with many researchers providing detailed analytical and experimental treatments of a wide variety of fan types. These types include, among others, the *centrifugal, mixed flow* and *axial fans* used in most air cushion craft. Most of the work in air cushion craft fan design has its origins in the work by Osborne[11] (1966) and Shipway[12] (1968) for centrifugal fan systems and by Wallis[13](1961) and Shipway[14] (1968) for axial fan systems. Earlier more detailed treatments were developed by Baljé[15] (theoretical) (1960) and by Cordier[16] (empirical) (1955).

While there are many considerations in fan design concerning blade shaping, inlet design and other shape variables, it is usually found that from standard theory

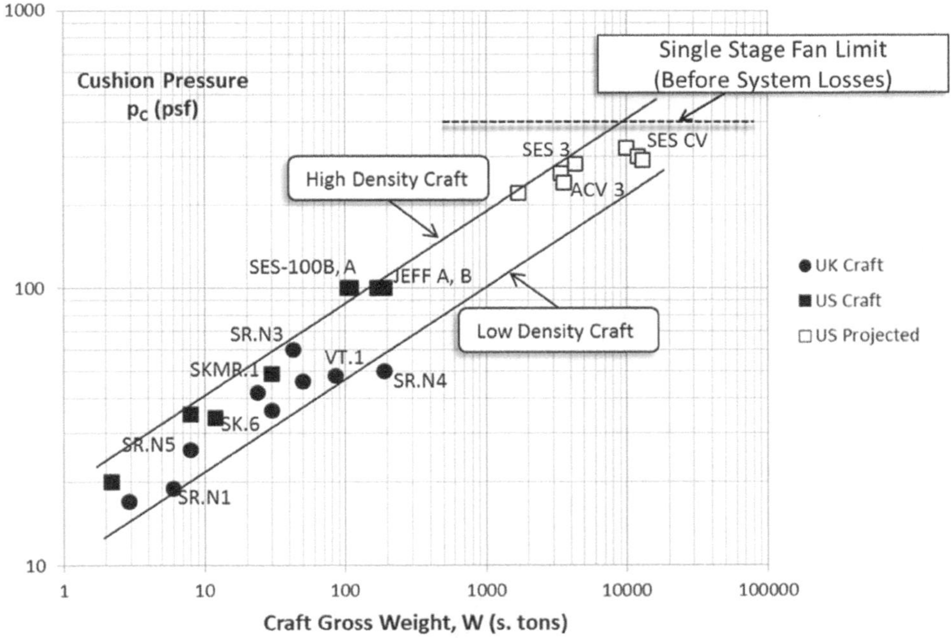

Figure 10.17: Air Cushion Craft Cushion Pressures and Single Stage Fan Limit

of turbomachines that the main parameters can be captured in the use of the non-dimensional terms of *specific speed* and *specific diameter* defined by:

$$\text{Specific Speed, } n_S = \frac{n\sqrt{Q}}{\left(p/\rho\right)^{3/4}} \tag{10.14}$$

$$\text{SpecificDiameter, } D_S = \frac{D\left(p/\rho\right)^{1/4}}{\sqrt{Q}} \tag{10.15}$$

These non-dimensional parameters contain the essential fan characteristics (in English units);

n	fan rotational speed in radians/sec
Q	air flow through the fan in ft^3/sec
p	total pressure rise across the fan in lb/sq.ft
ρ	air density in slugs/ft^3
D	fan diameter in ft

A collection of fan data from a wide selection of air cushion craft expressed in these non-dimensional parameters is shown in Figure 10.18. It is seen that the data of centrifugal, mixed flow and axial fans collapse to a common form. More complete variations on the theoretical predictions may be found in the cited references, particularly Baljé and Cordier, but the mean lines through the data in Figure 10.18 show a simple relationship between specific diameter and specific speed that can be expressed as:

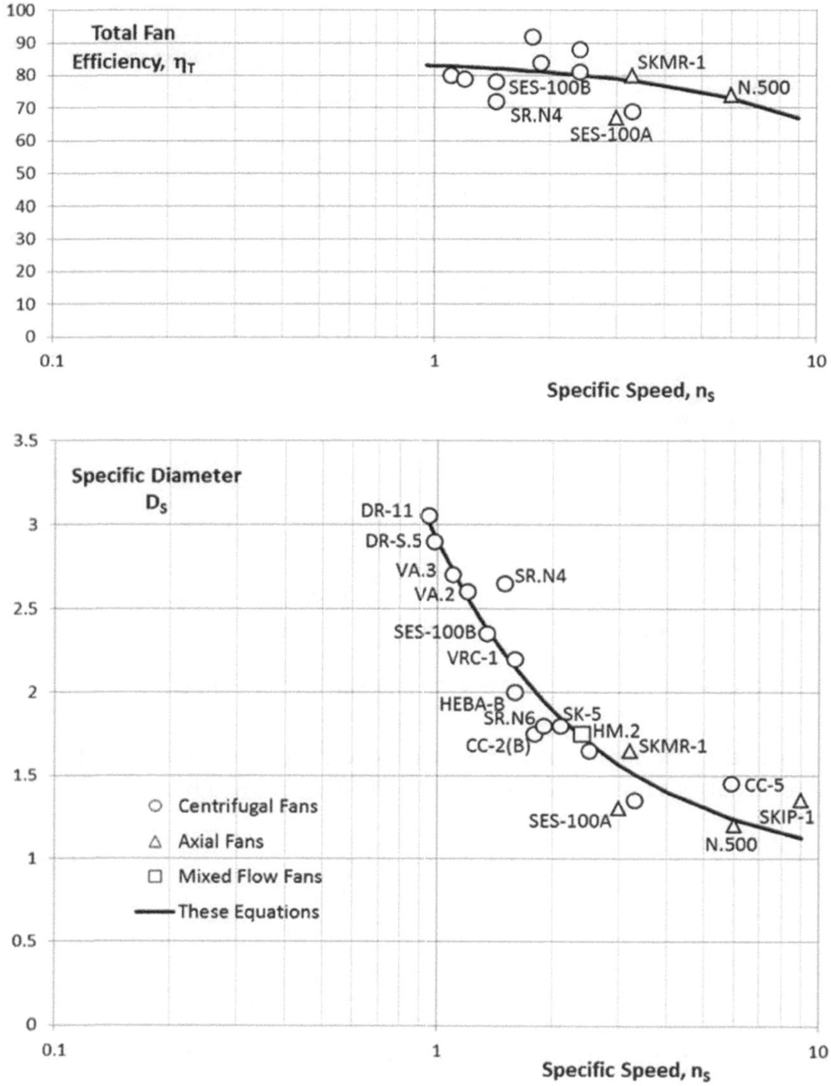

Figure 10.18: Generalized Single Stage Fan Sizing

$$D_S = 0.90 + \frac{2}{n_S} \tag{10.16}$$

The total fan efficiency (η_T) is more variable but can be represented by:

$$\eta_T = 0.85 - 0.02n_S \tag{10.17}$$

Refinements of the fan design to optimize the efficiency includes careful design of the blade shape to improve the air flow over the blade and this was done for the SES-100B centrifugal fan as discussed in Chapter 6. The particular shaping of the blade was shown in Figure 6.39. This is important to minimize the fan losses together with the pressure losses through the ducting system that varies from craft layout to craft

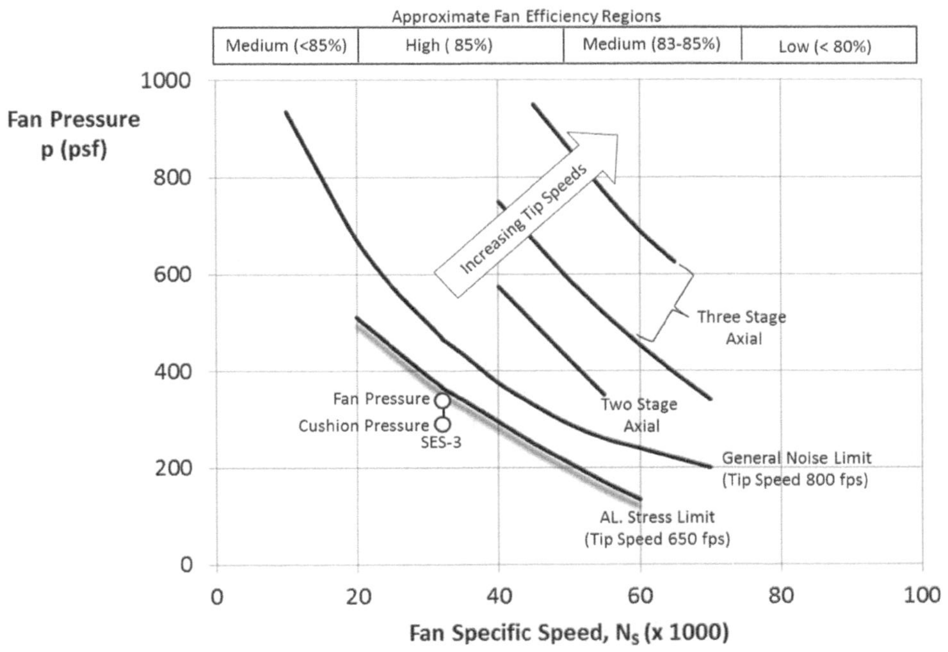

Figure 10.19: State of the Art in Single & Multi-Stage Fans

layout. The losses through typical air cushion systems are discussed in detail in the 1980 report. The significance of the lift system fan design in different craft sizes was explored in the ANVCE Project and Figure 10.19 shows the general characteristics and the need for multi-stage fan systems in the larger craft sizes.

Multi-Stage Fan Design An examination was made of current fan technology and the single stage fan limit shown in Figure 10.17 is shown more graphically in Figure 10.19 where the performance characteristics of centrifugal fans (as used on the SES-100B and proposed for use on the frigate size 3KSES or SES-3), together with mixed flow fans and axial fans are displayed. It is common practice in the fan and pump industry to depart from the non-dimensional forms of *specific speed* and *specific diameter* defined by Equations 10.14 and 10.15 and to use instead various forms of dimensionally impure forms using easy-to-use values for their customers. For example, it is found more convenient to quote the fan speed (N) in RPM rather than radians/sec ($n = \frac{2\pi N}{60}$) and the total pressure rise across the fan in inches or feet of water (H) rather than in (p) lb per sq.ft. There is no common standard in the industry and the particular units used by the fan manufacturer providing the study results in Figure 10.19 for the ANVCE Project used a convenient and popular dimensional form of the specific speed given by:

$$N_S = \frac{N\sqrt{Q}}{H^{3/4}} \tag{10.18}$$

In this form, N is the fan rotational speed in RPM, the fan flow Q is now given in CFM and the pressure rise across the fan, H is measured in inches of water. This form is not non-dimensional but used in common practice and is directly related numerically to the non-dimensional form used in Figure 10.18.

The cross hatched line in Figure 10.19 indicates the best pressure (static) versus fan speed for current (at time of the ANVCE Project) single stage fans. Any further increase in pressure, regardless of whether a centrifugal, mixed flow or axial flow single stage fan is used, results in problems of noise or material stress limits as shown. Therefore, multistage fans must be developed before the size of air cushion craft can be increased into the large ocean going ship sizes. As an example, the centrifugal fan for the SES 3 frigate sized SES modeled after the SES-100B fan is shown on Figure 10.19. The cushion pressure for the design is 290 psf. With typical losses through the lift fan system this would require a fan with a capability of producing a fan total pressure of about 340 psf or more depending on the craft design and system losses. As shown from the study this meant that the single stage fan was close to a stress limit that would not be safe. Figure 10.19 shows that consideration for large SES (frigate size or larger) would require consideration of multi-stage fan systems with the added weight and complexity of design.

Flexible Skirt System State of the Art During the design phase of the SES-100B in the late 1960s, considerable effort was expended on the design of the flexible skirt system. Chapter 6 describes this effort especially as it pertained to choice of materials as well as to the understanding of the dynamic motion in response to wave action. Laboratory tests of materials under a wide range of operating conditions were conducted and the results are included in Chapter 6.

As the ANVCE Project got underway in 1976, this exploration was expanded as the considerations for much larger ships were contemplated. The ongoing operational experience in the UK was continuing and advantage was taken of that valuable real world data and information. At the time of the ANVCE Project, the largest air cushion craft was about 300 tonnes displacement and the question of producibility of skirts into the larger ship sizes (3000 tonnes and higher) prompted an evaluation to be included in the work. Four major companies involved in skirt design and manufacture were contracted with on this important subject. They were Aerojet General Corporation[17] and Bell Aerospace[18] in the US; and Vosper Thornycroft Ltd.[19] and Hovermarine Corporationn[20,21] in the UK. The results of that work are summarized in Figure 10.20 and Figure 10.21. The work explored the various schools on light weight and heavy weight materials and especially on the effect on skirt wear.

From a simple weight scaling analysis, the skirt weight would be proportional to the cushion pressure times skirt height, all else being equal. This is borne out by the high density cushion pressure designs (e.g. JEFF (B) and JEFF (A)) that have higher skirt weights that the low density cushion pressure designs such as on the SR.N4. Other factors that determine skirt weight are use of "deep skirts" (see later) or use of compartmental dividers for improved stability etc. Such variations cause the spread in the data shown in Figure 10.20. One generally agreed upon view from the different companies was that only moderate development would be required to meet the requirements of the larger sizes over the next 15 years (1975–1990). It was concluded by the companies listed above, in consort with material manufacturers, that strong, durable materials would have to be developed before such craft would be economical and well-suited for open ocean operations. The weights of the flexible skirt

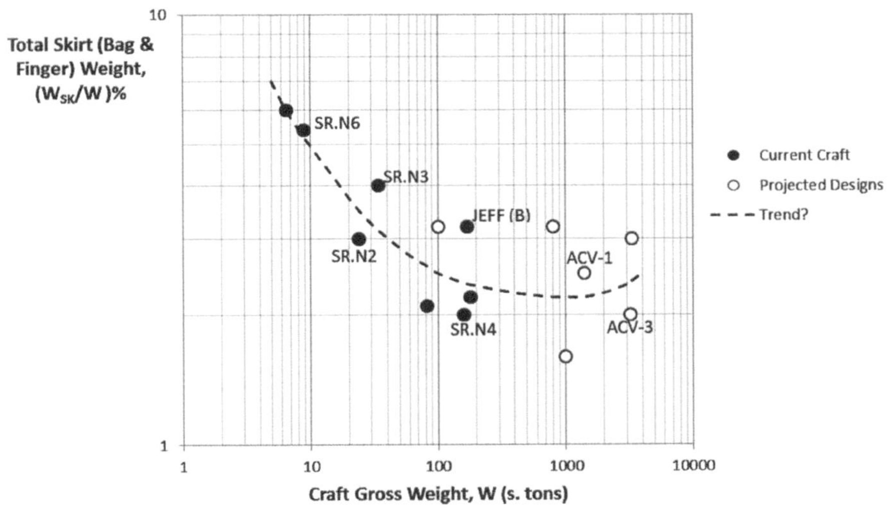

Figure 10.20: Flexible Skirt Systems Weights

systems varied widely from about 1.5% to 3.5% of the craft gross weight. The life of the skirts, particularly at the lower edge in contact with the water was steadily improving through operations in the UK on routine commercial (and military service in Vietnam) and through well directed R&D of new materials.

At the time of the ANVCE Project (circa 1976) a finger life of 450 hours was easily achievable (as shown in Figure 10.21) when operating at speeds of 50–80 knots on a continuous basis. This life expectancy was confirmed by the sea trials of the SES-100B *sidehull SES* with similar demonstrated skirt life. Projections were being made at that time of 2000 hours of life at those speeds but this would need verification from actual operational experience. After the ANVCE Project was completed, more work on high speed SES continued and in 1982–1983 the SES 200 demonstrated a skirt (finger) life of 1500–3000 hours but this was with a lower operational speed of 20–30 knots in various sea states. Chapter 6 discusses the life of flexible skirts in more detail obtained from the SES-100B program. Figure 6.32 in Chapter 6 provides details on finger life with different developmental materials from both Goodyear and Aerojet General Corporation from various laboratory and full scale operations.

Because the *rigid planer seal* of the CAB concept on the SES-100A had no significant operational test time there was no data on the life of such seals that could be evaluated by the ANVCE Project.

In addition to the exploration of types of materials and life of flexible skirt materials, the ANVCE Project explored the effect of "deep skirts" for the larger ship displacements expected for ocean going craft. The concern was related to the flexibility of such large structures and the effect on the drag and seakeeping characteristics. There are several different types of flexible skirt systems that have been used with success in the UK hovercraft industry and in the US Navy SES-100B Program and in the US Marine Corps Air Cushion Landing Craft (LCAC) Program. Some of the key differences are displayed in Figure 10.22.

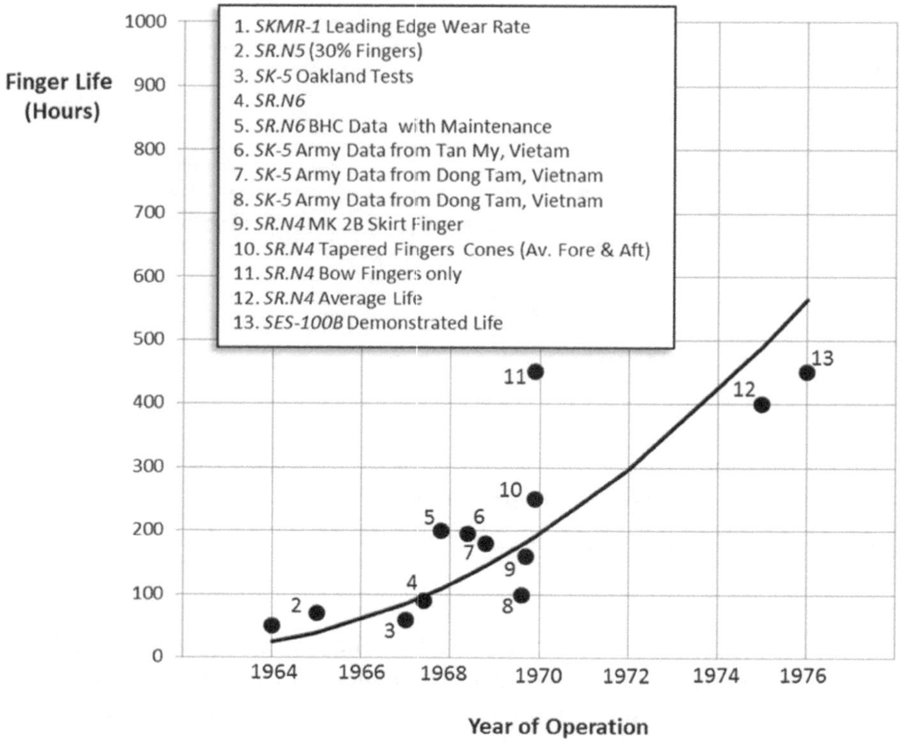

Figure 10.21: Demonstrated Life of Flexible Skirt Fingers

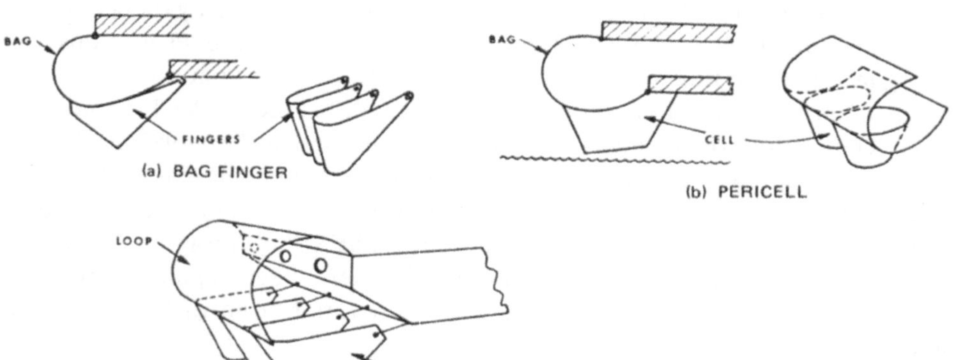

Figure 10.22: Successful Flexible Skirt Systems Used on Operational Air Cushion Craft

The (a) *"bag-finger"* skirt system shown in Figure 10.21 has been used on many British Hovercraft Corporation air cushion craft. It was also used successfully on the SES-100B. The (b) *"pericell"* used on many French air cushion craft and on the developmental prototypes in the US Marine Corps air cushion craft is shown in Figure 10.22. The (c) *"loop-segment"* type was used successfully on the Vosper-Thornycroft VT-2 air cushion craft in the UK is also shown in Figure 10.22. Each

(a) Deep Skirt Model Undergoing Test US Navy photo

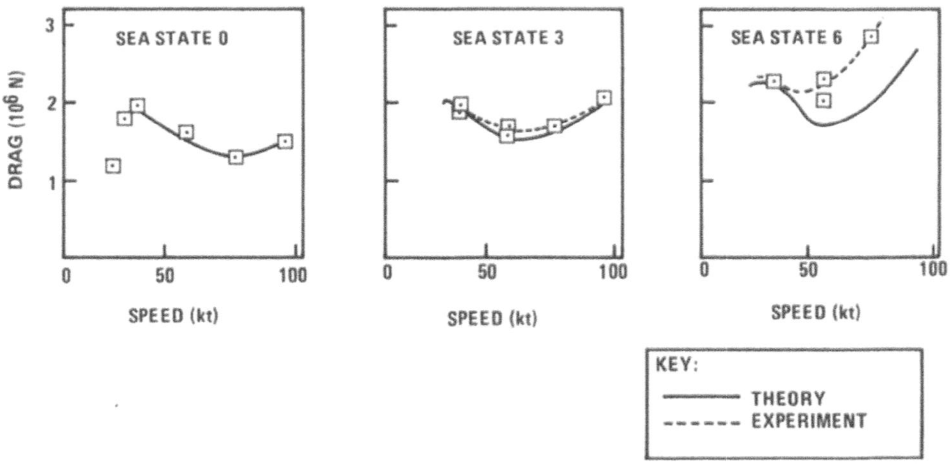

(b) Deep Skirt Model Drag Results in Rough Seas

Figure 10.23: ANVCE Model Test on Deep Skirt Configuration

of these types of flexible skirt systems have their own unique features that give different drag and seakeeping characteristics that provide a wealth of data on air cushion craft flexible skirt designs. A model using the *"pericell"* flexible skirt system as used on the US Marine Corps JEFF (A) AALC but re-configured for a deep cushion depth was tested as part of the ANVCE Project in various Sea States as shown in Figure 10.23.

The drag measurements compared to predictions are shown in the lower half of Figure 10.23. It was found that the agreement between prediction and model test was good in calm seas and for a modest Sea State 3 but that as the Sea State increased to

Sea State 6 the agreement was not as good. The ability to predict the drag at high Sea States continued to be a problem for amphibious air cushion craft (as tested here) and for the sidehull air cushion craft such as the SES-100B as discussed in Chapter 8 (see Figure 8.56) and surrounding text.

There are several contributing factors for this departure from prediction. The very high cushion pressures (and cushion densities) coupled with the relatively stiff skirts (especially the *pericell* configuration) can give rise to scooping action not easily predicted. The large size of the deep skirts gives different flexibility characteristics (and moment of inertia characteristics) than for the smaller sizes of current craft for the same geometry. This difference will cause different dynamic motion of the skirt and hence different drag characteristics. While the model test was conducted for the fully skirted amphibious air cushion craft, one would expect similar issues with the non-amphibious sidehull SES type. Such tests are a reminder that as ship size is scaled up the characteristics of even geometrically similar (Froude Number scaled) craft will behave differently. Hence the ANVCE Project general recommendation to proceed in smaller steps in the development process of such craft. The different response characteristics in pitch, heave, roll and yaw while not discussed here were also explored as part of the ANVCE model tests of deep skirted air cushion craft.

10.4 Ride Quality of High Speed Hydrofoils and Air Cushion Craft

A third key technology area explored in the ANVCE project was the ride quality of high speed ships when operating at high speeds in rough seas. This proved to be a difficult subject on which to obtain agreement because of the subjective nature of the topic. While "old sea dogs" can tolerate outrageous sea states while going about their sea duties, the "fare paying public" is much more critical and quite capable of selecting alternative modes of transportation if the ride is not comfortable. Further, the ability to operate high technology weapon systems and obtain "lock-on" for firing sequences also severely impact the design and acceptance of high speed military ships pounding through the waves.

A key issue therefore in the design of craft operating at high speed over waves is the determination of the criteria to be used on this subjective subject.

10.4.1 Ride Quality Criteria

The level of acceleration, and the encounter frequency at which it occurs, that would be acceptable by passengers and crew is in most cases highly subjective. There is the classification of levels by duration, that is, high acceleration can be tolerated for short periods of time but the level must be much lower for long duration travel. Also, the psychological factors must be included in determining levels of acceptance. People will normally accept a higher level of acceleration in a personal sport boat than in a commercial (or military) means of transportation that subjects them to even moderate sea roughness. The level of acceptable acceleration is also affected by whether the person is a passenger sitting in a warm, comfortable lounge or whether the person is a crew member working and reading vibrating gauges in the engine room.

A considerable effort was expended in the ANVCE Project to put together a reasonable set of criteria based on the extensive efforts in the aircraft and space industry and in the truck industry on proper standards. Unfortunately, there has not been a comparable effort in the high speed maritime industry on setting usable standards and most of the existing criteria have been adapted from the other industries even though it is a very different environment.

There are two main groupings in "ride quality" determination that are characterized by (a) motion sickness and (b) work efficiency. Typically, these are grouped very broadly as:

(a) Motion sickness (at low frequencies)
(b) Work efficiency (at high frequencies)

The transition from "low" to "high" frequency is in the region 0.60–1.0 Hz approximately.

At the time of the ANVCE Project, the criteria was based on laboratory experimentation that at that time was questionable as to its applicability but in the absence of better information was used in many advanced marine vehicle programs. Since the days of the ANVCE project, this situation has not improved significantly in the marine world especially in the context of high speed ships.

Most of the experimentation has been conducted under conditions of single (sinusoidal) frequency oscillation. Very little testing has been accomplished using broad band, random oscillations which characterize the air cushion craft, hydrofoil and other high speed craft motion. The literature is voluminous on the subject but two particular references are pertinent. The first is the work by O'Hanlon and McCauley[22] on motion sickness, and the second is the Guide[23] issued by the International Standard Organization (ISO) on whole body vibration more applicable to work efficiency. The key results from these two sources are shown in Figure 10.24.

The set of curves in the left hand half of Figure 10.24 are taken from the O'Hanlon & McCauley work on motion sickness and the set of curves on the right hand side of Figure 10.24 are taken from the ISO Standard for fatigue decreased proficiency (FDP) due to vibration.

The curves shown were being generated during the time of the development of the 100 ton SES Test Craft Program and were adopted as reasonable criteria at that early stage of SES development even though it was known that the criteria were tentative and suspect for the use intended. It is interesting to note that these curves have been reproduced in many technical publications since that time as being definitive and without debate!

The originators of the criteria emphasized the tentative nature of their data as follows:

Motion Sickness Criteria The motion sickness criteria curves in Figure 10.24 show curves for different percentages of the people getting sick in a continuous 4 hour duration of being subjected to the acceleration level shown. The criteria were obtained by testing untrained, unadapted subjects (college students). The subjects were enclosed in a motion generator "room" that was oscillated in a vertical direction only and the motions were sinusoidal. All subjects were seated

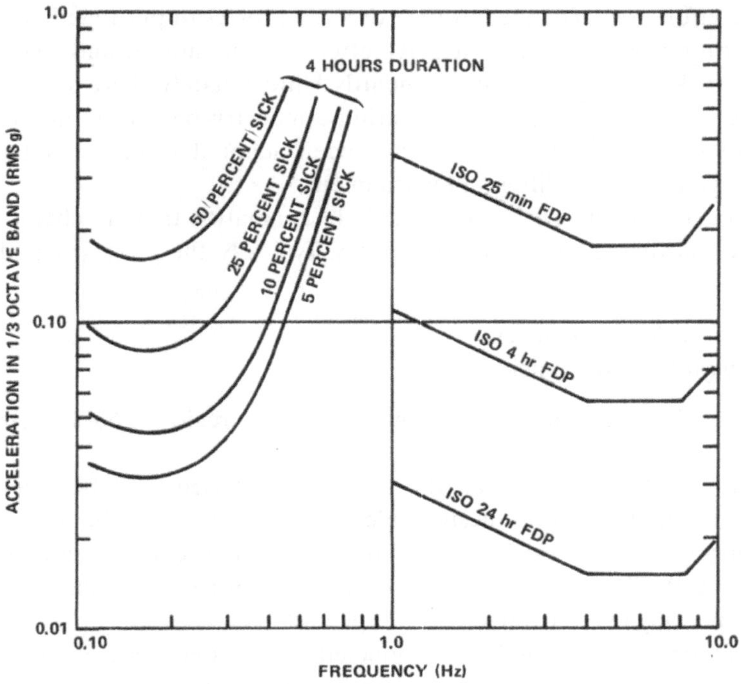

Figure 10.24: Composite Ride Quality Criteria

on conventional shipboard chairs (i.e. no special padding). Other variables such as complex waveforms, body orientation, visual reference, temperature and smell were not varied. Tests were conducted using the narrow band frequency only and is not representative of the broad band sea spectrum energy. The criterion used for determining motion sickness was initial vomiting. No assessment was made of the subject's ability to continue on a task or mission after his or her initial vomiting. No available information has been obtained as to any difference between male and female from these tests. No available information has been obtained indicating a difference between "untrained, unadapted subjects" and seasoned naval personnel. This is an area that clearly needs more work if large high speed ships are envisaged that will spend hours pounding through waves at high speed and where the crews are expected to conduct complex missions at high efficiency.

Work Efficiency Criteria The International Standard Organization (ISO) also was aware of the tentative nature of their data and provisional criteria. The referenced ISO Standard (ISO 2631) states that there are three kinds or categories of vibration in the frequency range of interest, they are:

1. Whole body vibration
2. Vibration transmitted through the feet (of a person standing) or buttocks (of a person seated)
3. Vibration applied to particular parts of the body (such as head or limbs)

The ISO Standard applies principally to Vibration Category 2 but all conclusions are provisional. Also, the criterion is considered (by ISO) to be applicable only in the 1–80 Hz range. The general limits, analyzed by ISO, are for:

1. Comfort	"Reduced Comfort Boundary"
2. Working Efficiency	"Fatigue Decreased Proficiency (FDP)
3. Safety of health	"Exposure Limit"

During the SES Program (and later adopted for use in the ANVCE Project) the Fatigue Decreased Proficiency (FDP) limit was selected and these are the limits shown in Figure 10.24. The 4hr FDP curve is considered to be most applicable to US Navy applications as it corresponds to a typical "4 hr watch" for onboard crew members. This was issued to all ANVCE Point Design teams as a common standard to be used. It should be pointed out, however, that the ISO standards were developed for passenger acceptance and the applicability to Navy crews conducting military operations is not clear.

A cautionary note is that the ISO curves are envelopes of results obtained by vibrating the subjects at each frequency separately, i.e. the subjects were subjected to simple frequency (sinusoidal) vibration and not to a spectrum of frequencies typical of high speed marine craft motion. The ISO (in 1972) considered their work very tentative and urged more complete testing be done. Unfortunately, as the decades have passed no new comprehensive criteria has appeared – but at least it is a starting point, awaiting more college students eager to be vibrated to exhaustion for the good of science!

10.4.2 Ride Quality of SES-100B

By way of illustration of the use of the ANVCE Ride Quality Criteria, just one set of data from the many test runs is shown in Figure 10.25 for the SES-100B operating (a) with the ride control system "on" and (b) with the ride control system "off". This case is for the SES-100B operating at 50 knots in Sea State 2. The Ride Quality Control boundaries shown by the shaded line corresponds to the "4 hr watch" selection. It is seen that even with the Ride Control System turned "off", the SES-100B satisfies the criteria except briefly in an isolated mid

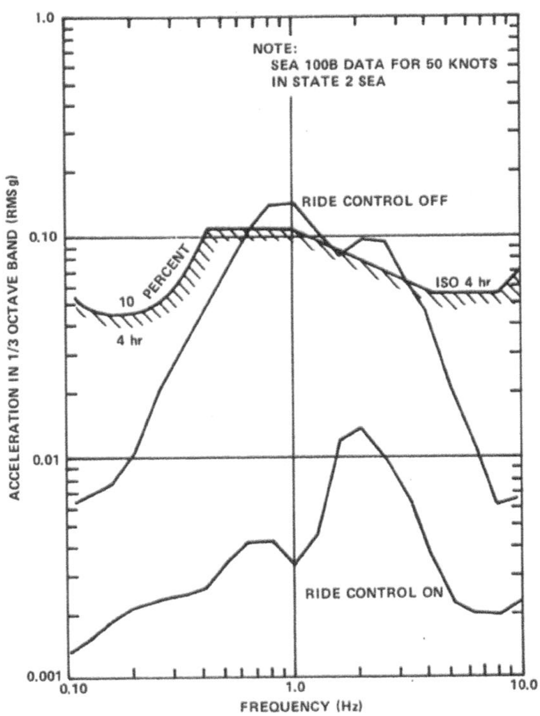

Figure 10.25: SES-100B Ride Quality

frequency range. Acceleration levels of about 0.10 RMSg expressed in the 1/3 octave band is a comfortable level for a reasonable duration. With the Ride Control System activated that acceleration level is reduced significantly by about one order of magnitude. The crew members report that the ride quality under these conditions is "quite acceptable".

Extensive work has been done on determining ride quality criteria for high speed ships and much of that work has been summarized in the ANVCE Final Report; and in the author's 1980 book; and elsewhere in the literature and will not be repeated here.

While the above treatment of ride quality is necessary and was implemented in the ANVCE Project for assessing the various candidates, it is useful to consider some more simple criteria for quick assessments. It is informative to give general observations based on many thousands of operating hours of high speed craft as provided by the experts in the various fields of hydrofoils and air cushion craft. An inexact but convenient parameter for comparing hydrofoils and air cushion craft is a "sea roughness parameter" that relates the strut length or skirt height to the wave height. While somewhat simplistic it does capture the main features of how high off the water the craft needs to be to achieve a reasonable ride quality. This simple parameter is defined as:

$$\text{Sea Roughness Parameter} = \frac{h_W}{h_S} \tag{10.19}$$

This inexact but simple measure is indicative of how high the main structure is above the waves, where (h_W) is the significant wave height and (h_S) is the depth of the skirt (for air cushion craft) or length of the foil strut (for hydrofoils). This is certainly not an exact measure but it does provide some indication of where the main structure is in relation to the oncoming waves.

Figure 10.26 (left hand side) shows some collected data of strut lengths and skirt heights of some representative hydrofoils and air cushion craft. The open symbols are the strut lengths of hydrofoils and the closed symbols are air cushion craft skirt depths. Earlier air cushion craft had lower cushion depths (shown by the dotted line)

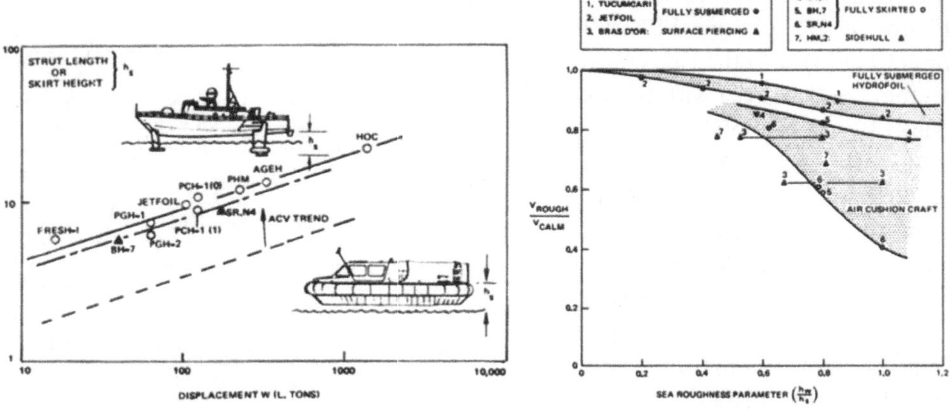

Figure 10.26: Strut & Skirt Lengths and Hydrofoil & Air Cushion Craft Speeds

but as the technology of skirts improved, deeper cushions were more attainable and as can be seen are similar to the strut lengths of (fully submerged) hydrofoils. The same intention is at work in that the aim is to keep the hard structure clear of the waves with the "suspension system" of both types of craft[f]. This simple parameter has held over a wide range of craft sizes up to about 1000 tons of displacement. If now the speed in rough seas is measured against this same sea roughness parameter a plot such as that shown in Figure 10.26 (right hand side) appears.

As might be expected, the speed fall-off in rough seas is much less for the fully submerged hydrofoil when compared to that of surface piercing hydrofoils and air cushion craft. For the fully submerged hydrofoil, as the wave height (h_W) approaches the strut length (h_S), the speed loss in waves is 20%. For the surface-piercing hydrofoil and the air cushion craft the speed loss in waves is much greater. While not exact, the surface piercing hydrofoil and the air cushion craft tend to have similar fall-off in speed in rough seas. More specific descriptions of speed of air cushion craft in rough seas is given in Chapter 6 where the performance of the SES-100B is shown to follow a similar pattern of speed at sea.

10.4.3 Variable Geometry Hydrofoil

Because the goal was to seek advanced marine vehicles capable of 100 knots at sea, the ANVCE Project included a hydrofoil with advanced technology features embodied in the foil that appeared to offer that possibility or close to it. The advanced foil design was being explored jointly by NASA and the US Navy. It was basically a conventional *subcavitating* foil (albeit with different cross section shape) that could, through simple flap angle change, convert into a *supercavitating* foil section capable of operating at higher speed without the associated drag penalty. Figure 10.27 illustrates the concept in a diagrammatic sketch form and is not intended to represent the specific geometry of an optimum variable geometry foil efficient in both sub– and super-cavitating flow.

The foil section resembles a conventional foil in that the upper surface is shaped to give the desired pressure (and lift) characteristics in the normal subcavitating mode. The shape of the underside is designed to be efficient in a supercavitating mode of operation.

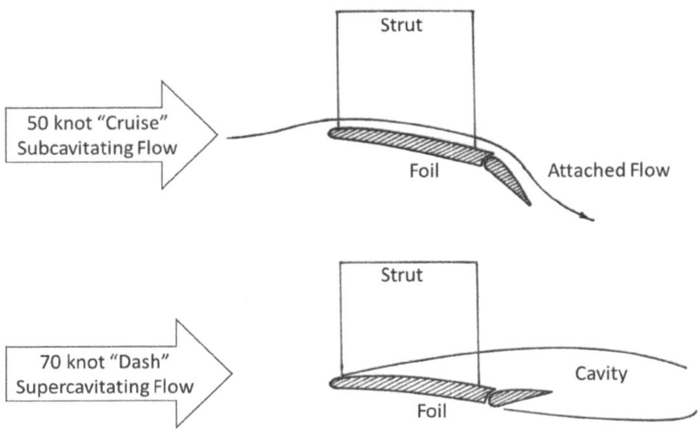

Figure 10.27: Variable Geometry Foil

The foil is subcavitating up to

[f] Chapter 1 provides descriptions of the two types of hydrofoils and their lift mechanisms.

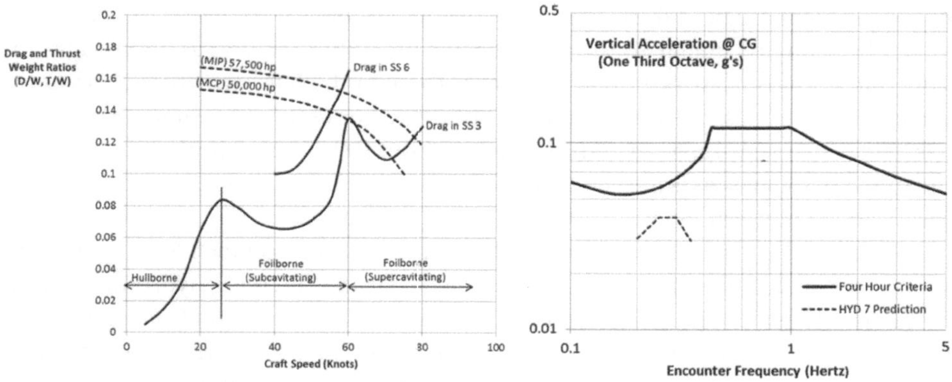

(a) Thrust & Drag Performance (b) Ride Quality

Figure 10.28: Drag Performance and Ride Quality of Variable Geometry Hydrofoil

approximately 55 knots; at that point the trailing edge flap is raised as shown whereby the flow separates from the upper surface and a cavity occurs. The lift is now generated by the positive pressure on the underside of the foil. This feature was integrated into a Point Design called "HYD 7" which is included in Figure 10:6 as one of the candidates for the "1000 tonne class of Non-Air Capable Surface Combatants". This variable geometry feature was intended to give the hydrofoil a cruise speed of 55 knots in an efficient subcavitating mode but be capable of a higher "dash speed" to 70 knots or more if the mission required that capability. Both drag performance and ride quality analyses were performed for such a novel concept as part of the ANVCE Project search for alternative ways of obtaining high speed in a practical manner.

Figure 10.28 shows the predicted drag performance and the predicted ride quality of the HYD 7 Point Design using this variable geometry feature. The left hand chart (a) in Figure 10.28 shows that the craft is hullborne up to about 25 knots, at which point it becomes foilborne until it needs to "dash" to the higher speed that is then limited by the thrust of the propellers. For this particular design the maximum speed is approximately 71 knots in Sea State 3 (the common design point set for all the ANVCE Point Designs). It is seen that engine power must be set to its Maximum Intermittent Power (MIP) level rather than the Maximum Continuous Power (MCP) to get over the "second hump" to achieve this dash speed.

While not critical to the concept, a comment on thrust margin is provided. Over the years of development of high speed craft such as air cushion craft and hydrofoils, it has become common practice to set the hump thrust margin at 0.025, that is:

$$\left|\frac{T-D}{W}\right|_{hump} = 0.025 \tag{10.20}$$

In terms of acceleration this can be written as:

$$\left|\frac{dV}{dt}\right|_{hump} = 0.025g \tag{10.21}$$

Not all craft have this built in "requirement" but it has become a useful criterion to avoid designing craft that are "sluggish" in getting over hump and taking a long time to accelerate to the desired speed. An analysis of many craft (hydrofoils and air cushion craft of many different types) showing the various drag characteristics and hump thrust margins can be found in the author's 1980 book. In the particular design of the HYD 7, it is seen that there is a large thrust margin (approx = 0.07) at the "first hump" in transiting from hullborne mode to (subcavitating) foilborne mode but when transiting from subcavitating mode to supercavitating mode the margin is low (approx 0.016) and the craft will be sluggish in getting to its "dash" speed of 71 knots. This does not appear to be a good feature from a military operation mission profile viewpoint. In any actual design program, a higher thrust level would have to be integrated into the design, especially to attain any "dash" capability for military purposes.

The right hand chart (b) in Figure 10.28 shows the predicted heave acceleration characteristics of the design for the cruise condition of 50 knots in Sea State 3. The curves have been superimposed on the ride quality criteria shown earlier in Figure 10.23. It is seen that the ride quality is well inside the criteria boundaries.

This particular design concept is included here to show that it is possible to achieve higher speeds for hydrofoils in a manner that appears to be a safe extrapolation of such craft without the earlier historical mishaps of high speed hydrofoil designs. While it does not appear to be capable of "100 knots", at a speed of 71 knots it is a valid projection for future consideration. This particular point design was for a 980 tonne displacement craft. Further analyses within the Project indicated that it would be feasible up to about 1350 tonnes displacement. For the frigate size craft (see Figure 10.5) it was found that the size could increase to about 2400 tonnes but the foil system would drop back into the subcavitating mode and a speed of 53 knots. These attractive designs were included in the data set for ANVCE evaluation.

Data for the ride quality was collected for both the hydrofoil and the air cushion craft when operating in rough seas and the data measured as RMSg of the heave acceleration is shown in Figure 10.29 as a function of the simple sea roughness parameter $\left({}^{hw}/_{hs}\right)$. The particular data shown for the air cushion craft is without any ride control systems operational.

Figure 10.29 is useful in that it shows the *general* comparison of the ride quality of air cushion craft (in its basic form without ride control) with that of hydrofoils, both surface piercing and fully submerged, without the analytical complexities described above. It provides a quick overview for discussion of the various types of craft designed to operate at high speed in waves. Note that the acceleration levels are in the same range as those determined by the more correct analytical methods previously discussed. Note also that with the ride control system activated (see Figure 10.24), the air cushion craft can meet the same low levels as the hydrofoil, although this proved to be a subject of much debate among the designers of these types of craft. There is little spread between sidehull air cushion craft and amphibious air cushion craft in terms of heave acceleration and the band of data shown shows both types at the upper and lower boundaries of the data. In the case of hydrofoils, there is a tendency for the surface piercing type to have higher acceleration values than for fully submerged foils although improvements in surface piercing hydrofoil ride control schemes is closing that gap.

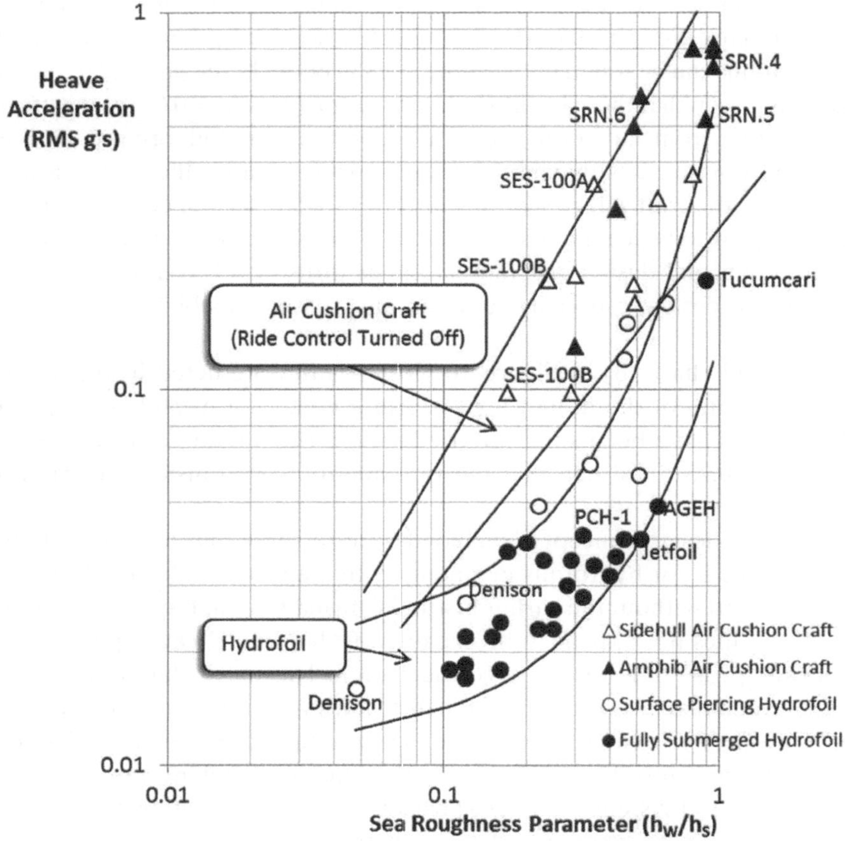

Figure 10.29: Ride Quality of Cushions and Foils

10.4.4 What Price Ride Quality?

A natural question to ask is: *Is the increased cost and complexity of the fully submerged hydrofoil over that of the simpler design of surface piercing hydrofoil commensurate with the degree of improved ride quality achieved by such designs?* The ANVCE Project was charged with determining the costs of the various advanced marine vehicles to aid the decision as to which candidate might be a better choice than that selected by the US Navy. To that end an extensive cost analysis was done of all candidates and included in the 1979 ANVCE Final Report. As part of that analysis, costs were collected of "today's" representative craft and a typical chart is that shown in Figure 10.30. The ANVCE work was done in 1976–1979 and the costs in Figure 10.30 are expressed in 1976 dollars. As a quick "update" to today for this chart one can use the inflation factor as a first pass. The inflation factor (actually the Consumer Price Index) between 1976 and 2012 is 4:1. That is, an item that cost $1 in 1976 would cost $4 in 2012! To keep the discussion in context of the ANVCE Project, the costs in 1976–1977 dollars will be used.

Included in Figure 10.30 is a shaded line representing estimates for the cost of a 1000 ton commercial SES[24] which is in line with the actual costs of air cushion craft at

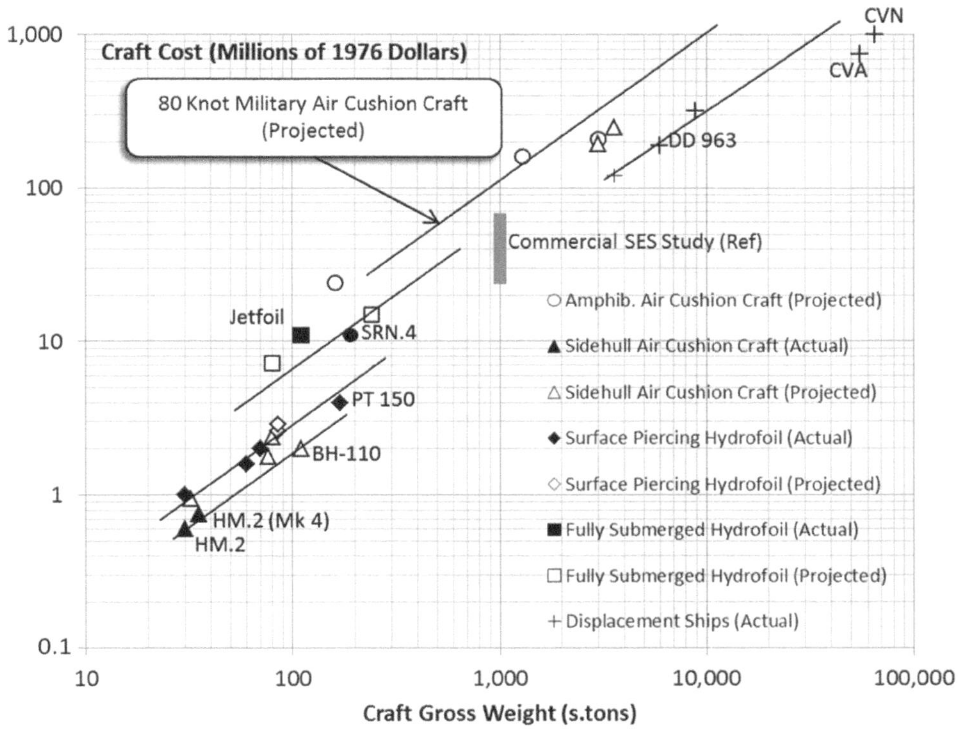

Figure 10.30: ANVCE Collected Costs (1976 dollars)

that time. Also included are the acquisition costs of US Navy ships (DD and CV) for comparison. The ANVCE Project analyzed all costs (R&D Costs, Acquisition Costs and Operating Costs) to give the Life Cycle Costs and the general trends found were as shown in Figure 10.31. The costs represent the many variations studied to determine the general trend.[25] It is seen that both the Life Cycle Costs (LCC) and the Platform (Ship Acquisition) Costs follow a general upward trend although as displacement increases, the platform cost becomes a smaller percentage of the LCC costs. These charts and others allow an appreciation to be obtained as to the contributors to the large costs associated with high speed craft. It is to be emphasized that the many possibilities in design and manufacturing techniques by different manufacturers preclude precision in any such costing but what is achieved is the ability to compare costs of different types of craft using a common costing methodology for all.

One question facing the decision makers was: *In addition to the speed increase, was the ride quality difference between the different types (see* Figure 10.29*) worth the cost difference (see* Figure 10.30*)?* There were many well experienced manufacturers around the world[g] of high speed craft at the time that seriously questioned the increased cost as being not acceptable by the customer (either military or commercial). In addition to the experts brought on the ANVCE Project from the UK, the

[g] There were no commercially operational high speed hydrofoils or air cushion craft at that time in the US but an extensive industry of such craft in the UK, Italy, USSR (at that time) and elsewhere that provided a wealth of background information on such craft.

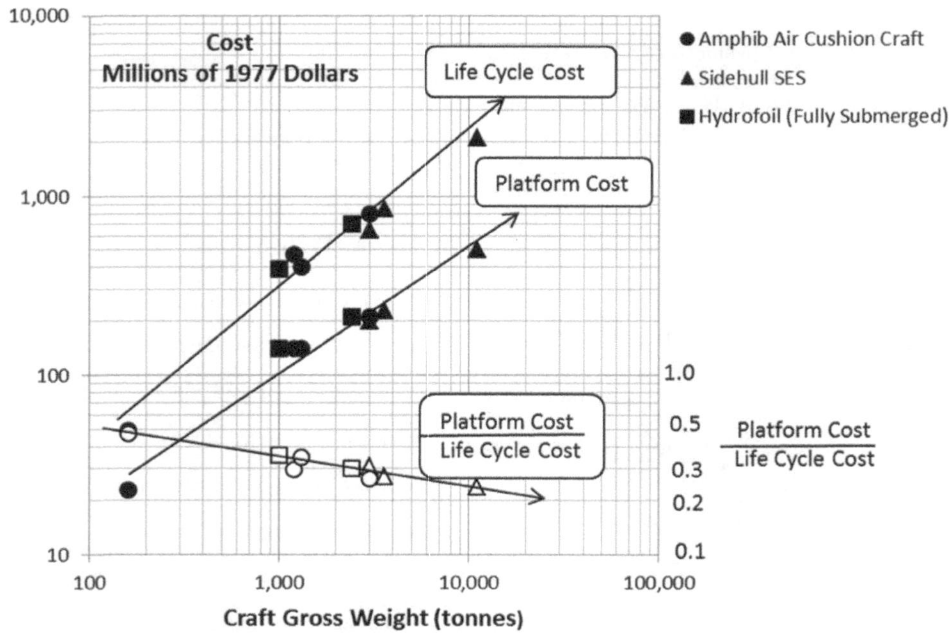

Figure 10.31: Cost of High Speed Marine Vehicles

author sought the help of two pillars in the European hydrofoil community to help shed some insight into this question of which concept was likely to be the most cost-effective. These pillars (Figure 10.32) were Baron Hanns von Schertel from Supramar AG, Lucerne, Switzerland and Dr. Leopoldo Rodriquez from Cantiere Navali Spa (née Leopoldo Rodriquez Shipyard), Messina, Italy.

One of this author's favorite memories is meeting Baron von Schertel for the first time in his offices in Switzerland in 1976. After being ushered through the general drawing office where a large number of engineers were seated on stools like elves in front of their almost vertical drafting tables industriously at work on some latest hydrofoil system, the secretary brought me into the end office where I expected to see the "the Baron" at his managing director's large mahogany desk. Instead, he was seated on a stool in front of an almost vertical drafting table working on his latest idea for a stabilization system to improve the ride quality of yet another Supramar hydrofoil! Baron von Schertel believed strongly that the product was the thing and not the trappings usually associated with people in his position. He believed in keeping his mind active on a daily practice of new improvements of his designs. The history of the Schertel-Sachsenburg hydrofoils is well documented. Supramar AG was founded in 1952[h] when the Sachsenburg Shipyard and von Schertel teamed up to develop hydrofoils of superior capability. In Chapter 1, a brief description is given of Supramar's air stabilization system for surface piercing hydrofoils. More description of von Schertel's work is provided in Chapter 2. Baron von Schertel provided much valuable advice to the ANVCE project[26].

[h] Baron von Schertel's original work starting in 1927 was on fully submerged foils but he switched to surface piercing foils in 1935 and all development since that time has been on surface piercing foil systems

Baron Hanns von Schertel
1902 - 1985

Dr. Leopoldo Rodriquez
1925 - 1998
Photo courtesy
Leopoldo Rodriquez

Figure 10.32: Two Distinguished Hydrofoil Pioneers

Based on discussions with the Baron, paraphrasing some general comments:

1. Speeds of 50 knots are easily attainable. Above that speed the onset of cavitation is a problem for all types (fully submerged and surface piercing). Improvements are possible to achieve supercavitating foil shapes but the issue is to keep them functioning in a practical sense in actual operations. The maintenance of the shapes can be an issue and is it worth the extra cost of maintenance and safety hazard of supercavity implosion? (Author: This needs more R&D).
2. The improvements in ride quality of *surface piercing foil* systems are occurring on a regular basis (especially with the Supramar air stabilization system) and it is believed that the ride quality as shown on Figure 10.29 is adequate and the large cost increase incurred by switching to *fully submerged hydrofoils* (see Figure 10.30) is of questionable value.

With the cancelation of all fully submerged hydrofoil programs in the US Navy in 1993 (the last program, the PHM, was canceled for being "not cost-effective"), were Baron von Schertel's comments prescient? Meanwhile today, surface piercing foil hydrofoils are in regular commercial service throughout the world at *low cost* and with *good ride quality* with slowly increasing speed capability.

Dr Leopoldo Rodriquez was an effervescent proponent of the hydrofoil. Leopoldo was the grandson of his namesake Leopoldo Rodriquez who formed Rodriquez Shipyard in 1887 in Messina, Italy as a repair yard that developed into a major shipyard (under the leadership of the elder Leopoldo's son Carlo Rodriquez) for hydrofoils, and in 1990 has expanded into other high speed ships under the new name of Rodriquez Cantieri Navali SpA. In 1976 when the author befriended Leopoldo (the grandson) he was the Technical Director of Rodriquez Shipyard

Figure 10.33: Rodriquez RHS-160 Surface Piercing Hydrofoil (with improved ride quality)
Courtesy Leopoldo Rodriquez

(under Carlo) which had recently changed its name to Cantieri Navali SpA. In 1976, Leopoldo as Technical Director had just led the shipyard into the new and improved electronic control system (developed by Hamilton Standard in the US) to control the trim tabs to automatically control pitch, heave and roll motions in a seaway that was applied to the latest Rodriquez surface piercing hydrofoil, the RHS-160 to give it superior ride quality in high sea states. The RHS-160 was launched in early 1976.

The RHS-160 (Figure 10.33) is an 85 ton ship with up to 200 passengers and operating at 36 knots with "good ride quality" based on the Leopoldo led improved ride control system. With the new ride control system, the RHS-160 was said to be able to transit in higher sea states than previously done with "good ride quality". Unfortunately, no data was available to include in this book on this promising feature of the Rodriquez' designs. While Leopoldo was a fierce proponent of the surface piercing foil type there was a brief excursion into the fully submerged foil type with development, in consort with Boeing in 1992, on the so-called "Maximum Efficiency Craft (MEC)" which had a fully submerged foil system. But this brief encounter did not go beyond the prototype stage and Rodriquez returned to the successful surface piercing foil system. Leopoldo was brought on board the ANVCE project to provide advice on ride control systems and other features of surface piercing foil geometries including the W-foil version.

The author developed a better understanding of surface piercing hydrofoils while inspecting and being led through the Rodriquez craft dockside in detail by an informed Leopoldo, then riding the craft on various runs between Messina and Taormina on the Island of Sicily and from Messina to Reggio di Calabria on the mainland of Italy. Observing the marine manufacturing techniques in the Messina yard and mentally comparing them with the aerospace manufacturing techniques of the fully submerged hydrofoils (Jetfoil and PHM) in the Boeing facilities in Seattle in the US was particularly enlightening. With charts like Figure 10.29 and Figure 10.30 in hand, spread out on an outside table in a restaurant in Milazzo overlooking the Tyrrhenian Sea with the smoking Stromboli in the distance, and discussing vigorously with Leopoldo the merits of surface piercing hydrofoils versus fully submerged hydrofoils was an enlightening discussion to have! Leopoldo's advice to the ANVCE Project was of great value.

Unfortunately, time and budget did not allow a more in depth discussion of the competing hydrofoil concepts or the advantages and disadvantages of the aerospace and marine manufacturing practices for high speed craft, especially for the 50 to 100 knot range. Now that the US Navy has discontinued its development of the fully submerged foil concept, a definitive answer to "*what price ride quality?*" may not be answered anytime soon. What would the Baron and Leopoldo have answered?

10.5 Other Novel Forms Evaluated by ANVCE

As part of the technical exploration of various ways to improve speed or ride quality or both of the surface vehicle candidates, the ANVCE Project included examination of several promising ideas. Many of these examinations were not as extensive in terms of time and effort as for the main candidates but were important enough to include in the work so that the US Navy would have sufficient material to make proper decisions and possibly outline future more in depth evaluation. A selection from the 1979 ANVCE Final Report is provided here in a concise form.

10.5.1 Supercritical Planing Hull

The natural extension of the conventional *displacement hull* (that made up the "blue ocean" US Navy Fleet) to attain higher speeds would be the *planing hull.* While the displacement hull of the Fleet could be characterized by 25 knot speeds in rough seas, the planing hull offered speeds twice that at 50 knots. A limiting factor is that the technology of these hull forms allowed the displacement hull to provide their speed capability up to very large sizes but that the planing hull was limited to the smaller displacements. These smaller sizes usually meant that the planing hull was not a candidate for open ocean ship sizes compatible with US Navy Fleet needs. The main thrust of the ANVCE Project was to seek solutions for open ocean capable ships (at least of frigate size around 3000 tonnes) which meant that emphasis on the planing hull was not as great as for the other candidates.

The main missions envisaged for the planing hull craft because of the technological size limitation was for *non-air capable ships* and for *coastal patrol craft* of around 1000 tonnes displacement. Even for those missions there was concern that while the speed could be increased from 25 knots to 50 knots[i], the ride quality deteriorated as the conventional V-shaped hull form (both round bilge and hard chine) pounded on the waves. It is known that the response of ships (displacement or planing) in a seaway depend on how the ship's natural frequency compares to the wave encounter frequency. A tuning factor, Ω is defined such that:

$$\Omega = \frac{\text{Wave Encounter Frequency}}{\text{Ship Natural Frequency}} \tag{10.22}$$

In 1955, Prof Edward V. Lewis at Stevens Institute of Technology[j] first described[27] the types of motion called "subcritical" and "supercritical" as zones referenced to the tuning factor as follows:

[i] The speeds of "25 knots" and "50 knots" are only used here as general guidelines to keep the discussion bounded. Speeds greater and lower than these values occur in particular designs as shown in the Point Designs.

[j] Later Prof Lewis became Director of Research at Webb Institute of Naval Architecture, New York

Subcritical Zone of Operation	$\Omega \leq 0.75$
Critical (Resonance)	$\Omega = 1$
Supercritical Zone of Operation	$\Omega \geq 1.2$

While there appears to be missing values of tuning factors in the above characterization, the intention is clear in that the design of any ship should be that it "stays away" from $\Omega = 1$, the resonance condition. In ship motion, the amplitude of the pitching motion increases as it approaches the critical zone ($\Omega = 1$) and the slamming into the waves gives rise to high accelerations and poor ride quality. This resonance with the waves can be avoided by increasing the speed of the ship such that it cuts through the waves rather than rising on to the oncoming waves.

Prof Lewis recognized this characteristic in all hull forms but in particular noted that in planing hulls where attempts to go faster ended up with the dynamic forces dominating giving rise to the high loads through wave impact, derived recommendations for possible solutions. In a later paper (1961)[28], Lewis summarized the problems of conventional planing hull forms and made recommendations on what needed to be done to alleviate the problem in his words:

"In general, planing craft have been poor performers in heavy weather. Although they are often very highly powered, their hull form has prevented their attainment of true supercritical operation. The reason appears to be that the bottom flatness and considerable flare forward prevents the hull from cutting through the waves. Instead, at high speed there is a tendency for the bow to be forced up by the oncoming waves, with heavy bottom slamming following. If it were possible to devise a hull that would both plane and cut through the waves, then we could attain high speed supercritical operation in a small craft. Several requirements should be noted:

1. *Little buoyancy forward*
2. *Relatively small flare at the bow*
3. *No bottom flatness forward*
4. *Extra cockpit protection from boarding seas*
5. *Ample reserve power"*

Figure 10.34: Peter Rowland Payne (1927–1997)
Courtesy Peter Payne

In 1971, Peter Rowland Payne in Annapolis, Maryland (see Figure 10.34) set about devising such a hull form. Peter Payne was a brilliant engineer and a prolific inventor in many fields including air cushion craft, hydrofoils, unique propulsion systems and in many other areas.

His many technical papers presented at aeronautical and marine professional society conferences attest to his ability to delve into complex subjects and emerge with practical solutions. He was always fascinated with speed at sea and has designed and built

Figure 10.35: Payne *Dynafoil* (1994)
Courtesy Peter Payne

many craft to develop his findings and to promote his concepts. Born in England, he emigrated to Canada in 1956 where he did seminal work on autogyros at Avro (Canada) before it closed its doors. He founded Payne Inc in Annapolis, Maryland in 1964 and proceeded in designing and building many unique craft that went under intriguing names such as *FICAT* (Favorable Interference CATamaran) (1964); the *Gayle Boat*, an early planing boat (named after his wife)(1969); the *Sea Knife* – a supercritical planing hull (1971); the *Wavestrider*, a foil-supported catamaran (1982); and the *Dynafoil* (1994), a hydrofoil with a sprung foil to give the boat "automobile-like ride comfort".

The *Dynafoil* (Figure 10.35) has a sprung *fully submerged* foil suspended beneath the boat hull on a simple mechanical hinge. It can operate with the foil at the surface where it has a strong heave stability. It has a strong pitch stability because of the fixed fully submerged foils aft. It also has stubby *surface piercing* foils on the foil struts, which can be seen just above the water surface in Figure 10.35. During take-off, the stubby winglets clear the surface and the main foils ventilate such that the craft stabilizes automatically. The claim is that this design is basically stable without the need for a costly electronic control system that is essential for the fully submerged foil system as used on the US Navy fully submerged foil systems. A technical analysis (in 1996) of the *Dynafoil* and its characteristics is included in the references.[29] This unique combination of a surface piercing/fully submerged foil system was introduced by Peter Payne in 1994 some fifteen years after the ANVCE Project was completed (1979) so no evaluation of that unique concept was possible. It is included here, as a "space holder", for some future development to indicate the hydrofoil system has some ideas left!

Returning to the supercritical planing hull and the 1961 "Lewis design requirements" laid down above, Payne set about designing such a hull form and devised the *Sea Knife*, a name that aptly captures Lewis' definition of a supercritical planing hull form. The generic shape of Payne's supercritical hull form is shown in Figure 10.36 taken from his 1973 patent[30]. Examining the patent drawing and comparing it with

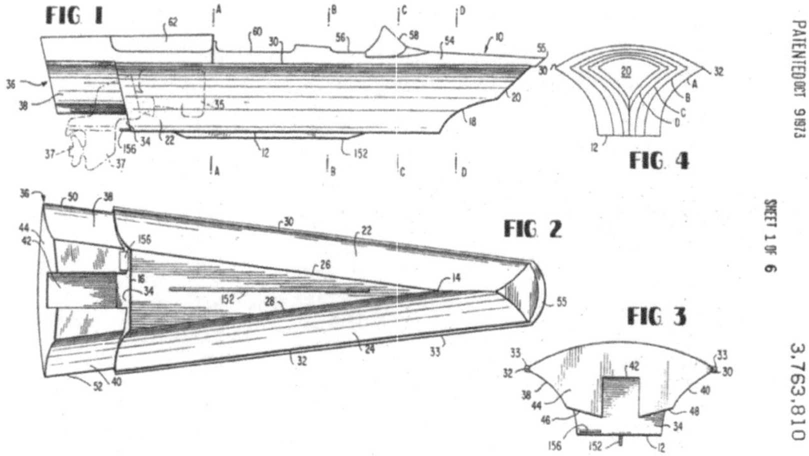

Figure 10.36: Supercritical Planing Hull Form

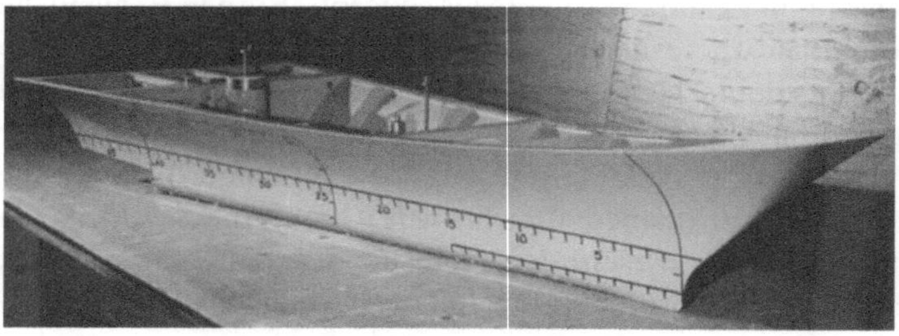

(a) *Sea Knife* Round Bilge Model

(b) *Sea Knife* Hard Chine Model with Spray Suppressors

Figure 10.37: Supercritical Planing Hull Model in Test
UK Admiralty photos

the Lewis "specifications", one can see the fine entry bow shape and the flat bottom at the stern.

This innovation in planing hull geometry that broke away from the conventional hard chine V-shape was intriguing enough to subject to analysis and test. Both a hard chine and a round bilge model were made and tested as part of the ANVCE Project exploration of possible new hull types.

In addition to the basic hull forms tested, variations to tackle various characteristics such as spray drag were also evaluated. Figure 10.37 (a) shows a round bilge version sitting in the model shop and Figure 10.37 (b) shows the hard chine model under test with the added spray deflectors.

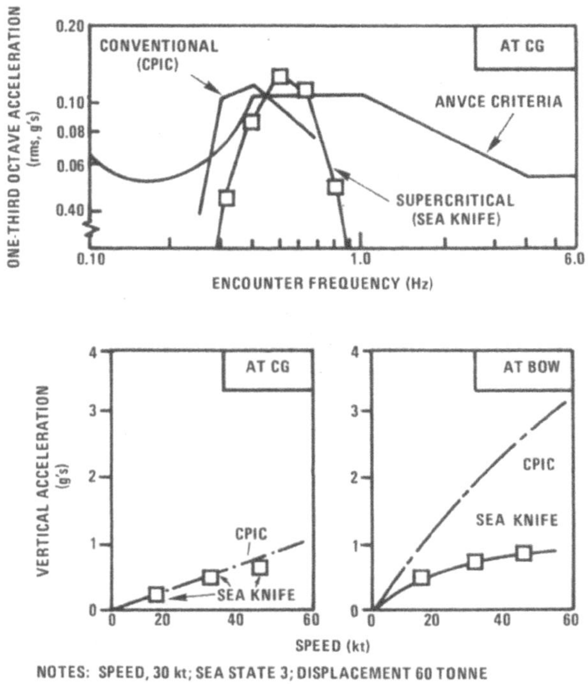

Figure 10.38: Supercritical and Conventional Planing Hull Ride Quality

These models were tested at the Admiralty Experiment Works, Haslar, Gosport, Hampshire, UK and at the US Naval Academy (USNA) Annapolis for the ANVCE Project in 1976. Tests were conducted on these hull forms and their variations in both calm water and in simulated rough sea conditions. The drag performance of this hull shape was disappointing in the blunt transom version (shown) and test reduction facility problems precluded analysis of tests with a planned boat-tail transom. When the hull drag was compared with an equivalent conventional planing hull modeled after the CPIC, it was found that in the blunt transom version, the supercritical hull exhibited a 32% higher drag at a speed of 40 knots than for the CPIC at an equivalent displacement of 60 tonnes. Difficulties with the test reduction facility precluded the analysis of the rough water tests but it appeared that the two hulls (conventional and supercritical) were comparable but this could not be verified. On the other hand, the data from the ride quality measurements (heave acceleration) however were encouraging. Figure 10.38 shows the comparison between the supercritical hull and the conventional (CPIC) hull and the ANVCE ride quality criteria discussed earlier.

Measured at the craft center of gravity (C.G.) it is seen that the ride quality of both the CPIC and *Sea Knife* were about the same and that both were in reasonable agreement with the ANVCE Criteria.

The difference at the bow, however, was quite striking in that the pitching motion of the supercritical hull was much reduced in agreement with the predictions by Prof Edward Lewis. This is because of the low buoyancy and flare that allowed the craft to "slice through" the waves without imparting large pitch motion or high accelerations. At the test speed of 40 knots, the accelerations at the bow were reduced by a factor of about 3:1 over the conventional hard chine V-shaped planing hull (see Figure 10.38).

Considerably more testing and analysis needs to be done on this novel hull shape and time and budget did not allow any further work other than the model tests. The results are provided here for future consideration. The complete data can be found in the OP 96 files from the ANVCE Project, including preliminary designs for a 1000 tonne displacement version.

In the public literature can be found instances of where various groups have built and raced the *Sea Knife* hull form in small craft sizes. A Russian built version of the *Sea Knife* was tested on the Neva River near Leningrad. A larger version was built and tested in Norfolk, Virginia by the US Naval Sea Combat Systems Engineering Department with US Coast Guard participation. The craft displaced 16,400lb and was powered by 2x625hp engines and reportedly achieved a speed of 71 knots. These various trials have not been well publicized or documented but unofficial reports say that the maneuverability of the craft is excellent with tight turning capability and with good ride quality. One would expect a softness in roll static stability but this does not appear to be a limiting factor. The available reports outside of the ANVCE Project include the 1976 UK Admiralty report[31] on seakeeping and a later detailed treatment by Peter Payne in 1984[32].

In a sense, the supercritical planing hull is to *dynamic craft* that the SWATH is to *displacement craft* in that the "lifting surfaces" are moved away from the disturbing circular motion of the waves to improve speed and ride quality in rough seas and is a germ of an idea that requires further exploration.

10.5.2 Double Propeller Transom Configuration

A concern in the development of the sidehull SES was the high power per propeller shaft required and how that would impact the width of the sidehulls. There were the concerns about propeller size and the incurred high stresses and also the increased drag cause by wider sidehulls to accommodate the larger propellers. To this end, the ANVCE Project explored overlapping of smaller propellers in thinner sidehulls and also staggering the propellers along the sidehulls. The concern then would be the efficiency of such configurations. The complete set of results from the model tests are included in the ANVCE Project files but a selected result is shown in Figure 10.39 and Figure 10.40.

Figure 10.39 shows one configuration tested. The photographs are of one concept with overlapping propellers at the stern (transom) of the port sidehull. The slope of the outer shaft can be seen angling toward the inside cushion area of

(a) Fish Eye View of Stern Propellers (b) Stern View of Overlapping Propellers

Figure 10.39: Double Propellers On Sidehull SES Model Tests
US Navy photos

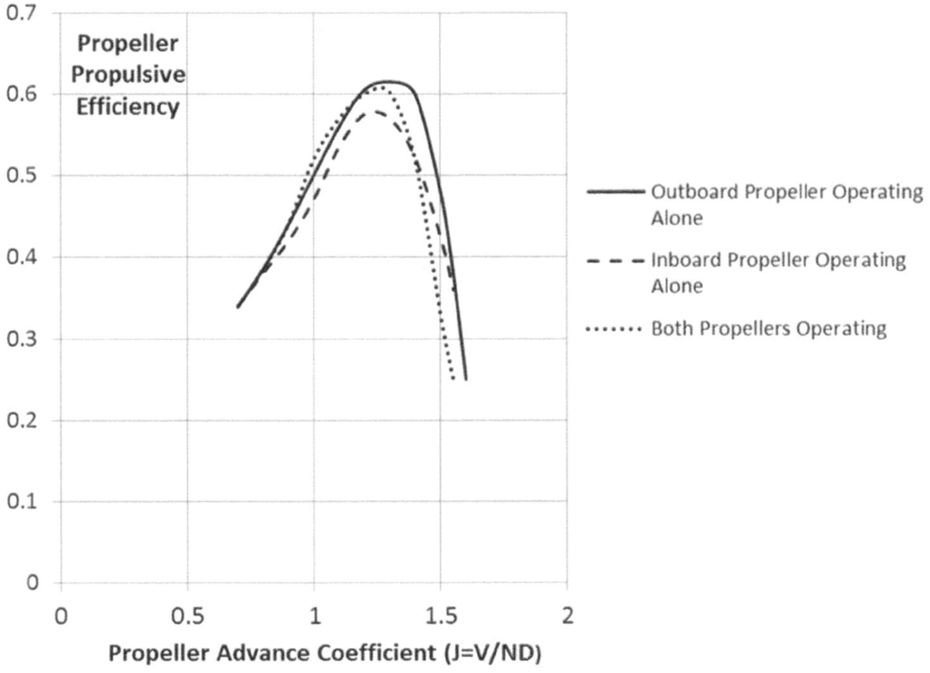

Figure 10.40: ANVCE Model Tests on Overlapping Propellers

the craft. The degree of overlap can be seen from both photographs. Other concepts tested included staggered propellers along the sidehulls instead of at the stern. There were other variations all designed to determine if multiple propellers could operate without undue interference. This was driven by the realization that a smaller number of propellers would be individually large and a strain on the fabrication capability of most propeller manufacturers.

One set of model test results can be seen in Figure 10.40 which shows the propulsive efficiencies of the propellers when operating singly or together. The results are expressed as propulsive efficiency (η) versus the advance coefficient $J = {}^{V}/_{ND}$ where V is the forward speed, N is the propeller RPM and D is the propeller diameter. Based on several tests with different configurations it was determined that propulsive efficiencies of about 60% were attainable and the overlapping (in the case of Concept 1) did not appear to be a problem. These ANVCE tests gave confidence that as the size of the sidehull SES increased, the need to accommodate propulsive schemes that could handle the large power requirements could be handled without incurring propeller scaling problems.

10.6 Sidehull or No Sidehull?

Another concern in the choice of type of air cushion craft: sidehull SES or amphibious air cushion craft, was where in the speed spectrum should the choice of concept be made? As speed is increased above those speeds associated with displacement hulls and planing hulls the drag of the sidehulls becomes a problem. The trade between sidehull drag and air cushion momentum drag depends on several

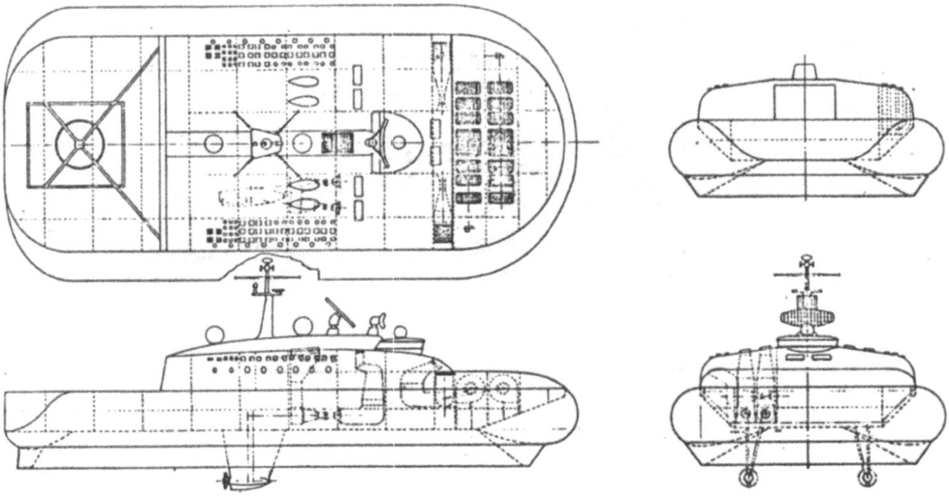

Figure 10.41: ANVCE ACV 3 Point Design General Arrangement

characteristics. It was suspected for frigate size ships, the crossover from sidehulls to amphibious would be around 100 knots. Coupled with that trade space was the cross over point for air and water propulsion. Again, many different considerations come into play in that determination. For example, in the choice of propulsion scheme, the consideration of the two design speeds of importance to air cushion craft namely hump speed (V_{hump}) and maximum speed ($V_{\max}$) help determine the number of propulsors required. Once these requirements have been satisfied the third design speed called cruise speed (V_{cruise}) will automatically be satisfied[k]. To this end, the ANVCE project conducted design studies on a range of configurations of "amphibious" air cushion craft using either water or air propulsion in different arrangements. The baseline frigate size amphibious air cushion craft labeled ACV 3 was identified in the collected air capable Point Designs in Figure 10.5. This design is repeated with more detain in Figure 10.41.

After the many trade studies it was concluded that for the frigate size, that water propeller propulsion was the best choice. This particular design, done[33] for the ANVCE project, had a gross displacement of 3000 tonnes and a speed of 110 knots in calm seas (and 94 knots in Sea State 3) with a total (lift and propulsion) installed power of 240,000 hp. While the fully skirted designs of air cushion craft are normally associated with true amphibious capability, it was reasoned that there is no need for a frigate size ship to come up onto the beach and so advantage was taken of water propulsion which was integrated into the design as shown in Figure 10.41. This frigate size craft was designed with a payload of 350 tonnes made up of a multi-mission combat suite for AAW, ASW and ASUW. The combat suite included two LAMPS Mk-XX helicopters as part of its "air capable" features.

[k] This key point in dynamic craft design is often misunderstood in some treatments of vehicle design in the literature by assuming that it is the design (cruise) speed (V_{cruise}) that determines the vehicle size and design, which would be an erroneous conclusion (see Chapter 8).

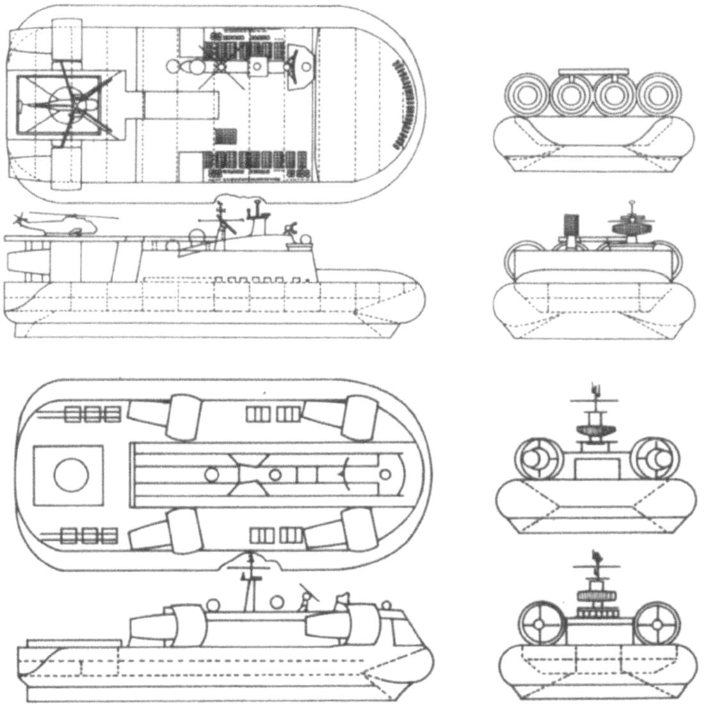

Figure 10.42: Air Propelled Air Cushion Craft (Two Versions)

Not obvious from the design shown is the problem of designing water propelled amphibious air cushion craft to transit the hump speed between subcavitation and supercavitation flow conditions. The thrust curves for the ACV 3 show a drop off of thrust at speeds below about 40 knots because ventilation (or cavitation) of the propeller coupled with the existence of the primary and secondary wave drag humps diminish the available accelerating force for "getting over hump". This particular problem is relatively easy to solve for the sidehull SES through use of the flow control devices such as that incorporated into the SES-100B design (see Chapter 6, Figure 6.45 and Figure 6.46) but this feature is difficult (but not impossible) to incorporate into the ACV 3 design.

Variations were then made around this baseline design (designed to the ANVCE standards discussed at the beginning of this Chapter) to examine the impact of using air propulsion of different types. Free air propeller; shrouded propeller and Q-fan propulsion schemes were examined. Figure 10.42 shows the general arrangement of two particular configurations. As can be seen, air propulsion is a problem for craft designed to be air capable. Not only is the desk space limited, the hazards of aircraft (helicopters) negotiating such a platform operating in rough seas with such exposed propulsion means and their turbulent air efflux can be difficult to accommodate in any design.

These configurations were explored for ships from small craft (non-air capable) up to 5000 tonne displacement (air capable) ships. The number of propulsors required not only varied due to size but also varied due to the relative drag predicted between hump speed and maximum speeds. Sometimes the number of propulsors required for attaining maximum speed was sufficient to provide thrust to get over hump, yet for other configurations extra engines and propulsors were found necessary. It is a complex trade of ship size, speed and propulsor characteristics that is required before the actual number of propulsors can be determined. It is not a linear or easily scalable design characteristic and only general guidelines can be provided in

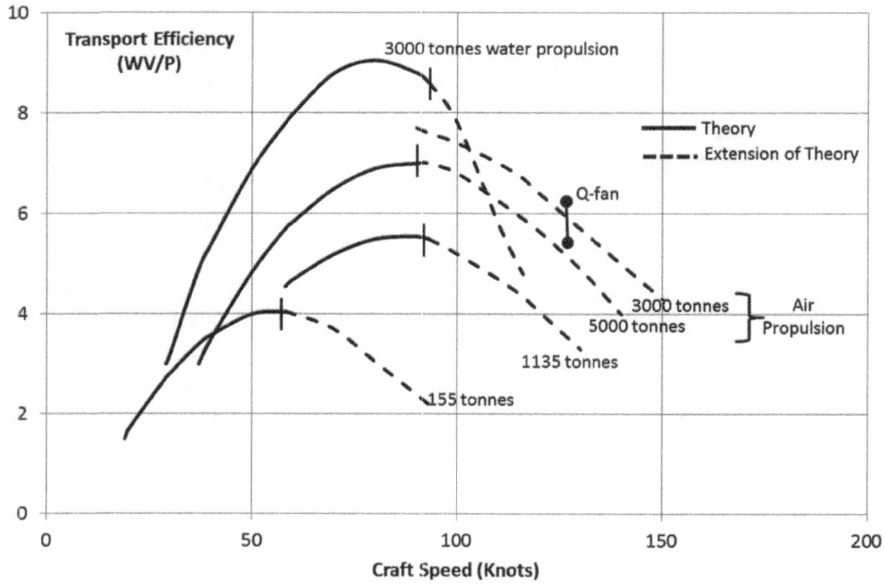

Figure 10.43: Air and Water Propelled Air Cushion Craft Performance

any particular case. An indication of the choice to be made as to which type of propulsion is to be used can be seen from Figure 10.43 where the Transport Efficiency ($^{WV}/_{P}$) of several candidates is displayed.

An exhaustive study was made in the ANVCE Project to determine which type of propulsion schemes would be best for craft designed to operate at speeds of 100 knots. The study encompassed amphibious air cushion craft sizes from 155 tonnes to 5000 tonnes displacement and for cruise speeds of between 65 and 200 knots. In addition to water propulsion systems, air propulsion including free-air propellers, shrouded propellers and high density, axial-fan propulsors (Q-fans) were included in the analysis. It was found that free-air propellers were a feasible proposition for all speeds of interest and for craft sizes up to about 2000 tonnes. The use of shrouded air propellers allowed that size to increase up to about 5000 tonnes if the speed was kept below 90–100 knots. In the analysis it was found that the size of the propulsion system was determined by the maximum speed condition; in other cases, the requirement to generate sufficient thrust to get over hump dictated propulsion system size. It was also found that the use of water propulsion schemes (propeller or waterjet) rapidly lost efficiency as speeds were pushed beyond 100 knots (see Figure 10.43). Hence, In the context of seeking alternative designs to best achieve "100 knots at sea", it is seen from plots such as Figure 10.43 that the choice between water propulsion and air propulsion is not clear cut and must be the result of design studies that include consideration of propulsor efficiency, craft displacement, deck arrangements (with consideration of air capable platforms) and performance considering both maximum speed and acceleration requirements to get through hump.

The above brief outline results from the ANVCE Project provide an overview of the depth of the evaluation in making selections and recommendations of the "best" way to achieve a 100 knot capability to the Fleet. One particular candidate that received an in depth evaluation was the introduction of the wing-in-ground-effect vehicle. Because of the choices made by the US Navy, this deserves especial treatment. The choices made were quite controversial as outlined in what follows.

10.7 ANVCE Project Evaluation of Aerodynamic Air Cushion Craft

Chapter 1 laid down the four basic types of aerodynamic air cushion craft (see Chapter 1, Figure 1.13). The directive from the Office of Secretary of Defense was quite general in that it expected all forms of "wing in ground effect aircraft" to be evaluated, but the US Navy only considered the type initiated by the Russian *ekranoplan* type of aerodynamic air cushion craft. This limited the ANVCE Project effort somewhat to provide a broader treatment. Some variations were introduced however to at least examine other possible types. A wider treatment of aerodynamic air cushion craft of all types is covered in Chapter 11 and only the ANVCE Project results are presented in this Chapter.

It is convenient to identify two basic types evaluated. They are:

Ekranoplan The Russian design of aircraft flying in surface effect and assisted in take-off by air ingestion from forward mounted jet engines. The *ekranoplan* flew a few feet above the waves.

PAR-WIG Essentially the same as the Russian *ekranoplan* except for the addition of thin endplates to achieve higher lift. The PAR-WIG flew with the endplates touching the waves (for greater air sealing efficiency).

It was this difference between the Russian *ekranoplan* design and the US Navy PAR-WIG design that proved to be an issue in the evaluation. Before discussing the evaluation of these basic types of aerodynamic air cushion craft, a short discussion is provided on the ability to describe the "waves" in any sea state. Some other types of aerodynamic air cushion craft will be mentioned in passing as necessary as they relate to the discussion.

10.7.1 The Vagaries of the Sea

The sea is not a well behaved environment and any ship designer would be foolish to ignore its "fickleness". While displacement ships can be placed in extreme danger of being hit by large waves at sea; for the high speed, light weight aerodynamic vehicles, the impact would be catastrophic. Despite this reality, there has been no systematic treatment of sea states and how to design for them in the design work for high speed (aerodynamic) vehicles such as the Wing-In-Ground-Effect (WIG) craft in its many forms. The several crashes by the Russian *ekranoplan* craft do not give rise to comfort in this area. It has been shown earlier in this Chapter on determining the drag on the high speed (aerostatic) sidehull SES-100B in various Sea States, of the difficulty in measuring and treating analytically the nature of the oncoming waves. For aerodynamic air cushion craft planned to operate at speeds *two to four times* that of the SES-100B, the problem becomes acute. Before discussing the different design philosophies used by the many WIG designers, a brief outline of Sea States and how they are defined and treated is given.

Most of the work on sea state description concentrates on wind generated waves. The nature of the sea will depend on the distance (called fetch) that the wind must blow over to attain fully developed seas. Out in the ocean, the craft will experience a range of conditions from "starting seas", "developing seas", "fully developed seas" and "decaying seas". The craft may experience different combinations of these sea formations along any particular sea route depending on where it is in the world. The wave spectra will be different for each of these stages of sea formation. Other determining factors include the effect of storms, local currents and sea floor topography. Each of these effects will contribute to different wave heights and wave lengths.

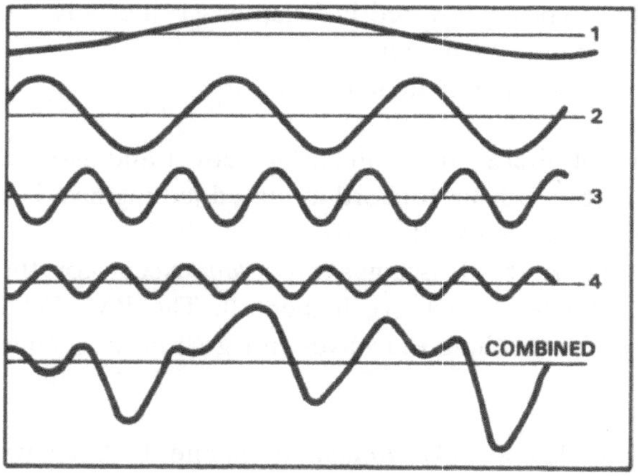

Figure 10.44: Wave Pattern Combining Four Regular Waves

As an example of the complexity of analyzing the sea state consider the hypothetical example of just *four* regular waves combining into *one* irregular sea pattern. Figure 10.44 shows the combined effect of the four regular waves into one sea state. As can be seen by this simple example, an irregular sea (a normal everyday occurrence) cannot be characterized by its pattern or shape[1] and other approaches must be taken.

The description of random seas is not simple and is still being pursued by mathematicians, oceanographers and others and new forms of describing Sea States appear regularly in the literature.

In most of the treatments used thus far and, in particular, those used in this book, random seas are considered to be two dimensional and created by a broad-scoped continuously blowing wind. The wave crests are continuous in a breadth direction and all waves move in the same direction. In actual seas, of course, the fetch or the distance over which the wind has been blowing is of varying length and the seas become confused due to the effects of wind strength and duration, the presence of storm centers, proximity of land masses and other effects that cause the sea to take on definite three-dimensional characteristics. This alone raises questions on ship designs based on model tests conducted in linear tow tanks and especially where the model is constrained to move in a given direction.

The characterization of an irregular sea comes from the simple assumption that its total energy is equal to the sum of the energies of all the small, regular waves assumed to make up the sea. The wave energy of a regular sea is given by:

[1] A fact that cautions the designer on basing his design using simple representations of sinusoidal or regular seas for both drag and stability analyses – even though they are done by most writers (including this author!)

$$E = \frac{\rho_W g}{\lambda} \int h_W(x)dx = \frac{\rho_W g}{8} h_W^2 \tag{10.23}$$

In Equation 10.18 it is assumed that the wave pattern is sinusoidal as before, with wave height (h_W) and wavelength (λ). In which case, the total energy in each square foot of surface of the random sea is given by:

$$E = \frac{\rho_W g}{8} (h_1^2 + h_2^2 + h_3^2 + \ldots\ldots\ldots) \tag{10.24}$$

If each of the single frequency waves are of different frequencies ($\omega_1, \omega_2, \omega_3, \ldots\ldots\ldots$) then this energy can be thought of as being distributed over these frequencies in an energy spectrum. Without getting into too much detail, it can be shown that the energy (E) is the area under the curve of spectral density, $S(\omega)$, measured in "energy seconds" against the wave frequencies (ω).[m]

The form of this spectral density has been found to be closely given by:

$$S(\omega) = \frac{A}{\omega^5} e^{-B/\omega^4} \tag{10.25}$$

The values for the "constants" A and B have been the subject of continuing research in the meteorological literature. One of the simplest empirical representations of the sea spectrum is the Pierson-Moskowitz spectrum which was developed in 1964 for fully developed seas in the North Atlantic. It is a single parameter spectrum and used in many meteorological charts. It expresses the spectral density in terms of the *significant wave height*, $H_{1/3}$. This "wave height" has its origins during WW II, when it was determined that the waves can be characterized by these heights as estimated by a "trained observer". It corresponds to the wave height that is the average of the highest third of the waves seen by the observer. After WW II, it was found that such heights corresponded to being proportional to the standard deviation using a Rayleigh distribution of the wave heights. The significant wave height ($H_{1/3}$) has tended to be the standard in measuring wave heights in irregular seas. Later, other forms appeared in the treatment of the sea spectra. The Bretschneider Spectrum was introduced to overcome the simplicity of fully developed seas as needed in the Pierson-Moskowitz spectrum. The Bretschneider Spectrum is a two parameter description using both the significant wave height ($H_{1/3}$) and the modal frequency (ω_m).

Another sea spectral density formulation is the JONSWAP (**JO**int **N**orth **S**ea **WA**ve **P**roject) spectrum that was developed for the short wavelength or limited fetch North Sea and adds a characteristic wave height to the Bretschneider spectrum. There are many more mathematical treatments found in the literature. The Ochi sea spectral density is a further expansion of the Bretschneider spectrum bringing in the effects of wide and narrow developing seas. It is a three parameter description of the sea. Still other spectra bring in combinations of storm and swell on the sea, corrections for uni-directional and three-dimensional wave direction. This short summary has been given merely to emphasize that the description of the sea is not a simple matter. The reader is referred to the ample literature on this complex subject.

[m] The reader can go to standard texts in the meteorological literature for a complete derivation or see an overview in the author's 1980 report, Appendix D for sea keeping formulae and derivation.

A good description of all of the sea state descriptions and their characteristics is provided in the 23rd International Towing Tank Conference (ITTC) Proceedings of 2002[34]. This is a continuing ongoing process to understand sea states.

The energy (E) can then be expressed as:

$$E = \int_0^\infty S(\omega)d\omega = \frac{\rho_W g}{8} H^2 \tag{10.26}$$

In this expression for the energy, (H) is the root mean square amplitude over the entire area. Using a Rayleigh distribution it is common to express the various wave heights in the following representations of the sea as shown in Table 10.5

Table 10.5: *Random Sea Wave Heights (Double Amplitude)*

	Random Sea Wave Heights	Common Notation
Root mean square amplitude, RMS	H	
Average amplitude	$2.50H$	
Average of highest 1/3 amplitude (significant wave)	$4.00H$	$H_{1/3}$
Average of highest 1/10 amplitude	$5.10H$	$H_{1/10}$
Highest expected amplitude in 100 successive amplitudes	$6.06H$	
Highest expected amplitude in 1000 successive amplitudes	$7.44H$	
Highest expected amplitude in N successive amplitudes	$\sqrt{2\ln(N)}H$	

These are the values that appear in the Pierson-Moskowitz description of Sea States. As an example, the following values of wave heights (in feet) appear for Sea State 4 in the Pierson-Moskowitz tables that occurs about 30% of the time in the North Atlantic and North Pacific Oceans:

Table 10.6: *Pierson Moskowitz Sea State 4 Wave Heights*

	Low	Medium	High
Average Wave Height	3.76	4.19	4.65
Significant Wave Height ($H_{1/3}$)	6.01	6.70	7.42
Average 1/10 Highest Wave Height ($H_{1/10}$)	7.64	8.54	9.43
Average 1/1000 Highest Wave Height	11.18	12.46	13.80

Use of the other forms of spectral density (Bretschneider, JONSWAP, Ochi, etc) will give different values because of the calculated different energy levels. Because of this it must be recognized that these wave heights (h_W) are only approximations useful for consistent analyses but the values should NOT be used in a prescriptive sense and certainly not to the second decimal place! Also, the Pierson-Moskowitz definition of Sea States assumes that the wind generating the waves is about 19.5m (approx 64feet) above the sea surface. The wave heights would be different for other wind speed assumptions. Because the *ekranoplan* flies at low altitude (the KM flies with its tailplane at about the height where the Pierson-Moskowitz assumes the generating wind is blowing) this is an important assumption.

Not detailed here is the additional problem of defining the different sea systems from other locations as they pass through the same sea route being used by the WIG. There will be long waves and short waves in the same spectrum that will interfere with the above detailed set of wave heights. This combination of long and short waves and swells will produce temporarily (but nevertheless very real!) build up of even larger waves than those computed by the various methods cited for describing wave heights. Such combinations can produce giant waves called Rogue Waves. Such waves have been known to reach heights of 100 ft or more. These are not related to tsunamis that can be predicted and measured from some movement of the ocean floor (such as from an earthquake) and thus can be avoided by the high speed WIG by appropriate course correction.

The Rogue Wave, on the other hand, is a phenomenon caused by the unpredictable confluence of high winds, strong currents and other environmental conditions already described. The literature from such sources as the Ocean Prediction Center, National Weather Service[35] provides documentation of the many incidents at sea that has caused the loss of large ships and coastal structures due to Rogue Waves. There is no specific prediction method available for computing the heights of Rogue Waves but factors of 2:1 to 3:1 times the significant wave height appear frequently in the literature – although the statistics show much higher values have been recorded on a fairly frequent basis. How does a WIG designed to fly within inches of the wave tops avoid such monsters? Such vagaries of the sea have not been lost on those designers who have designed and built operational forms of aerodynamic air cushion craft and their variants. Three main types of aerodynamic craft and their design philosophy are described as to how they treat this problem.

10.7.2 German School of WIG Craft Design

For want of a better title, this class of craft refers to the various designs by the German aeronautical engineer Dr Alexander Lippisch[n]. In Dr Lippisch's designs, there are three pertinent characteristics that relate to the operation over waves. In addition to the use of the reverse delta wing form for stability reasons, his designs incorporated a dihedral angle that lifted the main body higher than the wing tips. At the wing tips are incorporated float structures such that the wing can float on the water without being swamped. The wing tip floats also serve to alleviate the loads on the wing for the occasional impact with waves. Figure 10.45 shows two such craft, the experimental X-112 (1963) and the later craft X-113 (1970). The particular photo shown of the X-113 was taken on the Weser River near Bremenhaven, Germany in November 1972.

In the case of a later craft, the X-114 (1977) (See photo in Chapter 11, Figure 11.52), the wing tip floats incorporated various novel features such as shock

[n] See Chapter 1 for a description of Dr Lippisch's history as it relates to wing-in-ground-effect craft

X-112

X-113

Figure 10.45: Lippisch Design Philosophy for WIG "End Plates"
Photos permission from Special Collections Department of Iowa State University Library

absorber for impact with the waves, and other features such as ladder foils as landing and take-off aids. While these were exploratory features in the designs, Dr Lippisch's main approach was simply to NOT impact the waves at all during cruise and to accept a lower value of "ground effect" advantage (see Chapter 11) and to *fly higher to avoid the waves* for safety reasons because of the safety reasons associated with flying over a surface that has unpredictable characteristics of wave heights.

10.7.3 Russian School of WIG Craft Design

A very similar design philosophy was used by the Chief Designer of the former Soviet Central Hydrofoil Design Bureau (CHDB) Rostislav Alekseyev[o] in the design of the *ekranoplan* series of wing-in-ground-effect craft. Two of the CHDB designs, the "Caspian Sea Monster, KM" and the missile carrier "Lun" are shown in Figure 10.46.

In similar fashion to the Lippisch designs, the Alekseyev designs have float-like structures at the wing tips that aid in containing the air pressure build up under the wing at take-off from the air injection from the engines but the essential design feature is to *fly higher than the waves* and avoid wave impact.

[o] See Chapter 1 for description of Rostislav Alekseyev and his work

Figure 10.46: *Ekranoplan* Design Philosophy for "End Plates"

A close up of the KM port wing end plate is shown in Figure 10.47. In the cruise mode, the *ekranoplan* flies close to the sea surface but with the wing end plates clear of the surface (note the shadow of the end plate on the surface of the water) such that no part of the craft touches the waves. The water disturbance behind the end plates, seen in Figure 10.46, is from the interaction of the end plate air vortices striking the water and not the physical end plates themselves.

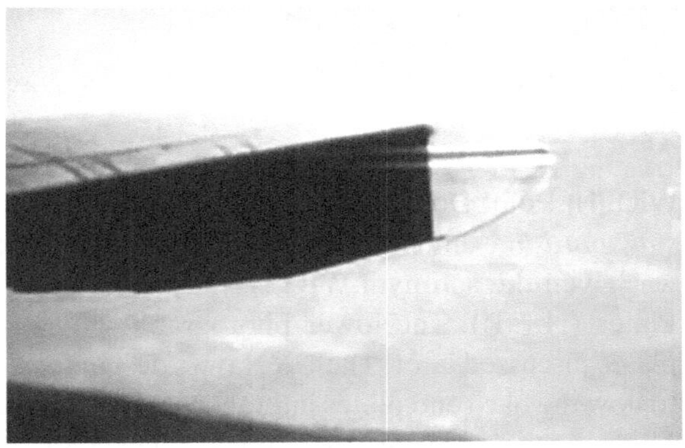

Figure 10.47: *Ekranoplan "KM"* End Plate Clear of Water in Normal IGE Flight

10.7.4 US Navy Design Philosophy for WIG Design

The US Navy, in its adoption of the Russian WIG design, departed from both the German and Russian approach and elected to retain the benefits of thin "hard end plates" at the wing tips, for increasing the ground effect and designed their models[p] to *fly THROUGH the waves and not ABOVE them.* This is shown in

[p] The US Navy only built models and not any operational flying machines based on the USN PAR-WIG adaptation of the Russian *ekranoplan* design.

Figure 10.48: US Navy Design concepts for PAR-WIG "End Plates" (Hard and Soft)
US Navy photos

Figure 10.48 for two variants of the US Navy adaptation of the Russian air injection system.

WIG (H) Point Design The upper photograph in Figure 10.48 shows the hard *thin sideboard-like* end plates to contain the pressurized air. In the ANVCE Project this was given the acronym WIG (H) to designate the WIG craft fitted with Hard End Plates (H.E.P.). The lower photograph shows the WIG(S) design using soft end plates discussed later. The US Navy's design for the WIG (H) with specific reference to how the hard end plates interact with the waves is shown in Figure 10.49.[q] Both the 1/10th highest wave and the 1/1000th highest wave is shown with the WIG flying at its cruise height. The wave heights selected in Figure 10.49 correspond to the "medium" wave height for the 1/10th highest wave and the 1/1000th highest wave is for the "high" wave height from the earlier Table 10.6.

In this selected cruise position, the WIG (H) is flying with the under surface of the main body of the craft approximately 14½ feet above the mean wave height level (ignoring craft motion in pitch or heave). The end plates impact the 1/1000th wave to a depth of about 1½ feet. If it is assumed that the wave period is about 10 secs[r] then the end plate will strike this wave about once every 3 hours and the 1/10th highest wave every 15 minutes (if the higher wave is chosen instead of the medium value). If the WIG (H) were crossing the Atlantic Ocean under these conditions at an average

[q] The chord is shown at 91 feet in this sketch. The final design selection by the US Navy was 108 feet.
[r] It is actually a shorter time according to Pierson-Moskowitz

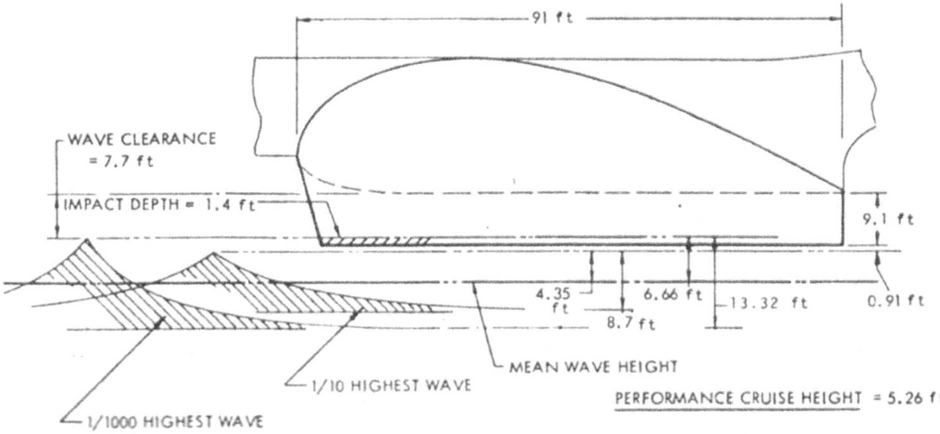

Figure 10.49: US Navy WIG (H) Flying in Sea State 4

speed of 200 knots, then it will strike the 1/10th highest wave approximately 60 times and the 1/1000th highest wave approximately 6 times in each crossing.

If the WIG (H) comes in contact with a Rogue Wave, that wave height would be between 15 and 22 feet (if the 2:1 and 3:1 times the significant wave height estimate is correct). That means such a wave will strike the main body of the WIG (H) which would be catastrophic.[s]

The ANVCE Project Office considered the US Navy WIG (H) design to be unsafe. A set of calculations were made by independent structural designers on the suitability of the *thin sideboard end plates* as shown in the sketch Figure 10.49 and also in the design concept drawings for the WIG (H) as shown in Figure 10.50.

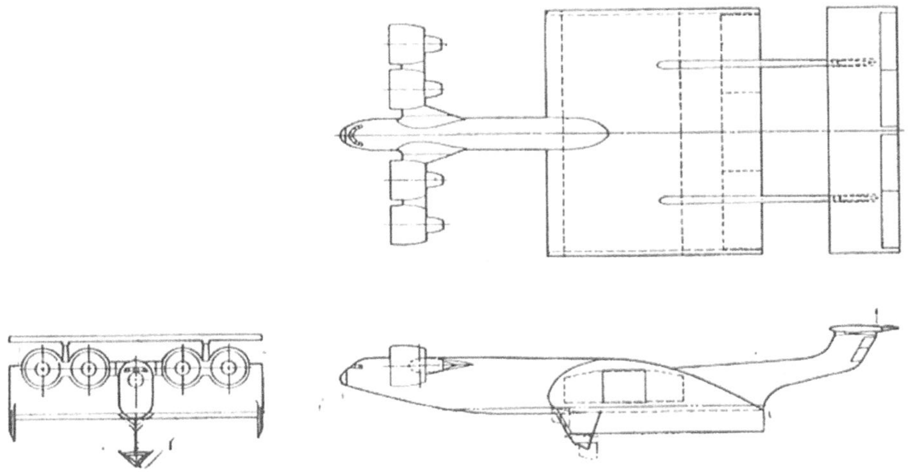

Figure 10.50: US Navy WIG (H) Design Concept

[s] Chapter 11 describes several instances where the *ekranoplan* has struck waves unexpectedly with extreme levels of impact accelerations and serious mechanical failures to the craft, and injuries to the crews.

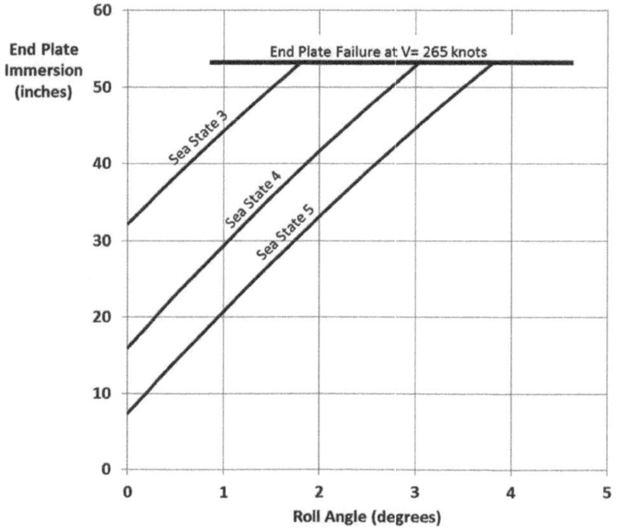

Figure 10.51: WIG (H) End Plate Immersion

The features of drag plates and forward mounted hydrofoil shown (lowered into the water at 90 knots!) are to aid the WIG (H) to slow down from high speed. The results of the stress analysis on the end plates are shown in Figure 10.51. The end plate immersion of 1.4 feet (17 inches) at zero roll angle is shown on the chart. As the craft rolls, the immersion increases for the different Sea States.

In Sea State 4, it was found that complete structural failure would occur after about a modest 3 degrees of roll. The structural analysis showed complete structural failure at 265 knots and an immersion of about 50 inches This analysis was done under the assumption of a rigid structure. If structural deformation would have been included in the analysis, the end plate would have deformed and failed at a lower value of roll angle and speed.

No calculations were done on the yaw stability as a result of the end plate digging in and deforming but this would have shown a roll-yaw interaction with catastrophic consequences for the craft. Much of the evaluation done for the WIG (H) motion in rough seas was left to conjecture because of the complexities of response characteristics to the forcing function of the waves. For example, as the WIG (H) continues on its cruise it will tend to rise as it passes wave crests and sink as it passes over the troughs. This motion will lag the rise and fall of the waves themselves which could induce the WIG (H) to counter the rise by trimming down which would be out of phase with the waves causing it to plunge into the next oncoming wave.

There has been no treatment of the flight control mechanism necessary to overcome these normal tendencies and such work would need to be done for high speed aerodynamic vehicles operating over (irregular seas) wave action especially at high speed. The wing span of the US Navy WIG (H) design in Figure 10.50 is approximately 108 feet.

After a glance at Figure 10.52, the reader might question whether the assumed immersion depth of 1.4 feet for the port end plate would be the same on the starboard end plate; some 108 feet further away! In reality, there would be (large?) differences on an assumed common immersion depth on both port and starboard end plates which would manifest itself in large drag changes in an asymmetric manner causing yaw forces and motion that would unlikely to be controlled by normal flight control systems in any practical manner. Again, craft roll motion is not included.

From this US Navy design work, it is seen why the German and Russian design approach of *"flying above the waves"* is the more prudent design philosophy for operation of WIG craft in rough seas. The ANVCE Project considered the USN approach of allowing *hard* end-plates to interact with the oncoming waves at high speed was NOT a safe design and other approaches were pursued in the ANVCE Project.

Figure 10.52: NOAA Ship in Rough Seas Courtesy NOAA

WIG (S) Point Design Because of the concern of incorporating such structures on a flying platform operating close to (in?) the water; at the urging of the ANVCE Project Office, the US Navy also included a variant where the end plates were of *soft and compliant* structure to still retain some form of end plate effect but at the same time be compliant to wave action. This variant was given the acronym WIG(S) to designate the WIG fitted with Soft End Plates (S.E.P.) shown in the lower photograph of Figure 10.48. Unfortunately, these soft endplates used on the USN models were simply inflated cloth tubes dragging in the water. In the model tests[36], the soft end plates were made up of hollow unsupported neoprene tubes (inflated with laboratory air). These tubes or hollow fingers were unconstrained and the front and rear fingers were of larger diameter than the rest so as to avoid wrap-around of the first finger with the second and adjacent finger! As might be expected, this gave a higher drag version of the end plate than that exhibited by the hard rigid end plate. This configuration of inflated neoprene tubes also contributed to higher vertical accelerations to the model. This was a disappointing configuration for the model tests. It was originally intended by the ANVCE Project Office that the *soft end plates* be more like *hovercraft skirts* to negotiate any waves, inflated by fan air or bleed air from the engines, and not dragging in the water, but cost and time precluded this refinement. More on this possibility is given later in this chapter and in Chapter 11.

If the soft end plates were similar to hovercraft skirts, the WIG(S) would look more like the sketch given in Figure 10.53. It is seen that the use of "soft end plates", or more correctly hovercraft-like skirts, would also give the craft an amphibious capability as well as a sea wave compliant capability. ANVCE Project calculations showed that the power required to inflate the skirts and seal the air cushion (with engine air bleed) would be less than the power required to drag the inflated tubes

Figure 10.53: USN PAR WIG modified with Hovercraft-Like Soft Skirts

Figure 10.54: Variable Geometry WIG with integrated Lift and Propulsion

(shown in Figure 10.48) through the water. The amphibious feature of such a WIG has been included in Figure 10.53. Another variation (included in the ANVCE Report) is shown in Figure 10.54 that shows that the PAR WIG feature has been replaced with engines that are buried within the body of the craft and providing direct lift and propulsion. The vulnerability of the forward mounted PAR engines of the *ekranoplan* and USN PAR-WIG designs is of concern. A concept, called Variable Geometry WIG or VG WIG (pronounced "vig wig") removes many of the concerns about a PAR WIG, that loses its amphibious capability with its use of rigid end plates, that will inhibit the ability to traverse most beach or shoreline terrain likely to be encountered in military missions (see Chapter 9).

More complete descriptions of the Variable Geometry WIG (VG WIG) including the integration of retractable skirts is given in Chapter 11 where an expanded treatment on aerodynamic forms of WIG craft is provided. The variations shown in Figure 10.53 and Figure 10.54 from the ANVCE Report are shown here to highlight that there are many other possibilities for an air cushion supported WIG vehicle that could solve many of the problems of earlier WIG craft, including the *ekranoplan* type, that were highlighted in the ANVCE Project Evaluation but not pursued in the US Navy work.

WIG(O) Point Design The third variant studied in the ANVCE Project was the variant called WIG(O) to designate a WIG design that had the capability to occasionally fly out of ground effect when the mission so demanded.

The most promising version of such a wing-in-ground-effect vehicle studied in the ANVCE Project was that which could operate in ground effect over long distances (and take advantage of the power reducing effects of ground effect) but have the capability to "hop up" out of ground effect for limited periods of time to (a) avoid high sea wave conditions or (b) prosecute some military maneuver. For convenience, this type of vehicle was labeled WIG(O). This "occasional" OGE capability had been incorporated in some of the Russian *ekranoplan* (e.g. *Orlyonok* and *Lun*).

To keep the analysis within the bounds of state of the art technology, the WIG (O) was modeled after the Russian *ekranoplan* shown in Chapter 1, (see Figure 1.15). As will be discussed further in Chapter 11 there are other possible variants of this concept that might offer even more attractive capabilities. Figure 10.55 shows the configuration and the leading particulars for analysis in the ANVCE Project for WIG(O) sometimes referred to as the "Jumping WIG". While not essential, the design philosophy was that the WIG(O) would be designed to be most efficient in the IGE mode and that the climbs to OGE to clear obstacles (or mission needs) could be done with a surge in power, and efficiency was not the driving feature for such bursts of "jumping".

For analysis purposes, a "representative" mission profile was assumed such that the feasibility of such a variant of the WIG designed could be evaluated.

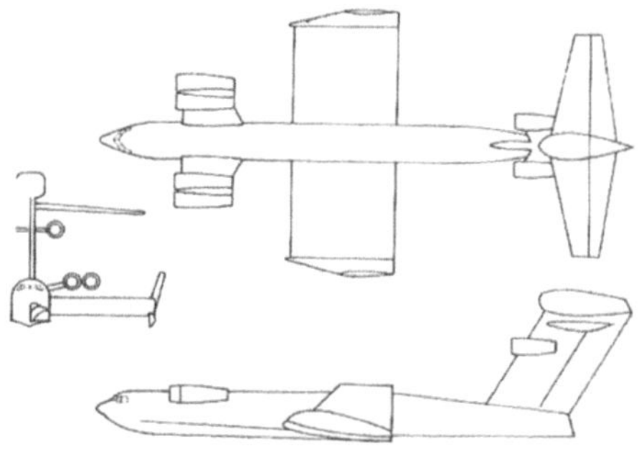

LEADING PARTICULARS

Gross Weight (tonnes)	536
Payload (tonnes)	48
Maximum speed (knots)	
At 20,000 ft	417
At 10,000 ft	445
At Sea Level	285
Time on station (hours)	22
Radius of Action (nm)	1,000

Figure 10.55: ANVCE WIG(O) Point Design

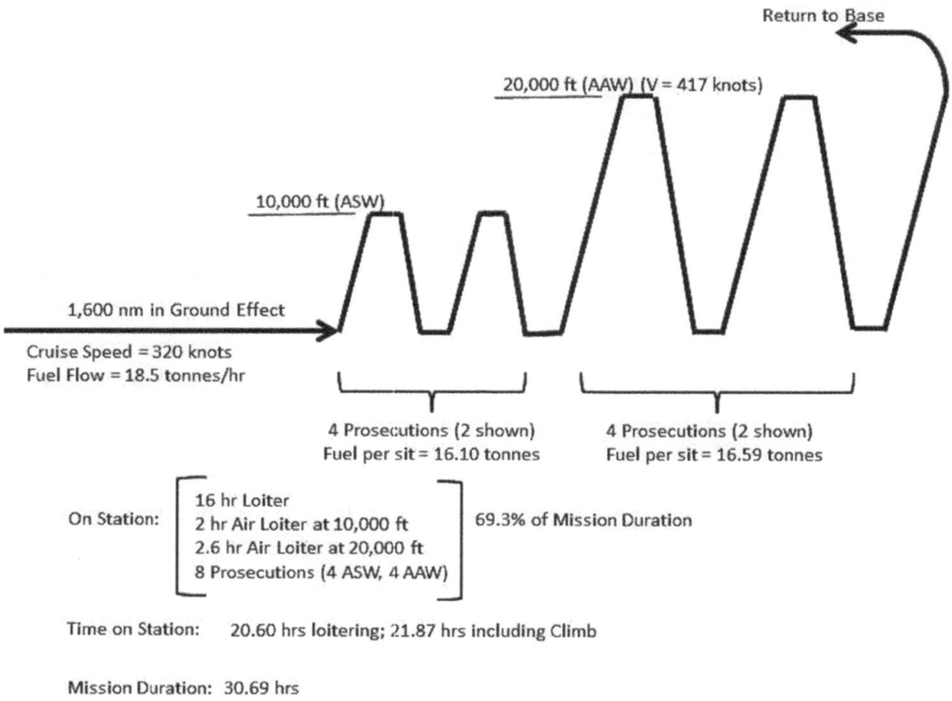

Figure 10.56: Out-of-Ground Effect Capable WIG (Typical Mission Profile)

That mission profile is shown in Figure 10.56. While the numerical values in the example WIG(O) shown in Figure 10.56 are not prescriptive of the concept they do demonstrate a consistent set of values that was used for analysis, and costing and mission treatment. The basic intent was to show that this particular wing-in-ground-effect vehicle would be capable of cruising over the ocean for some distance in ground effect; that could climb to 10,000ft for ASW mission prosecution (an arbitrary 4 such climbs was selected); the WIG(O) could then prosecute say 4 additional climbs to 20,000ft for surveillance and then return to base. This was a needed mission profile to include in the vehicle comparisons rather than the simple range requirement of 3,000 nm used for the other surface vehicles (see Figure 10.5) that did not have this unique capability.

The driving force behind the ANVCE project was to determine an alternative (if there was one) to the US Navy design for the 3,000 ton SES concept. This meant that, while other missions and capabilities were examined in the ANVCE Project, the main emphasis was on *large tonnage ships* carrying *large payloads* over *long distances.*

The SES 3 design (see Figure 10.5) was a frigate size ship (3,658 tonnes); capable of carrying a military payload of 391 tonnes; and a range capability of 3,208nm. It is seen that none of the WIG concepts examined could meet this capability in any directly comparable analysis. The "logistics" WIG(H) and WIG(S) offered to carry about half that payload over 4,000nm or more, at 2–3 times the speed. This emphasis tended to mask many of the advantages of the WIG concept in other mission areas.

The amphibious mission discussed in detail in Chapter 9 is just one military mission where it is believed that the WIG in some improved variant form would show distinct benefit.

The real issue uncovered during the ANVCE Project was the technical maturity of the PAR-WIG and *ekranoplan* concepts that dominated the US Navy proposed use of the WIG concept. This showed up in examination of some of the basic performance determinants as outlined below.

10.8 Empty Weight Trends for Landplanes, Seaplanes and WIGs

Designing an "aircraft", whether seaplane or WIG craft to withstand the various loads on its structure as it comes into contact with the water is a difficult task if the structural weight and empty weight is to be kept low enough such that there is room left for the payload and fuel. Historically, this has been a major problem in the design of seaplanes and this has contributed in many cases to their demise except for special mission needs. Figure 10.57 shows data for Empty Weight trends[t] of landplanes and seaplanes.

As can be seen from Figure 10.57, the general trend is that generally, seaplanes exhibit higher values of Empty Weights (W_e) than landplanes because of their higher structural weights (W_S) incurred by having to design for high impact loads during landing and take-off from the water. There are, of course, cases where by specific design niches it is possible to overcome this trend but the general rule that those light weight structures characteristic of aircraft structures must be strengthened and made heavier than their land based counterparts in order to operate at the interface between the air and sea surface.

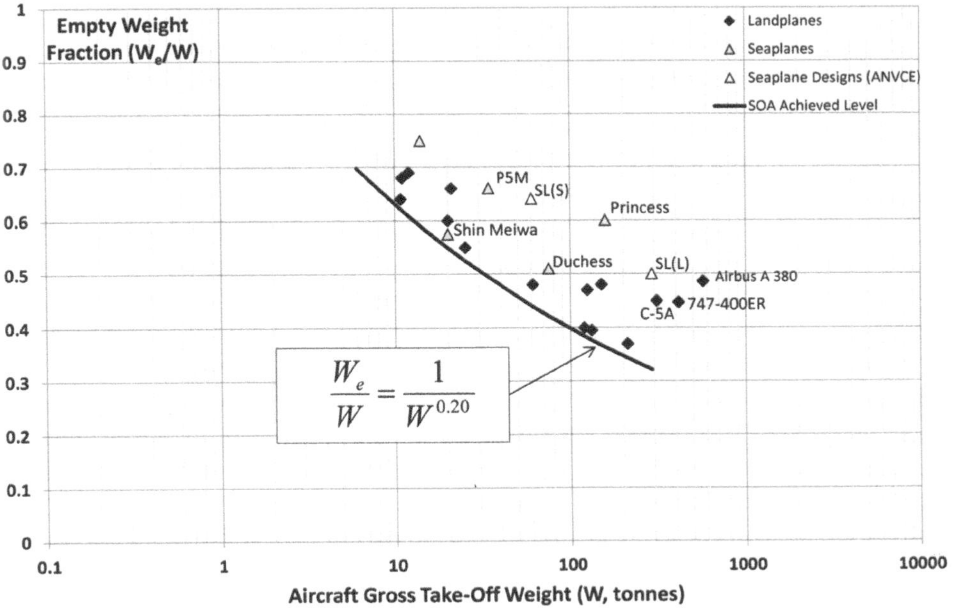

Figure 10.57: Empty Weight of Landplanes and Seaplanes

[t] See Chapter 7 for the standard breakdown of subsystem weights for all craft

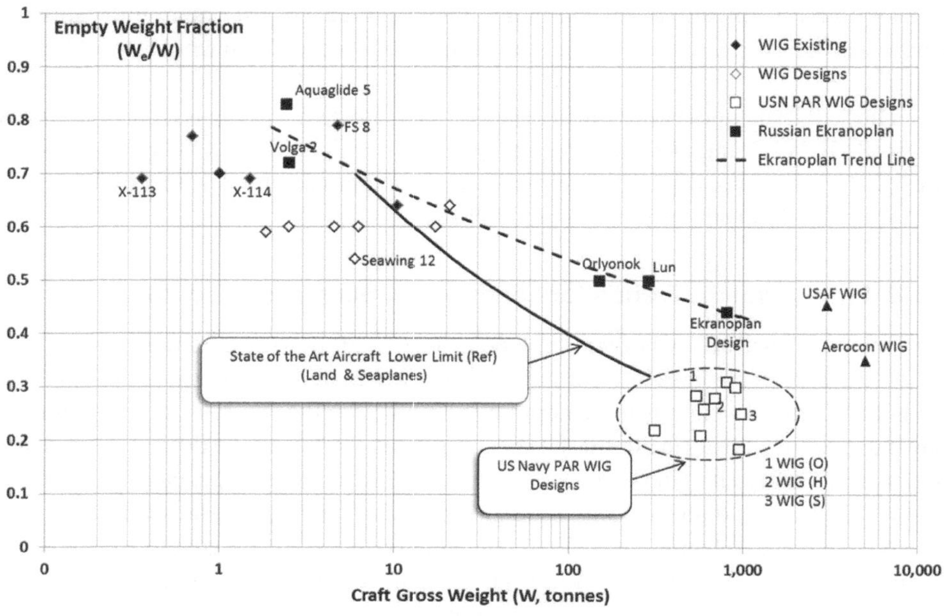

Figure 10.58: Empty Weight of WIG Craft

Figure 10.58 collects some statistics of WIG craft of various types. Specifically it shows:

1. Conventional WIG (of which the Lippisch type craft is a good example, see Figure 10.46)
2. Russian *Ekranoplan* type of WIG using air ingestion (see Figure 10.47)
3. USN PAR-WIG, an adaptation of *ekranoplan* but with deep endplates (see Figure 10.48)

The data is for both operational craft (solid symbols) and paper designs (open symbols).

The statistics show some startling comparisons. For reference, the lower limit (best) values of both land and seaplanes empty weight trend line, taken from Figure 10.57, has been added to Figure 10.58. During the ANVCE Project work it was strongly believed, based on independent analysis, that the US Navy PAR-WIG designs were optimistic as to the estimates for the Empty Weight Fraction (W_e/W). The USN PAR-WIG designers were predicting Empty Weight values, at 20% to 30% weight fractions (W_e/W), which is actually *lower* than the structural weights (W_S/W) of land and seaplanes (see dotted ellipse in Figure 10.58)!! Seaplanes have to be designed to withstand impact with waves during landing at speeds of 100 knots to 200 knots that contribute to their higher structural weights (and empty weights).

The WIG, even more dramatically, must be designed for *continuous* contact with the waves at speeds of 300 knots or more – much higher than seaplane TO & Landing speeds. As shown earlier, the impact pressures on the structure increases rapidly with speed (see Figure 10.12) and any load alleviation due to air ingestion would be minimal. The USN PAR-WIG designers believe that the more compact shape (lower Aspect Ratio) of the WIG would contribute to lower structural weights[37]. This is certainly true for the smaller craft as indicated in Figure 10.58 but would not be true for the larger sizes contemplated for military missions. The non-compact geometries of the *ekranoplan (*Figure 10.59) and the USN PAR-WIG (Figure 10.50 and Figure 10.55) where the engines are cantilevered away from the hull involve a structural penalty. The heavy propulsion systems for *ekranoplan* and PAR-WIG also contribute to high Empty Weight Fractions. While the analyses conducted during the ANVCE Project showed that the optimistic weight estimates by the PAR-WIG designers were not credible, it was not until later when the Russian *ekranoplan* data became available where the actual data showed dramatic increases in the Empty Weights of such craft as predicted by ANVCE. In 1994, DARPA[38] examined the Russian *ekranoplan* craft on location in Russia.

Figure 10.59: *KM* (above) and *Lun* (below) *ekranoplans*

The reports from that group provide much valuable information that substantiates many of the concerns expressed by ANVCE Project in 1979 on the US Navy PAR-WIG designs. The data on *ekranoplans* from the DARPA report is included in Figure 10.58.

According to the 1994 ARPA Report, the structural weight (W_S) of the Caspian Sea Monster (*KM*) is "greater than 32%" of the craft Take-Off Weight (W). The exact value was not available from the Russian sources. The *Lun ekranoplan* shown on Figure 10.58 is reported to have (DARPA data) a structural weight of 34% and an Empty Weight (shown) of 50%. An indication why the high values of structural weight and empty weight can best be illustrated by the photographs shown in Figure 10.60.

The dominant reasons for these high structural and empty weight values are driven by the fact that the *ekranoplan* derives its lift by dynamic flight. Unlike hovercraft or amphibious air cushion craft, the *ekranoplan* cannot hover and must move forward at speed in a "take-off" mode to generate lift. The upper photograph in Figure 10.60 shows the trailing edge of the *KM* as it accelerates to take-off speed which is approximately 244 knots and requires a take-off run of more than a kilometer. During that time and at such high speeds the trailing edge flaps are subjected to extremely high hydrodynamic loads. Earlier in this chapter, it was reported by the Russian designers that impact pressures as high as 250psi and more were experienced on the craft. This is reminiscent of the findings by the SESOC[39] Committee work in 1965 on the hydrodynamic loads on the captured air bubble (CAB) design for *hinged rigid planer seals* of similar geometry in interaction with the water. A more complete summary is provided in Chapter 4 but a quote from SESOC on the topic is germane:

> *"... the potential water loads on a hinged, but otherwise rigid, ski are enormous..."*

The structural analysis behind that conclusion was based on speeds of 100 knots for the CAB craft. The *ekranoplan* designs are for speeds of over 250 knots. For this reason, the structure of all the *ekranoplans* makes extensive use of thick gauge, welded aluminum structures, which contribute directly to high values of structural weight. Photographs of the *KM* structure are not available; the sole KM crashed in 1980 during take-off; but the trailing edge flaps of the *Lun* can be seen in the lower photograph of Figure 10.60 where the craft is up on blocks in a vacant naval station in Kaspiysk on the Caspian Sea where it now resides. The main wing structure is of welded aluminum and the flaps are a combination of riveted and bolted aluminum structure. The ARPA group reported large (compared to aircraft structures) thicknesses of plating throughout the craft (up to 1 inch thickness on the lower planing surfaces).

A further contributor to the high structural weights can be seen from Figure 10.59 lower photograph that shows the complex shaping in the bow of the craft and all along the fuselage made up of steps and spray strips, of all welded structure. Not visible in the photograph are the (seaplane-like) steps along the hull to improve "unsticking" of the hull familiar to all seaplane designers to reduce hydrodynamic drag during take-off. All these features contribute to increased weight[u].

[u] The reader is cautioned that higher values than given here for the Empty Weights on *ekranoplans* may be found in the literature and the internet; e.g. *Lun* has been quoted as 44% for W_S /W and >70% for W_E/W. If correct, the discrepancy between the Russian *ekranoplan* "as built" craft and the US Navy PAR-WIG projected designs would be even greater. It is believed that the ARPA data as used here is more reliable.

(a) Trailing Edge Flaps of *KM ekranoplan* During Take-Off

(b) Trailing Edge Flaps of *Lun ekranoplan* up on storage blocks

Figure 10.60: Trailing Edge Flaps of two *ekranoplans*

More details on the various features of the *ekranoplan* and other aerodynamic air cushion craft are given in Chapter 11 but these few points illustrate the concern over such low weights given by the US Navy PAR WIG designers during the ANVCE Project (see Figure 10.58) that led to the criticism of the designs, which was unfortunate because it is believed that some form of aerodynamic lift in the ship design is essential if such high speeds at sea are sought, and this needs to be given fair evaluation.

10.9 Techno-Economic Parameters

Included in the analysis by ANVCE of the various US Navy proposed concepts and designs, was the consideration of those technical characteristics that pertain most directly to the economic viability of the concepts. These considerations start with the range capability and the ability to transport payload over those ranges with economic fuel consumption. Some highlights from the ANVCE Final Report are given here.

The range of the vehicle is determined by the type of propulsion (and lift) system used. As derived in Chapter 8, the Transport Efficiency (TE) for those vehicles that do not use an auxiliary lift system is:

$$\text{Transport Efficiency (TE)} = \frac{WV}{P} = \eta_P \frac{L}{D} \tag{10.27}$$

In this expression, the propulsion efficiency is (η_P) and the aerodynamic efficiency is ($^L/_D$).

It was shown that for those vehicles using some form of pressurized air and flow for lift, the Transport Efficiency (TE) is modified to include the lift system, such that:

$$\text{Transport Efficiency (TE)} = \frac{WV}{P} = \eta_P \left(\frac{W}{D + \frac{\eta_P}{\eta_L} \frac{p_C Q}{V}} \right) = \eta_P \left(\frac{L}{D} \right)_{EFF} \tag{10.28}$$

In this expression for the Transport Efficiency (TE), the lift system is represented by the air cushion pressure (p_c), the air flow (Q) and the lift system efficiency (η_L). These efficiencies are then seen in the Breguet Range Equation that includes the effect on range of the fuel consumption and the amount of fuel available. A slight complication arises in the Breguet range equation in that the range depends on *how* the power or thrust is generated on the vehicle.

The original derivation of the Breguet Range Equation by Louis-Charles Breguet (see Chapter 7) was for reciprocating engines, or more generally, for those engines that provided the *power* that is then transmitted by different means (transmission, propellers, etc) to produce *thrust*. In this instance the fuel consumption is measured by the *specific fuel consumption (sfc)*, defined as:

$$\text{Specific Fuel Consumption } (sfc) = \frac{\text{lb of fuel}}{(\text{ft} - \text{lb/sec})(\text{sec})} \tag{10.29}$$

Expressing the specific fuel consumption (*sfc*) as lb of fuel/BHP-hour is typical in the engine manufacturing industry and represents the normal characteristic of reciprocating engines that the fuel consumption rate depends on the *power* level of the engine independent of speed.

With this form of engine, the Breguet Range Equation can be written:

$$\text{Range (nm)} = 325 \frac{1}{sfc} \frac{WV}{P} \ln \left(\frac{1}{1 - {^{W_f}}/_{W}} \right) \tag{10.30}$$

In this form, the units have been accommodated such the specific fuel consumption is expressed in lb/BHP-hour and the range in nautical miles (nm). The total fuel consumed (W_f) is shown as a ratio to the gross weight of the vehicle (W). Alternatively, this can be written in terms of the Effective Lift Drag Ratio $\left({}^{L}/_{D}\right)_{EFF}$ in the following form:

$$\text{Range (nm)} = 325\frac{\eta_P}{sfc}\left(\frac{L}{D}\right)_{EFF}\ln\left(\frac{1}{1-{}^{W_f}/_{W}}\right) \tag{10.31}$$

With the introduction of the *ekranoplan* form of air cushion craft where the engines produce *thrust* directly through the use of *turbojet engines*, the fuel consumption takes on a different form and is expressed by a *thrust specific fuel consumption* (*tsfc*), where:

$$\text{Thrust Specific Fuel Consumption } (tsfc) = \frac{\text{lb of fuel}}{(\text{lb of thrust})(\text{sec})} \tag{10.32}$$

In this case, the Breguet Range Equation takes on a form that includes craft speed and given by:

$$\text{Range (nm)} = \frac{2.85}{tsfc}\left(V\frac{L}{D}\right)\ln\left(\frac{1}{1-{}^{W_f}/_{W}}\right) \tag{10.33}$$

As before, the units have been accommodated in Equation 10.28 such that the *thrust specific fuel consumption* (*tsfc*) is in the more familiar unit of lb fuel/lb thrust-hour and the new term speed (V) is expressed in knots.

The propulsive efficiencies are also different for these two forms of craft. An analysis of all the various types of propulsion systems used in air cushion craft may be found in the paper by Sheets and Mantle[40]. Figure 10.61 taken from that reference shows the ideal efficiencies for the various functional forms of propulsive efficiency. The classical Froude theory used for these efficiencies considers the thrust (T) generated by a "magic disc" of area (A_j) that ejects rearward a stream of fluid at velocity (V_j) thus propelling the craft forward at a velocity (V). Since all the real fluid effects are neglected in the Froude theory, the theory provides an ideal comparative upper bound for the efficiency of the propulsive device for any given disk loading or thrust coefficient C_T, defined as:

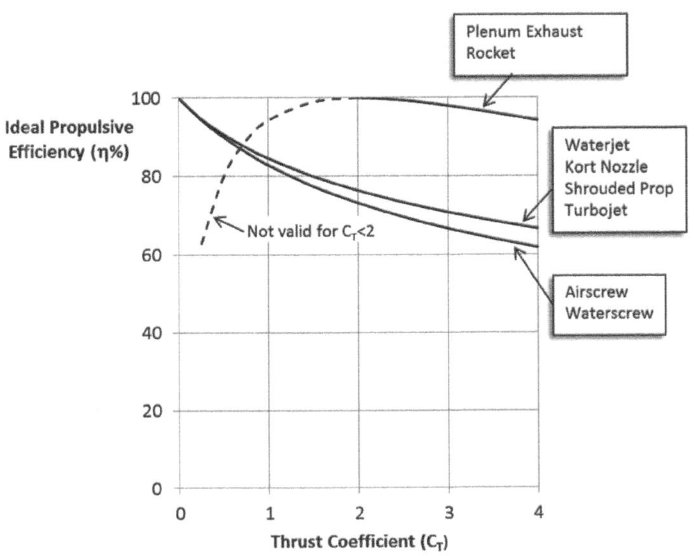

Figure 10.61: Ideal Propulsive Efficiencies

$$C_T = \frac{T}{\frac{1}{2}\rho V^2 A_j} \tag{10.34}$$

Table 10.7 shows the classical (Froude) ideal efficiencies for the various forms of propulsion used. Unfortunately, the ideal theory predicts the same efficiency for different propulsive devices, and it is only the real fluid effects that separate one from the other.

Table 10.7: *Maximum Efficiencies (Ideal)*

Propulsive Device	Froude (Ideal) Efficiency	Thrust Coefficient	Valid Region
• Unshrouded propeller (water & air)	$\frac{2}{1+(1+C_T)^{1/2}}$	$2\frac{V_j}{V}\left(\frac{V_j}{V}-1\right)$	$C_T \geq 0$
• Shrouded Propeller (water & air) • Kort Nozzle • Waterjet • Turbojet	$\frac{4}{3+(1+2C_T)^{1/2}}$	$2\frac{V_j}{V}\left(\frac{V_j}{V}-1\right)$	$C_T \geq 0$
• Airjet • Rocket	$\frac{2(2C_T)^{1/2}}{2+C_T}$	$2\left(\frac{V_j}{V}\right)^2$	$C_T \geq 2$

10.9.1 Speed for Maximum Range

For those craft that use *power* that is then transmitted through some propelling device (air propeller, water propeller or water jet) via a transmission system, the speed for maximum range is that speed for which the following is a maximum, i.e.:

$$\frac{\eta_P}{sfc}\left(\frac{W}{D + \frac{\eta_P}{\eta_L}\frac{p_C Q}{V}}\right) \text{ is a maximum} \tag{10.35}$$

This would apply to such concepts as the ACV, SES, hydrofoil, planing hull, etc in most cases depending on their form of propulsion.

For those craft that generate *thrust* directly (such as the WIG craft studied), the speed for maximum range would occur when the following parameter is a maximum:

$$\frac{1}{tsfc}\left(V\frac{L}{D}\right) \text{ is a maximum} \tag{10.36}$$

To summarize the speed for maximum range in terms of $\frac{L}{D}$ and other craft characteristics:

For Power Generating Engines (e.g. reciprocating and gas turbine engines providing shaft power)

$$\text{Speed for maximum range is when } \frac{\eta_P}{sfc}\left(\frac{L}{D}\right)_{EFF} \text{ is a maximum} \tag{10.37}$$

For Direct Thrust Producing Engines (e.g. turbojet engines)

$$\text{Speed for maximum range is when } \frac{1}{tsfc}\left(V\frac{L}{D}\right) \text{ is a maximum} \tag{10.38}$$

Complications occur in these relationships when other forms of aerodynamic craft are considered. One example discussed in Chapter 11, is the case when the compressor air is bled off the thrust engines to pressurize the air cushion beneath the wing. In the ANVCE work such complications were not pursued.

It is seen from Equation 10.30 and Equation 10.31, it is important to consider all the terms shown in determining the speed for best range and not just the hydrodynamic or aerodynamic terms of the craft. That is, just determining at what speed the $\left(\frac{L}{D}\right)$ is a maximum can be misleading when determining the correct speed for maximum range or economic advantage. This is why it is important to determine the Transport Efficiency $\left(\frac{WV}{P}\right)$and under what conditions it is a maximum because all the contributors $\left({}^{L}/_{D}, p_C, Q, \eta_P, \eta_L, sfc, tsfc\right)$ are automatically taken into account. This was done for all the projected designs in the ANVCE Project such that they could be compared on common ground. Note that the speed for maximum Transport Efficiency is not necessarily the maximum speed for the craft under consideration, although that is possible for craft designed for long range transport as will be shown in the following example.

To illustrate how the various parameters impact the final design, one example from the ANVCE set of Point Designs will be given. Figure 10.5 showed the set of designs for the "air capable" high speed ships. In particular, the "3000 tonne class" sidehull SES called the "SES 3" illustrates how the designer sought to maximize the speeds while operating at high efficiency. Figure 10.62 shows the three determinants: the "hydrodynamic" ${}^{L}/_{D}$; the Effective $\left({}^{L}/_{D}\right)_{EFF}$; and the Transport Efficiency $\left({}^{WV}/_{P}\right)$.

As seen in Figure 10.62, the maximum hydrodynamic efficiency $\left({}^{L}/_{D}\right)$ is about 22 and occurs at a speed of 70 knots. When the lift system characteristics (p_c, Q, η_L) are

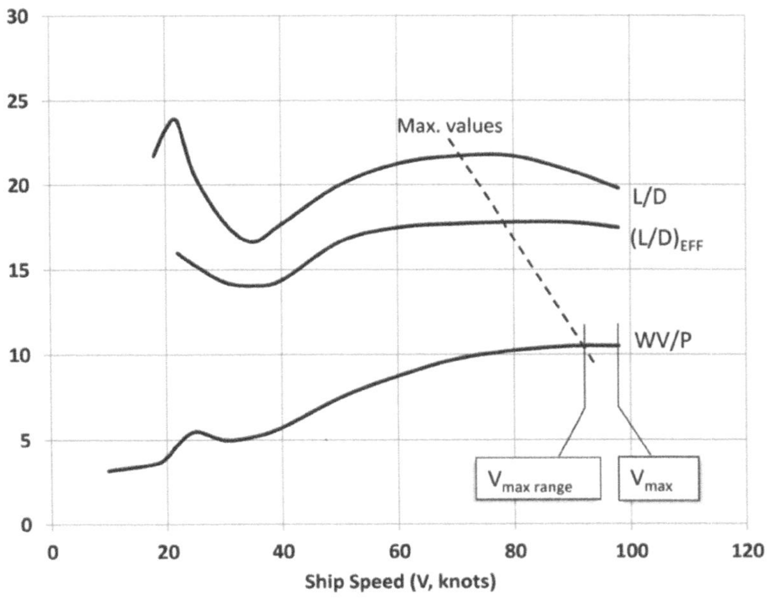

Figure 10.62: ANVCE Point Design "SES 3" Efficiency Characteristics

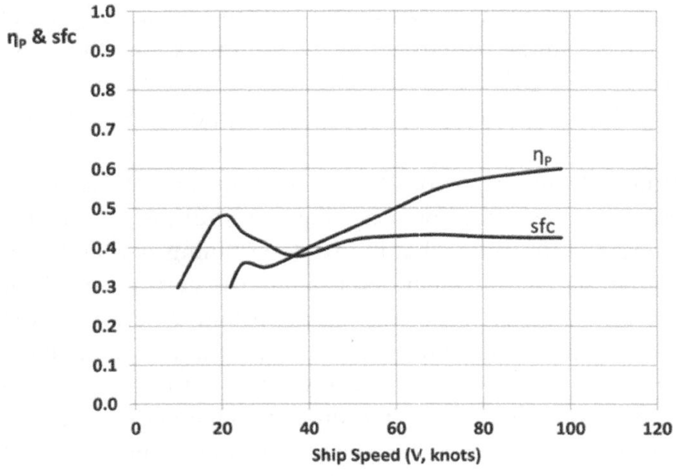

Figure 10.63: SES 3 Propulsion System Characteristics

taken into account, this reduces the efficiency to give the effective lift-drag ratio $(^{L}/_{D})_{EFF}$ of about 18 but at a higher speed of 79 knots. When now the propulsion system characteristics (η_P, *sfc*) are taken into account this gives the final value, the Transport Efficiency $(^{WV}/_{P})$ which shows a maximum value of greater than 10 but at an even higher speed of 92 knots. The maximum speed of the SES 3 is shown at the higher speed of 97 knots in Figure 10.62 and earlier in Figure 10.5.

The driving force and reason why the speeds increase as each parameter is included can be seen from Figure 10.63 that shows (for this design) how the propulsion parameters of propeller efficiency (η_P) and engine specific fuel consumption (*sfc*) variation with speed effect the final values. In this case the fuel consumption remains constant but the propeller increase in efficiency as speed is increased causes the increase in Transport Efficiency $(^{WV}/_{P})$ as speed is increased.

10.9.2 Maximum Speed and Speed for Maximum Range

The above example calculation shows how important it is to understand the different mechanisms at work in each design concept that determine the performance characteristics. The speeds of the ANVCE 3000 tonne class of vehicles (taken from Figure 10.5) are collected below in Table 10.8. These speeds are for Maximum Range and for Maximum Speed.

Generally speaking, for those vehicles that use some form of *dynamic* lift, it is possible to design the vehicle such that the speed for maximum range is close to the maximum speed of the craft. For those ships that operate in the displacement mode and rely on *buoyant* lift, there is a wider spread between the two speeds because of the nature of the hydrodynamic drag rise with speed. For the SWA 4 (a small water plane displacement hull) and the MONO 3 (a conventional displacement monohull), it is seen that the best range speed is much lower than the maximum speed of the ship. This highlights why one must be careful in comparing different vehicles that operate on different lifting physics for their performance. The details, and reasons for their differences, on all the various craft examined by ANVCE in this regard can be found in the 1979 Final Report.

Table 10.8: *Speeds for 3000 tonne class ships*

ANVCE Point Design	Speed for Maximum Range ($V_{max\ range}$)	Maximum Speed (V_{max})
SES 3	92	97
ACV 3	93	93
SWA 4	19	37
HYD 2	45	53
MONO 3	15	40

10.9.3 Transport Efficiency

Much of the emphasis on high speed ships over the previous decades has been on seeking ships that operate as *transport* vehicles operating over *long distances* (see discussion for maritime commerce in Chapter 4 and for military missions in Chapter 5). This leads directly to examination of the ability to generate a high value of Transport Efficiency $\left({}^{WV}/_{P}\right)$ in the designs. The ANVCE Project included such an examination in its work. After the intensive design effort by all the companies who were well versed in the design and building of air cushion craft and high speed ships, the ANVCE Project produced the designs summarized in the earlier Figure 10.5 and Figure 10.6. The Transport Efficiencies (at maximum speed) of the surface vehicles from that set are displayed together with an historical set of existing air cushion craft on Figure 10.64. It is seen that they represent the most advanced

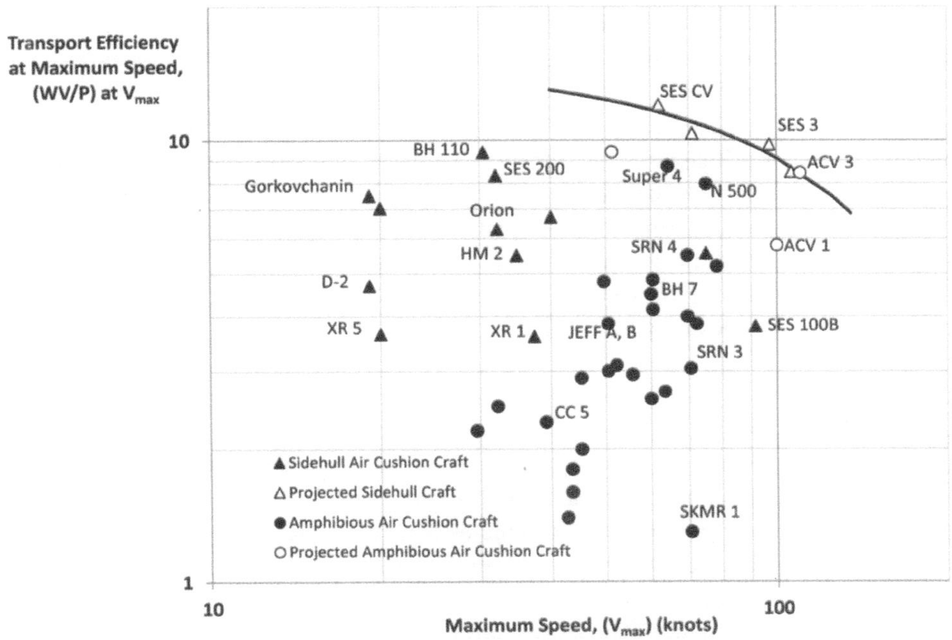

Figure 10.64: ANVCE Point Designs Transport Efficiencies of Sidehull SES *et al.*

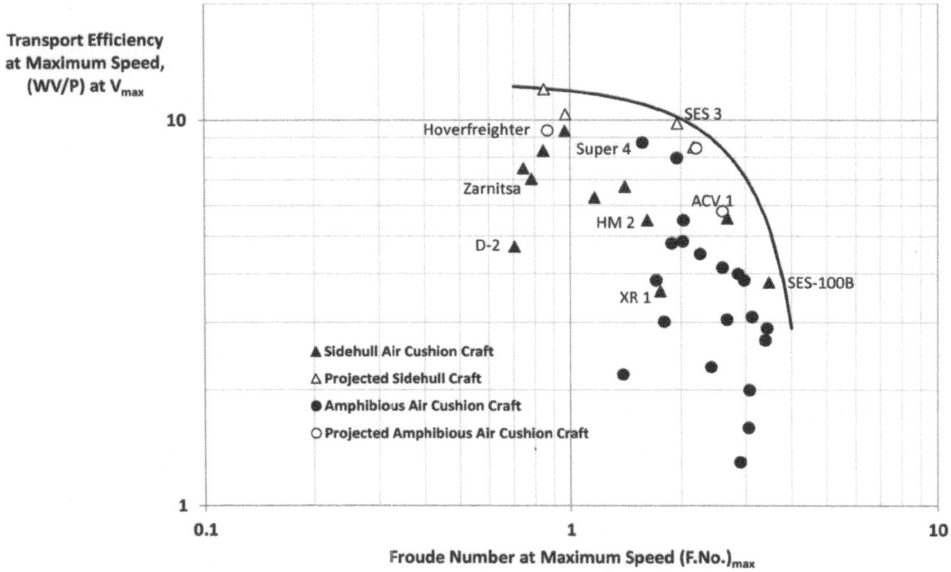

Figure 10.65: ANVCE Point Designs Transport Efficiencies as Function of Froude Number

designs with Transport Efficiencies of the order of $(\frac{WV}{P} = 10)$. There is a little less spread in the data, if the same data is plotted against Froude Number as shown Figure 10.65. There is a marked fall-off in efficiency as speed is increased.

It would be a little more difficult to plot the Transport Efficiency versus the speed for maximum range (maximum value of Transport Efficiency) because such data would require the details of each of the craft shown, and of its individual power versus speed curves. But in any case, regardless of the lowered power level required for *cruise speed*, each craft must be designed to install the maximum power required for *maximum speed* as noted by Gabrielli and Theodore von Karman[41] in their 1950 seminal work on efficiencies of transport vehicles, and thus the display shown in Figure 10.64 and Figure 10.65 is correct.

If that same data (from Figure 10.64) is displayed against the backdrop of all ships and all aircraft, the comparison is as shown in Figure 10.66. A discussion of all the data and display of Transport Efficiency is given in more detail in Chapter 7 on economic considerations where the treatment of the envelopes of all the data is provided.

Throughout the early development of high speed craft, a significant emphasis was placed on seeking (transport) vehicles that operate at the interface between the sea surface and the air to also have an efficiency between "displacement ships" and "conventional aircraft". It was called seeking to "fill the gap" between the two envelopes of conventional types of vehicles shown on Figure 10.66. Many reports, papers, books, etc. over the years are replete with proposals of various forms of high speed ships that promised Transport Efficiencies of $({}^{WV}/_{P} > 20)$ and at speeds $(V > 100 knots)$. As Figure 10.66 shows, after intensive development by experienced builders this has proven to be not the case. The ANVCE designs covered a range of

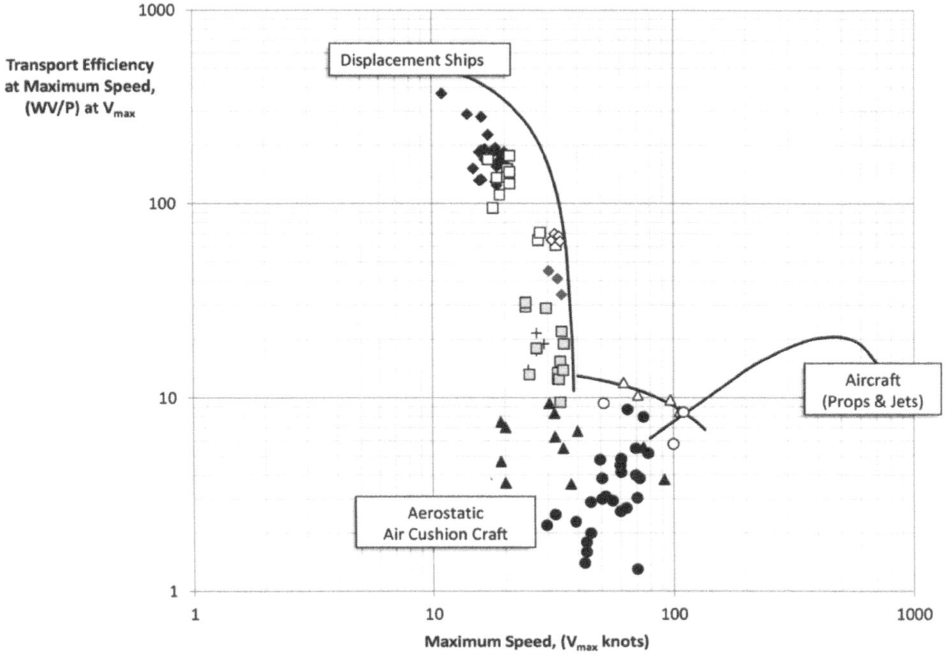

Figure 10.66: ANVCE Point Designs Transport Efficiencies on Gabrielli-von Karman Chart

ship displacements from 1000 tonnes to 11,000 tonnes and "achieved" values of Transport Efficiency of about *half* those early projected efficiencies. The intensive US Navy effort in 1976 on the ANVCE Project came one decade later than the first evaluation in 1965 by the SESOC[42] Committee that also challenged the projections for a CAB ship of 10,000 tonnes to have a Transport Efficiency $\left({}^{WV}/_{P}{\approx}20\right)$ at a speed of 129 knots. The ANVCE Point design for an 11,000 tonne sidehull SES (called "SES CV" in Figure 10.66) had a value $\left({}^{WV}/_{P} = 12\right)$ at 62 knots. This is a significant difference in predicted performance and confirms the earlier 1965 SESOC work and committee concerns.

Another view of the same data of Transport Efficiency $\left({}^{WV}/_{P}\right)$ but versus a dimensional form of the maximum speed Froude Number $\left({}^{V}/_{W^{1/6}}\right)$ is shown in Figure 10.67. This particular chart shows just the surface bound craft of both advanced designs (for ACV, SES, hydrofoil and planing craft) together with actual operational craft (hydrofoil, monohull and planing craft).

The characteristic rapid fall-off of efficiency with speed is again illustrated in this plot. Except for some optimistic projections for large (>10,000 tonnes) nuclear power designs none of the advanced designs exhibited values of Transport Efficiency greater than 10. This is a significant departure from earlier projections (vintage 1965 and later) found in the literature by advocates of various proposed concepts.

A similar set of findings was found for the air borne vehicles, especially for the PAR-WIG designs. This can be seen in Figure 10.68 that shows the Transport Efficiency for the PAR-WIG designs (WIG(H), WIG(S) and WIG(O) concepts);

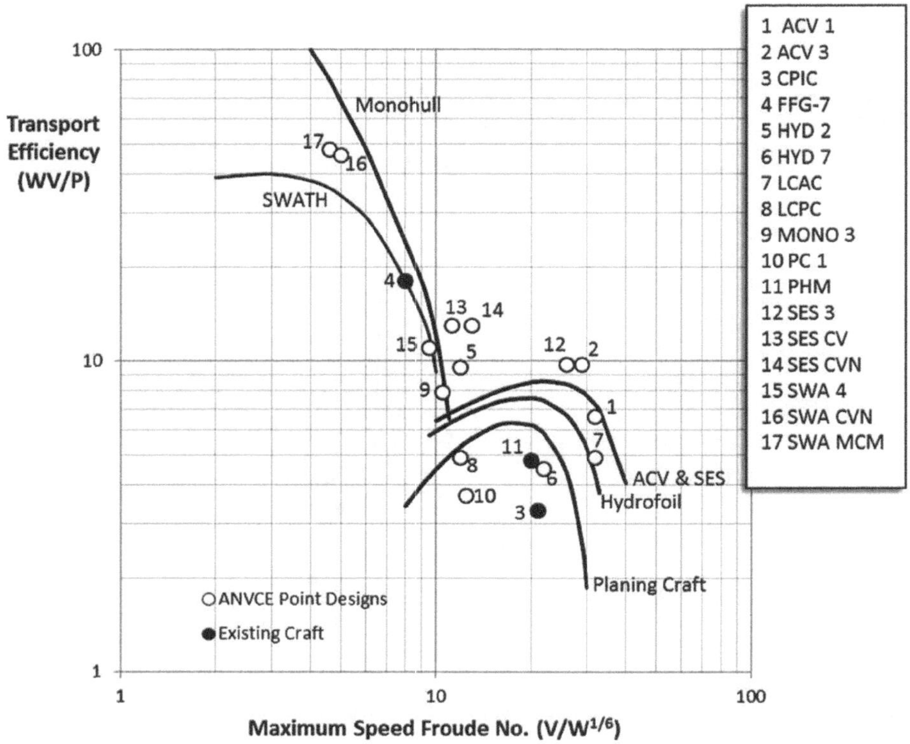

Figure 10.67: Transport Efficiencies of ANVCE Point Designs and Existing Surface Craft

the sea loiter aircraft (SL); and the air loiter aircraft (AL), both conventionally powered and nuclear powered, AL(N).

Again, as shown in Figure 10.68, except for the nuclear powered aircraft, none of the advanced concepts showed Transport Efficiencies any greater than the ongoing state-of-the-art of conventional air craft. The reader is reminded that the efficiencies given for the PAR-WIG craft shown on Figure 10.68 are based on the optimistic performance estimates already cited.

In the ANVCE Final Report other design measures such as productivity (Payload Tonne Miles/Tonne of Fuel) and other measures were evaluated and again, no advantage over more conventional means of transport were found. While the data bases and maturity of designs were different between the 1965 SESOC Report and the 1976 ANVCE work many of the governing conclusions were the same with general agreement that the performance capabilities of the concepts questioned had been overstated. More detail discussion of the general class of aerodynamic air cushion craft, including the Russian *ekranoplan* and US Navy PAR-WIG concept is provided in Chapter 11, especially as it applies to the goal of achieving "100 knots at sea" in a practical platform.

Many of the differences found in the 1976 work by the ANVCE Project are covered in the Final Report but the Conclusions and Recommendations from the ANVCE work are worth summarizing.

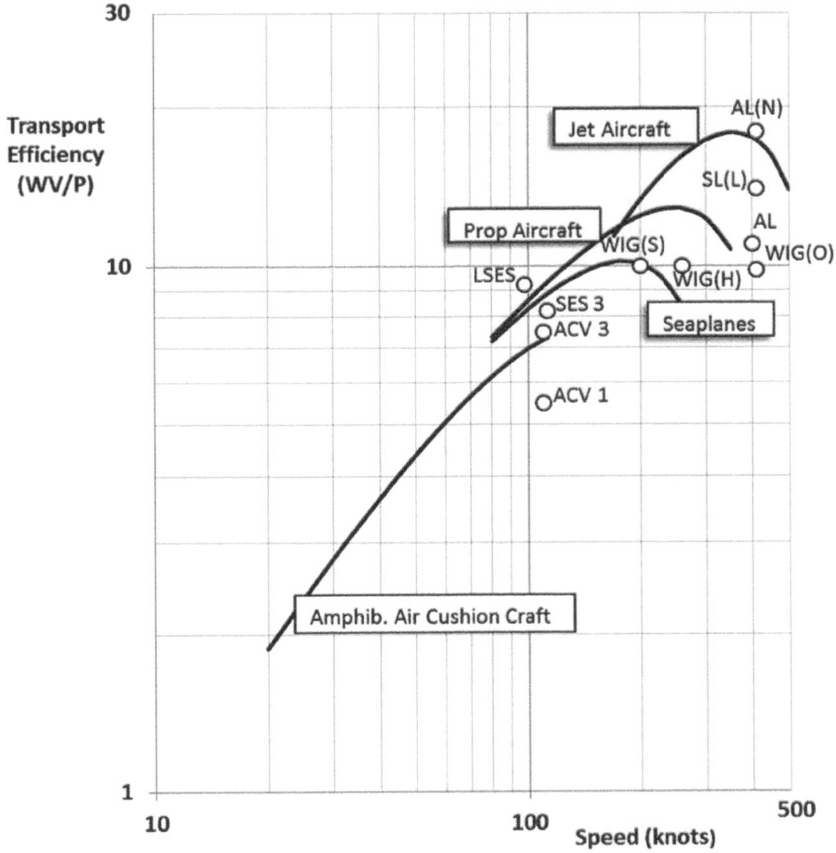

Figure 10.68: Transport Efficiencies of ANVCE Point Designs and Existing Aircraft

10.10 Conclusions and Recommendations from ANVCE Final Report

The analysis, design and testing work accomplished under the ANVCE Project was extensive. Similarly, the conclusions and recommendations were also expansive and laid down for a program plan to accomplish the much needed Research & Development called for to answer the questions posed by the US Navy, the Office of Secretary of Defense, Congress and GAO. Much of this material is out of context of "100 knots at sea" to repeat here.

There were some basic recommendations related to the goal of a "100 knot Navy" that can be summarized here to guide future work. Again attention will be restricted to those concepts that come closest to "surface vehicles" that would plan to operate close to the sea surface at the high speed of around 100 knots. This eliminates further discussion of all air vehicles (except certain versions of wing-in-ground-effect vehicles) and concentrates on; SWATH ships; mono-hull displacement ships; conventional planing hulls and certain forms of hydrofoils.

Of most direct interest were the ANVCE findings relative to the "3KSES" program and related issues. They are self-explanatory and summarized here as:

Surface Effect Ships

". . . . severely restrict the operating envelope to no more than 50 knots in any sea state until all engineering tests in that portion of the planned operating envelope are complete. . . . ."

". . . . . . study a 50 knot SES in the 3000-tonne class for the same missions as the 80 knot SES to determine cost effectiveness. The results of the analyses performed by the ANVCE Project indicate that by using high L/B ratio (more slender) platforms, the investment cost of the platform can be cut in half. . . . ."

Air Cushion Vehicles

". . . . . . skirt sizes for ACV of more than 3000 tonnes displacement would strain the capability of manufacturing looms. . . . . ."

". . . . . air propulsion, using conventional air propellers (shrouded or unshrouded), becomes impractical for ACV of more than 4000 tonnes; at this point, water propulsion (screw propeller or waterjet) becomes less of a risk. . . . ."

Hydrofoils

". . . . . once a 50 knot PHM[v] class is in operation, its ships should be fitted with variable geometry foils to give them dash capability of 70–80 knots. The tactical, open-ocean use of such a foil is best explored under actual sea conditions. . . ."

". . . . . hydrofoils in the 1000 to 2000 tonne class should be explored first, to avoid the greater technical risk and higher costs which are associated with heavier hydrofoils. . . ."

". . . . The surface piercing hydrofoil should be evaluated particularly in terms of the PHM hull form. The surface piercing hydrofoil is relatively lower in cost than a hydrofoil with fully submerged foils and also has a self stabilizing capability. . . . . ."[w]

Supercritical Planing Hulls

It is generally thought that planing craft is the realm of "small boats". This is driven by (a) the high power requirements and (b) the unacceptable "pounding" in rough water for equipment and personnel. During the ANVCE Project, one new concept called the *"supercritical planing hull"* offered the opportunity to "re-open" the planing hull discussion based on full scale test and analyses. Chapter 8 on Technical Considerations discusses this novel hull form that could prove a high speed hull form that could extend ship size into competitive regions of greater than 50 knots without the need for the "dynamic features" embodied in hydrofoil, ACV or SES designs. The specific recommendation from the ANVCE Project was:

[v] Unfortunately, all six PHM hydrofoils were retired and scrapped in 1993 having being declared by the US Navy as *"not cost-effective"*. The closing down of much of the advanced ship work after the 1980 program cancelation did not allow the cost reduction proposals by ANVCE summarized here to be put into place soon enough.

[w] See Chapter 1 for descriptions of the surface piercing hydrofoil ship. Again the scrapping of the PHM ships in 1993 precluded any such cost reduction activities to go into effect.

"..... Further development of the supercritical hull is highly recommended. Research is needed on design parameters for supercritical hulls that will allow sufficient internal volume to meet payload requirements without jeopardizing the hydrodynamic advantages of the hull form. Research should be conducted toward reducing its relatively high calm-water resistance while still retaining its good rough sea performance....."

Wing-in-Ground-Effect Vehicles

Of all the high speed concepts evaluated, the wing-in-ground-effect (WIG) vehicle offered the greatest promise for sustained high speed operation (> 100 knots) at sea. Unfortunately, this concept did not receive the broad or in-depth analysis of all the possibilities of this novel form of "surface vehicle" that has a history that almost predates all the other dynamic ship forms (see Chapter 1) up to the extensive efforts by the Russian designers and builders (see Chapter 1) that were underway in the same period as the ANVCE Project was being conducted. The fact that the WIG can take different forms complicated the analysis and objective treatment. The various WIG forms reflect whether the vehicle operates solely "in ground effect" all the time or in other forms that can operate for different amounts of time "out-of-ground-effect". This was an unexplored mission issue for the US Navy.

Enough analysis and model testing and design work was done by the ANVCE Project, to make the following conclusions and recommendations on the WIG concept:

".....the WIG vehicle operating only in ground effect does not compare favorably with ships and aircraft in the logistics role: ships and aircraft can perform the role better. Similarly, the "out-of-ground-effect capable" WIG did not fare well compared to an aircraft in the given naval missions. However, applications for the truly unique capability of a WIG vehicle, that of operating (fairly efficiently) in ground effect and being able to "hop up" into the air for a short period of time to perform a task were not fully explored....."

".... R&D should concentrate on solving the rough water stability and performance problems. The use of ACV-like skirts should be explored through R&D efforts...."

"...... hard end plates for the WIG vehicle were shown to be structurally unsound in any practical operation. The use of hovercraft skirt-like appendages to alleviate the dynamic loading was explored...."

"..... solutions examined for stability involved water skiis, hydrofoils, and similar devices but all lacked a degree of practicality for operation in rough water. Thus these issues remain unresolved...."

Chapter 11 goes beyond the work of ANVCE on this attractive form of high speed vehicle and shows other variations for future consideration for achieving high speed at sea and advantages in other missions for the US Navy.

10.11 Assessment of the ANVCE Project

The ANVCE Project was the last systematic study made by the US Navy into possible candidates for a design of a ship that could be the basis for the "100 knot Navy". The value of the study was that all candidates were evaluated by a common methodology with common requirements and common mission needs. The costing methodology was also a consistent method so that all candidates were treated on a common basis. This setting up of a common comparative method was specifically done so as to not repeat the problems experienced by the SESOC work (see Chapter 4) where the evaluators were hampered by the many different designs all done to different standards, requirements and technology levels. Because of the difficulties of obtaining exact costs from existing craft for comparison (different manufacturing methods, different accounting methods, different time frames, etc) the value of the ANVCE costing method was in using a consistent accounting to help in the relative costs between concept types which aided in the comparison and decisions to be made between the various candidates.

As was suspected by other experts, it was confirmed during the ANVCE Project that many of the claims by early advocates for high speed ships were highly exaggerated (e.g. speeds higher by a factor of 2:1 and transport efficiencies higher by 3:1) leading to lengthy analyses and costs to place all the results, designs and analyses on a fair footing for final evaluation. The key results have been summarized here. Out of the two year study came a wealth of data and designs put together by the best designers in the land from both industry and the Government, that offer a solid basis for further work on seeking a practical high speed platform. Many of the suggestions have been highlighted throughout this chapter and form a firm basis for further study, even if the speeds sought are less than 100 knots. It was found as a result of the technical analyses and the military operations analyses that in many cases the designs were found to be better suited for other missions than for open ocean frigate size ship missions. With the need to develop a navy with a greater flexibility of operations for future undetermined threats, the use of high speed much faster than the current Fleet speeds of 30+ knots is within the practical realm of high speed ship design. The ANVCE Project verified that such designs are possible.

References

1. Meeks, Capt Thomas L. USN and Peter J. Mantle, *"The Advanced Naval Vehicles Concepts Evaluation"*, AIAA/SNAME Advanced Marine Vehicles Conference, Paper No. 76–846, 20–22 September 1976
2. Mantle, Peter J., *"Advanced Concepts for Sea Control"*, SAE Paper No. 770966, SAE Aerospace Meeting, Los Angeles, California, 14–17 November 1977
3. Meeks, Capt Thomas L., Cdr Clark Graham USN and Richard C. Hu, "*Combat Systems for the Advanced Naval Vehicles Concepts Evaluation"*, Naval Engineers Journal, Vol 90, No. 5; October 1978
4. *"Advanced Naval Vehicles Concepts Evaluation (ANVCE) Project"*, Prepared by Chief of Naval Operations (OP-96), Report No. CNO LTR 96/C194136, Vols 1,2,3,4; December 1979
5. Crewe, Peter R. et al, *"Analysis of Acceleration and Pressure Measurements on the SR. N1"*, Hydrodynamics Note PRC/DGG/MFKi/3823, Saunders Roe Ltd, November 1960
6. *British Hovercraft Safety Requirements*, published by Civil Aviation Authority, April 1972
7. *"Wingship Investigation"*, Final Report, Vol 1,2,3 published by Advanced Research Projects Agency (ARPA), 30 September 1994
8. Smiley, Robert F., *"Water Pressure Distributions During Landings of Prismatic Model Having an Angle of Deadrise of 22.5°"*, NACA Tech Note 2816, November 1952
9. Heller, S.R. and D.J. Clark, "*The Outlook For Lighter Structures in High Performance Marine Vehicles*", American Institute of Aeronautics and Astronautics/Society of Naval

Architects and Marine Engineers Advanced Marine Vehicles Conference, AIAA Paper 74–330, 25–27 February 1974

10. *"Ship Work Breakdown Structure (SWBS)"*, by the Naval Ship Systems Command, NAVSHIPS 0900-039-9010, March 1973
11. Osborne, W.C., *"Fans"*, Pergammon Press, London, England, 1966
12. Shipway, J.C., *"Estimation of Head-Flow Curves for Centrifugal Fans"*, National Physical Laboratory Report, Hovercraft Report 5. October 1968
13. Wallis, R.A., *"Axial Flow Fans"*, Newnes Academic Press, London, England (1961)
14. Shipway, J.C., *"Aerodynamic Design of Axial Lift Fans"*, National Physical Laboratory Report, Hovercraft Report 4, July 1968
15. Baljé, O.E., *"A Study on Design Criteria and Matching of Turbomachines, Part A: Similarity Relations and Design Criteria of Turbines"*, American Society of Mechanical Engineers Paper 60-WA-230 (1960)
16. Cordier, O., *"Ahnligkeitsbedingungen Für Strömungsmaschinen"*, VDI Berichte, Vol 3, pp 85–88 (1955)
17. "*Study of Skirt Design Concepts for Large ACV's with High Speed Potential*", Report No. AGC-T-596, Aerojet General Corporation, December 1976
18. *"Study of Large ACV Skirt Design Concepts"*, Report No. 7588–950048, Bell Aerospace Textron, 31 January 1977
19. *"Air-Propelled Air Cushion Vehicles"*, Report No. P61/7701, Vosper-Thornycroft Ltd, March 1977
20. *"SES Lift Systems"*, Hovermarine Corporation, Report to ANVCE Project Office, December 1976
21. "*SES Skirt Systems*", Hovermarine Corporation, Report to ANVCE Project Office, March 1977
22. O'Hanlon, J.F. and M.E. McCauley, *"Motion Sickness Incidence as a Function of the Frequency and Acceleration of Vertical Sinusoidal Motion"*, Human Factors Research, Inc., Technical Report 1733–1, September 1973
23. *"Guide for Evaluation of Human Exposure to Whole Body Vibration"*, International Standard Organization (ISO), Document ISO/DIOS 2631, 1972
24. Eggington, W. and Nat Kobitz, *"The Domain of the Surface Effect Ship"*, Society of Naval Architects and Marine Engineers Annual Meeting, Paper 11, November 1975
25. Mantle, P.J., *"Advanced Naval Surface Vehicles for AD 2000"*, Joint DDR&E – Intelligence Community Technology Trends Colloquium, US Naval Academy, Annapolis, 29 March–1 April 1978
26. Baron von Schertel, *"Surface Piercing Hydrofoils and Other Hydrofoil Systems*, Supramar, AG, Report to ANVCE Project Office (1976)
27. Lewis, Edward V., *"The Influence of Sea Conditions On the Speed of Ships"*, American Society of Naval Engineers (ASNE) Journal, Vol 67, Issue 2, pp 303–319, May 1955
28. Lewis, Edward V., "*Research Towards More Efficient Transportation by Sea"*, Trans. Society of Naval Architects and Marine Engineers (SNAME), Vol 69, 1961
29. Payne, Peter R., *"A New Type of Hydrofoil"*, Paper presented at Small Craft Marine Engineering Resistance & Propulsion Symposium, Ann Arbor, Michigan, 15–17 May 1996
30. *"High Speed Boat with Planing Hull"*, inventor Peter R. Payne, US Patent No. 3,763,810 Filed 24 March 1972, Issued 9 October 1973
31. Hammond, M.A. and K. Nicholson, *"Sea Knife Analysis of Selected Seakeeping Data, Vol 1"*, Admiralty Experiment Works, Haslar Report, Gosport, Hants, UK, November 1976
32. Payne, Peter R., *"Supercritical Planing Hulls"*, Ocean Engineering, Vol 11, No.2, pp 129–184, 1984
33. *"Point Design for 3000-Ton Air Cushion Vehicle"*, Bell Aerospace Textron Report 7588–950049, 31 January 1977
34. *"The Specialist Committee on Waves"*, Proceedings of the 23rd International Towing Tank Conference (ITTC), Venice Italy, September 2002

35. National Weather Service, Ocean Prediction Center, www.opc.ncep.noaa.gov
36. Krause, Fred H., *"Evaluation of a Power-Augmented-Ram Wing Operating Free in Heave and Pitch Over Water"*, US Navy DTNSRDC ASED Report 385, August 1977
37. *Appendix: Wing-in-Ground-Effect Vehicles* by Capt James W. Kehoe to *Chapter VI: The Surface Effect Ship* by Edward A. Butler, "Modern Ships and Craft, Naval Engineers Journal, Special Edition, February 1985
38. *"Wingship Investigation"*, Final Report, published by Advanced Research Projects Agency (ARPA),now Defense Advanced Research Projects Agency (DARPA), Arlington, Virginia, September 1994
39. *"Surface Effect Ships for Ocean Commerce (SESOC), Final Report on a Study of the Technological Problems*", by the SESOC Advisory Committee, issued by US Department of Commerce and published by US Government Printing Office, February 1966
40. Sheets, H.E. and P. J. Mantle, *"Air Cushion Craft Propulsion"*, AIAA/US Navy Second Marine Systems and ASW Conference, AIAA Paper 66–731, 8–10 August 1966. See also AIAA Journal of Aircraft, Vol 4, No. 3, pp 237–244, May-June 1967
41. Gabrielli, Giuseppe and Theodore von Karman, *"What price Speed?"*, Mechanical Engineering, October 1950 Issue.
42. *"Surface Effect Ships for Ocean Commerce (SESOC), Final Report on a Study of the Technological Problems*", by the SESOC Advisory Committee, issued by US Department of Commerce and published by US Government Printing Office, February 1966

11 Aerodynamic Air Cushion Craft

"Speed, bonnie boat, like a bird on the wing"
Skye Boat Song

Harold Edwin Boulton (1859–1935)

Before proceeding with this Chapter, the reader might pause for a moment and glance again at the photograph in Figure 2.1 in Chapter 2 and imagine a wing-in-ground-effect (WIG) craft (or "ship") flying close to the water to obtain the lift advantage from "ground or surface effect" in those same sea conditions. One might also wonder why all the photographs that have appeared in the literature from the small personal aircraft using the WIG principle to the largest sizes in the Russian, Chinese and German work on WIG, and with the US PAR WIG and other variants of wing-in-ground-effect, are all taken in calm waters or in relatively calm seas. This is a problem that needs to be addressed and will be referred to in this chapter on this promising concept to achieve high speed over the seas of the world.

11.1 Aerodynamic Underpinnings

The basic aerodynamic principles that apply to *aerodynamic air cushion craft* can be traced to before the Wright brothers first flew at Kitty Hawk on 17 December 1903. In the late 1800s several scientists and renowned engineers were developing the theory of flight and building practical man-powered gliders. Sir George Cayley (1773–1857) in Scarborough, Yorkshire, England is frequently referred to as "the father of aeronautics" because of his extensive work in determining the four basic forces for flight: *lift, drag, thrust* and *weight* and building the first glider to carry a man aloft. The first flight was in 1849 with a young boy as pilot. A much larger glider that flew across Brompton Dale (near Scarborough, Yorkshire) was in 1853 with a grown man as pilot. Historical records are not clear as to who the pilot was but possibly an employee or probably Sir Cayley's grandson. A history of Sir George Cayley and his work may be found in Richard Dee's excellent 2007 book.[1] Sir George Cayley was a prolific inventor and well versed in basic research methods. An interesting and prescient set of short articles written by him appears in the 1809 and 1810 Journal of Natural Philosophy[2] where he outlines the governing principles for flight. These three articles are a remarkable discourse on the theory of flight and recommended reading where he develops all the necessary forces for sustained

Figure 11.1: Otto Lilienthal flying one of his many gliders (1895)

flight, including the requirements for stability. His analysis of bird flight during both take-off and in cruise flight laid the groundwork for many a subsequent aeronautical engineering analysis.

After Sir George Cayley, Otto Lilienthal is regarded as one of the most important figures in aeronautics at the end of the nineteenth century. Otto Lilienthal (1848–1896) born in Anklam, Prussia moved to Berlin, Germany with his mother and brother Gustav in 1870. Lilienthal studied George Cayley's work and set about (with help from his brother) to develop gliders based on the principles laid down by Cayley. Lilienthal made over 2000 flights of his many designs of both monoplane and biplane gliders, exploring the many features that were found essential for sustained flight. Figure 11.1 shows just one of his many flights off a manufactured hill in Lichterfelde (in Berlin[a]) that he had constructed in 1894 for launch of his many experimental flights. As can be seen in the photograph, Lilienthal was well aware of the advantages of *dihedral* and *camber* (a significant development over Cayley's "flat plate" surface gliders) in developing lift. In all of the earlier work by many engineers and researchers and by Lilienthal himself, the idea of "aerofoil thickness" to give lift to the wing had not yet been developed. While not discussed in the historical records, it appears (from Figure 11.1) that Lilienthal was approximating an elliptic wing in his designs. See later for discussion on the significance of this shape.

Lilienthal based much of his research (using Sir George Cayley's pioneering work) on the flight characteristics of birds and published his findings later (1889)[3]. Otto Lilienthal was particularly intrigued how, after the "powering mode" of wing flapping, birds could fly or glide long distances in level flight without flapping their wings. His research led him to experiment with the four forces identified by Cayley. As a dedicated experimenter, Lilienthal considered such an approach essential practice before considering the application of powered flight.

[a] Later (in 1932) made into a monument to Lilienthal's life's work.

Lilienthal was working on an engine powered craft – strangely chosen to be an ornithopter – in 1894 but the engine froze and the flight curtailed. He was working on a larger version in 1896 but before it could be tested, Lilienthal crashed one of his unpowered gliders and sustained a broken back among other injuries. He died the next day in hospital on 10 August 1896.

The idea of exactly *how* the wing developed lift started to take on a firm engineering basis when an English aeronautical engineer, Frederick W. Lanchester (1868–1946) began his work at the Royal College of Science (now the Imperial College of Science and Technology) in London, England. Earlier, Lilienthal had suggested in his book *Birdflight as the Basis for Aviation*, that if the airfoil was curved to be more in line with the streamlines flowing over the wing, much greater lift would be experienced. Lanchester studied this observation and postulated that some type of circulatory flow superimposed on the oncoming free stream flow would generate the lift experienced by the wing. This deceptively simple idea became the basis for much of the understanding of *lift* and *drag* of wings to this day. Lanchester developed these thoughts starting in 1891 and published his basic analyses in 1907.[4] He believed that this "circulation" could be explained by vortex filaments spread along the span of the wing culminating in a vortex stream emanating from the span tips. Lanchester theorized this arrangement based on his reading of the work by Hermann L. Helmholtz in 1858[5] on vortex characteristics. Sketches of these vortices along the span of the wing may be seen in Lanchester's 1907 book. Historians[6] generally agree that Lanchester can be credited with putting forward two fundamental ideas on *lift* and *drag*, which are:

(a) Circulation is the cause of *lift*
(b) Tip vortices are the cause of *drag*

Lanchester conceived these seminal results in the period of 1891–1894 and based on historical records he presented his ideas, albeit unsuccessfully, to unresponsive professional societies at the time. There is an unfortunate historical footnote in that Lanchester was known for obtuse explanations of his work using terms that were unfamiliar to his listeners. This made acceptance of his ideas difficult. In 1908, Lanchester met with Ludwig Prandtl (Director of Institute for Technical Physics at the University of Göttingen, Germany) together with Prandtl's student, Theodore von Kármán[b], and outlined his findings and conclusions on *lift* and *drag* but historians report that the meeting did not go well because of language problems and Lanchester's style of explaining his work. Immediately following this meeting, Prandtl started developing a mathematical model that put Lanchester's theories on to a sound basis. This became known as the "Prandtl Lifting Line Theory" around the world – although in England this is more often referred to as the "Lanchester-Prandtl Lifting Line Theory"!

These two giants in the field, Frederick Lanchester (Figure 11.2) and Ludwig Prandtl (Figure 11.3) provided much of the foundation of the theory of flight used today.

This greatly simplified historical snapshot does not include the many other major contributions along the way by such giants in aerodynamics as Hermann Glauert (England), Wilhelm Kutta (Germany), Nicolai Joukowski (Russia),

[b] See Chapter 7 for discussion on Theodore von Kármán's work in the context of this book's subject matter

Figure 11.2: Frederick Lanchester (1868–1946)

Samuel Langley (USA), Albert Betz (Germany) and Max Munk (Germany); all of whom contributed significantly to the understanding of the lifting line theory. In the interest of space, the reader is referred to the many texts on biographical histories of these men and standard texts on aerodynamics on their theoretical and experimental efforts in the understanding of lift and drag of wings.

11.1.1 Lanchester-Prandtl Lifting Line Theory

Lanchester had concluded that there was a linear relationship between *lift* and *circulation* and this was put on a mathematical basis by Wilhelm Kutta in 1902 and independently by Nicolai Joukowski in 1906 to yield what is referred to as the Kutta-Joukowski Law:

$$L = \rho V \Gamma \tag{11.1}$$

Figure 11.3: Ludwig Prandtl (1875–1953)

In this relationship, L is the lift generated by the wing; ρ is the air density and V is the velocity of the oncoming airstream to the wing. The circulation, Γ, is defined as the line integral of the airflow velocity along the curve of the flow over the wing. For a two-dimensional wing, the circulation would be constant along the span. Prandtl postulated that for a three dimensional wing this would cause a discontinuity at the wing tips and accordingly assumed an elliptical distribution as shown in Figure 11.4.

The original theory assumed a single line of vortex flow but Prandtl quickly switched to a set of "horseshoe" vortices emanating from the wing as shown in Figure 11.4 and integrated along the span to get the lift and drag forces. From dimensional analysis, it is known that the lift and drag can be expressed as:

$$L = C_L {}^1\!/_2 \rho V^2 S \tag{11.2}$$

$$D = C_D {}^1\!/_2 \rho V^2 S \tag{11.3}$$

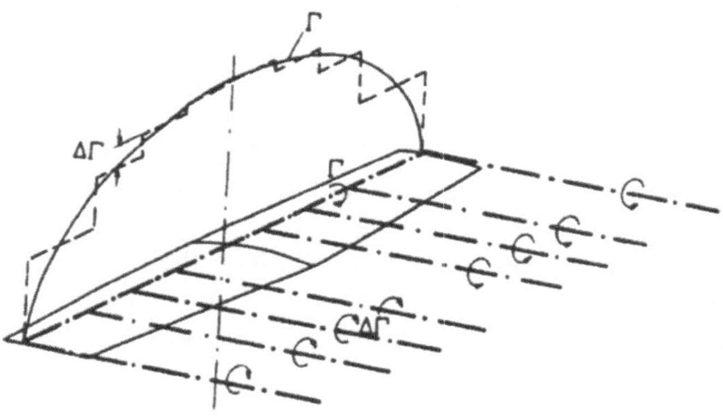

Figure 11.4: Sketch of Prandtl's Lifting Line Model

The lift (L) and drag (D) forces are referenced to the wing area (S). The lift coefficient (C_L) and drag coefficient (C_D) are functions of the characteristics of the wing geometry and other factors as determined in the lifting line analysis. A complete description of Prandtl's analysis took some time to appear in the technical literature and probably the first definitive description appeared in 1918–1919 by Prandtl himself[7] in internal notices at Göttingen (although different articles by Prandtl in technical journals circa 1914 contained bits and pieces of the theory) but a more complete description (in English) appeared later in 1926 by Hermann Glauert[8] who was summarizing Prandtl's contributions to aerodynamics.

The key elements of the Lanchester-Prandtl lift line theory are:

$$C_L = \frac{\partial C_L}{\partial \alpha}\alpha \tag{11.4}$$

In two dimensional flow, the lift curve slope ($^{\partial C_L}/_{\partial \alpha}$) is linear (up to the point of stall) and is given by:

$$\frac{\partial C_L}{\partial \alpha} = 2\pi \tag{11.5}$$

For a three dimensional wing, the lift curve slope is given by:

$$\frac{\partial C_L}{\partial \alpha} = \frac{2\pi A}{2 + A} \tag{11.6}$$

The final result from the lifting line theory is that the (induced) drag is related to the lift as:

$$C_{D_i} = \frac{C_L^2}{\pi A} \tag{11.7}$$

There are several important conditions or provisos in the results from the original lifting line theory that must be understood. They can be summarized as:

1. The circulation (and pressure distribution) is an elliptic distribution over an elliptic wing.
2. The profile drag must be added in order to obtain the total drag for the wing, such that:

$$C_D = C_{Do} + \frac{C_L^2}{\pi A} \tag{11.8}$$

The terms "*profile drag*" and "*induced drag*" took some time to be accepted terms in the aeronautical community. Prandtl and his assistants and other researchers originally referred to *induced drag* as "edge drag" and "additional drag" and it was not until much later that the term "induced drag"[c] was used to reflect the fact that the downwash from the wing caused by the vortex action caused the lift vector to tilt back by half the downwash angle to "induce" a drag given by Equation 11.7.

11.1.2 Modifications to Lifting Line Theory for Finite Wings

Immediately following the publication of the Lanchester-Prandtl Lifting Line Theory, several researchers began to tackle the problem of different wing planforms and other

[c] A term attributed to (Michael) Max Munk, one of Prandtl's renowned students.

geometric properties such as wing airfoil shape. Albert Betz (1885–1968), a close assistant to Prandtl at Göttingen, was the first to develop expressions for rectangular wings. Later, Glauert proposed changes to the lifting line theory to handle both rectangular and trapezoidal wings in his 1926 book based on earlier publication in 1921[9]. Glauert's results showed that the lift curve slope would *decrease* slightly and the induced drag would *increase* depending on the aspect ratio ($A = b^2/S$) over that given by the elliptic wing. Glauert's results were expressed in graphical terms.

It became common practice to rewrite the drag equation (sometimes called the *drag polar*) as:

$$C_D = C_{Do} + \frac{C_L^2}{\pi A e} \tag{11.9}$$

In this expression, a factor "e" later called the *span efficiency factor* has been included in the drag equation. This factor allows the aspect ratio (A) to be modified to give an *effective aspect ratio* to reflect the difference between actual wings pressure distribution and the assumed elliptic distribution in the lifting line theory. Various researchers have provided experimental values of "e" for their different wing shapes and this will not be expanded upon here in the context of examining WIG craft. Generally it is found (for OGE flight) that the value of "e", which has a value of $e = 1$ for elliptic wings with elliptic circulation (and lift) distribution and thus minimum induced drag, has values that cover a range $0.70 < e < 0.90$ depending on the geometry of the rectangular or trapezoidal wing and on aspect ratio.

In 1941, Robert T Jones[10] at Langley Aeronautical Laboratory (now Langley Research Center) in Virginia, US developed a theory to modify the Prandtl lifting line theory to give a better comparison with experimental results for elliptical wings and in a separate publication in 1941 for rectangular wings[11]. This correction (for rectangular wings) was to the lift curve slope, such that[d]:

$$C_L = \left(\frac{2\pi A}{2 + \xi A}\right)\alpha \tag{11.10}$$

Jones suggested that the correction factor (ξ) could be approximated by simply relating the total combined length of the wing span and chord ($b + c$) to the span (b) itself. This can be written as:

$$\xi = \frac{b + c}{b} = 1 + \frac{1}{A} \tag{11.11}$$

This then gives the revised form for the lift coefficient:

$$C_L = \left(\frac{2\pi A}{A + 3}\right)\alpha \tag{11.12}$$

In which case, the lift curve slope (for rectangular wings)[e] is given as:

[d] This result also appears in Chapter 8 when discussing Jones results applied to hydrodynamic planing hulls but with different notation to that used in this chapter.

[e] The original theoretical work by R T Jones showed that the number "3" in Equation 11.13 was "2.5" but Jones found that a better agreement with experimental results was achieved with the form as shown here.

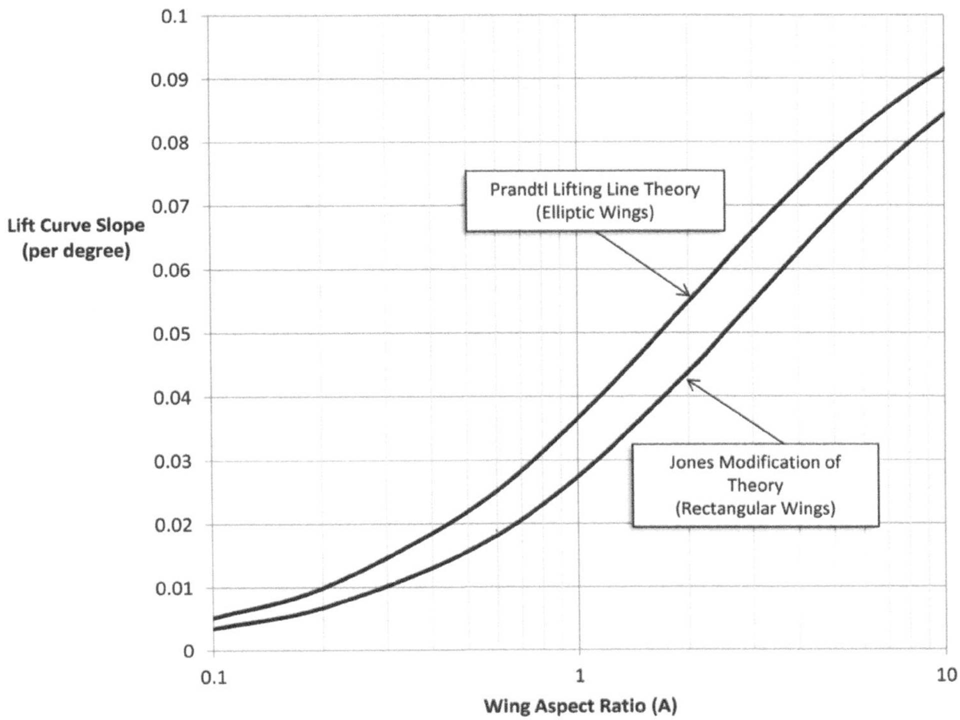

Figure 11.5: Lifting Line Theories for Lift Curve Slope (OGE)

$$\frac{\partial C_L}{\partial \alpha} = \frac{2\pi A}{A+3} \tag{11.13}$$

A comparison between Prandtl's result and Jones modification for the lift curve slope ($^{\partial C_L}/_{\partial \alpha}$) is shown in Figure 11.5:

All of the original work tended to be concentrated on the higher aspect ratios ($A>6$) more applicable to aircraft flying out-of-ground effect (OGE). Because of the interest here in wing-in-ground effect craft (WIG), interest is more in lower aspect ratios ($0.50<A<4.0$) and the original curves by Prandtl and Jones have been extended to the lower values in Figure 11.5. Also, for convenience, the lift curve slope is given here "per degree" rather than "per radian" as given in the original theoretical work.

Earlier, Glauert also had tackled adapting Prandtl's lifting line theory to rectangular wings and produced slightly different results to those given by R.T.Jones. Glauert produced[12] his results in graphical form and in a different form than Jones, modifying both the *lift curve slope* and the *span efficiency factor*. Glauert's result for the lift curve slope (for rectangular wings) was not that different numerically from Prandtl's result (for elliptical wings), especially for the lower aspect ratios ($A<6$) and is not reproduced here. There was a difference, however, for the span efficiency factor (e), as shown in Figure 11.6.

It is noticed that for wings of lower aspect ratio, more applicable to WIG craft, the effect of plan form on the span efficiency factor (e) is not that marked. Both results by Glauert and Jones show that for geometries different than the elliptic wing

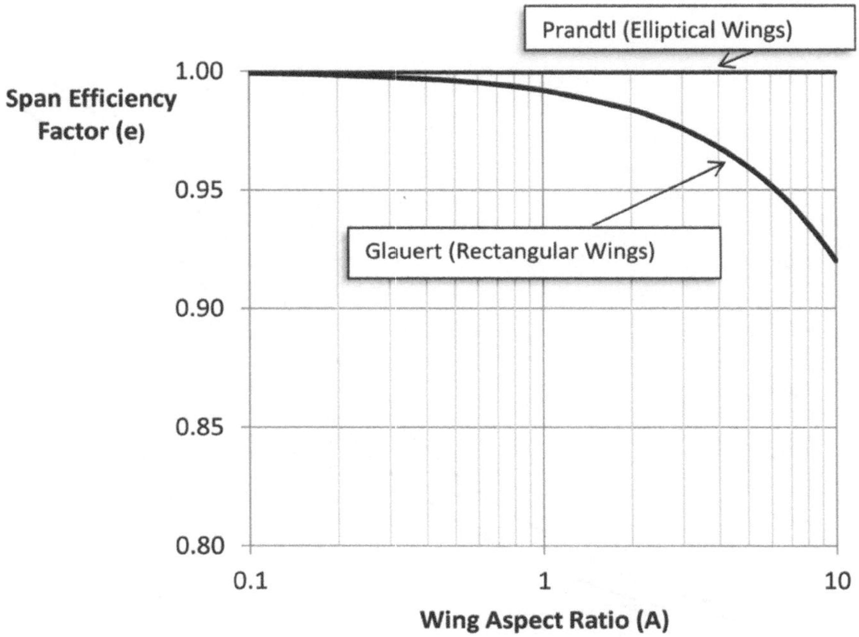

Figure 11.6: Lifting Line Theory for Span Efficiency Factor (OGE)

show *less* lift (per degree) and *more* induced drag. As a numerical example, from Figure 11.5, for a wing of aspect ratio ($A = 1$) the lift curve slope of a rectangular wing is approximately 75% of that for the minimum induced drag elliptic wing. Based on the theoretical results, the span efficiency factor is not that affected by aspect ratio (for $A < 3$). However, experimental data would indicate that a large reduction in efficiency would be experienced. Because of the large data base developed by NACA[f] and others, this is not reproduced here but variations in rectangular and trapezoidal wings with different thickness wing section geometries, etc., give values of the span efficiency factor between $0.70 < e < 0.90$ and accordingly, much higher induced drag (C_{Di}).

11.1.3 Adaption of Lifting Line Theory to Wing-in-Ground-Effect (WIG)

About the time that Prandtl and his colleagues were developing the lifting line theory, powered flight was becoming more common. Pilots were reporting that they experienced *increased lift* and *reduced drag* as their aircraft (wings) approached the ground. Prandtl assigned two of his most able assistants (Albert Betz and Carl Wieselsberger) to analyze this phenomenon. Betz concentrated on monoplanes and Wieselsberger worked on both monoplanes and biplanes (the preferred geometry at the time). Betz published his findings in 1912[13] and Wieselsberger published his work in 1921[14]. The reader is referred to the original work by Wieselsberger but some general results are summarized here. Wieselsberger used several results taken

[f] NACA: National Advisory Committee for Aeronautics (now, since 1958, the National Aeronautics and Space Administration (NASA).

from Prandtl's 1919 findings, and included in his 1921 report. Specifically, working from Wieselsberger's paper (but with current notation), the induced drag coefficient (C_{Di}) can be written:

$$C_{Di} = (C_{Di})_{OGE} + (\Delta C_{Di})_{IGE} \tag{11.14}$$

In this expression[g]:

$$(C_{Di})_{OGE} = \frac{C_L^2}{\pi A} \tag{11.15}$$

The *change in induced drag* due to flying in ground effect (IGE) is given by Wieselsberger as:

$$(\Delta C_{Di})_{IGE} = -\sigma \frac{C_L^2}{\pi A} \tag{11.16}$$

This then gives for the *induced drag in ground effect* of the wing as:

$$C_{Di} = (1 - \sigma) \frac{C_L^2}{\pi A} \tag{11.17}$$

The modifier (σ) in Equation 11.16 is defined as an *influence coefficient* for which Prandtl had developed an approximation, based on the height of the wing above the ground (measured at the quarter chord point), to be:

$$\sigma = \frac{1 - 1.32^h/_b}{1.05 + 7.4^h/_b} \tag{11.18}$$

In Wieselsberger's 1921 paper, the equation for the influence coefficient (σ) was expressed in terms of a value of (h) that is *twice* the actual height because of the method of using mirror images in the derivation. In Equation 11.18, the height is the *actual* height of the wing above the ground. Prandtl cautioned that this result could not be extrapolated to values out-of-ground effect (OGE) and also recommended that the approximation be limited to the height above the ground to be no lower than values of $^h/_b$ to be 1/30–1/4. This recommended limit was due to consideration of the viscous effects from the boundary layer not accounted for in the inviscid theory by Wieselsberger.

It is important to note that Wieselsberger was developing a relationship for the *change* in induced drag (ΔC_{Di}) and not the *induced drag* (C_{Di}) itself. The minus sign in his paper should be the clue, but this is frequently missed in later papers that quote Wieselsberger's work. This becomes clear if one plots Prandtl's approximation for the influence coefficient (σ) from Equation 11.18, as shown in Figure 11.7.

Upon inspection of Figure 11.7 and Equation 11.18, it is immediately obvious that the induced drag would not *decrease* as the height above the ground is *increased*. Hence, the more usual form is to show the *induced drag factor*, as shown in Figure 11.8. For reference, the two recommended limits for the applicability of the *influence coefficient* (σ) and *induced drag factor* (1–σ) have been superimposed on both Figure 11.7 and Figure 11.8. It is this latter form of the influence of the ground

[g] Variations on the OGE form of induced drag have already been discussed earlier with the work by R. T. Jones, Glauert and others and may be substituted here for work on specific wing planform shapes.

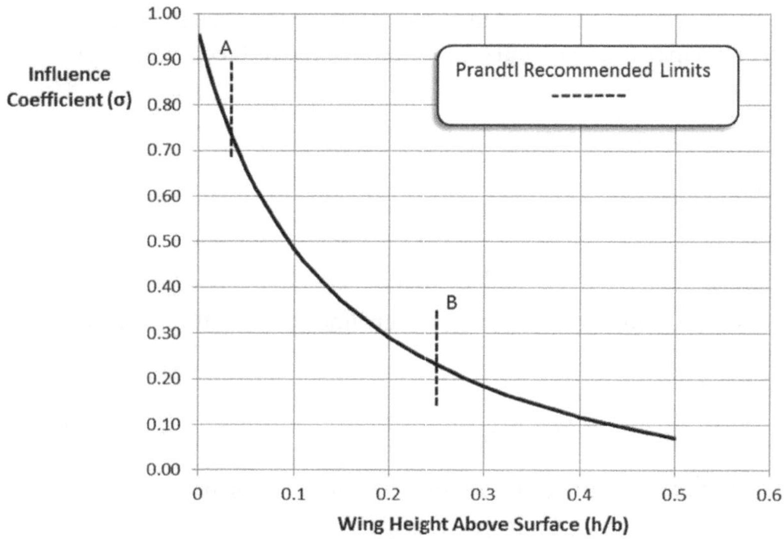

Figure 11.7: Prandtl's Approximation for Influence Coefficient (σ)

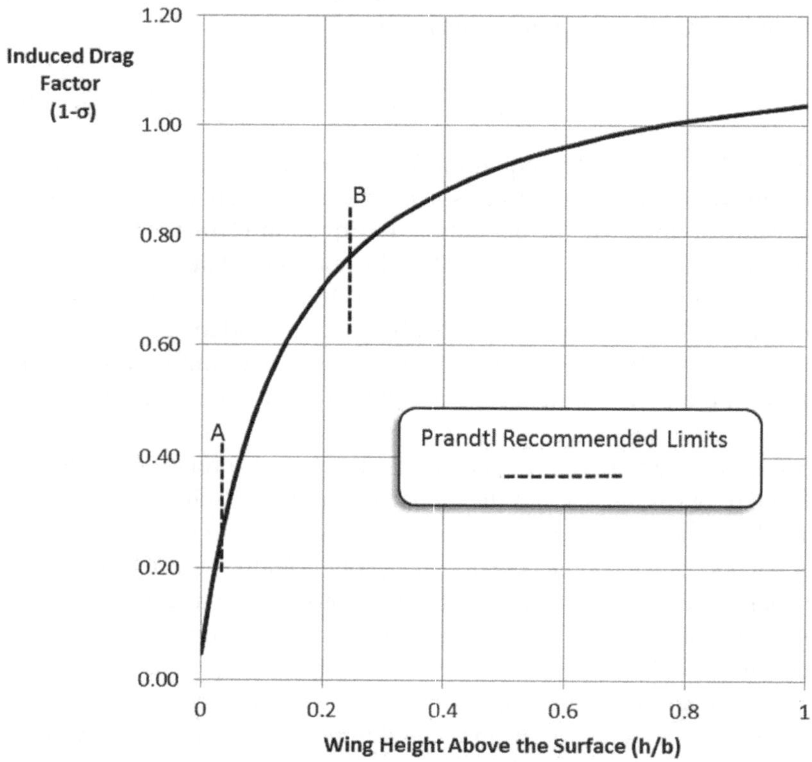

Figure 11.8: Prandtl's Induced Drag Factor ($1-\sigma$) in Wieselsberger Theory

on induced drag that appears most often in the literature, although frequently mislabeled as noted.

Wieselsberger took this result and applied it to a monoplane wing in the Göttingen wind tunnels and compared the drag polar for the model both in ground effect (IGE) and out of ground effect (OGE). This drag polar, Figure 11.9, is shown for the particular case of A=9.09 and $^h/_b = 0.242$ (using Wieselsberger's values). This result showed a significant reduction in induced drag (C_{Di}) under these conditions. An indication of the reduction in induced drag experienced by a wing approaching the ground can be obtained using Prandtl and Wieselsberger result for the induced drag as a function of height above the ground shown in Figure 11.8 using Equation 11.17 and Equation 11.18

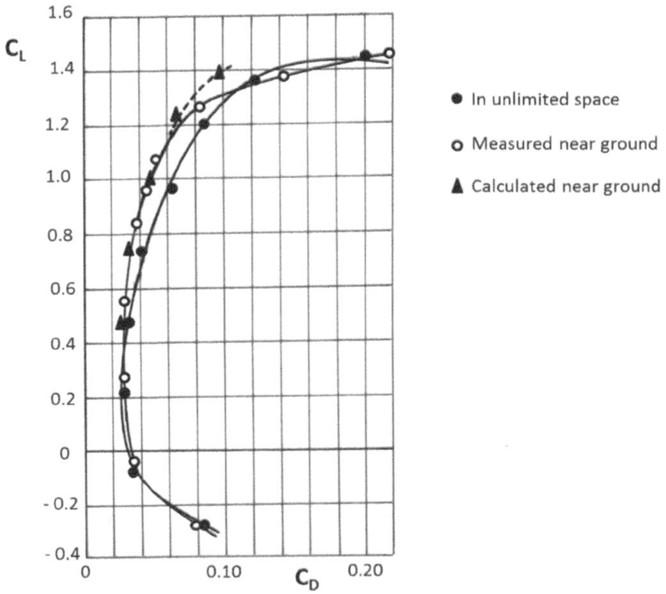

Figure 11.9: Wieselsberger's Drag Polar of a Wing operating in Ground Effect (1921)

By way of a numerical example, the Wieselsberger theory would predict that a wing close to the ground with a height to span ratio ($^h/_b = 0.04$) would experience an induced drag coefficient (C_{Di}) some 30% of the induced drag value out of ground effect. For height values above ($^h/_b = 0.5$) the theory suggests very little reduction in induced drag and the influence of the ground on lift and drag characteristics.

The work by Prandtl and Wieselsberger circa 1921 is probably the first to quantify the beneficial effects of ground effect but the data was limited and not all the conditions from the wind tunnel work were documented in this original work. It was not until the 1960s did more complete documentation appear on verification of Prandtl's and Wieselsberger's original theories of wings in ground effect (see later).

11.2 Simple Theory for Lift and Drag of Wings In Ground Effect (IGE)

There has been no published rigorous theoretical treatment of wings operating in ground effect (IGE) to provide guidance in the choice of airfoil shape, wing shape or operating height; but, with some simplifying assumptions, it is possible to derive a simple theory suitable for design use. It is reasonable to assume that the characteristics of such wings would asymptotically approach the out of ground effect (OGE) values as the wing moved higher above the ground. It is also reasonable to assume that the basic aerodynamic characteristics of the airfoil would remain relatively unchanged between operating IGE and OGE. With these assumptions, the lift and drag (coefficients) can be written:

$$C_L = C_{Lo} + \frac{\partial C_L}{\partial \alpha}\alpha \tag{11.19}$$

$$C_D = C_{Do} + C_{Di} \tag{11.20}$$

The choice of airfoil section (Glenn Martin, Clark, Göttingen,. . ..) will determine the basic lift and drag coefficients (C_{Lo}, C_{Do}) shown in Equation 11.19 and Equation 11.20. The lift curve slope $\left(\partial C_L/\partial\alpha\right)$ is then assumed to be the OGE value plus a term inversely proportional to the height above the ground so that it approaches the OGE value at some height determined by the specific wing characteristics. In which case, the equation for the lift curve slope can be written:

$$\frac{\partial C_L}{\partial \alpha} = \left.\frac{\partial C_L}{\partial \alpha}\right|_{OGE} + \frac{K_1}{\left(h/c\right)^n} \tag{11.21}$$

The constants (K_1, n) would be determined by experimentation and choice of airfoil. If it is taken that the wing planform would be rectangular in shape then it would reasonable to assume that the RT Jones value for lift curve slope given earlier would be applicable, such that Equation 11.21 can be written:

$$\frac{\partial C_L}{\partial \alpha} = \frac{2\pi A}{A+3} + \frac{K_1}{\left(h/c\right)^n} \tag{11.22}$$

It is also postulated, based on the work by Wieselsberger and Prandtl discussed earlier, the drag coefficient of the wing can be written:

$$C_D = C_{Do} + K_2(1-\sigma)_{NEP}\frac{C_L^2}{\pi A} \tag{11.23}$$

After examination of Wieselsberger results (see Figure 11.9) a constant (K_2) has been included in Equation 11.23 to retain the same functional form but also to account for the minor differences between experimental data and theory. Fortunately, there are some reliable model tests conducted on wings in ground effect by NASA in 1961 to test these simple theory predictions for lift and drag in ground effect.

11.3 NASA Wind Tunnel Tests of Wing-in-Ground-Effect with No End Plates

NACA had conducted considerable systematic research into the characteristics of wings operating close to the ground throughout the 1940s and into the 1960s and later. In 1961 (after NACA had become NASA) two key sets of data were published that laid the foundation for much of the wing-in-ground-effect (WIG) research that followed. The two most relevant reports are by Marvin P. Fink and James L. Lastinger[15] who researched the characteristics of wings in ground effect in a wind tunnel; and Arthur W. Carter[16] who conducted tow tank tests on wings operating IGE over water both with and without end plates. Some indication of the effect of the ground on wing lift and drag can be extracted from the exhaustive research in the 1961 Fink & Lastinger work in the Langley wind tunnel. All model

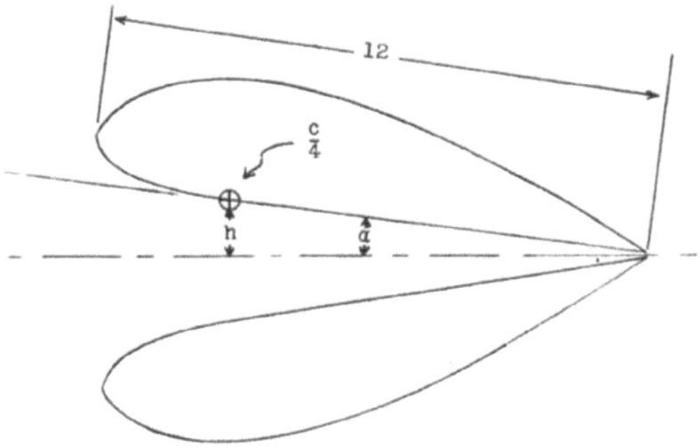

Airfoil coordinates

Station	Upper	Lower
0	8.86	8.86
1.25	12.21	6.03
2.50	13.81	4.79
5	15.98	3.33
7.50	17.65	2.35
10	18.92	1.72
15	20.72	0.75
20	21.68	0.28
30	22.13	0.00
40	21.13	0.00
50	19.20	0.00
60	16.64	0.00
70	13.35	0.00
80	9.43	0.00
90	5.00	0.00
95	2.57	0.00
100	0.00	0.00

Station and ordinates are in percent of airfoil chord

Figure 11.10: Model Wing Section for NASA Wind Tunnel Tests

work was conducted using a standard airfoil section Glenn Martin 21[17] wing modified to have a flat bottom over most of the chord from the 30% chord point to trailing edge (TE). A sketch of the model (with 12 inch chord) and section coordinates (from Fink & Lastinger's report) is shown in Figure 11.10. This is a 22% thick airfoil, which would be useful for a WIG designed to carry a large payload. The model test results while specific to this Glen Martin wing are representative of most wings likely to be used for WIG craft.

Later results for an 11% airfoil section will also be shown in the discussion of WIG with end plates. The model test set-up was to use an image wing as shown, to avoid the usual boundary layer problems of testing wings over a non-moving ground board. The height above the ground was measured at the quarter chord point to be close to the aerodynamic center. The tabulated ordinates are included in Figure 11.10. Before the model tests for the wing operating in ground effect are shown, it is important to understand that experimenters do not always use the same techniques. In this case, Fink and Lastinger, had chosen to express their heights above the ground board as a ratio to the chord (c) and not to the span (b) as used by Prandtl and Wieselsberger in their work. The relationship between the two expressions is:

$$\frac{h}{c} = \frac{h}{b}A \tag{11.24}$$

That is, the height to span $\left({}^{h}/_{b}\right)$ values used earlier must be multiplied by the wing aspect ratio (A) to get the height to chord $\left({}^{h}/_{c}\right)$ values used by Fank & Lastinger in their tests. Also, the height (h) is measured at the quarter chord point, which means that as the height (h) and angle of attack (α) are varied, the air gap at the trailing edge of the wing (TE) varies. To keep the comparison to be the least complex, the test results for the aspect ratio ($A = 1$) wing when tested in ground effect is selected as shown in Figure 11.11. For this particular case, $\left({}^{h}/_{c}\right)$ is numerically the same as $\left({}^{h}/_{b}\right)$.

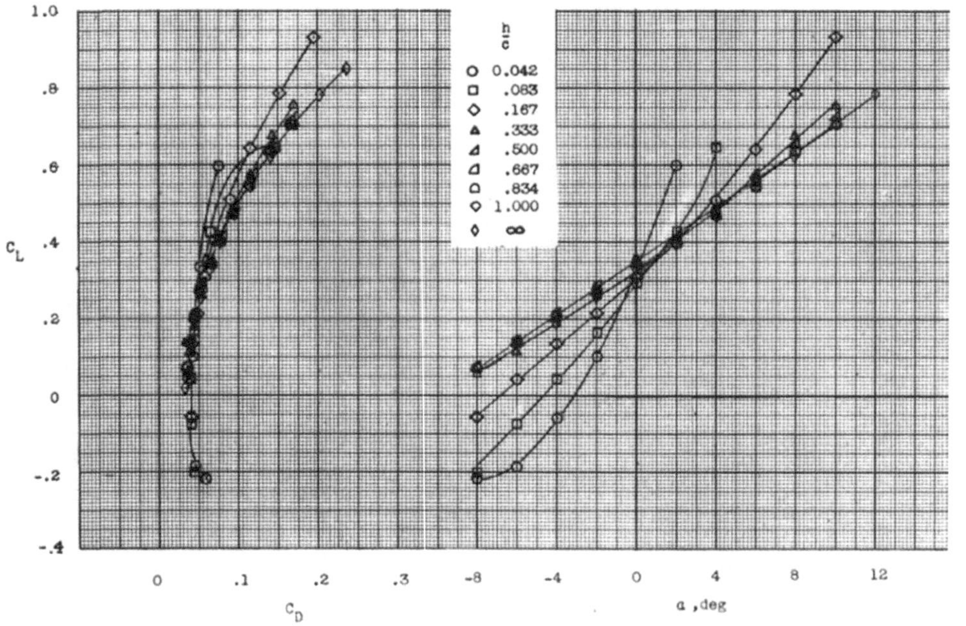

Figure 11.11: Lift and Drag of Wing-In-Ground-Effect (IGE) (A=1) (Fink & Lastinger)

Because most wing-in-ground-effect (WIG) craft tend to have aspect ratios around this value (although the *ekranoplan* used higher values) this is not seen as a detraction from the conclusions.

The complete set of results for all wing aspect ratios can be found in the original work by Fink and Lastinger in the cited reference. Some representative results are reproduced here. Figure 11.12 shows the lift and drag measurements for a set of aspect ratios (A=1, 2, 4 & 6) with the wing operating *out-of-ground-effect*. The drag polar, or C_L versus C_D, is shown on the left and lift (C_L) versus angle of attack (α) on the right of Figure 11.12. For reference, the lift curve slopes $\left(\partial C_L / \partial\alpha\right)$ as predicted for elliptic wings (Prandtl) and rectangular wings (Jones), given earlier, have been added in the lower right hand corner of Figure 11.12. It is seen that the comparison between these early predictions (1921) and the 1961 model test results is good.

While the data of the wing *in ground effect* (Figure 11.11) is not as simple as the data shown for *out-of-ground-effect* (Figure 11.12), the drag polar (on left of Figure 11.11) confirms that the induced drag is *reduced* over that for flight out of ground effect and the data on the right hand side of the Figure shows that the lift curve slope is *increased* for flight in ground effect. The form of the drag polar in the Fink and Lastinger results is seen to follow the original form given by Wieselsberger (see Figure 11.9). Both the effects contribute to increasing the lift-drag ratio ($^L/_D$) of the wing in ground effect.

Fink and Lastinger cross plotted their data to show this effect on the lift-drag ratio for all the models tested. Keeping wi1th the illustrative case of the wing of aspect ratio ($A = 1$), the results for lift-drag ratio are shown in Figure 11.13. In this particular case, the lower curve in Figure 11.13 shows the lift-drag ratio in flight out of ground effect (OGE) with the maximum value seen to be $^L/_D = 5$ approximately. Referring back to

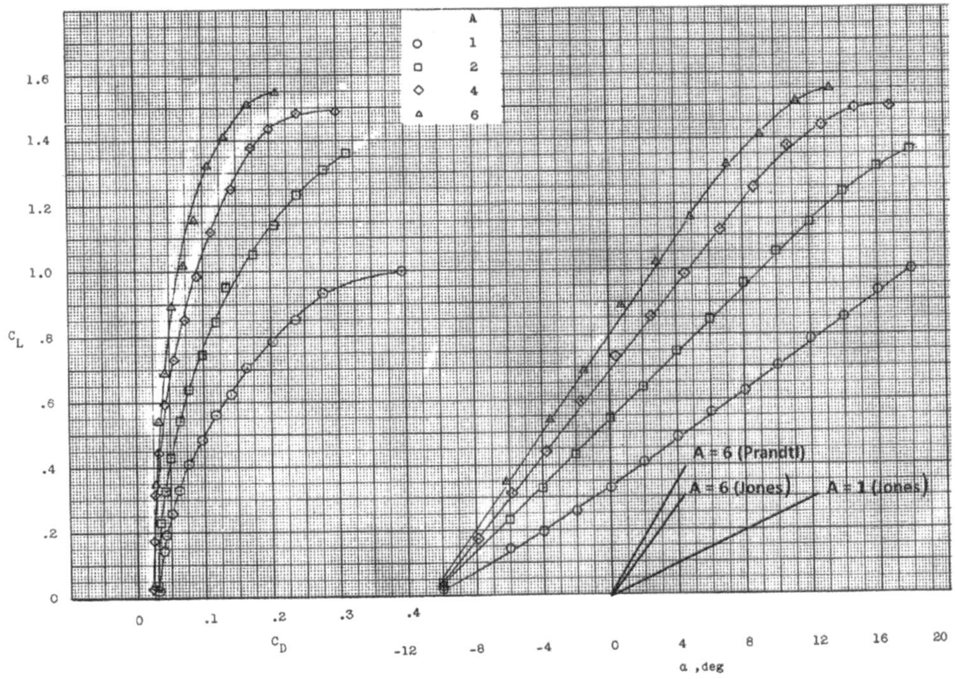

Figure 11.12: Lift and Drag of Wing Out-of-Ground-Effect (OGE) (A = 1–6) (Fink & Lastinger)

Figure 11.12, this occurs at a value of lift coefficient $C_L = 0.35$, a drag coefficient of $C_D = 0.07$ and at an angle of attack, $\alpha = 0.5$ degrees. As the wing approaches the ground, the lift-drag ratio increases and occurs at different wing conditions of lift coefficient and different angles of attack. By way of numerical example, if the height to chord ratio is $^h/_c = 0.083$, the lift-drag ratio increases to a value, $^L/_D = 6.8$ which occurs at a lift coefficient $C_L = 0.56$, a drag coefficient, $C_D = 0.08$, and at an angle of attack, $\alpha = 3.8$ degrees (see Figure 11.11). Using the same data for this aspect ratio, ($A = 1$) wing, if the height is further reduced to $^h/_c = 0.042$, the lift-drag ratio is even further increased to a value, $^L/_D = 7.8$, which occurs at a lift coefficient value of $C_L = 0.60$, a drag coefficient $C_D = 0.077$, and at an angle of attack of $\alpha = 3.9$ degrees.

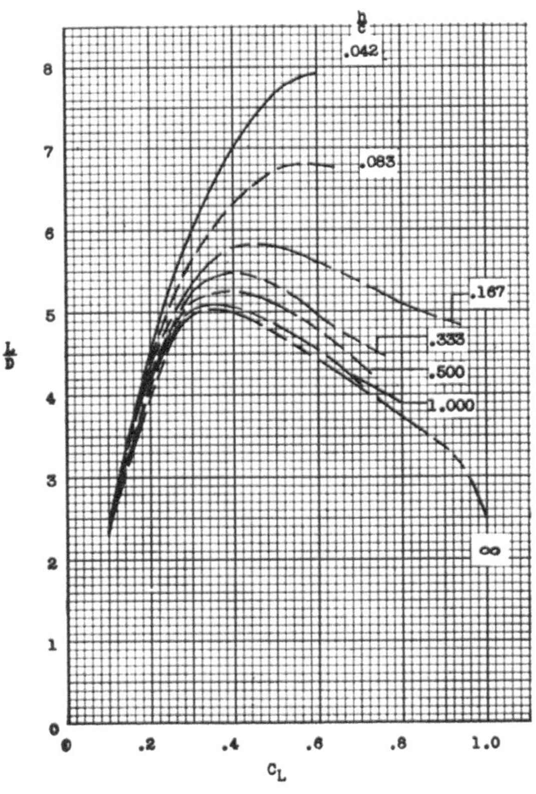

Figure 11.13: Lift-Drag Ratio of Wing In Ground Effect (A=1) (Fink & Lastinger)

It is noticed that these changes in lift and drag as the ground is approached are in general agreement with the predictions by Prandtl, Wieselsberger and others discussed earlier. In tabular form these results become clearer (see Table 11.1) for the operating conditions for $(^L/_D)_{MAX}$.

Given this backdrop of some of the NASA results for wings operating in ground effect, it now becomes possible to show the application of the earlier derived simple theory. First, it is recognized that the results (both theory and experimental) must start with the basic airfoil characteristics being used for the wing. The thickness of the wing affects the camber and the lift coefficient at zero angle of attack (C_{Lo}); the drag coefficient at zero lift (C_{Do}); and the lift curve slope (C_{La}) of the basic air foil. The lift curve slope of the two sets of data (t/c = 11% and 22%) is only lightly affected as shown in Figure 11.14.

Table 11.1: *Increased Lift-Drag Ratio in Ground Effect*

$^h/_c$	C_L	C_D	α (degrees)	$(^L/_D)_{MAX}$	L/D increase (IGE/OGE)
OGE	0.35	0.070	0.5	5.0	–
0.083	0.56	0.080	3.8	6.8	36%
0.042	0.60	0.077	3.9	7.8	56%

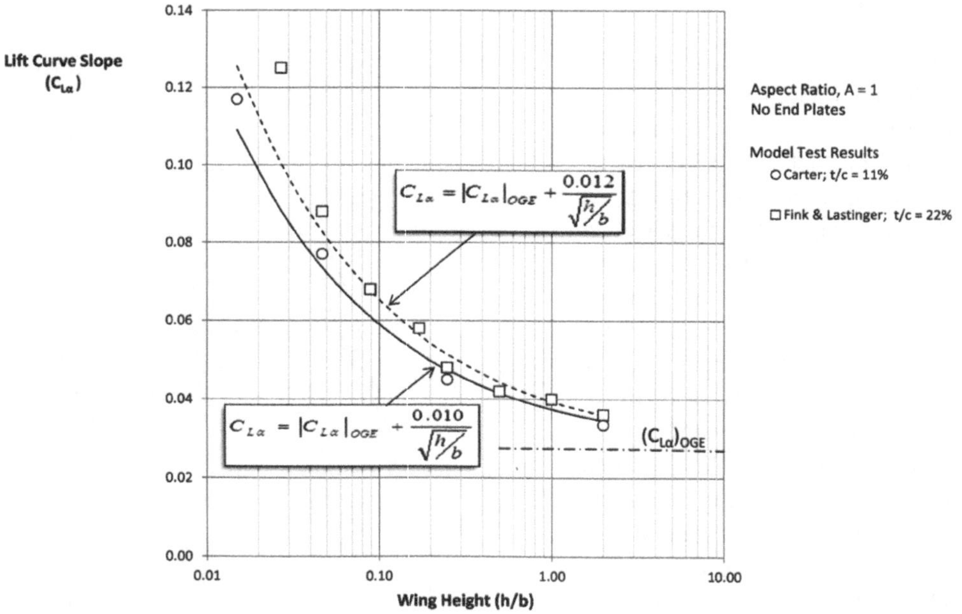

Figure 11.14: Effect of Airfoil Thickness on Lift Curve Slope In Ground Effect

The lift curve slope asymptotically approaches the Out-of-Ground-Effect (OGE) value as seen and the lift curve slope can be expressed as:

$$\frac{\partial C_L}{\partial \alpha} = \left|\frac{\partial C_L}{\partial \alpha}\right|_{OGE} + \frac{0.010}{\sqrt{h/b}} \qquad \text{for t/c} = 11\% \tag{11.25}$$

and

$$\frac{\partial C_L}{\partial \alpha} = \left|\frac{\partial C_L}{\partial \alpha}\right|_{OGE} + \frac{0.012}{\sqrt{h/b}} \qquad \text{for t/c} = 22\% \tag{11.26}$$

For the two airfoils used in the NASA tests, it is seen that there is about a 20% increase in the lift curve slope increment over the OGE value for this example case of a WIG with an aspect ratio A = 1. Such differences need to be monitored carefully when comparing data from different sources for the lift and drag characteristics of WIG craft. As stated earlier, the OGE value of lift curve slope will depend on the Aspect Ratio (A) and the planform shape of the wing (rectangular, trapezoidal, elliptical, etc) and can be calculated using the various formulae cited here. For the aspect ratio (A=1), it was found that the dependence on height was found to be in good agreement with the assumption that $n = 0.50$ in Equation 11.22. Figure 11.15 collects some of the NASA results for an aspect ratio (A = 1) wing together with the simple theory results. The prediction for the lift coefficient, see Figure 11.15(a), follows the test data well for the two selected angles of attack ($\alpha = 4$ and 10 degrees), i.e.:

$$C_L = C_{Lo} + \frac{\partial C_L}{\partial \alpha}\alpha \tag{11.27}$$

where

$$\frac{\partial C_L}{\partial \alpha} = \frac{2\pi A}{A+3} + \frac{0.012}{\sqrt{h/c}} \tag{11.28}$$

In this case, since A=1, the height parameters $h/_c$ and $h/_b$ are identical numerically. The lift coefficient at zero angle of attack (C_{Lo}) for this airfoil is taken to be $C_{Lo} = 0.33$, (see Figure 11.11).

The drag polar, see Figure 11.15(b), shows good agreement between the simple theory and test where the drag coefficient is given by Equation 11.23. The profile drag coefficient (C_{Do}) for the NASA wing model is taken as C_{Do} = 0.035 (see Figure 11.11). It was found that in comparing the simple theory with test, a constant K_2 = 0.80 gives a better agreement with the model data as shown. The influence coefficient (σ) is taken to be the value given in the Prandtl approximation (see Equation 11.18). The last set of curves in Figure 11.15, give the comparison between theory and test for the lift drag ratio:

$$\frac{L}{D} = \frac{C_{Lo} + {}^{\partial C_L}/_{\partial \alpha}\,\alpha}{C_{Do} + 0.80(1-\sigma)^{C_L^2}/_{\pi A}} \tag{11.29}$$

Note that the maximum value of the lift drag ratio ($^L/_D$) is given by:

$$\left(\frac{L}{D}\right)_{MAX} = \sqrt{\frac{\pi A}{4C_{Do}K_2(1-\sigma)}} \tag{11.30}$$

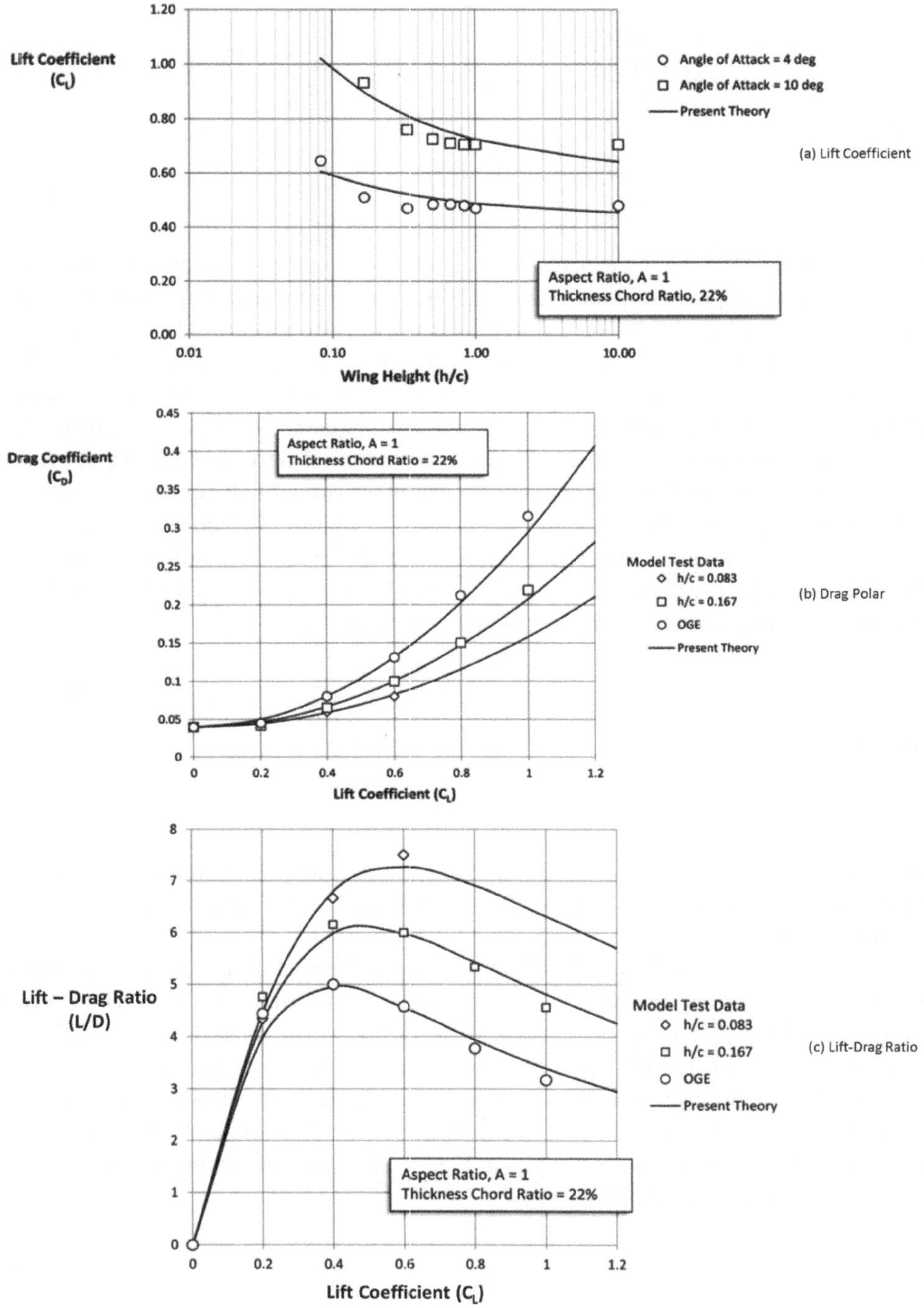

Figure 11.15: Lift and Drag Characteristics in Ground Effect (IGE) (Theory and Test) (A = 1) No End Plates (NEP)

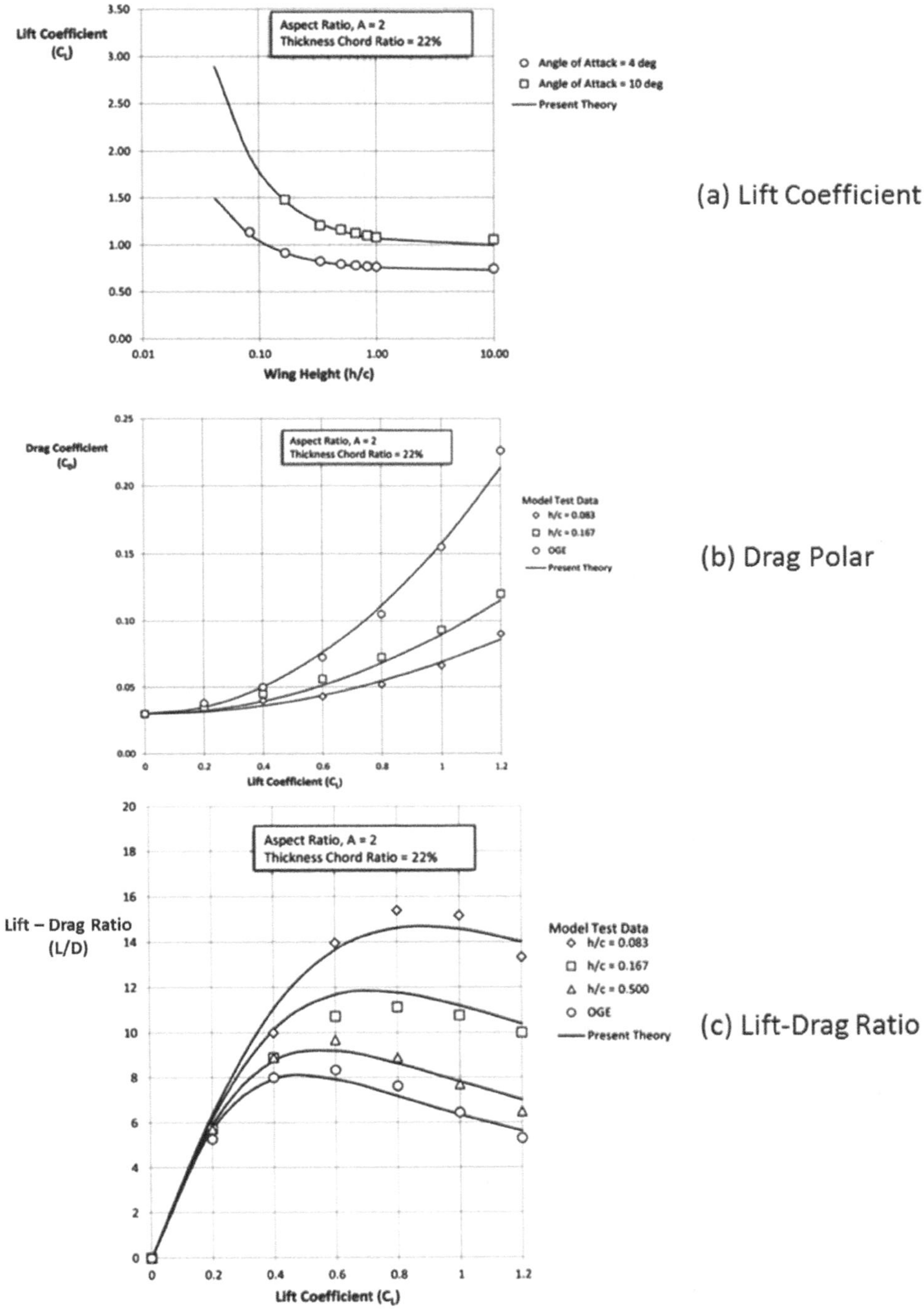

Figure 11.16: Lift and Drag Characteristics in Ground Effect (IGE) (Theory and Test) (A = 2) No End Plates (NEP)

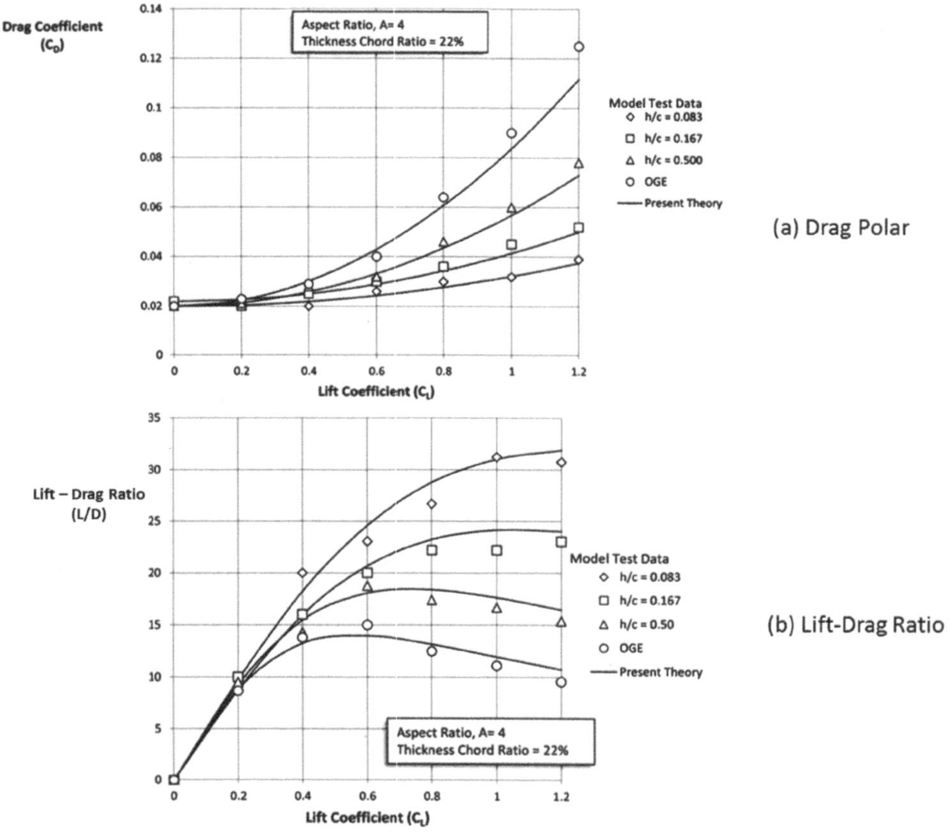

Figure 11.17: Lift and Drag Characteristics IGE (Theory and Test) (A = 4) (NEP)

The values of $C_{L\alpha}$, C_{Lo}, C_{Do} used in the calculation for $L/_D$ are as already provided earlier. It is seen that the simple theory provides a good agreement with the NASA data for this particular low aspect ratio wing (A = 1).

A similar comparison was done for a higher aspect ratio wing (A = 2) between theory and test and is summarized in Figure 11.16. As for the aspect ratio (A = 1) wing, a good agreement is found between the simple theory and the NASA model tests. From the model tests, it was noticed that the basic airfoil characteristics of (C_{Lo}, C_{Do}) also varied with wing aspect ratio and these variations are included in the simple theory results and again these variations of basic airfoil features must be taken into account when comparing the lift and drag characteristics of different vehicles and their wing geometries.

The simple theory developed here was exercised on all of the Fink & Lastinger data for a wide range of heights and aspect ratios and good agreement was found in all cases. An additional sample set for an aspect ratio A = 4 wing is included in Figure 11.17. Just the drag polar and lift drag ratio results are included here for illustration. Again, it is seen that the agreement is good. Some of the scatter in the test results displayed in Figure 11.15, Figure 11.16 and Figure 11.17 is caused more by the difficulty of reading the small charts in the original reference on test results and the faired nature through the original points rather than any fundamental aerodynamic effect. The trends are clear and well within the accuracy of the theory.

11.4 Effect of Aspect Ratio on Lift Curve Slope

One result found in the original NASA data that was not explained in the presentation of model test results was the non-linearity between the lift curve slope of the wing of aspect ratio A = 1 and the rest of the (higher) aspect ratio wings tested. This non-linearity can be seen in Figure 11.18 where the results of the various theories and the model tests on the lift curve slope are compared. The results are for the case of angle of attack, $\alpha = 0$ degrees. While the results are quite pronounced, it is possible to express the trends in two distinct groupings, as follows:

$$\frac{\partial C_L}{\partial \alpha} = \left|\frac{\partial C_L}{\partial \alpha}\right|_{OGE} + \frac{0.012}{\sqrt{h/_c}} \text{ for } A \approx 1 \tag{11.31}$$

$$\frac{\partial C_L}{\partial \alpha} = \left|\frac{\partial C_L}{\partial \alpha}\right|_{OGE} + \frac{0.008}{h/_c} \text{ for } A \geq 2 \tag{11.32}$$

This variation in the lift curve slope does not appear immediately upon examination of the results shown in Figure 11.15, Figure 11.16 and Figure 11.17 it is quite pronounced as seen in Figure 11.18. This difference in lift curve slope increases as the gap between the wing and the ground is decreased to very low values. Both Prandtl and Wieselsberger allude to simple theories not being applicable at very low heights because of viscous effects and other flow disturbances not reported by any of the later experimenters. Pending a more in depth research into the flow patterns around wings of low aspect ratio ($A \approx 1$), the simple theory results given by Equation 11.31 and Equation 11.32 provide a good approximation to the characteristics of such wings.

From Figure 11.18, it is seen that the lift curve slope for A=1 follows the general trend from OGE down to wing heights about h/c ≈ 0.40, after which it follows the result given by Equation 11.31. The lift curve slope for wings of $A \geq 2$ follows more closely the result in Equation 11.32.

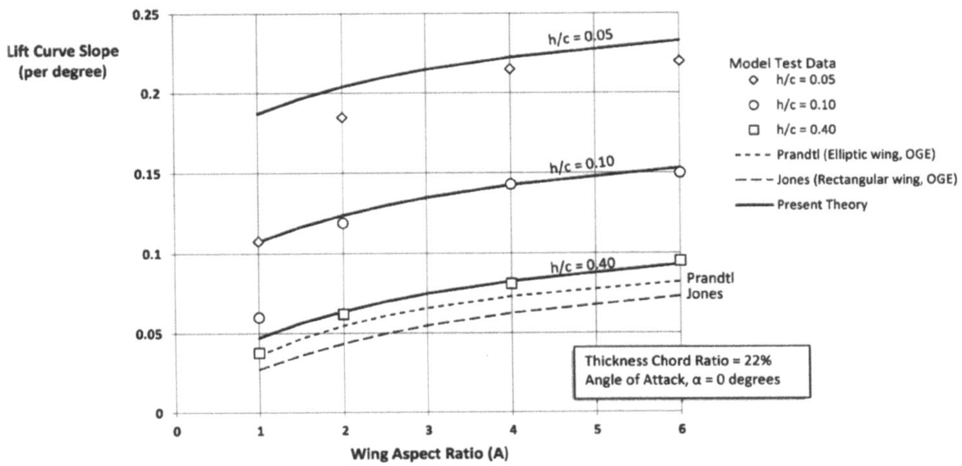

Figure 11.18: Effect of Aspect Ratio and Height on Lift Curve Slope In Ground Effect

11.4.1 Effect of Aspect Ratio on Lift Drag Ratio in Ground Effect

In Figure 11.15, Figure 11.16 and Figure 11.17, it is seen that the maximum value of lift drag ratio for the wing's $(L/D)_{MAX}$ increases as the ground is approached and further occurs at increasing values of the lift coefficient. If a cross plot is made of all the Fink & Lastinger test results for $(L/D)_{MAX}$ for all the aspect ratio wings tested and for all the heights tested, a chart like Figure 11.19 is obtained. Such a chart provides an opportunity to guide the designer in the choice of wing aspect ratio and operating height. Figure 11.20 is a close up view of Figure 11.19 to aid in an example. Consider a base design of a wing has been set with an aspect ratio $A = 1.5$ and an operating height of, $h/c = 0.167$. This WIG is labeled as **WIG "A"** in Figure 11.20 and is seen to have a maximum value of L/D of, $(L/D)_{MAX} = 9.5$. If the designer changes the operating height to be $h/c = 0.070$, which is a 58% reduction in operating height, the maximum value of the L/D rises to a value of $(L/D)_{MAX} = 12$, which is a worthwhile performance improvement. This WIG is labeled **WIG "B"**. Alternatively, if the design is changed to have an aspect ratio, $A = 2$, which is a 25% increase in span, the same value of $(L/D)_{MAX} = 12$ can be achieved at the same original operating height of $h/c = 0.167$. This might be a better design choice for operating over waves at sea. This design is labeled as **WIG "C"** in Figure 11.20.

Before the WIG designer can make a final decision on the correct geometry, however, the beneficial effects of "end plates" must be considered.

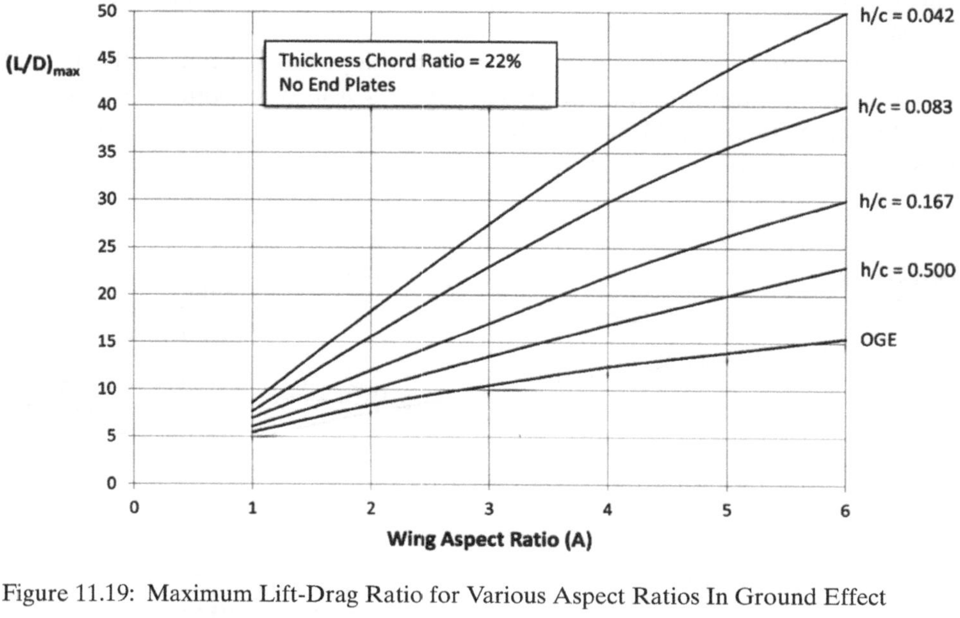

Figure 11.19: Maximum Lift-Drag Ratio for Various Aspect Ratios In Ground Effect

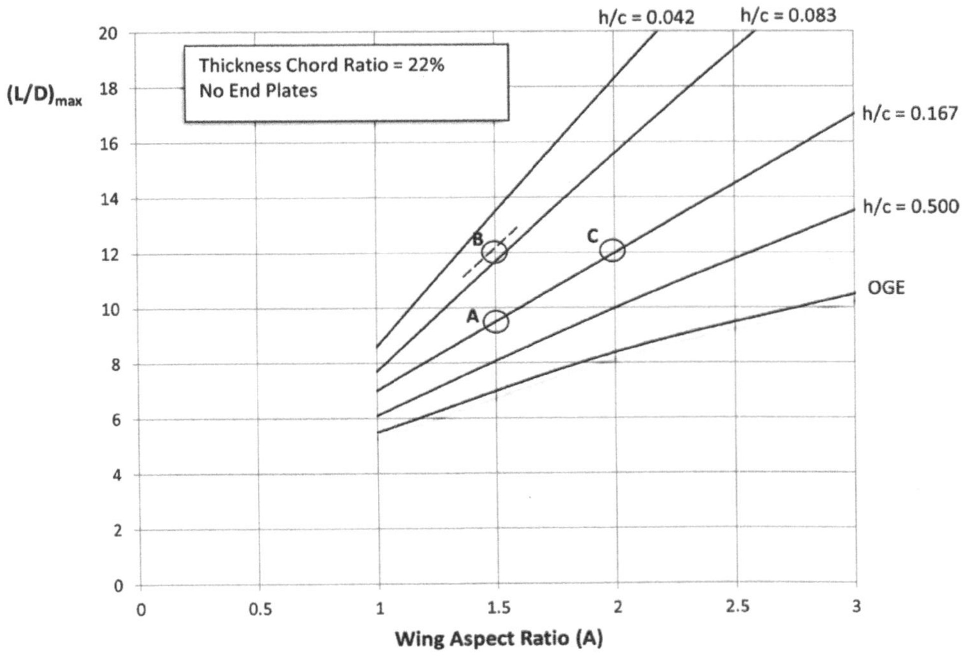

Figure 11.20: Example Design Trade to Achieve High Lift Drag Ratio in Ground Effect

11.5 Lift and Drag in Ground Effect with End Plates

About the same time (1961) that Fink & Lastinger were conducting their experiments in the Langley wind tunnel, another researcher, Arthur W. Carter (also from Langley), cited earlier, was conducting similar tests of wings in ground effect with end plates (shown as ram wings in Chapter 1, Figure 1.13b). These particular tests used the same airfoil sections (but with four times larger scale models) as used by Fink & Lastinger, and tested them over water in the Langley Tow Tank No. 1, instead of in a wind tunnel. Instead of the image method used by Fink & Lastinger with static models, Carter towed his models over the water, so again the issue of boundary layer interaction was minimized. An additional difference between to the two sets of model testing is that Carter used an 11% thick airfoil of the Glenn Martin 21 section when end plates were added. The ordinates were thus the same as used in the Fink & Lastinger tests (see Figure 11.10) except divided by 2 to give the thinner section.

A further difference between the two sets of testing is that Carter set the heights to be measured at the trailing edge (TE) of the wing, instead of at the quarter chord point used by Fink & Lastinger. This difference meant that the height to chord ratios tested were different from those used by Fink & Lastinger. While these are frustrating differences that hamper direct comparison, the general results are not obscured directly. The model with end plates is shown in the tow tank in Figure 11.21.

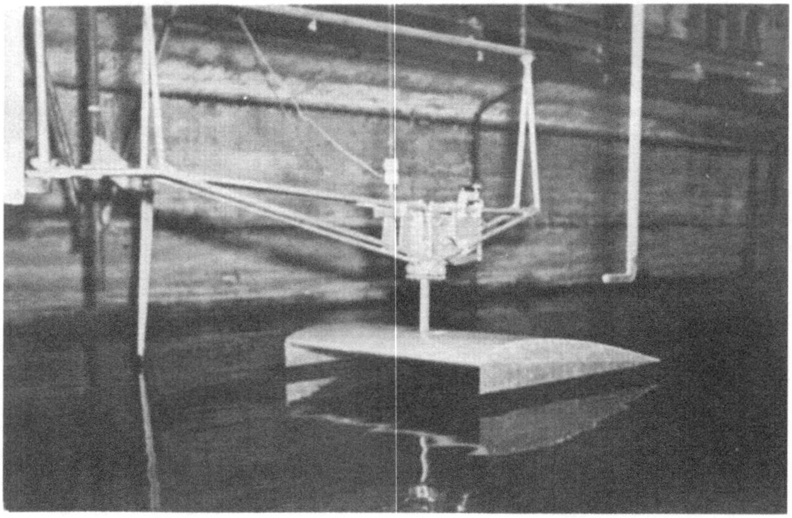

Figure 11.21: WIG Model with End Plates in Langley Tow Tank (1961)

The transformation between measuring the height at the trailing edge $\left({}^{h_{TE}}/_{c}\right)$ and measuring at the quarter chord point $\left({}^{h}/_{c}\right)$ is simply:

$$\frac{h}{c} = \frac{h_{TE}}{c} + {}^{3}/_{4} \sin \alpha \qquad (11.33)$$

As before, α is the angle of attack of the wing relative to the water. In each test, as the angle of attack was varied, a different end plate was used that ensured that the lower edge of the end plate was always flush with the trailing edge (TE) and parallel to the water as seen in the photograph. This transformation meant taking each data point at each angle of attack and then using Equation 11.33 to get the equivalent plot to compare with Fink & Lastinger's test results. This understandably introduced small errors when comparing the two sets of data. Figure 11.22 shows the geometric relationship between wing height, wing chord and angle of attack using Equation 11.33.

The reader is encouraged to read the full report to see test results for lift coefficient (C_L), drag coefficient (C_D) and also the moment coefficient (C_M)[h]. Because of the interest in the increased lift-drag ratio $\left({}^{L}/_{D}\right)$ for WIG operation, discussion has been concentrated here on this characteristic from the model tests. The model tests for the Fink & Lastinger work used the 22% thick airfoil section, and the work by Carter on models with end plates was done using the 11% thick airfoil section, it is important to see how these two versions of the airfoil section compare when operating in ground effect. Figure 11.23 compares the two sets of data.

[h] Discussion in this section has concentrated on the lift and drag characteristics. The aerodynamic moment, which aids in the determination of the center of pressure on the wing, and impact on stability, is considered later.

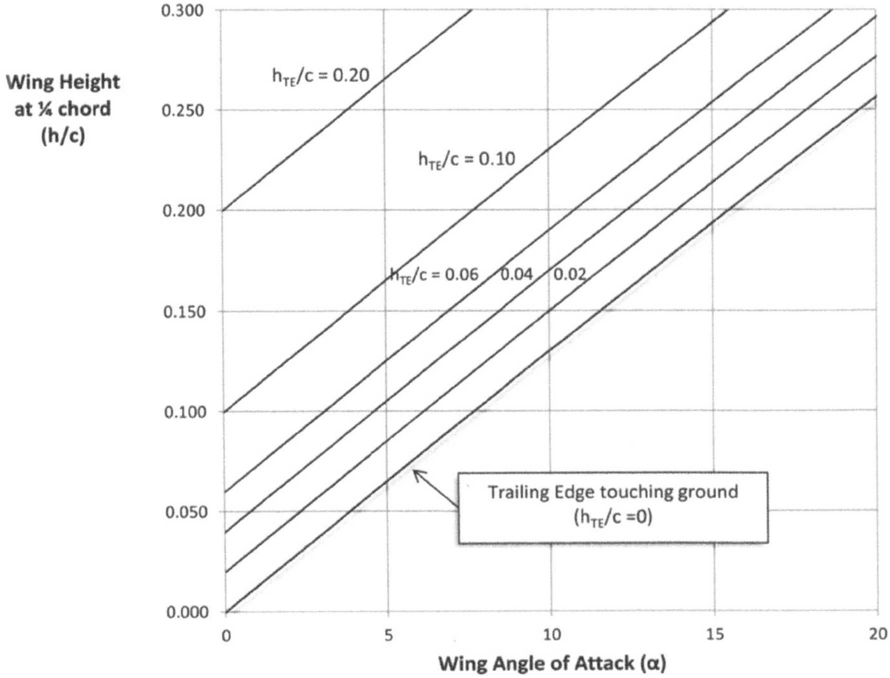

Figure 11.22: Wing and Trailing Edge Height at Angle of Attack

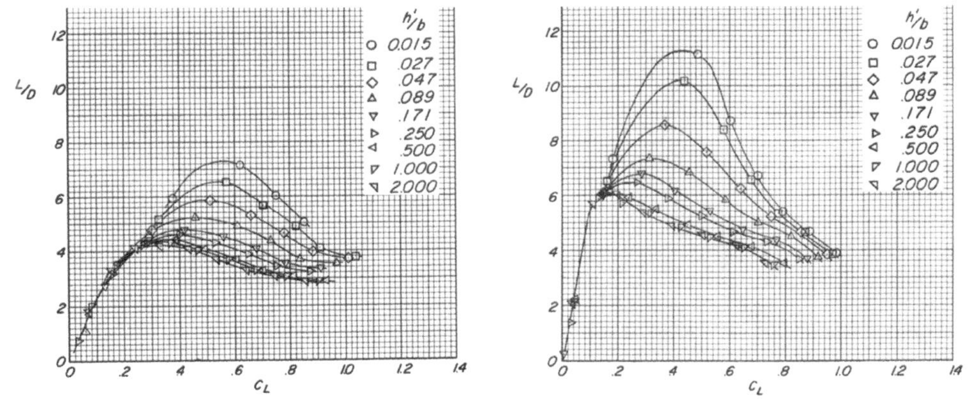

(a) Lift-Drag Ratio for 22% Thick Wing (Carter)

(b) Lift-Drag Ratio for 11% Thick Wing (Carter)

Figure 11.23: Comparison of 22% thick and 11% thick airfoils in Ground Effect ((A = 1) (No End Plates)

It is seen from Figure 11.23, that the thick wing (22%) has a lower lift-drag ratio ($^L/_D$) than the thin wing (11%). At the lowest ground clearance, the thin wing has about 55% higher lift-drag ratio. This distorts the comparison for the effect of end plates but it is believed that this difference can be rationalized in the final results. The difference between the results shown for the 22% wing measured in the tow tank (Carter) from that measured in the wind tunnel (Fink & Lastinger) was recognized by the researchers and attributed to the different Reynold's Number between the two sets of data and also the model set up. The difference was attributed to the increase in the profile drag (C_{Do}) and not the induced drag (C_{Di}).

The additional difficulty in comparing the results is that already mentioned in that the measurements for the height above the ground in the tow tank tests were taken at the trailing edge while in the tests in the wind tunnel, the heights were measured at the quarter chord point. The required correlation between the two height measurements can be obtained by Equation 11.33. These were frustrating differences in the comparison between the two sets of test data conducted by the same Research Center in the same year but the total set of data is highly valuable nevertheless.

The model test results for the 11% thick wing with end plates attached (see Figure 11.21) have been collected in Figure 11.24. The data in the left hand side of Figure 11.24 is for the measured heights taken at the wing trailing edge ($^{h_{TE}}/_{b}$). The conversion to heights at the quarter chord point ($^{h}/_{b}$) are shown on the right hand side of Figure 11.24. Note that these tests are for a wing of aspect ratio $A = 1$ such that values of $^{h}/_{b}$ are numerically equal to values of $^{h}/_{c}$.

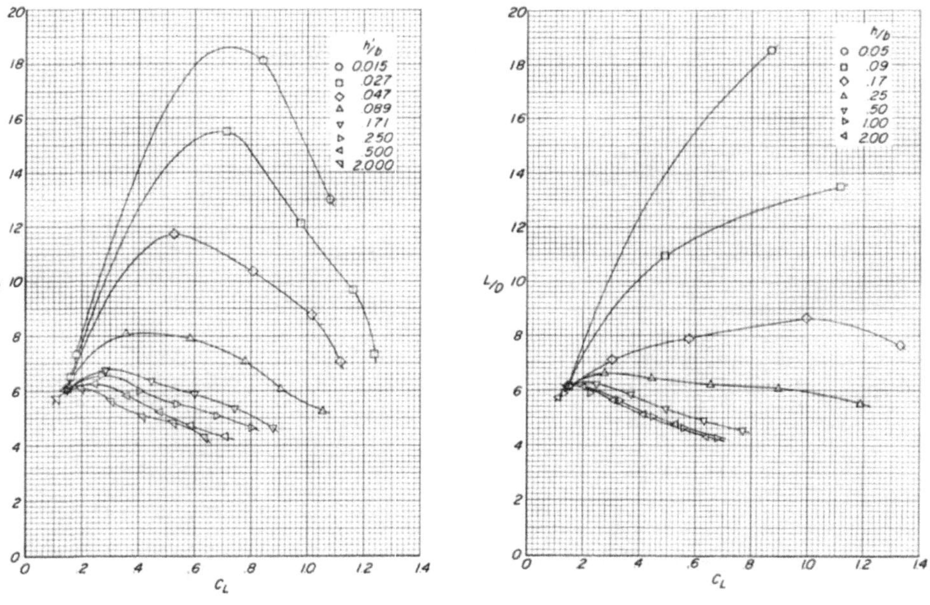

(a) Heights measured at Wing TE (b) Heights measured at Quarter Chord

Figure 11.24: Lift-Drag Ratio for 11% Thick Wing (With End Plates) ($A = 1$) (Carter)

A direct comparison showing the effect of the end plates can be made by comparing the data shown in Figure 11.23(b) and the data shown in Figure 11.24(a) for the 11% thick wing of aspect ratio $A = 1$, both with and without end plates. For ease of comparison, Table 11.2 shows the increase in lift-drag ratio with the addition of end plates using the data from Figure 11.23 and Figure 11.24.

Table 11.2: *Measured Increase in Lift-Drag Ratio with Ground Effect & End Plates*

Height at Wing TE ($^{h_{TE}}/_{b}$)	$(^L/_D)_{max}$ Without end plates	$(^L/_D)_{max}$ With end plates	% Increase in $(^L/_D)_{max}$
0.0015	11.2	18.6	66%
0.027	10.2	15.5	52%
0.047	8.6	11.7	36%
0.089	7.3	8.1	11%
0.171	6.8	6.8	-
0.250	6.4	6.5	-
0.500	6	6.2	-
2.000	6	6.1	-

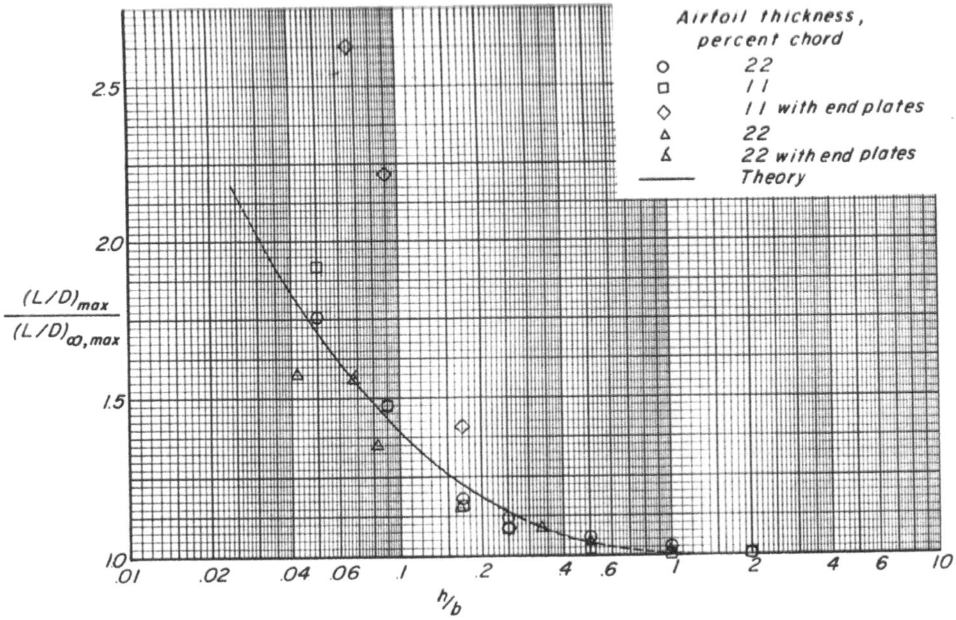

Figure 11.25: Maximum Lift-Drag Ratio as Ground Is Approached (Carter)

Carter cross-plotted his data together with those by Fink and Lastinger to show how the effect of ground proximity increases the lift drag ratio, as seen in Figure 11.25 for different wings and ground clearance.

This increase in lift-drag ratio as the ground is approached and shown in Table 11.2 is in general agreement with the theory by Wieselsberger (without end plates). It can be shown that the Wieselsberger theory, using Parndtl's approximation for the influence coefficient (σ), gives the ratio of the maximum lift-drag ratio in ground effect (IGE) to that out of ground effect (OGE) as:

$$\left[\frac{(^L/_D)_{IGE}}{(^L/_D)_{OGE}}\right]_{MAX} = \frac{1}{\sqrt{(1-\sigma)}} \tag{11.34}$$

Note that the data in Table 11.2 is expressed as a function of the height at the wing trailing edge $\left({}^{h_{TE}}/_{b}\right)$ while the data in Figure 11.24(b) is expressed as a function of the wing height at the quarter chord point $\left({}^{h}/_{b}\right)$ to be compatible with the theory by Wieselsberger and Prandtl. The height at the trailing edge would be less (using Equation 11.33). These results are significant as they represent the only consistent set of experimental evidence conducted under controlled conditions into ground effect.

One characteristic in the model testing technique used that is not immediately apparent in the Carter tests and that is the relationship between the end plate and the wing itself as the height was varied. As described above and seen in Figure 11.18, the end plate was always set to be flush with the wing trailing edge. This meant that as the height was varied, so was the air gap beneath the end plate, which would change the leakage or "sealing" effect at the wing tips to affect the wing tip vortices and thus the induced drag. No tests were conducted with the wing flying higher or lower and the air gap beneath the end plates maintained at some minimum value. The model results of the 11% wing with end plates at very low wing heights $\left({}^{h}/_{b} < 0.10\right)$ shown in Figure 11.24(b) show a strong increase in the lift-drag ratio. What is not known is how much that increase is due to the wing flying lower and how much is due to the air gap under the end plate being minimized. It would be useful to have a series of model tests done where this separation of the two effects could be measured. This was a point of contention in the US Navy ANVCE Project evaluation (see Chapter 10) where the USN PAR-WIG design extended the end plates into the water to achieve high sealing. The issue is: *Is the increased lift-drag ratio achieved by low (or zero air gaps) offset by the increased hydrodynamic drag incurred by dragging the end plates through the water?* In the Russian *ekranoplan* designs, the decision had been made that this was not a good design feature and ensured that the end plates on the *ekranoplan* designs were minimal (with high air gaps as the wing approached the water) and that it was more efficient to simply increase the span a little to increase lift (see Figure 11.20) and avoid the unsafe practice of dragging end plates through the water at high speed. A discussion on operating heights of WIG is given later.

11.6 Wings in Ground Effect with Air Seals

In seeking a design of a craft to travel at high speed over the water safely, it was recognized in 1959 during the US Maritime Administration program on an aerodynamic air cushion craft (see Chapter 3), that dragging any structure through the water was not a good idea. The *Columbia* design of a channel flow wing in ground effect (see Chapter 1, Figure 1.13d) recognized this problem and integrated the idea of an "air seal" instead of hard structural end plates to seal the wing tips vortex flow and minimize induced drag without any part of the hull or other structure dragging in the water. A complete description of that program and details of the design are given in Chapter 3 but some specific features of the *air seal* are repeated here in the context of the above described efforts on the use of *end plates*.

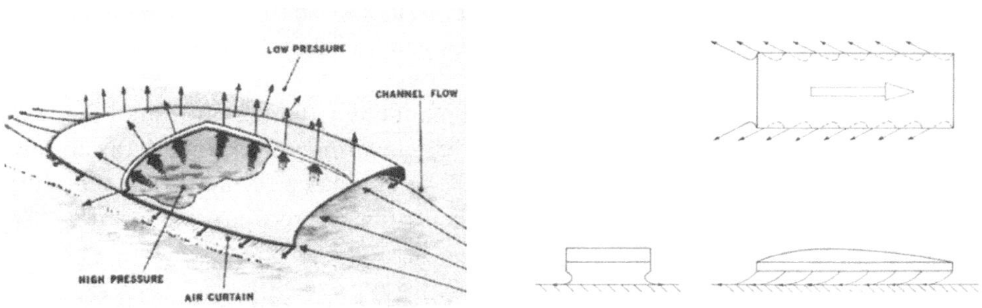

(a) Channel Flow System (b) Channel Flow Analytical Model

Figure 11.26: Channel Flow Concept

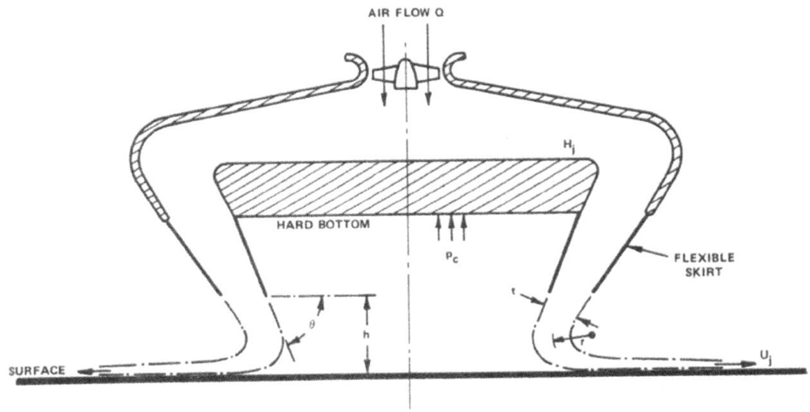

Figure 11.27: Peripheral Jet (Air Seal) Geometry

The channel flow air cushion craft is described in detail in Chapter 3 and the basic concept is reproduced from Chapter 3, Figure 3.2 in Figure 11.26. In cruise mode, the front and rear air seals are retracted and the side air seals remain "on" to provide the necessary sealing to ensure minimal induced drag.

The use of an air seal, called a "peripheral jet", was the original form used in the early hovercraft to seal the cushion air beneath the craft. Sir Christopher Cockerell's[18] original experiments in hovercraft design used the peripheral jet. There have been several developments (both theoretical and experimental) on the formulation of the peripheral jet characteristics and these have been summarized in the author's 1980 report[19] and only some of the key derivations will be included here. Richard Stanton-Jones presented his theory at the Princeton Symposium in 1959[20] but unfortunately the Appendix to his paper that contained the derivation was not included in the proceedings or circulated. Stanton-Jones presented the work again in 1961[21] in the Institute of Aerospace Sciences (IAS)[i] journal.

Figure 11.27 shows the governing geometry of the peripheral jet taken from the author's 1980 report. Flexible extensions or skirts, as used in the early hovercraft,

[i] IAS was the forerunner of the current AAIA (American Institute for Aeronautics and Astronautics).

have also been shown in Figure 11.27 for later discussion. In the sketch, the air flow is shown to be from a lift fan, but it could also be generated by engine bleed or similar air flow generation.

In the peripheral jet theory, the air jets emanating from the bottom of the craft (or flexible extensions if fitted) create an increase in air pressure beneath the craft by virtue of the centrifugal forces generated by the curvature of the jets as they turn away from the craft at the surface. These centrifugal forces are balanced by the pressure gradient across the jet, such that:

$$\frac{\partial p}{\partial r} = \frac{\rho u^2}{r} \tag{11.35}$$

In this expression, r is the local jet radius of curvature (see Figure 11.27) and u is the local jet velocity. Then it can be shown[22], upon integration, that the cushion pressure becomes:

$$p_c - p_o = (H_j - H_o)(1 - e^{-2x}) \tag{11.36}$$

In this relationship, p_c is the air cushion pressure under the craft, p_o is the ambient static pressure. H_j is the pressure source total head, and H_o is the total head of the free stream. The pressure source total head (H_j) could be generated either by lift fans as in the conventional hovercraft or it can be generated by directing propulsion engine air (compressor bleed or turbofan air) into the plenum. The jet geometry term in the exponential of Equation 11.31, is given by:

$$x = \frac{t}{h}(1 + \cos\theta) \tag{11.37}$$

This non-dimensional value of x includes the physical thickness of the jet (t), the height of the jet nozzle above the ground (h). The jet inclination is given by the angle θ. Note that the height h is the actual clearance beneath the jet nozzle and *not* the height or clearance of the hard bottom of the craft. This is an important consideration when determining power requirements of the craft with flexible extensions as shown in Figure 11.27.

The total head in the air jet, which is assumed to be equal to the total head at the source, is:

$$H_j = \frac{1}{t}\int_o^t p(r)dr + {}^1/_2 \rho u_j^2 \tag{11.38}$$

The variation of the static pressure $p(r)$ and the velocity distribution across the jet are functions of the jet geometry. Peter Payne provided a detailed treatment of these characteristics of an air jet in 1965[23] but the results are not summarized here of the pressure and velocity distribution across the jet.

While it is recognized that the pressure and velocity vary across the jet, Elsley[24] and Stanton-Jones[25] made the simplifying assumption that the average values would give a good approximation to experiment. This simplified momentum theory resulted in:

$$\frac{p_c - p_o}{H_j - H_o} = \frac{2}{1 + x} \tag{11.39}$$

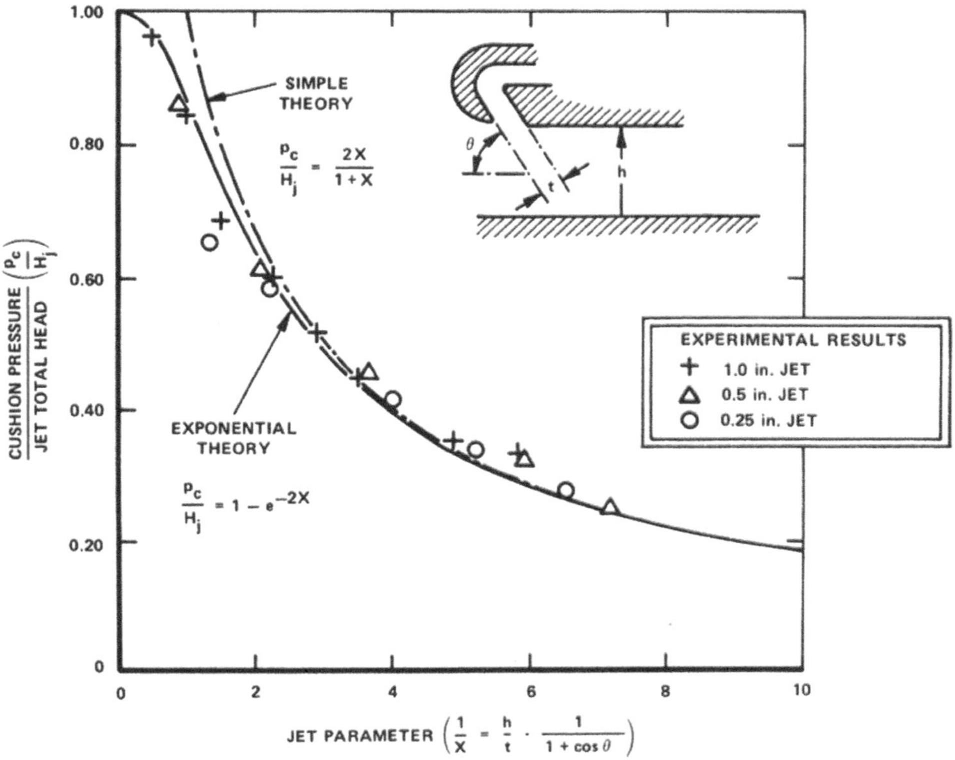

Figure 11.28: Theory and Test for Peripheral Jet Air Seal

A comparison between experimental results conducted by Saunders-Roe and the exponential theory (Equation 11.36) and the momentum theory (Equation 11.39) is shown in Figure 11.28.

It is seen that the agreement is good for determining the pressure in the air seal. The flow in the jet is:

$$Q = C\int_{o}^{t} u(r)dr \tag{11.40}$$

In this expression, C is the total peripheral length of the jet. For conventional hovercraft this would be:

$$C \approx 2(b + c) \tag{11.41}$$

For the channel flow wing in cruise mode the front and rear jets would be retracted, such that:

$$C \approx 2c \tag{11.42}$$

Upon substitution of the pressure terms and performing the integration, the air flow can be written:

$$Q = \left(\frac{2}{\rho}\right)^{1/2} (H_j - H_o)^{1/2} \frac{Ch}{1 + \cos\theta}(1 - e^{-x}) \tag{11.43}$$

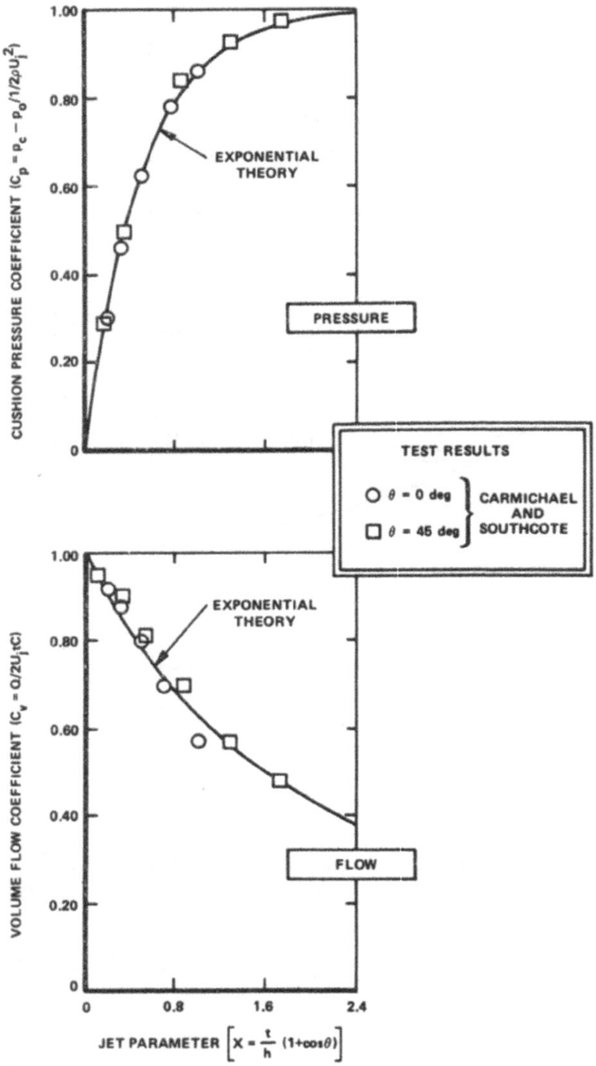

Figure 11.29: Pressure and Flow in an Air Seal (Theory and Test)

In 1960, two researchers at Ford Aeronutronic[26] conducted experiments to compare theory and test for both pressure and flow. The result is shown in Figure 11.29. Given these results, it becomes a simple matter to determine the power requirements to generate the air seal with knowledge of the pressure (Equation 11.36 or Equation 11.39) and the air flow given by Equation 11.43.

While clearly dependent on the configuration in any particular case, it is generally found that the power to generate the air seal is far less than the power required to drag endplates through the water especially at the high speeds envisaged for WIG craft.

The author[27] conducted model tests in 1962 of the *Columbia* channel flow craft and calculated the lift and drag based on model tests conducted in the California Institute of Technology (CALTECH) wind tunnels. The lift coefficient (C_L) determined from one test at a height of the wing $\left({}^{h}/_{c} = 0.02\right)$ is shown compared to the earlier data by Fink & Lastinger and Carter in Figure 11.26. The wing used in the *Columbia* work is a 10% thick airfoil section with a flat bottom not unlike the Glenn Martin 21 airfoil section. It is seen that the increased lift in ground effect achieved with a wing with *air seals* is of the same magnitude as the wings with *hard end plates.* In addition to the theories discussed earlier, the theory developed by Braunss and Lincke[28] for a two-dimensional wing has been included in Figure 11.30 as an additional data point to show general agreement with the trends in ground effect on lift and drag as the wing approaches the ground.

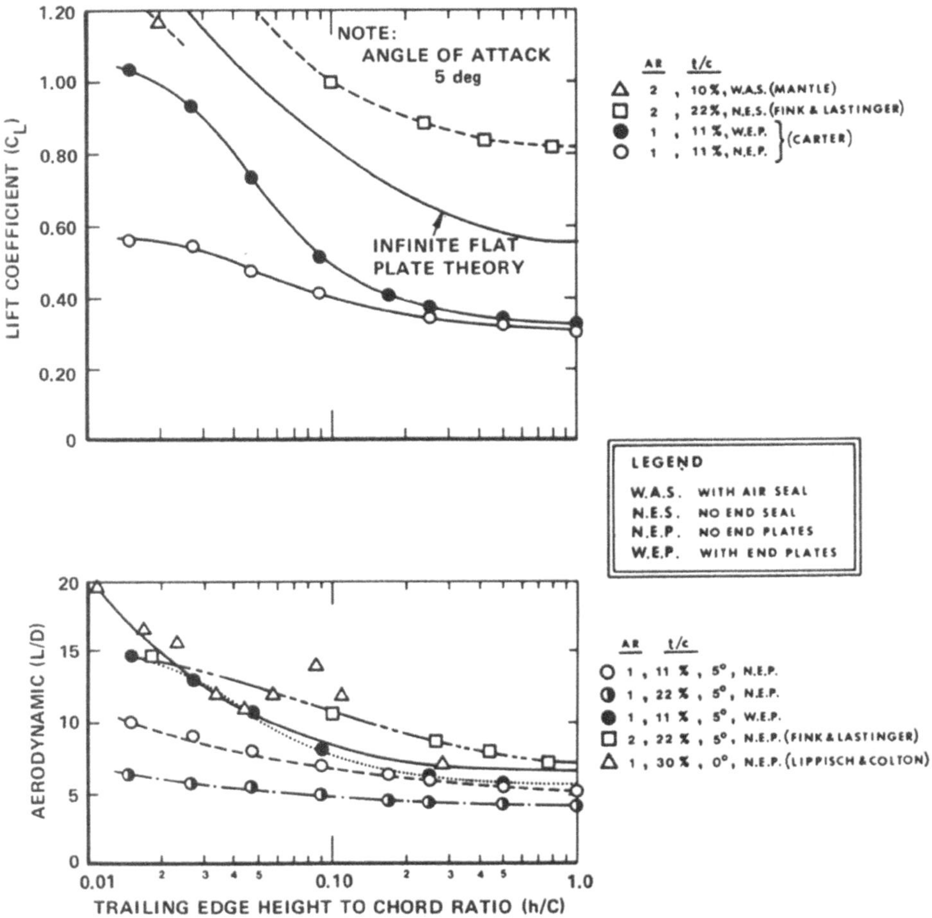

Figure 11.30: Collected Lift and Drag Results with *End Plates* and with *Air Seals*

11.7 Theory for Wings in Ground Effect with End Plates

In 1970, Peter Ashill (Cranfield Institute of Technology[j]) made an important contribution to wing–in-ground-effect technology by developing a theory for handling the induced drag for wings in ground effect, both with and without end plates.[29] Ashill developed an approximate theory for the minimum induced drag of a planar wing (flat plate) with end plates in ground effect. What was useful in this analysis is that both the height of the wing above the ground *and* the depth of the end plate were taken into account in the theory. Figure 11.31

[j] In 1946, the *Cranfield College of Aeronautics* was established on the Royal Air Force (RAF) Airfield in Cranfield near Bedford, England. In 1969, it became the *Cranfield Institute of Technology*, and upon expansion of its curriculum to fields other than aeronautics in 1993 it became the *Cranfield University*. The Alumni Association still includes "*College of Aeronautics*" in its name. Hence, depending on the time and subject under discussion, reference may be made to any of these names for the same education establishment.

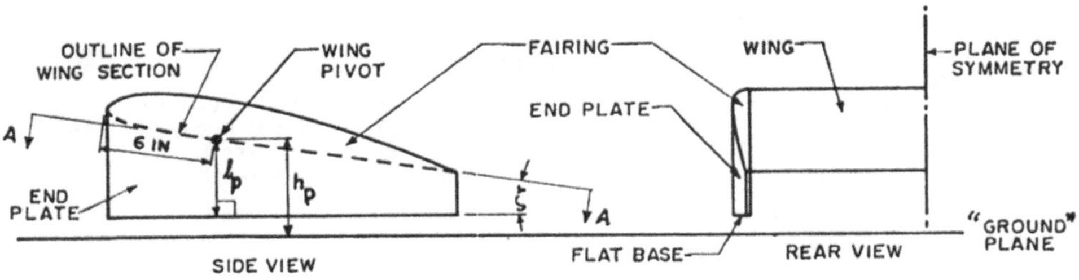

Figure 11.31: Planar Wing in Ground Effect (With End Plates) (Ashill)

shows a sketch of the model analyzed. The height of the wing (h) is measured at the wing quarter chord point and the depth of the end plate (l) is measured at the same point as shown.

In similar vein to the earlier geometries tested at Langley, the wing height is given by:

$$\frac{h}{b} = \frac{h_{TE}}{b} + {}^{3}/_{4}\frac{\sin\alpha}{A} \tag{11.44}$$

Except that now an additional geometric parameter is included $\left({}^{l}/_{b}\right)$ for the end plate depth that extends below the trailing edge (TE) of the wing. While this was an excellent contribution to the theory of wings in ground effect, there are some confusing use of terms in the paper that could hamper comparison with other work. For example, Ashill chose to use the symbol "σ" to represent the same function as "1-σ" as used by Prandtl and Wieselsberger in their work.

A comparison between Ashill's theory for wings with end plates and Wieselsberger and Prandtl's theory for wings without end plates is shown in Figure 11.32. In this book, the use of "σ" as the *influence coefficient*, and (1-σ) for the *induced drag factor* as defined by Wieselsberger and Prandtl, has been retained, and the notation used by Ashill has been adjusted accordingly to be in line with this use.

With this recognition of the symbol change, it is seen that the two theories for wings without end plates are in close agreement over the range of wing height to span ratios $\left({}^{h}/_{b}\right)$ studied by Ashill. The Ashill results for a wing with no end plates show a *lower* value of induced drag of between 4% and 10% from that predicted by Wieselsberger and Prandtl. The complete set of the Ashill predictions for the induced drag of WIG craft *with end plates* is shown in Figure 11.32.

In Figure 11.32, the ordinate is expressed as:

$$\frac{(\text{Induced Drag})_{IGE}}{(\text{Induced Drag})_{OGE}} = (1-\sigma) \tag{11.45}$$

As stated above, the symbol "σ" as used here is in agreement with the Wieselsberger and Prandtl's use of the influence coefficient. The upper curve in Figure 11.32, labeled $\left({}^{l}/_{b} = 0\right)$ corresponds to the wing without end plates (NEP) and may be compared to the result by Prandtl and Wieselsberger. Other notation changes from the Ashill work is that the data has been re-plotted in terms of $\left({}^{h}/_{b}\right)$ and $\left({}^{l}/_{b}\right)$ rather

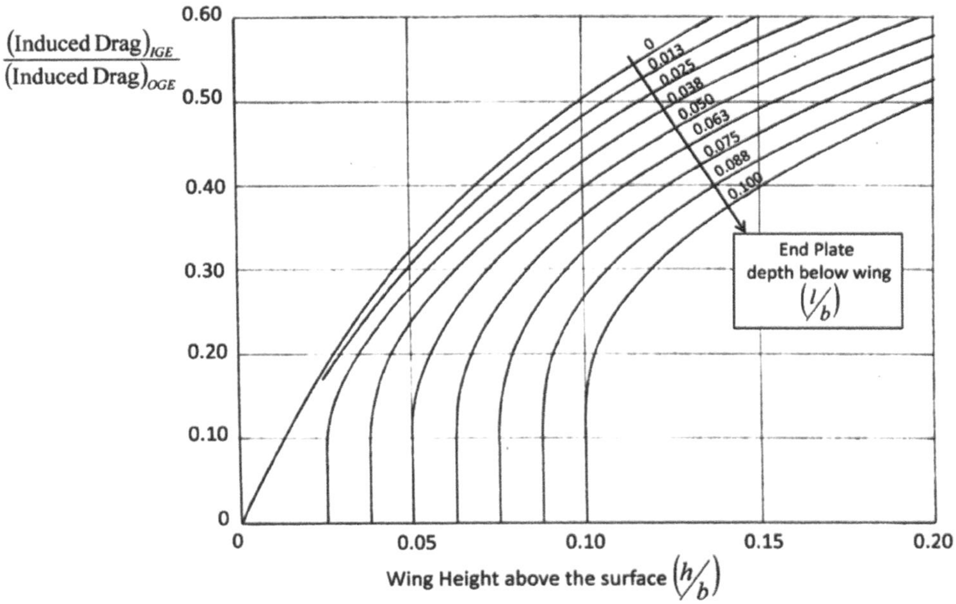

Figure 11.32: Ashill Theory for WIG (With End Plates)

than in terms of $\left(^{2h}/_{b}\right)$ and $\left(^{2l}/_{b}\right)$. It is believed that this makes it easier to compare the results with those shown earlier by Fink & Lastinger, Carter and others, and use in design work. Unfortunately, the theoretical prediction for induced drag for WIG without end plates given by the Ashill theory is not an easily understood form. The equation given for the *induced drag factor* $(1 - \sigma)$ for the wing with *no end plates* (upper curve in Figure 11.32) is a complex equation containing elliptic integrals, modulus of elliptic integrals, a transformation parameter and other terms requiring table look up from standard mathematical tables but contain no explicit information on wing geometry (span, chord, aspect ratio, etc). Additionally, the results given in the Ashill theory for WIG *with end plates* (WEP) are even more complex and further, contain terms that are undefined. The theoretical results are displayed graphically in Figure 11.32 for a range of end plate depths from $\left(^{l}/_{b}\right) = 0$ to 0.10. To provide a more tractable form for design use, the author re-plotted the data from Figure 11.32 to express the results in terms of the air gap, $^{(h-l)}/_{b}$, under the end plates. This is shown in Figure 11.33.

The early researchers (Wieselsberger et al.) in the 1920s and the later researchers in the 1960s, (Fink, Lastinger, Carter et al.) all recognized the effect of end plates in disturbing the vortex flow from the wing tips and its benefits in reducing the induced drag. It was also recognized that in addition to the end plates themselves, was the nature of the vortex flow that moved from the wing tips to the lower edges of the end plates depending on wing geometry and operating conditions. The air gap beneath the end plates was as important as the height of the wing above the surface in the determination of the best conditions for WIG craft. To this end, the author took the results by Ashill and constructed curves of constant air gap for each

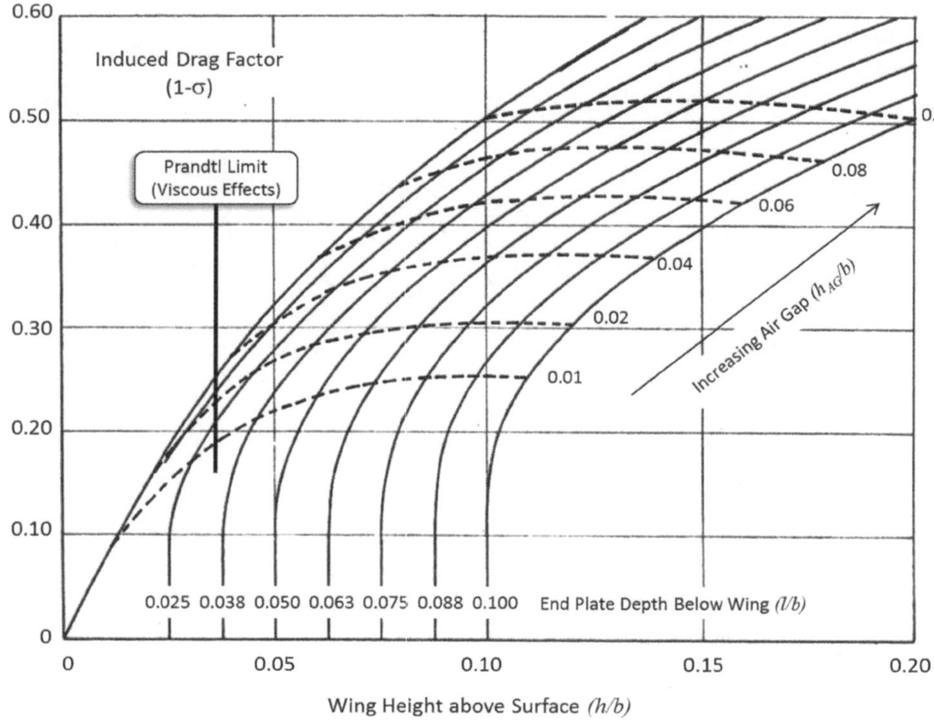

Figure 11.33: Cross Plot of Ashill Theory to Show Effect of Air Gap on Induced Drag

combination of wing height above the surface $\left({}^{h}/_{b}\right)$ and the depth of the end plate $\left({}^{l}/_{b}\right)$. The end result was that the complex representation of Ashill's numerical solution for induced drag theory for wings in ground effect can be augmented by the following simpler closed form equation containing direct and measurable geometric features of a wing with end plates, viz:

$$(1-\sigma)_{WEP} = 1.5\left(\frac{h_{AG}}{b}\right) + 5\left(\frac{h}{b}\right) - \frac{10}{\left({}^{h_{AG}}/_{b}\right)^{1/5}}\left(\frac{h}{b}\right)^{2} \tag{11.46}$$

In this relationship; the subscript WEP means "with end plates". The air gap $\left({}^{h_{AG}}/_{b}\right)$ is given as:

$$\frac{h_{AG}}{b} = \frac{h-l}{b} \tag{11.47}$$

Curves of these values have been superimposed on the original curves by Ashill for six typical air gap values of $0.01 < \frac{(h-l)}{b} < 0.10$ in Figure 11.33. Note that the approximation given by Equation 11.47 is not valid for zero air gap. This is not seen as a problem because the inviscid theory is also not valid for very low values of air gap because of the impact of viscosity and the boundary layer. These particular values were chosen to bracket some of the limits recommended by Prandtl in the Wieselsberger theory. The value of the induced drag factor without end plates, $(1-\sigma)_{NEP}$, can be taken either graphically from Ashill's plot (Figure 11.33) or it

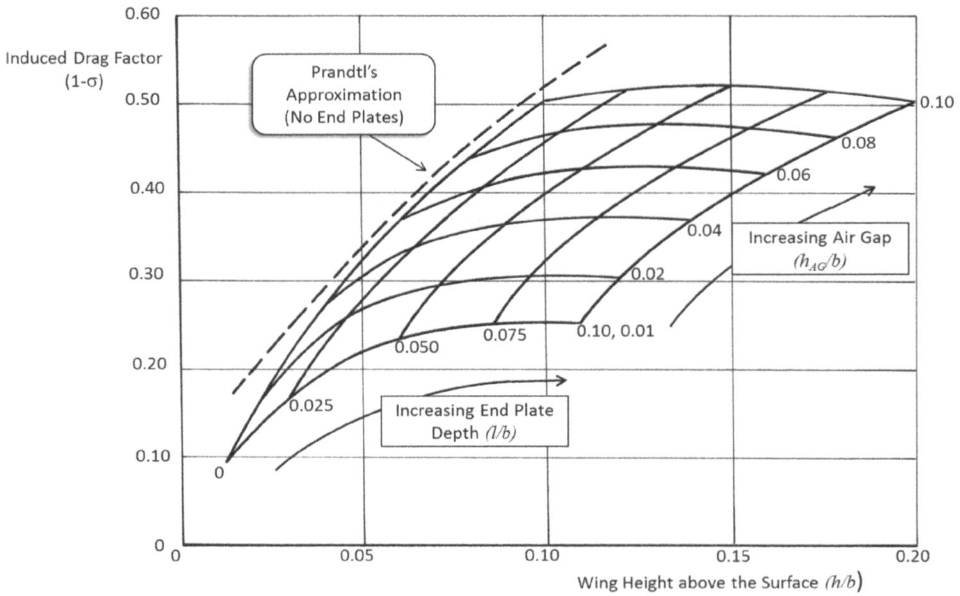

Figure 11.34: Cross plot of Theory to Show Effect of Air Gap With End Plates

can be calculated using Prandtl's approximation for the influence coefficient (σ), discussed earlier:

$$\sigma_{NEP} = \frac{1 - 1.32\left({}^{h}/_{b}\right)}{1.05 + 7.40\left({}^{h}/_{b}\right)} \tag{11.48}$$

In this relationship, the subscript *NEP* means "no end plates". Also included on Figure 11.33 is the recommended lower limit by Prandtl $\left({}^{h}/_{b} = 0.033\right)$, because of the impact of viscous effects from the boundary layer not included in any of the theories to date for inviscid flow on ground effect. In the Fink & Lastinger work discussed earlier, they found that the effect of end plates became negligible when the trailing edge (TE) of the wing was 15% of the span above the ground $\left({}^{h_{TE}}/_{b} = 0.15\right)$or when the quarter chord was 25% of the span above the ground$\left({}^{h}/_{b} = 0.25\right)$. This upper "limit" is not included in Figure 11.33 but is noted here for completeness. It now becomes possible, with Equation 11.46, to see the effects of both air gap and end plate depth on induced drag. Figure 11.34 is a cross plot of the theoretical results to show the effect of the air gap directly using Equation 11.47. The Prandtl approximation for the case of no end plates is included in Figure 11.34 for comparison. For clarity, just five values of end plate depth $0 < {}^{l}/_{b} < 0.10$ have been shown. Also, six values of the air gap $0 < {}^{h_{AG}}/_{b} < 0.10$, beneath the end plates have been shown. A numerical example helps clarify the approach. Figure 11.35 is another plot of the same results but with an example set of cases superimposed for discussion. Assume a design for a WIG *without end plates* is designed to operate with a wing height of $\left({}^{h}/_{b} = 0.04\right)$ and thus has an air gap also of $\left({}^{h_{AG}}/_{b} = 0.04\right)$. This example is labeled **"WIG (A)".** Figure 11.35 would show that the induced drag is approximately 28% of the value of the induced drag out of ground

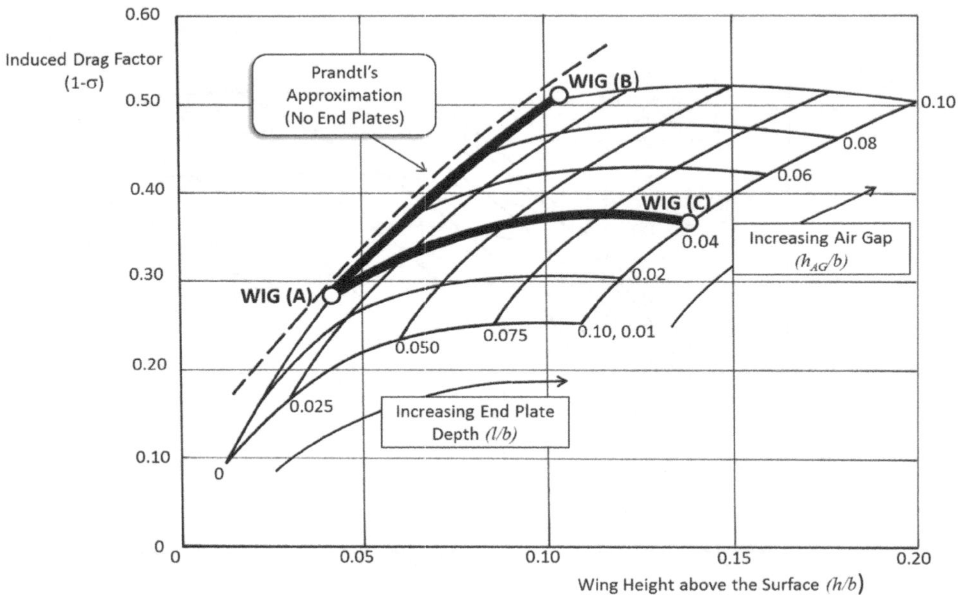

Figure 11.35: Example of WIG Induced Drag With and Without End Plates

effect. If, say for operational safety, the design point is raised to be at an operating height of $\left({}^{h}/_{b} = 0.10\right)$. This would be a craft labeled **"WIG (B)"** and would have an increased induced drag to be approximately 50% of the out of ground effect value, approximately double the value of **WIG (A).** If, on the other hand, *end plates* are added of depth $\left({}^{l}/_{b} = 0.10\right)$ this would produce a new design, labeled **"WIG (C)"** that would now have an induced drag significantly reduced from the "no end plate" **WIG (B)** design. It is seen that such a craft, **WIG (C)** could operate at an even greater wing height $\left({}^{h}/_{b} = 0.10 - 0.14\right)$ for approximately the same value of induced drag of 39% of the out of ground effect value which is 78% of the induced drag of **WIG (B).** The closed form relationship (Equation 11.46) approximates the Ashill numerical results values as shown below:

For **WIG (A)**: $({}^{h}/_{b} = 0.04, {}^{h_{AG}}/_{b} = 0.04, {}^{l}/_{b} = 0)$

$$\frac{(\text{Induced Drag})_{IGE}}{(\text{Induced Drag})_{OGE}} = (1-\sigma)_{WEP} = 1.5(0.04) + 5(0.04) - \frac{10}{(0.04)^{0.20}}(0.04)^2 = 0.24 \tag{11.49}$$

For **WIG (B)**: $({}^{h}/_{b} = 0.10, {}^{h_{AG}}/_{b} = 0.10, {}^{l}/_{b} = 0)$

$$\frac{(\text{Induced Drag})_{IGE}}{(\text{Induced Drag})_{OGE}} = (1-\sigma)_{WEP} = 1.5(0.10) + 5(0.10) - \frac{10}{(0.10)^{0.20}}(0.10)^2 = 0.49 \tag{11.50}$$

For **WIG (C)**: $({}^{h}/_{b} = 0.14, {}^{h_{AG}}/_{b} = 0.04, {}^{l}/_{b} = 0.10)$

$$\frac{(\text{Induced Drag})_{IGE}}{(\text{Induced Drag})_{OGE}} = (1-\sigma)_{WEP} = 1.5(0.04) + 5(0.14) - \frac{10}{(0.04)^{0.20}}(0.14)^2 = 0.39 \tag{11.51}$$

The agreement between the numerical results and the closed form relationship increases for the higher values of end plate depth ($l/b > 0.025$), wing height ($h/b > 0.03$) and air gap ($h_{AG}/b > 0.03$). This is not seen as a problem as these lower values fall into the realm where Prandtl suggests the theory is not valid because of viscous effects. The closed form equation gives good agreement for the likely practical values of the values shown. Unfortunately, the theory does not explain why, as the wing height is lowered, the induced drag first *increases* slightly before it *decreases* in the expected fashion as experienced in all WIG craft operating without end plates. There is insufficient experimental evidence to confirm these results and must await more systematic model testing and theoretical treatments. As a design tool, Figure 11.34, Figure 11.35 and Equation 11.47 can be used to obtain different combinations of wing height $(^h/_b)$, end plate depth $(^l/_b)$ and air gap $(^{h_{AG}}/_b)$ to achieve the desired performance of lowered induced drag in ground effect. The choice of which is the *best* operating height for WIG craft is discussed later in this section.

What is not obvious from these design trades in wing and end plate combination geometry is the overall effect on total aerodynamic drag and on craft stability in flight. Figure 11.36 illustrates the considerations that must be included before final determination of the WIG plus end plate geometry design is made.

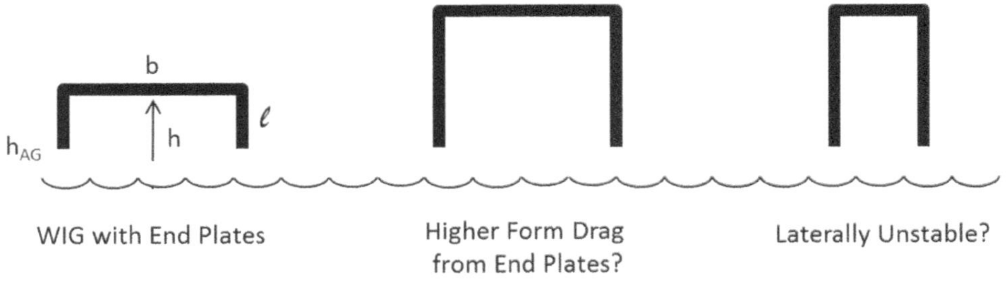

Figure 11.36: Other Considerations in Determination of Wing End Plate Geometries

As illustrated schematically in Figure 11.36, there are two areas of concern. First, if larger (deeper) end plates are considered to further lower the induced drag, the accompanying increase in the profile or form drag of the end plates themselves must be included in the final drag count. It could be that any decrease achieved in induced drag could be offset by the associated increase in end plate form drag (C_{Do}). To date, this has not been included in the various theories. The second consideration is the lateral stability of a geometry that has a shorter span (lower aspect ratio) but with an increased end plate depth. The roll or lateral stability of such a consideration would be of concern for certain geometries and must be included in any final analysis and choice of wing and end plate geometry.

These considerations are in addition to the problem of too small an air gap causing end plate contact with waves at high speed already discussed earlier in this chapter and in the ANVCE Project discussed in Chapter 10 and will not be expanded on further here.

11.8 Available Experimental Evidence of WIG with End Plates

Despite the interest in this novel form of achieving high speed at sea, there has been a paucity of systematic model testing of the various proposed configurations to support the various theories. The Fink & Lastinger data at NASA summarized earlier provided a good data base to compare with theory for wings *without end plates.* In their 1961 report, they also included one test of an aspect ratio, A=1 wing of thickness chord ratio 22% with end plates but unfortunately did not record the air gap beneath the end plates, so a direct comparison could not be made with theory. The reported data by Carter (see Figure 11.21) did provide data for a wing *with end plates.* However, the data was for a different end plate configuration than used by Fink & Lastinger and also for a thinner wing (11% thickness chord ratio).

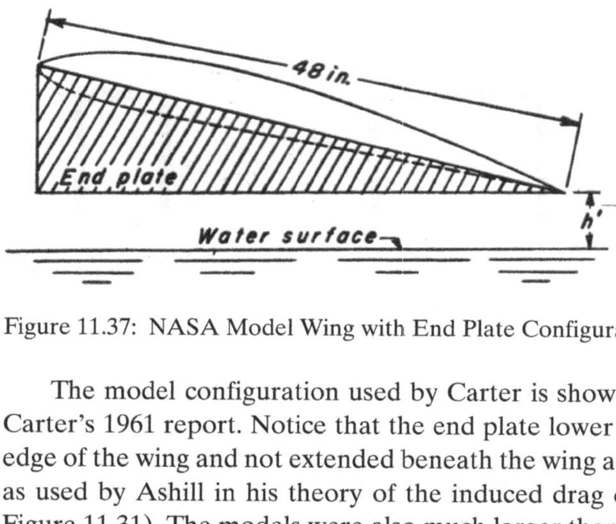

Station, percent chord	Airfoil ordinates, percent chord	
	Upper	Lower
0	4.43	4.43
1.25	6.10	3.02
2.5	6.90	2.40
5	7.99	1.66
7.5	8.82	1.18
10	9.46	.86
15	10.36	.38
20	10.84	.14
30	11.06	0
40	10.56	0
50	9.60	0
60	8.32	0
70	6.68	0
80	4.72	0
90	2.50	0
95	1.28	0
100	0	0

Figure 11.37: NASA Model Wing with End Plate Configuration (Carter)

The model configuration used by Carter is shown in Figure 11.37, taken from Carter's 1961 report. Notice that the end plate lower edge is flush with the trailing edge of the wing and not extended beneath the wing as used by Fink & Lastinger, or as used by Ashill in his theory of the induced drag of wings with end plates (see Figure 11.31). The models were also much larger than the Fink & Lastinger models (48 inch compared to 12 inch chord) meaning that the Reynold's Number was different and contributory to slight changes in drag characteristics between the two sets of data.

Despite these differences, it is possible to at least gain some insight as to the effect of end plates from a consistent set of model testing. Figure 11.38 shows the model data for the lift coefficient (C_L) versus the angle of attack (α) for the wing both with and without end plates from the Carter model tests at Langley.

The data was taken using the height of the wing at the trailing edge as the governing parameter. This means that with the notation used here,

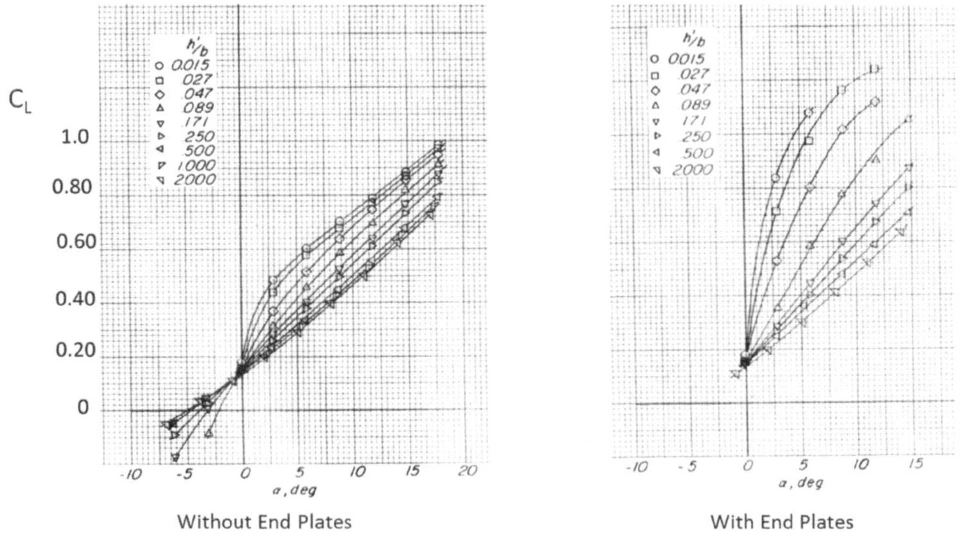

Figure 11.38: Lift of Wing in Ground Effect With and Without End Plates (A = 1)

$$\frac{h'}{b} \equiv \frac{h_{TE}}{b} \equiv \frac{h_{AG}}{b} \tag{11.52}$$

Also, because the data was only obtained for an aspect ratio, A = 1 wing then, $b = c$ for all the comparison of data and theory for the Carter model tests. The model tests both for the Fink & Lastinger and the Carter tests showed a non-linear relationship for the aspect ratio, A = 1 wing, that was not evident for the higher aspect ratios. This indicates that the vortex flow patterns for this short span wing were not the same as for the other wings but this observation was not commented on by the researchers and must await more detailed testing to confirm as to why this would be the case.

From Figure 11.38, it is seen that the lift curve slope is linear for the small angle of attack range $0 < \alpha < 3$ degrees, after which there is a distinct fall-off in the lift coefficient at the higher angles of attack. For the case with end plates, this linear range is increased to approximately $0 < \alpha < 5$–10 degrees depending on the wing height $\left({}^{h_{TE}}/_{b}\right)$.

The lift coefficient can be expressed by:

$$C_L = C_{Lo} + \frac{\partial C_L}{\partial \alpha}\alpha \tag{11.53}$$

The lift curve slope is written, as before, for the linear range, as:

$$\frac{\partial C_L}{\partial \alpha} = \left|\frac{\partial C_L}{\partial \varepsilon}\right|_{OGE} + f\left(\frac{h}{b}\right) \tag{11.54}$$

For the aspect ratio wing (A = 1) *without end plates*, the lift curve slope can be written:

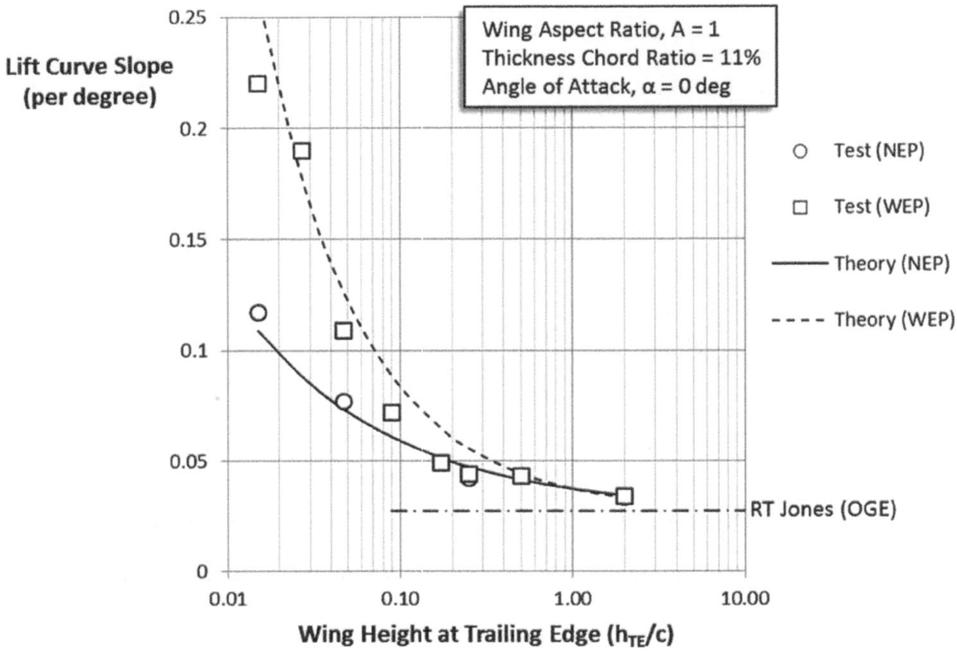

Figure 11.39: Effect of End Plates on Lift Curve Slope in Ground Effect

$$\frac{\partial C_L}{\partial \alpha} = \frac{2\pi A}{A+3} + \frac{0.01}{\sqrt{h_{TE}/b}} \tag{11.55}$$

For the wing *with end plates*, the lift curve slope increases to:

$$\frac{\partial C_L}{\partial \alpha} = \frac{2\pi A}{A+3} + \frac{0.01}{\left(h_{TE}/b\right)^{3/4}} \tag{11.56}$$

In the above expressions, it has been taken that the RT Jones result for the OGE value of lift curve slope is valid for both sets of wings (NEP and WEP). A comparison between these expressions and the data is shown in Figure 11.39.

It is seen from Figure 11.39, that the derived expressions for the lift curve slope is a good approximation over the full range of wing heights tested. If these expressions are now applied to the case of the wing with end plates for the lift and drag of the wing, the results shown in Figure 11.40 and Figure 11.41 are obtained. Allowing for the normal problems of reading small graphs for the experimental data and smoothing of the test curves, it is found reasonable agreement for the wing heights likely to be used in practice. Figure 11.40 shows good agreement for trailing edge wing height of $h_{TE}/b = 0.089$ and with slightly less accuracy as the wing approaches very small gaps $\left(h_{TE}/b \approx 0.015\right)$ above the water surface and high lift coefficients ($C_L > 0.05$). As noted by Prandtl, any inviscid theory would not be valid at such low heights or immersion into the boundary layer at the water surface.

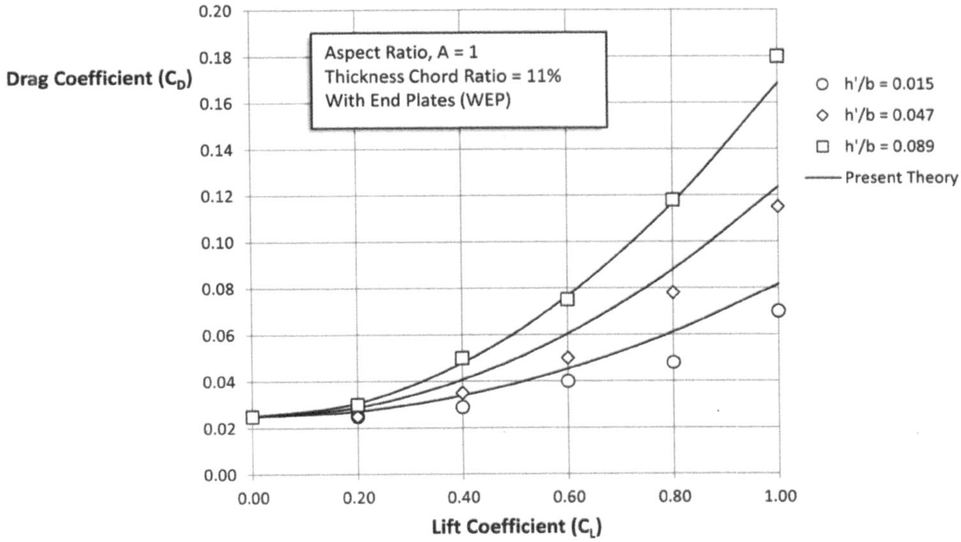

Figure 11.40: Drag Polar in Ground Effect (A = 1) (WEP)

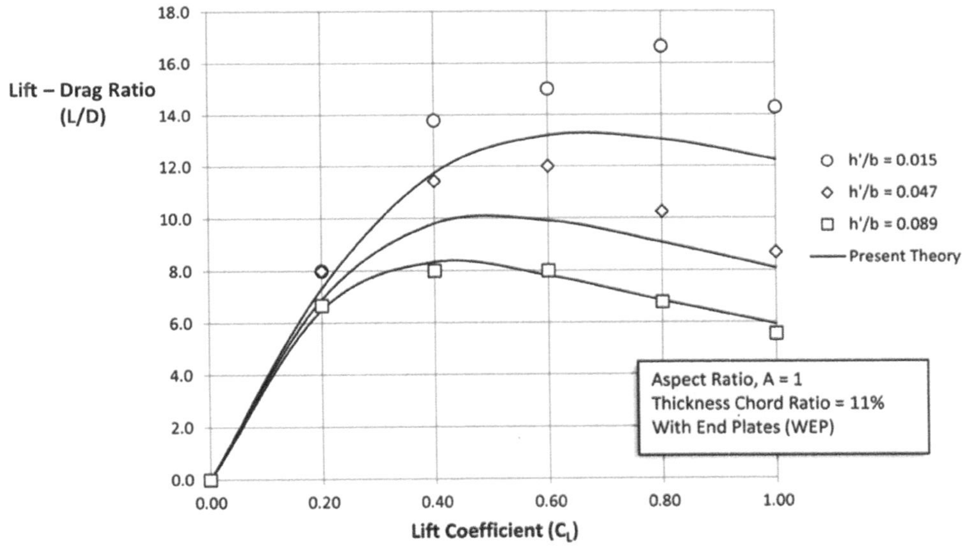

Figure 11.41: Lift Drag Ratio in Ground Effect (A = 1) (WEP)

For the drag polar shown in Figure 11.40, the drag coefficient in this simple theory is written as:

$$C_D = C_{Do} + K_3 f(h_{AG}, h) C_L^2 \qquad (11.57)$$

In Equation 11.52, the following parameters have been assumed:
C_{Do} is the profile drag coefficient of the wing (= 0.025 for the wing tested)
K_3 is a constant (assumed to be 0.90 for a better match of theory and test)

$$f(h_{AG}, h) = 1.5\left(\frac{h_{AG}}{b}\right) + 5\left(\frac{h}{b}\right) - \frac{10\left(^{h}/_{b}\right)^2}{\left(^{h_{AG}}/_{b}\right)^{1/5}} \tag{11.58}$$

This latter equation is that taken from the author's approximation to the Ashill theory for the induced drag of wings with end plates (Equation 11.42). Note that in this formulation, the constant K_3, absorbs the assumed "πA" in the expression for the OGE value of induced drag coefficient in the Prandtl and Wieselsberger's original formulation for the induced drag factor. Expressed in terms of Prandtl and Wieslesberger induced drag factor:

$$(1-\sigma)_{WEP} = K_3 f(h_{AG}, h)\pi A \tag{11.59}$$

This expression has the explicit assumption that the induced drag coefficient out of ground effect *with end plates* is the same as for a wing OGE *without end plates*, as used by Ashill in his theory, that is:

$$|(C_{Di})_{OGE}|_{WEP} = |(C_{Di})_{OGE}|_{NEP} = \frac{C_L^2}{\pi A} \tag{11.60}$$

It is seen that the Ashill theory (using the approximation Equation 11.47) provides a good agreement with the available experimental evidence for "medium" wing heights $\left(^{h_{TE}}/_{b}\right)$or air gaps $\left(^{h_{AG}}/_{b} \approx 0.089\right)$ but over predicts the induced drag slightly at the lower heights (see Figure 11.40). It is emphasized that the only available test data is for a wing of aspect ratio (A = 1) and also for the special case, where the end plate is flush with the trailing edge of the wing (see Figure 11.37). It is reasonable to assume that pending more systematic model testing that these trends are fairly representative of the likely savings in induced drag of wings with end plates operating in ground effect.

11.9 Summary of Key Equations for Lift and Drag in Ground Effect

For convenience and ease of reference, the various equations derived here together with the assumptions are summarized in Table 11.3 for the case of a wing operating in ground effect with *no end plates* (NEP).

A similar summary of the key equations for the case of a wing operating in ground effect *with end plates* (WEP) is provided in Table 11.4.

Table 11.3: *Summary of Key Equations for Wing in Ground Effect (No End Plates) (NEP)*

Item	Equation	Note
LIFT COEFFICIENT		
Lift Coefficient	$C_L = C_{L_o} + \frac{\partial C_L}{\partial \alpha}\alpha$	Linear assumption
Lift Curve Slope (IGE)	$\frac{\partial C_L}{\partial \alpha} = \left\vert\frac{\partial C_L}{\partial \alpha}\right\vert_{OGE} + \frac{0.008}{^{h}/_{c}}$ for $A \geq 2$	New
	$\frac{\partial C_L}{\partial \alpha} = \left\vert\frac{\partial C_L}{\partial \alpha}\right\vert_{OGE} + \frac{0.012}{\sqrt{^{h}/_{c}}}$ or $A \approx 1$	New
	$\left\vert\frac{\partial C_L}{\partial \alpha}\right\vert_{OGE} = \left(\frac{2\pi A}{A+3}\right)$	RT Jones (rect.wings)

Table 11.3: (*cont.*)

Item	Equation	Note
DRAG COEFFICIENT		
Drag Coefficient	$C_D = C_{Do} + C_{Di}$	
Induced Drag Coefficient (NEP)	$C_{Di} = K_2(1-\sigma)_{NEP}\frac{C_L^2}{\pi A}$	New
Influence Coefficient (NEP)	$\sigma_{NEP} = \frac{1-1.32^h/_b}{1.05+7.4^h/_b}$	Prandtl's approximation
LIFT DRAG RATIO		
Lift Drag Ratio	$\frac{L}{D} = \frac{C_{Lo} + \frac{\partial C_L}{\partial \alpha}\alpha}{C_{Do} + K_2(1-\sigma)_{NEP} C_L^2 / \pi A}$	New
Geometric Parameters		
Wing height (at quarter chord) h	Aspect Ratio: $A = \frac{b}{c} = \frac{b^2}{S}$	
Wing height at TE h_{TE}	Wing Area: $S = bc$	
Air Gap (beneath wing) h_{AG}	Height Ratio: $\frac{h}{c} = A\frac{h}{b} = \frac{h_{TE}}{c} + {}^3/_4 \sin\alpha$	
Wing span b	C_{Lo} Lift coefficient at zero angle of attack	
Wing chord (mean) c	C_{Do} Drag coefficient at zero lift coefficient	
Wing Angle of Attack α	$K_2 = 0.8$ for best fit with available test data	

Table 11.4: *Summary of Key Equations for Wing in Ground Effect (With End Plates) (WEP)*

Item	Equation	Note
LIFT COEFFICIENT		
Lift Coefficient	$C_L = C_{Lo} + \frac{\partial C_L}{\partial \alpha}\alpha$	
Lift Curve Slope (IGE)	$\frac{\partial C_L}{\partial \alpha} = \left\lvert \frac{\partial C_L}{\partial \alpha} \right\rvert_{OGE} + \frac{0.01}{\left(^{h_{AG}}/_b\right)^{3/4}}$	New
	$\left\lvert \frac{\partial C_L}{\partial \alpha} \right\rvert_{OGE} = \left(\frac{2\pi A}{A+3}\right)$	RTJones (rect.wings)
DRAG COEFFICIENT		
Drag Coefficient	$C_D = C_{Do} + C_{Di}$	
Induced Drag Coefficient (WEP)	$C_{Di} = K_3 f(h_{AG}, h) C_L^2$	New
Induced Drag Factor $(1-\sigma)_{WEP}$ ●	$f(h_{AG}, h) = 1.5\left(\frac{h_{AG}}{b}\right) + 5\left(\frac{h}{b}\right) - \frac{10}{\left(^{h_{AG}}/_b\right)^{1/5}}\left(\frac{h}{b}\right)^2$	New
LIFT DRAG RATIO		
Lift Drag Ratio	$\frac{L}{D} = \frac{C_{Lo} + {}^{\partial C_L}/_{\partial\alpha}\,\alpha}{C_{Do} + K_3 f(h_{AG}, h) C_L^2}$	New
Geometric Parameters		
Wing height (at quarter chord) h	Aspect Ratio: $A = \frac{b}{c} = \frac{b^2}{S}$	
Wing height at TE h_{TE}	Wing Area: $S = bc$	
Air Gap (beneath end plate) h_{AG}	Height Ratio: $\frac{h}{c} = A\frac{h}{b} = \frac{h_{TE}}{c} + {}^3/_4 \sin\alpha$	
Wing span b	C_{Lo} Lift coefficient at zero angle of attack	
Wing chord (mean) c	C_{Do} Drag coefficient at zero angle of attack	
End Plate Depth l	$K_3 = 0.90$ for best fit with test data	
Wing Angle of Attack α		
Air Gap $\frac{h_{AG}}{b} = \frac{h-l}{b}$		

11.10 Choosing Cruise Height for a WIG Craft

There are no established methods for determining the best cruise height for a wing operating in ground effect. If the wing flies *low* to achieve a low value of induced drag, it is in danger of crashing into the oncoming (and unpredictable) waves at sea. If it flies *high* to avoid the waves, it loses most of the advantage of ground effect. The only practical data available for large craft operating in real sea conditions are the Russian *ekranoplans*, specifically the original prototype, the 544 tonne *Caspian Sea Monster (KM)*; the 400 tonne *Lun*; and the 125 tonne *Orlyonok*. The leading particulars of these three craft are provided in Table 11.5. Figure 11.42 shows these craft operating in ground effect.

These three *ekranoplans* were designed and built by the Central Hydrofoil Design Bureau (CHDB) in Nizhny Novgorod (formerly Gorky), Russia. The Chief Designer was Rotislav Alekseyev, known for his work on the main hydrofoils in use in Russia.[k] The development of the *ekranoplans* grew out of the Soviet Navy interest in the 1960s in developing high speed marine missions (amphibious warfare and missile attack). This work received direct personal support from Nikita Kruschev (1894–1971) who led Soviet Russia from 1953–1964), first as First Secretary and then as Chairman of the Communist Party.

Table 11.5: *Leading Particulars of* ekranoplans

	KM	Lun	Orlyonok
TOGW (W)	544t (1,198,976lb)	400t (881,600lb)	125t (275,500lb)
Length Overall (LOA)	92.4m (303.15ft)	73.8m (242.12ft)	58m (190.29ft)
Span (b)	37.8m (124.01ft)	44m (144.36ft)	31.5m (103.35ft)
Mean Chord (c)	18.9m (62.01ft)	12.46m (40.88ft)	9.66m (31.69ft)
Wing Aspect Ratio (b^2/S)	2.00	3.53	3.26
Wing Area (S)	7,690 sq ft	5,900 sq ft	3,272 sq ft
Craft "density" ($W/S^{3/2}$)	1.78	1.95	1.5
Max Speed	270 knots	300 knots	253 knots
Cruise Speed	230 knots	256 knots	216 knots
Range	1852km (1,000nm)	3000km (1,620nm)	2000km (1080nm)
Power: Take-off (T.O.)	Ten Dobryin VD-7 turbojets 16,000kg (35,274lbf) thrust each	Eight Kuznetsov NK-87 turbofans 12,973kg (28,600lbf) thrust each	Two Kuznetsov NK-8 turbofans 10.5t (23,142lbf) thrust each
Power: Cruise	Only Two VD-7 engines for cruise	Split between T.O. and cruise not specified	One Kuznetsov turboprop NK-12, 11,000KW (14,751HP)
Payload (W_P)		100t (220,400lb)	15t (33,060lb)
Weight of fuel (W_f)			15.85 (34,933lb)
Empty Weight (W_E)			94t (207,176lb)

Note: The numerical values in this Table are considered to be the most reliable out of the many different values that appear in the literature.

[k] More background on Alekseyev's work on hydrofoils is given in Chapter 1.

(a) *Caspian Sea Monster (KM)* (1966) (Prototype)

(b) *Orlyonok* (1972) (Amphibious Warfare)

(c) *Lun* (1987) (Missile Attack)

Figure 11.42: Russian *ekranoplans*

With Kruschev's personal interest, Alekseyev enjoyed unprecedented support in developing these unique craft. The first of these, the "*Caspian Sea Monster(KM)*", see Figure 11.42(a), was spotted by US Intelligence in the mid-1960s. The *Orlyonok*, Figure 11.42(b), was launched in 1972, and later, the *Lun*, Figure 11.42(c), entered service in 1987[1]. None of these craft are currently operating. The prototype *KM* crashed in 1980. The single *Lun* is mothballed in Kaspiysk, Russia on the Caspian Sea and the sole remaining *Orlyonok* (of the three built) is now on display in Moscow, Russia. Reports appear from time to time about possible plans to build more *Lun* and *Orlyonok* but such plans have not yet transpired. Much smaller versions of the *ekranoplan* design have appeared over the last few years around the world, many being more of novelty small craft design rather than operational large craft suitable for the navy missions or commercial use.

Despite the demise of these large craft, these early original *ekranoplans* provide the only full scale data of operational craft that can provide practical guidance in design.

The design philosophy by the CHDB was that the *ekranoplan* should fly *over* the waves and not *through* them. That is, no contact with the water (waves) while cruising in ground effect. It accomplished this by setting a "rule of thumb" that the cruise height be such that the lowest part of the craft structure (the end plates) be at (a) an altitude equal to the wave amplitude of the design Sea State, plus (b) an additional "clearance" or "safety margin" built in, equal to 10% of the wing chord. That is:

$$\frac{h_{TE}}{c} = \frac{H_{1/30}}{2c} + 0.10 \tag{11.61}$$

In Equation 11.61, the Russian design practice of using the average of the 30th highest waves $(H_{1/30})$ to characterize the Sea State is used in the calculation. For the design Sea State 4, the Pierson-Moskowitz sea spectrum would give a wave height (double amplitude) equal to 2.61 times the average (RMS) wave height (see Chapter 10, Table 10.5). This would give then values for $H_{1/30}$ in Sea State 4, of between 9.9ft and 10.9ft. The *Lun* end plate shown in Figure 11.43 will be used as an example. As can be seen from Figure 11.43, the "end plate" in the *ekranoplan* designs are not the simple shapes assumed in the analyses and model tests shown earlier in determining the effect of end plates, so only general trends can be summarized here as to their effect on the induced drag of the *Lun* wing. The *ekranoplan* end plates are more like wing tip floats that are designed more to stabilize the wing when the craft is sitting in the water prior to take-off. These floats have typical hydrodynamic float shaping on the outboard surfaces with steps and rakes to eliminate water sticking during take-off but the inner surfaces are flat and vertical

[1] The photographs (taken by Soviet Navy photographers) of the *ekranoplans* throughout the book now appear liberally throughout the open literature and websites. All *ekranoplan* flights were during the Soviet Union era that ended on 26 December 1991. Many of the photographs are identical in each web site. The most detailed set is duplicated in at least two sites: http://englishrussia.com/ekranoplan and http://thevelvetrocket.com. Both these sites show both external and internal views of the *Lun* ekranoplan at the Kaspiysk(iy) shipyard on the Caspian Sea, Russia, which provide a useful set for understanding the design features as discussed here. Similar photographs appear in the1994 DARPA Report on the US Government investigation of *ekranoplans.*

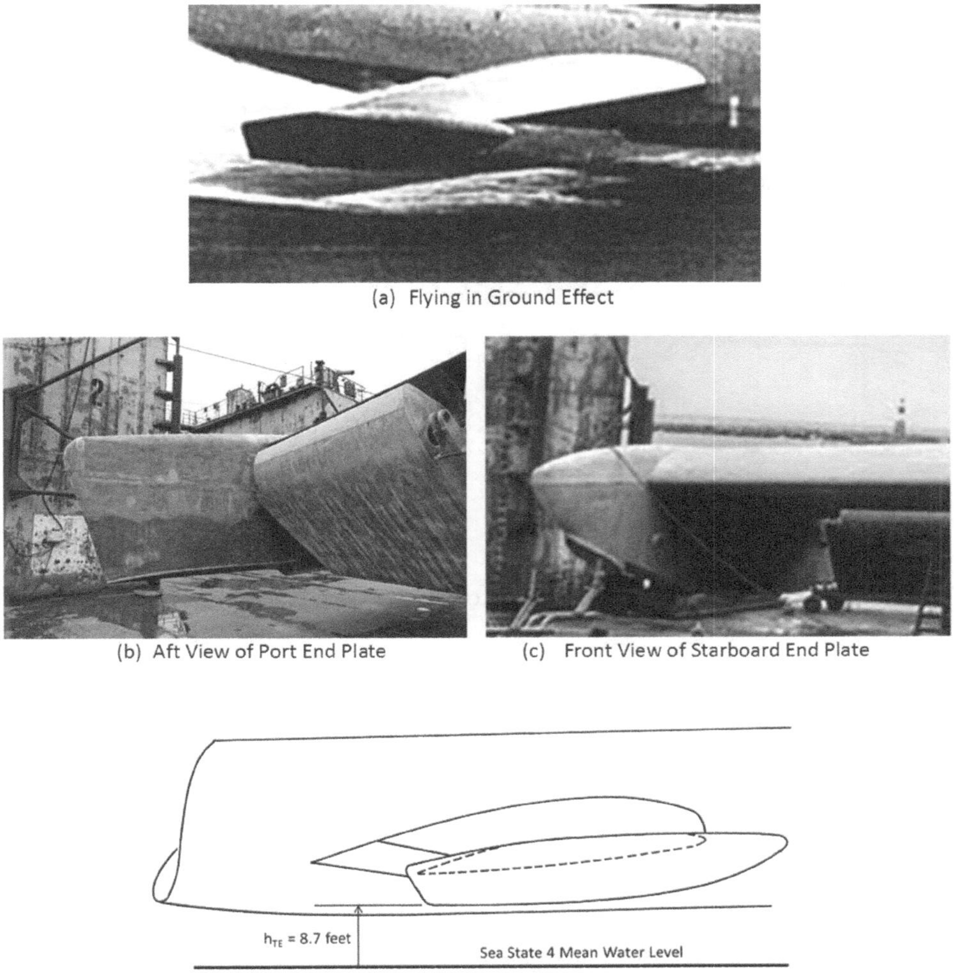

(a) Flying in Ground Effect

(b) Aft View of Port End Plate

(c) Front View of Starboard End Plate

(d) Sketch of End Plate in Cruise Mode

Figure 11.43: Wing Tip End Plates of Lun *ekranoplan*

and serve in an "end plate role". Figure 11.44 shows the *Lun* sitting in the water illustrating this need for wing tip floats to avoid the wings becoming immersed in green water when hullborne and at low speed. Another feature of the *Lun* end plate is the addition of air jets emanating from the end plate keel to further weaken the vortex flow from the end plates. The air jet is seen in Figure 11.43 (b) near the lowered flap trailing edge. There is no published data on the success of this technique.

If an average value for $H_{1/30}$ of say, 10ft is assumed, this gives for the "air gap" beneath the end plate:

$$\frac{h_{TE}}{c} = \frac{10}{2c} + 0.10 \tag{11.62}$$

Figure 11.44: *Lun* hullborne and at rest

The mean chord (c) of the *Lun* is taken to be 41ft approximately, such that the air gap is:

$$\frac{h_{TE}}{c} = 0.222 \tag{11.63}$$

This matches the height shown in Figure 11.42 and Figure 11.43 of approximately 9ft. From Table 11.5, the wing aspect ratio is $A = 3.53$ for the *Lun*, in which case, in terms of wing span, the air gap becomes:

$$\frac{h_{TE}}{b} = 0.063 \tag{11.64}$$

The other pertinent dimensions taken from the *Lun* drawings are:

$$\frac{h}{b} = 0.120, \quad \frac{l}{b} = 0.057, \quad \frac{h_{AG}}{b} = 0.063 \tag{11.65}$$

The wing height is at the quarter chord point and (because of the dihedral) also at the mid-half span. The keel of the *Lun* hull is at the same height as the lower edge of the (float) end plates approximately. Note that in all these calculations "approximate" and "average" values are used because of the imprecision in determining the critical dimensions of wave heights at sea and the non-simple geometries in the design of the wings and end plates. Also, of paramount importance is that all calculations have assumed that the wing does not roll or pitch in its motion over the sea. Such motions could change the heights and immersion depths assumed in a significant way. This is recognized by the CDHB designers in adding margins in their calculations for the cruise height of their *ekranoplans*. If the *Lun* geometric relationships from Equations 11.64 and Equation 11.65 are superimposed on the earlier developed plot Figure 11.34, it is seen that the induced drag would be expected to be about 45% of its out of ground effect value as shown on Figure 11.45. Similar measurements for the *KM* and *Orlyonok* were made and are included in Figure 11.45 along with the result for the *Lun*.

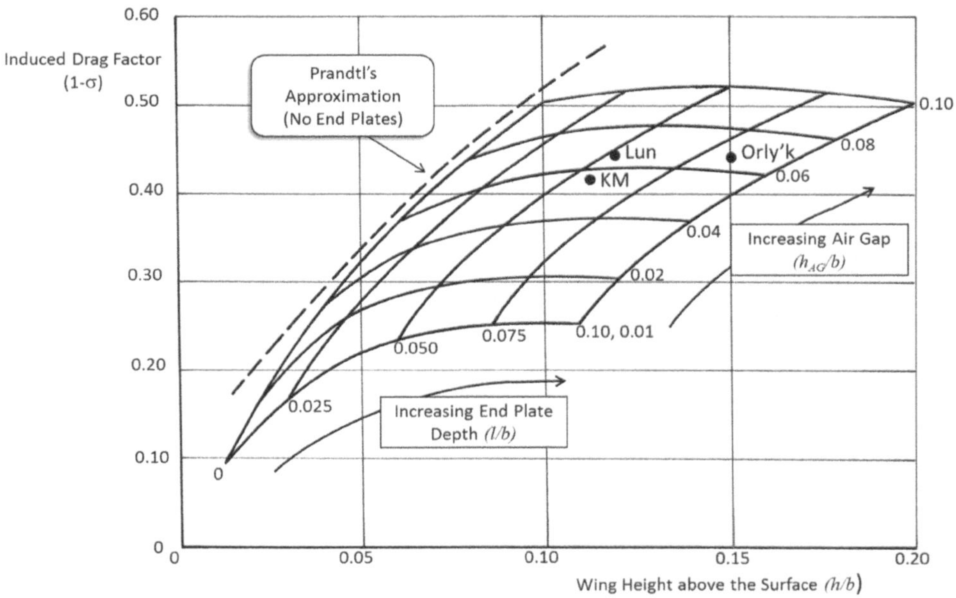

Figure 11.45: *Ekranoplan* Induced Drag Factors

Because of the lack of available precise measurements in the Russian literature, these results for the *ekranoplan* can only be taken as indicative of what the designers were seeking but they provide at least a guide as to the level of drag reduction expected. From CHDB documents made available from the DARPA report[30], the operating heights given for the *Lun* and *Orlyonok* were stated to be $0.10 < {}^{h}/_{b} < 0.30$. The heights shown in Figure 11.45 are at the mean values given for the craft between these operating limits. Because the *ekranoplans* were designed for the specific heights shown, the value of induced drag remains approximately constant at about 45% of their out of ground effect values of induced drag. If it is assumed that the *ekranoplan*s fly at a speed where the profile drag is approximately equal to the induced drag (to maximize the lift drag ratio) then it would be expected that the designs experience about a 22% reduction in total drag in their operational mode. This is confirmed from plots such as Figure 11.41 and earlier, where maximum values of lift drag ratio $({}^{L}/_{D})_{MAX}$ for "practical" cruise heights, showed an increase of approximately 20% when operating in ground effect.

11.11 Alternate Theories for Induced Drag in Ground Effect

Before considering how the lift and drag characteristics in ground effect manifest themselves into useful efficiencies and power consumption for such WIG craft, a brief summary is provided of the various alternative theories for the induced drag in ground effect as developed by other and later researchers in the early 1970s.

The work by Ashill, while very valuable in understanding the induced drag of wings with end plates, had the drawback of a lack of closed form analytic solutions for design use, relying instead on numerical solutions of complex formulations of circulation around the wings. What would be useful would be analytically derived expressions for lift and drag that would demonstrate the main contributing factors to reduction of induced drag of wings operating in ground effect.

About the same time that Ashill was developing his theory based on linearized lifting surface theory in 1970, Dr. Sheila E. Widnall[m] and Timothy Barrows, at Massachusetts Institute of Technology[31] developed an elegant theory based on a sophisticated system of replacing the wing (*without end plates*) with a distributed set of sources and sinks. This theory produced the expression for lift coefficient as follows:

$$C_L = \frac{\alpha}{\varepsilon} + \frac{2\alpha}{\pi}\left(1 + 2\ln\left(\frac{\pi}{\varepsilon}\right)\right) + \frac{2\varepsilon}{\pi^2}\left(4\ln\frac{\pi}{\varepsilon} + 2\left[\ln\frac{\pi}{\varepsilon}\right]^2\right) \qquad (11.66)$$

In Equation 11.66, Widnall and Barrows give α as the angle of attack of the wing and the height of the wing above the ground is expressed as $\varepsilon = {}^{h}/_{c}$, where c is the chord of the wing. Included in Widnall and Barrows theory were expressions for circulation and the effects of finite span, all of which provide an analytic basis of understanding the characteristics in ground effect of the pressure distributions both on the upper and lower surfaces of the wing. The theory was specifically applicable to elliptic planforms. This work concluded that minimum induced drag occurred for elliptic wings with straight trailing edge (TE) but with a parabolic lift distribution.

Shortly after Ashill developed his theory in 1970 for wings in ground effect *with end plates*, two groups of researchers in Japan tackled the problem of deriving closed form solutions, with an emphasis on determining the induced drag of wings in ground effect with end plates (and other geometric variations). The first group, Kida and Miyai of the Osaka Prefecture University, Japan published[32] their approximate theory in 1974 and compared their results with those of Ashill. Kida and Miyai used a similar technique to Ashill but for a set of wing geometries that apply to possible design configurations. These configurations include: semi-circular cross section wings; semi-elliptic wings[n] and rectangular wings with end plates.

For the particular case of rectangular wings *with end plates*, Kida and Miyai provide their results in a slightly different form to that of Ashill, in that they express the reduction in induced drag in terms of the *span efficiency factor* (e) rather than its inverse *induced drag factor* (1-σ) and the air gap beneath the end plates $\left({}^{h_{AG}}/_{b}\right)$directly instead of the depth of the end plate $({}^{l}/_{b})$ where:

$$\varepsilon = \frac{2h_{AG}}{b} \qquad (11.67)$$

[m] Later, Dr Widnall was the Secretary of the Air Force in the period (1993–1997)

[n] Kida & Miyai use the term "semi-elliptic" while Widnall & Barrows use the term "elliptic" for the same planform.

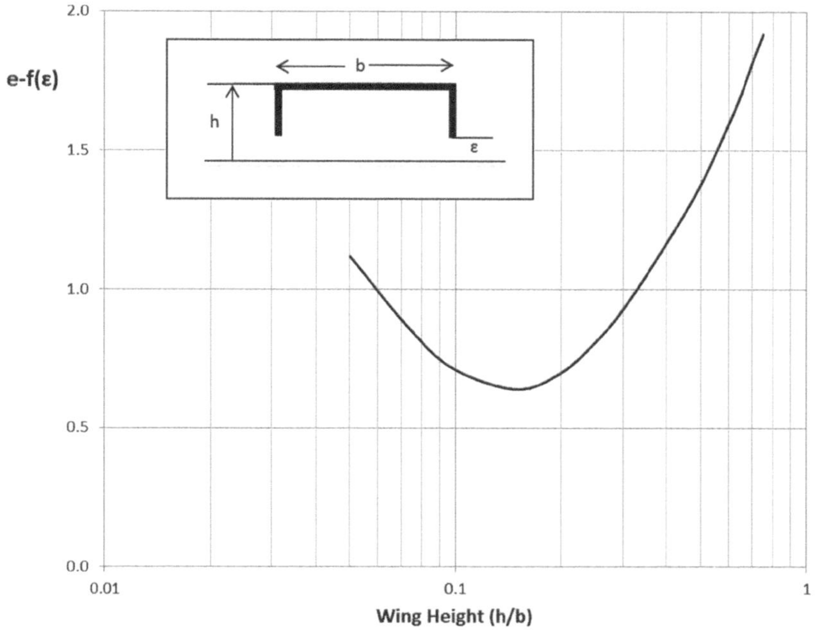

Figure 11.46: Span Efficiency of Wings With End Plates (WEP) (Kida & Miyai)

Like Ashill, an explicit formulation is not provided but given in graphical form as a result of a numerical computational method and graphical treatment of the elliptic integrals in their derivation. Figure 11.47 shows the result, where:

$$e - \frac{8}{\pi^2}\ln\left(\frac{1}{\varepsilon}\right) = f\left(\frac{h}{b}\right) \tag{11.68}$$

For clarity on Figure 11.46, the left hand side of Equation 11.68 is shown as:

$$f(\varepsilon) = \frac{8}{\pi^2}\ln\left(\frac{1}{\varepsilon}\right) \tag{11.69}$$

As stated above, the functional form $f\left({}^{h}\!/_{b}\right)$ is not given by Kida and Miyai explicitly as it is the result of both graphical and numerical computations and would be a sixth order polynomial of the dependent variable $\left({}^{h}\!/_{b}\right)$ not easily used in design work. It is of interest to note that below a wing height ${}^{h}\!/_{b} \approx 0.15$, the span efficiency function begins to increase. To this end, Kida and Miyai treat the special case of $h \rightarrow 0$ and determined that, in the limit:

$$e \approx \frac{1}{\pi\left({}^{h}\!/_{b}\right)} + O\left(\frac{1}{{}^{2h}\!/_{b}}\right) \tag{11.70}$$

For a more direct comparison between the theory by Ashill and the theory by Kida and Miyai, the author calculated the induced drag factor from the Kida and Miyai result (Figure 11.47) for two cases of air gap as shown in Figure 11.47. In this form, it is seen that the prediction by Kida and Miyai, for the selected

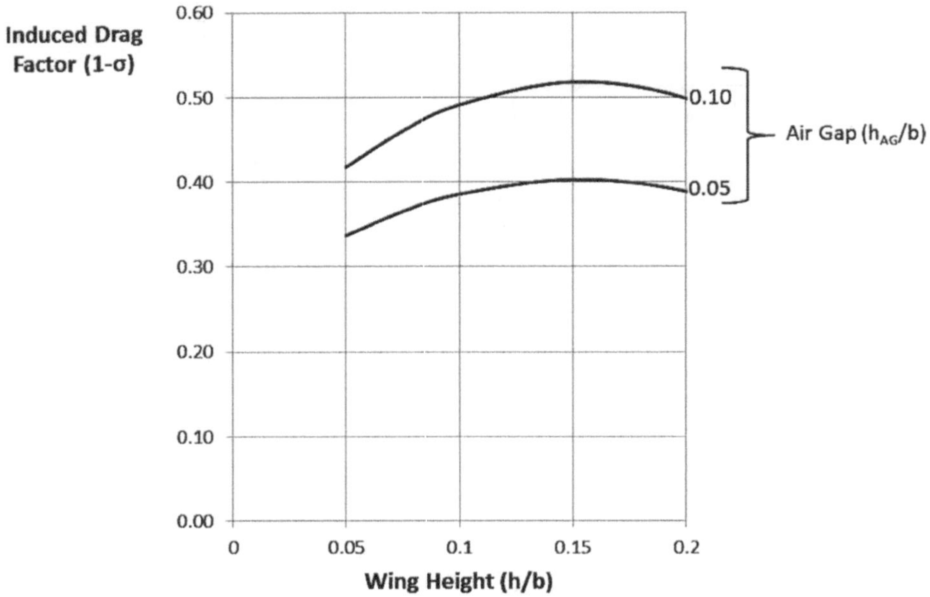

Figure 11.47: Induced Drag Factor for Wings With End Plates (Kida & Miyai)

values of air gap are similar to the results given by Ashill (see Figure 11.45 and earlier).

In 1976, two other researchers from Japan, Shigenori Ando and Haruki Yashiro from Nagoya University, Aichi, Japan[33] went one step further and developed a closed form solution for wings with end plates in ground effect. The cross section of these wings were semi-elliptic; rectangular and triangular. For the specific case of the rectangular wing with end plates as treated by Kida and Miyai, Ando and Yashiro derived a different expression for the span efficiency factor to be:

$$e \approx \frac{2\pi h}{3} + \frac{8}{\pi^2} \ln\left(\frac{1}{\varepsilon}\right) \tag{11.71}$$

This expression is an elegant closed form solution for the span efficiency factor that expresses the effect of both wing height and air gap beneath the end plates. There is a point of concern however in that while there is a similarity between this result by Ando and Yashiro (Equation 11.65) and that of Kida and Miyai (Equation 11.62 and Equation 11.64) the first term expressing the dependence of the span efficiency factor (e) on the wing height (h) is of a fundamentally different functional form at low wing heights than that given by Ando & Yashiro. Kida & Miyai show the height dependency as $h \to 0$, as being $\propto {}^{1}/_{h}$, while Ando & Yashiro give the result $\propto h$.

Ando and Yashiro state specifically in their paper that the difference is the result of an error by Kida and Miyai in their estimation of the elliptic integrals in the basic expression for the circulation. This causes a dilemma, in that both the Kida & Miyai prediction and the matching Ashill prediction provide an intuitively correct form of *reduced* induced drag as both the wing height (h) and the air gap (ε or h_{AG}) are reduced. By inspection, Ando and Yashiro's result in Equation 11.65, with the "corrected" term, gives a non-intuitive result of the induced drag factor *increasing* as the wing height is reduced.

Also, such a result does not agree with the results for a wing in ground effect (no end plates) as predicted by Prandtl, Wieselsberger, Ashill and Widnall and Barrows (see Figure 11.46). Intuitively one would expect the induced drag of wings with end plates would tend to give the same induced drag of wings without end plates as the depth of the end plate reduced to zero ($^l/_b \to 0$). Noting that the span efficiency factor (e), is the inverse of the induced drag factor, the results can be compared with the results for wings without end plates directly. For comparison with the prediction by Ashill, also for the case of no end plates, the author developed an approximation to Ashill's theory (shown in Figure 11.34) in terms of the *induced drag factor* (1-σ) to be:

$$1 - \sigma = 0.03 + 6.5\left(\frac{h}{b}\right) - 18\left(\frac{h}{b}\right)^2 \qquad (11.72)$$

It was shown earlier, that the 1961 Langley test data by Fink & Lastinger closely followed the Prandtl approximation with the inclusion of a constant $K_2 = 0.80$, such that the test data could be represented by a curve for the induced drag factor as:

$$(1 - \sigma)_{Test} = 0.80\left[\frac{0.05 + 8.72^h/_b}{1.05 + 7.4^h/_b}\right] \qquad (11.73)$$

These differences become clear when all results are compiled into one chart shown in Figure 11.48.

Each of the above theories for the case of a wing in ground effect but without end plates gave slightly different results and Figure 11.48 compares the theories of Prandtl, Ashill and Widnall & Barrows together with the test data obtained by Fink & Lastinger discussed earlier. As can be seen, generally, the theoretical results tend to have a similar functional form and all tend to predict a *smaller* induced drag reduction than the available test data as shown. Because of the many possible variations in wing shapes and geometries and the difficulty of including the viscous effects, it is believed that the differences between the various theories and the available test data are not sufficiently different to affect basic design decisions prior to detailed treatment in any particular case and that some general guidelines can be presented. It should also be pointed out that none of the theories include the effect of any changes in the wing profile drag as the wing approaches the ground that will also influence the final measure of the total drag ($C_{Di} + C_{Do}$) in these treatments.

For the heights envisaged for wings operating in ground effect and using the values found appropriate in the *ekranoplan* series of craft, it is not unreasonable to assume that for the likely operating heights, the reductions in induced drag can be expected to be:

$\frac{h}{b} \approx 0.15$ IGE induced drag approx 50–60% of OGE value
$\frac{h}{b} \approx 0.05$ IGE induced drag approx 25–35% of OGE value

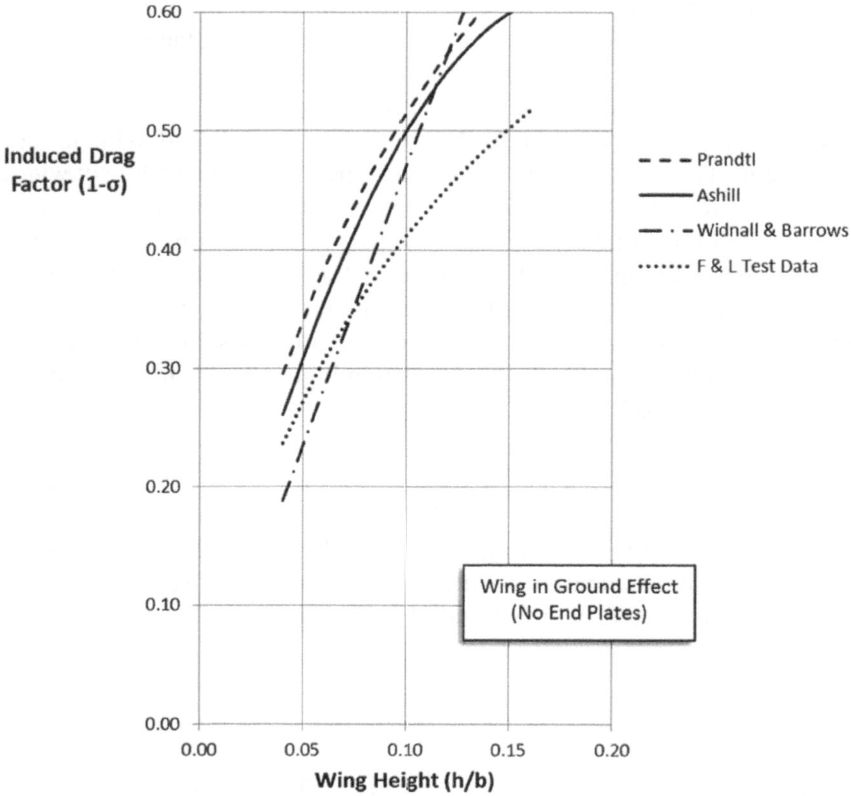

Figure 11.48: Comparison of Theory and Test for Wing In Ground Effect (No End Plates)

For the case of wings *with end plates*, the theoretical predictions by Ashill, Kida & Miyai, and Ando & Yashiro have been compiled in Figure 11.49 for the particular cases of air gaps beneath the end plates of ${}^{h_{AG}}/_{b} = 0.04$ and ${}^{h_{AG}}/_{b} = 0.10$ to be compatible with the earlier examples shown for the *ekranoplan* values. For reference, the prediction by Prandtl for wings in ground effect without end plates, together with the prediction by Ashill also for the no end plate case have been added to Figure 11.49.

It is seen from Figure 11.49 that the predictions by Ashill and Kida & Miyai have the same functional form and only differ numerically slightly, particularly at the lower wing heights $\left({}^{h}/_{b} < 0.15\right)$. The prediction by Ando & Yashiro, however, has a different functional form as noted earlier, because of the disagreement over the solution to the elliptic integrals. Fortuitously, over the range of wing heights envisaged for practical wings in ground effect, i.e. $0.10 < {}^{h}/_{b} < 0.20$ based on the *ekranoplan* experience, the numerical differences between all the theories are relatively minor and any differences would probably be inside any changes caused by inclusion of viscous effects and geometric wing variations. It is left as an exercise for the reader to determine which solutions to the troublesome elliptic integrals are correct!

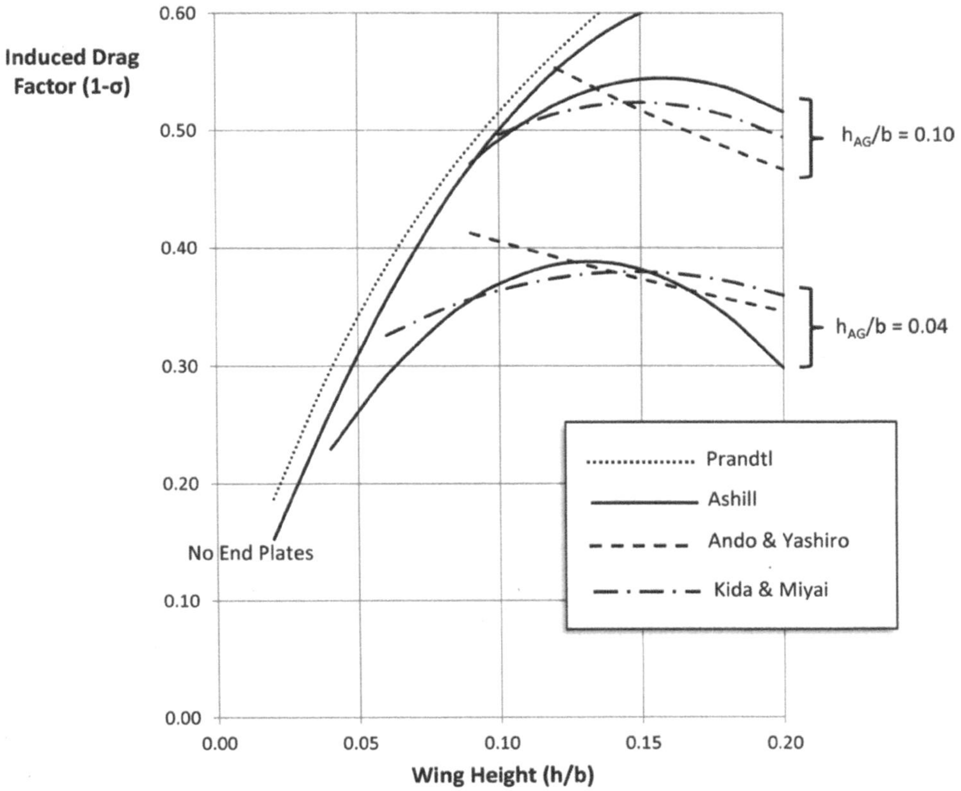

Figure 11.49: Comparison of Induced Drag Factors of Wings With End Plates

For economic value, the question now is *"how much power must be expended to achieve this higher aerodynamic efficiency and lower induced drag?"*. The data available for the *ekranoplan* suggests the power required is quite high and comes with a significant weight penalty.

11.12 Transport Efficiency of *Ekranoplan*

Chapter 8 on Technical Considerations, provides a detailed treatment of the key parameters in the economic equation for any vehicle, with an emphasis on deriving the power expenditure. It is shown in Chapter 8 that the range of the vehicle depends strongly on three vehicle parameters as expressed in the Breguet Range Equation[o]:

$$R = 325\left(\frac{1}{sfc}\right)\left(\frac{WV}{P}\right)\left(\log_e \frac{1}{1 - {}^{W_f}/_{W}}\right) \tag{11.74}$$

[o] For the purposes of this immediate discussion, the Breguet Range Equation for propeller driven craft will be used. Modifications to this equation for turbojet and turbofan engine propelled craft will be introduced later as necessary.

The three terms in the brackets of Equation 11.68, show the dependence on the fuel consumption (*sfc*) and the Transport Efficiency($\frac{WV}{P}$), and the ability to provide enough volume and weight for the on board fuel (W_f). The Transport Efficiency was shown to be:

$$\text{Transport Efficiency (TE)} = \frac{WV}{P} = \eta_P \left(\frac{L}{D}\right) \tag{11.75}$$

The propulsion system efficiency (η_P) is the *explicit* modifier to the aerodynamic efficiency ($^L/_D$) that determines how much power (P) must be installed in the craft. There is however, an additional *implicit* requirement not immediately obvious in the determining of the total power (P) that must be included, and that is the power required to give the very feature incorporated into the design for it to function in the manner described. A similar situation occurs for the *aerostatic air cushion craft* discussed more in Chapter 8 where it is shown that the Transport Efficiency must include the power required for lift, such that:

$$\text{Transport Efficiency (TE)} = \frac{WV}{P} = \eta_P \frac{W}{D + \frac{\eta_P}{\eta_L}\left(\frac{pQ}{V}\right)} \tag{11.76}$$

This required form for the Transport Efficiency shows where "how the lift was generated" is now included in the determination of the total power required. For the aerostatic air cushion craft, the lift system can be characterized by the lift system efficiency (η_L), the supporting air pressure in the cushion (p), and the air cushion flow (Q).

The difference between the conventional aero *static air cushion craft* and the *ekranoplan* is that for the air cushion craft the lift system must be "on" throughout its operational cruise in order for the craft to function to achieve its performance. The *ekranoplan*, on the other hand, only uses its "air ingestion" for its take-off and landing (TO&L) modes and operates in cruise with its lift system "off". For the *ekranoplan*, the air ingestion or blowing under the wing is essentially nothing more than a dynamic take-off & landing device. That is why, in the earlier discussion of the aerodynamic characteristics of the *ekranoplan* in cruise, it was only necessary to determine the aerodynamic performance of the wings flying in ground effect, regardless of how they achieved that mode of flight – BUT, the *ekranoplan* would not function unless it carried those "lift" engines throughout its entire operation and so the power for those "lift" engines must be included in the final determination of the efficiency of the design. The question is: *"should that extra power be included in the calculation of the Transport Efficiency and increased fuel consumption or as a "dead weight" that will detract from the craft's payload carrying capability?"*

In one sense, the ekranoplan concept is a version of the Jet Assisted Take-Off (JATO) designs used in WW II by both Allied and German aircraft to boost heavy aircraft during take-off to reach cruise speeds. In the JATO designs, the rocket or jet engines were jettisoned after take-off to improve the range capability without having to carry the dead weight of the used JATO engines.

US developments of JATO aircraft began with experiments by Theodore von Karman[p] at GALCIT at CALTECH, California in the period 1939–1941. The JATO concept was used on several military aircraft in the US, Germany, USSR and the UK both during and after WW II. It was not used for commercial aircraft.

In the concept of the *ekranoplan*, there is the disadvantage in that the weight of the take-off assisting engines (8 out of 10 engines in the case of the KM craft) must be carried throughout the flight compromising the payload and range capability. On the other hand, the advantage of the *ekranoplan* concept was that the same engines could be used again for the "return flight" which was not possible with the JATO concept.

Table 11.6 shows the Transport Efficiency (TE) for the three ekranoplans; KM, Lun and Orlyonok (using the available data from Table 11.5) under the assumptions (a) all engines are operating and (b) only the cruise engines are operating. Because different types of engines and propulsion systems are used by the craft, a representative propulsive efficiency of ($\eta_P = 85\%$) has been used for clarity in Table 11.6. A more detailed treatment could assign different values once the details of each particular configuration (adiabatic efficiency, propeller efficiency, transmission efficiency, etc) were made known.

Table 11.6: *Transport Efficiency of* ekranoplans

Operating Mode	KM	Lun	Orlyonok
All engines operating	2.9	3.3	2.9
Only cruise engines operating	14.4	13.1	12.3

As can be seen from Table 11.6, the use of power to ingest air under the wings of the ekranoplan pays a heavy price for the Transport Efficiency of the design.

In Chapter 8, a review is provided of vehicles' Transport Efficiency, with an emphasis on air cushion craft, ships and conventional aircraft. If the above values for *ekranoplan* are superimposed on such a plot, the "competitiveness" of such a transport vehicle can be assessed. This is shown in Figure 11.50. Both values of Transport Efficiency for each *ekranoplan* has been included in Figure 11.50. Compared to conventional aircraft, it is seen that even with the most optimistic assumption of power consumption, the *ekranoplans* did not exhibit the expected or promised improvement in performance. The more correct value of Transport Efficiency where all contributing engines to the craft performance has been included show that particular method of blowing under the wing to achieve sufficient lift capacity to allow the craft to move away from stationary position to cruise speed has severely compromised the efficiency of the total design. If some combination of power inclusion is used to be between the two extremes it is seen that the Transport Efficiency is still well below competing concepts.

[p] See Chapter 7 for discussion of Dr von Karman's work.

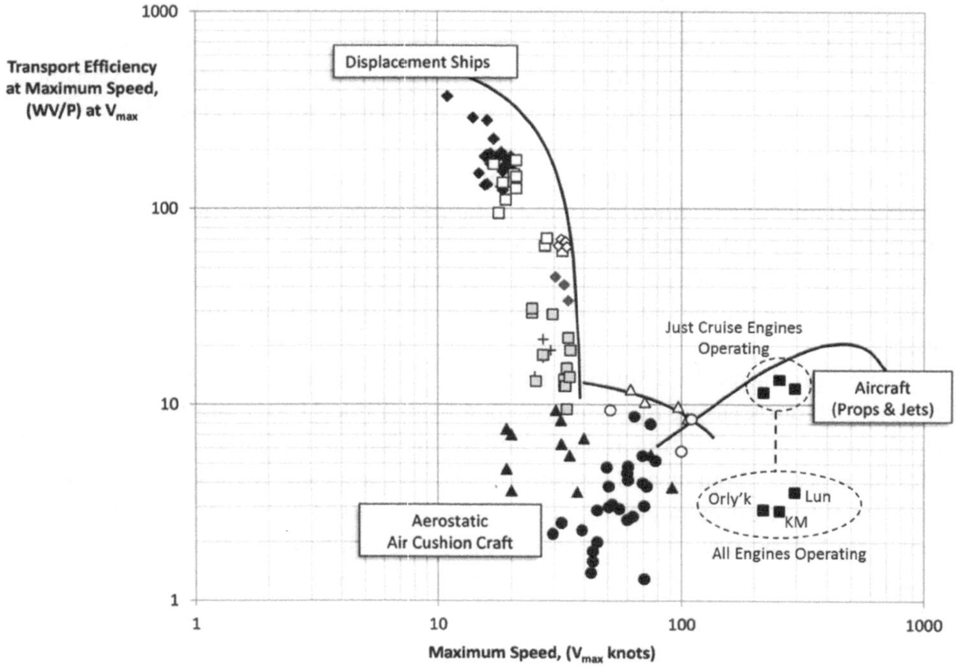

Figure 11.50: Transport Efficiencies of *Ekranoplans*

11.13 Alternative Forms of "End Plates"

While not documented in any detail, there are other forms of end plates that could have been tried in later *ekranoplan* and PAR-WIG designs to achieve the same benefit in lowered induced drag but without the danger of having hard structure fly very close to the waves. For conventional aircraft, the use of "winglets" is a well-known feature to improve the lift and drag of wings to improve their range characteristics, without the need to extend the "winglets/end plates" below the wing into harm's way.

Figure 11.51 shows a sample collection of winglet designs used on conventional modern day aircraft. These winglets allow for improved performance of the aircraft by reducing the induced drag. The upper two sketches in Figure 11.51 are the usual designs. The middle and lower sketches show more sophisticated designs for breaking up the vortex action at the tips of the wings. It has been found that even with high aspect ratio wings, further improvements or reductions in induced drag can be achieved with such devices. The reader is referred to standard texts on aircraft design to obtain specifics on the improvements achieved by winglets.

Alexander Lippisch took advantage of such "winglet" devices for his wing-in-ground-effect craft (WIG). Figure 11.52 shows the reverse delta wing *X-114* equipped with winglets. The first flight was on 15 April 1977. Lippisch[q] sold his patents to

[q] Actually sold by Collins Radio where Dr. Lippisch was employed at the time.

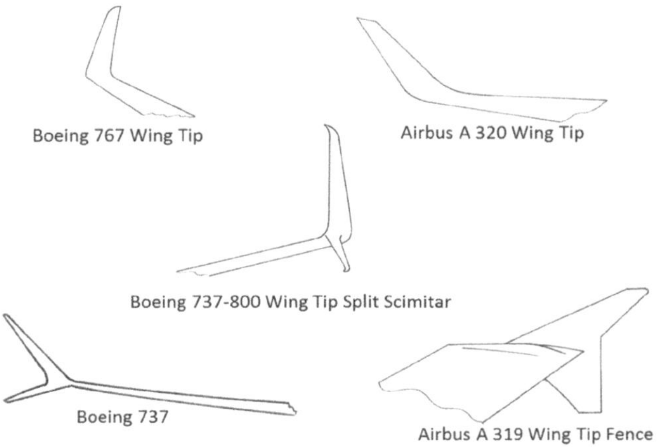

Figure 11.51: Conventional Aircraft Winglets

Figure 11.52: *X-114* Flying in Ground Effect
Photo permission by Airbus Group Corporate Heritage

Rhein-Flügzeugbau GmbH (RFB), Möchengladbach, Germany (a subsidiary of Fokker-VFW)[r] after the *X-112* craft design. Subsequent craft, the *X-113* and the *X-114*, were then built and tested by RFB after Lippisch passed away in 1976, although he did consult with RFB on these craft early in the design process. It is not clear if some of the additional features (such as shock absorbers, landing gear and ladder foils) were in the original designs of Lippisch or were added later by RFB. The *X-114H* ladder foils were mounted on the floats (not yet added in Figure 11.52). This feature was to decrease the take-off time and shorten the take-off distance, but this proved not to be the case. Unfortunately, during a "landing" approach, the pilot lowered the pitch angle too much such that the foils entered the water at a negative attitude and the craft crashed (the pilot survived without serious injury). The X-114 was discontinued after this accident. Such incidents highlight the difficulty of designing WIG craft to operate safely at the sea-air interface.

To complete the comparison of the *ekranoplan* concept with competing craft, the third parameter in Equation 11.74 must be examined. That term determines how well the design can accommodate a "useful load" which is a combination of fuel load

[r] Fokker-VFW was taken over by MBB in 1981, which became DASA in 1989 which subsequently has been become part of EADS (a multinational company formed by DASA, CASA, Aerospatiale-Matra et al.) in 2000. EADS North America is now part of Airbus Group.

(W_f), which is explicitly shown in Equation 11.74 and the empty weight (W_E)and the payload (W_P). That is:

$$W = W_E + (W_P + W_f) \tag{11.77}$$

In order to understand the nature of the Empty Weight (W_E) of the *ekranoplan* it is necessary to first review the governing design parameters that influence the two major sub-systems: structural weight and propulsion system weight as follows.

11.14 Impact of Wing Loading and Cushion Density on Craft Size

A key parameter that affects the design of air cushion craft is the *cushion density.* It affects both the drag and the structural weight. It is defined as:

$$\text{Cushion density} = \frac{p_C}{\sqrt{S}} = \sqrt{\frac{L}{B}}\left(\frac{p_C}{L}\right) \tag{11.78}$$

In this relationship, the supporting pressure for the craft can be written:

$$p_C = \left(\frac{W}{S^{3/2}}\right)(W)^{1/3} \tag{11.79}$$

Structural weight and overall sizing is determined by the wing loading $\left({}^{W}/_{S}\right)$ and the cushion density[s] given as $\left({}^{W}/_{S^{3/2}}\right)$. In Chapter 10, it is shown that the structural weight of the vehicle can be given as:

$$\frac{W_S}{W} = \frac{F(t, E, \ldots.)}{p_C} + \frac{n}{\sigma} G\left({}^{L}/_{B}, {}^{H}/_{L}, \ldots.\right)(W)^{1/3} \tag{11.80}$$

In this equation, it was shown that:

$F(t, E, \ldots.)$ includes the local properties of the panels and stiffeners and reflects the buckling criteria effect (through the Young's Modulus E) and the minimum gauge thickness (t) of the structural members.

$G\left({}^{L}/_{B}, {}^{H}/_{L}, \ldots\right)$ reflects some of the overall sizing or geometric effects on the design. It includes the overall shape through slenderness ratio $\left({}^{L}/_{B}\right)$ and the craft depth or height to length ratio $\left({}^{H}/_{L}\right)$ reflecting the beam strength.

n load factor dominated by the impact pressures caused by contact with the waves.

σ average stress level in the structure.

In Chapter 10, it is shown that this form for the structural weight was representative of many vehicle types and that the trends for air supported craft could be expressed in the form:

[s] In this discussion, it is convenient to use the terms developed in Chapter 8 and earlier chapters for the aerostatic air cushion craft. It should be clear from the context whether ${}^{W}/_{S}$ refers to "wing loading" or "cushion pressure" that supports the craft.

$$\frac{W_S}{W} = \frac{0.25}{W^{1/3}} + \frac{0.04}{\left[{p_C}/{\sqrt{S}}\right]^{1/3}} (W)^{1/3} \text{ (W in tonnes)} \tag{11.81}$$

It was shown that the sizing parameter, called "cushion density" or "loading density" depending on the context was crucial to the sizing of the craft. Equation 11.81 shows the strong influence of this parameter on the structural weight (W_S) which is the dominant component of the empty weight (W_E).

It was shown in Chapter 8, that this same sizing parameter played an important part in determining the hydrodynamic drag of the pressurized air cushion that forms beneath the wing. Such a drag was found to be directly proportional to this sizing parameter and directly impacted on the design of the propulsion system to ensure that the craft (*ekranoplan*) could accelerate from rest, "climb out" of the water depression and attain cruise speed. From Chapter 8, it was shown that this hydrodynamic wave drag can be expressed in the form:

$$\frac{D_W}{W} = \frac{4}{\rho_W g} \left(\frac{p_C}{L}\right) f\left(F.No.; {L}/{B}\right) \tag{11.82}$$

An expression developed for the maximum value of this Water Depression Drag was found to be:

$$\left|\frac{D_W}{W}\right|_{MAX} = \frac{4}{\rho_W g} \left(\frac{p_C}{L}\right) \left(\frac{1}{\sqrt{{L}/{B}}}\right) \tag{11.83}$$

The "loading density" can be written in different forms, as shown:

$$\frac{W}{S^{3/2}} = \frac{p_C}{\sqrt{S}} = \frac{p_C}{c} \left(\frac{1}{\sqrt{A}}\right) \tag{11.84}$$

The dimension "c" is the length of the water depression in the longitudinal direction of the vehicle, which would be the chord of the wing in the case of WIG or *ekranoplan* or cushion length "L" in the case of the aerostatic air cushion craft. Different researchers have used each of these variations of the same density parameter to express its importance. From a designer viewpoint, the craft density expressed in Equation 11.84 is instrumental in *lowering* the structural weight and the same parameter is instrumental in *increasing* the propulsion system power to overcome the hydrodynamic drag.

It was shown in Chapter 10 that the load factor "n" in Equation 11.80 was significantly higher for the high speed *ekranoplan* when compared to the load factors for aerostatic air cushion craft. Figure 10.12 in Chapter 10 showed that the load factor "n" for the *ekranoplans* was approximately 5 or 6:1 over that for the aerostatic air cushion craft because of the extreme difference in design speeds (250+ knots compared to 50+ knots). Each of these considerations has a major impact on the structural weight (W_S) and the empty weight (W_E) of the craft. It is not necessary to delve into the detail design of each craft to illustrate these effects on the final design of the craft. Fortunately, there is a sufficient data base of the many different types of craft built to show the major trends. Figure 11.53 shows the trends in the density parameter $\left({p_C}/{\sqrt{S}}\right)$ for a range of different craft: fighter aircraft (designed for high "g" maneuverability); conventional low speed aircraft designed for a less stringent

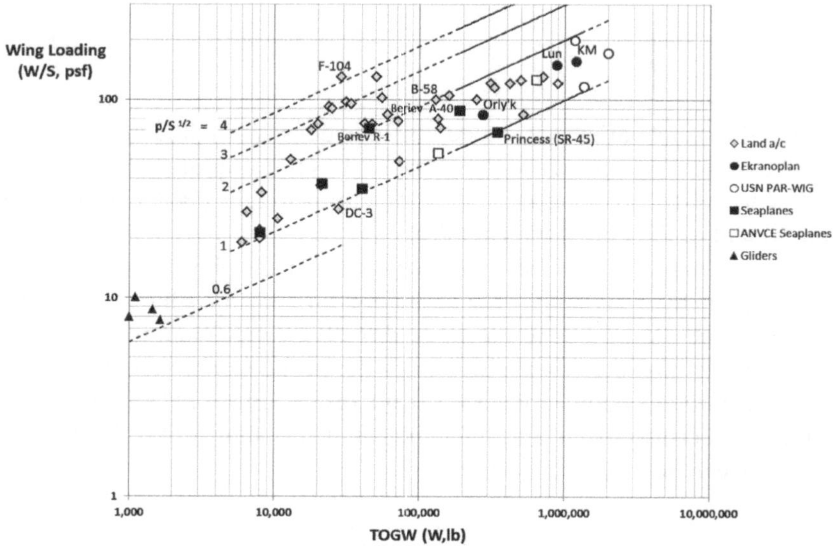

Figure 11.53: Wing Loading of Landplanes, Seaplanes, *ekranoplan* and WIG

environment; seaplanes (designed for high impact loads in waves); and for *ekranoplans* designed for high speed close to (and occasional impact with) the wave tops. Preliminary design data for the US Navy versions of similar craft (PAR-WIG) have also been included in Figure 11.53.

From this collected data, it is seen for *seaplanes* and for *ekranoplans*, that must deal with impact loads in waves, have density values $1.0 < p/\sqrt{S} < 2.0$. It is interesting to note that the wing loading for the aerodynamic air cushion craft (*ekranoplan*) and the aerostatic air cushion craft follow similar patterns with craft size. Figure 11.54 shows this comparison for a wide range of *aerostatic air cushion craft* (from Chapter 8) with the wing loading for the *ekranoplan* superimposed. Note that the comparison illustrates the design choice for high values of cushion density. These boundaries on the two schools of cushion density design can be expressed as:

$$\text{High density}: \; p_C = 1.45(W)^{1/3} \tag{11.85}$$

$$\text{Low density}: \; p_C = 0.75(W)^{1/3} \tag{11.86}$$

In these relationships, the craft weight (W) is expressed in tonnes.

11.15 Empty Weight of Aerodynamic Air Cushion Craft

Given the backdrop of the key parameters that determine the structural weight and propulsion system weight that are the largest contributors to the empty weight, the statistics of the many craft built can be reviewed. Figure 11.55 collects the empty weight (W_E) of landplanes and seaplanes. While there are instances of designs that do not follow the trend, it is normally found that seaplanes exhibit higher values of empty weight than landplanes because of the need to strengthen the hulls to withstand the high impact pressures from contact with waves as already discussed. The

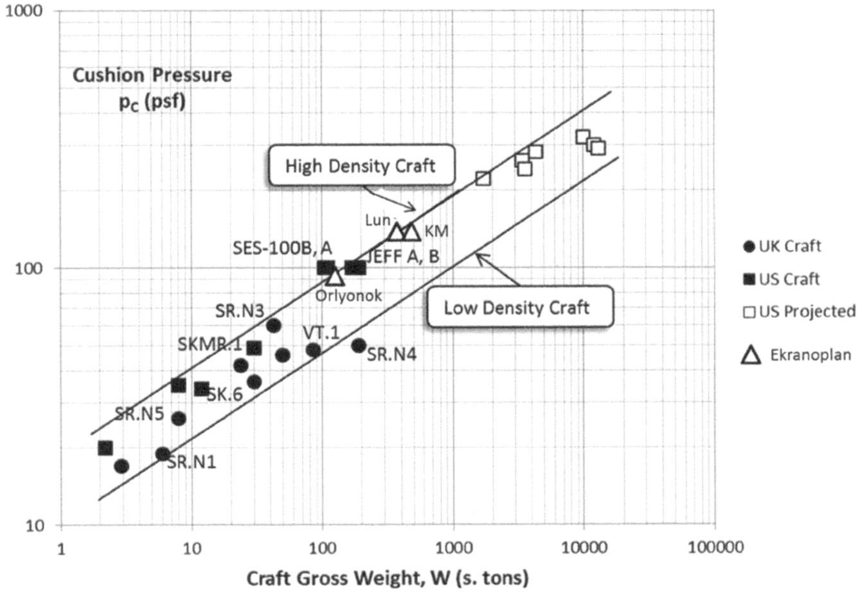

Figure 11.54: Cushion Density of Air Cushion Craft (including *ekranoplan*)

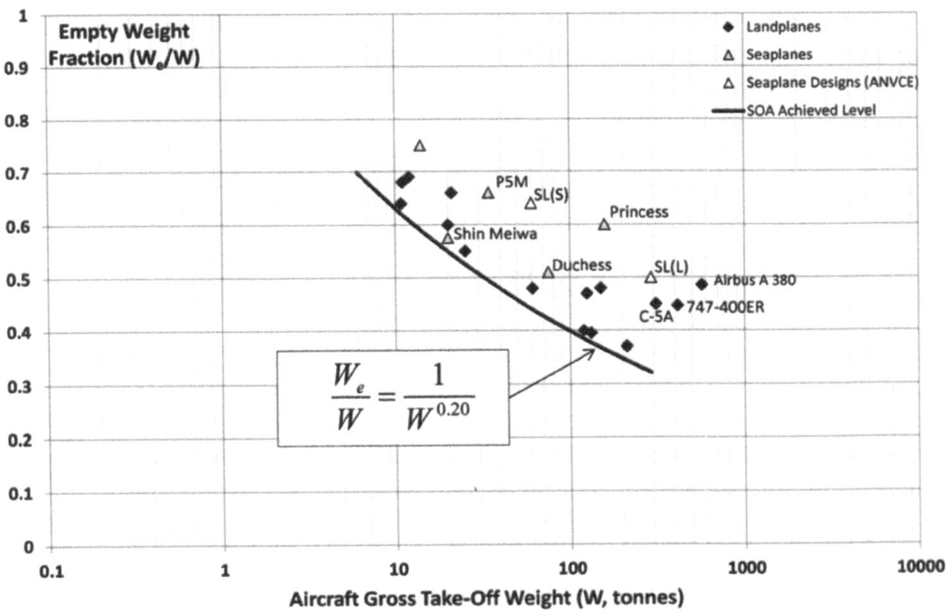

Figure 11.55: Empty Weight of Landplanes and Seaplanes

"best" values achieved by either landplanes or seaplanes are shown by the lower "limit" line in Figure 11.55.

In Figure 11.56, this same "limit" line for *landplanes* and *seaplanes* has been included for easy reference when comparing the empty weight of *aerodynamic air cushion craft* and in particular the *ekranoplan* both built and projected. Given the

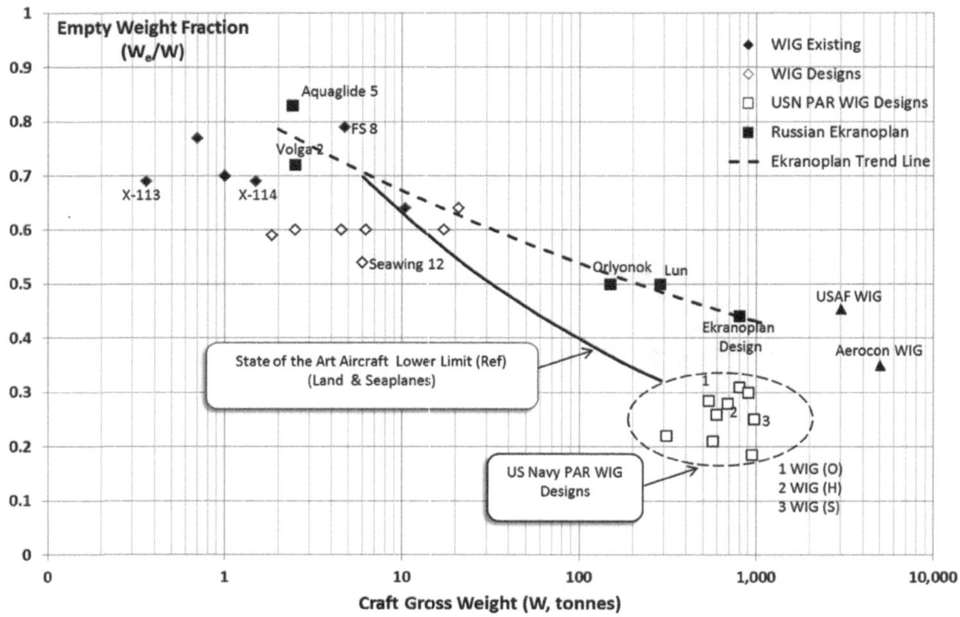

Figure 11.56: Empty Weight of *ekranoplan* and WIG

above discussion on the contributors to the empty weight, it should not be surprising to see the relatively high values of empty weight for *ekranoplan* as shown by the dotted line in Figure 11.56.

Some projected designs for PAR WIG done by the USAF and DARPA are included in Figure 11.56. It would be expected, based on the above discussion, that the weight trend would depend strongly on craft speed (V) and load factor (n), for a result similar to:

$$\frac{W_E}{W} = \frac{f(V,n)}{W^m} = \frac{0.85}{W^{0.10}} \tag{11.87}$$

Unfortunately, there is an insufficient number of actual craft to determine the function $f(V, n)$ but a constant value $f(V, n) = 0.85$ and an exponent $m = 0.10$ matches the dotted line shown for a range of ekranoplan gross take-off weights 2 < W <1000 tonnes. Because of the similarity between seaplanes and ekranoplans, three particular aircraft are shown for reference in Figure 11.57 (Glenn Martin P5M); Figure 11.58 (Saunders-Roe Princess); and Figure 11.59 (Lun ekranoplan) to illustrate the similarities.

The Glenn Martin *P5M-1G* entered service with the US Navy as a patrol aircraft in 1951. The original *P5M* series was built with a conventional tail, but in 1961 (or possibly earlier)[34] the design was changed to give the high "T-tail" as shown in Figure 11.57 for the US Coast Guard version of the *P5M-2G*. The high T-tail was installed primarily to avoid the spray generated by the main hull during landing and take-off. The *Marlin* was not amphibious and a beaching gear (see photo) was used to handle the aircraft on land. A unique design feature developed during extensive model testing was the shaping of the lower hull to provide a smooth take-off without porpoising. Note that the seaplane also was designed to have a high wing, again to

Figure 11.57: 43 tonne Glenn Martin *P5M-2G* (*Marlin*)
US Navy photo

Figure 11.58: 165 tonne Saunders-Roe "*Princess*"
Photo permission by GKN Aerospace©

Figure 11.59: 400 tonne *Lun ekranoplan*
http://thevelvetrocket.com

minimize spray and green water entering the propellers and engines. About 285 aircraft were built with the last plane retired from service in 1967.

In the same seaplane era (1950–1960+) a much larger seaplane appeared on the scene with the Saunders-Roe Ltd *SR.45, "Princess"*. This aircraft (called a "flying boat" in the UK) made its first flight on 22 August 1952. Figure 11.58 shows the only *Princess* flown (of three built)[t] flying at the air show in Farnborough, England in September 1953. Note the similarity in the lower hydrodynamic hull shaping between the *P5M-2G* and the *Princess*, with the unique shaping to achieve optimum spray and drag characteristics and stable flight during the critical TO & L modes.

The under hull shaping of the *ekranoplan* (Figure 11.59) followed similar rules for breaking the skin friction of the water as the craft accelerated during its take-off run to achieve its design speed.

The shaping of such hulls is a complex design problem and was part of the ongoing evaluation of high speed hull shaping for both seaplanes and *ekranoplan*-like craft in the ANVCE Project (Chapter 10). Details may be found in two pertinent reports by Peter Crewe of Saunders-Roe Ltd who was an advisor to the ANVCE Project in 1976–1977.[35,36]

The two examples (*Marlin* and *Princess*) are good examples of seaplane technology that came into being after World War II, but the seaplane era suffered several contributory causes to their eventual demise. These are (a) marine corrosion; (b) reduced payload capability; (c) rapid improvements in landplane technology, and (d) extensive availability of land plane airports. Without extensive (and costly) maintenance, seaplanes suffer from marine corrosion to both the aluminum structure and to the engines ingesting corrosive salt air. The two *Princesses* that were cocooned still suffered corrosion from the salt air and the hull structure actually crumbled to the touch when the cocooning was removed a few short years later. As can be seen from Figure 11.55, the *Princess* had a 50% increase in the empty weight over a comparable land plane, which is a considerable loss in useful load (payload + fuel load). The history of the *ekranoplans* has been discussed earlier but the stepped hull configuration for improved TO&L hydrodynamic performance of the *Lun* (1987) can be seen clearly in Figure 11.59 operating at its cruise height. This photo may be compared to other *Lun* photos in Figure 11.42 and later. Like the *Marlin* and the *Princess*, the *Lun* was also not amphibious requiring a beaching gear to handle the craft on dry land. From Figure 11.56, it is seen that the *ekranoplans* have not made any improvement in payload carrying capability over the earlier seaplane technology. The empty weight of the *ekranoplan Lun* is approximately the same as for the seaplane *Princess*. As an additional set of data, included in Figure 11.52, are the projected

[t] There is a sad footnote to history in that three *Princesses* were ordered (in May 1946) to be a transatlantic passenger service for British Overseas Airways Corporation (BOAC). All three aircraft were built but the order was cancelled when it was determined that the engines did not meet the contract performance requirements. Two *Princesses* were cocooned and the first *Princess* (Figure 11.58) operated for about 100 hours flying time. After several attempts to find alternative buyers, all three aircraft were broken up in 1967.

empty weights for the US Navy PAR-WIG variants of the *ekranoplan*. These projected weights have already been commented on in Chapter 10 in discussion of the ANVCE Project review of all the competing concepts for "high speed at sea". These values reported separately by the US Navy[37] and in the ANVCE Report[38] argue that such low values of empty weight can be achieved through use of compact structures (low aspect ratio wings) and through impact load alleviating use of the air cushion beneath the wing. As discussed in more detail in Chapter 10, the ANVCE Project concluded that such predicted low values of empty weight, in light of the available evidence of operational craft discussed above, were deemed not credible and the values shown in Figure 11.56 for the *ekranoplan* are more representative of what can be accomplished for *aerodynamic air cushion craft*.

11.16 Military Operations for *ekranoplan*

From the above analysis of (a) Transport Efficiency and (b) Empty weight, it would appear that the *ekranoplan* would not be a good choice for transport missions or long range transport across open seas. Its aerodynamic efficiency does not compete with other vehicles (conventional ships, transport aircraft, etc); neither does its transport efficiency, when the power cost to achieve cruise is considered. The eight out of ten engines installed required for the KM to get to cruise speed is a good example of this inefficiency and finally, the high values of

Figure 11.60: Typical Amphibious Landing (Omaha Beach 6 June 1944)
US Army photo

empty weight mean much less capability to carry meaningful payloads, which again mitigates against such designs for long range transport. The *ekranoplan* designers recognized this fact and accordingly concentrated on shorter range, special purpose missions for their craft. The *Orlyonok* was designed for short range *amphibious warfare* craft operating over relatively calm waters. The *Lun* was designed for similar missions with a *short range mission* equipped with just six missiles for limited attack military operations.With the emphasis on amphibious warfare, an issue not discussed by the designers is the *operational environment* for such missions. Figure 11.60 is a repeat shown in Chapter 9 for the amphibious landing on Omaha beach in 1944 during World War II. One is struck by the close proximity of all the varied fighting elements and the need to closely integrate the assets if the military objective is to be met.

The publicity photos circulated to show *Orlyonok* in an amphibious setting (see for example Chapter 9, Figure 9.7) are careful to show the craft sitting statically on a flat sandy beach. A more real depiction of the *Orlyonok* in an amphibious landing is shown in Figure 11.61. In the top photo (a) the *Orlyonok* is approaching the beach with its air injection system blowing high pressure jet exhaust beneath the wing generating large amounts of spray. The more dramatic photo is (b) where the *Orlyonok* has now turned and is leaving the beach where now the high pressure jet exhaust is churning up large amounts of wet sand into its immediate surroundings. At the power levels required for the *ekranoplan* to function, the jet exhaust blowing directly into the sand at over 23,000lb force from each of two engines, will most certainly gouge large holes into the beach and cause serious sand blast damage to structure, engines and more importantly into to the nearby military personnel who are on the same beach and part of the same amphibious invasion (Figure 11.60).

It should also be noted that the *ekranoplan* cannot reverse or maneuver in close quarters, further complicating amphibious landings of multiple craft (see later discussion on stability and maneuverability of aerodynamic air cushion craft.

In addition to the problem of spray and sand damage to nearby amphibious forces, is the problem to the craft itself. The *ekranoplan* designers used various techniques to protect the vulnerable engines to such treatment. As can be seen from Figure 11.61, the *Orlyonok* design buried the engines into the fuselage to minimize saltwater and sand ingestion. Also, the bow shaping of the *Orlyonok* includes upswept buttock lines and spray deflectors to minimize the green water intake of the engines. A similar shaping appears on the *Lun* craft (see Figure 11.42 and Figure 10.59 in Chapter 10). These improvements in design appeared after the initial design and construction of the *Caspian Sea Monster*. The issue of salt water ingestion limited the sea operation and the *Orlyonok* and *Lun* which were not qualified for open sea environment. The 1994 DARPA report[39] contains detailed assessment of this problem of the *ekranoplans* of saltwater and sand ingestion and the impact on military operations and craft availability.

(a) *Orlyonok* approaching a sandy beach . . .

(b) . . . *Orlyonok* leaving a sandy beach

Figure 11.61: *Ekranoplan* and Amphibious Warfare

11.17 Aerodynamic Air Cushion or Aerostatic Air Cushion?

While it must be left to later writings for detailed analysis, there are several solutions available to reduce the spray problem of *ekranoplans* and to ensure that such exposure during amphibious landings is minimized. What is sought is a solution closer to that experienced by the conventional *amphibious air cushion craft*. This can be seen in Figure 11.62 that shows the USMC *LCAC* traversing the same type of sandy beach and shoals as the *ekranoplan* but with minimal spray.

This was explored in the US Navy ANVCE Project in 1976–1978[40]. A sketch made for that report is shown in Figure 11.63 where all the features of the *Orlyonok*

Figure 11.62: *LCAC* traversing shoals near beach
US Navy photo

Figure 11.63: Combining *ekranoplan* and *hovercraft*

ekranoplan (burying the engines in the main hull/fuselage) are integrated with the conventional *amphibious air cushion craft* using side skirts.

The sketch in Figure 11.63 is meant not to be a definitive design but to illustrate that it is possible to take the features from one concept and combine them with another to produce a solution to one or more of the problems found in the original concept. The issue of stability and control can also be addressed by similar means. The difficulty of banking a *wing in ground effect* or *ekranoplan* because of the low surface clearance can be offset by the addition of a dorsal fin as shown in Figure 11.63. This was used successfully in the *Columbia* Project as discussed in Chapter 3 (see Figure 3.9). Different planforms can be used to control the aerodynamic center as used by the Lippisch designs (see for example, Chapter 1, Figure 1.16).

The world of aerodynamic *air cushion craft* contains many untapped solutions to date that could provide a practical "100 knots at sea" capability.

11.18 Mother Nature Knows Best

When the early pioneers were exploring manned flight, most efforts began by observing bird flight. The same observance of bird flight in ground effect will yield similar gems. After many, many millennia of Research & Development, Mother Nature concluded that attempting to achieve several very *different modes of operation for flight* into a *single geometry* was not a good idea. Birds that need to operate over a speed spectrum from zero speed on the ground (or water) to high speed in the air must have different geometries for each mode in order to function properly. The principle of *variable geometry* solves this problem. Figure 11.64 shows (as an example) the pelican in three modes of operation. Figure 11.64(a) shows the bird in a *hullborne* mode, where it can either float in a stationary mode or propel itself at low speed by use of its large webbed feet and strong legs (not shown). The pelican can also fly *in-ground-effect* mode once it has reached the necessary speed (by its webbed feet propulsion system) for sustained flight, as seen in Figure 11.64(b) to search for food (foraging) or migrate over long distances. The bird can also climb to altitude and reach higher speeds and attain the necessary altitude for flight over variable terrain, as seen in Figure 11.64(c) operating in *out-of-ground-effect* mode. Attempting to accomplish these different modes of operation and flight was simply not possible with a single geometry.

Of particular interest, is how the pelican, which is a big bird, can lift its large size through use of its light weight "structure" that encompass the use of air sacs, (a) in its skeleton and (b) under its skin, that ensure the necessary buoyancy for its "*hullborne*" mode, and light weight for its flying modes. Seen in both Figure 11.64(b) and Figure 11.64(c) is the clever use of its wing tips and its secondary feathers to disrupt the main vortex action and lower its induced drag. These feathers do not splay freely but under deliberate action by the pelican, as seen in Figure 11.64(b), where the secondary wing tip feathers are held together to provide "end plate" sealing *in ground effect*, and in Figure 11.64(c) are splayed forward like open fingers to achieve the low (induced) drag in *out of ground effect flight*, as necessary.

In the *ground effect* mode, the pelican lowers its wing tips to contain the pressurized air flow beneath its wings to lower the induced drag for sustained flight as it searches for its food. There is a familiar resemblance in this photograph to the many man-made designs using "end plate fingers" on WIG craft for the very same reason. The photo in Figure 11.65, taken at Laguna Beach, California, shows how this feature of creating "end plates" was used in unison with that familiar bird flight mode, of flying in a V-pattern formation to reduce the induced drag.

The combination of using (a) folded wing tips ("end plates") to reduce induced drag and (b) the use of V-pattern flight to further reduce induced drag gives birds a significant savings in drag and thus can sustain longer flights with reduced energy expenditure.

(a) Hullborne

(b) In Ground Effect (with "end plates")

(c) Out of Ground Effect Flight

Figure 11.64: Pelicans Operating in Three Modes

Figure 11.65: Brown Pelicans Flying in Ground Effect Mode (and V-formation)
Photo by author

11.19 Variable Geometry

This variable geometry feature used in bird flight led the author to postulate the same feature for variants of *wing-in-ground-effect craft* to achieve improved performance. In 1966, this led the author to put forward a variable geometry air cushion craft that would allow the craft to operate efficiently in all three modes of flight (hullborne, aerostatic air cushion mode, and aerodynamic air cushion mode). One example of the design appeared in the November 1966 issue of *Flight International (Air Cushion Supplement)*.[41] In that magazine different versions of possible variable geometries were discussed but Figure 11. 66 and Figure 11.67 show a display model in the General Dynamics Electric Boat Model Shop for one version of the *partial sidehull* design (see Chapter 8, Figure 8.36) integrated with the *wing-in-ground-effect* concept.

In Figure 11.66, the partial sidehulls are down for operation in (a) first a hullborne mode (to motor out of the harbor with minimum spray and noise) and also (b) for the aerostatic mode of operation where the craft is operating as a "conventional" air cushion craft. In Figure 11.67, the sideulls are raised such that the craft can operate in a wing-in-ground-effect mode and to be clear of the waves. This design concept was documented in more detail in the 1969 patent by the author.[42]

When analyzing a "multi-mode" craft operating at the interface between water and air, it must be recognized that different forces become dominant in each mode which affects the scaling laws. As discussed in Chapter 8, at low speeds the inertia forces acting on the craft are related to the gravitational forces inherent in wavemaking. This relationship is characterized by the Froude Number;

$$\text{Froude Number (F)} = \frac{\text{Inertia Forces}}{\text{Gravitational Forces}} = \frac{V}{\sqrt{gL}} \tag{11.88}$$

Figure 11.66: Variable Geometry WIG (Sidehulls down)
Photo Courtesy of General Dynamics Electric Boat

Figure 11.67: Variable Geometry WIG (Sidehulls up)
Photo Courtesy of General Dynamics Electric Boat

As the air cushion features are introduced into the craft concept, it was found that the air pressure and its interaction with the water began to play a significant role in the design of *aerostatic air cushion craft* (see Chapter 8). It was found early on in the development of hovercraft that the so called cushion density came into play, where:

$$\text{Cushion Density} = \frac{p_c}{S^{1/2}} = \left(L/B\right)^{1/2}\left(\frac{p_c}{L}\right) \tag{11.89}$$

The cushion density, expressed either in terms of the cushion area (S) or in terms of the cushion geometry (L,B) is a key parameter in determining the aerostatic forces acting on the craft. In the above expressions, V is the forward speed of the craft, "g" is the acceleration due to gravity, S is the cushion area, L is the cushion length and B is the cushion beam.

As the craft speed is further increased to speeds more appropriate to the *aerodynamic air cushion craft* the dependence on hydrodynamic and aerostatic forces diminishes and aerodynamic forces dominant with different scaling laws. In 1967, the author introduced the concept of the Pressure Number[43] defined by:

$$\text{Pressure Number (k)} = \frac{\text{Aerodynamic Forces}}{\text{Aerostatic Forces}} = \frac{1/2\rho V^2}{p_c} \tag{11.90}$$

The Pressure Number (k) expresses the ratio of the forces generated by the free stream to those generated by the cushion pressure. These "numbers" can be related through the simple relation:

$$F^2 = \left(\frac{2}{\rho g}\right)\left(\frac{p_c}{L}\right)k \tag{11.91}$$

Given these scaling parameters it becomes possible to show the results of model tests on a common scale, although the design of such a *multimode variable geometry air cushion craft* will require very different treatments in each mode. To illustrate how such a variable geometry craft could operate, a set of model test data non-dimensionalized by the Pressure Number (k) was presented in the 1967 paper and is reproduced here in Figure 11.68. Because of the emphasis on the final mode of operation where the aerodynamic forces dominate, the performance data, expressed as the Lift Drag Ratio (L/D), has been plotted against the Pressure Number (k).

In Figure 11.68, three types of air cushion craft have been collected to demonstrate the concept of a variable geometry air cushion craft. The shaded bold line, as an envelope of the data, would be representative of a craft that starts in the hullborne mode, then raises its sidehulls (in this example) to operate in its aerostatic air cushion craft mode and finally accelerates (with the skirts retracted) into the aerodynamic mode in surface effect. As shown in Chapter 8, the conventional hull forms lose their efficiency (L/D) as speed is increased. This is shown in Figure 11.68 with **Curve A**, using the test data from the partial sidehull model test discussed in Chapter 8 (see Figure 8.36).

In this constructed example, as the sidehulls are raised, the craft enters the aerostatic air cushion craft mode and now the lift-drag ratio (L/D) would exhibit values like those shown for different values of height above the surface (h/L). Finally, when the aerodynamic lift generated by the wing shape equals the aerostatic lift ($L = W = p_cS$), the skirts can be retracted (see Figure 11.68) and the aerodynamic forces dominate. The efficiency at high speed would then resemble the values shown by **Curve B**.

While the specifics for particular craft would vary, Figure 11.68 illustrates the three domains identified as "hydrodynamic", "aerostatic" and "aerodynamic". As discussed in more detail in Chapter 8, the hydrodynamic (and hydrostatic) forces are dominated by the Froude Number. These lose significance as speed is increased to higher values of the Pressure Number (k). In general terms, the dominance of the hydrodynamic forces occurs at speeds less than those given by $k<0.10$. The aerodynamic forces come into dominance as the Pressure Number (k) increases to values of $0.5 < k > 1.0$ and higher. Most *aerostatic air cushion craft* tend to be designed for a Pressure Number(k) of around $k = 0.30$–0.50 and a Froude Number (F. No.) of between 2 and 3.

An alternative variation of the variable geometry air cushion craft would be one more akin to the *channel flow craft* (see Chapters 1, 2, 3 and 4). In this case, there are no sidehulls and instead uses flexible hovercraft-like skirts. Figure 11.63 showed the design considered in the US Navy ANVCE Project (see Chapter 10). In this variable geometry design, the craft first operated in an aerostatic air cushion craft mode with complete peripheral skirts around the vehicle. As speed is increased to where the aerodynamic forces could support the craft (Pressure Number (k) $\approx$ 1.0)[u], the bow and stern skirts are retracted leaving just the skirts on the side edges of the craft. The purpose of those side skirts is now to function as "soft end plates" – like the Pelican's wing tip feathers – to aid in reducing induced drag.

[u] The actual value of "k" will depend on the lift coefficient (C_L) of the planform of the craft.

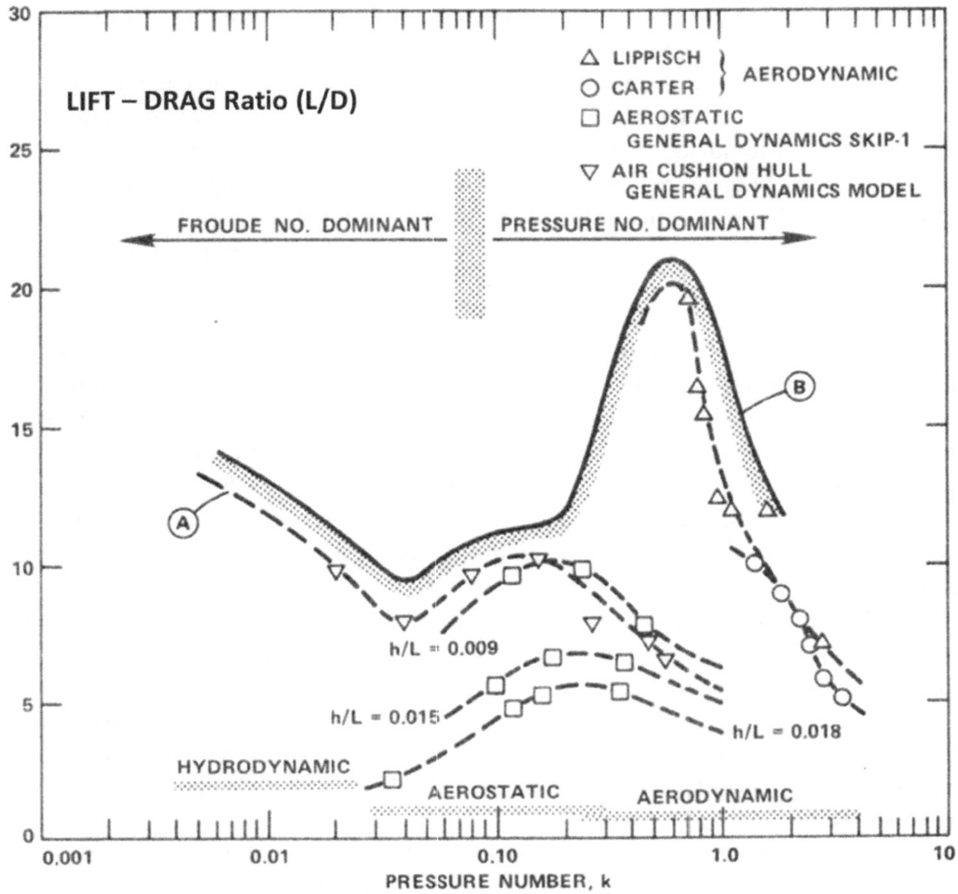

Figure 11.68: Lift Drag Ratio of Variable Geometry Air Cushion Craft

In Figure 11.63, it will be noticed that the propulsion engines are in the main body of the craft and the lift fans (to generate the cushion pressure) are not apparent. While earlier patent documents explain this feature, the sketches in the author's later 1979 patent[44] provide a good description of the feature employed. A set of sketches from that patent is provided in Figure 11.69.

In this instance, in a further nod to variable geometry, is to depart from the traditional practice in air cushion craft and surface effect ship design of having separate lift and propulsion schemes and to take advantage of available engine technology that allows separation of airflow for different functions. This can be seen in the "FIG 17" and "FIG 18" sketches in Figure 11.69 in that the incoming air flow to the engine (say a bypass turbofan) can be diverted to both "lift" and "propulsion" functions. At zero to low speeds, the low pressure compressor or fan air can be diverted through suitable nozzle arrangement into the air cushion for aerostatic lift. As speed is increased to higher and higher values, less compressor bleed air is used for the cushion and more to thrust. In the case shown in Figure 11.70 the compressor or fan air is diverted to inflating the side skirts for the induced drag reduction function. This engine technology is not new, of course, and has been successfully used in many applications. Figure 11.70 shows the Rolls-Royce

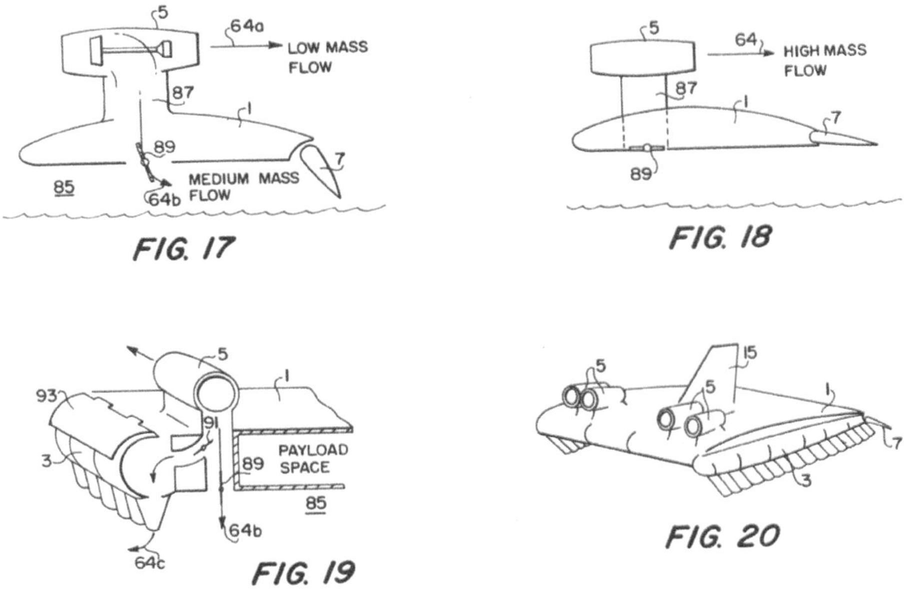

Figure 11.69: Integrated Lift-Propulsion for Variable Geometry Air Cushion Craft

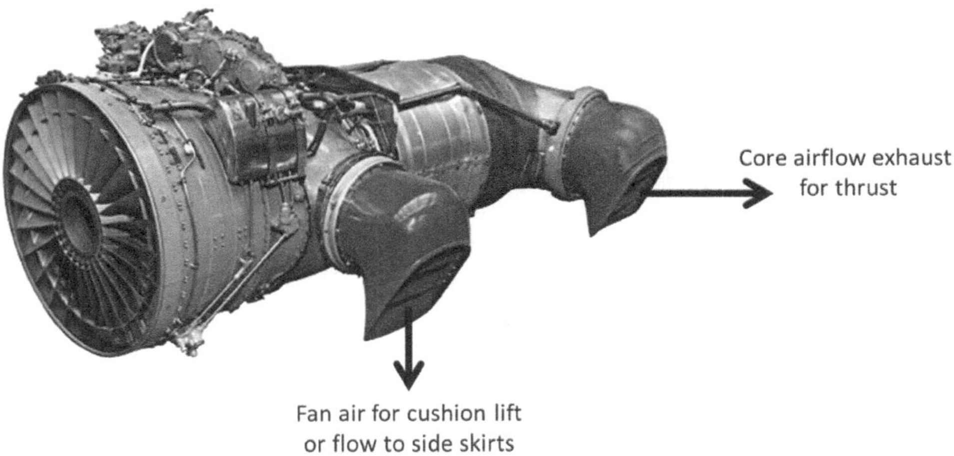

Figure 11.70: Illustrative possible application of *Pegasus* Turbofan Engine for Variable Geometry Air Cushion Craft
Photo permission by Rolls Royce plc©

Pegasus low bypass turbofan engine[45] shown in a diagrammatic manner for use on a variable geometry aerodynamic air cushion craft as shown in Figure 11.69.

The Rolls-Royce *Pegasus* turbofan is used successfully in such VTOL and STOVL aircraft as the *Harrier, AV-8B* and other multi-mode aircraft. The actual engine configuration will depend on the application and as shown in the above illustration has four rotatable nozzles to deflect both the fan air and the core exhaust air as needed for vertical lift-off and for forward flight. Without complicating the visualization it is clear that for application to the *variable geometry aerodynamic air cushion craft*, the nozzles shown would be arranged as necessary for the particular configuration under consideration.

The technology of turbofan engines and other variants (turboshaft, turboprop etc) and jet engines in general is well advanced and need not be explored here in this context. The choice of bypass ratio, pressure ratio and other important performance parameters would be selected as necessary for particular designs.

11.20 Wing in Ground Effect with Wing-Tip Jet Blowing

There have been other approaches put forward for the edge sealing of wings in ground effect to improve the lift-drag ratio. One approach was that tested by Raymond Mineck and Arthur Carter at NASA Langley in 1974[46]. In that research, Mineck and Carter ran model tests of a wing of low aspect ratio (A=1) flying over water in the Langley Tow Tank. The model was equipped with wing-tip blowing in different configurations. The configuration resembles the channel flow concept in the cruise mode (see Chapter 1, Figure 1.13d). The model set up and one set of model test results is shown in Figure 11.71. The air supply for the wing tip blowing was

(a) Model operating in Tow Tank

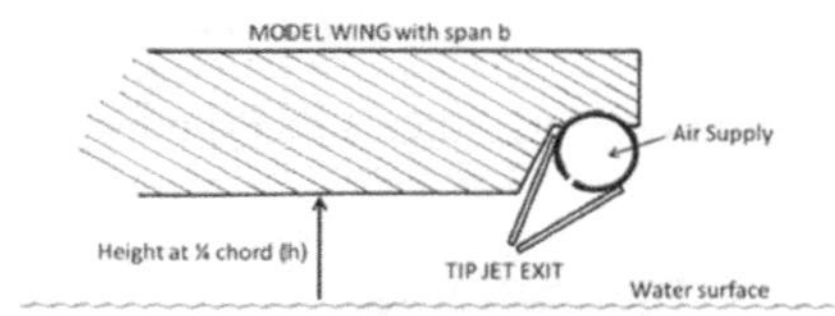

(b) Model Test Set Up

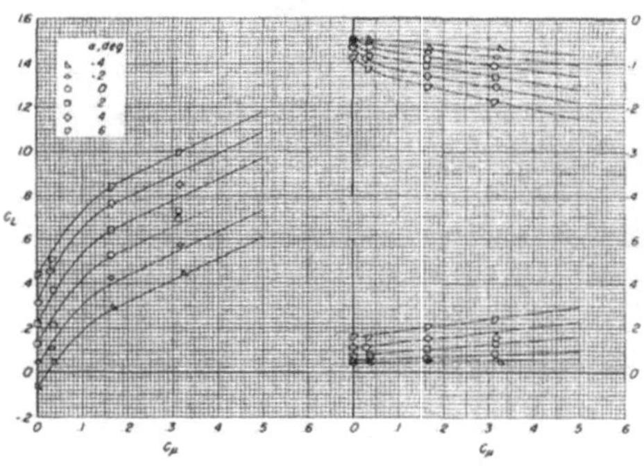

(c) Lift, Drag & Pitching Moment with Wing-Tip Jet Blowing

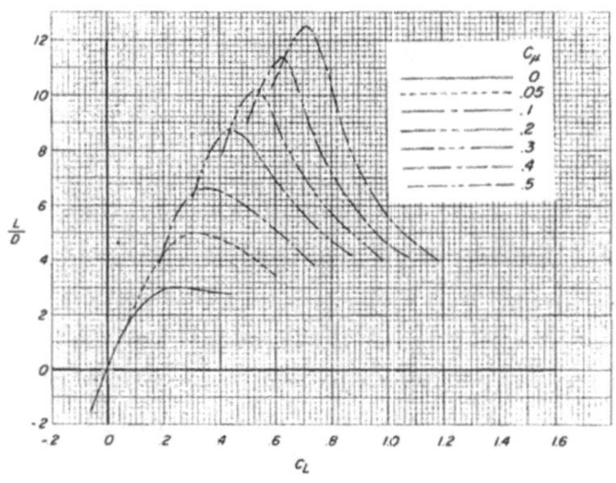

(d) Lift-Drag Ratio with Wing-Tip Jet Blowing

Figure 11.71: NASA Langley Model Tests of WIG with Wing-Tip Jet Blowing (Mineck & Carter) (1974) supplied by a compressed air source located on the towing carriage which meant it did not represent an actual real life configuration (the momentum drag of an actual craft air supply would not be represented) but it provided some valuable information. The tests covered three possibilities of wing-tip blowing, they were (a) spanwise blowing to provide an "equivalent span increase"; (b) wing-tip blowing straight down (deflection angle of θ = 90°); and an inward deflected jet with θ = 135°. The model speed was held constant at 72 ft/sec (Reynolds Number = 1.84 x 10^6). The model was an 11% thick Glen Martin airfoil (see Figure 11.37 for dimensions of the wing, no end plates). The wing chord (and span) measured 4 feet. The jet thickness at its exit measured (t = 0.031 inches). The model can be seen operating over the water at Langley in Figure 11.71 (a). The model geometry is as illustrated in Figure 11.71 (b).

The lift and drag measurements were defined by Mineck and Carter as follows:

$$C_L = C_{Lo} + C_\mu \sin\theta \cos\alpha \tag{11.92}$$

$$C_D = C_{Do} + C_\mu \sin\theta \sin\alpha \tag{11.93}$$

In these expressions for lift and drag; (C_{Lo}, C_{Do}) represent the basic lift and drag characteristics of the wing before wing-tip jet blowing. The inclination of the wing-tip jet is θ, and the angle of attack of the wing is α measured on the flat lower surface of the wing. There was no rearward deflection of the jets.

The wing-tip jet blowing coefficient, C_μ is defined as:

$$C_\mu = \frac{\text{Static Jet Thrust}}{\text{qS}} = \frac{\rho Q V_j}{qS} \tag{11.94}$$

In an actual case, the flow value Q could be that supplied from fan air or similar supply as shown in Figure 11.70 with the turbofan engine, but as already stated, was supplied "off model" by shop compressed air for simplicity. The free stream dynamic air pressure is (q) and the wing area is ($S = bc$).

There is an important observation in the formulation of the lift and drag equations (Equations 11.92 and 11.93) used by Mineck and Carter in their tests and that is that there was no provision made to deflect the jet air rearward to cancel any momentum drag of the wing air supply or to add thrust as a natural advantage to such a configuration (such as that used in the *channel flow wing* set up shown earlier in Figure 11.26). This was due to the simplicity of the model set up and would not be used in practice as was recognized by Mineck and Carter. It does mean however that the test results can be called conservative in terms of the values of lift-drag ratio (L/D) to be expected.

A wide range of test parameters were varied in the Mineck and Carter model tests, including angle of attack (α), wing-tip jet blowing coefficient (C_μ); height above the water (h) measured at the quarter chord point as shown in Figure 11.71 (b); and the deflection angle of the wing-tip jet (θ). For the case of direct wing-tip jet flow in the spanwise direction (θ=0°) there was very little improvement in lift-drag ratio; for downward deflection (θ=90°), there was a modest improvement, and for inward deflection (θ=135°) there was a good improvement in lift-drag ratio as might be expected from the original work on peripheral jet air cushion craft (see Figure 11.27).

To illustrate the results, just one particular set of data for a height to span value of h/b = 0.113 is shown in Figure 11.71 with an inward deflection of the wing-tip jet (θ=135°). This value of height was chosen to be reasonably close to the values used by the *ekranoplan* craft shown earlier, for the purposes of comparison. In Figure 11.71(c) it is seen that both the lift coefficient (C_L) and the drag coefficient (C_D) increase as the amount of blowing (C_μ) is increased, as expected in the light of no rearward jet deflection being incorporated. The pitching moment (C_M) is seen to be negative (pitching nose down) due to the effect of the increased drag. Despite this shortcoming, it is seen from Figure 11.71(d), there is a significant increase in the lift drag ratio (L/D) for the configuration. This significant increase in lift-drag ratio only occurs however at very high values of the blowing coefficient (C_μ). By comparison, it was shown in Chapter 3 in the discussion of the *channel flow* craft, much lower values of air flow are required to provide adequate sealing to reduce the induced drag. There was no separation of profile drag and induced drag from the Mineck and Carter results and it must be left to later writings to determine guidance on such features. The results of Mineck and Carter are intriguing enough, however, to stimulate more in depth treatments of edge sealing using onboard air supply and how best to use it.

11.21 Bréguet Range for Jet Engine Powered Craft

In Chapter 10 on the ANVCE Project (see 10.28 and surrounding text), it was shown that the Breguet Range equation for jet engines (such as the *Pegasus*) takes on a different form from that discussed mainly (for convenience) throughout this book. The calculation of *jet engine powered craft performance* is more complicated than that for craft that are powered by the engines through transmission or shaft driven propulsors. One essential difference is in the fuel consumption in that in the case of jet engines, the fuel flow is proportional to the drag (D) or thrust (T) of the craft and not the power (P). In this case, it is normal to express the fuel consumption as the *thrust specific fuel consumption* (*tsfc*), where the thrust (T) is the dominating force, such that the weight change throughout flight is given by,

$$dW = -tsfc.T.dt \tag{11.95}$$

The equation for the range (R) is then given by,

$$R = \frac{2\sqrt{2}}{tsfc\sqrt{\rho}}\sqrt{\frac{W}{S}}\frac{\sqrt{C_L}}{C_D}\left[1 - \sqrt{\frac{W_e}{W}}\right] \tag{11.96}$$

In Equation 11.96, air density is (ρ) and the wing loading is (W/S).

While Equation 11.96 is not quite as elegant as the form for the propeller driven craft, it contains the same features for the most efficient design in terms of maximizing the range of the craft.

In this case, for *maximum range*, the jet engine powered craft must be flown under the operating conditions of:

1. Minimun *thrust specific fuel consumption* (*tsfc*) (11.97)

2. Maximum value of $\left(\frac{\sqrt{C_L}}{C_D}\right)$ (11.98)

The lift coefficient (C_L) and drag coefficient (C_D) for the aircraft are defined by,

$$C_L = \frac{W}{{}^1/_2 \rho V^2 S} \tag{11.99}$$

$$C_D = \frac{D}{{}^1/_2 \rho V^2 S} \tag{11.100}$$

where under equilibrium conditions, L=W. The total resistance or drag of the aircraft is given by (D).

While Equation 11.96 is a good form for detailed analysis of any given craft, it is more useful in this context to express the range equation in terms of craft speed (V). In which case, after some substitution, the range equation can be written:

$$R = \frac{2}{tsfc} V\left(\frac{L}{D}\right)\left[1 - \sqrt{\frac{W_e}{W}}\right] \tag{11.101}$$

In this form, it can be seen that the design speed for maximum range (frequently called *cruise speed*) of a jet craft is not that speed for *minimum drag* or *maximum lift drag ratio* as for the propeller driven craft but at a slightly higher speed that corresponds to the speed for minimum $(\frac{T}{V})$ such that the product $(V\frac{L}{D})$ is a maximum.

The conditions for maximum range of the jet powered craft can then be rewritten as follows:

1. Minimun *thrust specific fuel consumption* (*tsfc*) (11.102)

2. Speed for maximum value of $\left(V\frac{L}{D}\right)$ (11.103)

These conditions will normally provide higher speeds that those required by propeller driven craft discussed above. It is the combination of these two key parameters that determine the speed for maximum range. If the jet engine powered air cushion craft is of the type proposed with a *Pegasus*-like jet engine shown in Figure 11.70 then the thrust can be expressed as:

$$T = \rho Q(V_j - V) \tag{11.104}$$

where ρQ is the mass flow through the engine and V_j is the jet exhaust velocity. This would apply in the cruise mode (where the craft is fully supported by aerodynamic lift). If the craft uses part of the flow to maintain some aerostatic lift or to use diverted fan air to inflate the side skirts (as in Figure 11.69) then the total flow (Q) through the engine is given as:

$$Q = Q_1 + Q_2 \tag{11.105}$$

In this expression, Q_1 is the core air flow used for thrust and Q_2 is the fan air diverted for air cushion use (cushion or side skirts). If part of the mass flow is diverted to provide lift (or sealing) to the craft then,

$$\text{Thrust (T)} = \rho Q_1 (V_j - V) \tag{11.106}$$

$$\text{Lift power, } (\text{P}_\text{L}) = p_c Q_2 \tag{11.107}$$

The expressions for Transport Efficiency (TE) used in earlier chapters would then be modified as necessary depending on the geometry of the design. A further modification would be made to the Breguet Range equation if another form of jet engine such as the turboshaft engine was used, where say the shaft would provide *power* to the lift (fan) system and the core air exhaust would provide the *thrust* to the craft. There would be many such variations to the performance equations not expanded upon here where the propulsion technology available from jet engines is integrated with the variations of aerodynamic air cushion craft geometries discussed in this chapter.

11.22 Summary of Aerodynamic Air Cushion Craft

The chapter has traced the historical beginnings and the technological developments of the promising technology of providing high speed over the sea using some form of aerodynamic lift. As shown throughout the chapter, there have been both elegant treatments and some not so elegant. The example of the *ekranoplan* KM is a case in point. That craft had ten engines of which eight were used only for the take-off mode! That feature alone showed the inefficiency of such a concept. The later designs (*Lun* and *Orlyonok)* were more efficient but still required large amounts of power for flight causing them to be much less efficient than other forms of air transportation.

This chapter has explored the various theories for ground effect and developed simple theories for the main functions that help discussions on basic design principles. To conclude the chapter, new ideas to be explored have been included that promise much more efficient craft that show a possible way toward achieving a "100 knot capability" in an efficient practical platform capable of conducting useful missions (commercial and military) in a rough sea environment. Areas requiring more experimental work offering improved performance of *aerodynamic air cushion craft* have been identified throughout the text.

References

1. Richard Dee, *"The Man who Discovered Flight: George Cayley and the First Airplane"*, published by McClelland & Stewart, Toronto, (2007)
2. Sir George Cayley, *"On Aerial Navigation"*, Nicholson's Journal of Natural Philosophy, Published in three issues (Part 1: November 1809; Part 2: February 1810; and Part 3: March 1810).
3. Lilienthal, Otto; *"Der Vogelflug als Grundlage der Fliegekunst"* pub by R. Gaertners Verlagsbuchhandlung, 1889. Later translated by I. W. Isenthal as *Birdflight as the Basis of Aviation*, and published by Longmans, Green; London 1911
4. Lanchester, Frederick W.; *"Aerodynamics"*, published by Constable & Co., London 1907
5. Helmholtz, Hermann L., "*On the Integrals of the Hydrodynamic Equations Corresponding to Vortex Motions*", published in Crelles Journal für die reine und angewandte Mathematik 60, pp 23–55, 1858
6. Giacomelli, R. and Pistolesi, E., *"Historical Sketch"* in *"Aerodynamic Theory"*, Vol 1 ed. W. F. Durand, pub by Springer, Berling 1934

7. Prandtl, Ludwig; "*Tragflächentheorie*" (Wing Theory), I Mitteilung (Note I) Nachrichten der K. Gesellschaft der Wissenschaften zu Göttingen, Math-phys. Klasse, 1918. Note II published in 1919.
8. Glauert, Hermann, "*The Elements of Aerofoil and Airscrew Theory*", Cambridge University Press, 1926
9. Glauert, Hermann, "*The Calculation of the Characteristics of Tapered Wings*", British R&M 767, 1921–1922
10. Jones, Robert T., "*Theoretical Correction for the Lift of Elliptical Wings*", Jnl Aeronautical Sciences, 9, No 1, 8, November 1941
11. Jones, Robert T., "*Corrections of the Lifting Line Theory for the Effect of Chord*", NACA Tech Note 817, July 1941
12. Glauert, Hermann, "*The Elements of Aerofoil and Airscrew Theory*", Cambridge University Press, 1926
13. Betz, Albert, "*Lift and Drag of a Wing near a Horizontal Surface (the ground)*", published in Zeitschrift für Flugtechnik und Motorluftachiffahrt", 1912
14. Wieselsberger, Carl; "*Über den Flugwiderstand in der Nähe des Boden*" (Wing Resistance Near the Ground) originally published in Zeitschrift für Flugtechnik und Motorluftschiffahrt, Vol No.10, p 145, 1921. Translated and published as NACA Technical Memorandum No. 77, April 1922
15. Fink, Marvin P. and Lastinger, James L., "*Aerodynamic Characteristics of Low Aspect Ratio Wings in Close Proximity to the Ground*", NASA Technical Note D-926, July 1961.
16. Carter, Arthur W., "*Effect of Ground Proximity on the Aerodynamic Characteristics of Aspect Ratio 1 Airfoils With and Without End Plates*", NASA Technical Note D-970, October 1961
17. NACA Report: "*Aerodynamic Characteristics of Airfoils–IV*", NACA Report 244, published 1926
18. Cockerell, C. S., "Ripplecraft", Report 1/55 (25 October 1955). See also Air Cushion Vehicle Supplement to *Flight International* Magazine, 28 February 1963.
19. Mantle, Peter J., "*Air Cushion Craft Development*", US Navy Report No. DTNSRDC-80–012, January 1980
20. Stanton-Jones, Richard, "*The Development of the Saunders-Roe Hovercraft SR.N1*", Symposium on Ground Effect Phenomena at Princeton, Paper 13 (21–23 October 1959)
21. Stanton-Jones, Richard, "*Some Design Problems of Hovercraft*", Institute of Aerospace Sciences Paper 61–45 (1961). Also Saunders-Roe Publication S.P. 404, January 1961
22. Elsley, Gordon H. and Anthony J. Devereux, "*Hovercraft Design and Construction*", published by Cornell Maritime Press, Inc., in 1968
23. Payne, Peter R., "*Curved Jet Flows*", United States Army Material Laboratory Technical Report 65–20, May 1965
24. Elsley, Gordon H., "*Two-Dimensional Hovercraft Lift*", Saunders-Roe Ltd, Report A/A/53, November 1958
25. Stanton-Jones, Richard, "*The Development of the Saunders-Roe Hovercraft SR.N1*", Symposium on Ground Effect Phenomena at Princeton, Paper 13, 21–23 October 1959
26. Carmichael, B.H. and M. P. Southcote, "*Air Cushion Feasibility Investigation, Performance and Stability Experiments, Configuration and System Evaluation Studies*", Ford Aeronutronic Co., Publication U-1066, November 1960
27. Mantle, Peter J., "*Thrust and Drag Estimate for the Channel Flow GEM*", Vehicle Research Corporation report, WP 36, September 1962
28. Braunss, G. and W. Lincke, "*Die Auftriebsverteilung einer Ebenen Platte in Bodennahe*" (The Lift Distribution on a Flat Plate in Ground Effect) published in Journal Zeitschrift für Flugwissenschaften, No. 10, 1962
29. Ashill, Peter R., "*On the Minimum Induced Drag of Ground-Effect Wings*", published in The Aeronautical Quarterly, Royal Aeronautical Society, Volume XXI, Part 3, pp 211–232, August 1970

30. *"Wingship Investigation"*, Final Report, published by Advanced Research Projects Agency (ARPA), now Defense Advanced Research Projects Agency (DARPA), Arlington, Virginia, September 1994
31. Widnall, Sheila E., and Timothy M. Barrows, *"An Analytic Solution for Two and Three-Dimensional Wings in Ground Effect"*, Jnl. Fluid Mechanics, Vol 41, part 4, pp 769–792, May 1970
32. Kida, T and Y. Miyai, *"Minimum Induced Drag of Non-Planar Ground Effect Wings with Small Tip Clearance"*, Journal of Aeronautical Quarterly, Vol XXV, pp 19–36, published by Royal Aeronautical Society, February 1974
33. Ando, Shigenori and Haruki Yashiro, *"Minimum Induced Drag of Ground Effect Wings"*, Journal of Hydronautics, Vol 10, No. 3, pp 106–112, published by AIAA, July 1976
34. Swanborough, Gordon and Bowers, Peter, *"United States Navy Aircraft Since 1911"*, pub by Naval Institute Press, Annapolis, 1990
35. Crewe, Peter R., *"A Review of the Seaplane Hydrodynamic Design Problems of Greatest Present Importance and Possible Solutions. Part I, Introduction and Water Performance of Streamlined Bodies"*, Saunders-Roe Ltd Report H/O/112/1, May 1950
36. Gerry, Dave G., and Peter R Crewe, *"Seaplane Design Problems"*, Final Report to ANVCE Project, Saunders-Roe Ltd Report SP 3920, June 1977
37. Capt James W. Kehoe: *"Appendix: Wing-in-Ground-Effect Vehicles"* to *Chapter VI: The Surface Effect Ship* by Edward A. Butler, "Modern Ships and Craft, Naval Engineers Journal, Special Edition, February 1985
38. Chief of Naval Operations, OP-96, *"Advanced Naval Vehicles Concepts Evaluation (ANVCE) Project, Vol I, II & III"*, Report No CNO LTR 96/C194136, December 1979
39. *"Wingship Investigation"*, Final Report, published by Advanced Research Projects Agency (ARPA), now Defense Advanced Research Projects Agency (DARPA), Arlington, Virginia, September 1994
40. *"Advanced Naval Vehicles Concepts Evaluation (ANVCE) Project"*, Prepared by Chief of Naval Operations (OP-96), Report No. CNO LTR 96/C194136, Vols 1,2,3,4; December 1979
41. *Flight International Magazine (Air Cushion Supplement)* issued November 1966
42. *Multi-Mode Variable Geometry Surface Effect Ship*, US Patent Number 3,481,297 by P. J. Mantle filed 13 Nov 1967, issued 2 December 1969
43. Mantle, Peter J., *"Some Design Aspects of Air Cushion Craft"*, paper presented at the International Congress of Subsonic Aeronautics, New York, NY, 3–6 April 1967. Published in the Annals of the New York Academy of Sciences, Vol 154, Article 2, pages 924–952, 22 November 1968
44. *Wing in Ground Effect Vehicle*, US Patent Number 4,151,893 by Peter J. Mantle, filed 8 September 1977, issued on 1 May 1979
45. http://en.wikipedia.org/wiki/Rolls-Royce_Pegasus
46. Mineck, Raymond E. and Arthur W. Carter, *"Effect of Ground Proximity on the Longitudinal Aerodynamic Characteristics of an Aspect Ratio-1 Wing With and Without Wing-Tip Blowing"*, NASA Technical Memorandum NASA TM X-3048, July 1974

12 Lessons Learned and Where to Next?

> *"Learn from yesterday, live for today, hope for tomorrow. The important thing is not to stop questioning"*
>
> Albert Einstein, (1879–1955)

A vast amount of experience in the technology of "*ground or surface effect*" for marine vehicles occurred during the half century 1950–2000. During that time, the *hovercraft* came into being in its birthplace in England and throughout the world. Both commercial and military uses benefited from these craft that typically operated at speeds up to 50 knots in littoral sea states. That history has been covered in the early chapters of this book (see Chapter 1 and Chapter 2 and throughout) as data from that era was used in the technical development chapters. A derivative of the hovercraft that became known as the *surface effect ship* enjoyed significant development during the two decades 1960–1980 most intensely in the US where speeds approaching 100 knots were achieved. That history has occupied many chapters in this book. The development of the *aerodynamic air cushion craft* has actually covered a much longer period from the original work by Prandtl and Wieselsberger in the 1920s to the present day. The most developed work in terms of operational craft was that done in Russia in the 1960–1990 period when such craft operated at speeds in excess of 200 knots (see Chapter 11).

Each of those key periods produced significant work in theories of ground effect, substantial experimental data and many operational vehicles. Copious references to that work appear in the various chapters and used in the development of the simple theories provided here in the understanding of a "100 knot at sea capability".

It is in the critical period, 1960–1980, when the most intense and fruitful development was undertaken to achieve a practical 100 knot capability, that the stumbles began to occur. They can be grouped under the two main headings of (1) unresolved technical problems and (2) neglecting to follow well-established R&D management practices. These issues have been addressed in the preceding chapters. It is believed that it would have been possible to have achieved a high speed marine craft for practical missions capability within the time and money expended if a few simple established procedures had been followed. In what follows, a summary is provided of "lessons learned" from that period to develop a capable "100 knots at sea", to provide a sound base for future development.

In today's world of uncertainty of threat, often referred to now as an "asymmetric threat" it is becoming increasingly clear that any weapon systems devised for national defense must now be flexible in both capability and mission use. The high speed marine craft will fit well in such scenarios. Flexibility of use arguments can also be put forward for commercial use of efficient high speed marine craft. This chapter concludes with some specific recommendations based on the developed findings on new avenues to explore to bring high speed into the marine world in a practical and useful manner.

12.1 The "Size-Speed-Mission" Triad

One of the underlying difficulties encountered during the decades of development of very high speed marine craft was the all too often mismatch between the type of high speed concept being put forward and its intended purpose or mission. Frequently, a concept came upon the scene and the tendency was to immediately postulate "bigger and faster" versions of the concept without proper regard to the mission or purpose of the craft and the inherent restrictions of the physics involved.

The seminal work in 1950 by Gabrielli and von Karman, entitled "*What Price Speed?*"[1] clearly showed the all-encompassing limits of size and speed where craft of all types are being examined as the transportation spectrum in air and water is examined. Chapter 7 explored that work (see Figure 7.18 as a summary chart) as it affected the choice of concept of high speed marine craft. The four underlying principles of *hydrostatic, hydrodynamic, aerostatic* and *aerodynamic* forces have a profound effect on the ability of any concept to perform efficiently at speed. It is indeed *possible* to push a conventional displacement hull through the water at high speeds but only at the expense of exorbitant amounts of power. For those concepts constrained to use water buoyancy for support, the size of the craft must be drastically reduced to achieve high speed as shown by the Gabrielli and von Karman "boundaries", hence the introduction of the modified hull form embodied in planing craft. This automatically affects the choice of mission, say from *large container (displacement) ships* operating at low speeds (say 10–30 knots) to smaller size (*planing*) *patrol boats* operating at higher speeds (approx. 50 knots). It has been found extremely difficult to overcome the physics of the problem of operating at speed in a medium that is 840 times denser than air! For the very high speeds, it is found that it is necessary to get the hull out of the water altogether and switch to *aerodynamic* support such as enjoyed by the aircraft concept. These two major divisions of craft designed to work either *in the water* or *in the air* with their different forms of "limits" are clearly displayed in the charts of Gabrielli and von Karman (Chapter 7 and 8).

The problems occurred at the interface between water and air where the many concepts of *hydrofoils, air cushion craft*, and *wing-in-ground-effect craft* were developed in that crucial period of 1960–1980. It was generally recognized that hydrofoils, relying on *hydrodynamic* support, were not scalable to large ocean going sizes (say 10,000 tons displacement or more) and missions had to be sought for smaller sizes (1,000 tons or less?) in more forgiving sea states such as in littoral waters. Similarly, the *air cushion craft* in its amphibious form was also subject to large hydrodynamic forces as it operated over the waves, and conventional aerodynamic means of propulsion (air propellers and the like) were not able to produce the necessary thrust for very large sizes (say 5,000 to 10,000 tons) and missions had to be sought in the smaller sizes (around 500 tons or less). Such air cushion craft could grow to *frigate size ships* (3,000+ tons) for open ocean missions but only with different propulsion schemes envisaged. Practically, however, smaller sizes suitable for *amphibious assault missions* were found to be most efficient in the development to date.

The non-amphibious form of *air cushion craft*, which became known as the *surface effect ship* (see Chapter 1 for a more descriptive classification of types) offered the promise of speed at the larger sizes because of the integration of water propulsion schemes in a more efficient manner (various forms of water propellers and waterjets are used) that led to projections for large high speed ships. Unfortunately, there was still the need to "push" the ship through water with its many issues associated with the large hydrodynamic forces (see Chapters 4 and 8) incurred. Early enthusiasts again projected "bigger and faster" versions, promising such water surface bound craft of 10,000 tons displacement with speeds of greater than 129 knots and a projected Transport Efficiency some three times greater than has been achieved before or since! This was the case of forcing a *ship concept* to meet a *mission need* to transport large payloads (approx. 5,000 tons or more) at high speed (greater than 100 knots) across the world's oceans in order to have an economical value in marine transportation. Such a mismatch between the concept and mission was quickly abandoned and much smaller ship sizes were sought with an associated change of mission from *commercial container ships* to *military frigate* size ship missions.

This mismatch of concept and mission occurred again in the 1990s when the high speed *ekranoplans* form of aerodynamic air cushion craft, which were designed for smaller sizes (150 tons to 550 tons) and speeds of greater than 200 knots) for *amphibious assault* and *missile attack* missions, were scaled (on paper) to sizes greater than 5,000 tons for *strategic lift* missions which was clearly a strained match of concept and mission and unattainable in any practical sense.

The matching of the concept in *size* and *speed* to the *mission* intended is a key step in any design analysis and must serve as a base point in any future design.

12.2 Lessons Learned

Without a retelling of the various issues that are adequately covered in the preceding chapters, it is possible to gather up some important lessons learned and to propose a positive way forward to achieve a high speed at sea capability in different mission areas both militarily and commercially.

What are some of the lessons learned?

1. Make the Mistakes in Small Sizes In many areas of new development, and especially so in the case of developing a capability to transit open ocean rough seas at high speed, it is better to proceed cautiously and in small steps. It is better to make mistakes in small sizes than to proceed imprudently into larger sizes with unresolved technical issues, where any mistakes will be expensive and dangerous. Technological work into new areas has frequently caught the inventor off guard. This was true in the development of the aircraft, the submarine and many other areas of vehicle development. There was no reason to believe that development of the *surface effect ship* and the other variants of *air cushion craft* explored in this pursuit of "100 knots at sea" capability would be any different.

The initial US Navy Program did follow a prudent development path up through the critical 100 ton displacement ship size. Sufficient model testing had been done together with sufficient analyses to justify arriving at that stage. Chapter 5 summarized this well-thought out plan with Critical Design Reviews (CDR) set up at Key Milestones in the development process. One concept, the SES-100B was successfully meeting all program objectives while the other concept the SES-100A was not meeting its contract objectives with several critical technologies unresolved. This was well documented at the time. It was at this critical stage that a fundamental flaw in the program management occurred. The next step in the original planned Program was to proceed to a 500 ton displacement ship but *only* if the 100 ton prototypes had successfully met their technical objectives.

The political enthusiasm at the time (1972) overruled this prudent approach in the program and direction was given to (a) move ahead to the next phase even though one prototype was not meeting the technical objectives and (b) to skip the 500 ton size altogether and jump to a much larger ship of 2000 ton displacement that grew, with political program enthusiasm, to greater than 3000 ton displacement (see Chapter 5, Figure 5.1 for the program schedule). The Achilles heel of this high speed ship development program can be traced directly to that fateful decision.

The hydrodynamics of hull forms designed to move at extremely high speeds through the water is still an unexplored technology. Understanding the mechanics in cavitating and aerating flow; to understand the dynamics of wave interaction with the hull (and seal systems) with the associated enormous loads; and to design new propulsion schemes and lift mechanisms all required careful development with controlled testing that must be followed scrupulously if any degree of success is to be expected. By "skipping" a step in the development program exacerbated the problem in the US Navy SES development program.

In any future program to develop such a high speed ship, all efforts must be made to "take small steps".

2. Meet Milestone Objectives Before Proceeding to the Next Step Research & Development is actually a well-structured discipline if followed properly. While there were good management schemes in place in the US prior to 1969, in that year following the Vietnam war, a major improvement in the Defense Acquisition Policies occurred with the establishment of the Defense Systems Acquisition Review Council (DSARC) to properly manage the Research & Development of programs and the transition into acquisition programs of operational systems to be introduced into the military inventory.[2]

While there have been various upgrades of the instructions over the intervening years, the latest versions of the directive itself[3] and the instructions[4] are referenced here[a]. The actual documents are somewhat cumbersome but the author has provided

[a] A new issue of the DOD instructions is expected to appear shortly but it will not contain any significant differences on how to properly manage acquisition programs and the associated R&D.

a succinct summary of the key subject matter (Mission Need Statements, Key Milestones, Milestone Decision Authority, Demonstration & Validation Phase, Alternative Concepts Evaluation etc) in his 2004 book on missile defense[5]. Whether talking about missile defense or ship design, the management disciplines for good Research & Development and System Acquisition remain the same. In the 2004 book, the management disciplines in other nations (UK, France, Germany and others) and in NATO itself are compared with the US system and while there are differences in detail, all follow a similar and *good management discipline* to ensure that Research & Development is conducted in a well-ordered manner with adequate reviews dispersed at Key Milestones in the program. These steps require technical reviews at each stage with required approval before moving to the next stage in the development process.

It is not intended to delve into the specifics of this management scheme (the cited reference contains those in detail) other than to say that such a disciplined approach calls for *peer reviewed* documentation of the key developments and established procedures for approval to progress through the management process. Unfortunately, in the *high speed surface effect ship* program many of the official Key Milestone Reviews either did not occur or were treated in a lax manner such that fundamental technological shortcomings were overlooked. The Mission Need Statements, for example, would have required review by Naval Operators familiar with military operations at sea and the stringent requirements for any ship to "go into harm's way".

In any future program to develop a high speed ship, all efforts must be made to adhere to well-structured R&D phases, milestones and an hierarchical approval process before moving to the next step. Moving forward without solving the technical issues in the ongoing phase can only lead to failure, as it did for the US Surface Effect Ship program.

3. Maintain Critical Peer Review as the Program Proceeds There is a tendency when working a program, for the engineers and program managers to concentrate on the program at hand using only the data extant within the program and within their libraries and focused data sources. This is an unfortunate but natural tendency that can lead to missteps in the development process. In Chapter 4, a discussion is given on the critical review that the US Maritime Administration (MARAD) made in 1965 on some of the candidates for a 100 knots at sea concept. It is a credit to the experts consulted by MARAD given the short time available (approximately 12 weeks) that most if not all the technological issues were properly identified and documented for MARAD at that time.

Their report could have been even more compelling if the ongoing work around the world, especially in the UK where commercially and military craft using the surface effect principle was well underway, would have been incorporated and the advice of these other (external) experts, protagonists of the art and peers had been included. The technology problems that caused the MARAD and the later US Navy program to stall had already been identified by those outside experts.

In any future program to develop such a high speed ship, all efforts must be made to ensure that peers and experts outside the immediate circle working the program are brought in to minimize the risk as the program proceeds.

4. Be Cognizant of Competing Concepts and Complementary Concepts at Each Stage There is a tendency in both small and large organizations alike (both government and private) for the engineers assigned to any project to become advocates about the merits of the concept on which they are working. This is a natural tendency caused by the organization, company, or division in which they work, to assign engineers into focused categories. These could be, for example, Ship Department, Air Cushion Craft Department, WIG Department and so on. While there are definite merits to such focused efforts (or departmental boundaries) in any one area and to create "centers of excellence", it does tend to cultivate "advocates" and mindsets for their concepts and synergism of effort of working together with their fellow workers in the same departments. Unfortunately, it also fosters a level of competition with other concepts (Departments) such that any possible value of combining the features of one concept with the features of another concept for an improved product tends to get lost.

Management of such organizational groups should become aware of the frequent need for cross-fertilization of ideas across groups at key stages of any particular project. Since it has been found and discussed in this book that it is the cross fertilization of ideas across competing or complementary concepts that offer the most promise, this should be considered as a standard management discipline in any future program on developing a high speed marine craft.

5. Do Not Isolate Engineers from Real World Experts in Complementary Areas As a complementary finding to that given above in **Lesson Learned No. 4**, it is essential that all the engineers *not* be isolated from other disciplines that will impact the design. It is common practice in many organizations, for example, to isolate the Cost Analysts from the Engineers. This has the unfortunate tendency to unwittingly include costly features into the design because engineers tend to strive for the "best" design (in terms of performance for example) that may also be, unfortunately, the most expensive.[b] A better interaction with cost analysts as the design unfolds could avoid this tendency. Another group that tends to be separated from the engineers and lab technicians are those experienced with actual operations of ships at sea.

In many Technical Conferences for example, where technical papers, designs, analyses are presented, military operators and users are seldom present. This separation of those involved has led to the cancelation of most high speed marine craft programs (see earlier chapters) on the basis that they were deemed not "cost effective".

In any future program on advanced ships, all efforts should be made to include these complementary groups for a "better" product with a higher potential for success.

12.3 Domains of High Speed Marine Craft

A considerable amount of research and development of the many different types of high speed marine craft candidates pursued over the previous decades has produced

[b] Francois-Marie Arouet Voltaire (1694 – 1778), French philosopher, is attributed with originating, in his 1772 poem, "*La Bégueule*", the saying: "*The best is the enemy of the good*".

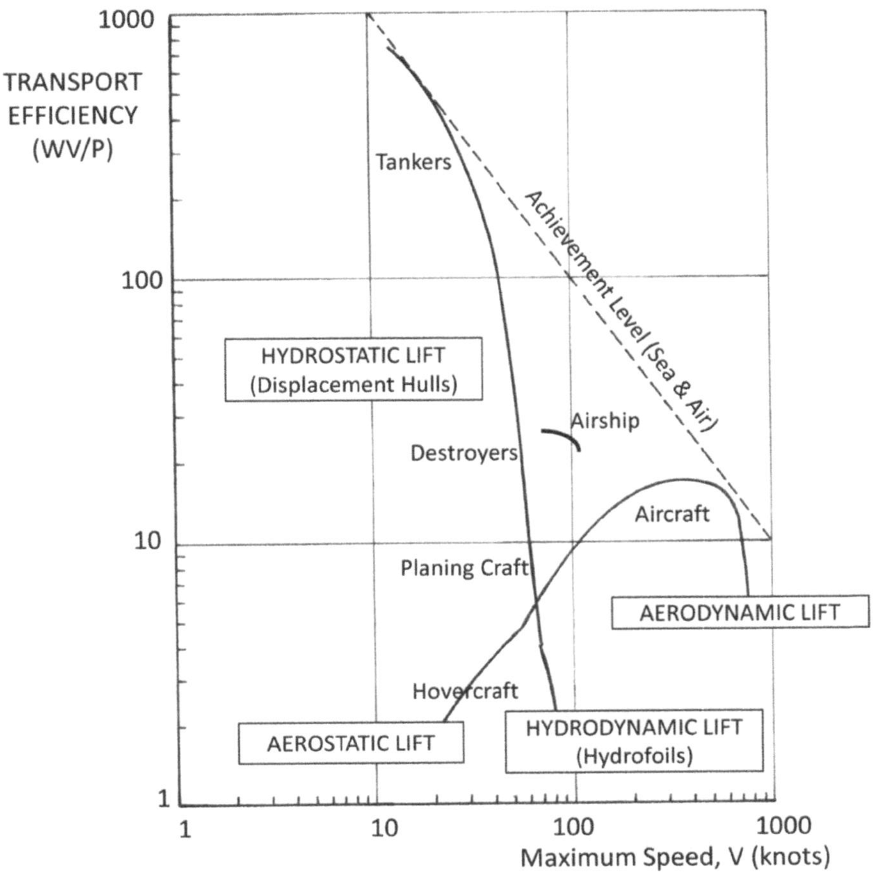

Figure 12.1: Domains of High Speed Marine Craft

a reliable view now of where in the speed spectrum such craft lay. Figure 12.1, reproduced from Chapter 8, provides a convenient display for discussion. As each concept was pursued, it became clear that those craft that are dominated by *hydrodynamic forces* and those dominated by *aerodynamic forces* coalesced into very distinct "classes" of craft. The impact of size was a secondary effect that also appeared from the development that emphasized that simply making the craft larger did not materially improve their efficiency, indeed it frequently resulted in less efficient craft. As seen in Figure 12.1, in the waterborne "class", the larger sizes (say, of several thousand tons of displacement) belonged in the class of the hydrostatic or buoyancy supported class of craft and provided the most efficient in terms of transport efficiency or "work done" by the craft but could not achieve high speed. Any attempt to go faster resulted in much lower efficiency and in much smaller sizes. The laws of physics simply precluded any other outcome.

By changing the design concept and bringing in lifting mechanisms that changed from hydrostatic shaping to hydrodynamic shaping, that resulted in *planing hull* forms, provided a marked improvement in speed, from say less than 30 knots (for the hydrostatic lift class) to greater than 50 knots, but at a great sacrifice in efficiency and size (down from thousands of tons displacement to hundreds of tons of displacement). In the 1960s and later, great effort got underway to develop two different forms of high speed marine craft: the *air cushion craft* (or *hovercraft* as it is known in its birthplace in the UK) and the *hydrofoil* to achieve high speed (50–100 knots). As seen from Figure 12.1 this increase in speed again incurred a significant drop in transport efficiency and size. For very high speeds, say 200 knots or more, leaving any part of the craft in the water was simply not practical as such craft dealt with a medium that was 840 times as dense as air. This led to the development of *aerodynamic air cushion craft* but as discussed in Chapter 11, this produced craft that achieved the high speed but again in the smaller sizes. The transport efficiency of such aerodynamic forms of air cushion craft was much greater than the hydrodynamic craft and much higher speeds but at efficiencies that did not rival conventional aircraft.

The data base that evolved from these various forms of air cushion craft showed that achieving transport efficiencies of about WV/P =10 was very difficult to achieve and even that value was *one third* of the initial claims by the US Maritime Administration and the US Navy as the development of a high speed ship at 100 knots and in the multi-thousand tons of displacement got underway. As a side note, it is noticed from Figure 12.1, based on actual vehicle data, that transport efficiencies of 20–30 were achieved at speeds around 100 knots but only in the "other" class of aerostatic craft – the airship. This is a demonstration of the effect of the forces acting on the vehicle that avoided any form of hydrostatic or hydrodynamic forces. In the case of the airship, of course, this higher efficiency was only obtained through the use of large craft volume, to achieve the lifting mechanism, which has hampered development of the airship in any major usage as a large transport vehicle.

In retrospect, the initial aim of the US Maritime Administration to seek a form of marine transport that could achieve high speed (around 100 knots) with efficiencies somewhere between ships and aircraft was indeed achieved with the US Navy program – just not at the level of the original ambitious claims for transport efficiency! A glance at Figure 12.1 will show that the intense development period in the 1960–1980 timeframe by the US Maritime Administration and the US Navy did actually achieve the original aim of a high efficiency ship capable of traveling at 100 knots. The domain of such ships was found to be characterized by the "WV/P =10 and V= 100 knots" achievement. Unfortunately, the pursuit beyond the demonstration of this capability at the prototype stage was hampered by technical and management issues that could have been resolved at the time of cancelation of the program in 1980. The details of the potential solutions have been discussed in detail in the preceding chapters.

As outlined above in the *"size-speed-mission"* triad section, the features not explored in the development of high speed marine craft was the potential for new types of missions that would materialize from use of such new forms of aerodynamic air cushion craft, such as improved amphibious assault missions and other new missions (see Chapter 9) that would capitalize on the unique features of such craft that would serve well in the need for a US Navy with new mission statements in a world needing flexible capabilities in an uncertain threat world. Exploring new missions was not well researched in the earlier time period of development of high speed marine craft.

12.4 Outgrowths of High Speed Marine Craft Development

It is encouraging to note that the intense efforts by several nations to seek solutions to the high speed marine craft, with speeds that ranged from 50 to 250 knots, did produce a wealth of data that has benefitted both the marine and aeronautical communities world-wide. In some cases, the developments stopped short, leaving the door open for future entrepreneurs to explore.

Some of these areas discussed in the book are summarized here.

12.4.1 High Speed Hydrofoils

The hydrofoil community historically had tended to diverge into two schools: (a) the *surface piercing hydrofoil* school and (b) the *fully submerged hydrofoil* school. Both schools however had explored the use of variable foil sections that would allow the craft to operate at speeds 35 to 50 knots in subcavitating flow and at speeds greater than 70 knots when operating in supercavitating flow regimes. The preceding chapters document the various programs and technical features of these two possibilities. Enough technical successes have been achieved to suggest that it might be possible to develop a high speed marine craft in ocean going sizes using some form of variable geometry foil system to achieve speeds of 70–100 knots. What has been lacking is a comprehensive R&D program in the laboratory on the hydrodynamics of such foil systems. Because the hydrofoil (mainly in the *surface piercing hydrofoil* school) has the longest operational experience of all the high speed marine craft and offers low cost (less moving parts?) this suggests a fruitful area for hydrodynamicists and naval architects to pursue. Some possible areas to explore are provided in the earlier chapters.

12.4.2 High Speed and Stealth

This is an area of research that offers a great opportunity for innovation in design. Operation at the interface of the sea and the air has proven to be both a difficult analytical problem to solve and an operational problem of significant magnitude. The law of conservation of energy is very visible in the wake generated by marine craft moving through (or near) the water. The energy required to push through the water manifests itself in the ship's wake that generates a highly visible and trackable wake (by enemy radar). The only marine craft that can claim essentially a non-visible wake on the water surface is the submarine. The submarine enjoys this feature by submerging to a depth where the wake energy combined with the circulatory energy of the waves is sufficiently minimalized to create a stealthy presence.

There have been some basic research & development into combining the features of submerged submarines with the non-submerged hull forms using hydrodynamic laws of drag cancelation to minimize the wake and its signature to enemy radars. These designs and applications are discussed in the earlier chapters (see Chapter 9). This is an unfinished area of research where hydrodynamicists would find fruitful avenues of new solutions. Additionally, there are possibilities being explored to give submarines a "100 knot" capability using technologies already used in advanced torpedo design but these possibilities have not yet materialized.

Some suggestions are given in the earlier chapters where this is left as a promising area for new high speed ship design with hitherto undefined hull forms.

12.4.3 Variable Geometry Craft

As a counterpoint to the design philosophy of selecting a design and then invoking "bigger and faster" versions is to recognize the different speed regimes where different concepts and geometries would be better suited and to seek use of the variable geometry design approach. This is a well-developed technique in the natural world and has seen many successes in many areas of vehicle design. The hydrofoil is an obvious case in point in using either hull support or foil support as needed in the mission. Other possibilities for the air cushion craft are discussed in detail in several chapters (see Chapters 3, 9, 10 and 11 for particular examples). Such features that have served well in other disciplines such as the variable wing sweep aircraft that allow aircraft to operate in both subsonic and supersonic modes efficiently are well known. Similar solutions discussed in the earlier chapters need to be explored in the hydrodynamic high speed marine craft world for a variety of missions and flexibility of operation.

12.5 Recommended Avenues to Pursue

The earlier chapters have explored in detail the technology and the design of high speed marine craft. The vast data base on those efforts has yielded a wealth of information to help guide future explorations. That data base tends to show the most likely avenue for success for "100 knots at sea" would be centered on two forms of air cushion craft, specifically:

1. *(Partial) sidehull surface effect ship*
2. Some new form of *aerodynamic air cushion craft*

With the two most likely choices, there is an important proviso that is tied to the mission and the environment. The initial impetus for development in the US was to seek large tonnage (say 15,000 tons displacement) *commercial* ships capable of operating across the oceans of the world. The second impetus was for *military* ships with missions that covered frigate size ships (approx. 3,000 tons) for transoceanic operation and for smaller ships and craft (say 100 to 500 tons displacement) for either littoral sea operation or for across the beach operation for amphibious warfare.

This mission and environment impact tends to direct the choice to the *sidehull surface effect ship* for the larger size, open sea operation and the *aerodynamic air cushion craft and variants* for the smaller sizes for littoral seas and across-the-beach missions. These are not hard and fast rules but it helps focus on the concept appropriate for its mission.

12.5.1 Large Ocean Going High Speed Ships (Next Step)

With this recognition of concept and mission, it is noticed that the US Navy had already accomplished a successful critical stage in the development for open ocean missions with its success in 1976 when the *sidehull surface effect ship, SES-100B* achieved 91.90 knots during Navy Trials. The SES-100B had met all the program objectives of performance, stability, rough sea operation and craft reliability (see for example Chapter 6, Table 6.6). These achievements satisfied all the Defense Acquisition Milestones for R&D as called for above in **Lessons Learned No.2**. Based on this, it is recommended that the "missed step" in the original US Navy Plan be reinstated and a 500 ton displacement ship should be pursued.

There is no hard and fast rule for setting the size of the "small step" (see **Lesson Learned No. 1**), but keeping the size increment or scale factor to be less than $\lambda = 2$ is a good "rule of thumb" to follow. Increasing the ship size from 100 tons to 500 tons corresponds to a scale increase to $\lambda = 1.71$, which is not an unreasonable value.

Such a ship size (500 tons) allows for open ocean testing and for testing of key components such as the flexible skirt life operating under real conditions. The demonstrated skirt life of 400 hours on the test craft SES-100B needs to be verified on the larger size and improved, and to allow time for fabric manufacturers to develop materials for the "100,000 mile ~~tire~~ skirt" that would be necessary for any open ocean ship to operate on a satisfactory maintenance schedule. Also, the increased power levels required of both the propulsion system and the lift system are not beyond the state-of-the-art provided the speed is not Froude scaled but kept to the 50 to 100 knot level, and require validation by testing.

All attempts to increase the ship displacement beyond 500 tons in the next critical phase to achieve "100 knots at sea" should be resisted for the reasons expanded upon earlier.

12.5.2 Aerodynamic Air Cushion Craft (Next Step)

Because the *aerodynamic air cushion craft* did not receive a well-disciplined R&D program as followed for the US Navy 100 ton SES Program, a different set of recommendations are necessary. As covered in Chapter 11, there are a wide variety of *aerodynamic air cushion craft* that have been explored in the past. These include *wing-in-ground effect craft* such as the Lippisch designs; the Rethorst *channel flow craft*; and the Alekseyev air ingestion *ekranoplan* craft, all described in Chapter 1. Other variations of *aerodynamic air cushion craft* such as *variable geometry air cushion craft* are described in Chapter 11. Each of these concepts offer promise but none have benefited from a formal Feasibility Phase or Demonstration & Validation Phase in a well-structured Acquisition Program (see **Lesson Learned No. 2**). The technology of these various types is presented in detail in Chapter 11. While there is much work still to be done, enough analysis is provided in Chapter 11 to indicate that even with advances in aerodynamics in the near future that craft sizes will be limited to the smaller sizes (say 100 to 500 tons) and shorter range missions than for the *sidehull surface effect ship* best suited for open ocean missions.

The advantages seen in the aerodynamic air cushion craft for amphibious warfare were recognized in the Russian *ekranoplan* craft but alternative versions considered in Chapter 11 should now be pursued to give a definite amphibious capability. The discussion on improved designs for amphibious warfare was included in the discussion on military operations in Chapter 9 (see for example Figure 9.6) It is recommended that an R&D program be set up in a formal Feasibility Phase program to include the *variable geometry aerodynamic air cushion craft.*

12.6 Summary

This book has explored both the history and the technology of possible ways to achieve a "100 knot at sea" capability. It has provided basic theories suitable for design use and for further mathematical and experimental development of fundamental concepts. The mathematical concepts require further development to understand hydrodynamic flow around high speed hulls with asymmetric waterlines. The theories developed here for wings operating in ground effect require more sophisticated treatments that encompass the various possible geometries of planform; end plate configuration; and for operating over irregular surfaces caused by sea waves. There is a need to explore further the stability and control of such platforms operating at high speed in such close proximity to the waves.

It is hoped that this book has provided both a well-rounded history of the development to date and a useful foundation for further exploration of such unique high speed craft for both commercial and military applications in, around and on the oceans of the world.

References

1. Gabrielli, Giuseppe and Theodore von Karman, *"What Price Speed?"*, Mechanical Engineering, October 1950 issue
2. Packard, David (Deputy Secretary of Defense), *"Establishment of a Defense Systems Acquisition Review Council"*, The Pentagon, Washington DC, 30 May 1969
3. *"The Defense Acquisition System"*, US Dept of Defense, DOD Directive 5000.1, issued 23 October 2000
4. *"Operation of the Defense Acquisition System"*, US Dept of Defense, DOD Instruction 5000.2, 5 April 2002
5. Mantle, Peter J., *"The Missile Defense Equation: Factors for Decision Making"*, published by American Institute for Aeronautics and Astronautics (AIAA) (ISBN 1-56347-609-6), published 2004

Index

Printed in the United Kingdom by TJ Clays Ltd.